Early Tertiary Volcanism
and the Opening of the
NE Atlantic

Geological Society Special Publications

Series Editor K . COE

GEOLOGICAL SOCIETY SPECIAL PUBLICATION NO 39

Early Tertiary Volcanism and the Opening of the NE Atlantic

EDITED BY

A. C. MORTON
British Geological Survey
Keyworth
Nottingham NG12 5GG

L. M. PARSON
Institute of Oceanographic Sciences
Deacon Laboratory
Brook Road
Wormley
Godalming
Surrey GU8 5UB

1988

Published for

The Geological Society by

Blackwell Scientific Publications

OXFORD LONDON EDINBURGH

BOSTON PALO ALTO MELBOURNE

Published for
The Geological Society by
Blackwell Scientific Publications
Osney Mead, Oxford OX2 0EL
  (*Orders*: Tel. 0865 240201)
8 John Street, London WC1N 2ES
23 Ainslie Place, Edinburgh EH3 6AJ
Three Cambridge Center, Suite 208, Cambridge,
  MA 02142, USA
667 Lytton Avenue, Palo Alto, California 94301, USA
107 Barry Street, Carlton, Victoria
  3053, Australia

First published 1988

Typeset, printed and bound in Great Britain by
William Clowes Limited, Beccles and London

DISTRIBUTORS

USA and Canada
  Blackwell Scientific Publications Inc.
  PO Box 50009, Palo Alto
  California 94303
  (*Orders*: Tel. (415) 965-4081)

Canada
  Oxford University Press
  70 Wynford Drive,
  Don Mills,
  Ontario M3C 1J9
  (*Orders*: Tel. (416) 441-2941)

Australia
  Blackwell Scientific Publications
  (Australia) Pty Ltd,
  107 Barry Street, Carlton,
  Victoria 3053
  (*Orders*: Tel. (03) 347 0300)

British Library Cataloguing
in Publication Data

Early tertiary volcanism and the opening of the NE
  Atlantic.—(Geological Society special publication,
  ISSN 0305–8719; no. 39)
  1. Geology—North Atlantic Ocean
  I. Morton, A. C.   II. Parson, L. M.   III. Series
  551. 46′ 08′ 0933   QE350.22.N65

ISBN 0-632-02171-3

Library of Congress
Cataloging-in-Publication Data

Early tertiary volcanism and the opening of the NE
  Atlantic. (Geological Society special publication;
  no. 39)
  Bibliography: p.
  Includes index
  1. Volcanism—North Atlantic Ocean.
  2. Geology, Stratigraphic—Tertiary. 3. Sea-floor
  spreading.
  I. Morton, A. C.   II. Parson, L. M.   III. Series.
  QE522.E27 1988   551.7′ 8′ 091821   87-32636

ISBN 0-632-02171-3

Contents

Review of Igneous Activity

Introduction

L. M. Parson & A. C. Morton

A conference dealing with 'Early Tertiary volcanism and the opening of the NE Atlantic' was held at the Geological Society of London, Burlington House, London, on 18 and 19 March 1987. The meeting was supported and promoted by three of the specialist groups of the Society (Volcanic Studies, Marine Studies and Petroleum groups), and carried the endorsement of the Norsk Petroleumsforening. This broad-based support reflects the truly multidisciplinary nature of the meeting, which saw contributions from geophysicists, geochemists, geochronologists, volcanologists, petroleum geologists, structural geologists, petrologists, sedimentologists and biostratigraphers. The conference was attended by over 180 delegates from throughout the N Atlantic borderlands, who enjoyed the presentation of 34 papers and five posters, and contributed to a number of lively and spirited discussion periods.

This volume contains 37 papers, four of which appear as extended abstracts, and some in forms very different to that presented at the meeting. For practical reasons some authors have combined manuscripts, some contributions presented at the meeting were not available for inclusion in this publication, and some papers have been included despite not being presented at the meeting. The papers have been grouped into six categories, essentially on the basis of structural and geographic setting, but these categories also serve to emphasize the multidisciplinary approach of the current research.

Volcanic and tectonic framework

The first group of papers deals with the volcanic and tectonic framework on which the subsequent regional studies are based. These papers deal not only with the regional plate-tectonic setting during the Palaeogene, but also with the Icelandic hot-spot, in particular its character and role in the inception of seafloor spreading, its relevance to the excessive volcanism, and its possible cause.

Dipping reflectors and NE Atlantic evolution

The large number of papers falling into this group reflects both the strong interest and perhaps the lack of agreement in these fields. Description and discussion of volcanic margins throughout the NE Atlantic provides a broad spectrum of the degree of development of volcanism and structural context. However, recent geochemical data from Deep Sea Drilling Project (DSDP) and Ocean Drilling Program (ODP) results have provided some of the most interesting finds. Powerful evidence is presented to show the existence of varying types of crustal contamination in the dipping reflector sequences and in the underlying associated extrusives. Without doubt, one of the greatest advances has been made in the field of geochemical assessment of the margin evolution, and the acquisition of further geochemical data from more NE Atlantic dipping reflector sequences must be considered to be a high priority.

E Greenland and the Faeroe Islands

The first geographical grouping of papers covers the E Greenland margin as well as the Faeroes block, and includes detailed accounts of the trace element and isotopic compositions of E Greenland lavas and dykes and valuable geochronological and palaeomagnetic studies that have further constrained timing of models of the development of the E Greenland margin.

Volcanism in basins to the N and W of the British Isles

Aspects of Tertiary volcanism in pre-existing Mesozoic rift basins that border the NE Atlantic to the N and W of the British Isles, such as the Faeroe Basin, Rockall Trough and Porcupine Seabight Basin, are covered by this group of papers. These studies depend heavily on results of hydrocarbon exploration, and include analyses of deep commercial borehole samples and studies of commercial seismic data, combined with data acquired by the British Geological Survey (BGS) during shallow drilling. Of particular interest is the association of peraluminous cordierite-bearing dacites overlain by tholeiites found in the northern Rockall Trough, because this sequence is remarkably similar to the association found on the Vøring Plateau margin. Prior to these discoveries, peraluminous extrusives had not been recorded in the NE Atlantic Igneous Province, and they may represent a volcanic association particular to these pre-existing deep sediment-filled Mesozoic rift basins. There is clearly a need for more deep drill sites in this structural setting.

From MORTON, A. C. & PARSON, L. M. (eds), 1988, *Early Tertiary Volcanism and the Opening of the NE Atlantic,* Geological Society Special Publication No. 39, pp. ix–xii.

British Tertiary Igneous Province

This section deals predominantly with stratigraphic aspects, as the petrogenesis of the extrusives and intrusives has already been studied in considerably greater detail than the offshore areas, or even the Faeroe–E Greenland province (e.g. Thompson 1982). Important constraints are put on some of the timing of volcanic events within the British Tertiary Volcanic Province by radiometric and isotopic data presented in the following set of papers. Regional stress patterns are assessed over the NW margin and can be compared with the larger scale tectonic review presented early in the volume.

The sedimentary record

The next section deals with the sedimentary record of the volcanic and tectonic events associated with the NE Atlantic opening, and in particular deals with the record of early Tertiary volcanism in the North Sea sedimentary record. Despite unresolved differences in interpretation of the provenance of the pyroclastic deposits found at particular levels within the Palaeogene sequences, these studies provide important information on the timing of activity in the NE Atlantic borderlands. It is clear that with continued work, particularly on the geochemistry of individual tephras, the sources of the volcanics will be further constrained, thus allowing precise biostratigraphic dating of volcanic activity.

Review of igneous activity

The volume is completed by an exhaustive and far-reaching review of igneous activity in the NE Atlantic borderlands, the first such attempted for well over a decade, since the review by Noe-Nygaard (1974).

Overview

It is impossible in a short introduction to comment on all the highlights of the meeting. However, important advances were made in several directions. Much of the meeting was devoted to the igneous activity at the continental margin itself, and helped to focus the attention of geochemists, petrologists and volcanologists familiar with the classic onshore localities of Scotland, Ireland, Iceland, Faeroes and E Greenland onto the less well-known but volumetrically and geographically more extensive offshore volcanics. Considerable effort was made to encourage presentations which would address general but long-standing problems of the 'rifting' to 'drifting' transition at volcanic passive margins, of which the NE Atlantic is arguably the best example world-wide. Many of the uncertainties concerning the structure and evolution of specific margins addressed during the early part of the meeting stem from a limited appreciation of the fundamental processes contributing to continental break-up and sustaining early seafloor spreading. These processes, involving localized magmatic activity on an enormous scale, need to be fully understood and modelled before more detailed scenarios can be objectively appraised. The timing and the mechanisms by which volcanic passive margins progress from subaerial rift phase to the almost inevitable submarine seafloor-spreading phase are controlled by deep crustal behaviour and magma supply, but locally perturbed by anomalies such as the Iceland mantle plume. Several papers address these aspects, and this move away from local studies of short sections of margins is to be welcomed.

The problems of evolution of specific volcanic passive margins, however, falls within the domain of geophysical and geological sampling. The seaward-dipping reflector sequences forming the blanket to many ocean–continent transitions world-wide are particularly abundant on the conjugate flanks of the E Atlantic N of 55°N. The debate over the timing of the onset of true seafloor spreading at these margins persists, and despite a significant increase in the quality of data available, both in terms of seismic reflection and refraction profiles and deep sampling results, the schools of thought still agree to differ. Although it is possible that some dipping reflector sequences described from passive margins elsewhere are non-volcanic in origin (Grow *et al.* 1983), deep drilling during DSDP Leg 81 and ODP Leg 104 has confirmed that this cover sequence is basaltic in the NE Atlantic at least. In many respects, these basalts resemble those generated at mid-ocean ridges, apparently providing support for proponents of a subaerial seafloor-spreading origin for the dipping reflectors. However, detailed geochemical and, particularly, isotopic studies show that the sources of the extrusive sequences drilled in the NE Atlantic have not only an oceanic component but also undoubted continental and/or subcontinental lithospheric components, and that continental (in its broadest sense) contamination of dipping reflector basalts themselves is not uncommon and can continue to high levels in the sequence. As White (1987) has recently pointed out, in these volcanic passive margins there must be an area between unstretched continental crust and true oceanic crust consisting of stretched and intruded crust, probably with an increasing igneous component oceanwards. The consequent futility of attempting to define a meaningful 'ocean–continent transition' on volcanic passive margins is reflected in the increasing use of the term 'transitional crust'.

The contributions also provide a fascinating insight into the nature of the magmas that were available immediately prior to and at the inception of seafloor spreading in the NE Atlantic. Basaltic rocks show a range from N-type mid-ocean ridge basalt (MORB) (e.g. in the dipping reflector sequence SW of Rockall), through T-type MORB (e.g. on the Vøring margin) to E-type MORB (as in E Greenland). The role of the subcontinental lithosphere is also evident: a contribution from an enriched subcontinental lithospheric source is postulated for early E Greenland extrusives, whereas the SW Rockall Plateau dipping reflector basalts are inferred to have a significant contribution from a depleted subcontinental lithospheric reservoir. Similarly, magmas derived by anatectic melts of continental crustal material are also in evidence, the most interesting perhaps being the peraluminous extrusives documented from the NE Rockall Trough and the Vøring Plateau.

The debate over the timing of the onset of volcanism in the NE Atlantic and the age relationships between distinct areas has continued for many years, largely because of the poor biostratigraphic control inherent with these predominantly terrestrial sequences. However, it is becoming apparent that the increased resolution of radiometric age dating, using the ^{40}Ar-^{39}Ar method, together with more critical evaluation of conventional K-Ar data, used in combination with palaeomagnetic data, is a powerful approach, and has been used to great effect in the British Tertiary Province in particular. However, the volcanic history can perhaps be best resolved when the effects of terrestrial volcanism are seen in the marine sedimentary record, and in this regard the North Sea and adjacent offshore Mesozoic basins provide valuable data. Thus, the record of Hebridean-derived basaltic volcaniclastics in the early part of the post-Danian Palaeocene of the North Sea confirms that the major part of the basaltic activity in the British Province can be constrained to the magnetic anomaly chron C26R. Similarly, the discovery of an early Selandian tuff sourced from the Faeroe/E Greenland Province confirms that there was basaltic activity in the Faeroes (and possibly, by inference, in E Greenland) as early as C26R, confirming the assignment of the two magnetic normal polarity zones in the Faeroes Lower Series to anomalies 26 and 25. Thus, the currently accepted view that

Tertiary volcanism in E Greenland began during C24R (Soper *et al.* 1976) now appears to be regionally anomalous, and it is perhaps now the time to re-evaluate the somewhat scanty data set on which this inference was made.

The increased understanding of the stratigraphic development of volcanism in response to the opening of the NE Atlantic might also prove useful in resolving the long-standing debate on the age of the Palaeocene–Eocene boundary, estimated at between 53 Ma (Curry & Odin 1982) and 57.8 Ma (Berggren *et al.* 1985).The sequences on the NE Atlantic borderlands have been, and will continue to be, of key importance in constraining the age of the boundary, and the combination of radiometric and palaeomagnetic studies of the terrestrial sequences with biostratigraphic studies of the volcaniclastics in the sedimentary record will undoubtedly prove invaluable here. The evidence presented in this volume would favour an age of ca. 58 Ma for anomaly 26, considerably younger than the 60.2–60.8 Ma range preferred by Berggren *et al*, but closer to its estimated age on the Curry & Odin timescale.

Summary

In summary, the results of this meeting clearly demonstrate the need to maintain a high degree of deep sampling and deep seismic work, together with integrated trace-element and isotope geochemistry. These approaches have resulted in major advances in understanding of the evolution of the NE Atlantic volcanic passive margin in the last decade, and deserve continued support. More precision is also available with regard to the timing of magmatic activity at the margin, and this has been derived mainly from the interpolation of data from adjacent land studies. Despite this, an integrated model of NE Atlantic margin formation, combining timing and detailed assessment of component volcanic and tectonic events, is still awaited. Until this is achieved, in what can rightly be termed an ideal natural laboratory for the study of volcanic passive margins, the origin and formation of such margins elsewhere will remain unclear.

ACKNOWLEDGEMENTS: A large body of people has worked long and hard to produce this work. It would be a lengthy and impersonal list if all the reviewers, typists, proofreaders and those involved with technical support were named individually, but we would particularly like to thank Moyra Scott who enabled us to bring these works to publication with such alacrity and at such a high standard. Special thanks must also be made to the office staff of the Geological Society, particularly Sian Roblings, for shouldering much of the organization of the meeting.

References

BERGGREN W. A., KENT D. V. & FLYNN J. J. 1985. Jurassic to Paleogene, part 2. Paleogene geochronology and chronostratigraphy. *In:* SNELLING, N .J. (ed) *The Chronology of the Geological Record.* Memoir of the Geological Society of London, **10**, pp. 141–198.

CURRY D. & ODIN G. S. 1982. Dating of the Palaeogene. *In:* ODIN, G. S. (ed) *Numerical Dating in Stratigraphy, Part 1.* Wiley, London, pp. 607–630.

GROW J. A., HUTCHINSON D. R., KLITGORD K. D., DILLON W. P. & SCHLEE J. S. 1983. Representative multichannel seismic profiles over the US continental margin. *In:* BALLY A. W. (ed) *Seismic expression of structural styles.* American Association of Petroleum Geologists Studies in Geology, **15 (2)**.

NOE-NYGAARD A. 1974. Cenozoic to Recent volcanism in and around the North Atlantic Basin. *In:* NAIRN A. E. M. & STEHLI F. G. (eds) *The Ocean Basins and Margins (Vol. 2).* Plenum Press, New York, pp. 391–443.

SOPER N. J., DOWNIE C., HIGGINS A. C. & COSTA L. I. 1976. Biostratigraphical ages of Tertiary basalts on the East Greenland continental margin and their relationship to plate separation in the Northeast Atlantic. *Earth and Planetary Science Letters,* **32**, 149–157.

THOMPSON R. N. 1982. Magmatism of the British Tertiary Volcanic Province. *Scottish Journal of Geology,* **18**, 49–107.

WHITE R. S. 1987. When continents rift. *Nature,* **327**, 191.

L. M. PARSON, Institute of Oceanographic Sciences, Deacon Laboratory, Brook Road, Wormley, Godalming, Surrey GU8 5UB, UK.

A. C. MORTON, British Geological Survey, Keyworth, Nottingham NG12 5GG, UK.

Volcanic and Tectonic
Framework

A hot-spot model for early Tertiary volcanism in the N Atlantic

R. S. White

SUMMARY: The initial stages of rifting of the N Atlantic were accompanied by massive extrusion of volcanic rocks and the accretion to the lower crust of even greater volumes of igneous rock. Emplacement of the bulk of the igneous rocks along the 2000 km long continental rift occurred in as little as 2–3 My. The pattern and timing of magmatic activity can be explained by partial melting of anomalously hot asthenosphere as it welled up to fill the space left by the stretched and thinned continental lithosphere. Asthenosphere temperatures were elevated regionally by 100–150°C over a 2000 km diameter region by the mushroom-shaped head of hot material fed by a mantle plume, which at present is centred beneath Iceland.

The purpose of this review is to show that the extensive early Tertiary magmatic activity which occurred during rifting of the continental lithosphere to form the N Atlantic Ocean can be explained by the elevated asthenosphere temperatures in the region caused by abnormally hot material fed by a mantle hot-spot. This model explains the distribution, volume and rapid production of the huge quantities of new igneous rock added to the crust at the time of rifting. It also explains why the volcanic margins of the N Atlantic remained above sea-level when they rifted, whereas the continental margins away from the influence of any hot-spot, such as those in the Bay of Biscay, subsided gently and did not exhibit extensive volcanism when they were formed.

The paper is split into two parts. In the first part observations from the N Atlantic are outlined, to demonstrate: (1) the area over which early Tertiary volcanism occurs; (2) the volume of extrusive and intrusive igneous rocks added to the crust; (3) the timing of the magmatism; (4) the elevation of the rifted margins; and (5) the subsequent evolution of the igneous crust in the region. In the second part it is shown how the hot-spot model can explain these observations.

Throughout this paper the term 'N Atlantic' refers to that part of the northern N Atlantic lying between Greenland and the NW European continent, comprising the Greenland Sea, the Norwegian Sea, the Iceland Sea and the Irminger Basin.

Observations on N Atlantic magmatism

Areal distribution

Significant early Tertiary volcanism occurs in two main settings: (1) the classic Tertiary Igneous Province of NW Scotland, northern Ireland, the Faeroes and Greenland; and (2) underwater along the rifted continental margins on both sides of the N Atlantic Ocean.

The onshore occurrences have been known about and studied for a long time and, indeed, some, such as the Skaergaard intrusion, have become classic areas for the detailed study of magmatic processes. Offshore, many regions of sill intrusion into continental shelf sediments have been recognized over the past few years from regional seismic reflection surveys (Fig. 1).

Perhaps the major change in our appreciation of the distribution of offshore volcanism has been over the past five years, with the recognition that characteristic patterns of seaward-dipping reflectors imaged on seismic reflection profiles across rifted continental margins are caused by stacked layers of extrusive lava flows. The dipping reflectors were first postulated by Hinz (1981) to represent volcanic flows. Although initially some people thought that the dipping reflectors might be sedimentary deltaic deposits, they have been shown to be basaltic from direct sampling by drilling (Roberts et al. 1984a; ODP Leg 104 Scientific Party 1986; Parson et al. 1986). The high seismic velocities they exhibit (Mutter et al. 1984; White et al. 1987a, b; Whitmarsh & Miles 1987; Spence et al., in press), are also consistent with an igneous origin.

Once it was realized that the dipping reflectors represented volcanic flows, it became easy to map the extent of syn-rift volcanism on the continental margins of the N Atlantic using seismic reflection profiles. Seaward-dipping reflectors have been recognized almost continuously along the rifted margins on both sides of the N Atlantic (solid black areas on Fig. 1), although they cease abruptly at the unnamed fracture zone at 57°N, and are not found further to the S. Younger dipping reflectors, postdating the initial continental split, are also found along the flanks

From MORTON, A. C. & PARSON, L. M. (eds), 1988, *Early Tertiary Volcanism and the Opening of the NE Atlantic,* Geological Society Special Publication No. 39, pp. 3–13.

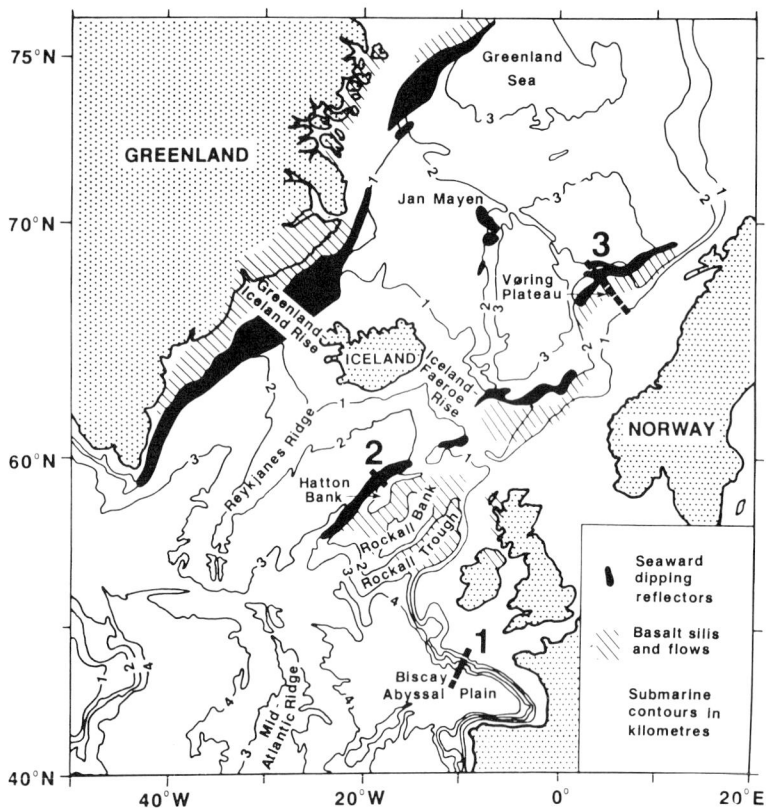

FIG. 1. Solid shading shows distribution of seaward-dipping reflectors caused by basalt flows around the rifted continental margins of the N Atlantic. Diagonal shading shows regions of Lower Tertiary basalt sills and dykes adjacent to the rifted margins. Seismic reflection profiles 1, 2 & 3 discussed in text and illustrated in Fig. 2. Distributions compiled from Nunns (1983), Smythe (1983), Larsen (1984), Roberts *et al.* (1984a), Uruski & Parson (1985), White *et al.* (1987a), Larsen & Jakobsdottir (1988), Skogseid & Eldholm (1987), Mutter *et al.* (in press).

of the Greenland–Iceland–Faeroes Volcanic Ridge.

Volume

The volume of igneous rock generated during rifting in the N Atlantic is huge. Typically, the pile of extrusive basalts which form the seaward-dipping reflectors is 3–6 km thick, although this varies considerably from zero at the landward feather edge to 8 km or more in the fully developed sections of oceanic crust (White *et al.*, 1987a; Larsen & Jakobsdottir 1988; Spence *et al.*, in press). The width of the zone of dipping reflectors is commonly 50–100 km, although it can be much more. The N Atlantic Rift between Greenland and NW Europe is about 2000 km long and dipping reflectors occur along the margins on both sides of the ocean basin (see Fig. 1), so the

total volume of extrusive basalts is in the region of $1–2 \times 10^6$ km^3. As Roberts *et al.* (1984a) point out, this is of about the same magnitude as that of the basalts in the Deccan Volcanic Province, to quote a well-known example.

The extrusive basalts on the rifted margins are, however, only the 'tip of the iceberg'. Beneath them lie huge volumes of igneous rock accreted to the lower part of the crust. Up to a 15 km thickness of new igneous material was added to the crust as the margins rifted. The thick lower crustal igneous section has been recognized on two high-quality deep seismic studies across continental margins in the N Atlantic, one across the Vøring Plateau (see Fig. 1, line 3), the other across the Hatton Bank margin (see Fig. 1, line 2). Both studies used two-ship multichannel expanding spread profiles to map the seismic velocity structure down to the Moho.

Across the Vøring Plateau transect (Mutter *et*

al., in press) the crust actually thickens beneath the rifted margin, with the Moho descending to a maximum depth of 24 km (Fig. 2). Seismic velocities of the igneous material in the lower crust are high, well in excess of 7 km/s. The continent–ocean boundary is considered by Mutter *et al.* to lie beneath the Vøring Plateau escarpment and to be almost vertical (see Fig. 2, dashed line on line 3), with the thick igneous section beneath the seaward-dipping reflectors produced by 'subaerial seafloor spreading' fed by a localized small-scale convection system under the rift. The author's interpretation of the structure of the margin (see Fig. 2), is of a wedge of pre-existing continental crust thinning seawards across the margin. The region beneath at least the landward portion of the dipping reflectors on this interpretation is probably a complex intermixture of blocks of old continental crust and newly intruded igneous crust. Whichever interpretation is correct, it is clear that there is a huge prism of new igneous material immediately beneath the margin.

Across the Hatton Bank continental margin (White *et al.* 1987a, b) the deep structure is similar to that beneath the Vøring Plateau, again with a prism of high-velocity lower crustal igneous material accreted beneath the dipping reflectors (see Fig. 2). The continental crust thins over a distance of perhaps 50–80 km, petering out beneath the dipping reflectors. The present Moho is a new feature formed at the time of rifting and lying beneath the thick prism of accreted igneous rocks in the lower crust. It is particularly noteworthy that nowhere else on the adjacent unstretched continental crust is high-velocity material, such as that beneath the rifted margin, found in the lower crust. Seismic velocity–depth determinations constrained by synthetic seismogram modelling from the Hebridean continental shelf (Hughes *et al.* 1984; Powell & Sinha 1987) and from the Irish Sea and the North Sea (Bott *et al.* 1985) never rise above 7.0 km/s in the lower crust. The rather less-well-constrained results from seismic refraction lines across adjacent continental crust, modelled using only travel-time information, also show lower crustal velocities of <7.0 km/s (Smith & Bott 1975; Bott *et al.* 1979; Barton & Wood 1984; Meissner 1986), with the sole exception of published LISPB results from northern Britain (Bamford *et al.* 1978); these assumed a velocity of 'about 7 km/s' for the lower crust, although without direct evidence of seismic arrivals from that layer.

In summary, the thick prism of high-velocity crust beneath the rifted margins is new material added during the rifting process. The volume of intrusive igneous rock is about five times that of the extrusive dipping reflectors, in total about 5–10×10^6 km^3 along the rifted margins of the N Atlantic. As we shall see below, this enormous volume was accreted in as little as 2–3 My.

By contrast, in that part of the Atlantic to the S of 57°N, where no dipping reflectors were developed, there was no significant addition of igneous material to the lower crust during rifting. A well-constrained cross-section across the Biscay continental margin (Ginzburg *et al.* 1985; Whitmarsh *et al.* 1986) shows that the crust thins gradually over a distance of 200–300 km (see Figs. 1 & 2, profile 1). The Moho lies at the base of the stretched continental crust.

Subsidence

A major difference between the volcanic continental margins of the N Atlantic and non-volcanic margins elsewhere is that at the time of rifting, the former remained above sea-level, whereas the latter subsided by 2 km or more. When lithosphere containing normal-thickness crust is stretched, as it is at continental rift zones, it responds by subsiding immediately to maintain isostatic equilibrium. This is followed over the next 100 My by exponentially decreasing subsidence as the perturbed lithosphere thermally re-equilibrates (McKenzie 1978). Such a pattern of subsidence, modified locally by the tilted fault blocks which accommodated the upper crustal extension, is indeed found on the non-volcanic Biscay continental margin.

The margins of the N Atlantic, by contrast, remained above sea-level and may even have been uplifted further during the main phase of continental rifting. This is documented by the extensive basaltic dipping reflector sequences which were erupted subaerially or in shallow marine conditions. Similar stacks of convex upwards-dipping basalt layers have been observed on land in Iceland. Pálmason (1980) has explained them by the loading effect of basalts fed from a retreating vent. By analogy, the seaward-dipping reflectors on the continental margins were probably formed by extrusion from vents to seaward of the continental crust (Mutter *et al.* 1982). The dipping reflectors feather-out landwards on stretched continental crust, so at the time of extrusion the feeder vents must have been higher than the surrounding areas across which the basalts flowed. Subsequently, there has been an inversion of topography to produce the present seaward-dipping configurations of the basalt flows. Since seaward-dipping reflectors continue oceanwards into oceanic crust, it follows that during the early stages of seafloor spreading,

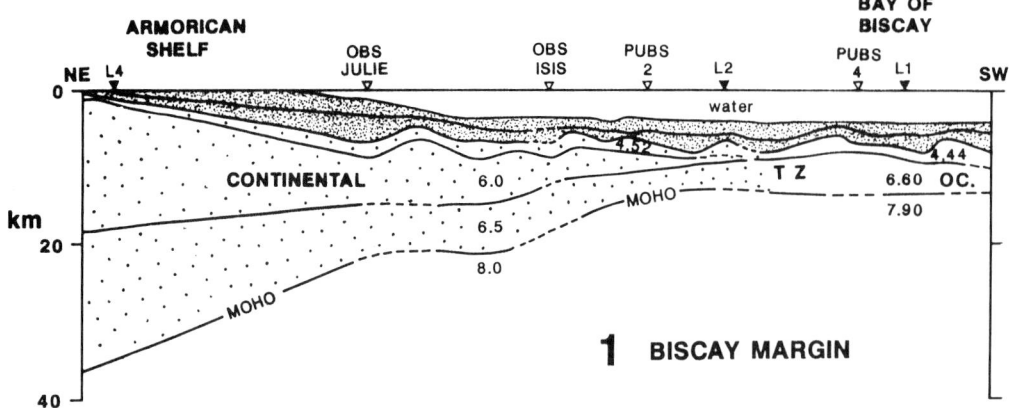

1 BISCAY MARGIN

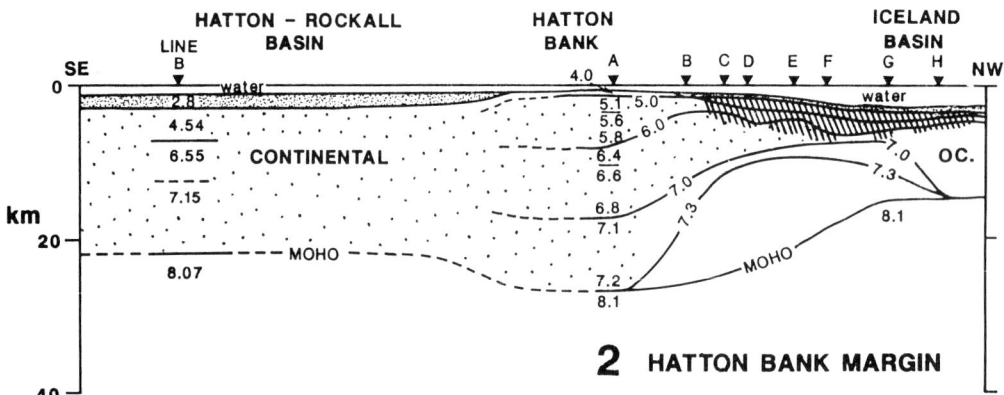

2 HATTON BANK MARGIN

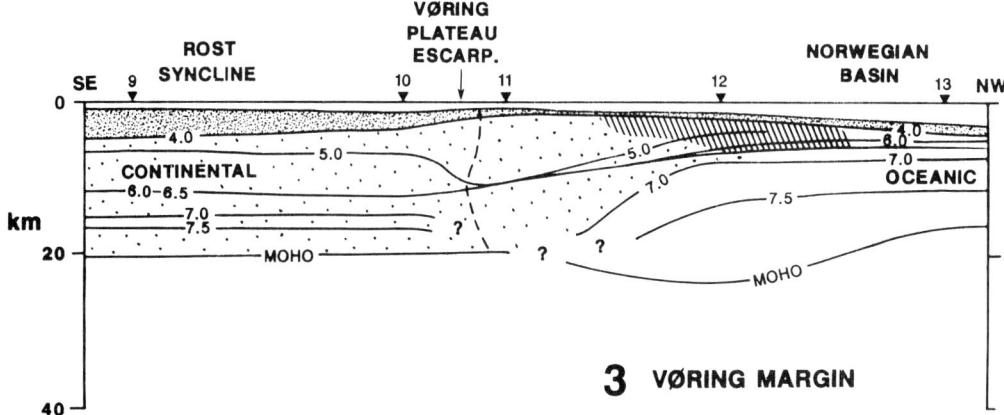

3 VØRING MARGIN

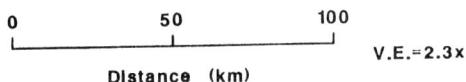

Distance (km)

V.E.=2.3x

as well as during the continental rift stage, the upper surface of the crust was near, or above, sea-level.

The dips of basaltic flows on conjugate rifted margins are in opposite directions (i.e. dipping towards the centre of the basin), showing that the feeder vents lay roughly in the centre of the evolving rift. Following the major outburst of igneous activity and elevation during rifting, the continental margins subsequently subsided at normal rates as they thermally equilibrated (Hyndman & Roberts, in press). At present the entire northern N Atlantic is about 1 km anomalously shallow compared with other ocean basins of similar age.

Timing

Perhaps the most remarkable feature of the early Tertiary igneous activity on the rifted continental margins and surrounding regions was its rapidity and intense concentration temporally.

The onset of separation between Greenland and NW Europe occurred throughout the N Atlantic between seafloor-spreading magnetic anomalies 25 and 24, over a period of only 2–3 My. The bulk of the extrusive basalts producing the seaward-dipping reflectors, and of the accreted lower crustal igneous section, was emplaced during this same interval during a period of rapid spreading. To the N of Iceland, Hinz *et al.* (1987) report that the margin split symmetrically, although there is clear evidence to the S of Iceland that the split was asymmetric during the early stages of rifting (White *et al.* 1987a).

The huge volume of igneous rock added to the crust during rifting, discussed above, was thus added along the entire margin of the N Atlantic in the remarkably short period of only 2–3 My. At any one location on the margin, the bulk of the igneous activity may have occurred over an even shorter interval.

In NW Britain and Ireland, radiometric dating of early Tertiary igneous rocks shows that most of the onshore igneous activity took place between 60 and 59 Ma, although the very earliest indications start at 64 Ma and magmatism continues sporadically until about 52 Ma (Mussett *et al.* 1988). In any one place igneous activity usually only lasted a few million years. The onset of basaltic volcanism in E Greenland and the Faeroes was probably a few million years later than in NW Britain (Noble *et al.* 1988; Tarling *et al.* 1988).

The precise temporal relationship between the onshore igneous activity and the continental-rift margin volcanism is somewhat uncertain, because the former relies largely on radiometric and biostratigraphic dates whilst the latter is constrained by seafloor-spreading magnetic anomalies, and there is disagreement amongst different workers of 2–4 My in the calibration of the magnetic reversal timescale during this period (e.g. Curry & Odin 1982; Harland *et al.* 1982; Berggren *et al.* 1985). Nevertheless, whichever timescale is adopted it is clear that there was some onshore volcanic activity in the region which preceded the onset of rapid spreading by 1–4 My (Gibson *et al.*, in press; Mussett *et al.* 1988; Knox & Morton 1988; Tarling *et al.* 1988). Following the onset of spreading, an enormous outburst of igneous activity occurred along the rifting margin within as little as 2–3 My. In volume the rift-margin magmatism is of far greater significance than the onshore igneous activity of the Tertiary Igneous Province.

Subsequent evolution of the N Atlantic

The most striking features of the evolution of the N Atlantic following the initial rift are the anomalously shallow bathymetry throughout the region and the anomalously thick crust of the Greenland–Iceland–Faeroes Ridge. The seafloor-spreading history of the region S of Iceland after the early stages of separation was straightforward, with continuous, symmetric spreading from the Reykjanes Ridge spreading centre continuing to the present. N of Iceland the

FIG. 2. Interpreted cross-sections showing crustal structure on the same scale across the non-volcanic Biscay continental margin (Line 1 on Fig. 1, redrawn from Ginzburg *et al.* 1985); the volcanic Hatton continental margin (Line 2 on Fig. 1, from White *et al.* 1987a and Scrutton 1972); and the volcanic Vøring Plateau (Line 3 on Fig. 1, redrawn and re-interpreted from Mutter *et al.*, in press). The Hatton and Vøring margins were both influenced by the hot-spot when they rifted, producing thick igneous crust, whereas the Biscay margin was away from the influence of any hot-spot and rifted with little igneous activity. Dense stipple shows syn- and post-rift sediments, open stipple shows inferred extent of stretched continental crust and diagonal shading shows extent of seaward-dipping reflectors. The dashed line beneath the Vøring Plateau escarpment on Line 3 shows Mutter *et al.*'s (in press) interpretation of the position of the boundary between continental and oceanic crust. Labelled triangles along the top of each profile indicate the location of velocity control points.

evolution was more complex, with a spreading centre jump towards the W (Nunns 1983).

The structure of the crust generated at the Reykjanes Ridge spreading centre is similar to that of normal oceanic crust (White 1984), although at 10 km thick (Bunch & Kennett 1980) it is rather thicker than normal. Beneath the Iceland–Faeroe Ridge the crust is up to 35 km thick (Bott & Gunnarsson 1980), and under Iceland it reaches 15 km or more in thickness. Bathymetrically, the whole N Atlantic is unusually shallow (see Fig. 1), reaching a depth anomaly of about 2.5 km above normal in the vicinity of Iceland.

Hot-spot model

In the second half of this paper it is shown how raised asthenosphere temperatures across a 2000 km diameter region caused by convection of hot material in a narrow, central, mantle plume can satisfactorily explain all the observations on volcanism in the N Atlantic discussed above.

Before discussing the model in detail, it is first necessary to remove a common misconception about hot-spots. This misconception is that a hot-spot only affects the relatively narrow region immediately above the rising plume. Thus a common argument against a hot-spot being responsible for the magmatism along the rifted margins of the N Atlantic is that it is implausible that a hot-spot should have dimensions of 2000 km (the length of the rift) by 100 km (the width of the rift volcanism) at the time of rifting. Indeed this is implausible, but it is also a quite unnecessary concern because, as is shown below, the hot material brought up by a mantle plume spreads laterally over an area of up to 2000 km diameter in the upper mantle.

Intraplate swells exhibiting significant bathymetric and geoid anomalies are common in the ocean basins. Recent detailed heat-flow measurements across the Hawaiian, Bermudan and Cape Verde swells have shown that they have a thermal origin. The model developed by Courtney & White (1986) for the mantle plume beneath the Cape Verde Rise is a good analogue for probable asthenosphere conditions beneath NW Europe just prior to the splitting which created the N Atlantic, so we will examine it in some detail. The Cape Verde model is useful to study because, firstly, the overriding Atlantic plate is moving only very slowly with respect to the hot-spot, so Courtney & White were able to use the surface heat flow, bathymetry and geoid measurements

to constrain the deep mantle flow using an axisymmetric convection model without the (at present computationally excessive) complications of 3-D convection modelling. Secondly, the Cape Verde hot-spot lies under old, thick lithosphere and is not generating large quantities of volcanic rocks; this is similar to the situation under the NW European–Greenland plate prior to rifting.

The temperature structure from the convection model for the Cape Verde Rise which best fits the observed bathymetric uplift of nearly 2 km at the centre, the geoid anomaly and the increased heat flow across the Rise is illustrated in Fig. 3. Volcanism occurs only above a narrow (150 km diameter) central plume, which is much hotter than the normal asthenosphere. The hot material from that plume is deflected laterally by the overlying plate, creating a circular region some 1500 km in diameter of anomalously hot asthenosphere beneath the plate. Almost all the bathymetric uplift is dynamically supported by this flow.

In the present N Atlantic the rising plume lies beneath Iceland, where it is generating copious quantities of volcanic rocks. The region of elevated bathymetry and geoid anomaly is somewhat larger than the Cape Verde Rise, extending over 2000 km diameter, with slightly greater maximum uplift of 2.5 km at the centre. Otherwise, the N Atlantic hot-spot is not dissimilar to the Cape Verde model. It is worth noting that the best resolution on the thermal structure of the model is in the region near the base of the plate (Courtney & White 1986; Parsons & Daly 1983). The observed bathymetry, geoid and heat flow give little information on the depth of the convection, although since the magmatism on the continental margins is produced by partial melting of the upper mantle, the depth of convection makes no difference to the predictions of the model.

In the remainder of the paper it will be assumed that a mushroom-shaped hot-spot similar to that illustrated in Fig. 3 developed beneath the NW European–Greenland plate prior to rifting, and I will investigate the consequences of splitting the N Atlantic above this region of anomalously hot asthenosphere.

When continental lithosphere is stretched, it responds by thinning. Asthenospheric material then rises to fill the space left by the thinning lithosphere. As the asthenosphere wells up and decompresses it generates partial melt. The volume of partial melt can be calculated, if the upwelling is assumed to be rapid, by requiring the entropy of the upwelling material to be constant and ignoring the movement between the

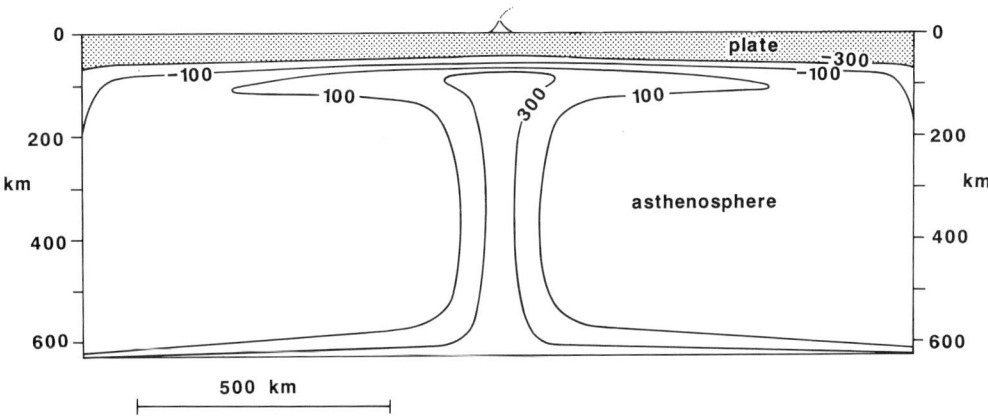

FIG. 3. Cross-section of the temperature structure through the mantle plume beneath the Cape Verde Rise based on the axisymmetric convection model of Courtney & White (1986) and constrained by geoid, bathymetry and heat-flow measurements. Temperatures are labelled in degrees Celsius with respect to the mean asthenosphere value. Note the narrow rising mantle plume at the centre and the broad, mushroom-shaped head of hot asthenosphere material deflected laterally by the overlying plate. Plume volcanism occurs only above the rising central plume.

melt and the residue. The hotter the asthenosphere, the more partial melt is generated. The curves for melt generation as a function of amount of stretching shown in Fig. 4 were calculated from an empirical relation between melt fraction, pressure and temperature, which accurately fits the experimental observations (McKenzie & Bickle, in press) and are abstracted from White *et al.* (1987a).

The amount of melt generated is a very sensitive measure of the asthenosphere potential temperature. Globally, the asthenosphere temperature beneath the oceans must vary little, because the thickness of igneous crust produced at oceanic spreading centres is consistently 6–8 km. From this we infer that the mean asthenosphere potential temperature is about 1280°C, with an uncertainty of approximately 20°C.

Areal distribution

When continental lithosphere lying above normal temperature asthenosphere is stretched, very little partial melt is produced by the upwelling asthenosphere. Even when the lithosphere is thinned to one-fifth of its original thickness ($\beta = 5$), then only 2 km of melt is produced (see Fig. 4). The lithosphere breaks to form new oceanic crust when the stretching continues above values for β of about five. However, when the asthenosphere temperature is only 100°C higher, almost 10 km

of melt are produced for the same amount of stretching (see Fig. 4).

The mushroom-shaped hot-spot (see Fig. 3) provides a ready explanation for the occurrence

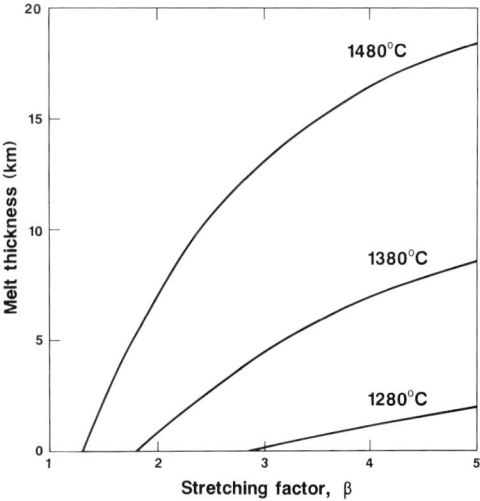

FIG. 4. Thickness of partial melt generated by decompression of upwelling asthenosphere for varying amounts of lithosphere stretching (assuming uniform stretching) from diagrams in White *et al.* (1987a), based on calculations by McKenzie & Bickle (in press). Curves are shown for three different representative asthenosphere potential temperatures of 1280°C (which is close to the normal value away from hot-spots), 1380°C and 1480°C, assuming an initial lithosphere thickness of 100 km.

of volcanics along the 2000 km long N Atlantic Rift. Where the rift ran across the 2000 km diameter of elevated temperature asthenosphere around the hot-spot plume, voluminous partial melt was produced by the upwelling hot asthenosphere. Away from the influence of any hot-spot, such as on the Biscay margin, asthenosphere temperatures were normal and no significant volcanism accompanied continental rifting. Rift margin volcanism would have occurred wherever the rift had crossed the region of hot asthenosphere fed by the mantle plume.

Volume

The thickness of igneous crust emplaced on the continental margin during rifting allows us to estimate the asthenosphere potential temperature using the curves in Fig. 4. Thus, the observed thickness of 10–15 km of igneous crust beneath the margins under the N Atlantic Rift requires 100–150°C elevation of the asthenosphere temperature at the time of rifting. This is in good agreement with the magnitude of the asthenosphere temperature increases predicted by hot-spot convection models (see Fig. 3). Where the rifting margin lay over normal temperature asthenosphere, such as on the Biscay margin, little extra partial melt was formed until the stretching factor exceeded three, when normal thickness oceanic crust was generated (Foucher *et al.* 1982).

Very great quantities of melt can be produced above the mantle plume at the centre of the hot-spot anomaly. Partial melting occurs continuously in the central rising mantle plume. Where the rift passed near the upwelling mantle plume in the vicinity of the Faeroes, huge volumes of igneous crust were generated. As the ocean continued to spread, the central mantle plume produced 25–35 km thick igneous crust beneath the Greenland–Faeroes Ridge.

Subsidence

There are two distinct reasons why the rifted margins under the influence of the hot-spot should be elevated above sea-level. The first is that the mantle hot-spot produces dynamic uplift across the entire region (Courtney & White 1986), so the whole area was probably uplifted prior to rifting.

The second reason is that as new igneous material is added to the crust, the elevation of the crust changes in order to maintain isostatic equilibrium. Stretched continental crust has very little flexural rigidity (Barton & Wood 1984; Fowler & McKenzie, in press) so the elevation

change for any given amount of stretching and asthenosphere temperature can be calculated easily assuming that local isostatic equilibrium is maintained, that lithosphere stretching is translated into thinning by pure shear, and using the volume of melt generated shown in Fig. 4. For normal temperature asthenosphere, as beneath the Biscay margin, stretching the lithosphere by a factor of five produces over 2 km of subsidence immediately. However, if the asthenosphere temperature is 100°C above normal, for the same amount of stretching the elevation caused by the addition of new material to the crust counterbalances much of the subsidence caused by lithosphere thinning and the net result is that the surface of the crust subsides by only 1.5 km (White *et al.* 1987a). If the asthenosphere temperature is 200°C above normal then the crust subsides by only 0.5 km.

The basaltic dipping-reflector sequences are known to have been extruded near or above sea-level at the time of rifting. Uplift caused by dynamic support from the hot-spot, together with the accretion of new igneous rock to the crust as the margin rifted, can explain this observation. The asthenosphere temperatures we infer are the same as those required to explain the volume of igneous rock generated.

Timing

Partial melting occurs as the asthenosphere wells up under the rifting margin. The partial melt segregates and rises rapidly upwards to the crust as soon as it is formed (McKenzie 1984). This explains the observed rapidity of the outburst of igneous activity on the rifting margins. Since it is the very process of lithosphere rifting that allows the asthenosphere to well up and therefore to partially melt, the igneous activity on the margins occurs at the same time as the lithosphere rifting.

The onset of igneous activity over the broad region of the Tertiary Igneous Province shortly before the start of separation between Greenland and NW Europe shows that the hot-spot was present before the Atlantic started to split in its present location. The trends of the widely spread early igneous activity derived, for example, from dyke swarms and from the inferred positions of explosive volcanic centres responsible for numerous ash layers deposited in the North Sea, are rather different to the NE–SW direction of the rift that developed shortly afterwards into the N Atlantic, suggesting that the stress field may have re-orientated prior to separation (Knox & Morton 1988). However, the occurrence of igneous activity throughout the Tertiary Igneous Province, wherever there was local tension, shows that the entire region was underlain by asthenosphere with abnormally elevated temperature. Wherever tension occurred, volcanism resulted. It is quite likely, indeed, that the lithosphere tension over a broad region, which allowed the volcanism to occur, was itself caused by uplift following initiation of the hot-spot mantle plume.

Subsequent evolution of the N Atlantic

The central mantle plume, approximately 150 km across, at present lies beneath Iceland where igneous crust of 15 km or more thickness is now being generated. As the N Atlantic opened, a trace of very thick igneous crust was left by the mantle plume, creating the 25–35 km thick Greenland–Faeroe Ridge (Vink 1984). The convecting hot-spot is also responsible for the abnormally shallow bathymetry throughout the N Atlantic. Most of the uplift is dynamically supported by the asthenosphere convection (Anderson *et al.* 1973; Courtney & White 1986).

The raised asthenosphere temperatures under the N Atlantic cause the oceanic crust now being generated to be a little thicker than on normal spreading centres away from hot-spots. However, the oceanic crust is not as thick as the accreted igneous section on the rifted margins. On a simple model with unchanged asthenosphere conditions we might have expected a rather thicker igneous crust than we observe to continue to be generated at the oceanic spreading centre. There are two factors which tend to decrease the thickness of igneous crust generated once a fully developed ocean has started to form.

Firstly, the process of stretching the continental lithosphere during the early rift stage causes a pressure perturbation which tends to focus the rising melt towards the margin (Spiegelmann & McKenzie, pers. comm.). So this focusing thickens the igneous crust under the margin.

Secondly, once the lithosphere has been fractured and large quantities of igneous rock start to be extracted from the central plume, as at present occurs below Iceland, the temperature structure of the residual mantle is modified. The upwards migration of large volumes of partial melt from the central plume advects a considerable amount of heat out of the asthenosphere. The asthenosphere left to flow laterally to feed the oceanic spreading centres will thus have a somewhat reduced temperature. Furthermore, the least refractory portion will be stripped off and extracted from the plume, leaving slightly depleted material to flow laterally beneath the ocean basin. So when the asthenosphere is finally tapped under the spreading centres it may have reduced

temperatures and be somewhat depleted compared to the material which lay beneath the continental lithosphere before rifting started and before plume volcanism commenced. Both these factors would tend to decrease the volume of partial melt generated under the present spreading centre, compared to that formed in the initial split, although maintaining a thicker than normal oceanic crust.

Conclusions

The distribution, volume and timing of early Tertiary volcanism during opening of the N Atlantic, and the elevation of the margins during rifting, can be explained by the presence of a convective mantle hot-spot in the centre of the region. Calculations of the volume of partial melt generated in the asthenosphere upwelling beneath the rifting continental lithosphere suggest that asthenosphere temperatures were 100–150°C above normal. Asthenosphere temperatures were increased regionally prior to rifting over a 2000 km diameter region by convection of hot material in a narrow central plume and its lateral deflection by the overlying plate.

The N Atlantic, with its currently active central plume and well-preserved conjugate rifted margins, which are not generally obscured by large thicknesses of post-rift sediments, is a superb natural laboratory in which to study igneous processes connected with the opening of a new ocean basin. A survey of rifted continental margins around the world shows that a similar model of proximity to an active hot-spot at the time of rifting can explain the occurrence of all the volcanic margins seen elsewhere (White & McKenzie, in press), although because of its excellent exposure the N Atlantic remains a classic example.

ACKNOWLEDGEMENTS: I am indebted to Dan McKenzie for his calculations of the effect of asthenosphere temperature on the amount of partial melt generated beneath rift zones. Dave Smythe and the organizers of the conference encouraged me to make this review of volcanism in the N Atlantic. I thank Sue Fowler, Hans Christian Larsen, Dave Smythe, George Spence and Graham Westbrook for helpful discussions. Thanks are due also to all those who sent me preprints of their work, and in particular to John Mutter for permission to use results from the Vøring Plateau in the compilation shown in Fig. 2. I am grateful to Barbara Dyson for typing the manuscript and to Phyl Fisher for drafting the figures. Department of Earth Sciences, Cambridge, contribution number ES1077.

References

ANDERSON, R. N., MCKENZIE, D. P. & SCLATER, J. G.
1973. Gravity, bathymetry and convection in the
Earth. *Earth and Planetary Science Letters*, **18**,
391–407.
BAMFORD, D., NUNN, K., PRODEHL, C. & JACOB, B.
1978. LISPB–IV. Crustal structure of Northern
Britain. *Geophysical Journal of the Royal Astronomical Society*, **54**, 43–60.
BARTON, P. & WOOD, R. 1984. Tectonic evolution of
the North Sea Basin: crustal stretching and
subsidence. *Geophysical Journal of the Royal Astro-
nomical Society*, **79**, 987–1022.
BERGGREN, W. A., KENT, D. V. & FLYNN, J. 1985.
Jurassic to Palaeogene; Part 2: Palaeogene geo-
chronology and chronostratigraphy. *In:* SNELLING,
N. (ed) *Geochronology and the Geological Record*.
Memoir of the Geological Society of London, **10**,
pp. 141–186.
BOTT, M. H. P. & GUNNARSSON, K. 1980. Crustal
structure of the Iceland–Faeroe Ridge. *Journal of
Geophysics*, **47**, 221–227.
——, ARMOUR, A. R., HIMSWORTH, E. M., MURPHY,
T. & WYLIE, G. 1979. An explosion seismology
investigation of the continental margin west of the
Hebrides, Scotland, at 58°N. *Tectonophysics*, **59**,
217–231.
——, LONG, R. E., GREEN, A. S. P., LEWIS, A. A. J.,
SINHA, M. C. & STEVENSON, D. L. 1985. Crustal
structure south of the Iapetus Suture beneath
Northern England. *Nature*, **314**, 724–727.
BUNCH, A. W. H. & KENNETT, B. L. N. 1980. The
crustal structure of the Reykjanes Ridge at 59°
30'N. *Geophysical Journal of the Royal Astronomical
Society*, **61**, 141–166.
COURTNEY, R. C. & WHITE, R. S. 1986. Anomalous
heat flow and geoid across the Cape Verde Rise:
evidence for dynamic support from a thermal
plume in the mantle. *Geophysical Journal of the
Royal Astronomical Society*, **87**, 815–867 (microfiche GJ87/1).
CURRY, D. & ODIN, G. S. 1982. Dating of the
Palaeogene. *In:* ODIN, G. S. (ed) *Numerical Dating
in Stratigraphy*. John Wiley & Sons, Chichester,
pp. 607–629.
FOUCHER, J-P., LE PICHON, X. & SIBUET, J-C. 1982. The
ocean–continent transition in the uniform litho-
spheric stretching model: role of partial melting in
the mantle. *Philosophical Transactions of the Royal
Society of London*, Series A, 27–43.
FOWLER, S. R. & MCKENZIE, D. P. (in press). Flexural
studies of the Exmouth and Rockall Plateaux using
Seasat altimetry. *Marine Geophysical Researches*.
GIBSON, D., MCCORMICK, A. G., MEIGHAN, I. G. &
HALLIDAY, A. N. (in press). The British Tertiary
Igneous Province: Young Rb–Sr ages for the
Mourne Mountains granites. *Scottish Journal of
Geology*.
GINZBURG, A., WHITMARSH, R. B., ROBERTS, D. G.,
MONTADERT, L., CAMUS, A. & AVEDIK, F. 1985.
The deep seismic structure of the northern conti-
nental margin of the Bay of Biscay. *Annales
Geophysicae*, **3**, 499–510.

HARLAND, W. B., COX, A. V., LLEWELLYN, P. G.,
PICKTON, C. A. G., SMITH, A. G. & WALTERS, R.
1982. *A Geologic Time-scale*. Cambridge University
Press, 131 pp.
HINZ, K. 1981. A hypothesis on terrestrial catastrophes:
wedges of very thick oceanward dipping layers
beneath passive continental margins. Their origin
and palaeoenvironmental significance. *Geologische
Jahrbuch*, **E22**, 3–28.
——, MUTTER, J. C., ZEHNDER, C. M. & the NGT
STUDY GROUP (1987). Symmetric conjugation of
continent–ocean boundary structures along the
Norwegian and East Greenland margins. *Marine
& Petroleum Geology*, **4**, 166–187.
HUGHES, V. J., WHITE, R. S. & JONES, E. J. W. 1984.
Additional constraints imposed on the velocity
structure of the northwest Scottish continental
margin by amplitude studies. *Annales Geophysicae*,
2, 669–678.
HYNDMAN, R. S. & ROBERTS, D. G. (in press). DSDP
heat flow and models of the rifted west Rockall
margin. *Journal of Geophysical Research*.
KNOX, R. W. O'B. & MORTON, A. C. 1988. The record
of early Tertiary N Atlantic volcanism in sediments
of the North Sea Basin. *In:* MORTON, A. C. &
PARSON, L. M. (eds) *Early Tertiary Volcanism and
the Opening of the NE Atlantic*. Geological Society
of London, Special Publication, **39**, pp. 407–419.
LARSEN, H. C. 1984. Geology of the East Greenland
Shelf. *In:* SPENCER A. M. *et al.* (eds) *Petroleum
Geology of the North European Margin*. Graham &
Trotman, London, pp. 329–339.
—— & JAKOBSDOTTIR, S. 1988. Distribution, crustal
properties and significance of seaward-dipping
sub-basement reflectors off E Greenland. *In:*
MORTON, A. C. & PARSON, L. M. (eds) *Early
Tertiary Volcanism and the Opening of the NE
Atlantic*. Geological Society of London, Special
Publication, **39**, pp. 95–114.
MCKENZIE, D. P. 1978. Some remarks on the develop-
ment of sedimentary basins. *Earth and Planetary
Science Letters*, **40**, 25–32.
—— 1984. The generation and compaction of partially
molten rock. *Journal of Petrology*, **25**, 713–765.
—— & BICKLE, M. J. (in press). The volume and
composition of melt generated by extension of the
lithosphere. *Journal of Petrology*.
MEISSNER, R. 1986. *The continental crust: a geophysical
approach*. Academic Press, London. 426 pp.
MUSSETT, A. E., DAGLEY, P. & SKELHORN, R. R. Time
and duration of activity in the British Tertiary
Igneous Province. *In:* MORTON, A. C. & PARSON,
L. M. (eds) *Early Tertiary Volcanism and the
Opening of the NE Atlantic*. Geological Society of
London, Special Publication, **39**, pp. 337–348.
MUTTER, J. C., BUCK, W. R. & ZEHNDER, C. M. (in
press). Convective partial melting I: a model for
the formation of thick basaltic sequences during
the initiation of spreading. *Journal of Geophysical
Research*.
——, TALWANI, M. & STOFFA, P. L. 1982. Origin of
seaward dipping reflectors in oceanic crust off the

Norwegian margin by "subaerial sea-floor spreading". *Geology*, **10**, 353–357.

——, —— & —— 1984. Evidence for a thick oceanic crust adjacent to the Norwegian margin. *Journal of Geophysical Research*, **89**, 483–502.

NOBLE, R. M., BROWN, P. E. & MACINTYRE, R. M. Age constraints on Atlantic evolution-timing of magmatic activity along the E Greenland continental margin. *In:* MORTON, A. C. & PARSON, L. M. (eds) *Early Tertiary Volcanism and the Opening of the NE Atlantic.* Geological Society of London, Special Publication, **39**, pp. 201–214.

NUNNS, A. G. 1983. Plate tectonic evolution of the Greenland–Scotland Rise. *In:* BOTT, M. H. P., SAXOV, S., TALWANI, M. & THIEDE, J. (eds) *Structure and Development of the Greenland–Scotland Ridge.* Plenum Press, New York, pp. 11–30.

ODP LEG 104 SCIENTIFIC PARTY, 1986. Reflector identified, glacial onset seen. *Geotimes*, March 1986, pp. 12–14.

PÁLMASON, G. 1980. A continuum model of crustal generation in Iceland: kinematic aspects. *Journal of Geophysics*, **47**, 7–18.

PARSON, L. M., MASSON, D. G., MILES, P. R. & PELTON, C. D. 1986. Structure and evolution of the Rockall and East Greenland continental margins. Report of work undertaken by IOS in the period up to April 1985. *Institute of Oceanographic Sciences*, Report No. **223**, 71 pp.

PARSONS, B. & DALY, S. 1983. The relationship between surface topography, gravity anomalies and temperature structure of convection. *Journal of Geophysical Research*, **83**, 1129–1144.

POWELL, C. M. R. & SINHA, M. C. 1987. The PUMA experiment west of Lewis, U.K. *Geophysical Journal of the Royal Astronomical Society*, **89**, 259–264.

ROBERTS, D. G., BACKMAN, J., MORTON, A. C., MURRAY, J. W. & KEENE, J. B. 1984. Evolution of volcanic rifted margins: synthesis of Leg 81 results on the west margin of Rockall Plateau. *In:* ROBERTS, D. G. SCHNITKER, D. *et al. Initial Reports of the Deep Sea Drilling Project.* US Government Printing Office, Washington, **81**, pp. 883–911.

SCRUTTON, R. A. 1972. The crustal structure of Rockall Plateau microcontinent. *Geophysical Journal of the Royal Astronomical Society*, **27**, 259–275.

SKOGSEID, J. & ELDHOLM, O. (1987). Early Cenozoic crust at the Norwegian continental margin and the conjugate Jan Mayen Ridge. *Journal of Geophysical Research*, **92**, 1147–1491.

SMITH, P. J. & BOTT, M. H. P. 1975. Structure of the crust beneath the Caledonian Foreland and Caledonian Belt of the north Scottish shelf region. *Geophysical Journal of the Royal Astronomical Society*, **40**, 187–205.

SMYTHE, D. K. 1983. Faeroe–Shetland Escarpment and continental margin north of the Faeroes. *In:* BOTT, M. H. P., SAXOV, S., TALWANI, M. & THIEDE, J. (eds) *Structure and Development of the Greenland–Scottish Ridge.* Plenum Press, New York, pp. 109–119.

SPENCE, G. D., WHITE, R. S., WESTBROOK, G. K. & FOWLER, S. R. (in press). The Hatton Bank continental margin: I. Shallow structure from two-ship expanding spread profiles. *Geophysical Journal of the Royal Astronomical Society.*

TARLING, D. H., HAILWOOD, E. A. & LØVLIE, R. 1988. A palaeomagnetic study of lower Tertiary lavas in E Greenland and comparison with other lower Tertiary observations in the northern Atlantic. *In:* MORTON, A. C. & PARSON, L. M. (eds) *Early Tertiary Volcanism and the Opening of the NE Atlantic.* Geological Society of London, Special Publication, **39**, pp. 215–224.

URUSKI, C. I. & PARSON, L. M. 1985. A compilation of geophysical data on the East Greenland continental margin and its use in gravity modelling across the continent–ocean transition. *Institute of Oceanographic Sciences*, Report No. **214**, 53 pp.

VINK, G. E. 1984. A hotspot model for Iceland and the Vøring Plateau. *Journal of Geophysical Research*, **87**, 10677–10688.

WHITE, R. S. 1984. Atlantic Oceanic Crust: seismic structure of a slow spreading ridge. *In:* GASS, I. G., LIPPARD, S. J. & SHELTON, A. W. (eds) *Ophiolites and Oceanic Lithosphere.* Geological Society of London, Special Publication, **13**, pp. 34–44.

——, SPENCE, G. D., FOWLER, S. R., McKENZIE, D. P., WESTBROOK, G. K. & BOWEN, A. N. 1987a. Magmatism at rifted continental margins. *Nature*, **330**, 439–444.

——, WESTBROOK, G. K., BOWEN, A. N., FOWLER, S. R., SPENCE, G. D., PRESCOTT, C., BARTON, P. J., JOPPEN, M., MORGAN, J. & BOTT, M. H. P. 1987b. Hatton Bank (northwest U.K.) continental margin structure. *Geophysical Journal of the Royal Astronomical Society*, **89**, 265–272.

—— & McKENZIE, D. P. (in press). Magmatism of rift zones: the generation of volcanic continental margins and flood basalts. *Journal of Geophysical Research.*

——, AVEDIK, F. & SAUNDERS, M. R. 1986. The seismic structure of thinned continental crust in the northern Bay of Biscay. *Geophysical Journal of the Royal Astronomical Society*, **86**, 589–602.

WHITMARSH, R. B. & MILES, P. R. 1987. Seismic structure of a seaward dipping reflector sequence southwest of Rockall Plateau. *Geophysical Journal of the Royal Astronomical Society*, **90**, 731–740.

R. S. WHITE, Bullard Laboratories, Department of Earth Sciences, University of Cambridge, Madingley Rise, Madingley Road, Cambridge CB3 0EZ, UK.

A new look at the causes and consequences of the Icelandic hot-spot

M. H. P. Bott

SUMMARY: New support for the hypothesis of a lower mantle plume rising near Iceland comes from recent developments in the theory of whole-mantle convection and is inferred from the large heat input required to support the major topographic swell which affects much of the N Atlantic and Greenland. The associated geoid and gravity highs are interpreted as due partly to the swell and its deep compensation and partly to the pressure anomaly caused by the plume. Rapid lateral migration of partially fused material from the plume in the asthenosphere prior to continental break-up caused the early Tertiary volcanism, with subsequent igneous activity concentrated at the ridge crest to form the Icelandic transverse ridge. Progressive expansion of the upper mantle thermal anomaly in response to pressure, in association with the complexities of plate evolution in this region, may account for the early uplift of S Greenland and later uplift of the Blosseville Kyst region. The low density upper mantle would also be expected to give rise to an anomalously high ridge-push force.

This paper discusses the causes and consequences of hypothetical mantle processes which gave rise to the early Tertiary continental volcanism, the opening of the north-eastern N Atlantic and the subsequent anomalous evolution of the region. The most coherent explanation stems from the suggestion by Wilson (1963) that such regions overlie mantle hot-spots. This developed into the mantle plume hypothesis of Morgan (1971). The plume hypothesis has been extensively applied to Iceland and the anomalous north-eastern N Atlantic by Vogt (e.g. 1971, 1974, 1976, 1983) and others (e.g. Schilling 1973). It assumes that a hot, narrow plume rises through the lower mantle beneath the region, possibly originating at the core–mantle boundary. The plume discharges hot, partially fused material into the asthenosphere, at several hundred degrees above ambient temperature. The earliest visible activity of the hypothetical plume (or plumes) was to cause the early Tertiary continental volcanism. Shortly afterwards, the new continental split was initiated between Greenland and Europe, and subsequently the volcanic activity concentrated at the ridge crest to produce the Icelandic transverse ridge. Most of the hot material from the Iceland plume and from a subsidiary plume beneath the Azores, possibly after removal of the main liquid fraction, continued to spread laterally in the asthenosphere. The present broad topographical swell of the N Atlantic ocean floor has been essentially produced in isostatic response to the hot, low density upper mantle beneath.

The lower mantle plume hypothesis has been criticized on two main grounds. First, Runcorn (1974) found mantle plumes difficult to reconcile with the principles of mechanics, preferring a broad cellular pattern of whole mantle convec-

tion. Second, geochemical evidence (e.g. O'Nions *et al.* 1979) indicates that the main source of igneous rocks has become depleted with respect to the whole-earth model. This can be taken to support the concept of convection in a depleted upper mantle with little interchange with the undepleted lower mantle. This has thrown further doubt on the concept of deep mantle plumes, suggesting that the igneous rocks of the Icelandic transverse ridge and the continental precursors come from the upper mantle. A further difficulty is the very wide lateral extent of the early Tertiary continental volcanism which is at first sight difficult to reconcile with an origin from an isolated narrow plume.

Recent work has provided some new insights which circumvent these difficulties. Theoretical modelling has validated the concept of plumes rising from a thermal boundary layer at the base of the mantle. Seismological observations show that lithosphere of the north-western Pacific subducts into the lower mantle. In this paper these and other new insights are briefly reviewed and some further speculations about the role of the hypothetical plume in the development of the N Atlantic region are presented.

Mantle convection and the plume hypothesis

The most convincing evidence that whole-mantle convection is taking place comes from the pattern of seismic residuals from Pacific deep-focus earthquakes, which demonstrates that subducting lithosphere is sinking into the lower mantle beneath the north-western Pacific to a depth of at least several hundred kilometres below the

From MORTON, A. C. & PARSON, L. M. (eds), 1988, *Early Tertiary Volcanism and the Opening of the NE Atlantic,* Geological Society Special Publication No. 39, pp. 15–23.

transition zone (Creager & Jordan 1986). This is strongly supported by interpretation of the belt of positive geoid and gravity anomalies over the circum-Pacific subduction zones in terms of the underlying cool sink for the subducting slabs (Rabinowicz *et al.* 1983), which suggests that recycling of cool slabs into the lower mantle applies to the whole circum-Pacific belt.

The idea that mantle plumes provide the main conduit for upward flow of hot material through the lower mantle and the mantle transition zone has been supported by recent interpretations of the D″ layer at the base of the mantle. This probably represents a thermal boundary layer where heat is transferred by thermal conduction from core to mantle. Stacey & Loper (1983) infer a downward increase of about 800 K across this 150 km thick layer. Viscosity in the mantle is highly temperature dependent, and a temperature increase of this order is calculated to reduce the viscosity by a factor of around 10^4; instability develops in such a hot, low-viscosity layer. Loper & Stacey (1983) concluded that the development of relatively narrow plumes to vent the heat through the higher viscosity lower mantle is inevitable in this situation. Numerical experiments by Christensen (1984) support the conclusion. The plume modelled by Loper & Stacey is about 10 km in radius and transfers material at about 800 K above ambient temperature into the asthenosphere at a rate of 4.8 m/y, transferring about 10^{11} W. The rising plume would heat the adjacent lower mantle, but according to the modelling this would be countered by lateral inflow towards the low pressure plume which would be constrained to a narrow chimney throughout its lifespan. These rather specialized advances in plume theory counter the previous mechanical objections.

A model of mantle convection which combines the plume hypothesis with the demonstrable sinking of linear subducted slabs into the lower mantle has been proposed by Loper (1985). A sketch of the hypothesis as applied to Iceland and the region westwards towards the circum-Pacific belt is shown in Fig. 1. Narrow axisymmetric plumes carry hot material up from the core–mantle boundary into the asthenosphere where the viscosity is significantly below that of the lower mantle. There may also be a complementary slow upward migration of displaced lower mantle material through the transition zone into the upper mantle to maintain balance between the rates of slab sinking and upwelling. Asthenospheric material rises passively at ocean ridges to form oceanic lithosphere, which cools and is recycled into the lower mantle at subduction zones, forming the return flow. According to

Loper's model, the sinking slabs are partly diverted sideways into the lower mantle, cooling it and displacing material upwards and downwards.

This model of mantle circulation fits the observations concerning subduction zones and hot-spots much better than the older concepts of cellular convection. As the sinking lithospheric material is at least partly recycled via the core–mantle boundary, the geochemical observations can be explained without need for layered convection. Although many aspects of the theory are uncertain, in broad terms it provides a basis for explaining the anomalous N Atlantic region in terms of the mantle–plume hypothesis.

The N Atlantic topographic swell

The early Tertiary continental volcanism and the production of the anomalously thick crust of the Icelandic transverse ridge are the most obvious indications of anomalous activity in the underlying upper mantle. However, by far the most extensive feature, but less conspicuous, is the broad swell which produces anomalously shallow bathymetry over most of the Atlantic N of 30°N. This uplift includes much of Greenland. The swell attains its maximum elevation of over 2.5 km in the vicinity of Iceland, with a subsidiary peak of 1.2 km over the Azores. This broad swell needs to be distinguished from the much more localized Icelandic transverse ridge which is superimposed on the swell and is the isostatic response to locally thickened oceanic crust of Icelandic type.

The ocean-floor swell was attributed by Haigh (1973) to an exceptionally hot underlying asthenosphere, with a temperature of up to 150 K above normal. Cochran & Talwani (1978) reached a similar conclusion, suggesting that a 75 K average excess temperature over 200 km depth range in the upper mantle could account for the isostatic support of 500 m excess elevation of the seafloor. There is also considerable evidence from heat-flow measurements that ocean-floor swells of the N Atlantic and elsewhere have a thermal origin (Langseth & Zielinski 1974; von Herzen *et al.* 1982; Courtney & White 1986; Detrick *et al.* 1986).

The swell covers an area of about 14×10^6 km². including the affected parts of the N Atlantic and Greenland. The average excess elevation would be about 750 m if entirely underwater, which is equivalent to 525 m entirely on land. This average elevation can be isostatically supported by an average reduction in density of 9 kg/m³ extending over a 200 km vertical extent in the underlying

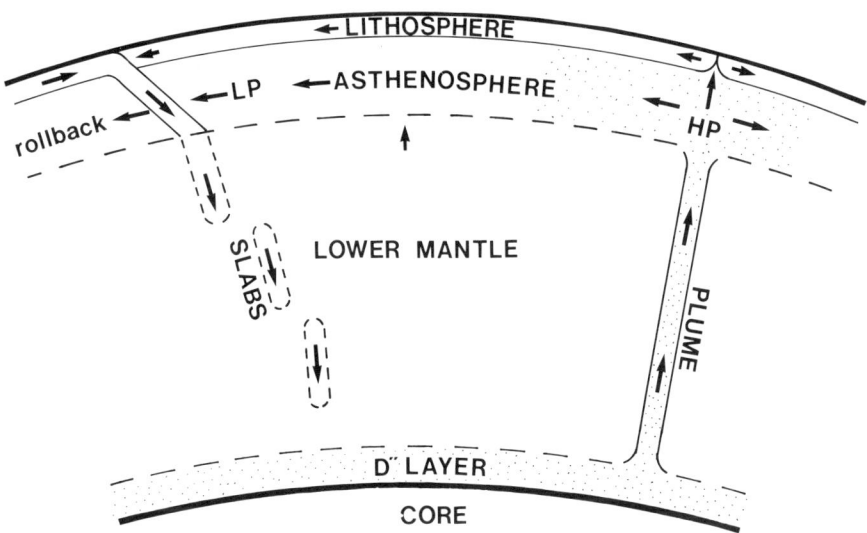

FIG. 1. A sketch of the model of whole-mantle convection as developed by Loper (1985), with application to the Icelandic region and the adjacent part of the circum-Pacific belt. HP: high pressure; LP: low pressure.

upper mantle. This reduction in density would be produced by a 100 K rise in temperature, taking the volume coefficient of thermal expansion as 2.5×10^{-5} K^{-1}. Taking the specific heat to be 1.25×10^3 J/kg/K and the density to be 3300 kg/m^3, the additional heat required to cause this temperature anomaly is calculated to be 1.15×10^{27} J. If this heat has been introduced at a constant rate since 60 Ma (the time since the first indications of the hot-spot) then the average rate of heat inflow amounts to 6×10^{11} W. This is equivalent to about 2% of the total global heat loss through oceanic regions.

The simple calculation above shows that a substantial amount of heat in global terms participates in the N Atlantic hot region. It is difficult to see how this could be produced on such a short time-scale by self-generation in the upper mantle, or by thermal conduction from deeper parts of the mantle, which would imply excessive geothermal gradients across the mantle transition zone. The simplest explanation is that the heat has been transferred upwards from the lower mantle by some sort of convective upwelling beneath the region, such as plumes. Other geochemical and topographic evidence suggests that this occurs beneath Iceland or nearby, with a subsidiary upwelling beneath the Azores. These considerations support the plume hypothesis, although they do not prove it.

The heat calculated above represents the present temperature excess relative to normal sub-oceanic lithosphere. The plume has introduced much more heat than this over its present lifespan. A similar amount of additional plume-derived heat will be lost in the future as the present asthenospheric material rises to form oceanic lithosphere and cools to a much lower average temperature before being subducted. Furthermore, a comparably large amount of heat has already been lost by oceanic heat flow over the last 55 My. Taking these factors into account, the plumes beneath Iceland and the Azores are probably bringing up heat which will eventually be lost to the surface at a rate of at least 8–10% of the total global oceanic heat loss. The upwellings beneath the N Atlantic are thus probably a substantial factor in the terrestrial heat flow.

The N Atlantic geoid and gravity high

The N Atlantic region coincides with a major global geoid high (see Fig. 2) which is centred near Iceland and E Greenland, and covers a similar area to the region of uplifted topography discussed in the previous section. A conspicuous elongated geoid low occurs to the W of the high, with minimum regions over Canada and E of the Caribbean. The N Atlantic high and the low to the W of it are also displayed on the global gravity map (Fig. 2) which is based on the same basic information as the geoid map but emphasizes higher harmonics relative to the lower ones. The long wavelength geoid and gravity anomaly maps are usually attributed to dynamically supported

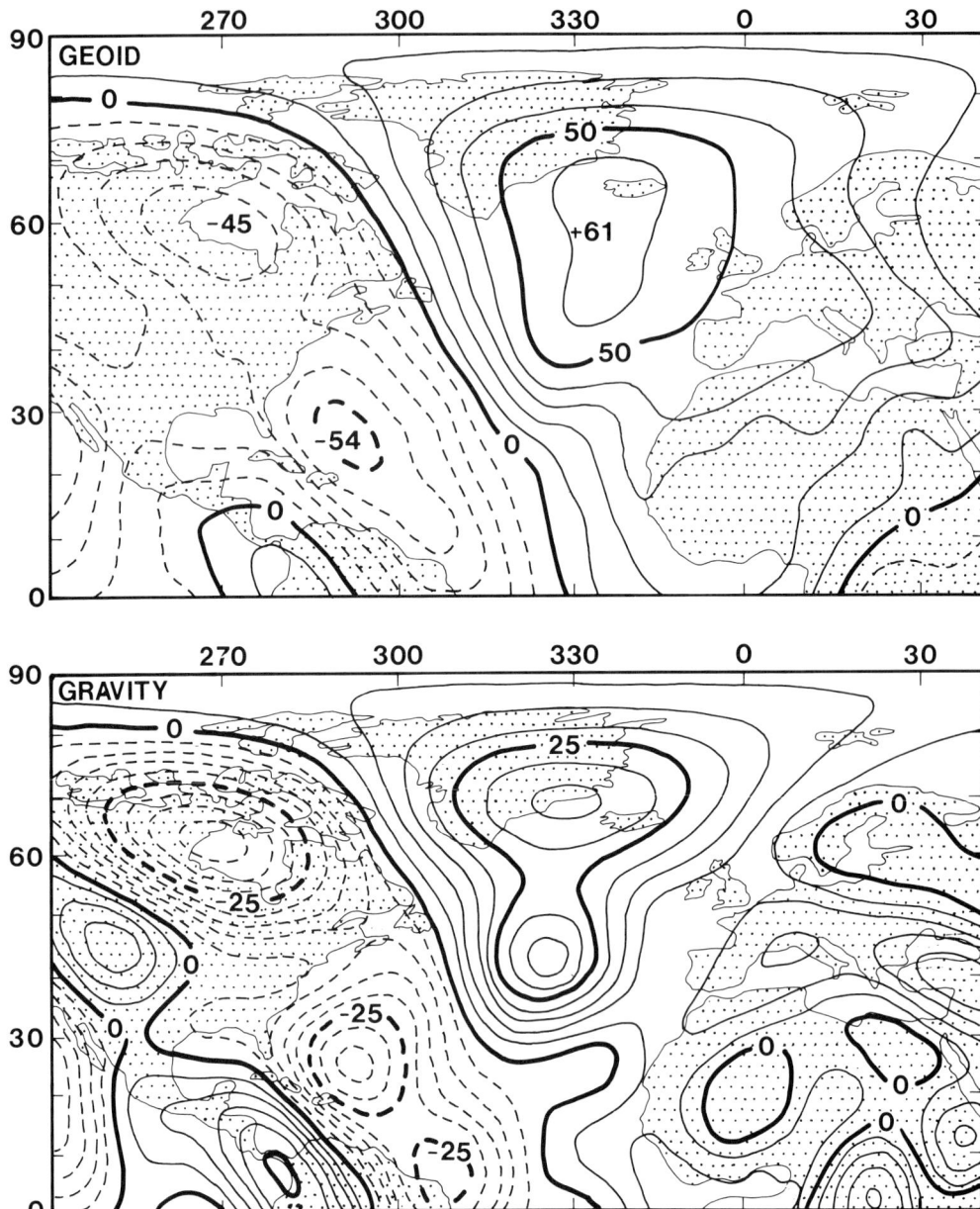

FIG. 2. The long-wavelength geoid and gravity anomalies of the N Atlantic and adjacent regions, redrawn from Lerch *et al.* (1979) and based on the GEM9 model. The geoid map includes harmonics up to $n = 22$, whereas harmonics above $n = 16$ have been truncated in the gravity map. Contours are at 5 m and 5 mGal intervals respectively.

mass anomalies associated with mantle convection, but their exact nature remains controversial.

Geoid and gravity anomalies of wavelength greater than about 2000 km are unlikely to be caused by lateral density variations in the lithosphere, as these are known to be flexurally compensated with negligible residual anomaly at this wavelength. As the underlying mantle can flow, the anomalies are probably caused by lateral density variations in the asthenosphere and surface deformation effects which are dynamically maintained. The following four potential sources can be identified:

(1) Isostatically supported lithospheric elevation related to lateral density variations of thermal origin in the underlying upper mantle, with significant residual anomalies resulting from deep isostatic compensation.
(2) Lateral density variation in the lower mantle and/or mantle transition zone associated with slow convective flow, such as sinking of cool subducted slabs.
(3) Dynamically produced (uncompensated) variation in depth of the core–mantle boundary.
(4) Deformation of the Earth's surface resulting from lateral pressure variations in the upper mantle associated with convection-related flow.

Geoid highs over hot-spots such as Iceland have been variously attributed to pressure variations associated with upper mantle convection (Sclater *et al.* 1975), to deep isostatic compensation of the associated topographic swell by hot, low density asthenosphere (Cochran & Talwani 1978) and to lateral variation of lower mantle density or depth to core–mantle boundary (Chase 1979; Crough & Jurdy 1980). The adjacent gravity low down the E side of the Americas, which forms part of a belt of negative anomalies on the continental side of the circum-Pacific belt, has been attributed by Rabinowicz & Lago (1984) to pressure variations associated with cellular mantle currents driven by the nearby subducting slab. A subsidiary part of the Canadian low is probably related to postglacial recovery.

The gradient of the geoid anomaly E of Iceland, where there are no other major disturbances (see Fig. 2), and of the long wavelength gravity anomaly, are much too steep to explain in terms of fluctuations of the depth of the core–mantle boundary. A large region of anomalously dense rocks in the underlying lower mantle is difficult to reconcile with a hot, uprising convection current or plume. We are left with deep isostatic compensation of the N Atlantic swell in the underlying upper mantle, and lateral pressure variation in the upper mantle, as possible explanations of the high. It is suggested that these two effects combine to produce the N Atlantic geoid and gravity high, and that the pressure low associated with asthenospheric flow from the hot-spot towards the circum-Pacific belt produces the adjacent geoid and gravity low.

As indicated by Cochran & Talwani (1978), the N Atlantic gravity high (see Fig. 2) shows excellent correlation with the topographic swell. Individual peaks occur in the regions of Iceland

and the Azores, and the 15 mGal contour approximately outlines the swell. The geoid shows a broader high without individual peaks because it emphasizes the longer wavelengths. Cochran & Talwani showed from surface gravity measurements that the gravity anomalies could be accounted for mainly by deep isostatic compensation. However, the following calculation based on the broader scale satellite geoid and gravity analysis suggests that only part of the anomalies can be explained by deep compensation. We can approximately estimate the amplitude of the anomalies caused by deep isostatic compensation by treating the surface swell as a spherical cap of height 750 m under water subtending 15° at the Earth's centre. The isostatic compensation takes the form of a similar cap of equal but opposite mass per unit area situated at 200 km depth. These together yield a maximum geoid anomaly of 18 m and a maximum positive gravity anomaly of 7 mGal. Even allowing for the gross approximation it seems unlikely that such deep isostatic compensation can account for more than 25–50% of the observed amplitude.

It is therefore suggested that the remaining part of the positive gravity and geoid anomalies, and the adjacent eastern American negative, are essentially caused by the pressure gradient in the upper mantle associated with asthenospheric flow from the high pressure hot-spot towards the low pressure region adjacent to subducting lithosphere.

According to Garfunkel *et al.* (1986), the subduction zones along the eastern Pacific margins are retreating westwards relative to the hot-spot frame of reference at about 6 mm/y as they converge on the Pacific region. There must, therefore, be an upper mantle flow from the Atlantic region at about this rate. It is suggested that material for this westward asthenospheric flow is provided by upwelling at the Iceland–Azores hot-spots, rather than by broad cellular convection as postulated by Rabinowicz & Lago (1984). The feasibility of this idea can be tested simply. An excess pressure P in the asthenosphere approximately produces a gravity anomaly of $\Delta g = 2\pi GP/g$ which is directly proportional to the pressure. Allowing for a deep compensation effect of 10 mGal, the remaining long-wavelength change in gravity between the Iceland–Azores high and the low to the W is about 50 mGal, corresponding to an average horizontal pressure gradient in the asthenosphere of 3.0 Pa/m. For an upper mantle viscosity of 10^{20} Pa s averaged over 300 km depth, this pressure gradient would produce an average channel flow of 6 mm/y. This would be supplemented by a smaller effect from lithospheric plate drag. Thus, the calculated flow using a realistic viscosity yields an estimated flow closely similar to that inferred from the completely independent observations of Garfunkel *et al.* (1986).

The high pressure in the N Atlantic asthenosphere can readily be accounted for by high pressure produced by one or more narrow low-density plumes, provided these have been active sufficiently long for the pressure to build up. Continuing plume activity is required, as otherwise the pressure anomaly would decay by identically the same process which causes postglacial recovery. The postglacial recovery of Fennoscandia, where the ice load had a radius of about 550 km, has a time constant of recovery of about 5000 years (Cathles 1975). The equivalent radius of the N Atlantic swell is about 2500 km. The time constant of decay of a load or pressure anomaly by asthenospheric channel flow is proportional to the square of the radius. Thus the N Atlantic pressure anomaly and its associated gravity and geoid anomalies would decay with a time constant substantially less than 1 My if not maintained by an active plume which provides a continuing pressure head. The thermal anomaly, however, would take much longer to dissipate.

It has been shown that the N Atlantic geoid and gravity highs can be interpreted as the combined effect of deep isostatic compensation associated with a hot asthenosphere, and high upper mantle (asthenospheric) pressure caused by upwelling from the lower mantle beneath the region. The geoid and gravity highs, as well as the topographic swell, thus receive a plausible interpretation in terms of the plume hypothesis.

Discussion

Some of the implications of the hot-spot and plume hypotheses as applied to the N Atlantic region are summarized as follows:

(1) The earliest detectable activity associated with the Icelandic hot-spot was the early Tertiary continental volcanism. This started several million years prior to the continental break-up between Greenland and Europe, and rapidly died out afterwards except at a few isolated localities. The Palaeocene igneous activity extended over more than 1000 km between E Greenland and Scotland on a Palaeocene continental reconstruction, and this prompted an earlier suggestion by the writer of a major convective overturn of the mantle as an alternative to the plume hypothesis (Bott 1973). However, the thermal anomaly in the upper mantle produced by a convective overturn would decay much too rapidly to account

for the subsequent developments continuing to the present time. Multiplicity or rapid migration of plumes also seems unlikely. The widespread continental volcanism can perhaps be best explained by very rapid lateral squirting of partially molten material from a single newly formed plume into the weakest level of the subcontinental asthenosphere. This would be expected to continue until continental splitting enabled the magma to be vented at the newly-formed ridge crest. Two geochemically distinct types of magma could be produced: (a) magma direct from the plume material; and (b) magma produced by diapiric upwelling of the existing heated asthenospheric material. These could rise together or separately through lines of weakness in the continental crust above (Fig. 3).

(2) Most of the continental volcanism demonstrably preceded the pre-anomaly 24B continental break-up, which probably occurred just after the development of the E Greenland coastal dyke swarm and flexure (Bott 1987). The hot-spot may thus have been a cause of the continental break-up, but cannot be a consequence of it.

(3) After early Eocene continental break-up, the main magma fraction from the plume appears to have been diverted to the nearest section of ocean

Pre-split

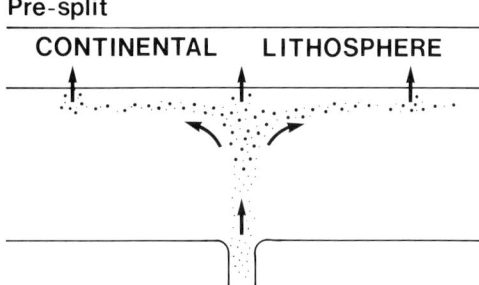

Post-split

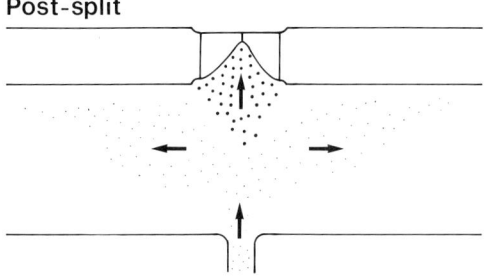

FIG. 3. Sketches showing the destination of plume material prior to (top) and subsequent to (bottom) continental break-up. Large dots represent the magma fraction and small dots represent the solid residuum.

ridge crest (see Fig. 3). The rate of volume production of plume-derived magma per unit length of ridge crest has continued to be much greater than that produced by normal upwelling at the ocean ridge crest. This has risen to form the anomalously thick Icelandic-type crust of the Icelandic transverse ridge. During the earlier stages of evolution, from 50–27 Ma, crust up to 30 km thick was produced beneath the Iceland–Faeroe Ridge, and probably also beneath the complementary Greenland–Iceland Ridge. The plume at this stage was presumably of smaller diameter than the 250 km width of the Iceland–Faeroe Ridge. Subsequently, the plume activity appears to have become more diffuse, producing the 10–15 km thick crust of Iceland over a longer section of the ridge crest; the diameter of the plume may have increased, or the plume may have become offset from the ridge crest, possibly towards the W. The continuity of production of Icelandic-type crust indicates continuing upwelling from 50 Ma to the present at approximately uniform rate.

(4) In order to maintain the thermal and pressure anomalies which affect a large area of the N Atlantic and Greenland, a volume of plume material far in excess of that diverted directly to the ridge crest must flow laterally into the upper mantle. As Vogt (1971) pointed out, the most rapid outflow would be expected to occur beneath the ocean ridge system. This is partly because of the lowered viscosity associated with the hot upper mantle beneath ridge crests, and partly because of low pressure associated with ongoing upwelling of the asthenosphere. In the N Atlantic, the outflow from the Iceland region has been concentrated towards the SW beneath the Reykjanes Ridge rather than towards the NE, as evidenced by the contrasting depths of the ocean floor to the S of Iceland and N of Jan Mayen. This may be because of blockage caused by the Jan Mayen microcontinent and associated fracture zones, which prevented a direct low-viscosity channel towards the NE during the earlier stages of evolution, as indicated by Vogt.

(5) A puzzling aspect of the topographic evolution of Greenland can possibly be explained by the developing pattern of asthenospheric outflow from the Iceland plume. The uplift of the Kangerdlugssuaq dome adjacent to the Greenland–Iceland Ridge took place at approximately the time of the break-up (52 Ma), and the uplift of S Greenland may have occurred at this time or shortly after. In contrast, the Blosseville Kyst region, where the early Tertiary continental volcanism was most pronounced, did not start to be uplifted until about 26 Ma. The Blosseville

Kyst uplift coincides approximately with the start of the last of three main stages of plate-tectonic evolution of the region (Bott 1987), when the complementary fan-shaped spreading both E and W of the Jan Mayen microcontinent ceased, and all spreading started to take place along the Kolbeinsey Ridge between Greenland and the microcontinent. This new regime probably expedited the flow of asthenospheric material towards the N from Iceland, allowing the N Atlantic thermal anomaly to spread westwards beneath this previously unaffected part of Greenland to produce the isostatic uplift which persists to the present.

(6) A further consequence of the N Atlantic thermal anomaly, which does not depend on the plume hypothesis but merely on the occurrence of deep isostatic compensation for the associated topographic swell, is its influence on the plate-boundary force. The strength of the ridge-push force depends on the elevation of the ridge and also on the density–depth profile in the lithosphere and asthenosphere beneath the ridge. The deeper the low density compensating material, the larger the ridge-push force. Thus the ridge-push force produced between the Azores and Jan Mayen is likely to be substantially greater than normal, with a maximum value in the vicinity of Iceland itself where the thermal anomaly is greatest. The N Atlantic thermal anomaly may thus have had a profound influence on plate evolution, including a significant role in the development of the Alpine mountain range.

Conclusions

The concept of whole-mantle convection has received new impetus from the seismological observations indicating subduction into the lower mantle and theoretical work suggests that narrow plumes may rise from the thermal boundary layer above the core–mantle boundary. The convection model of Loper (1985), in particular, appears to overcome the mechanical and geochemical objections to plumes and whole-mantle convection respectively. Two further considerations support the hypothesis of upwelling from the lower mantle beneath the Iceland region. The hot upper mantle underlying the N Atlantic topographic swell requires a major heat input which can best be explained by such upwelling. Also, the positive geoid and gravity anomalies which dominate the region can perhaps be best interpreted as the combined effect of the thermal and pressure anomalies caused by upwelling from the lower mantle. The compact nature of the Icelandic transverse ridge and the associated geochemical anomalies still provide the main indication that the upwelling takes the form of a relatively narrow plume. There are no direct observations of a lower mantle plume beneath the Iceland region, so the hypothesis remains unproved. However, there is a great deal of circumstantial evidence to support it.

Some of the consequences of the plume hypothesis are as follows:

(1) The Palaeocene continental volcanism may result from rapid lateral migration of partially fused plume material injected into the asthenosphere prior to Eocene continental break-up.
(2) The development of the hot, partially fused, low density asthenosphere may have triggered continental break-up in the early Eocene.

(3) Subsequent to continental break-up, the plume-derived magma became concentrated along a relatively short segment of the new ocean ridge where it produced the anomalously thick crust underlying the Icelandic transverse ridge.
(4) The upper mantle thermal anomaly has progressively expanded to underlie much of the N Atlantic and Greenland. The more pronounced development of the resulting topographic and bathymetric swell towards the SW, and the late uplift of the Blosseville Kyst region of E Greenland, can possibly be understood in terms of the effect on the outflow of the complexities of plate evolution N of Iceland.
(5) The anomalously low density upper mantle beneath the region would be expected to produce a significantly larger ridge-push force than at normal ocean-ridge crests, with the maximum centred on Iceland.

References

BOTT, M. H. P. 1973. The evolution of the Atlantic north of the Faeroe Islands. *In:* TARLING, D. H. & RUNCORN, S. K. (eds). *Implications of Continental Drift to the Earth Sciences.* Academic Press, London & New York, **1**, pp. 175–189.

—— 1987. The continental margin of central East Greenland in relation to North Atlantic plate tectonic evolution. *Journal of the Geological Society of London,* **144**, 561–568.

CATHLES, L. M. 1975. *The Viscosity of the Earth's Mantle.* Princeton University Press, Princeton, New Jersey. 386 pp.

CHASE, C. G. 1979. Subduction, the geoid, and lower mantle convection. *Nature,* **282**, 464–468.

CHRISTENSEN, U. 1984. Instability of a hot boundary layer and initiation of thermo-chemical plumes. *Annales Geophysicae,* **2**, 311–320.

COCHRAN, J. R. & TALWANI, M. 1978. Gravity anomalies, regional elevation, and the deep structure of the North Atlantic. *Journal of Geophysical Research,* **83**, 4907–4924.

COURTNEY, R. C. & WHITE, R. S. 1986. Anomalous heat flow and geoid across the Cape Verde Rise: evidence for dynamic support from a thermal plume in the mantle. *Geophysical Journal of the Royal Astronomical Society,* **87**, 815–867 (microfiche GJ 87/1).

CREAGER, K. C. & JORDAN, T. H. 1986. Slab penetration into the lower mantle beneath the Mariana and other island arcs of the northwest Pacific. *Journal of Geophysical Research,* **91**, 3573–3589.

CROUGH, S. T. & JURDY, D. M. 1980. Subducted lithosphere, hotspots, and the geoid. *Earth and Planetary Science Letters,* **48**, 15–22.

DETRICK, R. S., VON HERZEN, R. P., PARSONS, B., SANDWELL, D. & DOUGHERTY, M. 1986. Heat flow observations on the Bermuda Rise and thermal models of midplate swells. *Journal of Geophysical Research,* **91**, 3701–3723.

GARFUNKEL, Z., ANDERSON, C. A. & SCHUBERT, G. 1986. Mantle circulation and the lateral migration of subducted slabs. *Journal of Geophysical Research,* **91**, 7205–7223.

HAIGH, B. I. R. 1973. North Atlantic oceanic topography and lateral variations in the upper mantle. *Geophysical Journal of the Royal Astronomical Society,* **33**, 405–420.

LANGSETH, M. G. & ZIELINSKI, G. W. 1974. Marine heat flow measurements in the Norwegian–Greenland Sea and in the vicinity of Iceland. *In:* KRISTJANSSON, L. (ed) *Geodynamics of Iceland and the North Atlantic Area.* D. Reidel Publishing Company, Dordrecht, Holland, pp. 277–295.

LERCH, F. J., KLOSKO, S. M., LAUBSCHER, R. E. & WAGNER, C. A. 1979. Gravity model improvement using Geos 3 (GEM 9 and 10). *Journal of Geophysical Research,* **84**, 3897–3916.

LOPER, D. E. 1985. A simple model of whole-mantle convection. *Journal of Geophysical Research,* **90**, 1809–1836.

—— & STACEY, F. D. 1983. The dynamical and thermal structure of deep mantle plumes. *Physics of the Earth and Planetary Interiors,* **33**, 304–317.

MORGAN, W. J. 1971. Convection plumes in the lower mantle. *Nature,* **230**, 42–43.

O'NIONS, R. K., EVENSEN, N. M. & HAMILTON, P. J. 1979. Geochemical modeling of mantle differentiation and crustal growth. *Journal of Geophysical Research,* **84**, 6091–6101.

RABINOWICZ, M. & LAGO, B. 1984. Large-scale gravity profiles as evidences of a convective circulation. *Annales Geophysicae,* **2**, 321–332.

——, —— & SOURIAU, M. 1983. Large-scale gravity profiles across subducted plates. *Geophysical Journal of the Royal Astronomical Society*, **73**, 325–349.

RUNCORN, S. K. 1974. On the forces not moving lithospheric plates. *Tectonophysics*, **21**, 197–202.

SCHILLING, J.-G. 1973. Iceland mantle plume: geochemical study of Reykjanes Ridge. *Nature*, **242**, 565–571.

SCLATER, J. G., LAWVER, L. A. & PARSONS, B. 1975. Comparison of long-wavelength residual elevation and free air gravity anomalies in the North Atlantic and possible implications for the thickness of the lithospheric plate. *Journal of Geophysical Research*, **80**, 1031–1052.

STACEY, F. D. & LOPER, D. E. 1983. The thermal boundary-layer interpretation of D″ and its role as a plume source. *Physics of the Earth and Planetary Interiors*, **33**, 45–55.

VOGT, P. R. 1971. Asthenosphere motion recorded by the ocean floor south of Iceland. *Earth and Planetary Sciences Letters*, **13**, 153–160.

—— 1974. The Iceland phenomenon: imprints of a hot spot on the ocean crust, and implications for flow below the plates. *In:* KRISTJANSSON, L. (ed) *Geodynamics of Iceland and the North Atlantic Area.* D. Reidel Publishing Company, Dordrecht, Holland, pp. 105–126.

—— 1976. Plumes, subaxial pipe flow, and topography along the mid-oceanic ridge. *Earth and Planetary Science Letters*, **29**, 309–325.

—— 1983. The Iceland mantle plume: status of the hypothesis after a decade of new work. *In:* BOTT, M. H. P., SAXOV, S., TALWANI, M. & THIEDE, J. (eds) *Structure and Development of the Greenland–Scotland Ridge: New Methods and Concepts.* Plenum Press, New York, pp. 191–213.

VON HERZEN, R. P., DETRICK, R. S., CROUGH, S. T., EPP, D. & FEHN, U. 1982. Thermal origin of the Hawaiian swell: heat flow evidence and thermal models. *Journal of Geophysical Research*, **87**, 6711–6723.

WILSON, J. T. 1963. Evidence from islands on the spreading of ocean floors. *Nature*, **197**, 536–538.

M. H. P. BOTT, Department of Geological Sciences, University of Durham, Durham DH1 3LE, UK.

Palaeocene–Oligocene tectonics of NW Europe

J. F. Dewey & B. F. Windley

SUMMARY: During the Palaeogene in NW Europe, a Gallic subplate was very nearly created. This plate was almost entirely bordered by plate boundaries, although not of equivalent ages. To the N, the N Atlantic began to open about 56 Ma ago, the Alpine convergent boundary extended southwestwards to the Pyrenean subduction zone; on the NE side a N British Palaeocene/early Eocene fracture–dyke boundary appears to pass southeastwards across the North Sea into a Lutetian to Burdigalian rift system.

There is an age disparity between the fracturing and NW dyke intrusion in Britain and the fracturing and formation of the Rhine Graben. The last dykes in Britain (Ypresian) were intruded not long before the important subsidence of the Rhine Graben (earliest Lutetian), but their causes were different. The British dyke fractures propagated southeastwards via the Cleveland Dyke towards the North Sea, whereas the Rhine Graben propagated northwards towards the Zuider Zee depression. However, the structures from both sides converged in the Sole Pit Basin in the North Sea with the result that a continuous though not coevally-originated fracture system developed from NW Scotland to the Alps.

During the early Tertiary, there was a wide range of tectonic activity throughout NW Europe (Fig. 1). In NW Britain lavas were extruded and plutonic complexes and extensive dyke swarms were intruded. In the North Sea tuffs were deposited as a result of volcanism along the Rockall–Faeroe rift, reactivation of faults controlled magmatism, localization of gas fields, devolatilization of coal, and inversion of basins caused uplift, deformation and erosion. Along the Rhine Graben, doming, subsidence and rift propagation took place in association with alkaline volcanism and plutonism, coevally with the Meso–Alpine compressive deformations that occurred in the Alps. Faulting controlled sedimentation patterns during inversion of basins in southern England, the English Channel, northern France and Holland, old rift-faults were reactivated in the Western Approaches Basin, thrust-faults formed in the southern Bay of Biscay contemporaneously with thrusting in the Pyrenees, and Variscan massifs were tilted and uplifted along steep reverse-faults.

The main tectonic activity lasted from about 60 Ma to about 30 Ma, but it was diachronous, generally earlier in the NW and later in the SE. The igneous activity in northern Britain was related to the opening of the N Atlantic in the late Palaeocene, but the formation of the Rhine Graben and the inversion of sedimentary basins in the North Sea and near the English Channel were caused by major orogenic events in the Alpine region in the late Eocene to mid-Oligocene (Frisch 1981). Tectonic phenomena in the North Sea provide the link between those in northern Britain and those in the Rhine Graben. The aim of this paper is to demonstrate that these movements from the Atlantic and Alpine sides almost fragmented the lithospheric plates of NW Europe during the creation of a Gallic sub-block, an earlier northern boundary from NW Scotland to northern Germany being transected by a later eastern boundary along the Rhine Graben.

Previous views on the origin of structures

It is appropriate to consider, briefly, previous key ideas on the origin of the principal structures in the two main regions concerned—Northern Britain and the Rhine Graben.

British Igneous Province

The problem of the siting of the Tertiary central complexes and dyke swarms of northern Britain is long-standing. Richey (1937) suggested that the plutonic centres were located at the intersection of a N–trending lineament and NE–trending faults, but Vann (1978) pointed out that they occur at the intersection of the early basic dyke swarms and the NE–trending faults. The major lava piles are coincident with Mesozoic basins and the most intense dyke swarms are located in the basinal areas where the crust is relatively thin (Emeleus 1983). Brooks (1973) suggested control by an underlying mantle hot-spot. In contrast, geophysical studies show that the central complexes occur on pre-Mesozoic ridges (Binns et al. 1984; McLean & Deegan 1978). The magma bodies that gave rise to the central complexes and the dilation axes of the major dyke swarms have en echelon NW–SE alignments (Gass & Thorpe 1976; Speight et al. 1982). Some N–trending dykes extend for up to 100 km N of Scotland and

From MORTON, A. C. & PARSON, L. M. (eds), 1988, *Early Tertiary Volcanism and the Opening of the NE Atlantic,* Geological Society Special Publication No. 39, pp. 25–31.

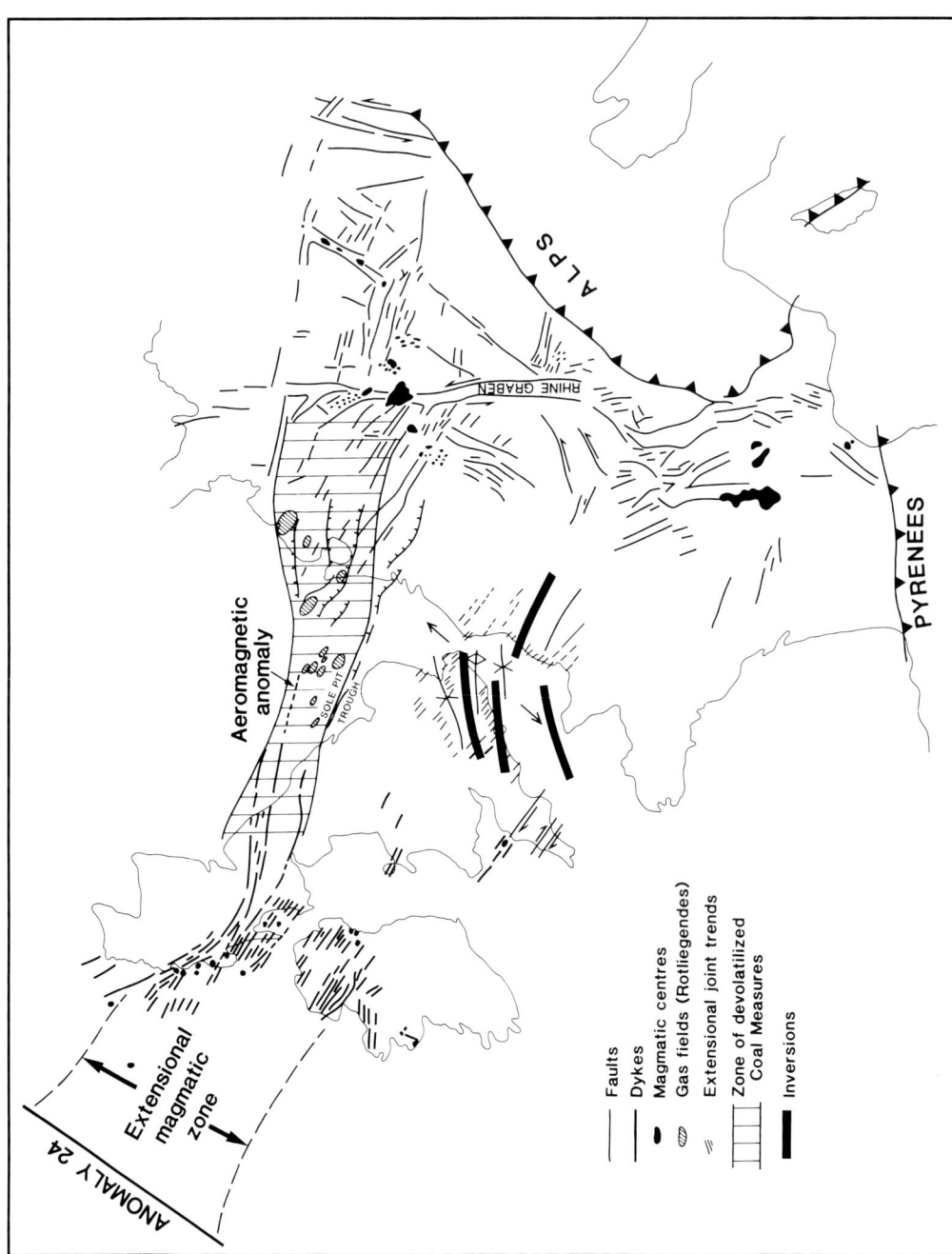

Fig. 1. Map of NW Europe showing the main tectonic features developed in the Palaeocene to Oligocene. Sources variable and mostly referred to in text. Aeromagnetic anomaly after Kirton & Donato (1985) and NW-trending mesofracture system in S England and N France after Bevan & Hancock (1986). Other data from Pegrum & Mountenay (1978) and Smalley & Westbrook (1982).

are oblique to the continental margin (Dunham 1972; Vann 1978). Thus, it is probable that the magma ridges developed as a result of simple shear stresses or compressive stresses oriented at a high angle to the continental margin of the northern Atlantic (Bell 1976; Vann 1978; Speight *et al.* 1982).

The Rhine Graben

Cloos (1939) argued, with respect to the Rhine Graben, that rifts form as a result of tensional stresses on domes, and Illies (1969) extended this idea to the E African Rift where many rift branches are sited on regional domes. In contrast, Illies (1975), Sengör (1976), Illies & Greiner (1978) and Ziegler (1982) pointed out that the Upper Rhine Graben did not start to form until the Eocene (Lutetian), was not superimposed on a dome and that the rifting events were contemporaneous with Alpine deformation phases, in consequence of which the Graben formed as a result of the Alpine Orogeny. Aulacogens at the start of the Wilson Cycle thus contrast with indentation rifts at the end of the Cycle (Sengör *et al.* 1978). Karner & Watts (1983) suggested that the Rhine Graben developed in a peripheral bulge, on the N side of the Alpine Belt, comparable to the bulge S of the Indo–Gangetic basin peripheral to the Himalaya.

The British Tertiary Igneous Province and its relation to the opening of the Atlantic

With the possible exception of 67 Ma lavas in northern Ireland, the igneous rocks of this province formed in the relatively narrow time-span of 61–52 Ma (Fig. 2) (Macintyre *et al.* 1975; Brown 1982; Emeleus 1983). There was a major phase of dyke intrusion at 59 Ma and a minor phase at 52 Ma. The Cleveland Dyke echelon, which extends to the coast of the North Sea, has a whole-rock K/Ar age of 55.8 ± 0.3 Ma (Fitch *et al.* 1978). The volcanism occurred during the period 61–55 Ma and the central complexes were intruded between 59 Ma and 52 Ma (Fig. 2).

The intrusion and extrusion of these rocks were tectonically controlled by events in the opening Atlantic (Bott & Watts 1971). The dykes were intruded into the faulted, thinned continental margin and the basaltic lava piles accumulated in areas of subsidence with Mesozoic sedimentary basins. Marine aeromagnetic maps have revealed the presence of mafic dykes in the Irish Sea offshore of Anglesey, the Lake District and the

Scottish borders (Kirton & Donato 1985). In the N Atlantic, magnetic anomaly 24 (56 Ma) developed at a time of maximum tectonic and magmatic activity, when new rifting assisted the opening of the Norwegian Sea and the final separation of Greenland from the Rockall–Faeroe microcontinent and Eurasia (Pitman & Talwani 1972; Pegrum & Mounteney 1978). The 60–58 Ma dykes and lavas in northern Britain formed in the fractured continental margin contemporaneously with this new rifting. Seafloor spreading was accompanied by subsidence of the Rockall Plateau on which sediments were deposited, the oldest of which have an age of 57 Ma. Minor dyke intrusion at 52 Ma (Macintyre *et al.* 1975) terminated the main igneous activity in the British Isles at a time of decreasing spreading-rate in the N Atlantic. At 48 Ma, spreading ceased in the Labrador Sea and the Davis Strait between Greenland and N America (Pitman & Talwani 1972).

The Rhine Graben and its relation to the Alpine Orogeny

The Upper Rhine Graben began to form during the Lutetian at about 48 Ma (see Fig. 2) (Sengör 1976). The initial downfaulting began in the S and was accompanied by deposition of freshwater sediments on Mesozoic basement and by extrusion of basaltic lavas that have an age of 48 Ma. By the late Eocene, accelerating subsidence (Villemin *et al.* 1986) in the S allowed the deposition of up to 900 m of fresh water *Limnea* marls, and rifts were propagated northwards in the early Oligocene. By the late Oligocene, extension had begun in the N and subsidence in the S had slowed down to such an extent that, by the onset of the Miocene, there was regression and disconformable deposition. In the early Miocene, subsidence in the N gave rise to thick freshwater deposits and, 18 Ma, new faulting controlled alkaline volcanicity in the Kaiserstuhl.

During the late Pliocene the extensional Upper Rhine Graben (4.8 km of post-Middle Eocene extension (Illies 1975) changed to a sinistral strike-slip zone (see Fig. 1) as a consequence of lateral extension and isostatic uplift of the Alpine range; this is a seismotectonically active zone today (Bonjer *et al.* 1984). Earthquake focal mechanisms indicate an annual seismic slip-rate parallel with the graben axis of 0.05 mm (Ahorner 1975). The same regional stress field gave rise to extension, dip-slip faulting and subsidence in the Lower Rhine Graben, where, during the Pleistocene, there was a dense network of active growth

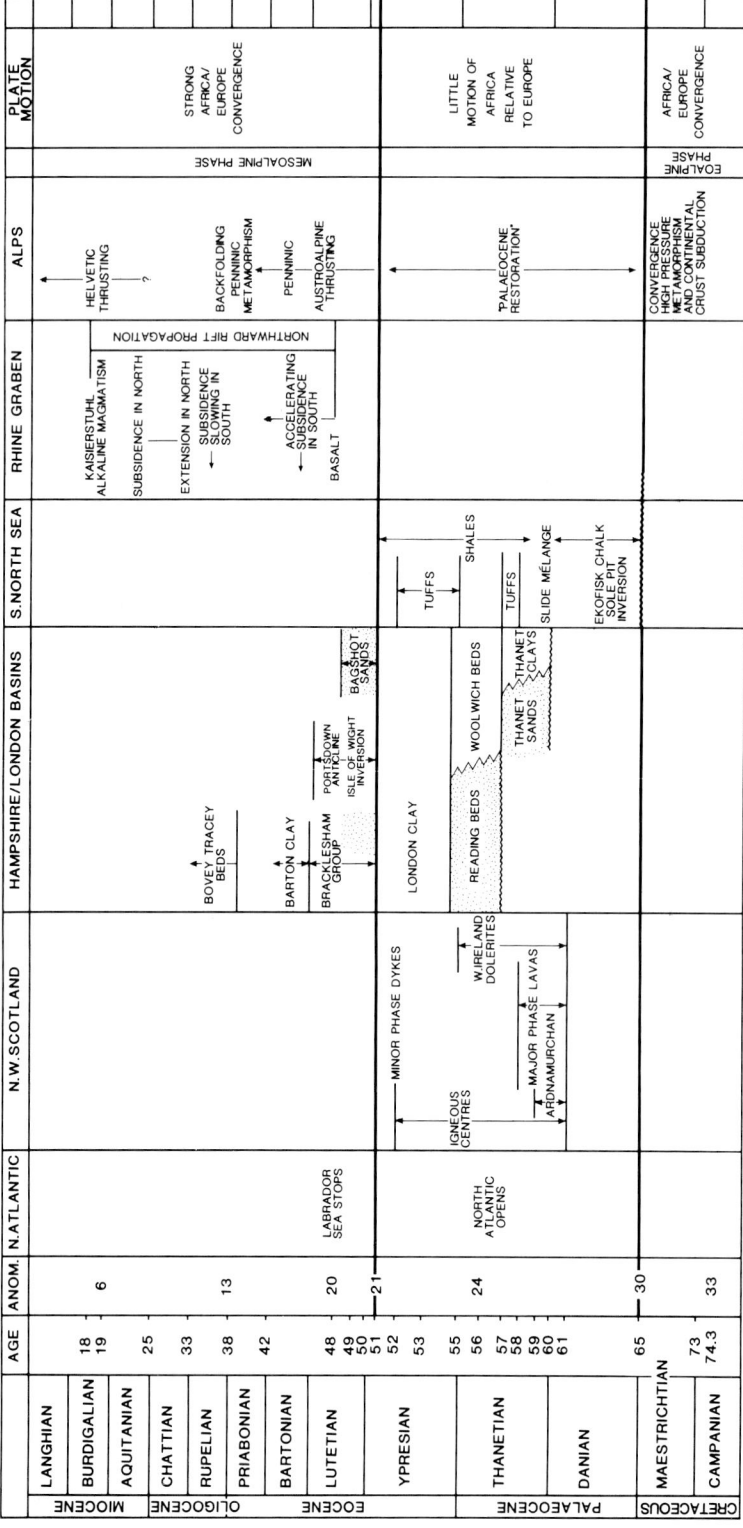

Fig. 2. Time chart of principal events in the evolution of NW Europe from the late Cretaceous to the Miocene. Data from Knox & Harland (1979), Knox (1984), Odin & Curry (1985), Aubry *et al.* (1986) and Mussett (1986).

faults and 380 m of sediments were deposited (Illies & Greiner 1978). The Rheinish Massif, situated between the Upper and Lower Rhine Graben, responded to these events by uplift and volcanism (Meissner *et al.* 1983). The uplift, which began at 30 Ma, accelerated at about 15 Ma and continues at present; the latest volcanism occurred 10 000 years ago (Fuchs *et al.* 1983). The northwesterly extension of the Lower Rhine Graben is expressed by the aseismic Zuider Zee depression, where Pleistocene sediments are more than 1 km thick. The Rhine Graben system has acted to the present day as an active subplate boundary linking the Western Alps to the Graben of the North Sea (Illies 1977).

The North Sea connection

Between northern England, where several mafic dykes continue the southeastward trend of the Scottish swarm, and the Lower Rhine Graben, a strong case can be made for a zone of extension crossing the southern North Sea. The southern North Sea gas fields in Lower Permian Rotliegend reservoirs contain gas that was derived from Upper Carboniferous source rocks and trapped beneath the Upper Permian evaporites. The Permian gas field zone lies within a zone of highly devolatilized coals that connect northern England to northern Germany. We believe that late Danian to Ypresian basic dyke intrusion in a zone of extension was the probable cause of the devolatilization.

Discussion and conclusions

During the Palaeogene in NW Europe, a Gallic subplate was created which was almost entirely bordered by plate boundaries. To the NW there was the opening Atlantic, the Alpine convergent boundary extended southwestwards to the Pyrenean subduction zone, and on the NE side there was a fracture–dyke boundary, which almost succeeded in splitting the subplate from its Eurasian parent to the E.

There is a disparity in age between the fracturing and NW-dyke intrusion in Britain and the fracturing and formation of the Rhine Graben. The last dykes in Britain (Ypresian) were intruded not long before the important subsidence of the Rhine Graben (earliest Lutetian), but their causes were entirely different. The common alignment of these structures was coincidental, because the British dyke fractures propagated southeastwards via the Cleveland Dyke towards the North Sea, whereas the Rhine

Graben propagated northwards towards the Zuider Zee depression. However, the structures from both sides converged in the Sole Pit Basin in the North Sea and thus the fracture system became continuous from NW Scotland to the Alps.

Fig. 2 shows that some important correlations exist for the Tertiary tectonics of northwestern Europe. From about 65 Ma to about 51 Ma, there was little motion of the African plate with respect to the Eurasian plate, as determined from magnetic anomaly finite difference filtering and fracture zone data in the Atlantic Ocean (M. Helman, pers. comm.). This period corresponds with a period of what Trümpy (1960) called the Palaeocene restoration in the Alps, a period separating the Eo–Alpine Cretaceous thrusting and high-pressure metamorphic phase from the Meso–Alpine Penninic thrusting phase. This period of little Africa–Europe motion contains the whole of the Thulean magmatic phase of northwestern Britain during which the N Atlantic opened, and the North Sea subsided rapidly with pyroclastic deposits (Knox & Morton 1983; Knox 1984). From 51 Ma, the renewed African/Europe convergence coincided with the northward propagation of the Rhine Graben, Alpine thrusting and the growth of inversion structures in the Wessex Basin with the attendant spread of the Bagshot sand-wedge. There was important Palaeogene intraplate deformation of two types within the Gallic subplate:

(1) Inversion tectonics were particularly pronounced. The main inversion axes trend E–W and extend from the Bristol Channel via the Weald and the English Channel to Holland (see Fig. 1). For example, the West Netherlands Basin is strongly inverted with steep reverse-faults and overthrusts (Ziegler 1978). The structural evolution of the Hampshire Basin during the Eocene (Cuisian–Lutetian) was largely controlled by uplift on bordering vertical fault-bounded blocks in the Palaeozoic basement, the superficial expression of which was the formation of the Isle of Wight and Purbeck monoclines (Plint 1982).

(2) On the continental margin SW of the British Isles there are two E–W-trending zones of late Eocene compressional deformation, one on the northern margin of the Bay of Biscay and the other in the southern Porcupine Bight (Masson & Parson 1983). The NW movement of Iberia gave rise to the Pyrenean orogenic belt and, offshore, to the overthrusting of Iberia onto the oceanic crust of the Bay of Biscay. Such movements may also have given

rise to NW–SE-trending strike-slip faults on the continental shelf in the Bristol Channel Basin (Kammerling 1979).

In situ stress measurements show that the uplifting and expanding Western Alps have high magnitude stresses up to 35 MPa in the NW–SE (140°) direction of maximum compression, that the Upper Rhine Graben has stresses in the same direction of 2.0 MPa, but the Lower Rhine Graben has low or negative stress values (Schmitt 1981). The N–S compression of NW Europe, evident from the extensional mesofracture pattern (see Fig. 1) (Bevan & Hancock 1986), probably resulted from N Atlantic ridge-push forces and Africa/Europe convergence.

References

AHORNER, L. 1975. Present-day stress field and seismotectonic block movement along major fault zones in Central Europe, *Tectonophysics*, **29**, 233–249.

AUBRY, M. P., HAILWOOD, E. A. & TOWNSEND, H. A. 1986. Magnetic and calcareous-nannofossil stratigraphy of the lower Palaeogene formations of the Hampshire and London basins. *Journal of the Geological Society of London*, **143**, 729–735.

BELL, J. D. 1976. The Tertiary intrusive complex on the Isle of Skye. *Proceedings of the Geologists Association*, **87**, 247–271.

BEVAN, T. G. & HANCOCK, P. L. 1986. A late Cenozoic regional mesofracture system in southern England and northern France. *Journal of the Geological Society of London*, **143**, 355–362.

BINNS, P. E., McQUILLAN, R. & KENOLTY, N. 1984. The geology of the sea of the Hebrides. *Report of the Institute of Geological Sciences*, **73/14**, 43 pp.

BONJER, K. P., GELBKE, C., GILG, B., ROULAND, D., MAYER-ROSA, D. & MASSINON, B. 1984. Seismicity and dynamics of the Upper Rhinegraben. *Journal of Geophysics*, **55**, 1–12.

BOTT, M. H. P. & WATTS, A. B. 1971. Deep structure of the continental margin adjacent to the British Isles. *Report of the Institute of Geological Sciences*, **70/14**, 89–109.

BROOKS, M. 1973. Some aspects of the Paleogene evolution of western Britain in the context of an underlying mantle hot spot. *Journal of Geology*, **81**, 81–88.

BROWN, G. M. 1982. Introduction to part 7: an appraisal of the igneous history; *In:* SUTHERLAND, D. S. (ed) *Igneous Rocks of the British Isles.* Wiley, Chichester, pp. 345–350.

CLOOS, H. 1939. Hebung–Spaltung–Vulkanismus, *Geologische Rundschau*, **30**, 405–527.

DUNHAM, K. C. 1972. *Aeromagnetic map of Great Britain, Sheet 1, 1:625,000.* 1st edition. Institute of Geological Sciences, Ordnance Survey, Southampton.

EMELEUS, C. H. 1983. Tertiary igneous activity. *In:* CRAIG, G. Y. (ed) *Geology of Scotland.* Scottish Academic Press, Edinburgh, pp. 357–397.

FITCH, F. J., HOOKER, P. J., MILLER, J. A. & BRERETON, N. R. 1978. Glauconite dating of Palaeocene–Eocene rocks from east Kent and timescale of Palaeogene volcanism in the North Atlantic region. *Journal of the Geological Society of London*, **135**, 499–512.

FRISCH, W. 1981. Plate motions in the Alpine region and their correlation to the opening of the Atlantic ocean. *Geologische Rundschau*, **70**, 402–411.

FUCHS, K., VON GEHLEN, K., MÄLZER, H., MURAWASKI, H. & SEMMEL, A. (eds) 1983. *Plateau Uplift: the Rhenish Shield—a Case History.* Springer Verlag, Berlin, 411 pp.

GASS, I. G. & THORPE, R. S. 1976. *Igneous Case Study: The Tertiary Igneous Rocks of Skye, NW Scotland.* Open University Press, Milton Keynes, 38 pp.

ILLIES, J. H. 1969. An intercontinental belt of the world rift system. *Tectonophysics*, **8**, 5–29.

—— 1975. Intraplate tectonics in stable Europe as related to plate tectonics in the Alpine system. *Geologische Rundschau*, **64**, 677–699.

—— 1977. Ancient and recent rifting in the Rhinegraben. *Geologie en Mijnbouw*, **56**, pp. 329–350.

—— & GREINER, G. 1978. Rhinegraben and the Alpine system, *Bulletin of the Geological Society of America*, **89**, 770–782.

KAMMERLING, P. 1979. The geology and hydrocarbon habitat of the Bristol Channel Basin. *Journal of Petroleum Geology*, **2**, 75–93.

KARNER, G. D. & WATTS, A. B. 1983. Gravity anomalies and flexure of the lithosphere at mountain ranges, *Journal of Geophysical Research*, **88 (B12)**, 10449–10477.

KIRTON, S. R. & DONATO, J. A. 1985. Some buried Tertiary dykes of Britain and surrounding waters deduced by magnetic modelling and seismic reflection methods. *Journal of the Geological Society of London*, **142**, 1047–1058.

KNOX, R. W. O'B. 1984. Nannoplankton zonation and the Palaeocene/Eocene boundary beds of NW Europe: an indirect correlation by means of volcanic ash layers. *Journal of the Geological Society of London*, **141**, 993–999.

—— & HARLAND, R. 1979. Stratigraphical relationships of the early Palaeogene ash-series of NW Europe. *Journal of the Geological Society of London*, **136**, 463–470.

—— & MORTON, A. C. 1983. Stratigraphical distribution of early Palaeogene pyroclastic deposits in the North Sea basin. *Proceedings of the Yorkshire Geologists Association*, **44**, 355–363.

MACINTYRE, R. M., McMENAMIN, T. & PRESTON, J. 1975. K-Ar results from western Ireland and their bearing on the timing and siting of Thulean magmatism. *Scottish Journal of Geology*, **11**, 179–192.

MCLEAN, A. C. & DEEGAN, C. E. 1978. The solid geology of the Clyde sheet *(55°N/6°W)*. *Report of the Institute of Geological Sciences*, **78/9**, 114 pp.

MASSON, D. G. & PARSON, L. M. 1983. Eocene deformation on the continental margin SW of the British Isles. *Journal of the Geological Society of London*, **140**, 913–920.

MEISSNER, R., SPRINGER, M., MURAWSKI, H., BARTELSEN, H., FLÜN, E. R. & DÖRSCHNER, H. 1983. Combined seismic reflection–refraction investigations in the Rhenish Massif and their relation to recent tectonic movements. *In:* FUCHS, K. *et al.* (eds) *Plateau Uplift*. Springer, Berlin, pp. 276–287.

MUSSETT, A. E. 1986. ^{40}Ar–^{39}Ar step heating ages of the Tertiary igneous rocks of Mull, Scotland. *Journal of the Geological Society of London*, **143**, 887–896.

ODIN, G. S. & CURRY, D. 1985. The Palaeogene timescale: radiometric dating versus magnetostratigraphic approach. *Journal of the Geological Society of London*, **142**, 1179–1188.

PEGRUM, R. M. & MOUNTENEY, N. 1978. Rift basins flanking North Atlantic ocean and their relation to North Sea area. *Bulletin of the American Association of Petroleum Geologists*, **62**, 419–441.

PITMAN, W. C. & TALWANI, M. 1972. Seafloor spreading in the North Atlantic. *Bulletin of the Geological Society of America*, **83**, 619–646.

PLINT, A. G. 1982. Eocene sedimentation and tectonics in the Hampshire basin. *Journal of the Geological Society of London*, **139**, 249–254.

RICHEY, J. E. 1937. Some features of Tertiary volcanicity in Scotland and Ireland. *Bulletin Volcanologique*, Series **2**, 13–34.

SCHMITT, T. J. 1981. The West European stress field: new data and interpretation. *Journal of Structural Geology*, **3**, 309–315.

SENGÖR, A. M. C. 1976. Collision of irregular continental margins: implications for foreland deformation of Alpine-type orogens. *Geology*, **4**, 779–782.

—— BURKE, K. & DEWEY, J. F. 1978. Rifts at high angles to orogenic belts: tests for their origin and the Upper Rhine graben as an example. *American Journal of Science*, **278**, 24–40.

SMALLEY, L. S., & WESTBROOK, G. K. 1982. Geophysical evidence concerning the southern boundary of the London Platform beneath the Hog's Back, Surrey. *Journal of the Geological Society of London*, **139**, 139–146.

SPEIGHT, J. M., SKELHORN, R. R., SLOAN, T. & KNAAP, R. J. 1982. The dyke swarms of Scotland. *In:* SUTHERLAND, D. S. (ed) *Igneous Rocks of the British Isles*. Wiley, Chichester, pp. 449–459.

TRUMPY, R. 1960. Palaeotectonic evolution of the central and western Alps. *Bulletin of the Geological Society of America*, **71**, 843–908.

VANN, I. R. 1978. The siting of Tertiary vulcanicity, *In:* BOWES, D. R. & LEAKE, B. E. (eds) *Crustal Evolution in Northwestern Britain and Adjacent Regions*. Seel House Press, Liverpool, 393–414.

VILLEMIN, T., ALVAREZ, F. & ANGELIER, J. 1986. The Rhinegraben: extension, subsidence and shoulder uplift. *Tectonophysics*, **128**, 47–60.

ZIEGLER, P. A. 1978. North-western Europe: tectonics and basin development, *Geologie en Mijnbow*, **57**, 589–626.

—— 1982. *Geological Atlas of Western and Central Europe*. Shell International Petroleum, Maatschappij B.V., 130 pp.

J. F. DEWEY, Department of Earth Sciences, University of Oxford, Parks Road, Oxford OX1 3PR, UK.

B. F. WINDLEY, Department of Geology, University of Leicester, University Road, Leicester LE1 7RH, UK.

Dipping Reflectors and
NE Atlantic Evolution

Deep crustal structure and magmatic processes: the inception of seafloor spreading in the Norwegian–Greenland Sea

J. C. Mutter & C. M. Zehnder

SUMMARY: An extensive investigation into the deep crustal structure of the conjugate Norwegian and E Greenland margins utilizing two-ship Expanded Spread and Wide Aperture CDP Profiling has led to the identification of a unique suite of crustal structures developed during the onset of seafloor spreading. On both margins, a region of low velocity crust was found proximal to marginal escarpments. Immediately seaward of the low velocity crustal block seaward-dipping reflectors occur within thick crust that is interpreted to be wholly igneous, a product of voluminous melt production enhanced by convective partial melting processes. This earliest formed oceanic crust is actually several kilometres thicker than the low velocity crust or the adjacent continental crust. Normal oceanic crustal thicknesses are not produced until about 5 My after the initiation of spreading when convective partial melting has abated and the magma budget has reduced. The extrusive unit reduces to normal thicknesses long before the intrusive and plutonic section, indicating that the thickening or thinning of oceanic crust is accomplished primarily by variations in the latter.

Only a few years after the original proposition of seafloor spreading by Dietz (1961), Vine & Matthews (1963), Hess (1965) and Vine (1966) Lamont–Doherty's research vessel *VEMA* entered the Norwegian Sea for the first cruise in an extended campaign of geophysical studies. Several earlier studies (Nansen 1904; Stocks 1950; Litvin 1965; Johnson & Eckhoff 1966; Johnson & Heezen 1967; Vogt *et al.* 1970) had defined the bathymetry and had broadly outlined some geophysical parameters such as the gravity and magnetic fields. During five surveys, each comprising several cruises, R/V *VEMA* mapped the magnetic lineation pattern, shallow seismic structure, gravity and bathymetry throughout the Norwegian–Greenland Sea from 60–80°N. Talwani & Edholm (1972, 1977) synthesized these data to provide a description of the plate-tectonic history of the Norwegian–Greenland Sea and the Norwegian continental margin. Nunns (1983) compiled all available ship-recorded magnetic data with airborne magnetic measurements to map the lineation pattern in the same area. He also redefined some of the rotation poles that describe the stages in the opening of the Norwegian–Greenland Sea and the separation of Jan Mayen Ridge from Greenland.

The seafloor-spreading history of the region is very well known in a kinematic sense. That is, plate-motion vectors and the timing of plate motions have been well constrained by the present data set and analyses, and are unlikely to change significantly were new data to become available.

Despite this, several problems have been identified regarding the initiation of seafloor spreading that demand that we examine mechanisms of rifting and plate accretion more closely. A kinematic description of the early opening provides little insight into why, for instance, we observe that the oldest oceanic crust in the Norwegian–Greenland Sea appears to be unusually shallow and anomalously thick (Talwani & Eldholm 1972, 1977; Mutter *et al.* 1984). This relationship has been established from the results of single and two-ship multichannel seismic reflection and refraction studies conducted over the last eight years (Hinz & Weber, 1976; Talwani *et al.*, 1981; Mutter *et al.*, 1984, 1985; Zehnder *et al.*, 1985; Zehnder & Mutter, 1986). Reflection transects across the margin (Figs 1 & 2) elucidate the structure and setting of the thick oceanic crust. Figure 2 describes the reflection seismic signature of a typical margin crossing (Fig. 1 gives the location). As Talwani & Eldholm (1972) previously noted, the Vøring Plateau Escarpment (VPE) divides the margin into two distinct structural provinces: a thickly sediment-filled basin landward, and an elevated basement block seaward which merges into the normal ocean crust further W. Formation of the sedimentary basin in the Upper Jurassic and Cretaceous considerably pre-dates the initiation of seafloor spreading in the lowermost Eocene (Bøen *et al.* 1984). In fact, the spreading event was not accompanied by renewed tectonism in the sedimentary basin; the stratigraphic level corre-

From MORTON, A. C. & PARSON, L. M. (eds), 1988, *Early Tertiary Volcanism and the Opening of the NE Atlantic,* Geological Society Special Publication No. 39, pp. 35–48.

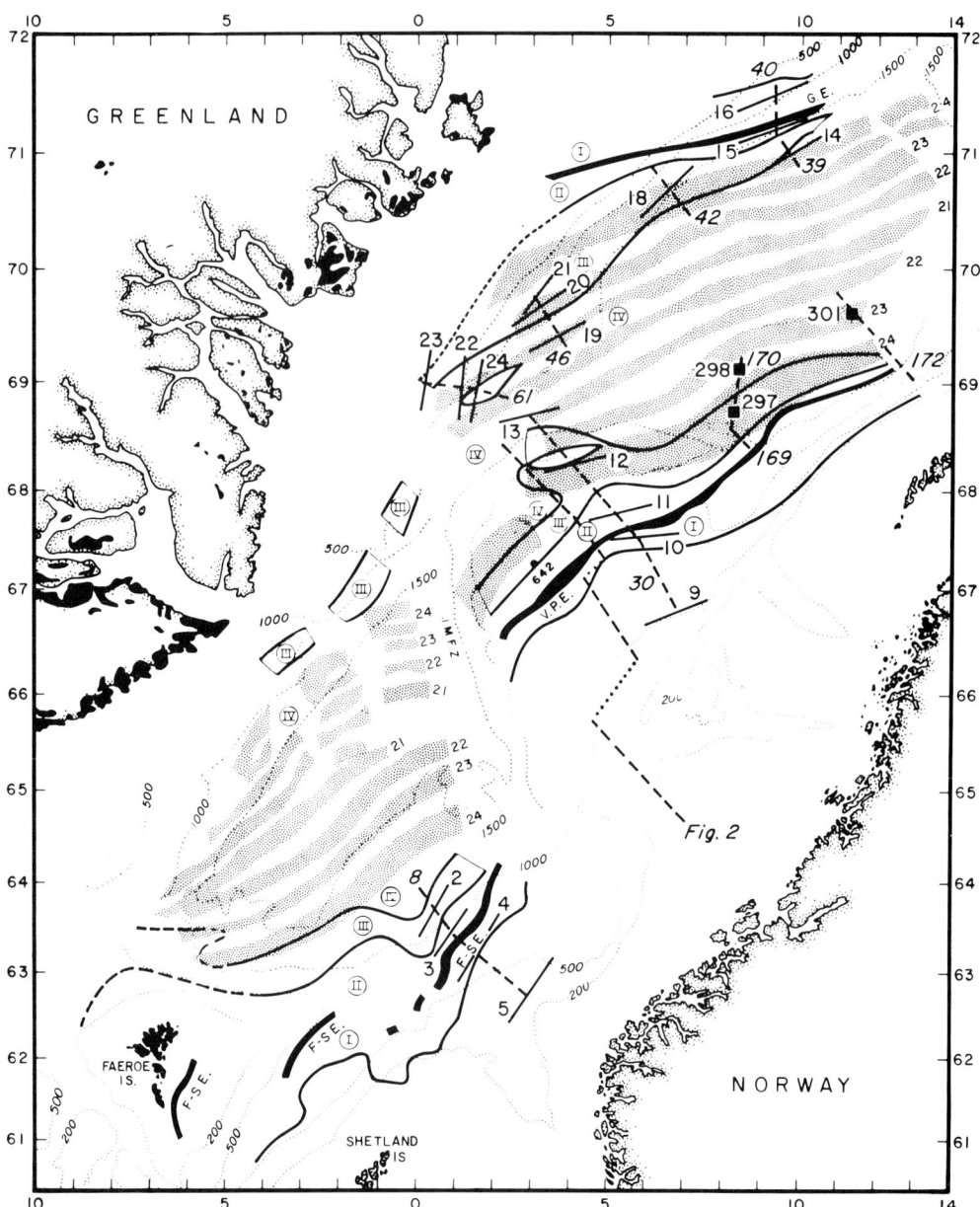

FIG. 1. The Norwegian–Greenland Sea at 50 Ma (reconstructed to anomaly 21 time using the poles of Talwani & Eldholm 1972). ESPs (numbered solid lines) are located along two-ship reflection transects (dashed lines, script numbers) that resolve continental basins, regions where the continental basins are capped by landward-dipping volcanic flows (I), regions characterized by a basement high landwardly terminated by an escarpment (II), regions where seaward-dipping reflectors are prominent (III), and the outlying oceanic crust (IV). Shading indicates magnetic lineations, which merge into the region of seaward-dipping reflectors on the Vøring Plateau, as well as the E Greenland margin.

sponding to break-up is about midway between horizons marked BT (base Tertiary) and MO (mid-Oligocene) in Figure 2 and is a smooth disconformity in the basin's sediment fill.

The basalt sequence has been drilled on two occasions (DSDP Leg 38, Talwani et al. 1976; ODP Leg 104, Eldholm et al. 1986). It is characterized by a seaward-dipping reflection

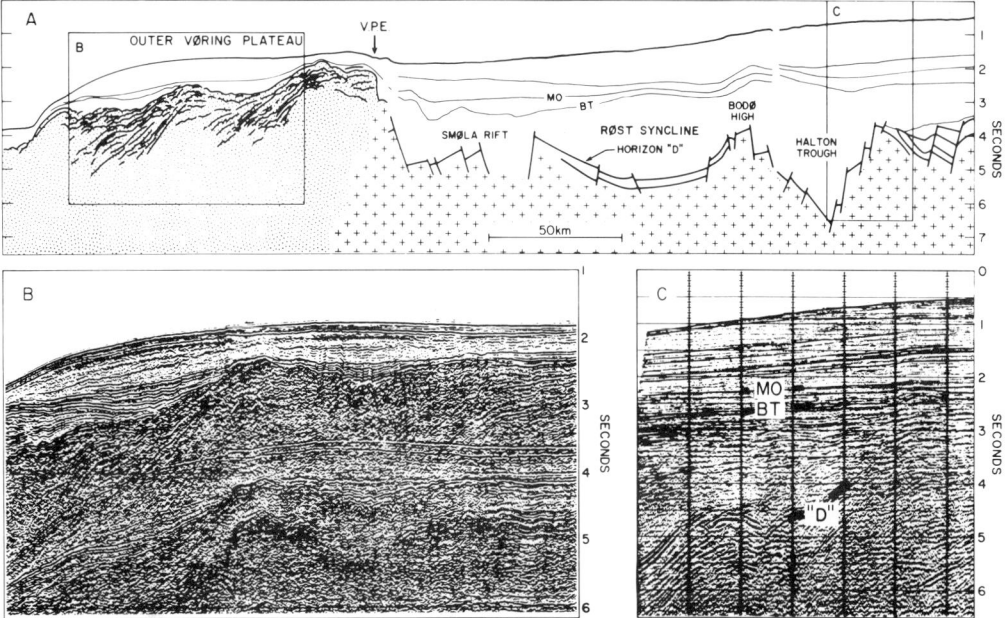

FIG. 2. The upper panel is a composite line drawing interpretation of the structure of the Vøring Plateau and the landward Norwegian margin defined by seismic reflection data, exploration, and scientific drilling. The section to the E of the VPE is taken from two illustrations in Bøen *et al.* (1984); that to the W is a line drawing of L-DGO, 24-channel seismic Line 164 (Mutter *et al.* 1984). The boxed regions on the upper section give the approximate locations of example reflection profiles shown beneath; the left is from L-DGO Line 194, the right is Line 162. The random dot symbol indicates igneous units, the crosses show continental crust.

pattern, the interpretation of which has led to several models for the emplacement of the igneous structures. Hinz (1981) has proposed that a highly distended continental crust may underlie parts of the basalt sequence, and that the igneous complex forms largely during late rift tectonism. Mutter *et al.* (1982) invoke 'subaerial seafloor spreading' to account for the seaward dips and arcuate, offlapping shapes to the reflectors within the basalt complex. They cite similarly shaped lava units in E Iceland as being analogous to the Norwegian margin units and therefore propose that the entire crust beneath the seaward-dipping units is an oceanic igneous complex. Roberts *et al.* (1984) have described a scheme for forming dipping reflector units that involves massive dyke injection into a distended continental crust. Smythe (1983) proposed a model akin to that of Hinz (1981), and Skogseid & Eldholm (1987) also advance a similar model.

While these various models may seem somewhat at odds with one another they are fundamentally similar in that they recognize that the inception of seafloor spreading was attended by a massive volcanic episode that built a thick igneous body between continental and normal thickness oceanic crustal sections. These structures, and the implied magmatic event, are not unique to the Norwegian margin (Hinz 1981). A general mechanism is therefore required that can account for a pulse of magmatic activity during passive margin formation. However, because not all margins exhibit these large igneous bodies, the mechanism must operate with variable strength, creating significant pulses of magmatism at some margins, while having no detectable effect at others. In this contribution we describe observations of the deep crustal structure of the conjugate Norwegian and E Greenland margins and a magmatic model developed therefrom.

Seismic structure of the conjugate margins

An extensive compilation of seismic reflection observations on the margins of the Norwegian–Greenland Sea (Hinz *et al.* 1987) allows the identification of regions where seaward-dipping reflectors occur (see Figs 1 & 2). They are developed seaward of marginal escarpments on the Norwegian margin and also on the E Greenland margin in places where seismic investigations have been able to penetrate the ice-covered shelf. In other locations on the E Greenland margin their association with marginal escarpments has not been determined. The seaward-dipping reflectors are frequently associated with crust of anomaly 24 age. They extend oceanward to include anomaly 23 age crust near the Vøring Plateau on the Norwegian margin, and the conjugate position on the Greenland margin. South of the Jan Mayen Fracture Zone (JMFZ) the dipping reflectors developed largely prior to anomaly 24 time. A temporal and spatial association of the seaward-dipping reflectors on the conjugate margins is established by a remarkably symmetric pattern that strongly implies formation by a seafloor-spreading process (Hinz *et al.* 1987). We describe here the deep seismic refraction data obtained by Expanded Spread Profiling (ESP) in 1983 (Zehnder *et al.* 1985; Zehnder & Mutter 1986), and discuss its implications.

The ESP technique utilizes the two-ship experimental configuration of refraction data acquisition described by Stoffa & Buhl (1979). The data were processed using standard techniques described in the NAT Study Group (1985), and Diebold *et al.* (1988). Demultiplexed data were reduced to 8 km/s and summed in 50 m bins at 50 m spacing in accordance with the shot density. Multiples were removed by time and distance windowing, and slowness stacks were performed on overlapping subarrays. Semblance statistics were calculated on each subarray and then applied as a filter. Filtered slowness stacks were linearly transformed and summed to produce tau-*p* representations of the data, which were inverted using the tau-sum method of Diebold & Stoffa (1981) to obtain trial $V(Z)$ solutions. Iterative forward modelling in X-T and tau-*p* produced final best-fit solutions incorporating amplitude information for the deep arrivals. The exceptional data quality resulted in well determined deep crustal structure at most of the ESP locations. Data are displayed in Figures 3 & 5. For reproduction purposes, the data were reduced to 8 km/s and summed in 150 m bins at 50 m spacing to enhance deep arrivals. Trace equalization based on 10 traces was applied for a more uniform background. The data were then filtered with a high cut of 15 Hz, 5 Hz ramp and a low cut of 6 Hz with a 3 Hz ramp, and gained appropriately.

To elucidate concisely similarities and differences in crustal structure across the margins, the solutions have been compiled into isovelocity sections. Isovelocity contours indicate the depth at which particular velocities are encountered, whether they occur at discontinuities or within gradients. We consider first the more familiar Vøring Plateau region and its conjugate on the E Greenland margin, then move northward to describe additional transects along both margins, and conclude with a transect across the Faeroe–Shetland Escarpment (FSE) of the Norwegian margin S of the JMFZ.

Vøring Plateau and conjugate Greenland margin

The transect (Line 30) across the Vøring Plateau (see Figs 1 & 3) has ESPs located in continental crust (ESP's 9 & 10), oceanic crust (ESP 13), crust associated with the seaward-dipping reflectors (ESP 12) and a complex intermediate structure (ESP 11). Within the continental sedimentary basin (ESP's 9 & 10) the 6.5 km/s discontinuity probably identifies the sediment–basement interface at approximately 12 km depth. The lowermost 3–4 km of crust has an average velocity in excess of 7.4 km/s, a velocity elsewhere interpreted as evidence for underplating by upper mantle partial melts ponded at the base of the crust (Drummond & Collins 1986; Furlong & Fountain 1986; LASE 1986; White *et al.* 1986). If the 7.4 km/s layer observed here is an underplating, it must have been emplaced during a pre-Tertiary rifting episode as it is not continuous with structures in the lower oceanic crust.

The oceanic crust at ESP 13 is anomaly 22–23 age, and is notably thicker than normal oceanic crust of anomaly 20 age in the nearby Lofoten basin (Mutter *et al.*, 1984). ESP 13 shows a typical oceanic, two-gradient basement structure with the upper layer thickness approximately equal to that of normal oceanic crust, but with the lower layer (> 6.5 km/s) approximately twice normal thickness. Moving landward into the anomaly 24 age crust at ESP 12, where dipping reflectors are observed (see Figs 1, 2 & 3), the crust can be directly correlated with that at ESP 13. It shows a two-fold thickening in the upper crustal structure and a further increase from 11 km to 17 km thickness in the lower crustal structure, as determined from X-T raytracing of pre-critical Moho arrivals which are apparent over a range of 70 km (Fig. 3, centre panel). The lower crust is nearly four times thicker at ESP 12 in the dipping reflector units than at anomaly 20 age crust in the Lofoten Basin.

Intermediate between oceanic and continental crust is the low velocity crust of ESP 11, located just seaward of the VPE (Fig. 3, lower panel). The upper crust shows several velocity discontin-

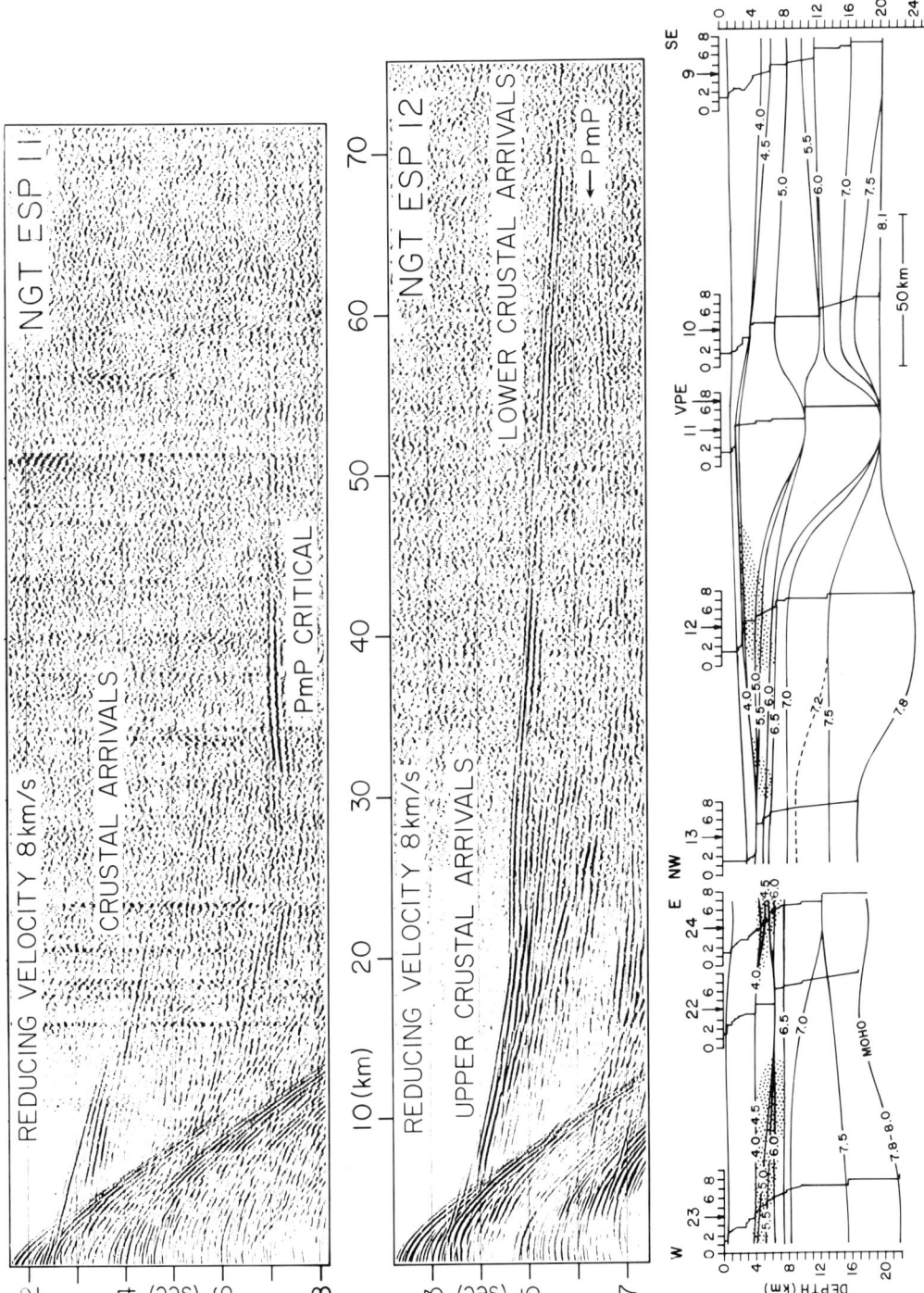

FIG. 3. ESP 11 (upper panel), ESP 12 (centre panel), and isovelocity contours (lower panel) for Greenland ESP's 23, 22 and 24 along Line 61 (left) and Norway ESP's 13, 12, 11, 10 and 9 along Line 30 (right); stipple denotes seaward-dipping reflectors. See text for further discussion.

uities and has an average velocity of only about 4.8 km/s. The lower crust is complex, apparently lacking internal reflecting horizons and has an average velocity of about 6.1 km/s. The lower boundary is based on brightening of a deep reflector interpreted as a critical Moho reflection from a discontinuity at 20 km depth (Fig. 3, upper panel). The crust at ESP 11 is strikingly different from either continental or thickened oceanic crust.

The conjugate Greenland margin (Line 61) has dipping reflector sequences that very nearly mirror those known on the Norwegian margin (Hinz *et al.* 1987). The associated deep crustal structure determined from ESP's 22, 23 and 24 (see Fig. 3, lower panel) demonstrates the same trends observed on the Norwegian margin. Thickening of oceanic crust first occurs by a two-fold thickening of the lower crustal structure (ESP's 22 and 24). The very thick crust associated with the main seaward-dipping reflector sequence (ESP 23) has an upper crustal structure that is nearly twice the thickness of that at ESP 24, in addition to a further thickened lower crustal structure.

E Greenland Line 46

Approximately 150 km NE of Line 61 (see Fig. 1) a transect of ESP's was located in oceanic crust of anomaly 23 age which displays seaward-dipping reflectors (ESP's 20 & 21) and oceanic crust of anomaly 22 age (ESP 19) which lacks reflecting events. Here, the isovelocity contours (Fig. 4) do not clearly indicate that crustal thickening occurs in association with the seaward-dipping reflectors (ESP's 20 & 21). Although there is a slight increase in the thickness of the upper crust where dipping reflectors are

observed, the lower crustal structure and depth to Moho are not well constrained. The crustal thickness at ESP 19 is very nearly identical to that known for anomaly 20 age crust of the Lofoten Basin (Mutter *et al.* 1984).

Greenland escarpment transect and the conjugate Norwegian margin

Analyses of ESP's collected along Lines 39/40 and 42 (Fig. 5) resulted in determinations of crustal thicknesses in continental crust (ESP 16), crust seaward of the Greenland Escarpment (GE) and landward of the dipping sequences (ESP 15), crust of anomaly 24 age associated with seaward-dipping reflectors (ESP 18) and oceanic crust just seaward of this region (ESP 14). The crustal thickness determinations are compiled into a single isovelocity section (see Fig. 5, lower panel) which highlights the location of the seaward-dipping reflectors known from reflection data along Line 39/40. Although ESP 18 is actually located within the dipping reflector sequence nearly 150 km to the SW, the thickness determination there is generally consistent with the trends known from ESP's 14, 15 and 16, and suggests that this margin has a fairly uniform along-strike structure.

The oceanic crust of ESP 14 is approximately 7 km thick. The crust associated with the seaward-dipping reflectors at ESP 18 shows a two-fold thickening in the upper crustal structure with respect to the crust at ESP 14, and an increase from 6 to 10 km thickness in the lower crustal structure (>6.5 km/s), accommodated primarily by the presence of crust with velocities in excess of 7.5 km/s (see Fig. 5, lower panel). The depth to Moho at ESP 18 is based on $X-T$ raytracing of pre-critical Moho reflections that appear through-

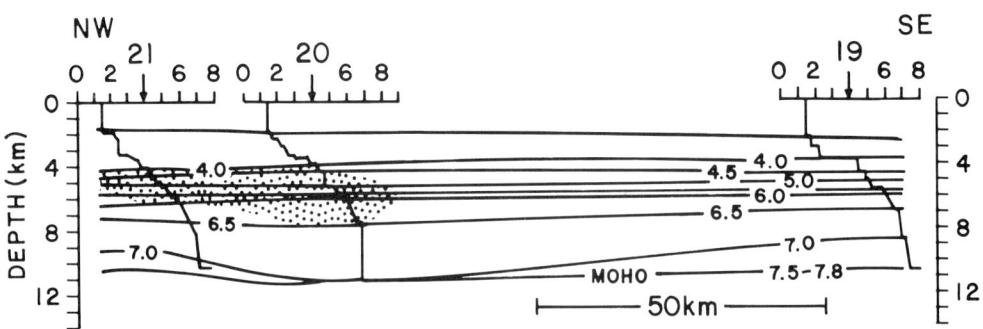

FIG. 4. Isovelocity contours for Greenland ESP's 21, 20 and 19 along Line 46; shading denotes seaward-dipping reflectors.

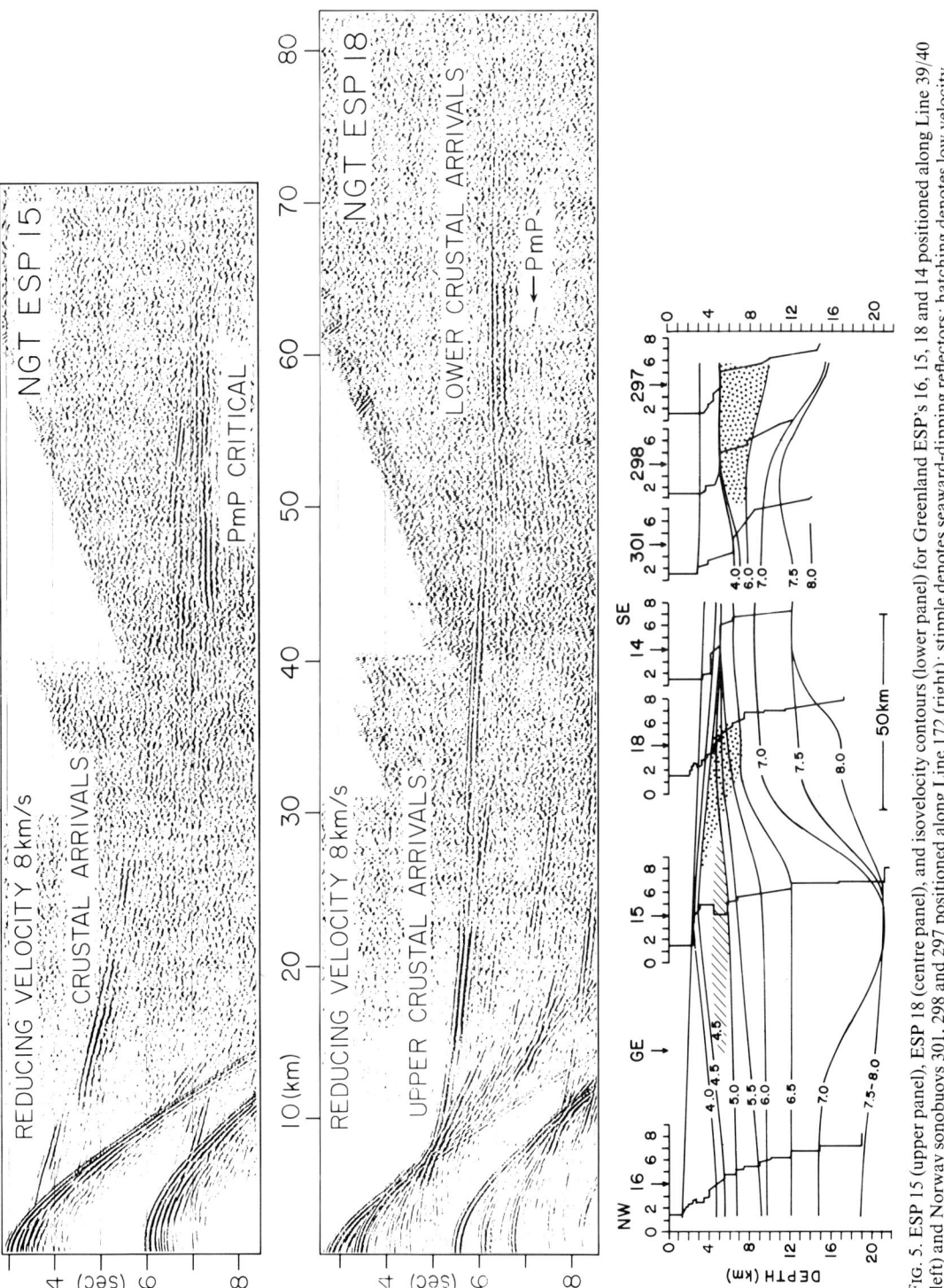

FIG. 5. ESP 15 (upper panel), ESP 18 (centre panel), and isovelocity contours (lower panel) for Greenland ESP's 16, 15, 18 and 14 positioned along Line 39/40 (left) and Norway sonobuoys 301, 298 and 297 positioned along Line 172 (right); stipple denotes seaward-dipping reflectors; hatching denotes low velocity crust overlain by higher velocity crust. See text for further discussion.

out the entire range of the ESP (see Fig. 5, centre panel).

The continental crust of ESP 16 is highly layered, the lowermost layer having an average velocity of 7.2 km/s. This layer does not extend into the intermediate region of ESP 15 where the lower crust is notably slower with an average velocity of about 6.8 km/s overlying Moho at 20 km depth (see Fig. 5, lower panel). This boundary is based on brightening of a deep reflector interpreted as a critical Moho reflection (see Fig. 5, upper panel).

The conjugate Norwegian margin structure is reconstructed from sonobuoy results of Mutter *et al.* (1984) and also indicates that thickening of oceanic crust associated with seaward-dipping reflectors occurs by a two-fold increase in upper basement structures and an unknown amount of thickening in the lower basement structure (see Fig. 5).

Faeroe–Shetland escarpment transect

South of the JMFZ a suite of ESP's was collected along Line 8 on the Norwegian margin (see Figs 1 & 6). The ESPs are located in the continental sedimentary basin 100 km SE of the Faeroe–Shetland Escarpment (FSE; ESP 5), crust 20 km landward of the FSE (ESP 4), crust 10 km seaward of the FSE (ESP 3) and seaward-dipping reflector crust formed prior to anomaly 24 time (ESP 2). The continental crust at ESP 5 is similar to that of ESP's 9 and 10 N of the JMFZ. However, isovelocity contours identify a complex crustal structure in the vicinity of the FSE. The

crust at ESP 4 displays a greater thickness of lower velocity material than that as ESP 5 (Fig. 6); at ESP 4 there is an 8 km thick unit with an average velocity of 5.8 km/s overlying another 8 km thick unit having a 6.7 km/s average velocity. This velocity structure is not a likely extension of the continental crust identified at ESP 5.

Seaward of the FSE, ESP 3 resolves a 5 km thick upper crust with velocities increasing from 4.0–6.0 km/s, overlying a 10 km layer with an average velocity of 6.1 km/s (Fig. 6). The upper crust is quite similar to that at ESP 2, located within the seaward-dipping reflector crust 30 km to the NW. The lower crust, however, differs greatly from oceanic or continental crust.

The crustal structure at ESP 2 differs somewhat from that of ESP 12 (located within the seaward-dipping reflectors on the Vøring Plateau) in that the upper crust (4.0–6.5 km/s) is thicker and contains more velocity discontinuities. The lower crust (>6.5 km/s) includes a 5 km thick zone with an anomalously low velocity of 6.3 km/s at 15 km depth. The absolute thickness of crust is not known, but our results indicate that it is in excess of 18 km.

Summary of observations

The isovelocity sections presented above show that the deep crustal structure of the Norwegian and Greenland margins changes dramatically near the margin escarpments; although the total crustal thickness remains about the same as that of the continental basins, crustal velocities are

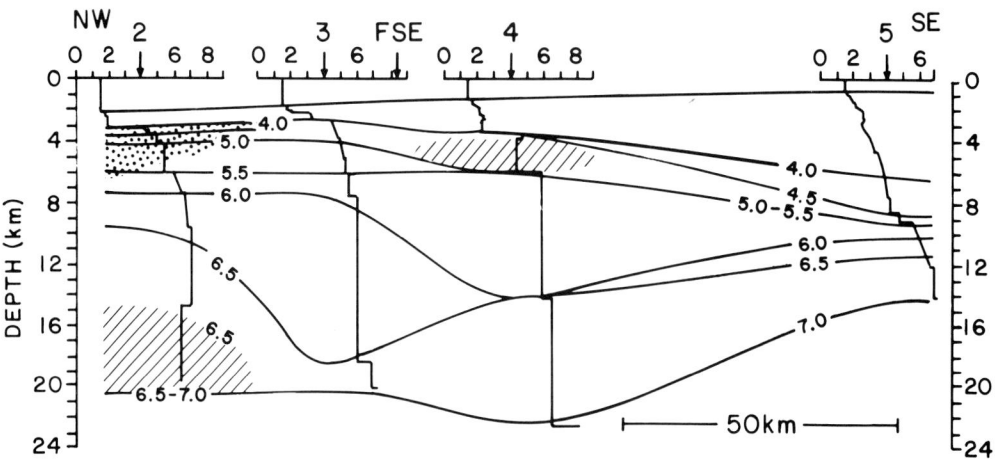

FIG. 6. Isovelocity contours for Norway ESP's 2, 3, 4 and 5 located along Line 8. Shading denotes dipping reflectors; hatching denotes the presence of low velocity crust overlain by higher velocity crust.

significantly lower than either continental or oceanic structures proximal to the escarpment. Seaward of the low velocity region crustal thickness increases to as much as 20 km where seaward-dipping reflectors occur. The two-gradient basement structure in the crust beneath the seaward-dipping reflectors is much like that of oceanic crust except that the upper and lower layer thicknesses are substantially enhanced. The crust thins seaward towards normal oceanic thicknesses and, as it does, the upper basement layer thins to normal oceanic thickness well before the lower basement layer attains normal thickness. The most prominent changes in crustal structure occur in the lower crust.

The Faeroe–Shetland transect is unusual in that low velocity crust occurs both seaward and landward of the escarpment. In addition, a low velocity zone occurs within the crust beneath the dipping reflectors.

Discussion

Igneous stratigraphy

The initiation of seafloor spreading in the Norwegian–Greenland Sea led to the development of passive margin structure profiles showing a suite of distinct characteristics. One of the most distinctive is a region of low velocity crust located on the seaward side of basement escarpments that separate the continental sedimentary basin from the thick crust where seaward-dipping reflectors are developed. The thickened crust thins seaward into oceanic crust primarily by a reduction in thickness of the lower crustal components, much like that observed in thinning of *bona fide* oceanic crust (Mutter & NAT Study Group 1985). There is no seismic evidence for a continuation of continental crustal structures into the region where seaward-dipping reflectors occur, even for the unusual crustal structure underlying seaward-dipping reflectors of the FSE transect, and we regard the crust underlying the seaward-dipping reflectors as wholly igneous.

The formation of the low velocity crust can be reconciled with events occurring during the onset of spreading activity. Igneous intrusions into continental crust cannot alone account for the low velocities we observe, unless the intrusions were substantially hydrated mafic sequences. Initial rifting and the inception of seafloor spreading occurred in a sedimentary basin covered by perhaps 1000 m of water (Mutter 1984; Eldholm & Mutter 1986). Hence, massive crustal fracturing that attended this tectonism was probably accompanied by substantial water influx that would considerably modify any associated magmatism. The low velocity crust might represent serpentinized mafic intrusive and plutonic rocks incorporating continental fragments at depth, overlain by intercalated sediments, hydromagmatic eruptives and basalts nearer the surface. The prevalence of low velocity crust seaward of the FSE (see Fig. 6) may indicate that deep fracturing and water penetration occurred over a greater area and was longer lived than in regions N of the JMFZ. The subsequent prolific basaltic eruptive activity is interpreted as indicating a well established magmatic system capable of building thick oceanic crust associated with seaward-dipping reflectors.

Drilling at Site 642 of ODP Leg 104 (Eldholm *et al.* 1986) revealed a volcanic stratigraphy that is apparently consistent with the events postulated above. Beneath a sequence of extruded flows forming the dipping events, an entirely different volcanic unit was encountered that included major components of explosive eruptive products such as ignimbrites. The lower unit is substantially less dense than the upper flows, and we suggest that it may mark the top of the low velocity crustal block we mapped from our seismic measurement.

The thinning of oceanic crust away from the margin reflects a diminishing magma budget (Fig. 7). Initially, the extrusive, intrusive and plutonic components diminish in thickness. The extrusive unit, however, approaches normal oceanic thicknesses much sooner than the intrusive and plutonic unit. The overall variation in crustal thickness occurs primarily within the intrusive and plutonic components of the oceanic crust. Increased extrusive layer thicknesses occur only when intrusive layer thicknesses are greatest. The production of crust with thicknesses as great as 20 km clearly requires an unusual magmatic flux during the earliest phase of plate accretion.

Convective partial melting

Beaumont *et al.* (1982), Foucher *et al.* (1982), McKenzie (1984), White *et al.* (1986) and Klein & Langmuir (1987) have discussed melting processes associated with rifting and seafloor spreading that include mechanisms by which unusual thicknesses of igneous crust could develop. During normal seafloor spreading and uniform extension, melt production occurs through passive upwelling along mantle adiabats. In the absence of elevated temperatures this process can yield a thickness of igneous rocks with a maximum equal to normal oceanic crust; approximately 6 km. However, if deep mantle

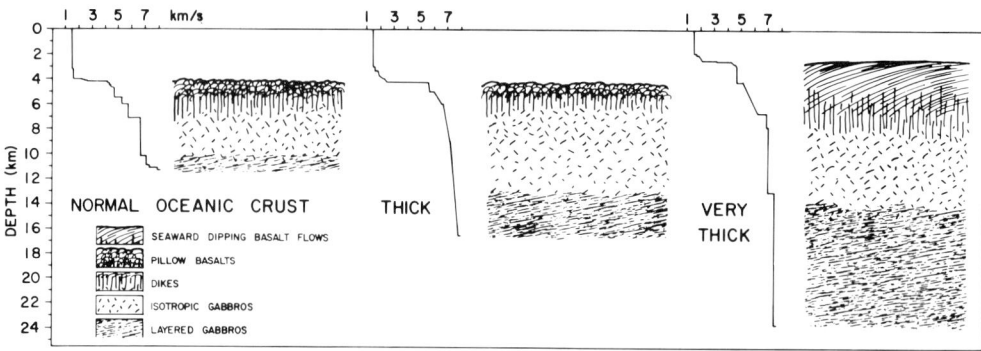

FIG. 7. Thinning of oceanic crust occurs primarily by the reduction in thickness of the intrusive and plutonic components and secondarily by a reduction in thickness of extrusive sequences. Very thick crust (e.g. ESP 12, right) is associated with seaward-dipping reflectors and has greatly enhanced thicknesses of intrusive, plutonic and extrusive components. As the crust thins (e.g. ESP 13, centre) the extrusive unit approaches normal thickness, whereas the intrusive and plutonic unit is double that of normal oceanic crust; 10–15 Ma after the initiation of seafloor spreading, normal thickness crust is produced (ESP 4 from Mutter, 1984, left).

temperatures are raised somewhat, the mantle material rising along an adiabat will intersect the solidus deeper than cooler mantle. Melting will then take place over a greater column of mantle, a larger total melt will be liberated, and a greater igneous crustal section will result. Klein & Langmuir (1987) have correlated major element chemistry of oceanic basalts with melt fraction and crustal thickness to show that this is a viable mechanism to account for crustal variability in the deep oceans. Beaumont *et al.* (1982) and Foucher *et al.* (1982) describe melting due to pressure release during lithospheric extension associated with rifting and note the need to invoke raised mantle temperatures to generate thick igneous underplating. The LASE Study Group (1986) have invoked that mechanism to explain the occurrence of an anomalous deep crustal layer off the US E Coast.

While we accept the general principle that enlarged igneous crustal thicknesses can be developed by the effect of increased mantle temperatures on the melting process we note that, to account for the structure profiles illustrated in Figures 3 to 6, mantle temperatures would need to be highest during the initiation of spreading and decline during subsequent spreading history. This requires the somewhat *ad hoc* notion that a hot-spot influenced spreading early in the separation history and was extinguished after a few million years, despite the fact that the most likely hot-spot source, Iceland, continues to be active to the present.

We have recently advanced a model to account for the extra melt production during the early separation history of a new accreting plate

boundary that does not require a transient heat input (Mutter *et al.*, in press) and can account for the observation of enlarged igneous crustal complexes that are present at many passive margins world-wide (as Hinz, 1981, has noted). Figure 8 summarizes the model.

With specific reference to the history of the Norwegian–Greenland Sea we note that a rifting event affected the region in the Late Jurassic,

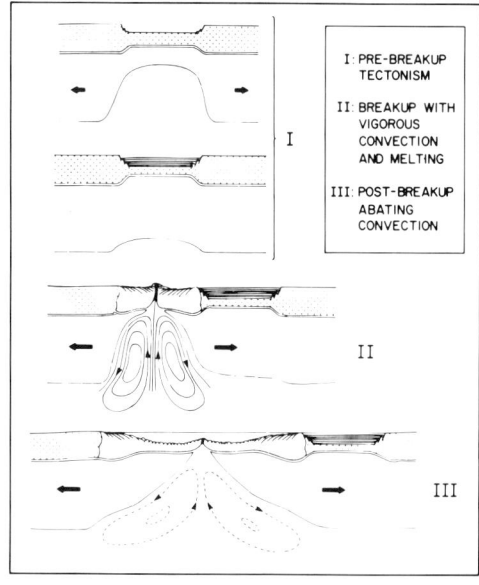

FIG. 8. The evolutionary scheme for the development of thick igneous crustal units adjacent to passive margins proposed by Mutter *et al.* (in press).

synchronously with that in the North Sea (Bøen *et al.* 1984). Sclater & Christie (1980) and Barton & Wood (1984) have shown that subsidence patterns in North Sea exploration wells can be adequately modelled as a passive rifting event from a theory developed by McKenzie (1978) to describe uniform extension of the lithosphere. While the subsidence history of the basins of the Norwegian margin are less well known (Buckovics *et al.* 1984), we assume that a dominantly passive mechanism of extension lead to thinning of the crust and lithosphere as shown in panel I of Figure 8. The thinning was probably accompanied by some melt liberation in the upper mantle, resulting in a late Jurassic underplating now manifest as the fast, lowermost layer of crust at ESP's 9 and 10 (see Fig. 3) of the Norwegian margin.

During Cretaceous and Palaeocene time the thinned crust subsided and acquired, in places, extremely thick sedimentary sequences (Eldholm & Mutter 1986). The lithosphere beneath the basin thermally equilibrated but, because the extension factors are unknown we cannot determine whether or not the lithosphere reacquired its original thickness. Panel II of Figure 8 shows, schematically, a deep sedimentary basin with somewhat thinned lithosphere, and pertains to a time immediately prior to the onset of seafloor spreading in the late Palaeocene.

The late Palaeocene–early Eocene spreading was characterized by two distinctive and related phenomena:

(1) The lack of renewed extension in the basins of the Norwegian margin; typical reflection profiles of the margin show no disruption to the sedimentary sequences at the stratigraphic level of the Palaeocene–Eocene units (see Fig. 2). Either renewed extension was highly localized and structural evidence for it has been subsequently buried beneath the ensuing volcanic sequences or, as we believe, spreading commenced following a relatively abrupt fracture of the whole lithosphere.

(2) The thickness of oceanic crust formed during the earliest spreading was unusually large and diminished as spreading continued.

To account for these two observations we invoke the effect of mantle convection in an extensional environment. Buck (1983, 1986) has shown that when lithosphere is extended by uniform stretching (McKenzie 1978) strong temperature gradients will be produced between the relatively cold continental lithosphere and the relatively hot upwelling asthenosphere beneath the extended crust and lithosphere. This thermal gradient causes density instabilities that can induce small-scale convection in the upper mantle beneath the extended lithosphere. Buck (1984) and Steckler (1985) have shown how this phenomenon can account for anomalous uplift of rift flanks (e.g. Gulf of Suez). Similarly, if the inception of seafloor spreading is brought about by the relatively abrupt juxtaposition of hot upwelling asthenosphere against relatively cold lithosphere, convection will result. The convection draws deep, hot mantle upward at a greater rate than purely passive upwelling can achieve (Panel III, Fig. 8). This material will cross the solidus, begin to melt and move to the surface to form an igneous crustal complex. The convective transport acts to enhance the amount of material passing upward and into the zone of melting. Thus a greater amount of melting will occur and a larger igneous crustal sequence will be formed.

As spreading continues (Panel IV, Fig. 8), the relatively cold, thick continental lithosphere separates. Thermal gradients attenuate and convection will slow. The additional melt production generated by the convective upward transport of mantle material will diminish also. A few million years into the history of spreading the input from convective melting will be negligible and a normal oceanic crustal thickness will develop. This progressive diminution of magma supply satisfactorily accounts for the progressive reduction in crustal thickness described above.

Conclusions

An extensive study of seismic refraction measurements reported here establishes that the transition from continental to oceanic crust does not occur by continuous, progressive thinning of stretched continental crust toward an oceanic section of normal thickness. Instead, we observed a deep, sediment-filled basin that formed well before the onset of seafloor spreading adjacent to a major basement structure displaying unusually low crustal velocities. This structure separates the basin from a massive igneous crustal complex having Moho depths greater than those beneath the basin. The thickest igneous crust is associated with seaward-dipping reflectors, and at some locations with magnetic lineation 24, the oldest recognized in the basin. The igneous complex thins progressively seaward, initially by a reduction in layer thickness of both upper and lower crust, then by loss of lower crustal layer thickness alone. The upper layers thus obtain normal oceanic thickness much sooner than the lower layers. Based on the observations we suggest a generalized igneous stratigraphy to describe thick igneous crustal sections. We regard the complete igneous sequences seaward of the low-velocity basement structure as accreted at a divergent plate boundary.

To account for the major pulse of magmatism that attended the initiation of seafloor spreading we invoke the mechanism of convective partial melting of Mutter *et al.* (in press). Initiation of spreading began by a relatively rapid and clean fracture of a continental lithosphere that may have been weakened by an earlier period of stretching which formed the deep Mesozoic sedimentary basin. Strong lateral thermal gradients then stimulated small-scale convection in the upper mantle which caused deep, hot mantle material to be transported upward into higher levels where melting was promoted at a rate that exceeds that associated with passive upwelling. The extra melt generated in this way is represented at the surface by a substantial increase in the thickness of extrusive igneous crust as well as the intrusive and plutonic sections produced during the earliest spreading.

As spreading progresses and the conjugate continental fragments retreat, convective melting abates and normal-thickness extrusive units occur. Crustal thicknesses produced at the spreading axis gradually reduce to those produced at mature spreading centres by a reduction in the intrusive and plutonic components of the crust. We believe that this mechanism accounts for the crustal profile mapped on the conjugate margins off Norway and E Greenland. We further emphasize that passive margins world-wide display a variable occurrence of seaward-dipping volcanic units and associated thick igneous crustal complexes, and we suggest that they may in general develop when seafloor spreading follows from a rapid rifting that leaves an abrupt transition between cold lithosphere and upwelling hot asthenosphere.

ACKNOWLEDGMENTS: Data presented here were acquired during a joint experiment with Lamont–Doherty Geological Observatory of Columbia University and Bundesanstalt für Geowissenschaften und Rohstoffe. United States participation was funded by the National Science Foundation, Department of Polar Programs, under Grant DPP82-16014 and DPP86-11825. German participation was funded by Deutsche Forschungsgemeinschaft. Prof. Karl Hinz led the German participation. The derivation of crustal structure from ESP data was greatly assisted by the input of P. Buhl, J. Diebold and J. Alsop. Solutions of ESP data from the Faeroe–Shetland margin were improved after comparison with solutions obtained for the same data set by Ingi Olafsson, University of Bergen. We thank Lindsay Parson and two anonymous reviewers for their comments, and Dennis Hayes for his constructive review and comments. We benefited from discussion on convection and melting with W. Roger Buck. N. Katz prepared the illustration; J. D. Graney and J. Guadagnini typed the manuscript. Lamont–Doherty Geological Observatory Contribution Number: 4170.

References

BARTON, P. & WOOD, R. 1984. Tectonic evolution of the North Sea basin: Crustal stretching and subsidence. *Geophysical Journal of the Royal Astronomical Society,* **79**, 987–1022.

BEAUMONT, C., KEEN, C. E., & BOUTILLIER, R. 1982. On the evolution of rifted continental margins: comparison of models and observations for the Nova Scotia margin. *Geophysical Journal of the Royal Astronomical Society,* **70**, 667–715.

BØEN, F., EGGEN, S. & VOLLSET, J. 1984. Structures and basins of the margin from 62° to 69°N and their development. *In:* SPENCER, A. M. *et al.* (eds) *Petroleum Geology of the North European Margin.* Graham & Trotman, London, pp. 253–270.

BUCK, W. R. 1983. Convection beneath continental rifts: the effect on cooling and subsidence. *EOS,* **64**, 838.

——, 1984. *Small-scale convection and the evolution of the lithosphere.* PhD thesis (unpublished). Massachusetts Institute of Technology, 256 p.

——, 1986. Small-scale convection induced by passive rifting: the cause for uplift of rift shoulders. *Earth and Planetary Science Letters,* **77**, 362–372.

BUCKOVICS, C., SHAW, N. D., CARTIER, E. G. &

ZIEGLER, P. A. 1984. Structure and development of mid-Norway continental margin. *In:* SPENCER, A. M. *et al.* (eds) *Petroleum Geology of the North European Margin.* Graham & Trotman, London, pp. 407–424.

DIEBOLD, J. B. & STOFFA, P. L. 1981. The traveltime equation, tau-p mapping and inversion of common midpoint data. *Geophysics,* **46**, 238–254.

——, —— & THE LASE STUDY GROUP 1988. A large aperture seismic experiment in the Baltimore Canyon Trough. *In:* SHERIDAN, R. E. & GROW, J. A. (eds) (1988) *The Atlantic Continental Margin (Vol. 1 & 2: The Geology of North America).* Geological Society of America.

DIETZ, R. S. 1961. Continent and ocean basin evolution by spreading of the sea floor. *Nature,* **190**, 854–857.

DRUMMOND, B. J. & COLLINS, C. D. N. 1986. Seismic evidence for underplating of the lower continental crust of Australia. *Earth and Planetary Science Letters,* **79**, 361–372.

ELDHOLM, O. & MUTTER, J. C. 1986. Basin structure on the Norwegian Margin from analysis of digitally recorded sonobuoys. *Journal of Geophysical Research,* **91**, 3763–3783.

——, THIEDE, J. & ODP LEG 104 SHIPBOARD SCIENTIFIC PARTY 1986. Ocean drilling at the Vøring Plateau in the Norwegian Sea. *Nature,* **319**, 360–361.

FOUCHER, J. P., LEPICHON, X. & SIBUET, J. C. 1982. The ocean continent transition in the uniform stretching model: role of partial melting in the mantle. *Philosophical Transactions of the Royal Society of London,* **A305**, 27–43.

FURLONG, K. P. & FOUNTAIN, D. M. 1986. Continental crustal underplating: thermal consequences and seismic-petrologic consequences. *Journal of Geophysical Research,* **91**, 8285–8294.

HESS, H. H. 1965. Mid-oceanic ridges and tectonics of the sea-floor. *In:* WHITFORD, W. F. & BRADSHAW, R. (eds) *Submarine Geology and Geophysics.* Colston Papers; Butterworths, London, **17**, pp. 317–333.

HINZ, K. 1981. A hypothesis on terrestrial catastrophes. Wedges of very thick seaward dipping layers beneath passive continental margins. Their origin and paleoenvironmental significance. *Geologisches Jahrbuch,* **E22**, 3–28.

—— & WEBER, J. 1976. Zum geologischen aufbau des Norwegischen kontinentalran des und der Barents See nach reflexionsseismischen Messungen. *Erdol-Erdgos Z,* **94**, 217–280.

——, MUTTER, J. C., ZEHNDER, C. M. & NGT STUDY GROUP 1987. Symmetric conjugation of continent-ocean boundary structures along the Norwegian and East Greenland margins. *Marine and Petroleum Geology,* **4**, 166–187.

JOHNSON, G. L. & ECKHOFF, O. B. 1966. Bathymetry of the north Greenland Sea. *Deep-Sea Research,* **13**, 1161–1173.

—— & HEEZEN, B. 1967. Morphology and evolution of the Norwegian Greenland Sea. *Deep-Sea Research,* **14**, 755–771.

KLEIN, E. M. & LANGMUIR, C. H. 1987. Ocean ridge basalt chemistry, axial depth, and crustal thickness and temperature variation in the mantle. *Journal of Geophysical Research,* **92**, 8089–8115.

LASE STUDY GROUP 1986. Deep structure of the U.S. East Coast margin from large aperture seismic experiments (LASE). *Marine and Petroleum Geology,* **3**, 234–242.

LITVIN, V. M. 1965. Origin of the bottom configuration of the Norwegian Sea. *Okeanologiya,* **5**, 692–700 (in Russian; translation in *Oceanology,* **5**, 90–96, 1966).

McKENZIE, D. 1978. Some remarks on the development of sedimentary basins. *Earth and Planetary Science Letters,* **40**, 25–32.

——, 1984. The generation and compaction of partially molten rock. *Journal of Petrology,* **25**, 713–765.

MUTTER, J. C. & NAT STUDY GROUP 1985. Multichannel seismic images of the oceanic crust's internal structure: evidence for a magma chamber beneath the Mesozoic mid-Atlantic Ridge. *Geology,* **13**, 629–632.

——, TALWANI, M. & STOFFA, P. L. 1982. Origin of seaward-dipping reflectors in oceanic crust off the Norwegian Margin by "subaerial sea-floor spreading". *Geology,* **10**, 353–357.

——, —— & —— 1984. Evidence for a thick oceanic crust off Norway. *Journal of Geophysical Research,* **89**, 483–502.

——, BUHL, P., ZEHNDER, C. M. & HINZ, K. 1985. Two-ship MCS investigation of the conjugate passive margins of Norway and East Greenland: part I-wide aperture CDP profiling. *EOS,* **6B(46)**, 106.

——, BUCK, W. R. & ZEHNDER, C. M. (in press). Convective partial melting I: a model for the formation of thick basaltic sequences during the initiation of spreading. *Journal of Geophysical Research.*

NANSEN, F. 1904. The bathymetrical features of the North Polar seas, with a discussion of the continental shelves and previous oscillations of the shore line; Norwegian North Polar Expedition, 1893–96. *Science Research,* **4**, 1–232.

NAT STUDY GROUP 1985. North Atlantic Transect: a wide-aperture, two-ship multichannel seismic investigation of the oceanic crust. *Journal of Geophysical Research,* **90**, 10321–10341.

NUNNS, A. G. 1983. Plate tectonic evolution of the Greenland–Scotland Rise. In: BOTT, M. H. P., SAXOV, S., TALWANI, M. & THIEDE, J. (eds) *Structure and Development of the Greenland–Scotland Ridge.* Plenum Press, New York, pp. 11–30.

ROBERTS, D. G., BACKMAN, J., MORTON, A. C., MURRAY, J. W. & KEENE, J. B. 1984. Evolution of volcanic rifted margins: synthesis of Leg 81 results of the west margin of Rockall Plateau. *In:* ROBERTS, D. G., SCHNITKER, D. *et al. Initial Reports of the Deep Sea Drilling Project.* US Government Printing Office, Washington, **81**, pp. 913–923.

SCLATER, J. G. & CHRISTIE, P. A. F. 1980. Continental stretching: an explanation of the post and mid-Cretaceous subsidence of the central North Sea basin. *Journal of Geophysical Research,* **85**, 3711–3739.

SKOGSEID, J. & ELDHOLM, O. 1987. Early Cenozoic crust at the Norwegian continental margin and the conjugate Jan Mayen Ridge. *Journal of Geophysical Research,* **92**, 11471–11491.

SMYTHE, D. K. 1983. Faeroe–Shetland escarpment and continental margin north of the Faeroes. *In:* BOTT, M. H. P., SAXOV, S., TALWANI, M. & THIEDE, J. (eds) *Structure and Development of the Greenland–Scotland Ridge.* Plenum Press, New York, pp. 109–119.

STECKLER, M. S. 1985. Uplift and extension of the Gulf of Suez—indications of induced mantle convection. *Nature,* **317**, 135–139.

STOCKS, T. 1950. Die Tiefenverhältnisse des europaischen Nordmeeres. *Deutsche Hydrographischen Zeitschrift,* **3**, 93–100.

STOFFA, P. L. & BUHL, P. 1979. Two-ship multichannel seismic experiments for deep crustal studies: expanded spread and constant offset profiles. *Journal of Geophysical Research,* **84**, 7645–7660.

TALWANI, M. & ELDHOLM, O. 1972. The continental margin off Norway: a geophysical study. *Geological Society of America Bulletin,* **83**, 3575–3608.

—— & —— 1977. Evolution of the Norwegian–Greenland Sea. *Bulletin of the Geological Society of America,* **88**, 969–999.

——, UDINTSEV, G. *et al.* 1976. *Initial Reports of the Deep Sea Drilling Project.* US Government Printing Office, Washington, **38**, 1256 p.

——, MUTTER, J. C. & ELDHOLM, O. 1981. The initiation of opening of the Norwegian Sea. *Oceanologica Acta*, **4**, 23–30.

VINE, F. J. 1966. Spreading of the ocean floor: new evidence. *Science*, **154**, 1405–1415.

—— & MATTHEWS, D. H. 1963. Magnetic anomalies over oceanic ridges. *Nature*, **199**, 947–949.

VOGT, D. R., OSTENSO, N. A. & JOHNSON, G. L. 1970. Magnetic and bathymetric data bearing on sea-floor spreading north of Iceland. *Journal of Geophysical Research*, **75**, 903–920.

WHITE, R. S., WESTBROOK, G. K., BOWEN, A. N., FOWLER, S. R., SPENCE, G. D., PRESCOTT, C., BARTON, P. J., JOPPEN, M., MORGAN, J. & BOTT, M. H. P. 1987. Hatton Bank (northwest U.K.) continental margin structure. *Geophysical Journal of the Royal Astronomical Society*, **89**, 265–272.

ZEHNDER, C. M. & MUTTER, J. C. 1986. The continent–ocean boundary at (volcanic) passive margins. *EOS*, **67(44)**, 1192.

——, —— & BUHL, P. 1985. Two-ship MCS investigation of the conjugate margins off Norway and East Greenland: part II-expanded spread profiling (ESP). *EOS*, **66**, 1106.

J. C. MUTTER, Lamont–Doherty Geological Observatory of Columbia University, Palisades, NY 10964–0190, USA.

C. M. ZEHNDER, Department of Geological Sciences, Columbia University, New York, USA.

Early Cainozoic evolution of the Norwegian volcanic passive margin and the formation of marginal highs

J. Skogseid & O. Eldholm

SUMMARY: The final continental break-up between Greenland and Norway which took place during the late Palaeocene/early Eocene was accompanied by extensive subaerial volcanism along parts of the plate boundary lasting for about 3 My. Subsequent subsidence and change to normal seafloor spreading have left behind marginal highs partly underlain by seaward-dipping reflector sequences composed of basaltic extrusives. We propose a model in which the formation of the marginal highs and the emplacement of the dipping series are related to the crustal configuration prior to the early Cainozoic plate break-up, causing an initial uplift if the rifting occurs within much thinned continental crust. This forms a 'volcanic' type rifted margin characterized by a small amount of crustal extension compared with a 'normal' margin segment that has experienced a large amount of extension and rapid initial subsidence. Recent drilling of the Vøring marginal high suggests that the plate boundary volcanism and the N Atlantic Volcanic Province may be manifestations of the same regional volcanic surge.

In the early Tertiary, continental break-up and generation of oceanic crust was achieved along a new plate boundary extending from the Hatton Bank region to the Eurasia Basin in the Arctic. This event initiated the formation of the Norwegian–Greenland Sea. The early Tertiary is also the time of major regional volcanism in the N Atlantic realm. Detailed geophysical surveys and scientific drilling along the margins of the plate boundary have yielded a number of geological features that were formed during the late rifting and early spreading stages. Moreover, it has become evident that volcanism played a major part during the history of the early opening of these ocean basins. This, in turn, suggests a relationship between the Brito–Arctic, or N Atlantic, Volcanic Province and the early evolution of the adjacent continental margins.

The objective of this study is to describe the key geological features and their associated geophysical parameters, and to suggest a scenario for the early Tertiary sequence of events that created the present margin. We focus on the Norwegian–Greenland Sea, in particular the Vøring Plateau continental margin which has been most thoroughly investigated both by geophysical techniques and by drilling. It is believed, however, that some of the inferences and the evolutionary model might also be applicable to other continental margins.

Characteristic features of the outer margin

The first seismic investigations off Norway revealed a characteristic configuration of the acoustic basement beneath the outer continental margin. Approaching the margin from the Lofoten and Norway basins, the irregular oceanic basement surface changes into a smooth, opaque acoustic basement reflector buried by a relatively thin sequence of sediments. The opaque reflector terminates abruptly at escarpment-like features that appear to mark the seaward extent of major sedimentary basins beneath the continental slope. Moreover, the smooth basement surface is elevated with respect to the adjacent oceanic crust, defining prominent structural highs between the early Cainozoic oceanic crust and the Mesozoic Møre and Vøring sedimentary basins (see Fig. 1). The Vøring Plateau and Faeroe–Shetland escarpments were first mapped by Talwani & Eldholm (1972) who suggested that they represent first-order geological features associated with the change from continental to oceanic crust. Furthermore, they noted that the escarpments mark significant changes in a number of geophysical parameters. In particular, the magnetic field becomes quiet on the landward side, a gravity high is located on the seaward side and there is a pronounced difference in the velocity–depth functions on either side of the escarpments.

More sophisticated seismic profiling techniques have revealed layering beneath the opaque horizon at the Faeroe–Shetland and Vøring Plateau marginal highs. The most spectacular observation in the seismic record is a series of seaward-dipping reflector sequences (Hinz & Weber 1976; Talwani et al. 1981; Smythe 1983; Smythe et al. 1983; Talwani et al. 1983).

Marginal highs have also been mapped elsewhere in the Norwegian–Greenland Sea. A

From MORTON, A. C. & PARSON, L. M. (eds), 1988, *Early Tertiary Volcanism and the Opening of the NE Atlantic,* Geological Society Special Publication No. 39, pp. 49–56.

prominent high exists along the western margin just S of the Greenland fracture zone. It terminates landward at the Greenland escarpment (Eldholm & Windisch 1974; Talwani & Eldholm 1977) and Hinz et al. (1987) have recently documented the existence of seaward-dipping reflectors beneath the high. Along the rifted part of the Barents Sea margin a marginal high described by Myhre & Eldholm (in press) is associated with early Eocene extrusive volcanism (Eldholm et al. in press; Faleide et al. 1988). Dipping reflectors are also described at the Norwegian margin N of the Vøring Plateau (Hinz & Weber 1976; Eldholm et al. 1979), along the eastern flank of the Jan Mayen Ridge (Fig. 1) (Skogseid & Eldholm 1987; Gudlaugsson et al. 1988), and at the margin off SE Greenland

(Featherstone et al. 1977; Larsen 1984) and the Hatton Bank (Roberts et al. 1979, 1984).

Deep Sea Drilling Project (DSDP) and Ocean Drilling Programme (ODP) drilling has demonstrated that the opaque acoustic basement reflector is comprised of extrusive basalts (Talwani, Udinstev, et al. 1976; Roberts, Schnitter, et al. 1984) and that the entire dipping sequence consists of flows and interbedded volcaniclastic sediments (Eldholm et al. 1987).

The above observations imply a relationship between dipping reflector sequences, smooth acoustic basement, the continent–ocean transition and the marginal highs. The fact that these features lie at or close to the oldest oceanic crust suggests an association with events taking place during the latest phase of rifting and/or the early

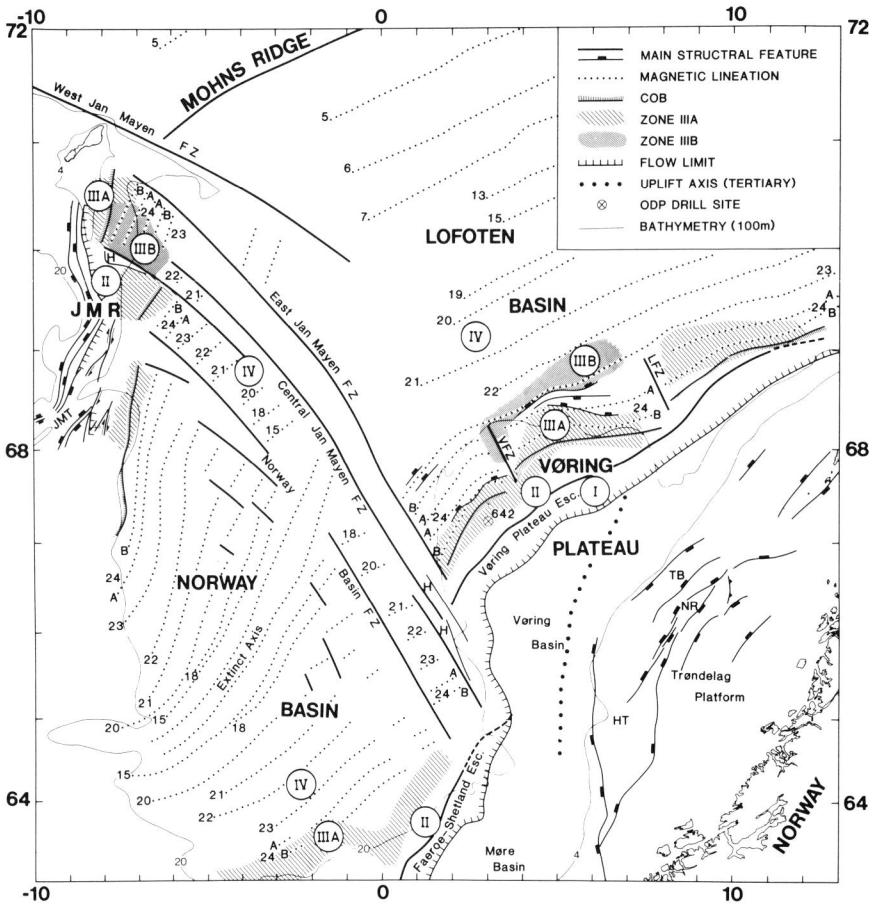

FIG. 1. Areas of early Cainozoic volcanism at the outer continental margin off central Norway and at the northern Jan Mayen Ridge (Talwani et al. 1983; Skogseid & Eldholm 1987). Included also are main structural features (Bøen et al. 1984; Bukovics & Ziegler 1985; Skogseid & Eldholm 1987) and magnetic seafloor-spreading anomalies (Talwani & Eldholm 1977; Hagevang et al. 1983; Nunns et al. 1983; Skogseid 1983). HT: Halten Terrace, JMR: Jan Mayen Ridge, JMT: Jan Mayen Trough, LFZ: Lofoten Fracture Zone, NR: Nordland Ridge, TB: Træn Basin, VFZ: Vøring Fracture Zone.

seafloor spreading. Furthermore, Hinz (1981) has shown that marginal features of this kind are not restricted to the N Atlantic realm. Thus, our contention is that they might actually represent a stage in the evolution of a special type of the rifted passive continental margin.

The Vøring Plateau marginal high

We now focus in some detail on the Vøring Plateau, noting that many of the inferences have also been shown to apply to the margin S of the plateau (Talwani *et al.* 1983), and the conjugate Greenland (Hinz *et al.* 1987) and Jan Mayen Ridge (Skogseid & Eldholm 1987) margins.

From the character of the opaque acoustic basement reflector (Fig. 2, EE) and its subsurface reflector configuration, Hinz *et al.* (1982) and Talwani *et al.* (1983) divided the outer margin into different zones (see Figs 1 & 2, I–IV). The areas comprising the wedge of seaward-dipping reflectors, zone III, lie between the normal oceanic crust, zone IV, and a structural high 10–35 km W of the Vøring Plateau escarpment. Between the apex of the dipping wedge and the escarpment, zone II, we observe some irregular sub-EE reflectors resting on a base reflector K (see Fig. 2). Reflector K (Hinz *et al.* 1982) continues into zone III but appears to terminate beneath the innermost dipping wedge. A region of volcanic extrusives landward of the Vøring Plateau escarpment, zone I or 'inner flows', is believed to be the continuation of reflector EE overlying a thick section of Mesozoic sediments in the Vøring Basin.

We have recently shown that zone III can be divided into two different sets of wedges indicated

as IIIA and IIIB in Figs. 1 & 2 (Skogseid & Eldholm 1987). Well identified seafloor-spreading type magnetic anomalies (23–24B) overlie these zones, whereas there is an irregular, indistinct magnetic pattern in zone II (Hagevang *et al.* 1983). In terms of seismic velocities the main wedges have been interpreted as overlying expanded oceanic crust with a more ambiguous velocity–depth distribution adjacent to the escarpment (Mutter *et al.* 1984; Hinz *et al.* 1987).

During ODP Leg 104 Site 642 (see Fig. 1) was drilled at the innermost dipping wedge. The drillbit penetrated the entire wedge which was composed of early Eocene flow basalts and interbedded sediments emplaced in a terrestrial environment. This Upper Volcanic Series is, in terms of biostratigraphy, petrology and magnetic polarity, indeed similar to the rocks of the N Atlantic Volcanic Province. The series might be characterized as tholeiitic basalts representing an Icelandic type of crustal generation. At the level of reflector K, a different series of flows and a few dykes were encountered. These flows include peraluminous andesitic and basaltic–andesitic rocks, and both the extrusives and the interbedded sediments show indications of continental contamination (Eldholm *et al.* 1987). The results from Site 642 are considered a key element in the interpretation of the marginal features described above, placing important constraints on evolutionary models.

The pre-opening basin

In a basin experiencing extension the relationship between lithospheric and crustal thickness influences the direction of the initial subsidence, and

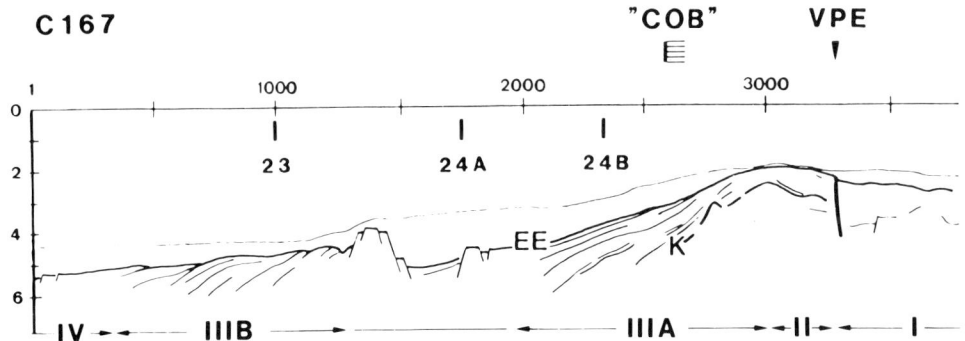

FIG. 2. Interpretation of L-DGO seismic line 167 across the Vøring marginal high revealing a double seaward-dipping reflector sequence (zones IIIA & IIIB). Note that identified seafloor-spreading anomalies overlie part of the dipping sequences. Reflectors EE and K represent the top of the Upper and Lower Volcanic Series respectively. The continent–ocean boundary (COB) is placed between the seaward termination of reflector K and the landward boundary of anomaly 24B. VPE: Vøring Plateau Escarpment.

the stretching factor determines its magnitude (McKenzie 1978; Royden *et al.* 1980). Therefore, we examine the geological setting prior to the opening of the Norwegian–Greenland Sea in earliest Eocene time when a series of rift basins had been established between Norway and Greenland. Off Norway, the large Møre and Vøring Basins might contain accumulations of more than 10–12 km of sediments (Eldholm & Mutter 1986).

Several phases of tectonic activity, during late Palaeozoic and Mesozoic times, have governed the structural and depositional patterns off Norway. Rifting occurred in the early Carboniferous to Permian, the Triassic to early Jurassic, middle to late Jurassic and the early Cainozoic times (Bøen *et al.* 1984; Bukovics & Ziegler 1985). However, the late Jurassic (late Kimmerian) rift episode has been most fundamental in structuring the margin. Subsequent rapid subsidence in the Cretaceous created the Møre and Vøring Basins, whereas the inner part of the present margin, the Trøndelag Platform, was not affected in a major way.

Estimates of the late Kimmerian crustal extension in the Møre and Vøring Basins are relatively uncertain, and stretching factors range from 2.0–3.5 (Bøen *et al.* 1984; Bukovics *et al.* 1984; Gravdal 1985). By assuming a pre-Kimmerian crustal thickness of 32 km and a McKenzie (1978) model of uniform extension and thinning we estimate a post-Kimmerian crustal thickness in the range of 9–16 km. However, these estimates might be too high in light of the pre-Kimmerian rift events. Subsequently, a thick sequence of Cretaceous sediments infilled the basins (Bøen *et al.* 1984) and we indicate a crustal thickness at the beginning of the Tertiary from 13–20 km. These results are comparable with the present 16–22 km depth to Moho in the basins (Eldholm & Mutter 1986; Hinz *et al.* 1987; R. Vially pers. comn.) when compensating for 2.5–4.0 km of water and Cainozoic sediments.

Evolutionary model

Two competing models for the emplacement of the Vøring Plateau dipping sequence have been advanced by Mutter *et al.* (1982; 1984), Hinz (1981) and Hinz *et al.* (1984), whereas Roberts *et al.* (1979, 1984), Morton & Taylor (1987) and Smythe (1983) have discussed the evolution of the Rockall and Faeroe–Shetland margins. The various models differ principally in terms of the location of the continent–ocean boundary and

the nature of the crust in zones II and III. Here, we focus on the entire marginal high, stressing that a geological model for the Vøring Plateau marginal high has not only to be compatible with the geophysical data and drilling results, but must also take into account the pre-rifting crustal configuration as well as the evidence for only a small amount of extension in the late Cretaceous/early Tertiary sediments in the Vøring Basin. We note that farther N, off the Lofoten–Vesterålen islands, the margin was down-faulted in the early Tertiary, suggesting major extension and rapid subsidence (Eldholm *et al.* 1979).

Many passive margins are associated with rapid fault-defined initial subsidence and onset of seafloor spreading at water depths of about 2.5 km. This situation is illustrated in Fig. 3 where the subsidence of the 'normal' margin is

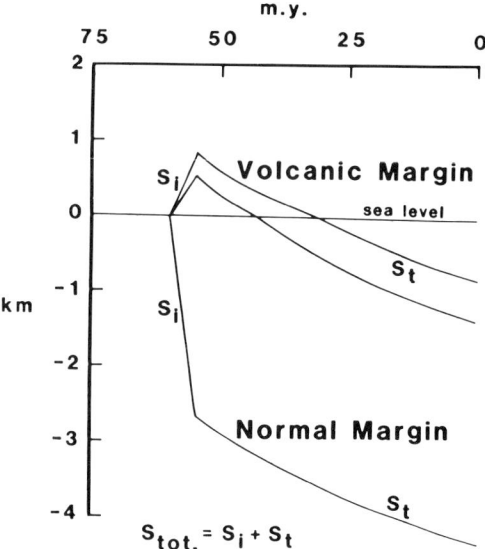

FIG. 3. Subsidence curves for passive continental margins assuming opening 56 Ma. The 'normal' margin curve is based on a uniform extension model (McKenzie 1978) ($\beta = 3.6$) and crustal parameters of Parsons & Sclater (1977). The 'volcanic' margin curve is calculated for a two-layer extensional model assuming low crustal extension ($\beta = 1.1$) and infinite lithospheric thinning (continental break-up), as well as 50% dyke intrusion during rifting. The upper curve is based on a complete restoration of the lithospheric thickness (125 km) since the late-Kimmerian rifting, whereas the lower curve is calculated with a 110 km thick lithosphere compensated for sediment loading. The total subsidence (S_{tot}) is the sum of the initial (S_i) and thermal (S_t) subsidence.

calculated according to McKenzie (1978) assuming initial crustal and lithospheric thicknesses of 32 and 125 km respectively and uniform stretching factor of 3.6. On the other hand, the amount and extent of the early Eocene extrusives that have been emplaced above or close to the sea surface argue for a different mechanism. Although we cannot rule out that there might be extensional structures hidden by the volcanic overprint in zones I and II, we infer that the upper crust underwent only a modest amount of stretching prior to opening. The extrusive nature and continental affinities of the Lower Volcanic Series at Site 642 imply, in our opinion, subaerial emplacement by feeder dykes through a thinned continental crust during the late rifting stage (Eldholm & Thiede 1987). Hence, the model should include dyke injection (Royden *et al.* 1980).

We have attempted to incorporate the above observations into a subsidence, or relative elevation, model as a function of time for a 'volcanic' rifted margin. The corresponding subsidence curves in Figure 3 are calculated for crustal extension, including dyke intrusion and differential lithospheric thinning (Royden & Keen 1980). The initial crustal thickness is 17 km. To compare with the 'normal margin', the lithosphere–asthenosphere boundary is back to its initial position with respect to the late-Kimmerian rifting in the upper curve. In the lower curve the lithospheric thickness is set to 110 km to indicate incomplete adjustment with respect to the late-Kimmerian rifting event. Other crustal parameters are shown in Figure 3. The modelling shows an initial uplift during the first few million years after break-up, and the plateau becoming submerged 15–25 My after opening. Although one might use other model parameters, our aim is primarily to show that initial uplift, or perhaps absence of rapid initial subsidence, might occur in certain geologic environments.

In the present model we have not included partial melting that might lower the mean density of the upper mantle resulting in reduction in the initial subsidence predicted from the uniform extension model (Foucher *et al.* 1982). The available melt is, however, insufficient to produce the observed amount of volcanics according to Mutter *et al.* (in press) who introduced the concept of convective partial melting of Buck (1986), to generate more material to replace thinned lithosphere than during passive upwelling. The increased thinning of the lithosphere inferred by Buck is not incompatible with our model of a two-layered lithosphere and the effect of lowered density in the melt would amplify the uplift.

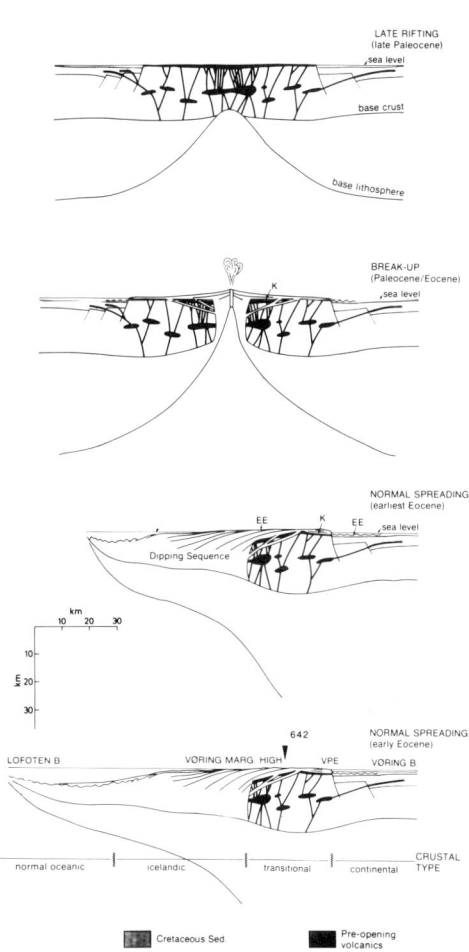

VØRING PLATEAU "VOLCANIC" PASSIVE MARGIN

FIG. 4. Schematic model for the formation of a volcanic-type passive margin (Skogseid & Eldholm, 1987). The structural position of ODP Site 642 is indicated.

The early Cainozoic evolution of the outer Vøring margin is schematically illustrated in Fig. 4, comprising the following stages:

(1) Late Cretaceous and Palaeocene lithospheric extension within the Mesozoic Vøring Basin underlain by previously thinned crust.

(2) Late Palaeocene dyke injection and subaerial extrusion of andesitic flows. The top of these flows represent reflector K.

(3) Break-up of the thinned lithosphere between anomaly 24B and 25 time without initial rapid subsidence. Generation of Icelandic-type tholeiitic crust above or close to sea

level. The improved flow properties of these basalts caused the emplacement of the dipping sequences and basalts also flowed over the adjacent thinned continental crust. Reflector K terminates at the line of break-up.

(4) About 3 My after break-up the intrusive centre subsided below sea-level initiating a transition from Icelandic to normal seafloor spreading.

(5) Final submergence of the most elevated parts of the basaltic edifice in the late Oligocene with subsequent thermal subsidence of the margin and smoothing of the sea-floor relief by sedimentation.

In this model the continent–ocean boundary is placed between the seaward termination of reflector K and anomaly 24B (see Figs 1 & 4). The region between the thinned continental and normal oceanic crust in the Vøring and Lofoten Basins respectively is characterized as Icelandic and transitional. The transitional crust is believed to be a strongly intruded part, possibly a fault block, of the continental crust. The term Icelandic describes the subaerial mode of seafloor spreading, including emplacement of long flow lengths and subsequent subsidence causing the characteristic seaward-dipping beds (Walker 1973; Palmason 1973, 1980). The existence of two separate dipping units, zones IIIA and B (see Fig. 1) has been ascribed by Skogseid & Eldholm (1987) to shifts in the position of the injection centre, as proposed in Iceland by Helgason (1984), or to relief along the strike of the high causing variation in the time of onset of normal seafloor spreading.

Conclusions

During a study of the most thoroughly surveyed marginal high in the Norwegian–Greenland Sea we have suggested a model for the evolution of a volcanic type of rifted passive continental margin. The sequence of events forming the margin appears to be related to the geological setting where the break-up occurs, as well as to physical properties of the lithosphere. Thus, in a young ocean like the Norwegian–Greenland Sea there are low-extension, volcanic, rifted margin segments and segments with evidence of major extension and rapid subsidence. Though many aspects of the margin history are still poorly understood, we believe that most of the inferences made with respect to the Vøring marginal high apply also to marginal highs elsewhere in the Norwegian–Greenland Sea and to other oceans as well.

The similarities in structures and rock properties of the early Cainozoic rocks drilled at the Vøring Plateau and those described elsewhere in the N Atlantic Volcanic Province suggest that large-scale mantle anomalies may have provided the conditions for increased volcanism, both at parts of the plate boundary and within the continental crust. At the plate boundary these conditions possibly amplified the effects of the pre-opening geology.

ACKNOWLEDGEMENTS: This work has been partially funded by the Norwegian Research Council for Science and the Humanities, Norwegian Petroleum Directorate and Statoil a.s. Norwegian International Lithosphere Project Contribution no. 33.

References

BUCK, W. R. 1986. Small-scale convection induced by passive rifting: the cause for uplift of rift shoulders. *Earth and Planetary Science Letters*, **77**, 362–372.

BUKOVICS, C. & ZIEGLER, P. A. 1985. Tectonic development of the mid-Norway continental margin. *Marine and Petroleum Geology*, **2**, 2–22.

——, SHAW, D. N., CARTIER, E. G. & ZIEGLER, P. A. 1984. Structure and development of the mid-Norway continental margin. *In:* SPENCER, A. M. *et al.* (eds) *Petroleum Geology of the North European Margin.* Graham & Trotman, London, pp. 407–424.

BØEN, F., EGGEN, S. & VOLLSET, J. 1984. Structures and basins of the margin from 62–69° N and their development. *In:* SPENCER, A. M. *et al.* (eds) *Petroleum Geology of the North European Margin.* Graham & Trotman, London, pp. 253–270.

ELDHOLM, O., FALEIDE, J. I. & MYHRE, A. M. (in press). Continent–ocean transition at the Western Barents Sea/Svalbard continental margin Geology.

—— & WINDISCH, C. C. 1974. Sediment distribution in the Norwegian–Greenland Sea. *Bulletin of the Geological Society of America*, **85**, 1661–1676.

—— & MUTTER, J. C. 1986. Basin structure on the Norwegian margin from analysis of digitally recorded sonobuoys. *Journal of Geophysical Research*, **91**, 3763–3783.

—— & THIEDE, J. 1987. Summary and preliminary conclusions, ODP Leg 104. *In:* ELDHOLM, O., THIEDE, J., TAYLOR E., *et al.* (eds) *Proceedings of the Ocean Drilling Program.* US Government Printing Office, Washington, **104A**, 751–771.

——, SUNDVOR, E. & MYHRE, A. M. 1979. Continental margin off Lofoten–Vesterålen, Northern Norway. *Marine Geophysical Researches*, **4**, 3–35.

——, THIEDE, J. & TAYLOR, E. *et al.* 1987. *Proceedings of the Ocean Drilling Program.* US Government Printing Office, Washington, **104A**, 783 pp.

FALEIDE, J. I., MYHRE, A. M. & ELDHOLM, O. 1988. Early Tertiary volcanism at the Western Barents Sea–Svalbard margin. *In:* MORTON, A. C. & PARSON, L. M. (eds) *Early Tertiary Volcanism and* the Opening of the NE Atlantic. Geological Society of London, Special Publication, **39**, pp. 135–146.

FEATHERSTONE, P. S., BOTT, M. H. P. & PEACOCK, J. H. 1977. Structure of the continental margin of Southeastern Greenland. *Geophysical Journal of the Royal Astronomical Society*, **48**, 15–27.

FOUCHER, J.-P., LE PICHON, X. & SIBUET, J.-C. 1982. The ocean–continent transition in the uniform lithospheric stretching model: role of partial melting in the mantle. *Philosophical Transactions of the Royal Society of London*, **A305**, 27–43.

GRAVDAL, N. 1985. *The Møre Basin*. Cand. Scient. thesis, University of Oslo, 121 pp.

GUDLAUGSSON, S. T., GUNNARSON, K., SAND, M. & SKOGSEID, J. 1988. Tertiary volcanic events and the evolution of the Jan Mayen Ridge microcontinent. *In:* MORTON, A. C. & PARSON, L. M. (eds) *Early Tertiary Volcanism and the Opening of the NE Atlantic*. Geological Society of London Special Publication, **39**, pp. 85–93.

HAGEVANG, T., ELDHOLM, O. & AALSTAD, I. 1983. Pre-23 magnetic anomalies between Jan Mayen and Greenland–Senja fracture zones in the Norwegian Sea. *Marine Geophysical Researches*, **5**, 345–363.

HELGASON, J. 1984. Frequent shifts of the volcanic zone in Iceland. *Geology*, **12**, 212–216.

HINZ, K. 1981. A hypothesis on terrestrial catastrophes. Wedges of very thick oceanward dipping layers beneath passive continental margins—their origin and paleoenvironmental significance. *Geologisches Jahrbuch*, **E22**, 3–28.

—— & WEBER, J. 1976. Zum geologischen aufbau des Norwegischen kontinentalrandes und der Barents-See nach reflexionsseismischen Messungen. *Petrochemie, Erdol und Kohle, Erdgas*, 3–29.

——, DOSTMAN, H. J. & HANISCH, J. 1982. Structural framework of the Norwegian Sea. *In: Offshore Northern Seas*. Norwegian Petroleum Society, 22 pp.

——, ——, —— 1984. Structural elements of the Norwegian continental margin. *Geologisches Jahrbuch*, **A75**, 193–221.

——, MUTTER, J. C., ZEHNDER, C. M. & NGT STUDY GROUP 1987. Symmetric conjugation of continent-ocean boundary structures along the Norwegian and East Greenland margins. *Marine and Petroleum Geology*, **4**, 166–187.

LARSEN, H. C. 1984. Geology of the North European Margin. *In:* SPENCER, A. M. *et al.* (eds) *Petroleum Geology of the North European Margin*. Graham & Trotman, London, pp. 329–339.

MCKENZIE, D. 1978. Some remarks on the development of sedimentary basins. *Earth and Planetary Science Letters*, **40**, 25–32.

MORTON, A. C. & TAYLOR, P. N. 1987. Lead isotope evidence for the structure of the Rockall dipping-reflector passive margin. *Nature*, **326**, 381–383.

MUTTER, J. C., TALWANI, M. & STOFFA, P. L. 1982. Origin of seaward dipping reflectors in oceanic crust off the Norwegian margin by "subaerial sea-floor spreading". *Geology*, **10**, 353–357.

——, ——, —— 1984. Evidence for a thick oceanic crust adjacent to the Norwegian Margin. *Journal of Geophysical Research*, **89**, 483–502.

——, BUCK, W. R. & ZEHNDER, C. M. (in press). Convective partial melting I: A model for the formation of thick basaltic sequences during the initiation of spreading. *Journal of Geophysical Research*.

MYHRE, A. M. & ELDHOLM, O. (in press). The Western Svalbard Margin (74°–80°). *Marine and Petroleum Geology*.

NUNNS, A. G., TALWANI, M., LORENTZEN, G. R., VOGT, P. R., SIGURDGEIRSSON, T., KRISTJANSSON, L., LARSEN, H. C. & VOPPEL, D. 1983. Magnetic anomalies over Iceland and surrounding seas. *In:* BOTT, M. H. P., SAXOV, S., TALWANI, M. & THIEDE, J. (eds) *Structure and Development of the Greenland-Scotland Ridge*. Plenum Press, New York, pp. 661–678.

PALMASON, G. 1973. Kinematics and heat flow in a volcanic rift zone, with application to Iceland. *Geophysical Journal of the Royal Astronomical Society*, **33**, 451–481.

—— 1980. A continuum model of crustal generation in Iceland; Kinematic aspects. *Journal of Geophysics*, **47**, 7–18.

PARSONS, B. & SCLATER, J. G. 1977. An analysis of the variation of ocean floor bathymetry and heat flow with age. *Journal of Geophysical Research*, **82**, 803–827.

ROBERTS, D. G., MONTADERT, L. & SEARLE, R. C. 1979. The western Rockall Plateau: stratigraphy and structural evolution. *In:* MONTADERT, L., ROBERTS, D. G., *et al, Initial Reports of the Deep Sea Drilling Project*. US Government Printing Office, Washington, **48**, 1061–1088.

——, BACKMAN, J., MORTON, A. C., MURRAY, J. W. & KEENE, J. B. 1984. Evolution of volcanic rifted margins: Synthesis of Leg 81 results on the west margin of Rockall Plateau. *In:* ROBERTS, D. G., SCHNITKER, D. *et al, Initial Reports of the Deep Sea Drilling Project*. US Government Printing Office, Washington, **81**, 883–911.

——, SCHNITKER, D., *et al.* 1984. *Initial Reports of the Deep Sea Drilling Project*. US Government Printing Office, Washington, **81**, 922 pp.

ROYDEN, L. & KEEN, C. E. 1980. Rifting process and thermal evolution of the continental margin of eastern Canada determined from subsidence curves. *Earth and Planetary Science Letters*, **51**, 343–361.

——, SCLATER, J. G. & VON HERZEN, R. P. 1980. Continental margin subsidence and heat flow: Important parameters in formation of petroleum hydrocarbons. *Bulletin of the American Association of Petroleum Geologists*, **64**, 173–187.

SKOGSEID, J. 1983. *Geophysical studies between the Vøring Plateau Margin and the Jan Mayen Ridge, and a plate tectonic model for the evolution of the Norway Basin*. Cand. Scient. thesis, University of Oslo, 126 pp.

—— & ELDHOLM, O. 1987. Early Cenozoic crust at the Norwegian continental margin and the conjugate Jan Mayen Ridge. *Journal of Geophysical Research*, **92**, 11471–11491.

SMYTHE, D. K. 1983. Faeroe Shetland Escarpment and continental margin north of the Faeroes. *In:* BOTT,

M. H. P., SAXOV, S., TALWANI, M. & THIEDE, J. (eds) *Structure and Development of the Greenland–Scotland Ridge*. Plenum Press, New York, pp. 109–120.

——, CHALMERS, J. A., SKUCE, A. G., DOBINSON, A. & MOULD, A. S. 1983. Early opening history of the North Atlantic—I. Structure and origin of the Faeroe-Shetland Escarpment. *Geophysical Journal of the Royal Astronomical Society*, **72**, 373–398.

TALWANI, M. & ELDHOLM, O. 1972. The continental margin off Norway: A geophysical study. *Bulletin of the Geological Society of America*, **83**, 3575–3606.

—— & —— 1977. Evolution of the Norwegian–Greenland Sea. *Bulletin of the Geological Society of America*, **88**, 969–999.

——, UDINTSEV, G. *et al.* 1976. *Initial Reports of the Deep Sea Drilling Project*. US Government Printing Office, Washington, **38**, 1256 pp.

——, MUTTER, J. & ELDHOLM, O. 1981. The initiation of opening of the Norwegian Sea. *Oceanologica Acta*, **4** (supplement), 23–30.

——, —— & HINZ, K. 1983. Ocean continent boundary under the Norwegian continental margin. *In:* BOTT, M. H. P., SAXOV, S., TALWANI, M. & THIEDE, J. (eds) *Structure and Development of the Greenland–Scotland Ridge*. Plenum Press, New York, pp. 121–131.

WALKER, G. P. 1973. Length of lava flows. *Philosophical Transactions of the Royal Society of London*, **A274**, 107–118.

J. SKOGSEID & O. ELDHOLM, Department of Geology, University of Oslo, Norway.

Dipping reflector styles in the NE Atlantic Ocean

L. M. Parson & the ODP Leg 104 Scientific Party

SUMMARY: The geometries and structure of volcanic rifted continental margins, that is, those with an ocean–continent transition masked by thick sequences of inclined tholeiitic basaltic flows, are examined. Small- and large-scale differences in sequence seismostratigraphy along strike, down-dip, and between conjugate configurations are discussed. Detailed appraisals of deep drilling results and seismic data for the Vøring Plateau and western Rockall Plateau are used to contrast margin architecture and demonstrate difficulties in interpolating data between sites of dipping reflector formation.

Certain structural and morphological characteristics of dipping reflector margins are now relatively well understood, and these aspects are described here. Thick sequences of inclined, oceanward-divergent seismic reflectors occurring in the vicinity of, and striking parallel to, the ocean–continent transition zone (OCT) are known to be present along many of the passive margins in the NE Atlantic, as well as many other margins throughout the world. On seismic reflection profiles orientated orthogonal to the continental margin, they can be characterized by a variety of geometries, ranging between arcuate and planar, and they may be imaged as either stacked, continuous seismic horizons extending for several tens of kilometres or as a complex of detached, subparallel, shorter events. In profile these sequences form oceanward-divergent wedges, locally up to 10 km in thickness, bounded at their upper surface by a smooth, prominent reflector, but almost invariably without a well-defined base. In plan view, they occupy a band adjacent to the OCT of variable width, generally between 40 and 120 km wide. With few exceptions, their landward boundary is on the continental side of the earliest oceanic seafloor-spreading anomaly, but in many cases may also overlap and extend oceanward beyond it.

The distribution of these reflectors in the NE Atlantic has been assessed by Roberts *et al.* (1984) and Parson *et al.* (1986). We here present a revised and comprehensive distribution based on both published and unpublished data. This distribution and the profile characteristics are speculatively interpreted in terms of the basement structure of the margins, and their formation and evolution are discussed by extrapolation from deep drilling results and deep seismic results from two margins, the Vøring Plateau and SW Rockall Plateau.

Within most of the dipping reflector sequence, conventional seismostratigraphic relationships such as offlap and onlap can be identified, although faulting and regional tectonic disruption locally disturb their overall continuity. An attempt to formalize a classification of individual dipping reflector profiles has been made by Roberts *et al.* (1984), who defined four structural units. Most volcanic passive margins have a complexity which varies across and along strike, and such a formal classification may prove of limited usefulness. In most cases the oceanward-thickening wedge of reflectors gives way to a complex zone of low signal strength, masked by a confused pattern of diffraction hyperbolae. The upper surface of this outer zone is commonly rough and may be slightly elevated with respect to the adjacent dipping package. This feature has been referred to as the 'outer high', although confusion over structural highs at passive margins which have developed some distance from the OCT (Skogseid & Eldholm 1987), and which may have no significance to them, may arise (Parson & Eldholm, in prep.). Beyond this disrupted zone, deeper into the ocean basin, a regular hummocky profile usually characterizes more typical oceanic crust with associated seafloor-spreading anomalies. Less continuous and more widely spaced inclined reflectors are frequently observed in this oceanic crust beyond the OCT (Uruski & Parson 1985; Roberts *et al.* 1984; Larsen & Jakobsdottir, 1988). These younger reflectors confirm that a layered structure of layer 2 upper crust continues oceanward to the mid-Atlantic ridge.

This summary description is highly generalized, however, and as has been observed on seismic profiles over dipping reflector margins throughout the NE Atlantic and elsewhere in the world (Hinz 1981), a wide variety of profile morphologies and styles of deposition exist. This range appears to be a result of variation in a number of factors, which include rate and extent of crustal attenuation, rate of supply and composition of volcanics associated with the margin,

From MORTON, A. C. & PARSON, L. M. (eds), 1988, *Early Tertiary Volcanism and the Opening of the NE Atlantic,* Geological Society Special Publication No. 39, pp. 57–68.

timing and degree of subsidence during or following separation, as well as gross structural framework of the continental basement. The importance of each of these factors is not well understood, although two successful drilling programmes (Deep Sea Drilling Project (DSDP) Leg 81 and Ocean Drilling Program (ODP) Leg 104) have gone some way in local interpretations (Roberts, Schnitker *et al.* 1984; Eldholm *et al.* 1987).

Dipping reflectors have been described and discussed by a large number of workers. On Rockall these have included Roberts *et al.* (1979), Roberts *et al.* (1984), Parson *et al.* (1986) and more recently White *et al.* (1988); in the Faeroes region, Smythe (1983); in the Møre Basin, Pelton (1985); on the Vøring Plateau, Mutter *et al.* (1982, 1984) and Eldholm *et al.* (1987); on the Jan Mayen Ridge, Pelton (1985), Skogseid & Eldholm (1987), and Gudluagsson *et al.* (1988); and E Greenland, Larsen (1984), Uruski & Parson (1985), and Larsen & Jakobsdottir (1988). These authors have summarized principally geophysical data for each of their study areas but have not synthesized data from a variety of regions. An attempt is made here to contrast and compare the principal geometrical styles of dipping reflector sequences.

Geometry of the reflectors

In simplest terms, there are four common configurations of the dipping reflector sequences, and these are illustrated in Figure 1. They are as follows: divergent-planar, divergent-arcuate, subparallel, flexured. These are each extremes of styles within a highly variable series, and any dipping reflector sequence generally incorporates a combination of these into a composite form. Within these broad subdivisions, detailed variations in seismic character, structural setting and degree of postdepositional tectonism further differentiate between types of dipping reflector sequence, and these may further reflect variations in composition, and supply rates of magma. Examples of each style of dipping reflector is illustrated and described here using a variety of seismic data and are located in Figure 2.

Divergent-planar (Norwegian Møre Basin)

A seismic reflection profile (a stacked 12 channel record) across the Møre Basin is illustrated in Figure 3. The prominent horizon Z at one second (two-way time) below seafloor marks the upper limit of the dipping reflector sequence, reflectors

a. Divergent—planar :

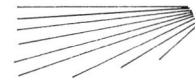

b. Divergent—arcuate :

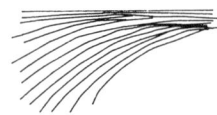

c. Sub—parallel :

d. Flexured :

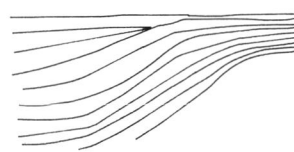

FIG. 1. Schematic illustration of the different styles of dipping reflector geometry discussed in the text: a) divergent-planar, b) divergent-arcuate, c) sub-parallel, d) flexured.

within which may be identified to a further two seconds depth. Common depth-point stacking velocities thus suggest an observed thickness of volcanics of at least 4 km. Within the dipping sequence there is a strong uniformity of planar reflectors disposed subparallel to each other in the SE, but weakly divergent in the central and northwestern section. There are local examples of syndepositional antithetic faulting. One of the most striking characteristics of the surface Z is the locally stepped nature of the composite reflector comprising the upper surface of the dipping sequence (A). Both onlap and offlap relationships are obvious, as well as seismic reflector 'outliers' (B–B') suggesting supply from out-of-section sources.

Divergent-arcuate (western Vøring Plateau)

One of the most intensely studied areas of dipping reflector occurrence lies to the W and NW of the

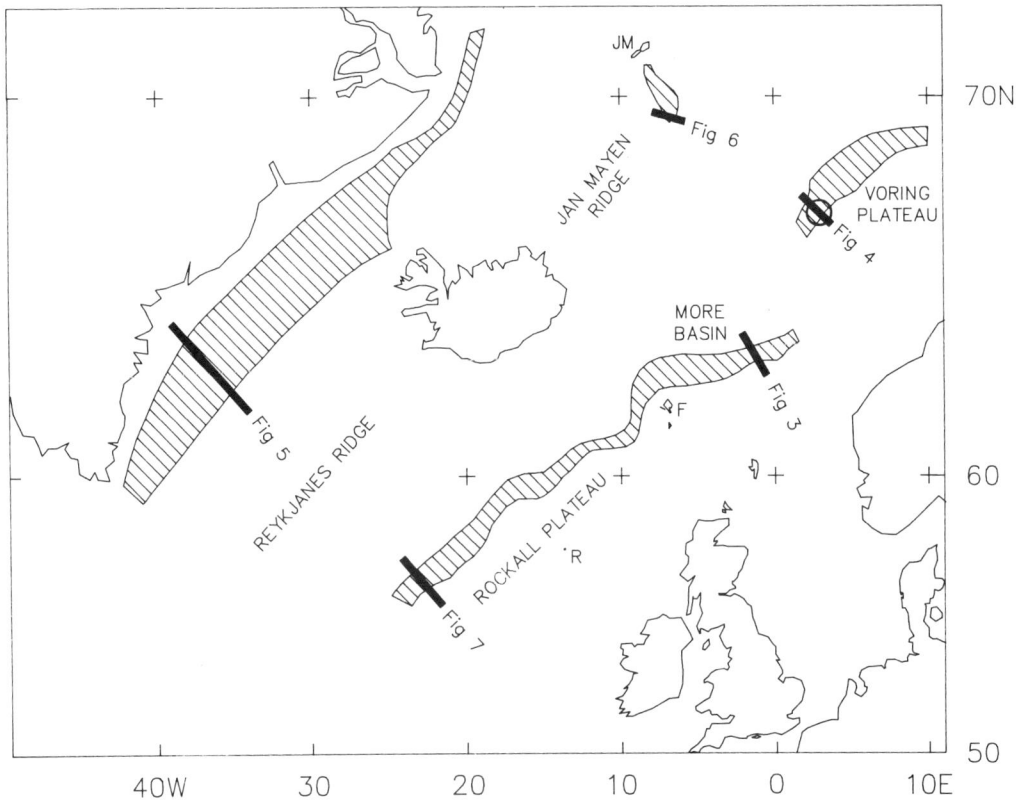

FIG. 2. Simplified location map of the NE Atlantic, including the location of seismic profiles discussed in the text, and ODP Site 642 (circled). The hatched area is the presently known distribution of dipping reflectors, not including horizontal stratigraphic equivalent sequences. VP = Vøring Plateau; R = Rockall Island; JM = Jan Mayen Island; F = Faeroe Islands.

Vøring Plateau, offshore northern Norway. The dipping reflector sequence has been described by numerous workers, including Talwani & Eldholm (1972), and was recently drilled during ODP Leg 104 (Eldholm *et al.* 1987). Figure 4, a depth-migrated multichannel seismic reflection profile adapted from Eldholm *et al.* (1986) illustrates part of the divergent-arcuate series of reflectors. Both the upper and lower surfaces of the dipping reflector sequence are well-defined between points B and C, but westwards of C, the lower surface (referred to as K by early workers, such as Talwani & Eldholm 1972) is less well resolved, and there is no clear base to the sequence. Here the upper series of reflectors appears to continue with depth in the section; it is suggested that this point marks the oceanward limit of the interme-diate volcanic suite identified during drilling at Site 642 on Leg 104. The continental affinities of these intermediate volcanics is unquestionable (Parson *et al.* 1988), as is the MORB-like character of the overlying dipping reflector

sequence (Viereck *et al.* 1988). It is tentatively suggested that this junction represents the closest approximation to an ocean–continent boundary possible, buried as it is beneath 3–4 km of basaltic flows. The differentiation between parts of the section has been emphasized by the geochemical and isotopic data from Leg 104, which differen-tiated a tholeiitic 'Upper Series' and an andesitic/dacitic 'Lower Series'. Seismically, the character of the Upper Series (the dipping reflectors) differs from the underlying 'basement' in its higher frequency, more rectilinear and curvilinear fabric and the more continuous nature of its seismic horizons. Even at the crest of the Vøring Plateau close to Site 642, where the Upper Series 'cover' is at its thinnest (<200 m), the Lower Series displays a weak reflector fabric. In addition, there are apparent seismostratigraphic relationships, such as onlap and offlap, both within the Upper Series and at its bounding surfaces (e.g. A & B on Fig. 4). This deep Lower Series type basement has not yet been resolved on any other volcanic

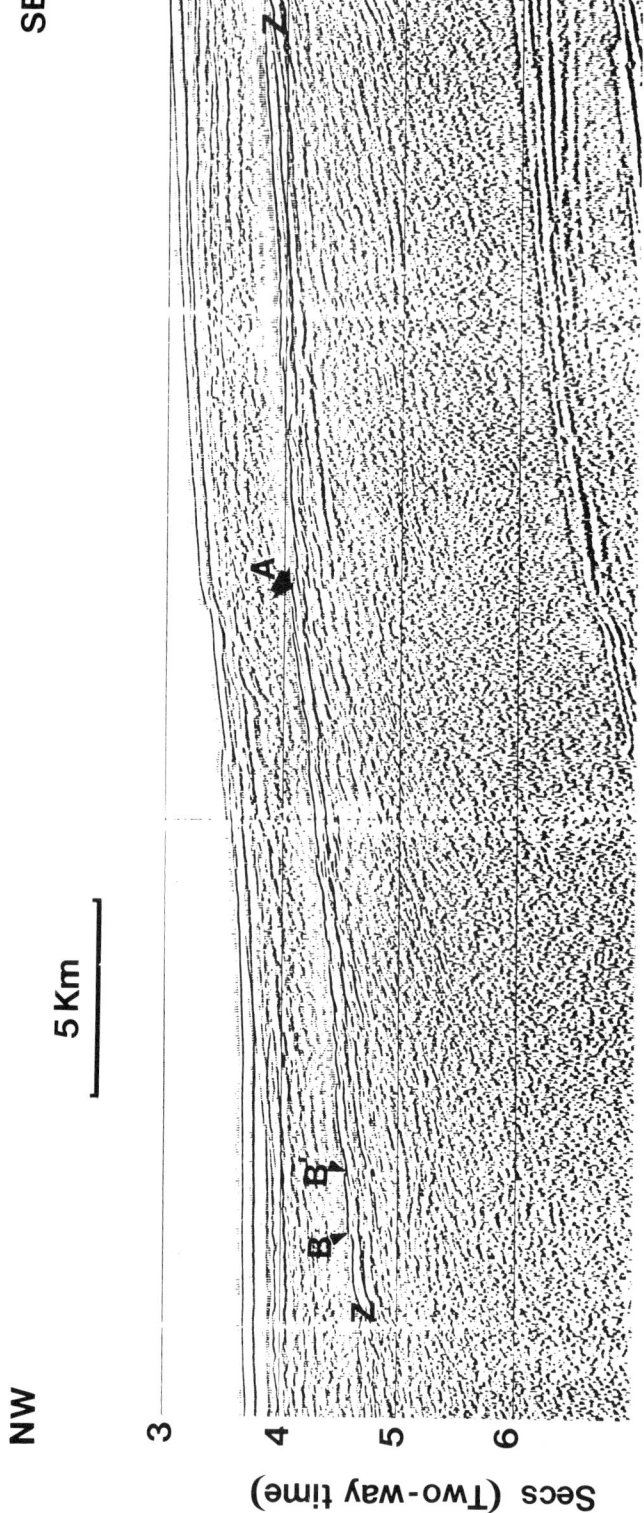

Fig. 3. Multichannel (12-channel) seismic reflection profile across the central Møre Basin margin. Subrectilinear seismic reflectors are present in the SE and NW sections of the sequence. Horizon Z represents the upper surface of the dipping reflector sequence. Offlapping reflectors at A are interpreted as compound lava fronts between 30 and 100 m thick. B–B' locates limits of seismic reflector 'outlier'.

(Figs 3–6 are single and multichannel seismic reflection records of dipping reflector sequences in the NE Atlantic, and are located in Fig. 2. NB vertical exaggeration varies between profiles as indicated, and levels of processing are not comparable)

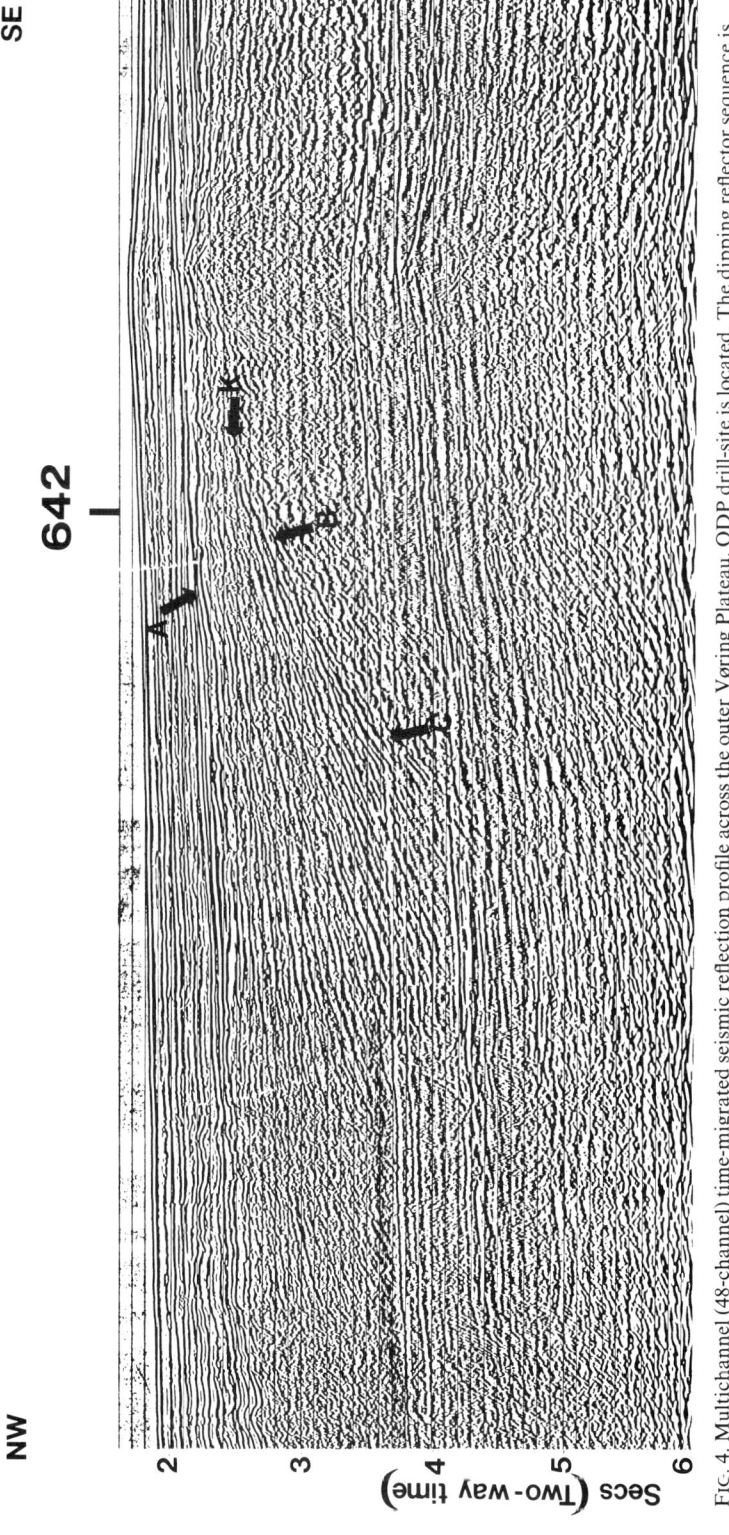

Fig. 4. Multichannel (48-channel) time-migrated seismic reflection profile across the outer Vøring Plateau. ODP drill-site is located. The dipping reflector sequence is dominated by an arcuate pattern. Reflector K is discussed in the text. A and B locate seismostratigraphic relationships discussed in the text. C marks the oceanward limit of reflector K.

passive margin, although similar lithologies to the unusual peraluminous dacites recovered from the base of Site 642 have been obtained from deep drilling within the Mesozoic rift of the Rockall Trough (Morton *et al.*, 1988).

Subparallel (southern E Greenland)

Figure 5 illustrates a single-channel seismic reflection profile interpretation orientated approximately orthogonal to the ocean–continent transition off south-eastern Greenland, and located in Figure 2. Here a series of subparallel, generally continuous, inclined intrabasement seismic horizons extends for about 40 km between the foot of the continental slope and the first identifiable oceanic magnetic anomaly (24B). Using approximate seismic velocities derived from comparable dipping reflector margins elsewhere (e.g. Mutter *et al.* 1984), the reflectors can be identified to a depth of only 2 km, although seismic penetration using this acquisition system (single gun, single channel) is limited. Unpublished multichannel seismic reflection data, however, (Larsen 1985) also fails to resolve a basal sequence such as that seen at the Vøring Plateau. The upper surface is well imaged, and describes a sharp, angular unconformable surface to the dipping horizons, which are continuous as indi-

vidual reflectors for up to 10 km. Seismic data covering the outcrop of the dipping sequences from beneath the shelf sedimentary sequence also show stepped surfaces indicating individual or composite lava flow terminations (Uruski & Parson 1985). Subparallel dipping reflectors occasionally occur as series of arcuate events, such as those off the eastern margin of the Jan Mayen Ridge, illustrated on the unmigrated seismic reflection profile in Figure 6.

Flexured (western Rockall)

Flexured series of dipping reflectors occur in which the most oceanward and the most landward sections of the reflector sequence are approximately horizontal, but the central section is inclined. This geometry is identified along much of the western and northwestern margins of the Rockall Plateau, W of the UK. A continuous, prominent reflector occurs everywhere at the upper surface to the reflectors, but no uniform basal reflector (e.g. K) is seen beneath the wedge (Fig. 7). The presence of a second, deeper sequence of dipping reflectors is detected on some seismic sections to the N and NW of Rockall Plateau (unpublished IOS data), where a strong seismic unconformity can be traced at the junction between the two sequences. This presum-

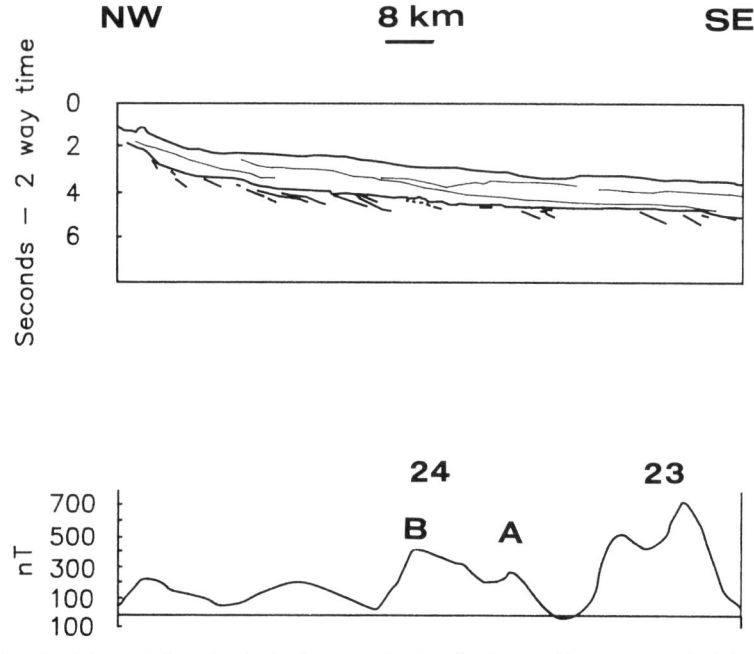

FIG. 5. Line-drawing interpretation of a single channel seismic reflection profile and magnetic data across the E Greenland margin, from Uruski & Parson (1985), illustrating subparallel inclined reflector geometry. Early seafloor-spreading magnetic anomalies 23 & 24 are indicated.

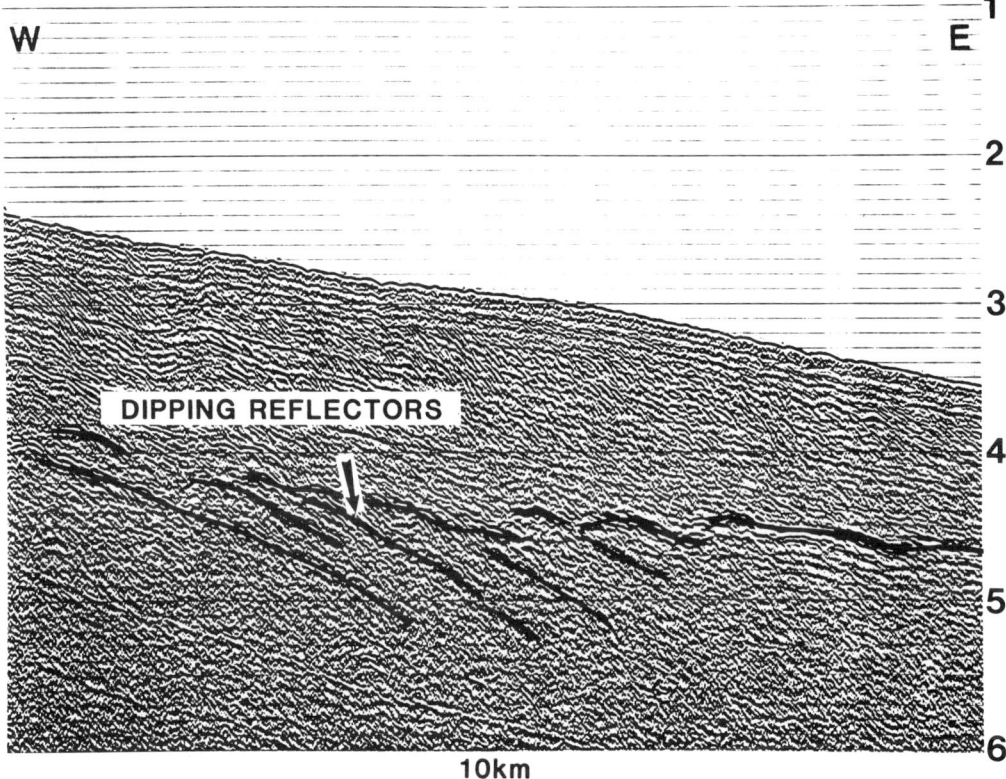

FIG. 6. Multichannel (12-channel) seismic reflection profile from the NE Jan Mayen Ridge, illustrating a subparallel sequence of arcuate dipping reflectors.

ably older sequence can be mapped throughout much of the area between the Faeroe and Rockall banks, and its seismic similarity to the overlying sequence suggests a (marginally) earlier episode of basalt extrusion rather than a volcanic 'Lower Series' equivalent such as is seen at the Vøring Plateau, (Parson & Eldholm, in prep.). Both landward and oceanward limits to the dipping sequence are confused by zones of diffraction hyperbolae and shorter, discontinuous seismic events. In the central section, however, a thick sequence of roughly parallel, divergent, rectilinear reflectors of various frequencies can be observed to a thickness of at least 5 km.

Two observations may be emphasized at this point:

(1) Precisely conjugate margins need not necessarily share styles of dipping reflector sequence. The NW Rockall and SE Greenland margins are a conjugate pair, but the style of sequence is different. The distribution and deposition of the dipping reflectors are undoubtedly controlled by basement topography and structure, as well as by the spreading rate and rate of crustal attenuation at the margin.

(2) Variations in the geometry of these reflector sequences may take place rapidly along strike. Seismic data from the Rockall margin (IOS unpublished data and Roberts, Schnitker *et al.* 1984) illustrate marked variations in seismostratigraphy and architecture of the dipping reflector sequence. The examples given above represent an attempt to illustrate a scheme for classifying the geometry of the reflector sequence as an aid to a comparison and contrast of volcanic passive margins. The second part of this paper serves to examine the nature of the reflectors, the possible reasons for the differences, and provide different models for their formation. Particular attention will be paid to the Vøring Plateau margin, where one of the most comprehensive databases exists. During the assessment of the different styles it is important to appreciate that the variation along strike of the styles may be very rapid,

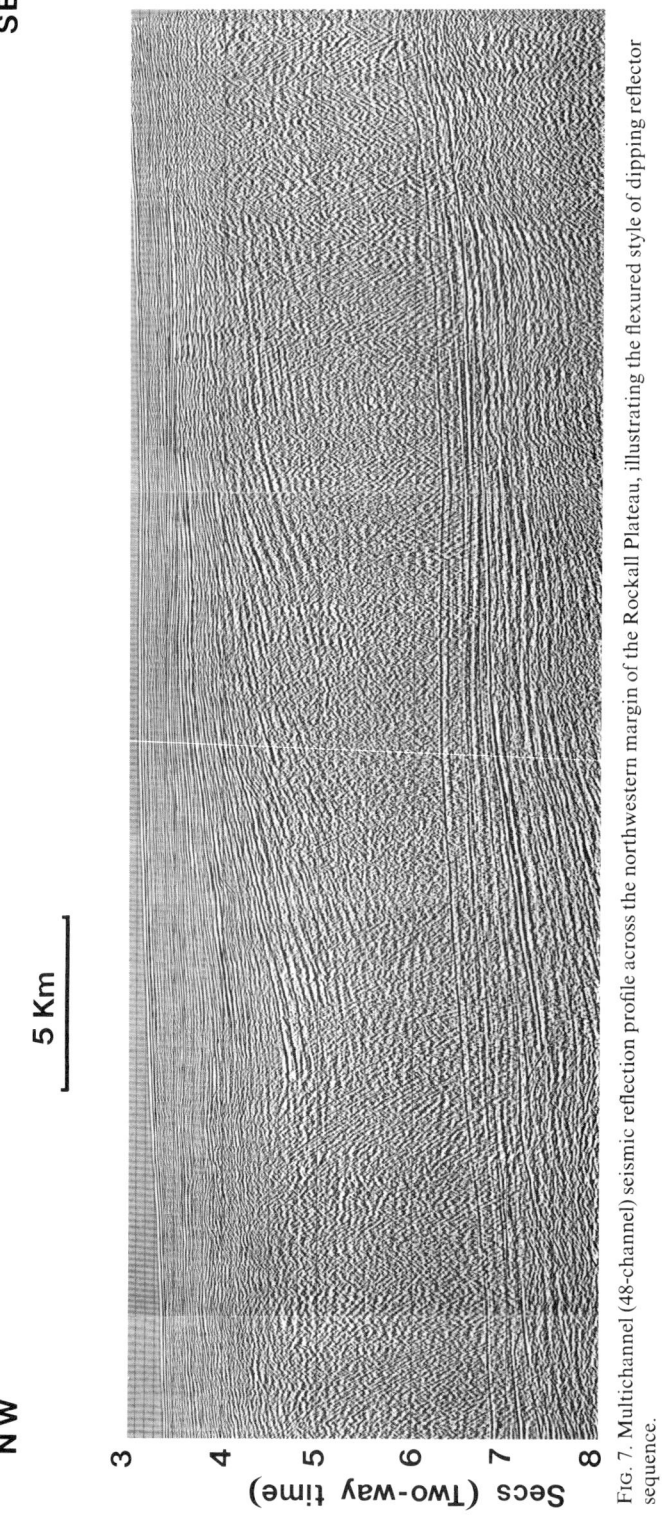

FIG. 7. Multichannel (48-channel) seismic reflection profile across the northwestern margin of the Rockall Plateau, illustrating the flexured style of dipping reflector sequence.

The nature of the reflectors

Both Leg 81 of the DSDP and Leg 104 of the ODP have successfully sampled dipping reflector sequences. However, the most comprehensive study available is from the Vøring Plateau, ODP Leg 104. This paper will re-examine the results from this site where arcuate reflector patterns predominate, and attempt to extrapolate the interpretation to sites of different geometry. A combination of deep drilling and wireline geophysical logs revealed over 900 m of upper basalt flows, which can be directly correlated with the entire dipping reflector sequence, in addition to a further 143 m penetration of the underlying volcanics which were not logged. The Upper Series (the dipping reflectors) comprise a series of T-MORB-type tholeiitic lava flows (as opposed to the N-MORBs recovered from Leg 81), numbering 120 in all, interleaved with 53 volcaniclastic units and cut by three dykes. Both the extrusive and the intrusive volcanics are comparable to many other Tertiary plateau basalt suites throughout much of the N Atlantic. More enriched basaltic vitric tuffs dominate the volcaniclastics. Analysis of the trace element and isotope geochemistry is in progress to understand the evolution of the magma chambers supplying these extrusives (Viereck *et al.*, 1988). The Upper Series is interpreted as a series of subaerial lavas, laid down entirely during the reversed polarity interval between magnetic anomalies 24 and 25, and probably over a period of only 2–3 My. By contrast, the underlying Lower Series flows comprise a series of peraluminous basaltic andesites and dacites, carrying cordierite as one of its phenocryst phases, and with initial $^{87}Sr/^{86}Sr$ ratios of 0.709–0.712 (Parson *et al.* 1988). The unit marking the base of the dipping reflectors was found to be characterized by volcaniclastic sediments of continental derivation (Eldholm *et al.* 1987).

A basal reflector to the Upper Series, referred to by previous workers as 'K' (Talwani & Eldholm 1972), corresponds to the upper surface of the intermediate volcanics, but has yet to be identified on dipping reflector margins elsewhere. This may be due to the shallower level of subcrop of K at the Vøring Plateau, or its absence elsewhere. In most other cases of volcanic rifted margins, primary reflectors become gradually less easily distinguishable from noise and artefact with increasing depth. The question remains as to whether the Vøring Plateau intermediate volcanics, as well as their elevation, are anomalous or even unique, or whether they are present, but deeply buried on other margins. Seismic refraction data across this type of margin are plentiful over Rockall and the Vøring Plateau, but sparse elsewhere. Lower crustal velocities of 6.3–6.5 km/s underly the reflectors off Rockall, (White *et al.* 1987; Whitmarsh & Miles 1987) but whilst implying a continental affinity, are not unambiguous.

The proportions and numbers of the volcaniclastic flows per extrusive flow unit, coupled with the relative numbers of each type of flow unit, can be further used to subdivide the Upper Series of the Vøring Plateau into a series of nine sequences of various thicknesses. The mechanism of how this occurs and its implications for the evolution of the margin are still poorly understood. In the seismic section which crosses this site only 12 individual seismic reflectors are observed. Although it has yet to be modelled successfully, it is considered likely that sampling and summing during processing of reflectors generated at the regular sediment/basalt interfaces produces a seismic signature such as that seen in Figure 4. Here, each single reflector does not correspond to a geological horizon, rather to a combination of several events. Alternatively, the grouping of several flow units into sequences as described from the ODP Leg 104 (Eldholm *et al.* 1987), may provide a zone of sufficient acoustic impedance contrasts to contribute to the reflector pattern. Therefore, apparent cut-off and/or onlap relationships must be viewed with some care, where doubt exists over the identification of individual horizons. Nonetheless, these relationships, however speculative, will prove important to an understanding of the direction of input of lava into the depositional basin. At the Vøring Plateau, the ODP Leg 104 drilling penetrated only the feather edge of the dipping reflector sequence. Here the dipping reflector sequence clearly overlaps a palaeosurface, referred to as K, which we now know to be composed of lavas with a continental affinity. Dykes of Upper Series composition clearly cut this lower sequence, and the assimilation of continental crust by some of the lower Upper Series flows is indicated by enriched Th levels and slightly higher initial $^{87}Sr/^{86}Sr$ ratios (up to 0.7035).

Higher and younger in the sequence there is less continental influence and lower initial ratios persist. K terminates at point C (see Fig. 4) beneath the central part of the volcanic sequence, at either a downfaulted junction or a depositional palaeoscarp. Skogseid (1983) has mapped the distribution and depth to K throughout the western Vøring Plateau and northwards towards the Lofoten Basin. As discussed above, in the absence of any other feature, the oceanward limit of K seems to be a suitably mappable approximation to the ocean–continent transition. Smythe (1983) argues for the position of the OCT at the '. . . lower end of the oldest of the oceanward-dipping series of intrabasement reflectors . . .'. It is clear, however, that with the exception of those margins where a base to the dipping reflector sequence is seen (to date, only the Vøring Plateau), the increased penetration, resolution and processing of existing and forthcoming seismic data sets will identify this position at increasingly deeper, and consequently more oceanward positions. Smythe's assessment of the continent–ocean boundary (COB) in the Faeroe–Shetland region is based on the technical inability to resolve seismostratigraphic or seismotectonic relationships at greater depths in the section. Where this *is* possible, the control on the deposition of the dipping reflectors is clearly seen as dominated by the 'K' palaeosurface. The transition from flow to feeder dyke is not resolved. Here, basement type alone provides a precise and easily located OCT. Elsewhere, until improved seismic data is available, a *zone* of transition must be defined between the lower limit of the oldest reflector as it laps onto the 'basement' surface *and* the junction between subaerial and submarine modes of deposition, recognizable from the distinct change in seismic facies.

The data available for the Vøring Plateau appears to support those models favouring continental or transitional crustal material below much of the dipping reflector accumulations. Oceanward of the outer limit of the Lower Series volcanics, feeder dykes may dominate the subdipping reflector sequence, as suggested by the increased dip of the seismic events. Figure 4, however, illustrates a time-migrated, rather than a depth-migrated section, and care must be exercised in comparing reflector gradients. Where a regional downturn of the reflector sequence is not evident, such as at the Møre and Rockall margins, a speculative interpretation is that attenuated continental crustal material dominates the floor to the dipping reflectors out to the site of the 'outer high', should it exist. No evidence is available to suggest attenuation of the lower crust by listric synthetic faulting, although some form of semi-brittle or brittle deformation must take place. One possibility is that the large number of feeder dykes contribute significantly to the extension. No downturn in the reflectors is suggested at their oceanward limits, as would be suggested by the Mutter-type model (Mutter *et al.* 1982).

Diffraction hyperbolae are locally observed within the thickest outermost part of the dipping reflector sequence, and suggest point sources at either younger feeder dyke margins cutting the older lava flows, or antithetic, normal dip-slip faults. The dykes at depth probably intrude the highly attenuated tongue of continental crust, since post-rift subsidence and margin collapse following the rupture and complete continental separation are unlikely to have preserved the stability and regularity of a plateau lava deposit. Immediately to the W (oceanward) of this 'intruded' zone, a chaotic seismic signature underlies an increasingly rough and hummocky upper volcanic surface. This change is interpreted as marking the change from subaerial/terrestrial dominated flow deposits to entirely or predominantly submarine, contemporaneous with the transition from rifting to drifting. The OCT here is less well constrained, but must lie between an oceanward limit marked by the initiation of persistent submarine conditions of extrusion, and continental limit, where attenuated and possibly listric-rifted continental basement underlying the dipping reflector sequence is so intruded by dykes that it is effectively oceanic basement. The mapping of this boundary is subjective, and its location is one which will require revision with the acquisition of further deep multichannel seismic data.

The flexured style of dipping reflector sequence is one that is relatively uncommon, and observed only from the western Rockall Plateau to date. It appears as a variation of the rectilinearly disposed reflectors, and suggests some structural control, presumably normal faulting of pre-existing palaeoscarp in the basement flooring the reflector sequence.

Concluding remarks

A range of geometries and structural styles of dipping reflector sequences at volcanic passive margins has been presented, and examples of these styles have been illustrated using a range of seismic reflection data of varying quality and sophistication. Four idealised end-member configurations of the reflector packages have been compared and contrasted between conjugate and non-conjugate margins. A detailed appraisal of the continent–ocean transition at the outer margin of the Vøring Plateau is compared with that off the western Rockall Plateau.

ACKNOWLEDGEMENTS: The author gratefully acknowledges the constructive criticism and reviews of a number of colleagues during the preparation of this study. In particular, thanks go to Doug Masson and Andy Morton and to many co-workers involved in ODP Leg 104. This work has been funded in part by the Department of Energy, UK.

References

ELDHOLM, O., THIEDE, J., *et al*. 1986. Dipping reflectors in the NE Atlantic—ODP Leg 104 drilling results. *Journal of the Geological Society of London*, **143**, 911–912.

——, ——, TAYLOR, E., *et al*. 1987. *Proceedings of the Ocean Drilling Program*. US Government Printing Office, Washington, **104A**, 783 pp.

GUDLAUGSSON, S. T., GUNNARSSON, K., SAND, M. & SKOGSEID, J. 1988. Tectonic and volcanic events at the Jan Mayen Ridge microcontinent. *In:* MORTON, A. C. & PARSON, L. M. (eds) *Early Tertiary Volcanism and the Opening of the NE Atlantic*. Geological Society of London, Special Publication, **39**, pp. 85–93.

HINZ, K. 1981. A hypothesis on terrestial catastrophes. Wedges of very thick oceanward dipping layers beneath passive continental margins—their origin and palaeoenvironmental significance. *Geologisches Jahrbuch*, **E22**, 3–28.

LARSEN, H-C. 1984. Geology of the North European margin. *In:* SPENCER, A. M. *et al*. (eds) *Petroleum Geology of the North European Margin*. Graham & Trotman, London, pp. 329–339.

—— 1985. *Project NAD—East Greenland. An integrated aeromagnetic and marine geophysical project off the east coast of Greenland*. Final Report (**No. 8**), 78pp (also unpublished multichannel seismic data).

—— & JAKOBSDOTTIR, S. 1988. Distribution, crustal properties and significance of seawards-dipping sub-basement reflectors off E Greenland. *In:* MORTON, A. C. & PARSON, L. M. (eds) *Early Tertiary Volcanism and the Opening of the NE Atlantic*. Geological Society of London, Special Publication, **39**, pp. 95–113.

MORTON, A. C., DIXON, J. E., FITTON, J. G., MACINTYRE, R. M., SMYTHE, D. K. & TAYLOR, P. N. 1988. Early Tertiary volcanic rocks in Well 163/6–1A, Rockall Trough. *In:* MORTON, A. C. & PARSON, L. M. (eds) *Early Tertiary Volcanism and the Opening of the NE Atlantic*. Geological Society of London, Special Publication, **39**, pp. 293–308.

MUTTER, J. C., TALWANI, M. & STOFFA, P. L. 1982. Origin of seaward dipping reflectors in oceanic crust off the Norwegian margin by 'subaerial seafloor spreading'. *Geology*, **10**, 353–357.

——, —— & —— 1984. Evidence for a thick oceanic crust adjacent to the Norwegian margin. *Journal of Geophysical Research*, **89**, 483–502.

PARSON, L. M. & ELDHOLM, O. (in prep). A comparative study of dipping reflector sequences. *Proceedings of the Ocean Drilling Program*, **104B**.

——, MASSON, D. G., MILES, P. R. & PELTON, C. D. 1986. Structure and evolution of the Rockall and East Greenland continental margins.

Institute of Oceanographic Sciences Report, **233**, 71 pp.

——, VIERECK, L., LOVE, D., MORTON, A. C., GIBSON, I. L. & HERTOGEN, J. (in press). The petrology of the Lower Series volcanics, ODP Site 642. *Proceedings of the Ocean Drilling Program*, **104B**.

PELTON, C. D. 1985. Geophysical interpretation of the structure and evolution of the Jan Mayen Ridge. *Institute of Oceanographic Sciences Report*, **205**, 38 pp.

ROBERTS, D. G., MONTADERT, L. & SEARLE, R. C. 1979. The western Rockall Plateau: stratigraphy and structural evolution. *In:* MONTADERT, L., ROBERTS, D. G. *et al*. *Initial Reports of the Deep Sea Drilling Project*. US Government Printing Office, Washington, **48**, 1061–1086.

——, BACKMAN, J., MORTON, A. C., MURRAY J. & KEENE, J. B. 1984. Evolution of volcanic rifted margins: synthesis of Leg 81 results on the west margin of Rockall Plateau. *In:* ROBERTS, D. G., SCHNITKER, D. *et al*. *Initial Reports of the Deep Sea Drilling Project*. US Government Printing Office, Washington, **81**, 883–912.

——, SCHNITKER, D. *et al*. 1984. *Initial Reports of the Deep Sea Drilling Project*. US Government Printing Office, Washington, **81**, 922 pp.

SKOGSEID, J. 1983. *A marine geophysical study of profiles between the Vøring Plateau and the Jan Mayen Ridge*. Cand. Sci. thesis. University of Oslo, Norway.

—— & ELDHOLM, O. 1987. Early Cenozoic crust at the Norwegian continental margin and the conjugate Jan Mayen Ridge. *Journal of Geophysical Research*, **92**, 11471–11491.

SMYTHE, D. K. 1983. Faeroe–Shetland escarpment and continental margin north of the Faeroes. *In:* BOTT, M. H. P., SAXOV, S., TALWANI, M. & THIEDE, J. (eds) *Structure and Development of the Greenland–Scotland Ridge*. Plenum Press, New York, pp. 109–119.

TALWANI, M. & ELDHOLM, O. 1972. The continental margin off Norway: A geophysical study. *Bulletin of the Geological Society of America*, **83**, 3375–3608.

URUSKI, C. I. & PARSON, L. M. 1985. A compilation of geophysical data on the East Greenland continental margin and its use in gravity modelling across the continent–ocean transition. *Institute of Oceanographic Sciences Report*, **214**, 53 pp.

VIERECK, L., TAYLOR, P. N., PARSON, L. M., MORTON, A. C., HERTOGEN, J., GIBSON, I. & the ODP LEG 104 SCIENTIFIC PARTY. 1988. Origin of the Palaeogene Vøring Plateau volcanic sequence. *In:* MORTON, A. C. & PARSON, L. M. (eds) *Early Tertiary Volcanism and the Opening of the NE Atlantic*. Geological Society of London, Special Publication, **39**, pp. 69–83.

WHITE, R. S. 1988. A hot-spot model for early Tertiary volcanism in the N Atlantic. *In:* MORTON, A. C. & PARSON, L. M. (eds) *Early Tertiary Volcanism and the Opening of the NE Atlantic.* Geological Society of London, Special Publication, **39**, pp. 3–13.

——, WESTBROOK, G. K., BOWEN, A. N., FOWLER, S. R., SPENCE, G. D., PRESCOTT, C., BARTON, P. J., JOPPEN, M., MORGAN, J. & BOTT, M. H. P. 1987. Hatton Bank (northwest UK) continental margin structure. *Geophysical Journal of the Royal Astronomical Society,* **89**, 265–271.

WHITMARSH, R. B. & MILES, P. R. 1987. Seismic structure of a seaward-dipping reflector sequence southwest of Rockall Plateau. *Geophysical Journal of the Royal Astronomical Society,* **90**, 731–739.

L. M. PARSON, Institute of Oceanographic Sciences, Deacon Laboratory, Brook Road, Wormley, Godalming, Surrey GU8 5UB, U.K.

Origin of the Palaeogene Vøring Plateau volcanic sequence

L. G. Viereck, P. N. Taylor, L. M. Parson, A. C. Morton, J. Hertogen, I. L. Gibson & the ODP Leg 104 Scientific Party

SUMMARY: During ODP Leg 104 a 900 m thick sequence of volcanic rocks was drilled at Site 642E on the Vøring Plateau. It was subdivided into an Upper and a Lower Series separated by 7 m of estuarine volcaniclastic and epiclastic rocks. The Upper Series comprises transitional-type mid-oceanic ridge tholeiites with chemical affinities to the Faeroes and E Greenland plateau basalts and to Recent basalts from transitional segments of the mid-Atlantic ridge such as Reykjanes Ridge. The magmas are interpreted as having been derived from a secondarily LIL-element enriched, primarily strongly depleted mantle source. Some flows and dykes show evidence of minor contamination by upper continental crustal rocks. Interlayered tuffs are more differentiated and ferrobasaltic, and are interpreted to have been derived from a LIL-element enriched plume-type mantle source. The Lower Series contains 13 peraluminous dacite flows representing melts of upper crustal metasedimentary rocks such as shales and greywackes. The dacites overlie five basaltic andesite flows formed by mixing of LIL-element depleted tholeiitic magma with upper crustal melts.

A 900 m thick volcanic section was drilled on the Vøring Plateau, NE Atlantic, at ODP Site 642E during Leg 104 (Fig. 1). It consists of 138 lava flows and 60 interlayered tuff beds and is cut by five dykes (Fig. 2). Core recovery was ca. 41%. The entire section was deposited in a terrestrial environment. A 7 m thick estuarine sequence of quartz- and white mica-rich volcaniclastic rocks and minor epiclastic sandy mudstones (unit S43) separates an Upper Series (US) and a Lower Series (LS), which differ significantly in textures, structures, mineralogy, and chemical composition. A detailed description of the volcanic section is given by Eldholm et al. (1987). Various manuscripts by the authors on the petrology of the Upper and Lower Series are presently in preparation and will be published in the Proceedings of the Ocean Drilling Program: Part B of Leg 104.

Lithology and petrography of Upper Series lavas

The Upper Series of the volcanic section corresponds to the seaward-dipping seismic reflector sequence observed in reflection profiles through the outer Vøring Plateau (Talwani et al. 1982; Hinz et al. 1984). An Eocene age is indicated by palaeomagnetic and microfossil data (Eldholm et al. 1987). The Series is approximately 760 m thick and comprises 120 aphyric to moderately phyric plagioclase, olivine and, rarely, clinopyroxene phyric lava flows. Core recovery was ca. 42%. The flows commonly show reddened tops and are thought to have been erupted subaerially, judging from the absence of

pillows or unequivocal hyaloclastites. Some flows, however, exhibit non-reddened, palagonitized, vesicular sideromelane rinds a few mm thick. They may indicate that a small number of flows were deposited in a shallow water environment. The common occurrence of neritic and paralic microfossils in the interlayered sediment units is taken as evidence for a general emplacement of the Upper Series in a near coastal environment and indicates that subsidence kept pace with extrusion of lavas (Eldholm et al. 1987).

The lava flows occur in medium- and fine-grained varieties that differ in crystallinity, internal flow fabric, physical properties and average thickness but are indistinguishable chemically (see Fig. 2, column 5). The fine-grained flows have an average thickness of 7 m (with a maximum of 18 m) with intense laminar fabric, thick top-breccias with common reddening and less than 5% of glassy mesostasis. In contrast, the medium-grained flows, with an average thickness of 4 m, lack internal laminar flow features and flow-top breccias; they rarely show reddening of flow tops, are vesicle-rich and contain between 5 and 10% of glassy mesostasis. These flows occur in sets of 4–8 flows and are thought to represent compound lava flows (Walker 1970). In analogy to observations made in Eastern Iceland by one of the authors (I.L.G.) and J. Helgason (pers. comm.), the medium-grained flows are thought to resemble proximal, small volume, high temperature/low viscosity, pahoehoe-type flow units. The fine-grained flows are believed to represent the distal facies of more voluminous aa-type flow units that were more strongly degassed, cooler and more viscous at the time of emplacement.

The two flow types are mineralogically uniform

From MORTON, A. C. & PARSON, L. M. (eds), 1988, *Early Tertiary Volcanism and the Opening of the NE Atlantic,* Geological Society Special Publication No. 39, pp. 69–83.

69

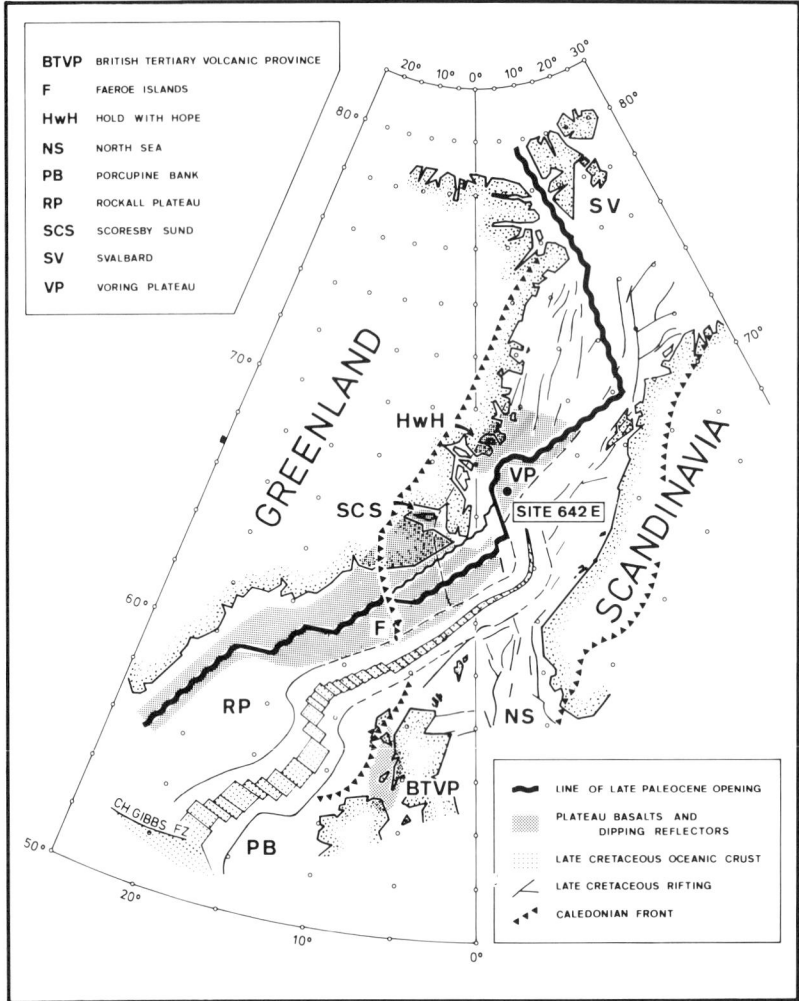

FIG. 1. Reconstruction of the NE Atlantic at Palaeogene time modified after Hanisch (1984), including data from Nunns (1982), Bott (1984), Eldholm *et al.* (1984), Larsen (1984), Parson *et al.* (1986), Mutter *et al.* (in press).

and consist of 55–60% plagioclase (phenocrysts An_{80-70}: groundmass and rims of phenocrysts An_{66-46}), 30–35% clinopyroxene (Ca_{41-29} Mg_{49-45} Fe_{12-26}), 3–4% saponite pseudomorphs after oxidized olivine, 2–3% Fe-Ti-oxides and traces of Mg-chromite inclusions (Figs. 3 & 4). Pigeonite, common elsewhere in NE Atlantic early Tertiary subaerial tholeiites, is absent.

Fe-saponite is the most common alteration product in the Upper Series lavas. Less common calcite and minor beidellite, nontronite, celadonite, analcite and heulandite also occur as vesicle and vein fills. Olivine is completely replaced throughout, but clinopyroxene and plagioclase are generally fresh. Surface weathering prior to sea water alteration resulted in considerable

replacement of primary mineral phases in the upper 70 m of the lava pile. Less intense weathering indicated by strong albitization of plagioclase and marked secondary chemical changes also affected intervals below 511 m, 607 m and 800 m, each of which is a few tens of metres thick. These depths coincide with boundaries between lithologic groups or chemical units (see Fig. 2, columns 3 & 11), indicating that the units defined on the basis of primary chemical changes may represent individual eruptive centres.

Petrology of the tholeiitic Upper Series basalts

All lava flows of the Upper Series have low-K, low-Ti transitional mid-oceanic tholeiite (T-MORB) composition with low initial $^{87}Sr/^{86}Sr$

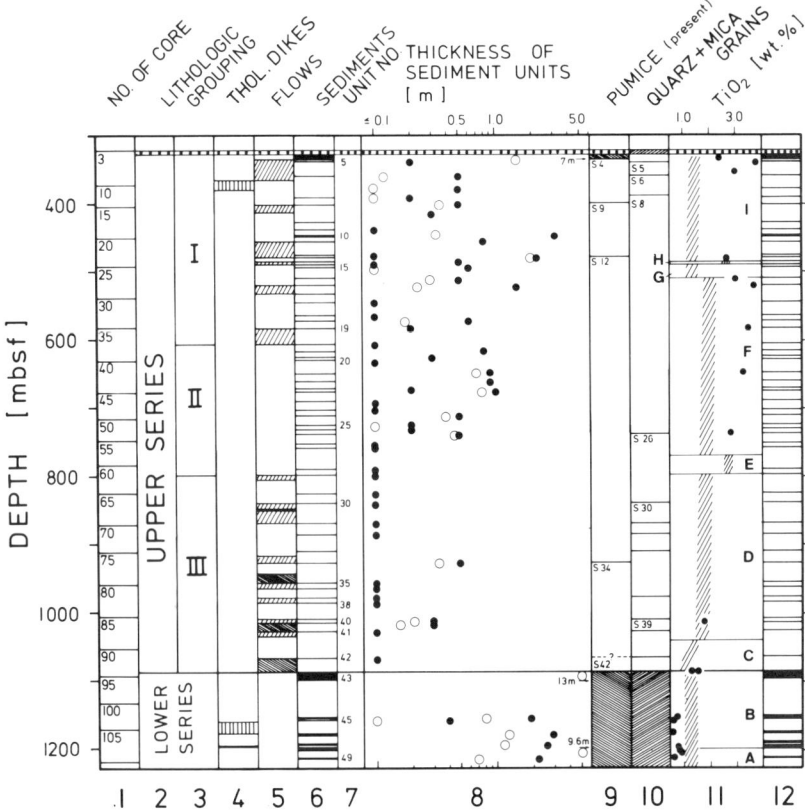

FIG. 2. Stratigraphic profile of ODP Leg 104 Site 642E. Symbols in column 5: wide shading = successions of medium-grained flows, unshaded = successions of fine-grained flows, densely shaded = mixed flows. Open circles in column 8 indicate recovered thicknesses; large dots represent sediment thicknesses deduced from geophysical logs. Column 11 lists the stratichemical grouping of the lavas based on their range of TiO₂ contents. Column 12 is the same as column 6. (mbsf = metres below sea level).

ratios (mean 0.7032) (Table 1; Figs. 5, 6, 7). They have rather flat chondrite normalized rare earth element (REE) patterns with $(La/Sm)_N < 1$ and $(La/Yb)_N > 1$ (Figs. 7, 8). They are moderately differentiated with Mg-values of 63–54 (MgO: 9–7 wt% except the lowermost flow F105 with 12–9 wt%; Ni: 180–70 ppm).

Flows with similar MgO contents range from low-Ti (1.0%) to moderately Ti-rich tholeiites (2.9%), generally with TiO₂ between 1.7 and 2.2%. Crystal fractionation of 5–12% olivine, 0–20% plagioclase and minor Mg-chromite may explain most of the variation of compatible elements, but can only account for about 15% of the incompatible element enrichment. There is no correlation between Cr and Sc in samples with Cr > 80 ppm, excluding fractionation of clinopyroxene as a major cause for most of the Cr variation observed (370–110 ppm; rarely 35–40 ppm). Positive Eu-anomalies in a number of flows in the uppermost

part of the section can be explained by the observed accumulation of plagioclase.

Different degrees of partial melting of a uniform, slightly depleted mantle source seem to account for most (ca. 85%) of the chemical differences. Simple batch partial melting calculations have been done based on established partition coefficients (Fujimaki *et al.*, 1984) The calculations suggest the basaltic magmas parental to the Vøring Plateau Upper Series tholeiites (Cr > 400 ppm, Ni > 300 ppm) to be derived by 5–16% of melting (TiO₂:2.25 wt% = 5–6% PM, 1.5 wt% = 11–12% PM, 0.9 wt% = 16% PM).

Several flows, especially near the base of the series, are selectively enriched in Th. This is accompanied by slightly higher initial $^{87}Sr/^{86}Sr$ ratios (>0.7035). Binary mixing calculations were done using appropriate Th values of uncontaminated Vøring Plateau tholeiites (0.2–5 ppm) and of upper continental crustal rocks

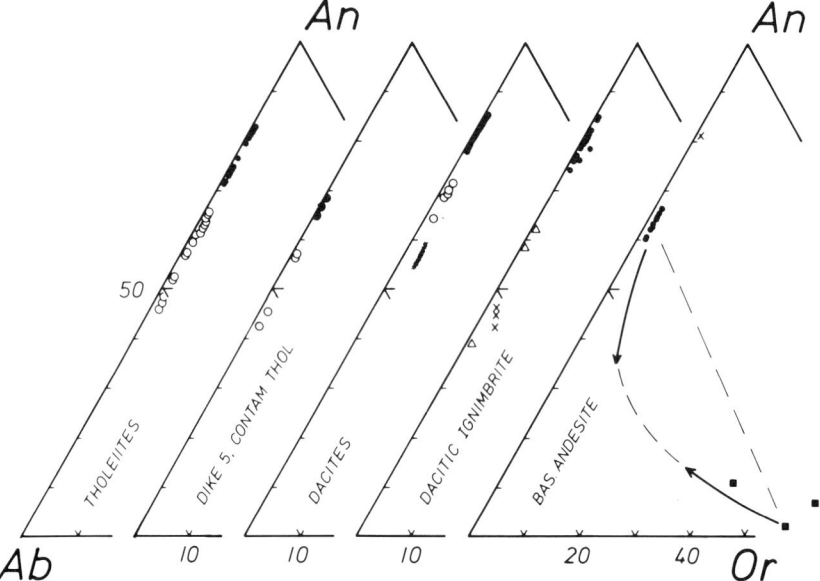

FIG. 3. Compositional range of plagioclase phenocrysts (dots) and microlites (circles). Black line in the dacite section at An_{55-60} is the composition of plagioclase intergrown with quartz in fasciculate bundles in glass of units F106–109. Crosses and open triangles in the dacitic ignimbrite field are compositions in accessory fragments. Tied arrows in the field of basaltic andesite indicate a range of compositions of coexisting groundmass plagioclase and alkali feldspar in shoshonitic rocks (Nicholls & Carmichael 1969).

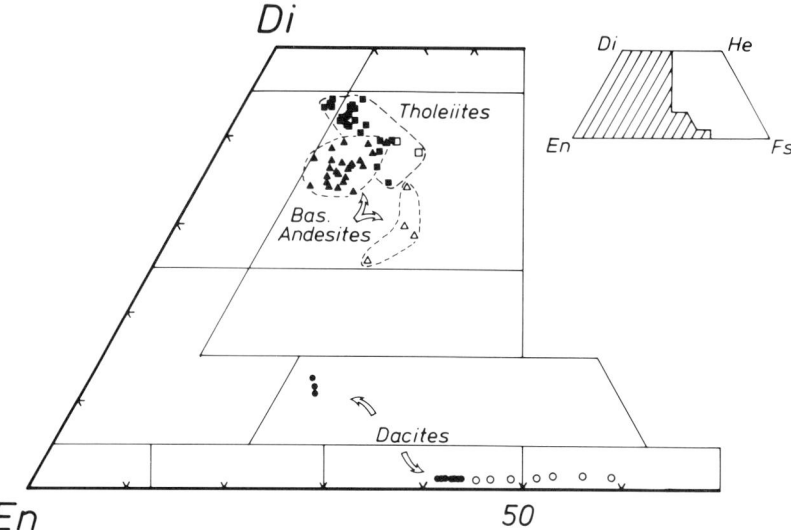

FIG. 4. Composition of pyroxenes (closed symbols = phenocrysts, open symbols = in matrix) in tholeiites (squares), basaltic andesites (triangles) and dacites (circles).

(non-metamorphosed to medium-grade meta-morphic) respectively (6–11 ppm; Wedepohl 1978). They indicate that assimilation of less than 2 wt% of crustal rocks (Viereck et al., in prep. b) can account for the enrichment observed.

The lava succession can be divided into seven stratichemical units (USC to USI) on the basis of their incompatible element characteristics (see Fig. 2, column 11; see Fig. 6). Flows of the lowermost unit USC are the least fractionated (Ni: 180–100 ppm) and are derived by the largest degree of melting (16% PM; Zr: 50–70 ppm,

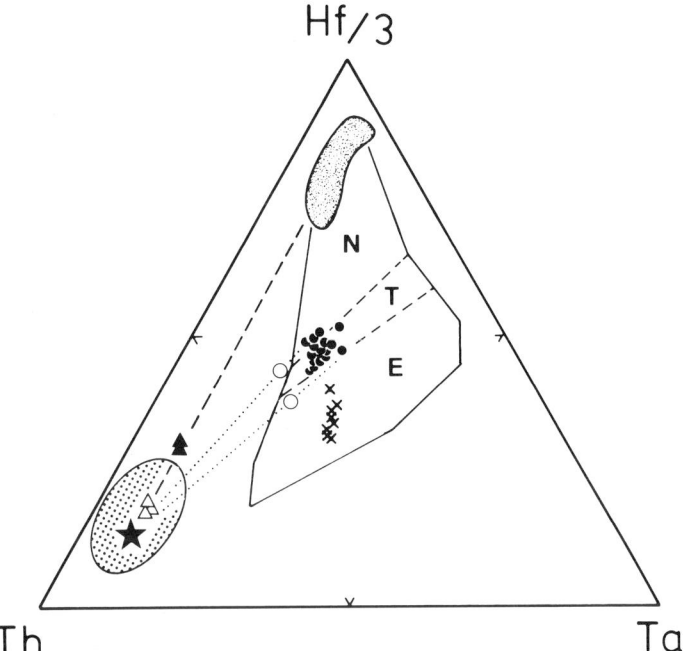

FIG. 5. Vøring Plateau lavas and sediments from Site 642E in Hf/3-Ta-Th plot (Wood 1980). Dots: Upper Series tholeiites, open circles: same but contaminated, crosses: Upper Series Fe-Ti-basaltic sediments, filled triangles: contaminated tholeiitic dykes D5 and D6 (Lower Series), open triangles: basaltic andesites (Lower Series), black star: dacites (Lower Series). Compositional field of crustal rocks from the Caledonides (ellipse, dotted) includes data for Moine schist, Torridonian sandstone, Palaeozoic slates, and Precambrian amphibolites from Wood 1980, Thompson 1982, and Thompson *et al.* 1986. Shaded field of strongly depleted N-MORB from DSDP Leg 81, Site 553 and 555 after data from Joron *et al.* 1984.

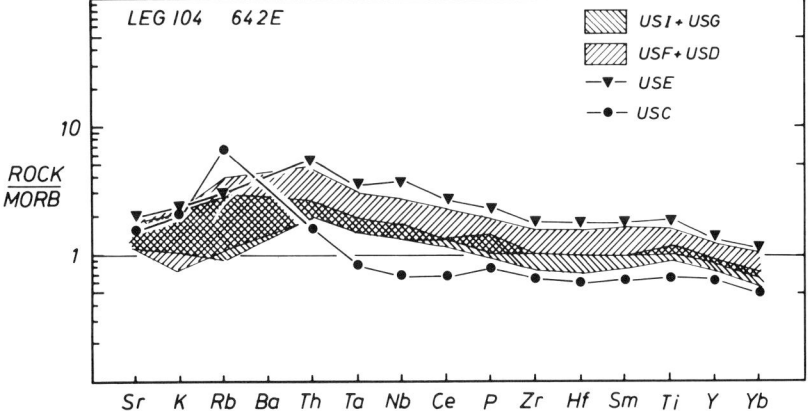

FIG. 6. MORB-normalized variation of trace elements in rocks of stratichemical units of the Upper Series (normalizing values given in Pearce *et al.* 1984).

TABLE 1. *Representative analyses of rock types drilled on Leg 104 at Site 642E. Tholeiites: F23, F90, F62; contaminated dyke: D5; ferrobasaltic volcaniclastic sediment: S19; dacite (glass handpicked): F106; basaltic andesite: F118; standard BHVO-1 given for comparison. Chemical groups as in Fig. 2 (LS: Lower Series, US: Upper Series); n.a.: not analysed. XRF analyses were done at Ruhr University, Bochum, Federal Republic of Germany, neutron activation analyses (NAA) were done at Leuven, Belgium.*

	F23		F90		F62		D5		S19		F105		F118		Standard BHVO-1
Lithol. Unit	F23		F90		F62		D5		S19		F105		F118		
Core-Section	22-4		76-2		59-3		105-2		35-2		97-1		108-2		
Interv. (cm)	148-150		32-34		69-71		84-86		108-110		97-100		51-53		
Depth (m)	475.3		937.7		775.0		1176.6		584.1		1110.8		1203.0		
Chem. Group	USI		USD		USE		LSB		USF		LSB		LSA		
Method	XRF	NAA	XRF	NAA	XRF	NAA	XRF	NAA	XRF	NAA	XRF	NAA	XRF	NAA	XRF
SiO_2	48.70		48.10		47.60		48.90		47.60		61.70		48.90		50.0
TiO_2	1.45		1.99		2.66		1.07		3.51		1.19		1.18		2.76
Al_2O_3	16.44		14.46		13.56		13.28		12.45		13.53		17.11		13.73
Fe_2O_3	4.62		5.86		6.03		4.56		18.60		7.97		6.35		2.74
FeO	5.72		6.24		7.34		7.66		1.36		n.a.		2.59		8.55
MnO	0.17		0.18		0.19		0.21		0.10		0.12		0.16		0.18
MgO	7.38		7.06		7.30		8.43		5.53		1.42		5.21		7.36
CaO	11.86		11.21		9.77		9.84		3.85		2.90		7.48		11.42
Na_2O	2.17		2.39		2.63		2.09		2.04		1.88		2.98		2.31
K_2O	0.07		0.10		0.23		0.11		0.96		2.43		1.66		0.53
P_2O_5	0.12		0.19		0.27		0.10		0.34		0.20		0.18		0.28
H_2O^+	1.23		1.14		1.28		1.46		4.24		5.66		2.39		0.20
CO_2	0.19		0.24		0.19		0.14		0.38		0.19		1.45		0.04
FeS	n.a.		n.a.		n.a.		1.35		n.a.		n.a.		1.23		n.a.
Sum	100.12		99.16		99.05		99.20		100.96		99.19		98.87		100.10
S	0.00		0.00		0.00		0.72		0.00		0.05		0.66		0.01
Cl	0.01		0.01		0.01		0.01		0.01		0.02		0.03		n.a.
Sc		39.8		44.2		43.9		52		30		24.7		51.5	
V	305		359		424		362		193		130		229		315
Cr	229	232	210	218	257	256	268	246	58	58	48	8.9	166	153	272
Co	41	45.2	37	47.1	39	47.3	53	59.5	31	35.1	8		35	6.76	42
Ni	93		91		95		163		66		15		11		129
Cu	106		149		275		65		222		13		30		122
Zn	87		110		136		111		79		125		129		102
Rb	3	<5	3	<7	6	<8	<2	<7	14	11	114	93	44	39	8
Sr	208	202	221	224	226	289	121	240	192	169	158	160	185	193	385
Y	22		31		39		34		78		44		36		27
Zr	71		117		157		67		264		261		127		162
Nb	4		6		11		1		27		18		6		17
Ba	15	<75	69	<85	141	<100	71	<75	162	<80	508	528	242	220	215
La		4.1		6.9		9.5		4.1		27.9		36.3		16.1	
Ce		11.8		19.7		26.9		10.5		67.8		81		38.2	
Nd		9.5		14.7		20.3		6.7		49		33.7		17.7	
Sm		2.93		4.54		5.86		2.44		12		7.19		4.45	
Eu		1.19		1.66		2.09		0.92		3.49		1.14		1.26	
Tb		0.64		0.96		1.22		0.75		2.18		1.21		0.98	
Ho		0.83		1.4		1.4		1.4						1.4	
Yb		2.15		3.04		3.9		3.93		7.4		4.4		3.63	
Lu		0.31		0.44		0.56		0.57		1.01		0.65		0.52	
Hf		1.88		3.16		4.22		1.81		6.8		6.4		3.33	
Ta		0.32		0.47		0.71		0.17		2.16		1.33			
Th		0.43		0.69		1.09		1.27		2.65		12.2		4.8	
U		<0.15		<0.15		<0.15		0.27		1.12		2.7		1.2	

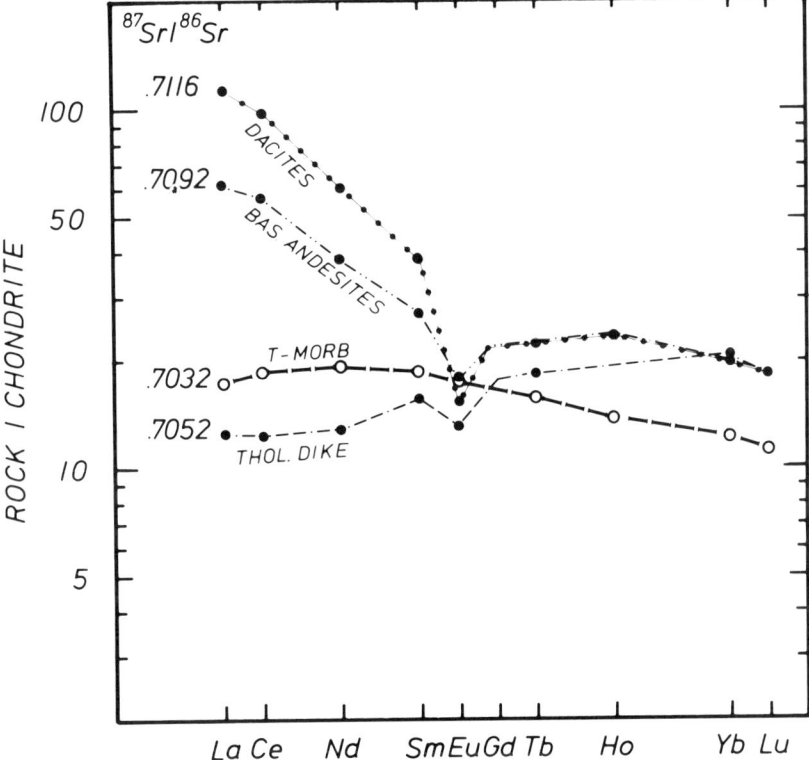

FIG. 7. REE and Sr-isotope variation of representative samples of each lava type of Upper and Lower Series.

TiO$_2$: 1.0–1.3%). The bulk of the Upper Series lavas (units USD and USF) are derived by 6–14% partial melting (TiO$_2$: 1.5–2.3%) while flows derived by lower degrees of melting are rare and constitute units USE (four flows) and USH (one flow) (TiO$_2$: 2.3–2.7%). The uppermost units USG and USI are mineralogically and chemically more variable, are, on average, less differentiated (Ni: 150–100 ppm), and are derived again by higher degrees of melting (12–14%, TiO$_2$ 1.3–1.5%).

Two tholeiitic dykes (D5 & D6) intrude the lower series (see Fig. 2). These have N-MORB character with (Sm/Yb)$_N$ = 0.7, although they have been strongly enriched in low-field-strength elements (LFS-) such as Th and light rare earth elements (LREE) (see Table 1; see Fig. 10). This can be explained by about 7–10% assimilation of continental crustal material (binary mixing calculations using Th-values) and show an initial ^{87}Sr/^{86}Sr ratio of 0.7052 (see Fig. 7). They were derived from a more depleted mantle source than the Upper Series lavas (see Fig. 7). This is also indicated in Figure 5, where the two dykes are displaced from the mixing line between the T MORB compositions of the Upper Series and

crustal rocks. They lie on a mixing line between crustal rocks and extremely depleted N-MORB, as drilled on Leg 81 at DSDP Site 553 on the western margin of the Rockall Plateau (Joron *et al.* 1984).

Comparison with NE Atlantic magmatic provinces

The lavas forming the Vøring Plateau dipping reflector sequence have chemical affinities to Palaeogene plateau basalts from the British Tertiary Volcanic Province, NE Greenland and the Faeroe Islands, as well as recent magmas from normal T-MORB segments of Mohns and Knipovich Ridge (see Fig. 8) (Mattey *et al.* 1977; Beckinsale *et al.* 1978; Wood 1980; Neumann & Schilling 1984; Upton *et al.* 1984a; Thompson *et al.* 1986). They are on the average less differentiated than lavas from Hold with Hope, e.g. the Lower Plateau Lava Sequence, or the Faeroes. Their REE-pattern is very similar to that of magmas from the Reykjanes Ridge between 62 and 63°N (see Fig. 8) (Schilling *et al.* 1983). The N-MORB dykes in the Lower Series on the

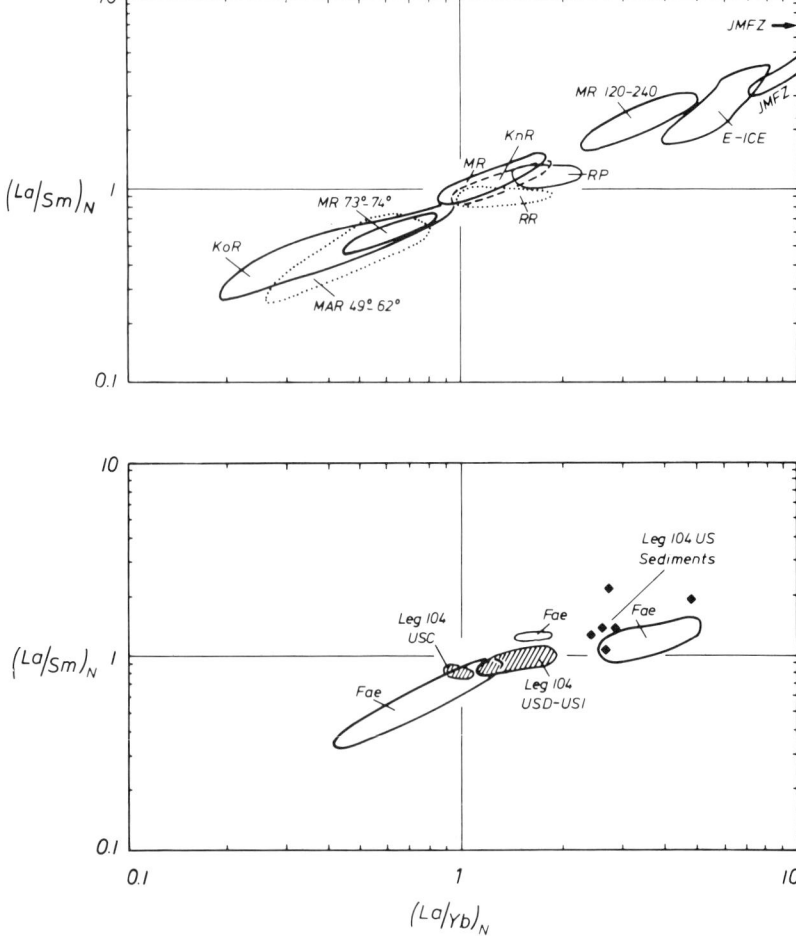

FIG. 8. Chondrite normalized ratios La/Sm versus La/Yb for Leg 104 Upper Series tholeiite lavas and Fe-Ti-basaltic volcaniclastic rocks (lower diagram). Values for Faeroe Islands (lower diagram) and Recent basalts from mid-Atlantic ridge between 49° and 80°N are given for comparison. KoR = Kolbeinsey Ridge, MAR 49–62 = segment of Reykjanes Ridge (given in degrees N), MR 73–74° = segment of Mohns Ridge, MR = Mohns Ridge between 240 km N of Jan Mayen and 74°N, KnR = Knipovich Ridge, RR = Reykjanes Ridge from 62.2°N to Reykjanes Peninsula (RP), MR 120–240 = Mohns Ridge between 120 and 240 km N of Jan Mayen fracture zone, E–ICE = Eastern Iceland (Miocene flows), JMFZ = < 100 km N and S of Jan Mayen fracture zone along Mohns and Kolbeinsey Ridge. Data are from Schilling & Noe-Nygaard (1974), Gibson et al. (1982), Schilling et al. (1983) and Neumann & Schilling (1984).

Vøring Plateau could only be derived from a depleted mantle source similar to that which presently underlies the Kolbeinsey Ridge. Their trace element enrichment pattern (see Fig. 10) is quite similar to that of basal lavas in the British Tertiary Volcanic Province (e.g. Staffa Magma Type, Isle of Mull) which are also thought to be strongly affected by assimilation of continental crust (Beckinsale et al. 1978; Thompson et al. 1986).

Vøring Plateau Upper Series sediments

Fifty-three dominantly basaltic lithic vitric tuff beds are interlayered with the lava flows, making up about 4% of the Upper Series (see Fig. 2, column 6). They are strongly altered, expressed by intense red colouration and almost complete replacement by secondary sheet silicates and, less often, zeolites (analcite and heulandite). Chemically, alteration led to a marked gain in oxidized iron, in water, Na, K, Rb, Ba and Cu as well as a loss in Ca and Sr (Fig. 9). Some beds show a peculiar irregular variation of P and Y that cannot be explained yet.

The tuff beds can be divided into an upper (interval 325–740 m) (see Fig. 2) and lower group (740–1189 m). In the upper group volcaniclastic units are thicker (up to 7 m) and more common with about one unit for every two flows (see Fig. 2, column 8). They may show repetitive fining-upward sequences, internal erosional surfaces and reworking. Pumiceous, presumably more differentiated, lapilli are rare but may form individual beds up to 10 cm in thickness in a few units of the upper group (see Fig. 2, column 9).

Incompatible element concentrations of the basaltic sediments are markedly higher than those of the lavas (e.g. TiO_2 : 2.7–3.8%) suggesting that they were derived from a different, more enriched mantle source (see Table 1; see Fig. 9). Their chondrite-normalized REE patterns are uniformly concave upward with $(La/Sm)_N = 1.5$–2.2 & $(La/Yb)_N = 2.0$–2.5 in contrast to the upwardly convex shape of all lava flow patterns $((La/Sm)_N < 1)$. These tuffs compositionally resemble slightly more differentiated (Ni: 90–60 ppm, Cr 200–40 ppm) ferrobasaltic magmas, similar to those of Eastern Iceland, the Jan Mayen Fracture Zone or those holes drilled on the Vøring Plateau at DSDP Leg 38 Site 342, that are derived from a LIL element-enriched plume-type MORB mantle source (Schilling 1976; Gibson *et al.* 1982; Neumann & Schilling 1984). They are chemically also similar to early Eocene tuffs of the Balder Formation and equivalents (e.g. tuffs of the Ølst Formation, Denmark; see

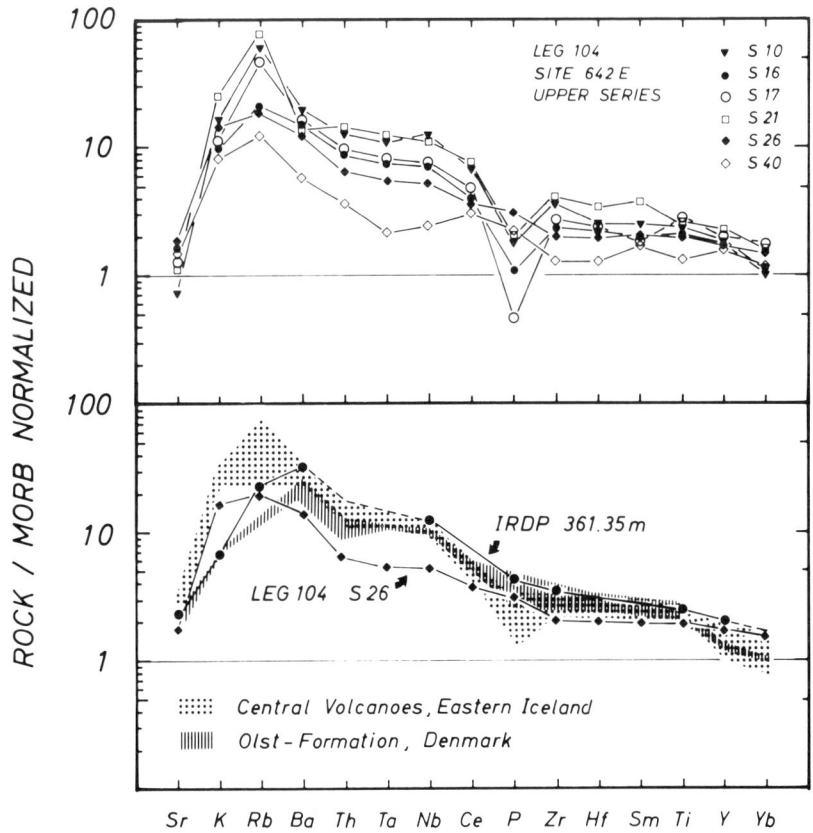

FIG. 9. MORB-normalized variation of hygromagmatophile elements in volcaniclastic units of the Upper Series (US). Least enriched sample S40 is a unit in the lower half of the US. Compositional ranges of Eocene ash layers in the Ølst Formation of Denmark and Miocene tuffs recovered in the IRDP drill hole in Eastern Iceland (Schmincke *et al.* 1982) are given for comparison.

Fig. 9) that are widespread throughout the North Sea and increase in thickness to the NW (Knox & Morton 1983).

The tuff beds in the lower half of the Upper Series are thinner (generally less than 10 mm) and petrographically more uniform (see Fig. 2, column 8). They are less differentiated (Ni 91 ppm, Cr 250 ppm) and less enriched in incompatible elements ($TiO_2 = 1.8\%$). White mica and tectonically deformed undulose quartz are present in most units while one unit also contains aegirine-augite (see Fig. 2, column 10). Correcting for the strong alteration that has affected all sediments, their major and minor element concentrations are similar to those of the lava flows they are interlayered with. However, like the tuffs of the upper group they also exhibit an enriched chondrite-normalized REE pattern with $(La/Sm)_N = 1.4$ and $(La/Yb)_N = 2.2$ and therefore cannot be derived from the same magma source as the lavas. They may resemble quartz-mica- and, rarely, also aegirine-bearing tuff beds of the Sele Formation that underlies the Balder Formation (cf. Knox & Morton 1983). However, any correlation on the existing data base is scientifically unjustified.

The mode of eruptive origin and emplacement of these tuff beds is still a matter of debate. Repetitive fining-upward sequences, internal erosional surfaces and reworking are clear indications that these sediment units do not represent primary pyroclastic deposits. They have been reworked by shallow marine or fluvial (or aeolian?) activity. Chemical similarities suggest their correlation over several hundred kilometres with the tuff beds of the Balder Formation in the marine basins of the North Sea. Only eruptions of Plinian-type would be able to generate such widespread deposits (Fisher & Schmincke 1984). The few pumice layers of probably more differentiated composition are, with no doubt, of Plinian origin.

However, the petrographic data are not very convincing evidence for a Plinian origin of the ferrobasaltic layers. Most of the units contain formerly glassy sideromelane shards and tachylitic to microcrystalline clasts in approximately equal amounts. Arcuate-shaped shards dominate over angular ones but average vesicularity of the shards is rather small for Plinian-type eruptions ($< 50\%$ vol). A hydroclastic influence during eruption seems to be likely, perhaps similar to the 1979 eruption of Soufriere Volcano, St. Vincent (Brazier *et al.* 1982). A more local derivation (< 100 km) of the tuff beds should therefore be favoured. The internal structure of accretionary lapilli found in two of the tuff beds is evidence for subaerial distribution of the fine-grained tephra by drifting ash clouds, prior to aquatic redeposition (Schumacher, pers. comm.).

Lithology and petrography of the Lower Series lavas

The Lower Series is about 140 m thick and extends to the bottom of Hole 642E at 1230 mbsf (see Fig. 2). It consists of about 70% lava flows and 17% volcaniclastic sediments, plus 14% tholeiitic dykes (dykes D5 & D6) which have already been discussed. Two flow types are present: 13 perlitic to variolitic dacite flows (F106–F117, Lower Series B) are underlain by five microcrystalline basaltic andesite flows (F118–F121, Lower Series A). Core recovery was ca. 33%.

The dacite flows have a perlitic, brecciated base and top, and a variolitic central zone, each of which is about 0.4–1.0 m thick. The brownish to greyish-pink glass has been pervasively hydrated to perlite with a purple vitreous lustre in hand specimen. Alteration to montmorillonite/beidellite has occurred along fractures, causing the brecciated appearance of most perlitic zones. Most vesicles and fractures in the glass are filled with montmorillonite/beidellite, chalcedony, agate, quartz, pyrite, and, less often, calcite and mordenite. Vesicles are elongated at an angle of 20–40° from the horizontal.

Phenocryst content ranges from 1–5% with calcic plagioclase (An_{85-70}, see Fig. 3) dominating over hypersthene in microcrystalline flows and pigeonite in glassy varieties (F106–F109) (see Fig. 4). The variolitic zones are characterized by swallow-tailed plagioclase with subparallel orientation and acicular ferrohypersthenes. Additional skeletal ilmenite shows length/width ratios of between 30 and 100. Untwinned prismatic cordierite occurs in trace amounts in glomerophyric aggregates with plagioclase in four flow units. Its $(Fe + Mn)/(Mg + Fe + Mn)$ ratio of 0.3 is similar to that from intratelluric cordierites in contact metamorphic melts and dacitic to rhyolitic magmas thought to be derived by partial fusion of continental crust (Wyborn *et al.* 1981; Clemens & Wall 1984; Grapes 1986; Viereck *et al.* in prep. a).

The basaltic andesite flows of Lower Series A are microcrystalline and intersertal, with a quenched matrix of plagioclase (An_{67-60}) and clinopyroxene and minor amounts of plagioclase intergrown with alkali-feldspar, of ilmenite and poikilitic quartz (see Figs. 3 & 4). Plagioclase and pyroxene are replaced by trioctahedral Fe-saponites.

Petrology of the Lower Series lavas

The dacites of Lower Series B are compositionally very uniform (MgO 1.4%) and are peraluminous rocks with $^{87}Sr/^{86}Sr$ ratios of 0.712–0.713 and $^{208}Pb/^{204}Pb$ ratios of 38.72–38.77 (see Table 1). Like other cordierite-phyric dacites and rhyolites they are high-K-type magmas indicated by microprobe analyses of glass in flows F106–109 with K_2O ranging between 2.7 and 2.9% (Ewart 1979; Wyborn *et al.* 1981; Clemens & Wall 1984; Munksgaard 1984; Viereck *et al.* in prep. a). The basaltic andesites cover a larger compositional range, with MgO contents varying between 2.3 and 8.5% (see Table 1). They have $^{87}Sr/^{86}Sr$ and

$^{208}Pb/^{204}Pb$ ratios of ca. 0.709 and 38.63–38.73 respectively.

The chondrite- and MORB-normalized incompatible element patterns show uniform, strongly alkaline enrichment with superimposed negative anomalies of Sr, Ba, Ta, Nb, P and Ti (see Figs. 7 & 10). The dacites are more strongly enriched, and their trace element patterns resemble those of average shales, greywackes, granites, schists, gneisses and amphibolites, as well as average continental crust (Wedepohl 1978). In the discriminative Th-Hf-Ta diagram by Wood (1980) the dacite compositions plot well in the range of the most common Caledonian metasedimentary rocks such as Moine schists (see Fig. 5; Thompson

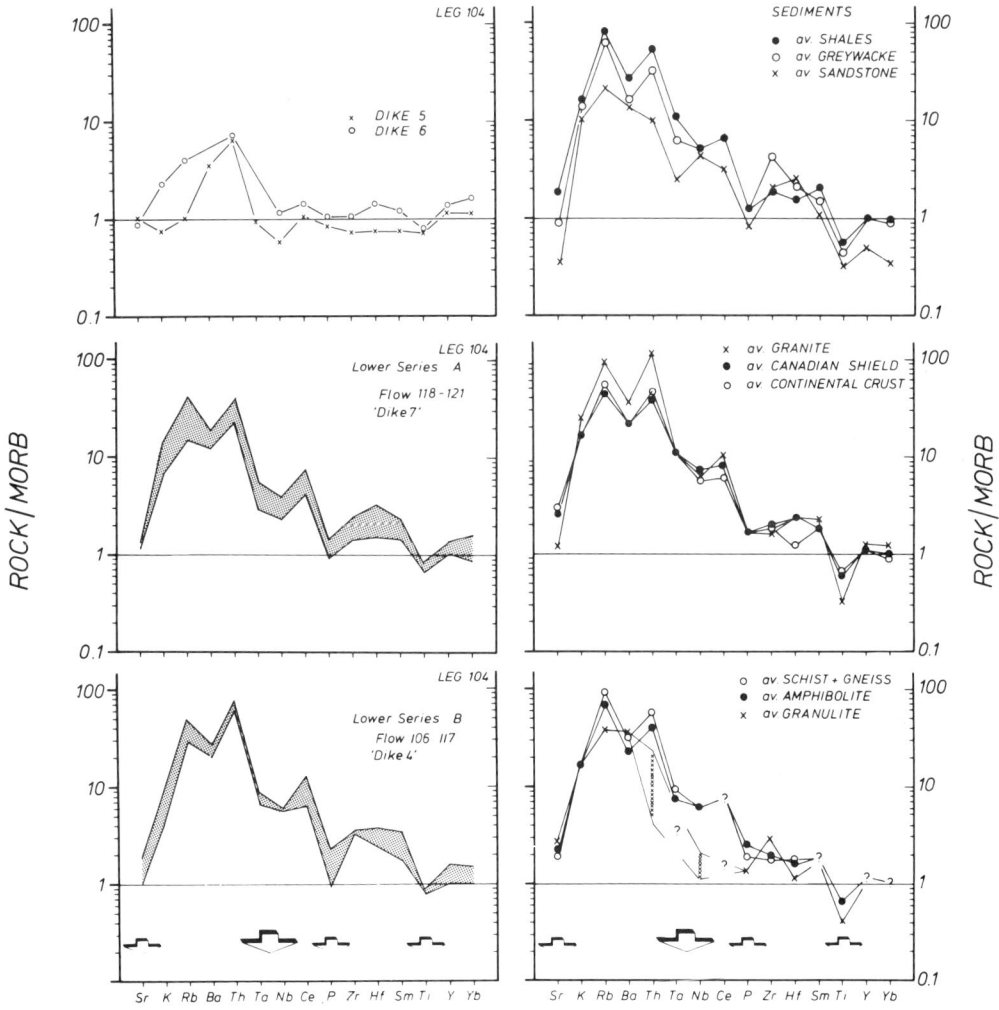

FIG. 10. MORB-normalized variation of dacites (Lower Series B), basaltic andesites (Lower Series A) and contaminated tholeiitic dykes D5 and D6. Average analyses for sediments, intrusive and metamorphic rocks are taken from Wedepohl (1978).

et al. 1986). Derivation by melting of granulite facies lower crustal rocks seems to be excluded by lack of Th depletion (Wood 1980).

The major and minor element and radiogenic isotope compositions of the dacites provide unambiguous evidence that these magmas represent melted upper continental crustal rocks, most likely of shale to greywacke composition (Miller 1985). The Vøring Plateau must therefore be underlain by basement of upper continental crustal composition. The basaltic andesites can be explained by mixing of 35–65% dacitic melts with strongly depleted N-MORB magmas, comparable to those of dykes D5 and D6, rather than with Upper Series T-MORB tholeiites (binary mixing calculations using Th; see Fig. 7).

Sediments of the Lower Series

Seven volcaniclastic sediment units make up 20% of the extrusive part of the Lower Series (see Fig. 2) and vary in thickness from 0.4–7 m. These sediments are grey to greyish green and moderately to well sorted with median grain diameters of about 0.1–0.5 mm; they are laminated or finely bedded, and display slump structures, cross-bedding, and erosional features suggesting fluvial or shallow water marine deposition. Altered glass shards and pumice fragments are the dominant constituents. Terrigenous components, such as light and dark mica, quartz, and rare fragments of quartz-mica schist, occur throughout the sediments, forming up to 10% by volume. The sediments are pervasively altered and thoroughly cemented by clinoptilolite, mordenite, carbonate and pyrite, indicating hydrothermal alteration temperatures of about 80–90°C for the Lower Series rocks (Iijima 1978).

Units S47 and S48 form a single, 9 m thick, unwelded ignimbrite unit that has been intruded by a later tholeiitic dyke. It consists of a 1 m thick basal breccia, a massive central part with normal size-grading of perlitic and aphanitic clasts (35–5 mm in diameter) accompanied by reverse size-grading of pumice fragments (1–20 mm). The upper 2.6 m are grey to light grey, well-sorted sandy to silty tuffs with parallel to increasingly inclined lamination upward. The relative thickness of the basal breccia zone indicates a proximal facies of emplacement for the ignimbrite (Fisher & Schmincke 1984).

The thorough alteration and the large admixture by terrigenous clasts makes an interpretation of the chemical composition of the sediments difficult. Their SiO_2 contents of 67–80% are in the range of rhyolitic compositions, but their incompatible as well as their compatible element contents (except for Ni) are lower than those of

the dacitic lavas they are interlayered with. The trace element pattern of the quartz-free ignimbrite is very similar to that of the dacites. Concentrations of Cr and Ni are 45–52 ppm and 15–20 ppm respectively. The incompatible element contents are generally lower by 10% rel., e.g. Zr 234 ppm and Th 11.0 ppm compared to 260 ppm and 12.5 ppm respectively in the dacites. The higher concentrations of the mobile low-field-strength elements Sr, K and Ba are probably due to enrichment during alteration. We therefore believe that the sediments are also of dacitic composition.

The sediment units of the Lower Series are thought to represent partly reworked volcaniclastic sediments that were derived from Plinian eruptions in the immediate vicinity of the Vøring Plateau, possibly within less than 10 km from Site 642E.

Conclusions

The Vøring Plateau, separated from Norway by an intracontinental rift system of Mesozoic age (see Fig. 1), is underlain by continental crust that became partially melted by ascending N-MORB tholeiitic magmas in the early Tertiary. The crustal magmas erupted as peraluminous dacites and, when mixed with tholeiitic magma, as basaltic andesite lavas, in a continental rift environment. We assume an environment much like the Pliocene Taos Plateau volcanic field (Rio Grande Rift, northern New Mexico) where tholeiites or transitional basalts with tholeiitic affinity are associated with less abundant high-K, calc-alkaline lavas of andesitic and dacitic composition that are thought to be derived by crustal fusion (Dungan *et al.* 1981; Dungan *et al.* 1984; McMillan & Dungan 1986). Analogies also exist with the early Palaeogene magmatic activity in the British Tertiary Volcanic Province, where the compositions of basal lavas (Staffa Magma Type) are strongly affected by crustal assimilation processes (Thompson *et al.* 1986).

After a time interval of unknown duration, final opening of the NE Atlantic to the W of the Vøring Plateau was accompanied by tholeiitic plateau basalt magmatism that built up a lava sequence ca. 800 m thick. The change to transitional MORB compositions may reflect a shift of the zone of magma generation from a depleted mantle region to a slightly enriched type, as it characterizes the Recent magma source of the NE Atlantic mid-ocean ridge N of the Jan Mayen Fracture Zone (Schilling *et al.* 1983; Neumann & Schilling 1984).

A correlation with the plateau basalts of NE

Greenland is strongly suggested from the palaeogeographic reconstruction for the early Tertiary (see Fig. 1). At Hold with Hope and Wollaston Forland, NE Greenland, a ca. 1 km thick plateau lava sequence has been subdivided into a lower division of T-MORB tholeiites (lower plateau lava sequence, LPLS) with TiO_2 contents of generally less than 2% and an upper division (upper plateau lava sequence, UPLS) of more variable (TiO_2: 1–4%), enriched tholeiitic to alkaline basaltic composition (Upton *et al.* 1980, 1984a). While the lower division may tentatively be correlated with the Upper Series lavas drilled on the Vøring Plateau at ODP Site 642E as well as DSDP Leg 38 Site 338, the upper division on the Vøring Plateau may be represented by enriched lavas drilled on DSDP Leg 38 at Sites 342 and 343 (Schilling 1976).

The volcaniclastic sediments of enriched ferrobasaltic composition interlayered with the plateau basalts at the Vøring Plateau most likely erupted from local (< 100 km) central volcanoes by phreatomagmatic eruptions. A possible source of less common pumiceous layers are more distal central volcanic complexes like those known from the continental margin of northern E Greenland by Larsen (1984), Upton *et al.* (1984b) and Nielsen (1987).

ACKNOWLEDGEMENTS: Thanks to Henny, Frank, Petra and Almut for their technical assistance during preparation of this work. An earlier version of the manuscript benefited from comments by H.-U. Schmincke, D. Barr and an anonymous reviewer. The project has been supported by the Deutsche Forschungsgemeinschaft (DFG) through grants Schm 250/28–12 and 28–13. A. C. Morton and L. M. Parson acknowledge the approval to publish of the Directors, B.G.S. and I.O.S. (NERC).

References

BECKINSALE, R. D., PANKHURST, R. J., SKELHORN, R. R. & WALSH, J. N. 1978. Geochemistry and petrogenesis of the Early Tertiary lava pile of the Isle of Mull, Scotland. *Contributions to Mineralogy and Petrology*, **66**, 415–427.

BOTT, M. H. P. 1984. Deep structure and origin of the Faeroe–Shetland Channel. *In*: SPENCER, A. M. *et al.* (eds) *Petroleum Geology of the North European Margin*. Graham & Trotman, London, pp. 341–347.

BRAZIER, S., DAVIES, A. N., SIGURDSSON, H. & SPARKS, R. S. J. 1982. Fall-out and deposition of volcanic ash during the 1979 explosive eruption of the Soufriere of St. Vincent. *Journal of Volcanology and Geothermal Research*, **14**, 335–359.

CLEMENS, J. D. & WALL, V. J. 1984. Origin and evolution of a peraluminous silicic ignimbrite suite: the Violet Town Volcanics. *Contributions to Mineralogy and Petrology*, **88**, 354–371.

DUNGAN, M. A., LIPMAN, P. W. & WILLIAMS, S. 1981. Continental rift volcanism. *In*: *Basaltic Volcanism on the Terrestrial Planets. Basaltic Volcanism Study Project*. Lunar and Planetary Institute, Houston, Texas, pp. 108–131.

——, MOORBATH, S., MCMILLAN, N. J. & HOEFS, J. 1984. Lead, strontium and oxygen isotopic evidence for the role of crustal assimilation in the genesis of andesites and dacites in the Taos Plateau volcanic field. *In*: DUNGAN, M. A., GROVE, T. L. & HILDRETH, W. (eds) *Proceedings of the Conference on Open Magmatic Systems*. ISEM, Dallas, Texas, pp. 40–42.

ELDHOLM, O., MYHRE, A. M., SUNDVOR, E. & FALEIDE, J. I. 1984. Cenozoic evolution of the margin off Norway and Svalbard. *In*: SPENCER, A. M. *et al.* (eds) *Petroleum Geology of the North European Margin*. Graham & Trotman, London, pp. 3–18.

——, THIEDE, J., TAYLOR, E. *et al.* 1987. *Proceedings of the Ocean Drilling Program*. US Government Printing Office, Washington, **104A**, 783 pp.

EWART, A. 1979. A review of the mineralogy and chemistry of Tertiary–Recent dacitic, latitic, rhyolitic, and related salic volcanic rocks. *In*: BARKER, F. (ed) *Trondhjemites, Dacites, and Related Rocks*. Developments in Petrology, **6**, pp. 13–121.

FISHER, R. V. & SCHMINCKE, H.-U. 1984. *Pyroclastic Rocks*. Springer Verlag, Berlin & Heidelberg, 472 pp.

FUJIMAKI, H., TATSUMOTO, M. & AOKI, K. 1984. Partition coefficients of Hf, Zr and REE between phenocrysts and groundmass. *Journal of Geophysical Research*, **89**, 858–860.

GIBSON, I. L., KIRKPATRICK, R. J., EMMERMANN, R., SCHMINCKE, H.-U., PRITCHARD, G., OAKLEY, P. J., THORPE, R. S. & MARRINER, G. F. 1982. Trace element composition of the lavas and dikes from a 3-km vertical section through the lava pile of Eastern Iceland. *Journal of Geophysical Research*, **87**, 6532–6546.

GRAPES, R. H. 1986. Melting and thermal reconstruction of pelitic xenoliths, Wehr volcano, East Eifel, West Germany. *Journal of Petrology*, **27**, 343–396.

HANISCH, J. 1984. West Spitsbergen Fold Belt and Cretaceous opening of the Northeast Atlantic. *In*: SPENCER, A. M. *et al.* (eds) *Petroleum Geology of the North European Margin*. Graham & Trotman, London, pp. 187–198.

HINZ, K., DOSTMANN, H. J. & HANISCH, J. 1984. Structural elements of the Norwegian continental margin. *Geologisches Jahrbuch*, **A75**, 193–211.

IIJIMA, A. 1978. Geological occurrences of zeolites in marine environments. *In*: SAND, L. B. & MUMPTON, F. A. (eds) *Natural Zeolites: occurrence, properties, use*. Pergamon Press, Oxford, pp. 175–198.

JORON, J. L., BOUGAULT, H., MAURY, R. C., BOHN, M. & DESPRAIRIES, A. 1984. Strongly depleted tholeiites from the Rockall Plateau margin, North

Atlantic: Geochemistry and mineralogy. *In*: ROB-
ERTS, D. G., SCHNITKER, D. *et al. Initial Reports of
the Deep Sea Drilling Project*, US Government
Printing Office, Washington, **81**, pp. 783–794.
KNOX, R. W. O'B. & MORTON, A. C. 1983. Stratigraphic
distribution of Early Palaeogene pyroclastic depos-
its in the North Sea Basin. *Proceedings of the
Yorkshire Geological Society*, **44**, 355–363.
LARSEN, H. C. 1984. Geology of East Greenland Shelf.
In: SPENCER, A. M. *et al.* (eds) *Petroleum Geology
of the North European Margin*. Graham & Trotman,
London, pp. 329–339.
MATTEY, D. P., GIBSON, I. L., MARRINER, G. F. &
THOMPSON, R. N. 1977. The diagnostic geochemistry, relative abundance, and spatial distribution
of high-calcium, low-alkali olivine tholeiite dykes
in the Lower Tertiary regional swarm of the Isle of
Skye, NW Scotland. *Mineralogical Magazine*, **41**,
273–285.
MCMILLAN, N. J. & DUNGAN, M. A. 1986. Magma
mixing as a petrogenetic process in the development of the Taos Plateau volcanic field, New
Mexico. *Journal of Geophysical Research*, **91**, 6029–
6045.
MILLER, C. F. 1985. Are strongly peraluminous magmas
derived from pelitic sedimentary sources? *Geology*,
93, 673–689.
MUNKSGAARD, N. C. 1984. High $\delta^{18}O$ and possible pre-
eruptional Rb-Sr isochrons in cordierite-bearing
Neogene volcanics from SE-Spain. *Contributions
to Mineralogy and Petrology*, **87**, 351–358.
MUTTER, J. C., BUCK, W. R. & ZEHNDER, C. M. (in
press). Convective partial melting I: a model for
the formation of thick basaltic sequences during
the initiation of spreading. *Journal of Geophysical
Research*.
NEUMANN, E.-R. & SCHILLING, J.-G. 1984. Petrology
of basalts from the Mohns-Knipovich Ridge; the
Norwegian-Greenland Sea. *Contributions to Mineralogy and Petrology*, **85**, 209–223.
NICHOLLS, J. & CARMICHAEL, I. S. E. 1969. A
commentary on the Absarokite-Shoshonite-Ban-
akite Series of Wyoming, U.S.A. *Schweizerische
Mineralogische und Petrographische Mitteilungen*,
49, 47–64.
NIELSEN, T. F. D. 1987. Tertiary alkaline magmatism
in E Greenland: a review. *In*: FITTON, J. G. &
UPTON, B. G. J. (eds) *Alkaline Igneous Rocks*.
Geological Society of London, Special Publication,
30, 489–515.
NUNNS, A. 1982. The structure and evolution of the Jan
Mayen Ridge and surrounding regions. *In*: WAT-
KINS, J. S. & DRAKE, C. L. (eds) *Studies in
continental margin geology*. American Association
of Petroleum Geologists Memoir, **34**, 193–208.
PARSON, L. M., MASSON, D. G., MILES, P. R. & PELTON,
C. D. 1986. Structure and evolution of the Rockall
and East Greenland continental margins. Report
of work undertaken by IOS in the period up to
April 1985. *Institute of Oceanographic Sciences
Report*, No. **233**, 71 pp.
PEARCE, J. A., HARRIS, N. B. W. & TINDLE, A. G. 1984.
Trace element discrimination diagrams for the

tectonic interpretation of granitic rocks. *Journal of
Petrology*, **25**, 956–983.
SCHILLING, J.-G. 1976. Rare Earth, Sc, Cr, Fe, Co, and
Na abundances in DSDP Leg 38 basement basalts:
some additional evidence on the evolution of the
Thulean Volcanic Province. *In*: TALWANI, M.,
UDINTSEV, G., *et al. Initial Reports of the Deep Sea
Drilling Project*. US Government Printing Office,
Washington, **38**, pp. 741–750.
—— & NOE-NYGAARD, A. 1974. Faeroe–Iceland
Plume: rare earth evidence. *Earth and Planetary
Science Letters*, **24**, 1–14.
——, ZAJAC, M., EVANS, R., JOHNSTON, T., WHITE, W.,
DEVINE, D. J. & KINGSLEY, R. 1983. Petrologic
and geochemical variations along the Mid-Atlantic
Ridge from 29°N to 73°N. *American Journal of
Science*, **283**, 510–586.
SCHMINCKE, H.-U., VIERECK, L. G., GRIFFIN, B. J. &
PRITCHARD, R. G. 1982. Volcaniclastic rocks of
the Reydarfjördur Drill Hole, Eastern Iceland 1.
Primary Features. *Journal of Geophysical Research*,
87, 6437–6458.
TALWANI, M., MUTTER, J. C. & HINZ, K. 1982. Ocean
continent boundary under the Norwegian conti-
nental margin. *In*: BOTT, M. H. P., SAXOV, S.,
TALWANI, M. & THIEDE, J. (eds) *Structure and
development of the Greenland–Scotland Ridge: New
methods and concepts*. Plenum Press, New York,
pp. 121–132.
THOMPSON, R. N. 1982. Magmatism of the British
Tertiary Province. *Scottish Journal of Geology*, **18**,
49–107.
——, MORRISON, M. A., DICKIN, A. P., GIBSON, I. L. &
HARMON, R. S. 1986. Two contrasting styles of
interaction between basic magmas and continental
crust in the British Tertiary Volcanic Province.
Journal of Geophysical Research, **91**, 5985–5997.
UPTON, B. G. J., EMELEUS, C. H. & BECKINSALE, R. D.
1984a. Petrology of the northern East Greenland
Tertiary flood basalts: evidence from Hold with
Hope and Wollaston Forland. *Journal of Petrology*,
25, 151–184.
——, ——, —— & MACINTYRE, R. M. 1984b.
Myggbukta and Kap Broer Ruys: the most
northerly of the East Greenland Tertiary igneous
centres (?). *Mineralogical Magazine*, **48**, 323–343.
——, —— & HALD, N. 1980. Tertiary volcanism in
northern E Greenland: Gauss Halvo and Hold
with Hope. *Journal of the Geological Society of
London*, **137**, 491–508.
VIERECK, L. G., FLOWER, M. F. J., HERTOGEN, J.,
SCHMINCKE, H.-U. & JENNER, G. A. 1988 The
genesis and significance of N-MORB sub-types.
Contributions to Mineralogy and Petrology. In press.
——, HERTOGEN, J. & HOEFS, J. (in prep. a) Melting of
continental crust during continental breakup:
geochemical and isotopic evidence from dacitic
glass flows of Eocene age drilled in the NE-
Atlantic. *In*: ELDHOLM, O., THIEDE, J., TAYLOR,
E., *et al. Proceedings of the Ocean Drilling Program*.
US Government Printing Office, Washington,
104B.
——, PARSON, L. M., MORTON, A. C., LOVE, D. &
GIBSON, I. L. (in prep. b) Chemical stratigraphy

and petrology of the Upper Series lavas and volcaniclastic sediments at Site 642E, ODP Leg 104. *In*: ELDHOLM, O., THIEDE, J., TAYLOR, E., *et al. Proceedings of the Ocean Drilling Program*. US Government Printing Office, Washington, **104B**.

WALKER, G. P. L. 1970. Compound and simple lava flows and flood basalts. *Bulletin Volcanologique*, **35**, 579–590.

WEDEPOHL, K. H. (ed) 1978. *Handbook of Geochemistry* (*Vol. II-1, II-2, II-3, II-4, II-5*). Springer-Verlag, Heidelberg.

WOOD, D. A. 1980. The application of a Th-Hf-Ta diagram to problems of tectonomagmatic classification and to establishing the nature of crustal contamination of basaltic lavas of the British Tertiary Volcanic Province. *Earth and Planetary Science Letters*, **50**, 11–30.

WYBORN, D., CHAPPELL, B. W. & JOHNSTON, R. M. 1981. Three S-type volcanic suites from the Lachlan Fold Belt, southeast Australia. *Journal of Geophysical Research*, **86**, pp. 10335–10348.

L. G. VIERECK, Institut für Mineralogie, Ruhr-Universität Bochum, Postfach 102148, 4630 Bochum, Federal Republic of Germany.

P. N. TAYLOR, Department of Earth Sciences, University of Oxford, Parks Road, Oxford OX1 3PR, UK.

L. M. PARSON, Institute of Oceanographic Sciences, Brook Road, Wormley, Godalming, Surrey GU8 5UB, UK.

A. C. MORTON, British Geological Survey, Keyworth, Nottingham NG12 5GG, UK.

J. HERTOGEN, Department of Geology, Katholieke Universiteit Leuven, Leuven, Belgium.

I. L. GIBSON, Department of Earth Sciences, University of Waterloo, Kitchener, Ontario N2L 3G1, Canada.

Tectonic and volcanic events at the Jan Mayen Ridge microcontinent

S. T. Gudlaugsson, K. Gunnarsson, M. Sand & J. Skogseid

SUMMARY: Two main tectonic phases were responsible for the formation of the Jan Mayen Ridge microcontinent: (1) the opening of the Norway Basin in late Palaeocene/early Eocene times, and (2) subsequent rifting within the Greenland margin by which complete separation was achieved in early Miocene times. During the first phase the eastern ridge flank developed as a volcanic passive margin. The initial break-up was associated with flexuring and the formation of sequences of eastward-dipping basalt flows, which are considered equivalent to similar features beneath the Vøring and Faeroe–Shetland marginal highs off Norway. Rifting along the Greenland margin during the second phase was accompanied by uplift, listric normal faulting and the formation of large extensional fault blocks. To the W and S of the ridge a flat volcanic marker of probable earliest Miocene age covers the subsided rift and masks the ocean–continent transition. It was formed by a volcanic event of large magnitude, either as submarine lava flows or as a sill complex.

In 1985 a detailed marine geophysical survey of the Jan Mayen Ridge area was carried out jointly by the Norwegian Petroleum Directorate and the National Energy Authority of Iceland. A total of 4000 km of multichannel seismic reflection data was obtained (Fig. 1) in addition to gravity, magnetic and sonobuoy measurements. The seismic data in particular have provided important information in terms of the tectonic and volcanic history of the ridge.

The main objective of this paper is to describe the nature of the igneous provinces at the Jan Mayen Ridge microcontinent and show how their formation relates to tectonic events at the ridge and the plate tectonic development of the Norwegian Sea.

Plate tectonics

The Jan Mayen Ridge (see Fig. 1) is a bathymetric ridge complex extending S from the volcanic island of Jan Mayen to about 67°N. The main northern ridge block is flat-topped with water depths increasing to about 1000 m in the S. It is separated from the still deeper southern ridge complex by the Jan Mayen Trough.

The Jan Mayen Ridge is a crustal fragment which split from the Greenland continental margin by a westward shift in the plate boundary at Oligocene/Miocene time. The evolutionary models for the Jan Mayen Ridge are based on a westward shift of the plate boundary and the observations of Talwani & Eldholm (1977) that the fan-shaped spreading pattern along the Aegir Ridge in the Norway Basin required that complementary spreading must have taken place further

W to account for the motion of Greenland relative to Eurasia. Talwani & Eldholm (1977) proposed that the Aegir Ridge was active until about anomaly 7 time and suspected that the complementary crust formed between the southern part of the Jan Mayen Ridge and the Norway Basin. At anomaly 7 time a westward jump to an 'intermediate' axis on the Iceland Plateau occurred. This axis supposedly became extinct when spreading started from the Kolbeinsey Ridge further W just before anomaly 5 time. Alternatively, Vogt et al. (1980) have rejected the existence of an intermediate axis and postulated spreading from the Kolbeinsey Ridge since anomaly 6C time.

The three-plate model has been further developed by several investigators (Unternehr 1982; Nunns 1983a, b; Bott 1985). According to Nunns (1983a, b) fan-shaped spreading formed two conjugate wedges of seafloor on either side of the Jan Mayen Ridge during the interval between the formation of anomalies 20 and 7. Similarly, Larsen (1988) suggests northward propagation of the Reykjanes–Kolbeinsey Ridge combined with gradual termination of spreading northwards along the Aegir Ridge during the 20–7 time interval.

Structure

The first order geological framework was reviewed by Myhre et al. (1984). The ridge is bordered by provinces of basaltic rocks on both sides. A flat volcanic marker, reflector F, characterizes the western province whereas irregular lava flows and seaward-dipping reflectors are

From MORTON, A. C. & PARSON, L. M. (eds), 1988, *Early Tertiary Volcanism and the Opening of the NE Atlantic,* Geological Society Special Publication No. 39, pp. 85–93.

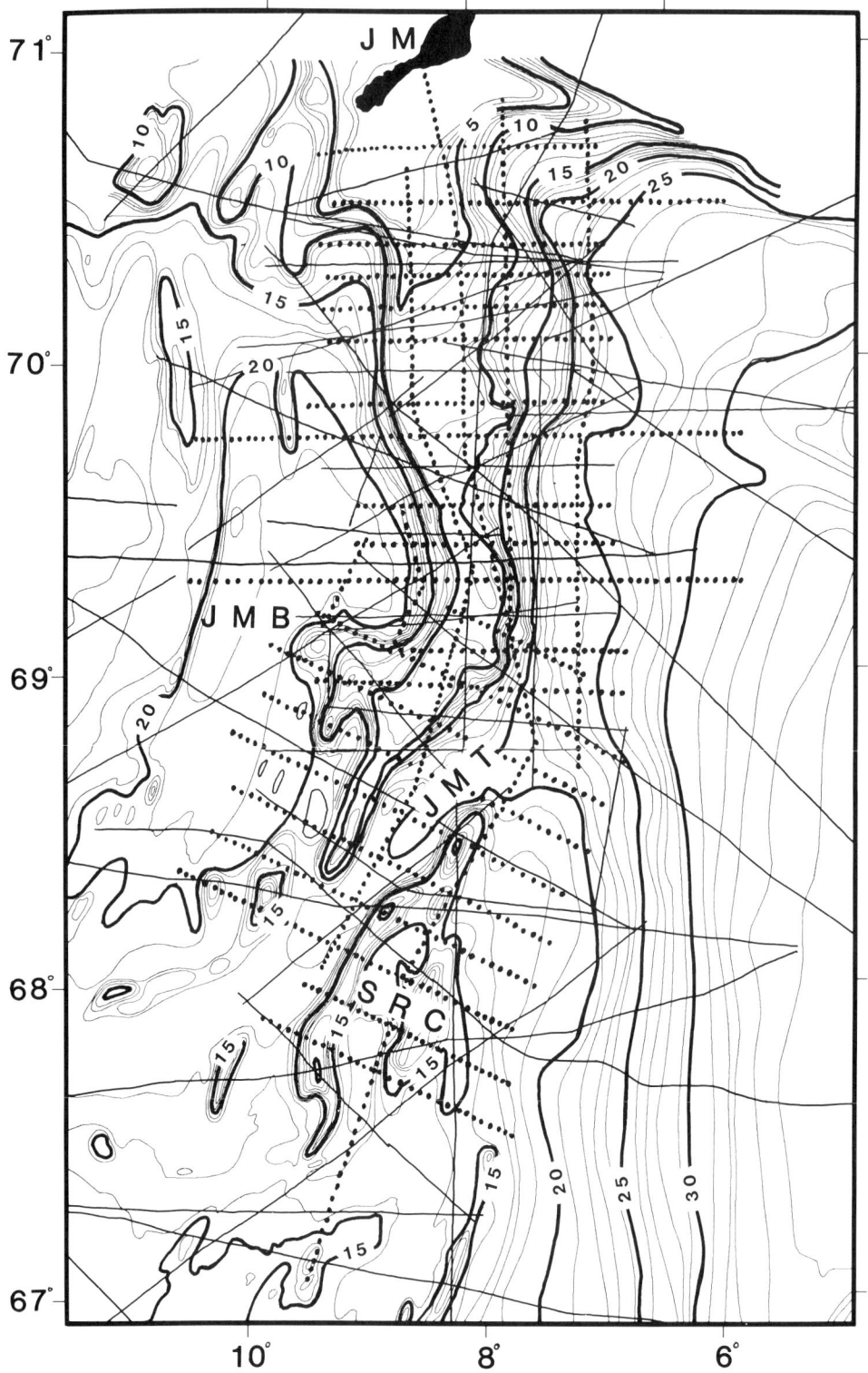

FIG. 1. The grid of multichannel seismic reflection lines used in this study superimposed on bathymetry. The new survey lines are shown as dotted lines. Contour interval 100 m with contours labelled every 500 m. JM = island of Jan Mayen, JMB = Jan Mayen Basin, JMT = Jan Mayen Trough, SRC = southern ridge complex.

found beneath the eastern ridge flank. The ridge is covered by a thick sequence of eastward-dipping sedimentary rocks, above a strong reflecting interface, which in early single-channel seismic records defined the acoustic basement (see Fig. 3, reflector JO). The sedimentary sequence comprises two main units separated by a prominent unconformity (reflector JA; J is prefixed to the commonly-used symbols O and A to indicate the Jan Mayen Ridge).

The structural map in Figure 2 is based on the interpretation of existing multichannel seismic reflection profiles including the new data. The seismic database is shown in Figure 1 and line drawings of three seismic profiles crossing the main geological provinces are shown in Figure 3.

The Jan Mayen Ridge is strongly affected by normal faulting. The number of fault blocks and the general structural complexity increase southward, as does the depth to the individual fault blocks (see Figs 2 & 3). The main ridge block in the N has a distinct asymmetric structure. The eastern flank dips steeply towards the Norway Basin and is almost undisturbed by faulting whereas the western flank of the ridge is down-faulted towards the Jan Mayen Basin. These faults form a listric fault complex in which individual faults can be seen to sole-out at depth. The ridges comprising the southern ridge complex (Pelton 1985) are also tilted fault blocks. The majority of the blocks, and almost all the large ones, have fault scarps facing W. Where the volcanic marker (reflector F) W of the ridge is absent a number of deep half-grabens are observed between the fault blocks.

Two main fault trends are observed at the Jan Mayen Ridge. The trends are also reflected in the bathymetry and intersect at the prominent bend in the ridge at 69°15′N. N of the ridge bend, the faults trend N–S. S of the bend, both on the main ridge block and in the southern ridge complex, the trend is nearly NNE–SSW. Complex structures are observed in the region where these trends intersect.

The system of eastward-rotated fault blocks at the Jan Mayen Ridge is interpreted in terms of crustal extension. There is also evidence of a later phase of compression. Reverse faults, similar to those reported by Skogseid & Eldholm (1987), have been identified in the new data, but their detailed correlation has yet to be worked out.

Between the eastern and western volcanic provinces there is a seismic window into the deeper crust. The new data show a number of reflecting interfaces beneath reflector JO, some of which are found at depths of 6–7 s two-way time (see Fig. 3).

Geophysical investigations and deep-sea drilling have not yet conclusively answered the question of whether the ridge is continental or oceanic in nature. In our opinion the balance of evidence favours a continental crust, probably thinned and modified by rifting processes. The seismic velocity structure (Johansen *et al.* in press) and the seismic reflector pattern below reflector JO are not compatible with normal oceanic crust.

The continent–ocean boundary is difficult to locate for two reasons: (1) the boundary is probably masked in many places by lava flows; and (2) the detailed location of the boundary is to a certain extent a matter of definition. We can, however, place limits on the zone where the crustal transition must occur. The outer limit is marked by the oldest seafloor-spreading anomalies on both sides of the ridge. At the eastern side the inner limit lies at the apex of the wedge of seaward-dipping reflectors. On the western side it follows the scarps of the westernmost fault blocks.

Volcanic provinces

The distribution of volcanic rocks on either side of the ridge is shown in Figure 2. On the eastern flank we differentiate between a wedge of eastward-dipping reflectors below reflector JO and a younger volcanic overprint. Reflector JO forms the top of the eastward-dipping reflectors. The wedge is most prominent S of the Central Jan Mayen fracture zone, where it is underlain by a sequence of parallel reflectors. On line C (see Fig. 3) this sequence may be traced up-dip towards the boundary fault without change in character, which shows that it predates the faulting and originally continued further W. On line B the same sequence may be continued a short distance beyond the apex of the wedge. The internal reflectors of the overlying wedge converge only to a certain point and then also become parallel. On the basis of these observations we suggest that an equivalent of the E Greenland plateau basalts may underlie the seaward-dipping reflectors and cover the southern part of the ridge. N of the Central Jan Mayen fracture zone the wedge has a different character. Both the dip and the divergence of the reflectors is less prominent and the reflector pattern is more irregular. The mapping of the extent of the dipping reflectors presented here is primarily based on the new survey, and their presence in Figure 2 is only shown where a well-developed dipping and divergent sequence is observed.

Reflector JO is overstepped from the E by a

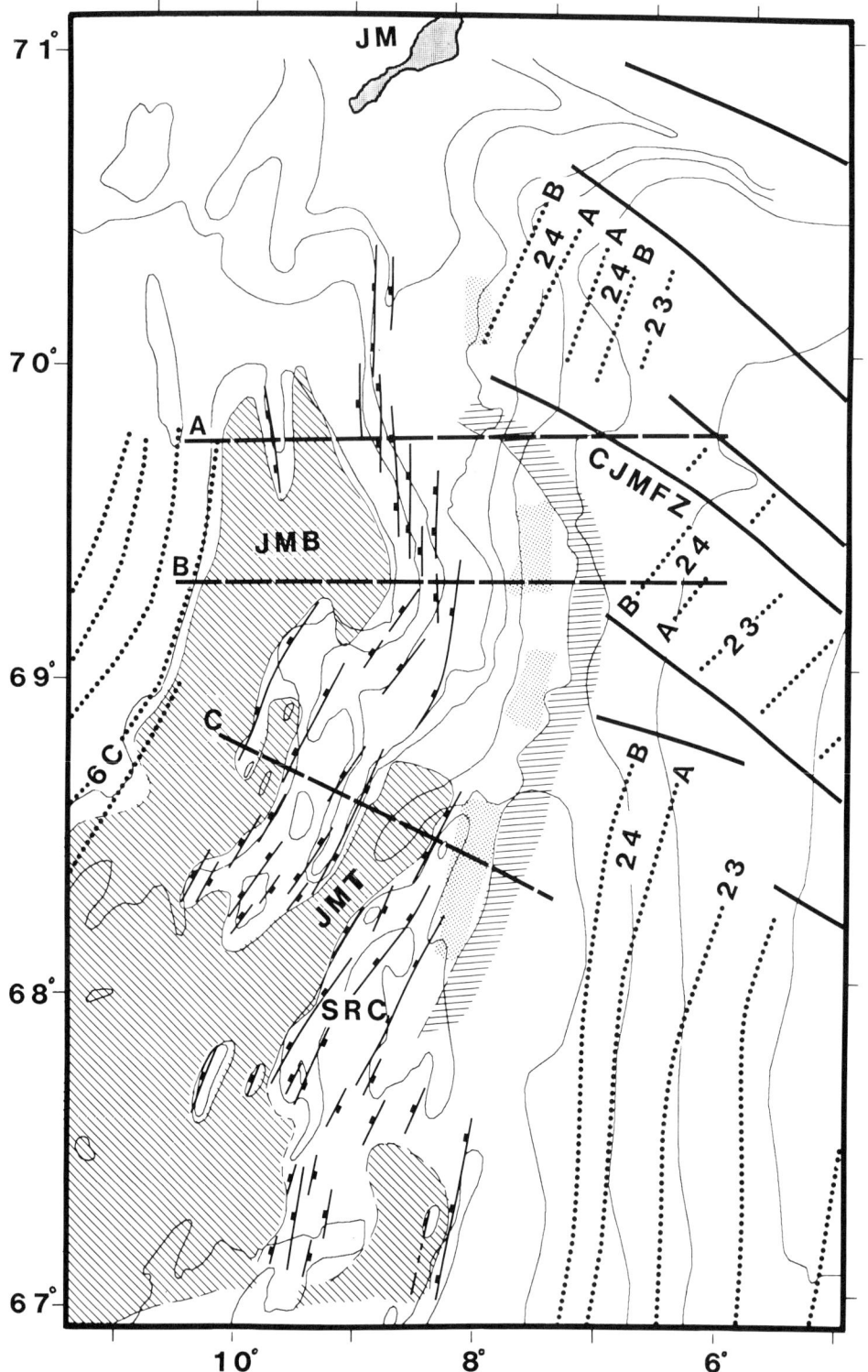

FIG. 2. Structural map of the Jan Mayen Ridge and the surrounding areas. Horizontal ruling = western front of volcanic overprint; diagonal ruling = area covered by reflector F; stippling = areas of seaward-dipping reflectors; thin continuous lines = bathymetric contours; thick continuous lines = fracture zones; dotted lines = seafloor-spreading anomalies (Skogseid & Eldholm 1987; Vogt *et al.* 1980; Grønlie *et al.* 1978). JMB = Jan Mayen Basin, JMT = Jan Mayen Trough, CJMFZ = Central Jan Mayen Fracture Zone. Fracture zones from Skogseid & Eldholm (1987). Dashed lines A, B and C show the location of the seismic profiles in Figure 3.

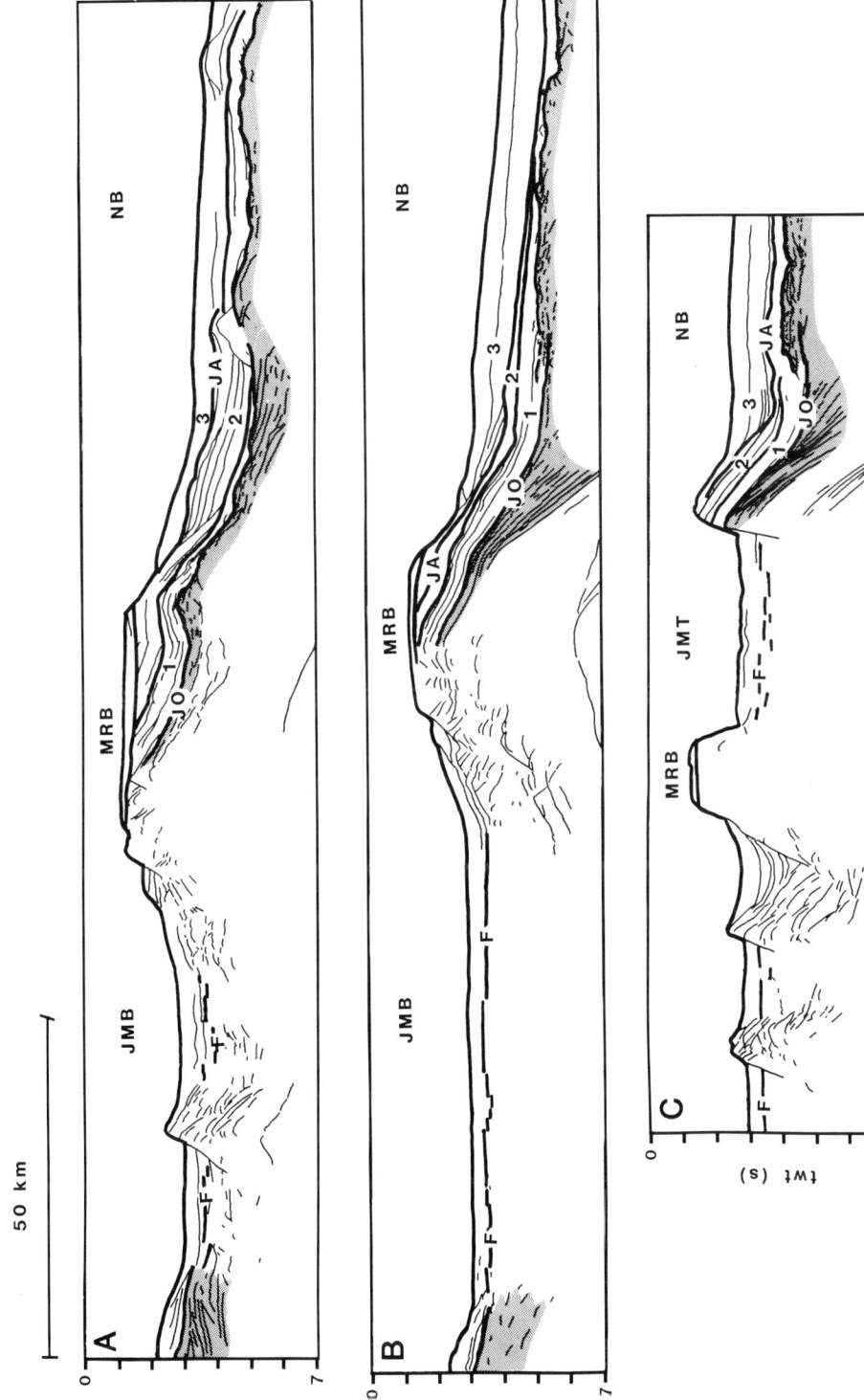

FIG. 3. Line drawings of seismic profiles A, B and C crossing the Jan Mayen Ridge. Location Figure 2. Dark shading denotes volcanic rocks. JA, JO, F and 1, 2, 3 are reflectors and seismic sequences discussed in the text. JMB = Jan Mayen Basin, JMT = Jan Mayen Trough, MRB = main ridge block, NB = Norway Basin.

strongly reflective horizon characterized by diffractions and both irregular and smooth reflector segments (see Figs 2 & 3). The surface can be traced from opaque oceanic basement westward, where it gradually becomes semi-opaque with irregular structure below before it terminates within the sediments above reflector JO. We interpret this layer as being caused by submarine lava flows, intrusions and pyroclastics interfingering with marine sediments. The western front of this younger volcanic overprint may be traced from the southern ridge complex, with decreasing stratigraphic gap with respect to reflector JO, N to the Central Jan Mayen fracture zone. At this point the front turns westwards and merges with reflector JO (see Fig. 2).

W and S of the ridge we have mapped a strong, flat-lying acoustic basement reflector beneath a thin sedimentary cover. This is the 'opaque horizon' of Eldholm & Windisch (1974) that covers a large part of the Iceland Plateau. The new data allow us to differentiate between an ultra-flat opaque volcanic marker (denoted F) in the area between the ridge and the oldest seafloor-spreading anomaly on the Iceland Plateau and a different more irregular type of basement further W.

Reflector F covers the area W of the southern ridge complex locally extending into the Jan Mayen Basin and the Jan Mayen Trough. On the eastern side it terminates abruptly at the fault scarps. The western boundary lies at the foot of an E-facing escarpment associated with the oldest seafloor-spreading anomaly. The height difference between the top of the escarpment and reflector F decreases towards the S. S of about 69°N the relationship between reflector F and the oceanic basement becomes unclear. Here the reflector seems to overstep the basement towards the W, but we cannot verify that it is continuous with the basaltic basement reflector which was drilled at Deep Sea Drilling Project (DSDP) Site 348 (Talwani *et al.* 1976). We note that W of the escarpment a number of sub-basement reflectors are observed on lines A and B. On line A the reflection pattern is westward-dipping and diverging similar to the pattern of a typical wedge of seaward-dipping reflectors.

Reflector F is very strong and normally no reflectors are observed below. The acoustic energy becomes trapped in the water layer and in the thin sedimentary layer. The reflector is not perfectly smooth but exhibits small-scale roughness. As Figure 3 shows, reflector F is composed of a number of offset segments. Because of the poorly developed stratification in the overlying sediments it is difficult to determine the nature of these offsets, but there are indications that some

of them are associated with high-angle normal faults. The most striking aspect of the reflector is, however, the extreme flatness of individual segments.

In our opinion, the possible interpretations of reflector F are: (1) high impedance sedimentary layer; (2) volcanic ash layer; (3) intrasedimentary sills; and (4) lava flows. In the sedimentary environment of the Norwegian Sea it is unlikely that such a high impedance contrast could occur within a sedimentary succession without the compaction effect of burial and later erosion. Reflector F does not show any evidence of submarine erosion. Volcanic ash may generate highly reflective flat surfaces. An ash layer must either have been pelagically draped over existing topography or redeposited as mass flows. The first mechanism is not viable as there is no draping effect over the blocks and the reflector terminates abruptly against fault blocks and escarpments. It is possible, however, that volcanic ash from an elevated source area close to Iceland may have been transported by turbidity flow northward onto the Iceland Plateau. A regional intrasedimentary sill is also a possible explanation, but seems unlikely from a mechanical viewpoint considering the areal extent of the reflector. However, if the reflector offsets are interpreted as the limits of individual sills and not as normal faults this argument does not apply and reflector F may represent the top of a sill complex. Nonetheless, we favour an interpretation of extensive submarine lava flows as suggested for the ultra-flat opaque reflectors on the Reykjanes Ridge (Vogt & Johnson 1973).

On line A (see Fig. 3), which crosses the Jan Mayen Basin just S of the northern termination of reflector F, the layer below the reflector is locally acoustically transparent. A number of reflections are observed beneath it and the reflection pattern suggests a continuity of sedimentary layers beneath the volcanic marker from the main ridge block to a fault block further W which rises above the floor of the Jan Mayen Basin. On seismic lines further S reflections from sedimentary layers and detachment planes in the fault complex at the western margin of the main ridge block are observed to dip under reflector F but are lost a short distance from its edge. In our interpretation, the listric fault complex continues underneath the reflector. A similar relationship is observed in the southern ridge complex where some of the fault blocks plunge beneath it.

Since the reflector is flat-lying and covers the basin fill in the half-grabens between some of the blocks, its formation postdates the block faulting by a considerable interval. It follows that most of the present elevation difference between the Jan

Mayen Ridge, and the lower areas to the W covered by reflector F, was already established at the time of its formation. The flat unconformity on top of the main ridge block represents the deepest level that wave erosion ever reached since the faulting. Therefore, the reflector was formed under submarine conditions.

Whether reflector F is interpreted as a layer of volcanic ash, submarine lava flows or sills, it was formed by a volcanic event of a large magnitude.

Timing of volcanic events and evolution of the microcontinent

No detailed chronology of events has been available for the Jan Mayen Ridge. The DSDP holes drilled during Leg 38 (Talwani *et al.* 1976) only sampled part of the sedimentary sequence and the stratigraphy of the ridge has not been well understood. We now propose a chronology for the main tectonic and volcanic events based on seismic stratigraphy and correlation with the geology of the conjugate Greenland and Norwegian margins.

We start by examining the age of the oldest well-defined rock sequence at the Jan Mayen Ridge, i.e. the sequence of seaward-dipping reflectors. Comparable sequences at the Norwegian margin are well known (Skogseid & Eldholm, 1987). In the light of their origin as extensive subaerial basaltic lava flows formed during the earliest spreading phase, their position relative to the Jan Mayen Ridge in plate tectonic reconstructions and the symmetry of the wedges on both sides, there can be no doubt that the sequence of seaward-dipping reflectors at the Jan Mayen Ridge has a similar origin and was formed at approximately the same time at the conjugate margin as proposed by Skogseid & Eldholm (1987). They interpret reflector JO as an equivalent to reflector EE of earliest Eocene age at the Vøring margin.

Gairaud *et al.* (1978) divided the cover of Tertiary sediments at the Jan Mayen Ridge into two sequences separated by a prominent unconformity, reflector JA. The unconformity was drilled at DSDP Sites 346, 347 and 349 and proved to represent a hiatus in the early Oligocene (Talwani *et al.* 1976).

Here, we prefer to divide the sedimentary series covering the Ridge into three sequences numbered 1, 2 and 3 from below (sequence no. 1 being the oldest, see Fig. 3).

The base of sequence 3 corresponds to reflector JA. Sequence 1 is mostly parallel-bedded and represents a widespread slope or shelf sequence.

It predates the block-faulting at the western margin of the ridge as does the parallel sequence below reflector JO and the wedge of seaward-dipping reflectors. Sequence 2 shows a more disturbed sedimentary pattern. On line A the sequence exhibits outbuilding and downlap followed by marine onlap. Over the entire ridge the apparent instability of the sedimentary environment increases upward and the uppermost part of the sequence consists of slumps and other mass-flow deposits. We interpret this as gradual uplift of the ridge flank associated with doming and rifting at the western margin. The uplift of the ridge culminated in subaerial exposure of the main ridge block, a strong erosional phase and the formation of a prominent submarine unconformity on the slope (reflector JA). Sequence 3 formed first by outbuilding of submarine fans and later by passive draping with much reduced sedimentation rates.

We now turn to the formation of the fault complex at the western margin of the ridge. Its conjugate part is the Liverpool Land margin, where Larsen (1984) describes a buried rift with an Eocene to early Oligocene graben-fill. He suggests the main tectonic episode of block-faulting to be early to mid-Eocene and that the rifting accompanied the initial formation of the Norway Basin, inferring that the block-faulting at the Jan Mayen Ridge dates from this time. However, the Jan Mayen Ridge seismic data do not support this model. In fact, the western margin of the Norway Basin is found at the eastern flank of the Jan Mayen Ridge, where it is associated with flexuring and formation of a wedge of seaward-dipping reflectors. No block faulting is observed. The only candidate at the Jan Mayen Ridge for a symmetric counterpart to the landwards-rotated system of fault blocks at the Greenland margin is the system of eastward-rotated listric normal faults at the western margin of the ridge. The two margin types reflect different tectonic events with different thermal and mechanical characteristics and must be separated in time. We associate the block-faulting at the Jan Mayen Ridge with the separation of the ridge from Greenland, thus suggesting a younger age for block-faulting at the Greenland margin. In this connection it is interesting to note that Larsen (1984) finds no evidence for a later tectonic episode at the Liverpool Land margin which might correlate with the separation of the ridge from Greenland. At the Jan Mayen Ridge the minimum time gap between the two events is the interval represented by sedimentary sequence 1. The upper boundary of the sequence represents the earliest possible date for the onset of rifting.

Thus, we propose that:

(1) Sequence 2 is time equivalent with the Liverpool Land rift basin-fill and is middle Eocene to early Oligocene in age. The deposition of sequence 2 was contemporaneous with the development of the rift beneath the Liverpool Land shelf. On the Jan Mayen side the sequence was deposited outside the rift, on its eastern flank. The sedimentary fill observed on the Greenland side was deposited within the rift during the same period.

(2) Sequence 1 at the Jan Mayen Ridge is early Eocene in age, possibly extending into mid-Eocene. It was deposited on the continental slope or shelf of Greenland on top of the seaward-dipping reflectors concurrently with extrusion of submarine lava flows in the Norway Basin. This sequence is probably found within the fault blocks in the Greenland Rift as well as at the Jan Mayen Ridge.

Reflector F, the volcanic marker W of the Jan Mayen Ridge, clearly postdates the block-faulting at the western margin. Since it lies flat on top of the half-graben-fill in the southern ridge complex and shows no evidence of being affected by movements on the boundary faults, neither there nor W of the main ridge block, the formation of the reflector postdates rifting between Greenland and the Jan Mayen Ridge by a significant time interval. W of the main ridge block it appears that normal seafloor spreading was not established until anomaly 6C time (24 Ma). Line A indicates that the final break-up was associated with the formation of a submarine sequence of westward-dipping reflectors W of the escarpment at anomaly 6C. Examining the western termination of reflector F at the escarpment we find it most likely that the escarpment was formed prior to the emplacement of the reflector. We propose an earliest Miocene age for reflector F.

ACKNOWLEDGEMENTS: We wish to thank the Norwegian Petroleum Directorate and the National Energy Authority of Iceland for their permission to publish this paper. S. T. Gudlaugsson and J. Skogseid were supported by a research grant from the Norwegian Petroleum Directorate, and K. Gunnarsson by a grant from the Nordic Council of Ministers.

References

BOTT, M. H. P. 1985. Plate tectonic evolution of the Icelandic transverse ridge and adjacent regions. *Journal of Geophysical Research*, **90**, 9953–9960.

ELDHOLM, O. & WINDISCH, C. C. 1974. Sediment distribution in the Norwegian–Greenland Sea. *Bulletin of the Geological Society of America*, **85**, 1661–1676.

GAIRAUD, H., JACQUART, G., AUBERTIN, F. & BEUZART, P. 1978. The Jan Mayen Ridge. Synthesis of geological knowledge and new data. *Oceanologica Acta*, **1**, 335–358.

GRØNLIE, G., CHAPMAN, M. & TALWANI, M. 1978. Jan Mayen Ridge and Iceland Plateau: origin and evolution. *Norsk Polarinstitutt Skrifter*, **170**, 25–48.

JOHANSEN, B., ELDHOLM, O., TALWANI, M., STOFFA, P. L. & BUHL, P. (in press). Expanding spread profile at the northern Jan Mayen Ridge. *Polar Research*.

LARSEN, H. C. 1984. Geology of the East Greenland Shelf. *In*: SPENCER, A. M., *et al.* (eds) *Petroleum Geology of the North European Margin*. Graham & Trotman, London, pp. 329–339.

—— 1988. A multiple and propagating rift model for the NE Atlantic. *In*: MORTON, A. C. & PARSON, L. M. (eds) *Early Tertiary Volcanism and the Opening of the NE Atlantic*. Geological Society of London, Special Publication, **39**, pp. 157–158.

MYHRE, A. M., ELDHOLM, O. & SUNDVOR, E. 1984. The Jan Mayen Ridge: present status. *Polar Research*, **2**, 47–59.

NUNNS, A. G. 1983a. Plate tectonic evolution of the Greenland–Scotland Ridge. *In*: BOTT, M. H. P., SAXOV, S., TALWANI, M. & THIEDE, J. (eds) *The Greenland–Scotland Ridge: New Methods and Concepts*. Plenum, New York, pp. 11–30.

—— 1983b. The structure and evolution of the Jan Mayen Ridge and surrounding regions. *In*: WATKINS, J. S. & DRAKE, C. L. (eds) *Continental margin geology*. Memoir of the American Association of Petroleum Geologists, **34**, pp. 193–208.

PELTON, C. D. 1985. Geophysical interpretation of the structure and evolution of the Jan Mayen Ridge. *Institute of Oceanographic Sciences Report*, No. **205**, 38 pp.

SKOGSEID, J. & ELDHOLM, O. 1987. Early Cenozoic crust at the Norwegian Continental Margin and the conjugate Jan Mayen Ridge. *Journal of Geophysical Research*, **92**, 11471–11491.

TALWANI, M. & ELDHOLM, O. 1977. Evolution of the Norwegian–Greenland Sea. *Bulletin of the Geological Society of America*, **88**, 969–999.

——, UDINTSEV, G. *et al.* 1976. *Initial reports of the Deep Sea Drilling Project*. US Government Printing Office, Washington, **38**, 1256 pp.

UNTERNEHR, P. 1982. *Etude structurale et cinématique de la mer de Norvége et du Groenland. Evolution du microcontinent de Jan Mayen.* Thesis. University of Bretagne, 227 pp.

VOGT, P. R. & JOHNSON, G. L. 1973. A longitudinal seismic reflection profile of the Reykjanes Ridge: Part II—Implications for the mantle hot spot hypothesis. *Earth and Planetary Science Letters,* **18**, 49–58.

——, —— & KRISTJANSSON, L. 1980. Morphology and magnetic anomalies north of Iceland. *Journal of Geophysics,* **47**, 67–80.

S. T. GUDLAUGSSON & J. SKOGSEID, Department of Geology, University of Oslo, Norway.

K. GUNNARSSON, Department of Geology, University of Oslo, Norway and National Energy Authority, Reykjavik, Iceland.

M. SAND, Norwegian Petroleum Directorate, Stavanger, Norway.

Distribution, crustal properties and significance of seawards-dipping sub-basement reflectors off E Greenland

H. C. Larsen & S. Jakobsdóttir

SUMMARY: We report in this paper the existence of seawards-dipping sub-basement reflectors along the entire E Greenland margin. The study is based on 8000 km of multichannel seismic data and a sonobuoy refraction seismic study providing information on the geographical and stratigraphical extension, internal geometry and crustal structure of the E Greenland dipping reflector sequence. A basaltic, subaerial seafloor-spreading origin of the reflector sequence is concluded from seismic stratigraphic analysis, including well information from the Rockall Plateau and the Vøring Plateau. Formation of the basaltic dipping reflector sequence off E Greenland took place within a period of a few million years along the axis of opening within the NE Atlantic. Duration of spreading above sea-level was relatively short (2 My) in areas of present-day deep basement, as opposed to 5–8 My in areas of present-day more shallow basement. On the highly elevated Iceland–Greenland Ridge, subaerial seafloor spreading continued into the Neogene and most likely into present-day subaerial spreading in Iceland. Following the mid-Tertiary westward shift of spreading towards Greenland, N of Iceland, spreading again took place above sea-level along this part of the Greenland margin until late Miocene, but this development only caused an erratic and shallow development of seawards-dipping reflectors.

Application of the kinematic model for crustal formation in Iceland (Pálmason 1980) onto the E Greenland dipping reflector sequence demonstrates a striking similarity between the two structures. However, volcanic productivity rate within the oldest part of the E Greenland dipping reflector sequence may be as much as three times the volcanic productivity rate recorded in Iceland with an original rift width equal to, or somewhat less than, that in Iceland. The high volcanic productivity rate caused the development of a thick extrusive upper crust (> 5–6 km) dominated by seawards-dipping reflections arising from lava flows or groups of flows which acquired their dip through postdepositional differential subsidence towards the rift zone. Refraction seismology defines a fairly flat-lying velocity zonation of the igneous crust with an anomalously thick layer 2 (3–5.5 km) in areas of well-developed dipping reflectors. The layer 2/3 boundary is seen to cut strongly across the dipping reflectors suggesting a metamorphic origin of this boundary. Initiation of seafloor spreading above sea-level is seen as a result of early upwelling of anomalously hot asthenospheric material that was able to ascend through a relatively mechanically unstretched crust and lithosphere and create a thick extrusive upper crust. Formation of a thick extrusive upper crust above sea-level only continued beyond the early spreading phase in the area of the Icelandic hot-spot (Iceland–Greenland Ridge).

The mapping and modelling of the Vine–Matthews seafloor-spreading magnetic anomalies of the oceans has, for a long time, been a very useful tool in outlining the extension and tectonic development of oceanic crust. However, in areas close to passive rifted continental margins, the magnetic lineation patterns of the oceans, if present at all, are frequently obscured by a complex initial development. Later obliteration of anomalies may follow as the basement of such regions becomes deeply buried below sediments advancing from the nearby continental margin. A so-called magnetic 'quiet zone' is accordingly often present along the outer continental margin and landwards of the oldest well-developed seafloor-spreading anomalies.

During the last few years, however, the existence of a widely distributed feature on passive margins, typically positioned within the magnetic quiet zone, has been demonstrated by modern multichannel reflection seismic profiling—namely the seawards-dipping sub-basement reflectors. It is more or less straightforward to interpret this feature in terms of the dynamic accretional model for oceanic crust formation proposed by Pálmason (1973, 1980), except that the quality and amount of data is generally not good enough to conclude a unique and common origin for the seawards-dipping reflectors at various places. Also, the reflectors are often found in magnetic 'quiet zones' and in regions of complex and poorly known tectonic development such as the Vøring Plateau (Mutter et al. 1984).

The present E Greenland study is primarily based on 8000 km of deep penetration multichannel seismic data acquired by the Geological

From MORTON, A. C. & PARSON, L. M. (eds), 1988, *Early Tertiary Volcanism and the Opening of the NE Atlantic,* Geological Society Special Publication No. 39, pp. 95–114.

Survey of Greenland in the region from SE Greenland northwards to the Jan Mayen fracture zone (Larsen 1983, 1984).

A magnetic quiet zone off SE Greenland between the outer shelf and seafloor-spreading anomaly 24 A/B was described by Featherstone *et al.* (1977) and by Larsen (1980). The present study demonstrates the presence within this quiet zone of a well-developed sequence of seawards-dipping sub-basement reflectors in an area of little post-drift sedimentation. The distribution of this dipping reflector sequence is, however, not limited to the quiet zone landwards of anomaly 24 A/B. In areas of relatively shallow basement, such as the Iceland–Greenland Ridge and the area NW of Iceland, our data show the continuous presence of dipping reflectors seawards into oceanic crust younger than anomaly 24 A/B.

This paper discusses the distribution and origin of the E Greenland seawards-dipping reflector sequence based on a seismic stratigraphic analysis of the reflection pattern and on a refraction seismic velocity study of the crust in which the sequence occurs.

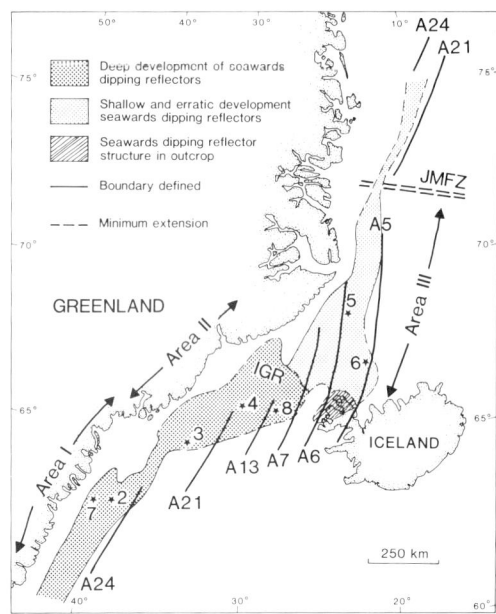

FIG. 1. Map of the seawards-dipping reflector sequence of E Greenland. Seafloor-spreading anomalies A24, A21, A13, A7, A6 and A5 shown. Numbered asterisks indicate positions of reflection displays from Figures 2, 3, 4, 5, 6, 7, 8. Extension of dipping reflectors N of JMFZ from Mutter (pers. comm.) and Mutter & Zehnder (1988). JMFZ = Jan Mayen Fracture Zone, IGR = Iceland–Greenland Ridge.

Distribution and geometry

Seawards-dipping sub-basement reflectors are found in a coast-parallel belt of variable width along the E Greenland margin from its very southern end northward to the Jan Mayen fracture zone (Fig. 1). Our data also indicates the presence of dipping reflectors N of the Jan Mayen fracture zone, and this is confirmed by an independent study (Mutter & Zehnder 1988). They are thus present along the entire rifted margin along the E coast of Greenland, but they show considerable variation in geometry and stratigraphical setting.

Distribution and stratigraphical setting

The age of the dipping reflectors can be indicated or determined from dated ocean-floor magnetic isochrons or from the interdigitating relationship between sediments and sub-basement reflectors. Using these criteria, some pronounced N–S variations in the stratigraphical setting of the dipping reflectors off E Greenland can be determined, even allowing for realistic uncertainty in our dating tools.

In the S, the dipping reflector sequence is well-developed in a broad zone landward of magnetic anomaly 24 A/B (see Figs 1 & 2). The emplacement of the dipping reflector sequence in this southern area thus took place prior to anomaly

24 A/B, most likely within a period of only a few million years.

The seaward boundary of the dipping reflectors, however, moves across anomaly 24 A/B further N. North of an area of poorly-developed dipping reflectors, well-developed dipping reflectors are situated in a crustal segment showing magnetic anomalies 24 A/B to 22 (see Figs 1 & 3) or perhaps even younger (seaward boundary of dipping reflectors not known exactly in this area).

In the Denmark Strait, precise magnetic anomaly dating is difficult (Nunns *et al.* 1983). A prominent elevated basement ridge is present here in the area between northwestern Iceland and E Greenland, and well-developed seawards-dipping reflectors (Figs 1 & 4) are found from close to Greenland (ca. anomaly 24 age) and seawards along most, if not the entire length of the basement ridge and towards north-western Iceland, where a very similar structure within the exposed lava pile has been observed by the first author along some of the deeply incised fjords. These basalts are of anomaly 5–6 age. Due to lack of deep seismic data, the likely existence of

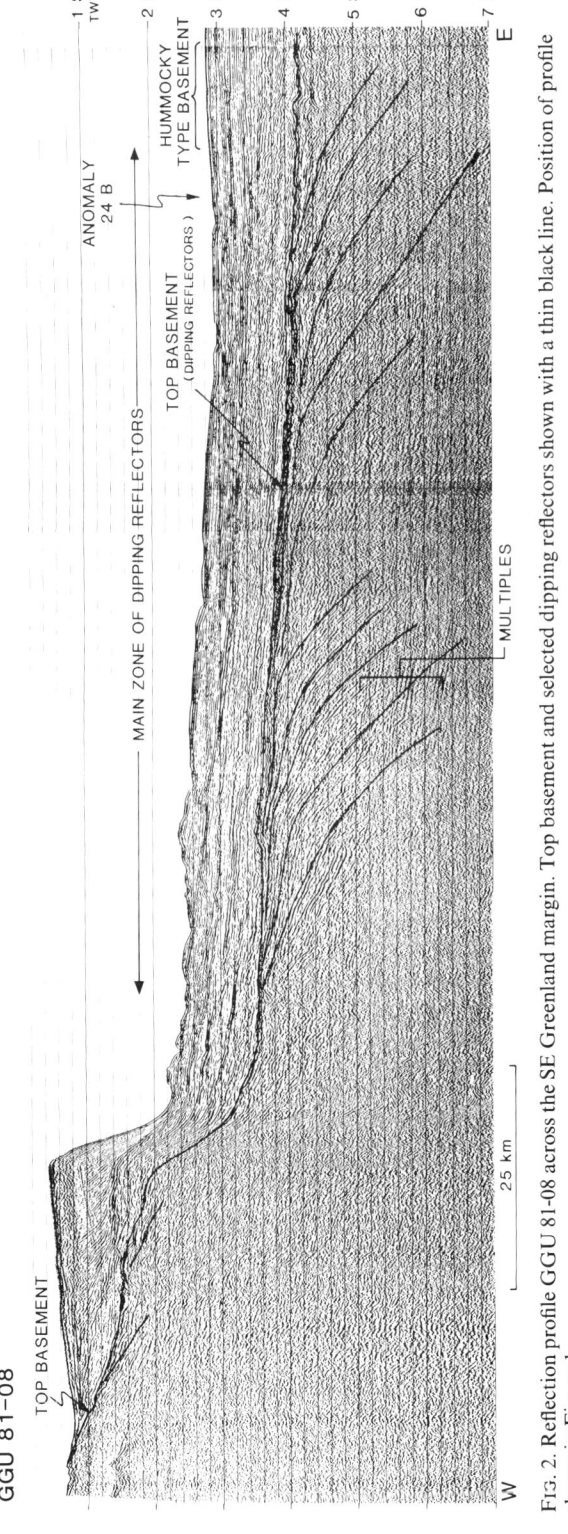

FIG. 2. Reflection profile GGU 81-08 across the SE Greenland margin. Top basement and selected dipping reflectors shown with a thin black line. Position of profile shown in Figure 1.

GGU 82-2A,2

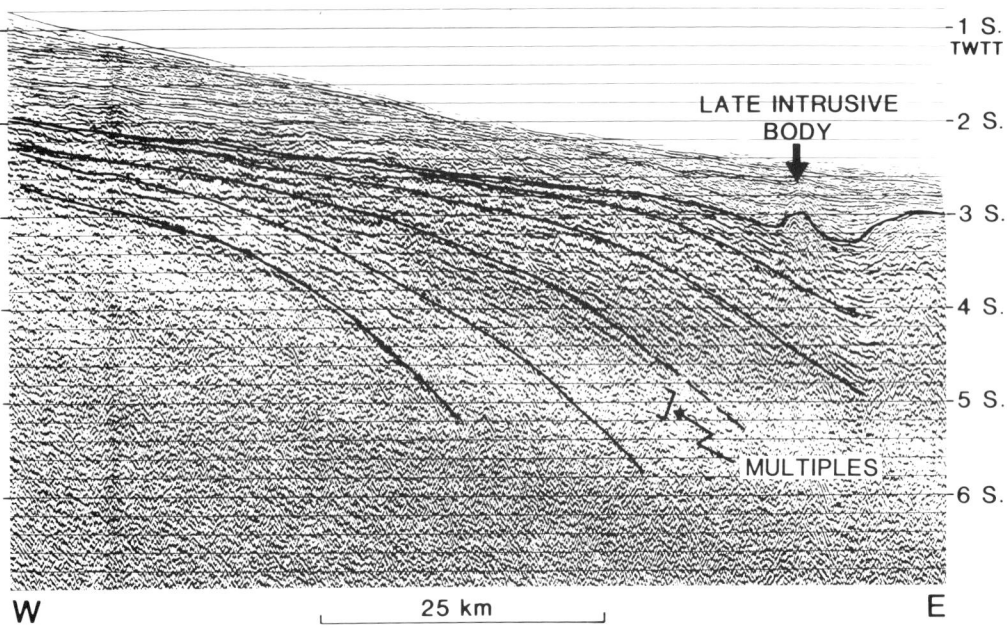

FIG. 3. Part of reflection seismic line GGU 82-2/2A showing the presence and development to large depth of seawards-dipping reflectors in anomaly 23–22 age crust S of the Iceland–Greenland Ridge (area II of Figs 1 & 11). Top basement and selected dipping reflectors shown with a thin black line. Note the late intrusive body and the basement escarpment at the eastern end of the profile. Transition into hummocky-type basement surface without dipping reflectors is indicated by other data to take place just E of this profile.

GGU 81-20

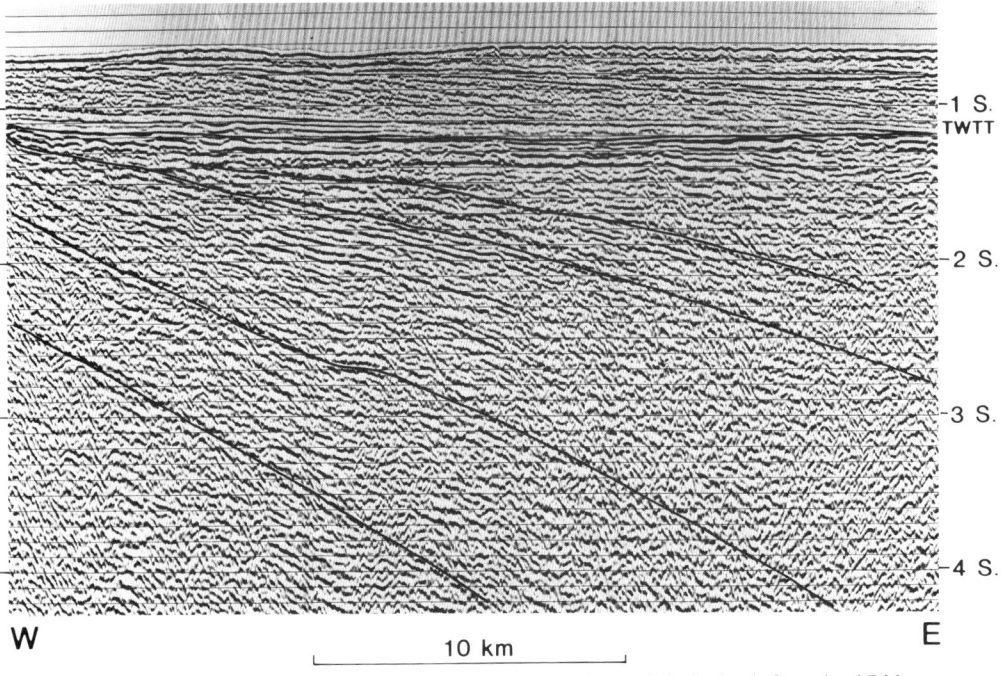

FIG. 4. Part of reflection seismic profile GGU 81-20 over the central part of the Iceland–Greenland Ridge (position shown in Fig. 1). Note the presence of dipping reflectors to great depth and the less convex pattern of dipping beds compared to Figures 2 and 3.

dipping reflectors on the Icelandic Shelf itself (ca. anomaly 7–6 age) cannot be confirmed yet.

Apparently, therefore, emplacement and formation of the seawards-dipping reflector sequence started at or somewhat prior to anomaly 24 A/B from the southern tip of Greenland and northwards to the Iceland–Greenland Ridge. Formation of dipping reflectors, however, continued longer in the northern regions than in the S. On the Iceland–Greenland Ridge, formation of dipping reflectors probably continued uninterrupted well into the Neogene. Therefore, deep basement to the S shows a stratigraphically short occurrence of dipping reflectors, whereas shallow basement in the N shows a stratigraphically much wider dipping reflector sequence (see Fig. 1).

Dipping reflectors are generally less well-developed and more erratic in occurrence N of the Iceland–Greenland Ridge compared to the ridge itself and the area S of the ridge. They occur in this region mainly within relatively young oceanic crust of anomaly 5–7 age due to the later initiation of spreading in this region, and are normally only developed to a very shallow depth (see Figs 1 & 5). The general shallow development of the dipping reflectors is discussed later. In general, dipping reflectors younger than anomaly 5A are not seen, except for the region of shallow basement just N of Iceland. In this region, non-exclusive commercial seismic data show the presence of dipping reflectors within young oceanic crust (see Figs 1 & 6).

Geometry

The geometry of the dipping reflectors and the fraction of the crust they occupy vary. Variations are mainly seen along the margin but also occur perpendicular to the margin. The geometry described here is based on seismic time sections (unmigrated and some migrated key sections). Depth conversions have been made in connection with modelling of refraction seismic data (see Fig. 14), as discussed later, and seem to confirm that the reflection geometry interpreted from the time sections is generally correct.

GGU 82-27

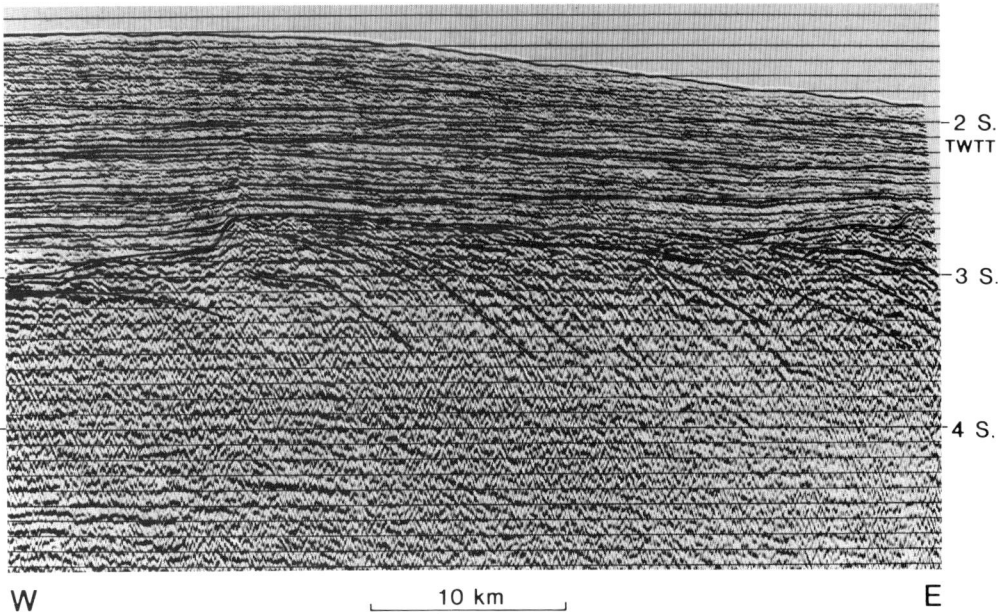

FIG. 5. Part of reflection seismic line GGU 82-27 showing the presence of dipping reflectors developed in anomaly 5–6 age crust N of the Iceland–Greenland Ridge. Dipping reflectors in this area (area III of Figs 1 & 11) only show erratic and shallow development. Note formation of the basement escarpment at the western end of the profile. The escarpment is interpreted as being virtually non-tectonic, arising out of co-existing shallow marine sedimentation sourced from the W (Greenland margin), and subaerial to perhaps shallow water volcanism sourced from the E. The escarpment represents a vertical build-up of the boundary zone between the two regions.

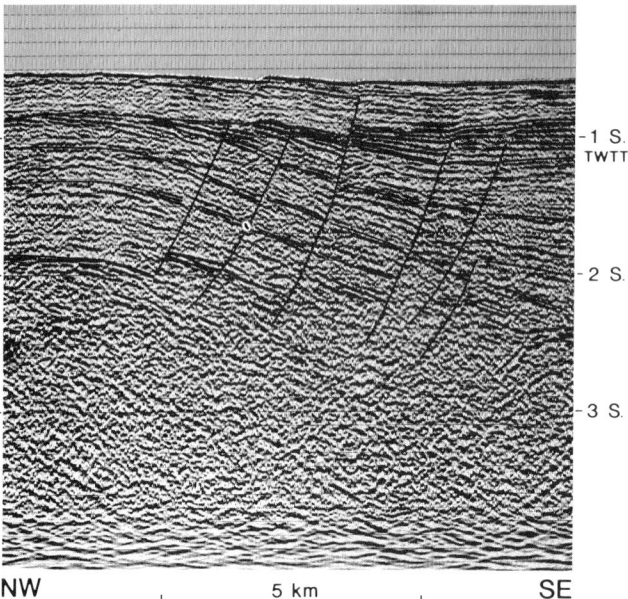

FIG. 6. Part of reflection seismic profile WGC ICE-C showing seawards (riftwards) dipping reflectors N of Iceland. Profile located within anomaly 5 age crust. Data have been migrated to highlight the faulting (some minor faulting can be followed to seabed). Typical dipping and divergent reflectors are seen to the SE. Dipping events in the central part of the section show a more parallel and flexure-like pattern which is associated with faulting. Data from Western Geophysical Co. of America Speculative Survey.

The sub-basement dipping reflectors typically comprise a sequence of seawards-dipping, divergent and upwardly convex reflection events (see Fig. 2). In general, the upper sequence boundary shows a seaward offlapping development somewhat similar to the offlapping configuration envisaged in seismic stratigraphy for a prograding sequence deposited during a slow relative sea-level fall (Vail *et al.* 1977). Erosion may have taken place locally at the upper sequence boundary and is quite widespread at the top of the Iceland–Greenland Ridge. In general, however, the upper sequence boundary is overlain conformably by a clearly distinguishable sequence of flat-lying and mainly parallel reflectors (see Figs 2, 3, 4, 5, 6 & 8).

The lower boundary of the dipping reflector sequence is generally not seen. Therefore, baselap relationships have not been observed. Dipping reflectors, irrespective of their penetration into the crust, show increased dip with depth and seem to disappear into a noisy or chaotic reflection pattern deeper in the crust. The dip range is from less than 1° at shallow crustal levels increasing to around 8° at 2 km and 20–25° at 5 km depth.

Systematic unconformity formation and faulting within the sequence is almost absent. The total thickness in place of the sequence may be as much as 6–8 km and total accumulated stratigraphical thickness much more than that. The sequence seems to wedge-out landwards into a thin sequence of apparently short stratigraphical length. In such marginal areas, a local lower boundary showing baselap configuration may be present (see Fig. 7). There is no indication of a seawards thinning of the sequence, but rather an abrupt change into a seismically opaque basement with a very irregular basement surface causing a hummocky-type reflection pattern (see Fig. 2).

The geometry of reflectors along strike is less well-controlled than in the dip direction. A sufficient grid-net of data exists, however, to enable individual reflectors or units of reflectors to be mapped in three dimensions in certain areas. This confirms that the dipping reflectors have a 3-D wedge geometry as opposed to a 3-D lobe geometry. In general, the reflector units (wedges) show good continuity along strike and over several tens of kilometres their strike

GGU 81-13

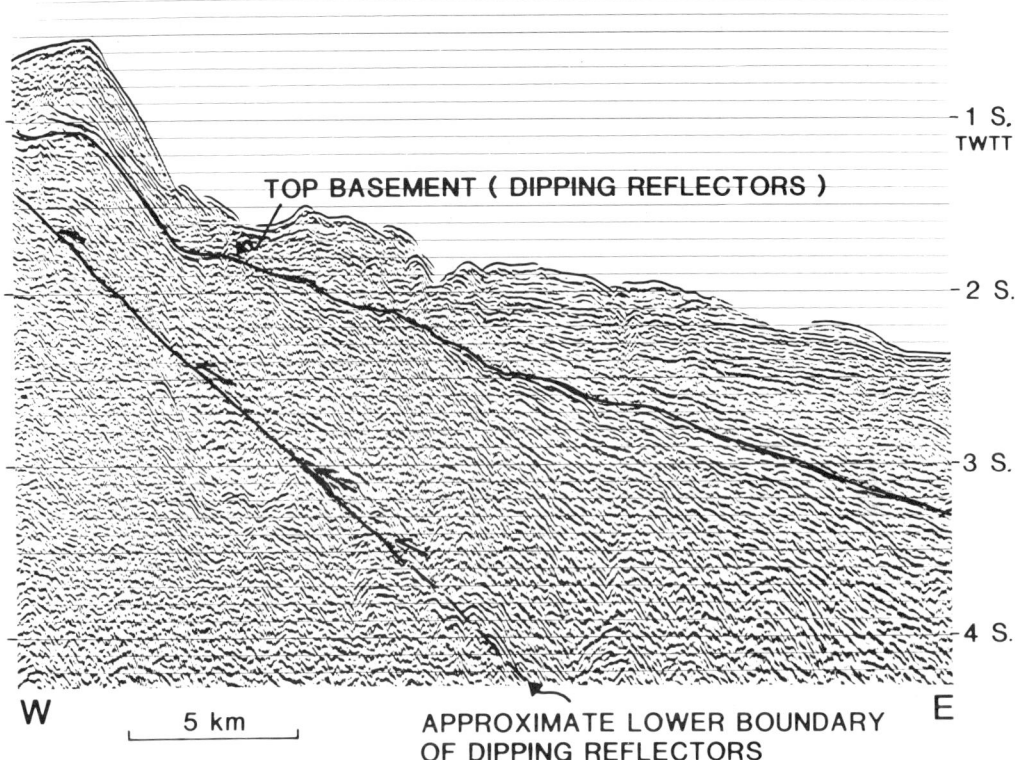

FIG. 7. Landward edge of dipping reflector sequence from reflection profile GGU 81-13. Approximate lower boundary of dipping reflector sequence indicated with dipping reflectors showing baselap (onlap). Note prominent landwards thinning of dipping reflector sequence. Position of profile shown in Figure 1.

directions are parallel or subparallel with the adjacent seafloor-spreading anomalies. Exceptions to both dip and strike geometry continuation are found below the Iceland–Greenland Ridge. The dipping reflector sequence on the Iceland–Greenland Ridge seems to lack the typical upward convex reflection pattern, sub-basement reflectors showing linear to weakly upward convex development and seawards dip being locally reversed into a landwards dip (see Figs 4 & 8).

North of the Iceland–Greenland Ridge, the dipping reflector sequence is generally insufficiently well-developed to allow an assessment of 3-D geometry.

Origin

The sequence has unusual seismic stratigraphic attributes and an apparently extreme rate of deposition (up to 10 m/1000 yr). Consequently, a sedimentary origin is highly unlikely. The reflection pattern, however, strongly suggests a layered structure formed by an exogene process. Gravity data, magnetic data and seismic velocity data (see later) all strongly indicate a volcanic origin for the sequence.

A sequence of basaltic flows erupted from long fissures and deposited subaerially is among the few, if not the only, known geological structures compatible with the seismic stratigraphic analysis. Strong support for this general origin was obtained through drilling off Rockall (Roberts *et al.* 1984) and recently further supported through drilling off Norway (Eldholm *et al.* 1986). The primarily volcanic and basaltic origin of the dipping reflector sequence is now widely accepted. It is, however, debatable as to whether the structure forms on attenuated continental crust in an advanced stage of rifting (Hinz 1981) or is the expression of seafloor spreading above sea-level (Mutter *et al.* 1984).

GGU 81-01 (01A)

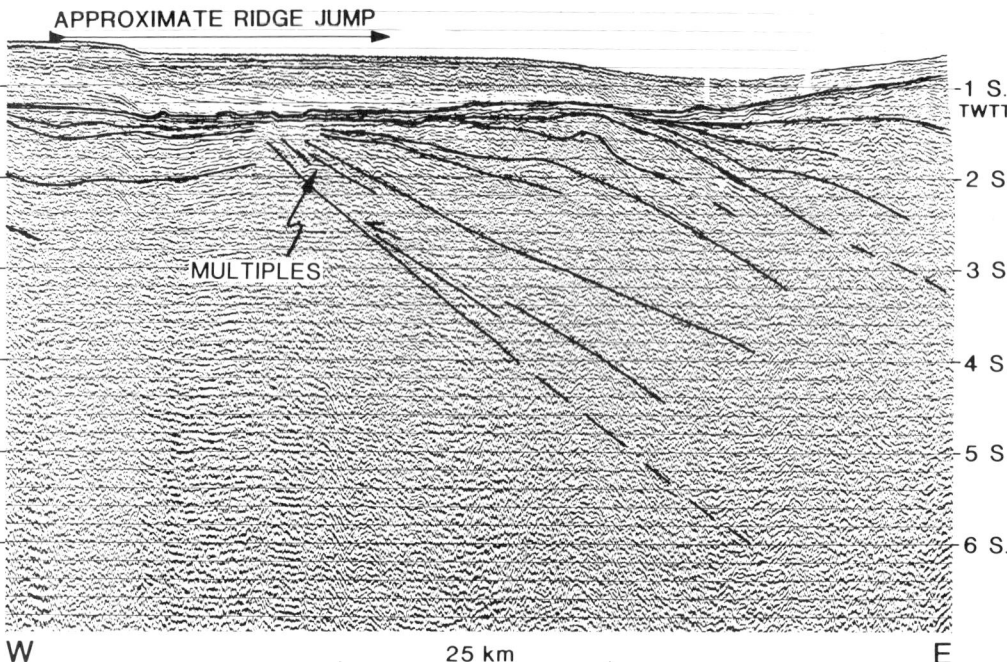

FIG. 8. Part of reflection seismic profile GGU 81-01/1A located on the central part of the Iceland–Greenland Ridge (location shown in Fig. 1). Top basement and selected dipping beds shown with a thin black line. Note the abnormal landwards dip at the western part of the profile and the change into the normal seawards (eastwards) dip at the eastern end of profile. The structure is interpreted as an approximately 30–40 km eastward jump of the spreading axis (see also Fig. 10). Profile location in Figure 1.

The occurrence off E Greenland of seawards-dipping reflectors seawards of anomaly 24 A/B in areas of evident oceanic origin strongly favours a subaerial oceanic origin following the 'Pálmason model' (Pálmason 1973, 1980) applied by Smythe (1983) and Mutter *et al.* (1984).

The origin of the structure can also be inferred from a seismic stratigraphic analysis. The seismic stratigraphic attributes (including recent well information) indicate that the dipping sequence is primarily made up of basaltic lavas extruded from linear sources (fissures) more or less parallel to the nearby seafloor-spreading anomalies. The volcanic source could either have been up-dip or down-dip. An up-dip source would imply deposition on a seawards-dipping palaeoslope and a seawards displacement of the volcanic line source with time (see Fig. 9). Down-dip deposition from an up-dip source should have caused a distinct baselap (downlap) and is likely to have caused destruction of the upper crustal layering by the seawards displacement, with time, of the volcanic line source and the associated eruptions through the volcanic wedge already in place (see Fig. 9).

The opposite is seen; there is no baselap and no reflection destruction. Also, an up-dip source could only be made compatible with the formation of a symmetric reflector wedge on the conjugated margin (Roberts *et al.* 1984; Andersen 1988; White 1988) by invoking great complexity.

UP - DIP SOURCE

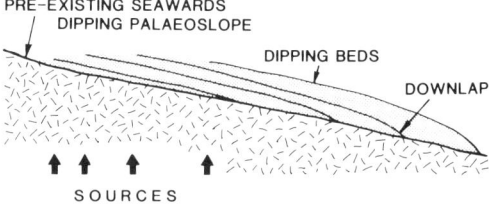

FIG. 9. Structure of dipping reflectors arising out of an assumed up-dip and linear volcanic source. See text for further discussion.

As opposed to the up-dip source, a down-dip source implies that present-day dip is a post-depositional feature caused by loading and differential subsidence. The down-dip source may cause a seawards–dipping reflector sequence of limited thickness and down-dip extension to form on a pre-existing basement (continental or oceanic) and a marginal landwards onlap relationship to the underlying basement can be expected in such a setting (see Fig. 10). However, more extensive down-dip continuity of the seawards-dipping reflector sequence can only be obtained from a down-dip source by creation of new crust and lithosphere, i.e. by seafloor spreading. The alternative, comprising a seawards and down-dip displacement of the volcanic source relative to a pre-existing crust and lithosphere, would interrupt the observed persistent seawards-dipping pattern and create repeated and frequent zones of reversed dip (see Fig. 10). Such local dip reversals can be observed in some places on the Iceland–Greenland Ridge (see Fig. 8) and are here interpreted as resulting from a small eastward jump of the spreading axis within oceanic crust.

It is concluded from the seismic stratigraphic analysis that the seawards-dipping reflector sequence has a basaltic, linear down-dip source and that down-dip extensive reflector sequences, such as those present along the southern and middle part of the E Greenland margin, are created by subaerial seafloor spreading. Marginal landward parts of the seawards-dipping reflector sequence may have been deposited on pre-existing continental crust, and are likely to show baselap (onlap) at their base as opposed to the seaward part of the dipping reflector sequence, where no baselap can be expected. Although the expected marginal baselap relationship to pre-existing crust may be difficult to locate on seismic data from passive margins, the change from such a baselap relationship seawards into a distinct lack of baselap and disappearance of dipping reflectors into a noisy or chaotic zone may be used to define the ocean-to-continent transition (see Fig. 10).

Comparison with Iceland

Subaerial seafloor spreading takes place in present day Iceland and a model explaining the structure of the Tertiary plateau basalts of north-western and eastern Iceland as a result of subaerial seafloor spreading was proposed by Pálmason (1973, 1980). In good accordance with this, the E Greenland dipping reflector sequence can be followed on top of the elevated basement ridge

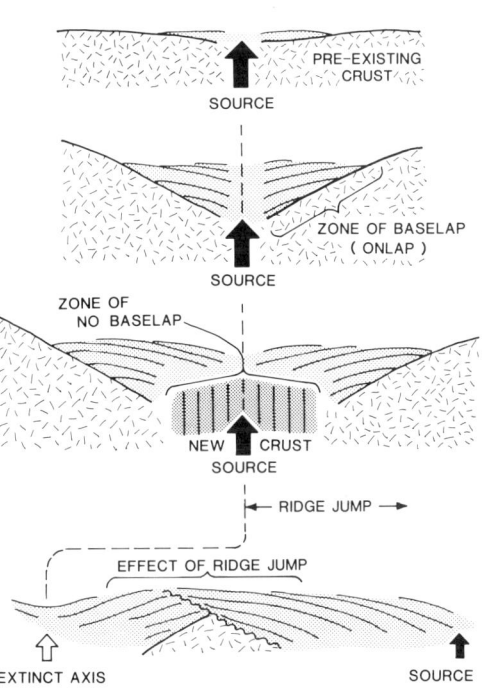

FIG. 10. Structure of dipping reflectors arising out of an assumed down-dip and linear volcanic source. This favoured mechanism of dipping reflector sequence formation implies that the seawards dip is a postdepositional dip acquired through rapid differential subsidence adjacent to the central rift zone as envisaged by Pálmason (1973, 1980). Note the effect of a ridge jump and compare with Figure 8.

between Iceland and Greenland to a position fairly close to Iceland. Lack of deep penetration seismic data from the insular shelf of Iceland prevents a complete tie with the plateau basalts of northwestern Iceland in which an eastward (=seawards) dipping structure very similar to the seawards-dipping reflectors can be observed (Pálmason & Sæmundson 1974).

Applying the kinematic model for subaerial seafloor spreading in Iceland (Pálmason 1973, 1980), one may interpret the downwards disappearance of seawards-dipping reflectors into a noisy or chaotic zone as a result of the change from an upper extrusive layer to a lower intrusive layer. According to Pálmason, the sharpness and the position of this change is a function of original rift width, volcanic production rate and spreading rate.

Using seismically and magnetically derived parameters for spreading rate, thickness of extrusive layer and the distribution and rate of volcanic production, an estimate of the original rift width can be obtained from the Pálmason model. We have used data from the extensively and deeply developed sequence of pre-anomaly 24 dipping reflectors off SE Greenland. We selected this area because of good data quality and because the simple steady-state conditions required by the model seem to be met in this area. According to Pálmason (1980), extension by normal faulting within the original rift zone mainly affects the thermal state of the crust and, hence, we have neglected the effect of normal faulting in our calculations.

The pre-anomaly 24 spreading rate is constrained by the lack of anomaly 25 within the approximately 80–90 km wide zone of oceanic crust landwards of anomaly 24 in this region. Seafloor spreading thus apparently started less than 2.5 My prior to anomaly 24 A/B and, hence, the minimum spreading rate is 3.2 cm/yr (half-rate) in accordance with rates obtained from modelling of the adjacent anomaly 24 A/B and 23 sequence (see Fig. 7 in Larsen 1980, 3.2 cm/yr).

The average minimum thickness of the extrusive layer is estimated to be 5–6 km as continuous dipping reflectors can generally be followed to a depth of 5–6 km (around 2.0 s, two way travel time (TWTT), below basement, with some reflectors penetrating to around 3.0 s TWTT). It is assumed that less than approximately 30% of intrusives are present at 6 km depth, but the model response in terms of original rift width only shows fairly small variations if this parameter is varied within the interval 20–35%.

The minimum volcanic productivity rate can be calculated from the volume of extrusives divided by the maximum time interval of 2.5 My The uncertainty of the time interval is not important for the model response, however, as the time interval also affects spreading rate and is equalized during the further calculations. The main problem is therefore to make a reliable estimate of the total volume of extrusives. The vertical zonation of lava versus intrusives in the crust can be estimated from Pálmason (Fig. 3 in Pálmason 1980). It appears that the total volume of extrusives is relatively insensitive to the exact distribution of extrusives with depth (Table 1). A value of 4.5–4.6×10^{-4} km^2/yr per rift length unit is used in our calculation. This figure is about three times greater than the volcanic productivity rate used by Pálmason for spreading in Iceland (Pálmason 1980, 1986), which is consistent with the high spreading rate involved (approximately three times the spreading rate of Iceland).

If we assume that the fraction of intrusives is

TABLE 1. *Approximate fraction in percentage of extrusives within the crust at two different values for the σ_1/σ_2 ratio according to Pálmason (1980). The integrated thickness of extrusives down to 16 km depth is 6.8 km for $\sigma_1/\sigma_2 = 0.3$ and 6.6 km for $\sigma_1/\sigma_2 = 0.6$. Thus, volcanic productivity rates in the present study only vary between 4.5 and 4.6×10^{-4} km^2/yr for a unit rift length, using the σ_1/σ_2 ratio range 0.3–0.6*

Depth km	$\sigma_1/\sigma_2 = 0.3$ %	$\sigma_1/\sigma_2 = 0.6$ %
0–2	100	95
2–4	95	80
4–6	85	60
6–8	50	40
8–10	10	25
10–12	0	15
12–14	0	10
14–16	0	5

within 20–35% over the depth interval of 5.5–6.0 km and that the volcanic depositional rate is between 4.5 and 4.6×10^{-4} km^2/yr per rift length unit, application of the Pálmason model (Fig. 3 in Pálmason 1980) provides a σ_1/σ_2 ratio of 0.17–0.40 ($\sigma_1 \approx$ zone of intense dyke injection or rift width; $\sigma_2 \approx$ zone of main lava deposition; see Pálmason 1980 for exact definition). From this range of σ_1/σ_2 ratios, an estimate of the original rift-zone width (σ_1) can be made for a given value of σ_2.

According to Pálmason (1980), σ_2 shows only slow response to variations in the spreading rate/volcanic productivity ratio. With the parameters given, we can expect only about 10% higher σ_2 values for the pre-anomaly 24 dipping reflector sequence compared to the σ_2 of Iceland. For Iceland, an average value of 20 km for σ_2 is assumed by Pálmason (1980, 1986) and a value of 22 km may therefore be applicable in our calculations. Pálmason (1980, Fig. 4) estimated σ_2 from observations on regional dip and lava deposition rate within the Tertiary lava pile in Iceland. As an independent alternative to the above figure (22 km) we have used seismically derived values on dip and lava deposition rate within the dipping reflector sequence. The dip at 2 km depth is around 8° and the estimated average lava deposition rate at 2 km depth is around 4–5 km/My. A σ_2 value of 20–25 km is obtained using these input values and relationships suggested by Pálmason (1980).

With the range of σ_1/σ_2 ratios given above and a σ_2 range of 20–25 km, we find a σ_1 range of 3.4–10 km. Using an average value of $\sigma_2 = 22.5$ km and the average σ_1/σ_2 value of 0.285, an average σ_1 value of 6.4 km is obtained. These calculations show very good correlation with data obtained from Iceland ($\sigma_1 = 5$–15 km, Pálmason 1980), although indicating a slightly narrower rift-zone width than in Iceland. Nevertheless, this is larger than on the mid-ocean ridges (Matthews & Bath 1967; Harrison 1968).

Application of the Pálmason model for spreading in Iceland to the pre-anomaly 24 dipping reflector sequence off SE Greenland thus offers a qualitative explanation of the deep development of dipping reflectors in this region. According to the model calculations, a relatively high spreading rate and a relatively high volcanic productivity rate in combination with a fairly narrow rift zone created an upper extrusive crust at least 5–6 km thick.

Velocity structure

The velocity distribution within the crust showing the presence of seawards-dipping sub-basement reflectors has been interpreted from 19 sonobuoy refraction profiles recorded within the area of seawards–dipping reflectors.

The sonobuoy registrations were carried out at selected positions while recording multichannel reflection seismic data, using an airgun array totalling 30–60×10^3 cm^3 as an energy source. The resulting 40 refraction profiles are of variable quality. Signals can only be seen to a distance of about 20 km on some profiles, while others show clear arrivals to a distance of 70 km. None of the profiles are reversed, but reflection data supply information on the upper part of the crust. Signals from the lower crust are observed on 19 profiles. These profiles were assessed using 1- and 2-D ray-tracing modelling. Velocity analysis and sediment thickness obtained from reflection data were used to constrain the upper part of the models. In most cases, the 1-D modelling provided acceptable results. Better fits were obtained for four profiles by using 2-D modelling.

Apart from one very long refraction profile, none of the 19 profiles show the strong critical reflections from the base of the crust that could be expected by comparison with the conjugate Rockall margin (White 1988). Weak, but rather systematic signals at a typical offset of 15–30 km have been used in an attempt to estimate the Moho depth on a number of profiles. As discussed below, the fairly shallow Moho position arising out of this attempt is probably in error, and a larger Moho depth estimated from the lack of strong signals at offsets 20–40 km and from comparison with the conjugate margin (Bott & Gunnarsson 1980; White 1988) has been incorporated as a likely alternative.

All but one of the refraction profiles used in this study are situated over locations of seawards-dipping reflectors. One profile is located at the seawards transition from the dipping reflector sequence into an opaque, hummocky-type basement.

Crustal changes along the margin

The refraction study reveals a variation in the crustal velocity structure along the margin and within the zone of dipping reflectors. This variation may be related to changes in the geometry and age of the dipping reflector sequence and seems to be associated with changes in the gravity and magnetic field data.

Based on the interpreted refraction data, a profile along the margin was constructed (see Fig. 11). This crustal profile is located within anomaly 24R to anomaly 22 crust along its southern part, around anomaly 20 crust in its central part across the Iceland–Greenland Ridge and within anomaly 7–5 crust in the N. The upper crustal structure including the top of layer 3 is well constrained from the available refraction data. The complete crustal structure including the position of Moho is only known with certainty on the northern end of the profile. Along the remaining part of the profile, two alternative Moho positions are shown, the shallow Moho position arising out of the study of weak signals at typical offsets of 15–20 km and a deeper Moho position estimated as discussed above.

The two different Moho positions show, to a large extent, the same general trend and, in combination with the upper crustal velocity distribution, define a clear three-fold division of the profile into a southern, a central and a northern segment (see Figs 11, 12, 13, 14 & 15); these correspond to the areas I, II & III of Figure 1. The southern and the central segments show about the same general layer 2 thickness (3–5.5 km), but the central segment is separated from the southern segment by a highly anomalous zone showing very large depth to typical layer 3 velocities (see Fig. 11). In addition, the total crustal thickness of the central segment is, irrespective of which Moho position is followed, significantly greater than the crustal thickness of the southern region. The change in crustal thickness is clearly reflected in the gravity data (see Fig. 11). Seawards-dipping reflectors are clearly and deeply developed in both segments penetrating into the upper part of layer 3, but are poorly developed or lacking in the anomalous transition zone between the two segments. Also,

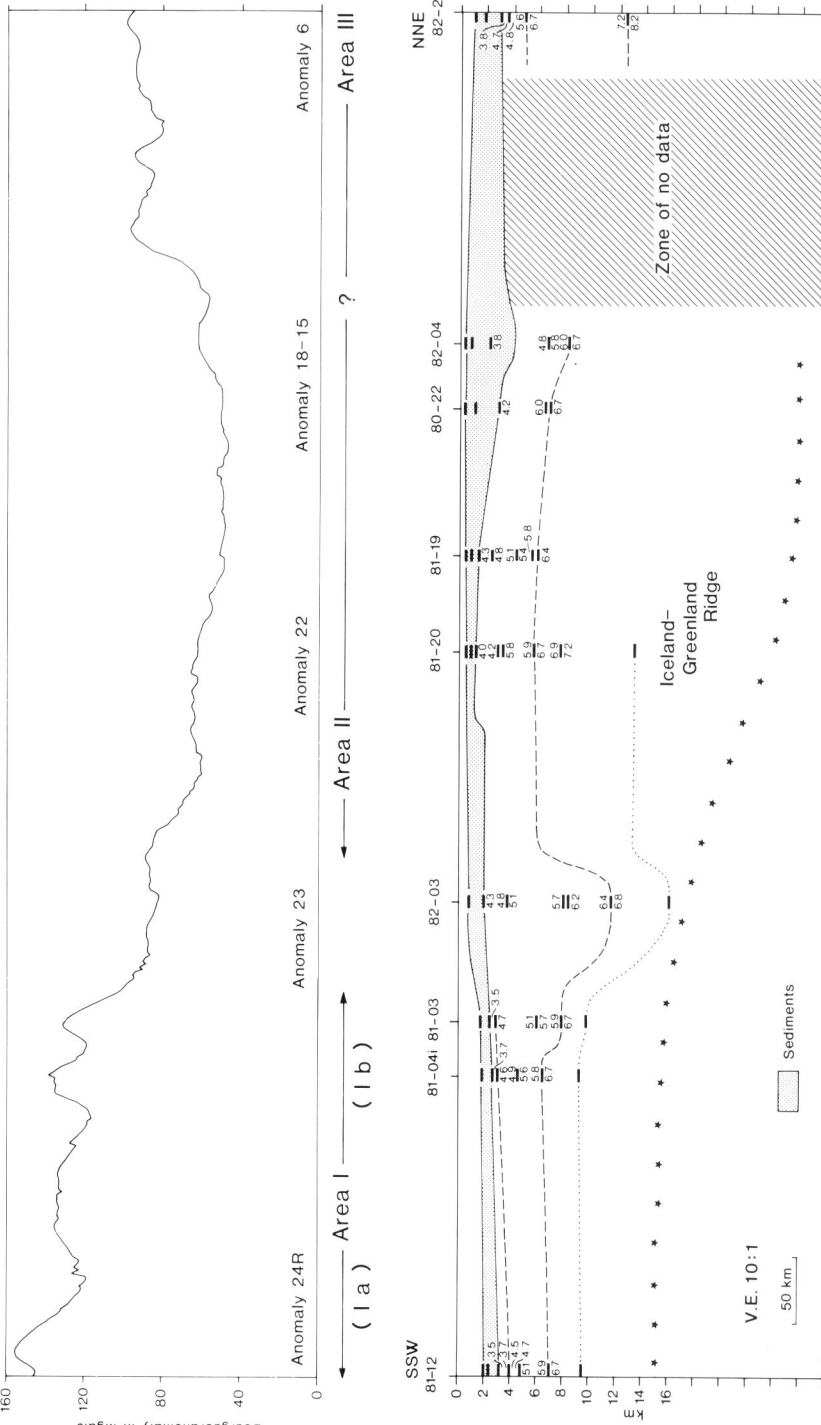

FIG. 11. Profile along the margin based on refraction data. The interpolation for the ocean-bottom and top basement (solid line) is based on multi-channel reflection seismic data. The dotted line represents Moho as indicated by weak signals close to or at noise level at offsets 15–30 km. Asterisks indicate a Moho position based on lack of strong mantle reflections at offsets 20–40 km and assuming symmetry with the Hatton Bank/Iceland–Faeroe Ridge margin (Bott & Gunnarsson 1980; White 1988). The Bouguer anomaly along the profile is drawn on top.

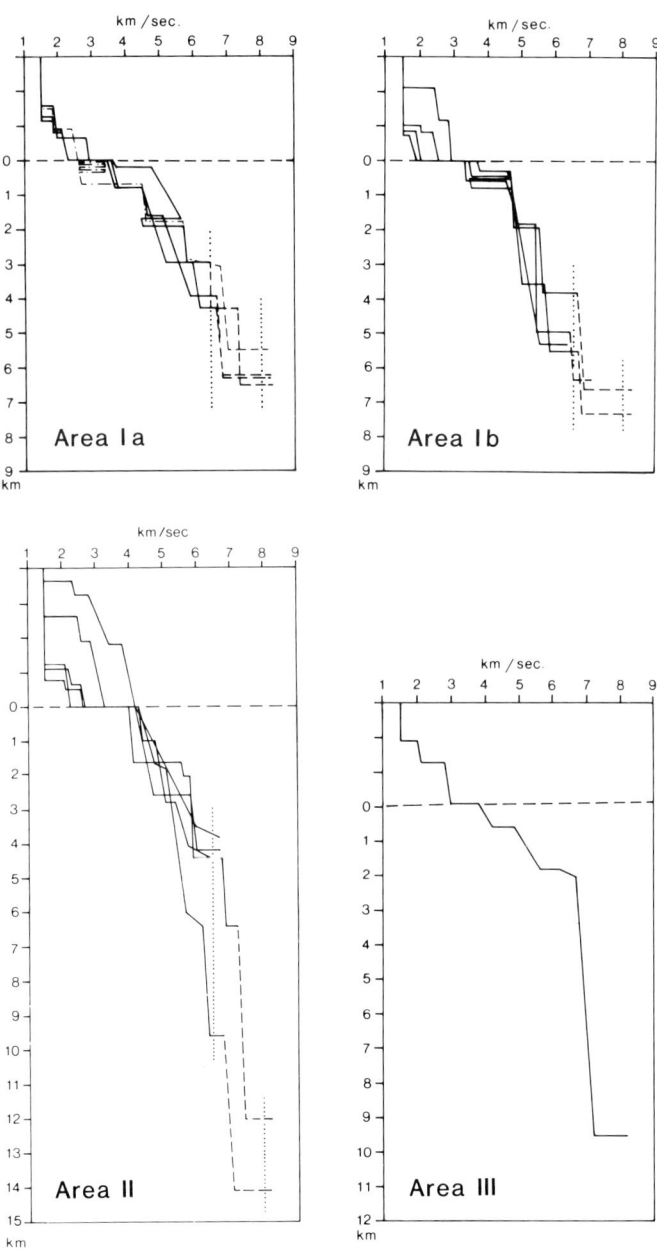

F<small>IG</small>. 12. Velocity distribution of refraction profiles from areas I, II and III. Velocity data from area I are further subdivided into area Ia (southern part) and Ib (northern part). Ia is in the conjugate position to the Hatton Bank study (White 1988). The zero line represents top basement. Water depth is not shown. Dotted lines mark the velocities 6.5 km/s and 8.0 km/s which indicate top Layer 3 and Moho respectively. The uncertain interpretation of Moho is shown by hatched lines (see also Fig. 11).

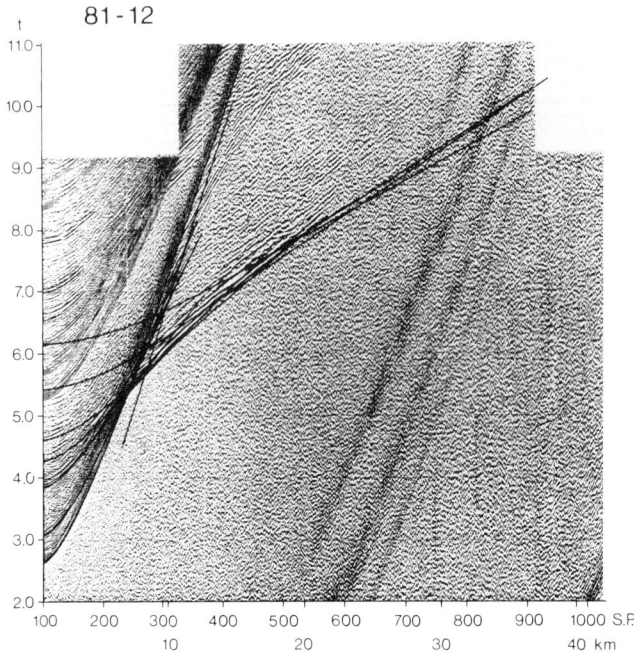

FIG. 13. Refraction profile 81-12 from area I with model response plotted in a thin black line. Data quality and offset range typical for area I and II.

there is a change from generally well-defined seafloor-spreading magnetic anomalies within the southern segment into a more disrupted, high amplitude pattern below the Iceland–Greenland Ridge situated within the central segment.

The northern segment shows a much thinner layer 2 thickness of only 1.5–2 km. The total crustal thickness at the northern end of the profile is well constrained and is around 11–12 km. Depending on whether a deep or a shallow position of the Moho below the southern and central segment is assumed, the crustal thickness of the northern segment is small to intermediate compared to the two other segments. The gravity data (see Fig. 11) might indicate that crustal thickness of the northern area is intermediate rather than thin.

The northern area shows seawards-dipping reflectors with only erratic and shallow develop-ment within the relatively thin layer 2, in marked contrast to the two other regions.

The exact position and nature of the transition from the central to the northern segment is not known due to lack of data (see Fig. 11).

Crustal changes across the margin

Due to existing data coverage, a study of the crustal changes perpendicular to the margin is only possible in the southern area (I). This section has the added advantage of being in a conjugate position to the Rockall–Hatton area (White 1988) in a pre-drift fit. In this area we have no firm evidence of the exact position of the Moho and the alternative positions discussed above are shown in Fig. 14. A possible subdivision of layer 3 into 3A and 3B is only observed directly on one line, but indicated by low upper layer 3 velocities on two other profiles.

The crustal thickness corresponding to the deep Moho position is similar to the crustal thickness found at the seaward end of the conjugate dipping reflector sequence off Rockall–Hatton Bank (White 1988). The strong landwards crustal thickening found below the Rockall–Hatton dipping reflector sequence is not likely to be found on the SE Greenland margin (see Fig. 14) according to the gravity data. Instead, the gravity data suggest a fairly constant crustal thickness across the dipping reflector wedge, and a relatively rapid crustal thickening at the landwards end of the wedge, corresponding to the inferred ocean to continent transition. The strong gradient in the Bouguer gravity profile associated with the landwards termination of the dipping reflector sequence (see Fig. 14) is seen along the entire SE Greenland margin.

The upper crustal velocity structure across the

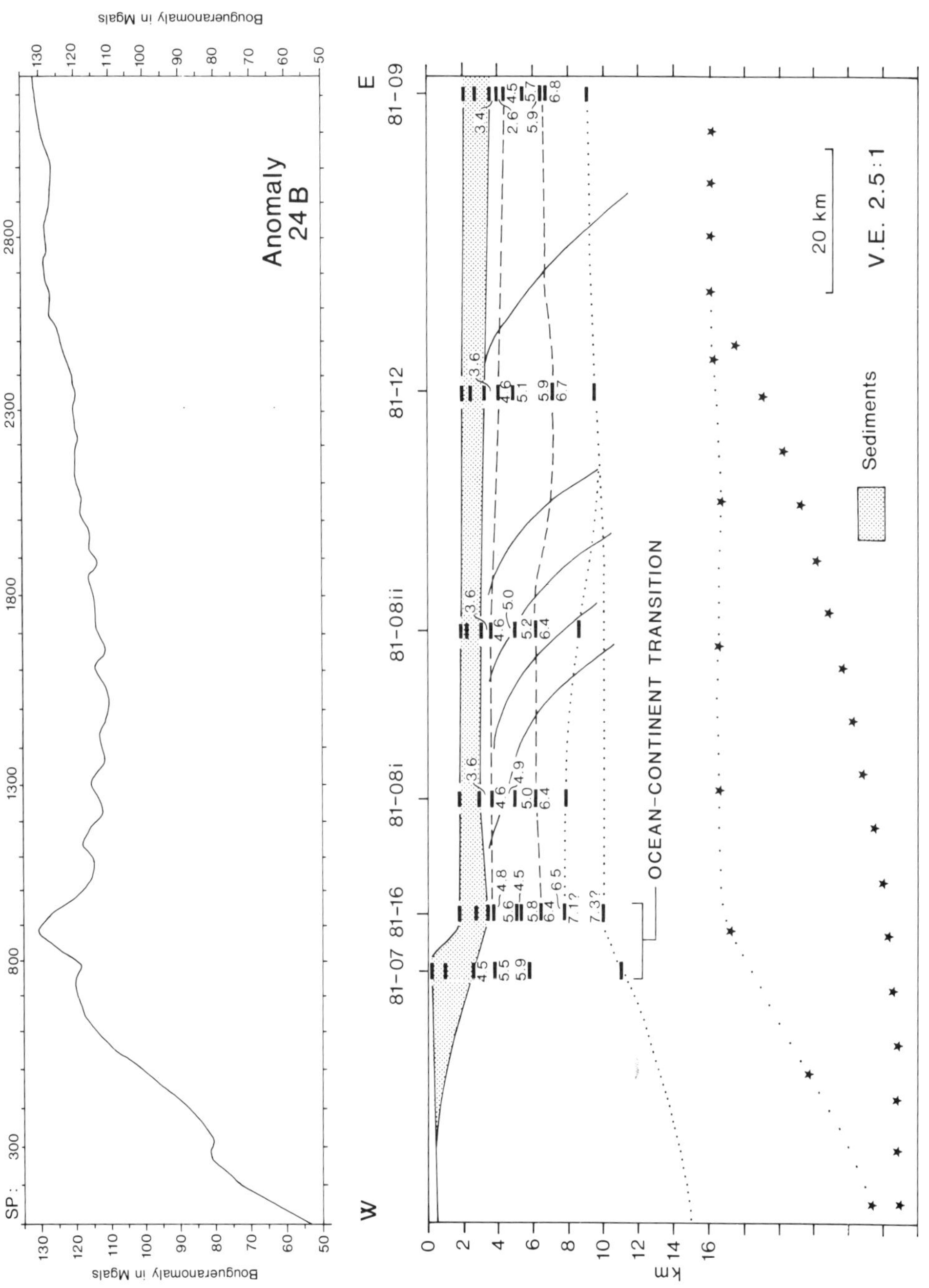

FIG. 14. Profile perpendicular to the margin and across area I. Solid lines as in Figure 11. Dotted lines show interpretation of Moho and Layer 3A/3B interface based on weak signals close to or at noise level. A deeper Moho position mirror imaging the Hatton Bank structure with a landwards increasing thickness of the crust below the dipping reflectors is shown by asterisks. The dot-asterisk line is suggested as the most likely Moho position. This interpretation is based on lack of strong mantle reflections at offsets 20–40 km indicating a rather thick crust and assuming that the weak signals of offsets 15–30 km are of intra-lower crust origin. The gravity data off SE Greenland does not indicate a strong landwards thickening of the crust below the dipping reflectors and therefore only a modest increase in crustal thickness is assumed for the Greenland margin. The apparent asymmetry between the Hatton and the Greenland margin in this aspect may be related to late volcanism and underplating within the Hatton Bank area (White, pers. comm.). The transition to Precambrian basement is suggested to take place within the first 30 km at the western end of the profile. The transition to the opaque, hummocky-type basement (pillow lavas) takes place at the eastern end of the profile (see also Fig. 2).

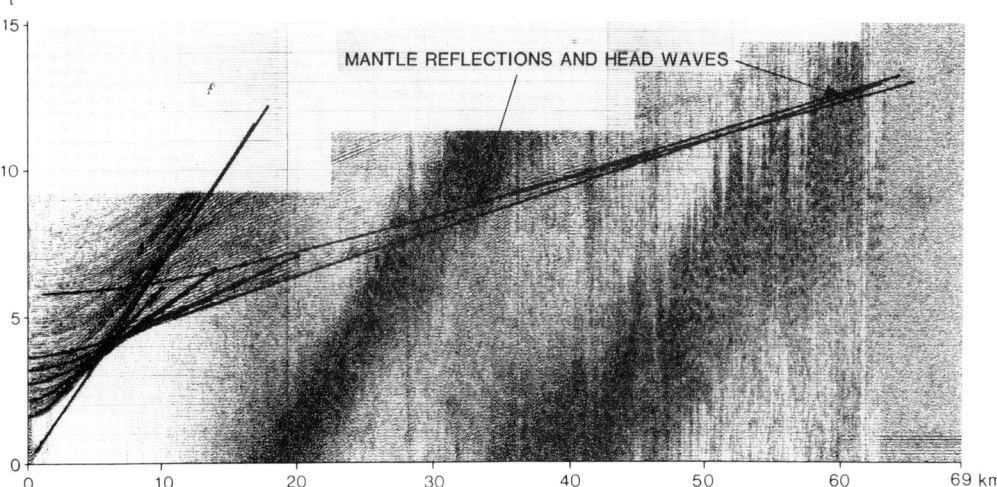

FIG. 15. Refraction profile 82-27 from area III with model response plotted in a thin black line. This is the only refraction profile where strong mantle reflections are observed. The only other profile with a similar offset range is 82-03 positioned at the anomalous transition zone between areas I and II (see also Fig. 11).

margin in this area seems rather uniform compared to the profile along the margin discussed above. The thickness of layer 2 is between 3 and 5 km across the entire wedge and shows no significant change at the seawards termination of the dipping reflectors into the hummocky acoustically opaque basement which we believe results from submarine volcanism. A slower velocity gradient in the upper part of layer 2 is, however, recorded within this opaque basement and may be attributed to zones of shallow-water volcanism (tuffs).

In general, the dipping reflector sequence can be followed to at least 5–6 km into the basement (around 2.0 s TWTT) and several prominent reflectors continue even deeper (around 2.5–3 s or 7–10 km). This observation implies that the majority of the dipping reflectors penetrate well into the depth level of layer 3 velocities.

Assuming that more than roughly 30% intrusives would destroy most of the dipping reflector pattern, this strongly suggests that the layer 2/3 boundary is not reflecting the interface between extrusives and intrusives, but is rather controlled by some secondary processes such as diagenesis or metamorphism. This is consistent with velocity measurements made on ophiolitic material, which show that increasing metamorphic grade to greenschist facies is 'capable of producing a seismic discontinuity similar to the layer 2/3 seismic boundary of refraction seismology' (Christensen 1978). Based on data from the Iceland Research Drilling Project, Robinson *et*

al. (1982) also suggest a metamorphic origin of the layer 2/3 boundary at the drill site location in eastern Iceland (approximate boundary between greenschist and amphibolite facies). If the layer 2/3 boundary is interpreted as the transition to greenschist facies, a layer 2 thickness of 3–5.5 km indicates a geothermal gradient of 40–80°C/km and about twice this gradient if correlation is made with the greenschist–amphibolite facies boundary. Our seismic data have no direct bearing on the exact nature of the suggested metamorphic origin of the layer 2/3 boundary, and different metamorphic history and grade may actually apply in different regions as discussed later.

Depending on which Moho depth is correct (shallow versus deep solution), a layer 3 thickness of around 3–3.5 km or around 7 km is suggested (see Fig. 14).

Adopting the shallow Moho interpretation would imply a strong asymmetric crustal structure of the Greenland and Rockall–Hatton margin and, perhaps even less likely, would suggest the presence of dipping reflectors to and perhaps across the base of the crust (see Fig. 14). If so, a Moho origin of similar nature to the inferred layer 2/3 boundary must be concluded. It is, however, difficult to envisage the formation of large quantities of sufficient olivine-rich extrusives that could provide mantle velocities through metamorphism or phase changes. We therefore consider the deeper Moho position to be the most likely.

Initiation of seafloor spreading above sea-level

It appears that initiation of seafloor spreading within the entire NE Atlantic took place above sea-level. The presence of seawards-dipping reflector sequences along other passive margins shows that a subaerial embryonic development of an ocean basin is not a unique phenomenon, whereas more mature spreading ridges are consistently situated well below sea-level.

In accordance with this observation, the classic occurrence of continental plateau basalts in India, N America and E Greenland shows that vast volumes of relatively dense magma can rise through the continental lithosphere, and through the relatively less dense continental crust and produce large and thick plateaux of basalts. In general, such continental basalt plateaux are elevated, flat-lying and relatively unfaulted. In the case of the E Greenland basalts, there is no indication of any substantial mechanical stretching and faulting of the crust immediately prior to the volcanism.

The fact that relatively heavy magmas can be erupted through and deposited on mechanically weakly stretched continental crust may be explained by a strong energy contribution from the hot asthenosphere, which may heat magmas well over their solidus temperature (Brooks & Nielsen 1982; White 1988). Unlike seafloor spreading, continental volcanism uses no magma, and thus no energy, for construction of new lithosphere.

It is thus not surprising that large-scale sub-aerial volcanism can take place along passive margins. However, Skogseid & Eldholm (1988) have pointed out that following the basin formation model proposed by McKenzie (1978), rifting and strong mechanical stretching will normally cause strong subsidence and, hence, one should expect seafloor spreading to start below sea-level, such as in the Red Sea. The McKenzie model, however, is aimed at describing sedimentary basin formation in relation to crustal stretching, and the basis of the model is violated if massive volcanism and non-conductive upward heat transfer takes place. Accordingly, the model cannot explain large-scale volcanic events.

It is therefore suggested that continental separation and initiation of seafloor spreading is not necessarily preceded by strong mechanical stretching, but can start with massive volcanism sourced from an abnormally hot asthenosphere (White 1988), emplaced through a relatively mechanically unstretched crust, and that the presence of seawards-dipping reflector sequences along a passive margin is indicative of this type

of continental separation. This type of 'active volcanic' separation in the NE Atlantic is supported by the following combined features: (1) no strong rifting or stretching episode immediately prior to spreading is recorded; (2) initial line of spreading was located in different tectonic realms (Archaean and Caledonian basement, Mesozoic rift basins); and (3) massive subaerial volcanism characterizing the entire conjugate margins of the NE Atlantic.

The two models of continental separation (mechanical stretching versus massive volcanism) may be two end-members of passive margin development as suggested by other studies (White 1988).

Conclusion and discussion

We conclude from a seismic stratigraphic analysis of the reflection pattern within the sequence of seawards-dipping sub-basement reflectors off E Greenland that down-dip extended occurrences of the dipping reflectors are unique indicators of subaerial seafloor spreading. We find this conclusion in agreement with the seismic velocity mapping included in our study and other geophysical data.

The distribution and geometry of the dipping reflector sequence off E Greenland show good correlation with Iceland. Application of the kinematic continuum model for subaerial crustal accretion in Iceland (Pálmason 1973, 1980) to the pre-anomaly 24 sequence of dipping reflectors off SE Greenland shows good consistency between the model and the seismically derived parameters. The modelling of this particular spreading interval (initial spreading) suggests that high-rate spreading (half-rate 3.2 cm/yr), high volcanic productivity along the rift (4.5–4.6×10^{-4} km^2/ yr per rift length unit) and a narrow rift zone (3.4–10 km) caused the formation of thick extrusive upper crust (> 5–6 km).

We conclude from the present study and other studies (Andersen 1988; Mutter & Zehnder 1988; White 1988) that seafloor spreading started within the NE Atlantic from the S of Greenland and progressed northwards to the Greenland–Senja fracture zone in a subaerial mode prior to or around anomaly 24 A/B. Spreading only stayed subaerial for a few million years except on the Iceland–Greenland Ridge where spreading was maintained above sea-level at least into the Neogene and probably continuously into present-day subaerial spreading on Iceland. Following the mid-Tertiary westward shift of spreading from the Norwegian Basin and onto the Icelandic Plateau N of Iceland (Larsen 1988), spreading along this part of the E Greenland margin again took place above sea-level until anomaly 5 time, at which time most of the spreading axis N of Iceland subsided below sea-level.

The early Tertiary seafloor spreading above sea-level has created a characteristic seawards-dipping, wedge-shaped, thick sequence of sub-basement reflectors. The seawards-dipping reflectors are believed to arise from bedding planes between basaltic lava flows or between groups of flows; sill complexes may be present locally. The lower boundary of the sequence may locally show baselap (onlap) towards the adjacent continental margin but, apart from such landward marginal areas, the seawards-dipping reflector sequence is diagnostic for oceanic crust formed by subaerial seafloor spreading.

The early Tertiary oceanic crust that formed subaerially along the SE Greenland margin is in its upper part (layer 2 and parts of layer 3) predominantly composed of extrusives (lava flows) and the layer 2/3 boundary evidently cuts across the layering of these extrusives. Layer 2 is normally quite thick in this region (3–5.5 km). We interpret the layer 2/3 boundary in this region as a metamorphic feature superimposed onto the pile of dipping extrusives. By comparison with experimental data and data from eastern Iceland (Christensen 1978; Robinson *et al.* 1982), we suggest that the metamorphic grade associated with the seismic layer 2/3 boundary may vary from greenschist facies to the greenschist–amphibolite facies boundary.

Unlike the early Tertiary seafloor spreading, the mid-Tertiary seafloor spreading above sea-level outside the central E Greenland margin only caused an erratic and shallow development of seawards-dipping reflectors. In this area, layer 2 is only 1–2 km thick, in contrast to the larger thicknesses found in the areas of well-developed dipping reflectors.

We can therefore conclude that the presence of dipping reflectors is not limited to layer 2. However, according to our and other studies (Mutter & Zehnder 1988; White 1988), areas of well-developed dipping reflectors show anomalous layer 2 thicknesses. If a correlation of the layer 2/3 boundary with a specific metamorphic grade is correct in areas of different layer 2 thickness and different dipping reflector development (deep versus shallow development), this suggests that an apparently high geothermal gradient characterizes the upper crust of weakly developed, normal layer 2 thickness dipping reflector sequences, and an apparently low geothermal gradient characterizes the upper crust of strongly developed, anomalously thick layer 2 dipping reflector sequences.

Such a systematic difference in the temperature history of the two kinds of crust is likely to be present. A thin extrusive layer implies that even upper crustal material originated from a position relatively close to the original rift zone and achieved its maximum temperature at an early stage due to its proximity to the hot rift zone. Upper crustal material within a crust that shows a relatively thick extrusive layer, however, originated in a rift-distal position and achieved its maximum temperature at a later stage due to burial within a zone of relatively low geothermal gradient. Data from the E Greenland continental plateau basalts suggesting a palaeogeothermal gradient of around 40°C/km may indicate the lower limit for the geothermal gradient of rift-distal crust, in good agreement with the estimates inferred by this study (Larsen *et al.*, in press).

We further conclude that initiation of seafloor spreading in the NE Atlantic started fairly abruptly in response to massive upwelling of volcanic material through crust and lithosphere that had undergone relatively little mechanical stretching, as opposed to other margins that may have developed through intense mechanical stretching, subsidence, rift-basin formation and eventually volcanism and continental separation.

Finally, we conclude that our findings of seawards-dipping sub-basement reflectors of E Greenland are generally in good agreement with information from the conjugate passive margin (Andersen 1988; Skogseid & Eldholm 1988; White 1988). Seawards-dipping reflectors showing a symmetric distribution to the E Greenland sequence are present on the NW European margin. Exceptions to this general symmetric occurrence are the lack of reported dipping reflectors on the Faeroe–Iceland Ridge and on the western part of the Jan Mayen Ridge but, to our knowledge, no deep penetration reflection seismic information is presently available from these areas.

ACKNOWLEDGEMENTS: The funding of the seismic programme through the EEC, Directorate of Energy and the Danish Ministry of Energy is greatly acknowledged. Useful comments on the manuscript and our interpretation of deep crustal information were obtained from A. C. Morton, R. S. White and G. D. Spence and are also gratefully acknowledged. Western Geophysical Co. of America is acknowledged for permission to use seismic data from their speculative seismic survey N of Iceland and for assistance through the seismic programme. V. Hermansen, B. S. Hansen and J. Lautrup assisted in the preparation of the manuscript. The paper is published with the permission of the director of the Geological Survey of Greenland.

References

ANDERSEN, M. S. 1988. Late Cretaceous and early Tertiary extension and volcanism around the Faeroe Islands. *In:* MORTON, A. C. & PARSON, L. M. (eds) *Early Tertiary Volcanism and the Opening of the NE Atlantic.* Geological Society of London, Special Publication, **39**, pp. 115–122.

BOTT, M. H. P. & GUNNARSSON, K. 1980. Crustal structure of the Iceland–Faeroe ridge. *Journal of Geophysics,* **47**, 221–227.

BROOKS, C. K. & NIELSEN, T. F. D. 1982. The E Greenland continental margin: a transition between oceanic and continental magmatism. *Journal of the Geological Society of London,* **139**, 265–275.

CHRISTENSEN, N. 1978. Ophiolites, seismic velocities and oceanic crustal structure. *Tectonophysics,* **47**, 131–157.

ELDHOLM, O., THIEDE, J., TAYLOR, E. & THE ODP LEG 104 SCIENTIFIC PARTY. 1986. Formation of the Norwegian Sea. *Nature,* **319**, 360–361.

FEATHERSTONE, P. S., BOTT, M. H. P. & PEACOCK, J. H. 1977. Structure of the continental margin of southeastern Greenland. *Geophysical Journal of the Royal Astronomical Society,* **48**, 15–27.

HARRISON, C. G. A. 1968. Formation of magnetic anomaly patterns by dike injection. *Journal of Geophysical Research,* **73**, 2137–2142.

HINZ, K. 1981. A hypothesis on terrestrial catastrophes. Wedges of very thick oceanward dipping layers beneath passive continental margins—their origin and palaeoenvironmental significance. *Geologische Jahrbuch,* **E22**, 3–28.

LARSEN, H. C. 1980. Geological perspectives of the East Greenland continental margin. *Bulletin of the Geological Society of Denmark,* **29**, 77–101.

—— 1983. Marine geophysical investigations offshore East Greenland. *Grønlands Geologiske Undersøgelse* Rapport, **115**, 93–100.

—— 1984. Geology of the East Greenland Shelf. *In:* SPENCER, A. M., *et al.* (eds) *Petroleum Geology of the North European Margin.* Graham & Trotman, London, pp. 329–339.

—— 1988. A multiple and propagating rift model for the NE Atlantic. *In:* MORTON, A. C. & PARSON, L. M. (eds) *Early Tertiary Volcanism and the Opening of the NE Atlantic.* Geological Society of London, Special Publication, **39**, pp. 157–158.

LARSEN, L. M., WATT, W. S. & WATT, M. 1987. Geology and petrology of the lower Tertiary Plateau Basalts of the Scoresby Sund region, East Greenland. *Bulletin of the Geological Survey of Greenland.* In press.

MATTHEWS, D. H. & BATH, J. 1967. Formation of magnetic anomaly pattern of mid-Atlantic ridge. *Geophysical Journal of the Royal Astronomical Society,* **13**, 349–357.

MCKENZIE, D. 1978. Some remarks on the development of sedimentary basins. *Earth and Planetary Science Letters,* **40**, 25–32.

MUTTER, J. C., TALWANI, M. & STOFFA, P. L. 1984. Evidence for a thick oceanic crust adjacent to the Norwegian margin. *Journal of Geophysical Research,* **89**, 483–502.

—— & ZEHNDER, C. M. 1988. Deep crustal structure and magmatic process: the inception of seafloor spreading in the Norwegian–Greenland Sea. *In:* MORTON, A. C. & PARSON, L. M. (eds) *Early Tertiary Volcanism and the Opening of the NE Atlantic.* Geological Society of London, Special Publication, **39**, pp. 35–48.

NUNNS, A. G., TALWANI, M., LORENTZEN, G. R., VOGT, P. R., SIGURGEIRSSON, T., KRISTJÁNSSON, L., LARSEN, H. C. & VOPPEL, D. 1983. Magnetic anomalies over Iceland and surrounding seas. *In:* BOTT, M. H. P., SAXOV, S., TALWANI, M. & THIEDE, J. (eds) *Structure and Development of the Greenland–Scotland Ridge.* Plenum Press, New York, pp. 661–678.

PÁLMASON, G. 1973. Kinematics and heat flow in a volcanic rift zone, with application to Iceland. *Geophysical Journal of the Royal Astronomical Society,* **33**, 451–481.

—— 1980. A continuum model of crustal generation in Iceland; Kinematic aspects. *Journal of Geophysics,* **47**, 7–18.

—— 1986. Model of crustal formation in Iceland, and application to submarine mid-ocean ridges. *In:* VOGT, P. R. & TUCHOLKE, B. E. (eds) *The Geology of North America, M, The Western North Atlantic Regions.* The Geological Society of America, pp. 87–97.

—— & SÆMUNDSSON, K. 1974. Iceland in relation to the Mid-Atlantic Ridge. *Annual Review of Earth and Planetary Science,* **2**, 25–50.

ROBERTS, D. G., BACKMAN, J., MORTON, A. C., MURRAY, J. W. & KEENE, J. B. 1984. Evolution of volcanic rifted margins: Synthesis of Leg 81 results on the west margin of Rockall Plateau. *In:* ROBERTS, D. G., SCHNITKER, D. *et al. Initial Reports of the Deep Sea Drilling Project.* US Government Printing Office, Washington, **81**, pp. 883–911.

ROBINSON, P. T., HALL, J. M. CHRISTENSEN, N. I., GIBSON, I. L., FRIDLEIFSSON, I. B., SCHMINCKE, H.-U. & SCHÖNHARTING, G. 1982. The Iceland research drilling project: synthesis of results and implications for the nature of Icelandic and oceanic crust. *Journal of Geophysical Research,* **87**, 6657–6667.

SKOGSEID, J. & ELDHOLM, O. 1988. Early Cainozoic evolution of the Norwegian volcanic passive margin and the formation of marginal highs. *In:* MORTON, A. C. & PARSON, L. M. (eds) *Early Tertiary Volcanism and the Opening of the NE Atlantic.* Geological Society of London, Special Publication, **39**, pp. 49–56.

SMYTHE, D. K. 1983. Faeroe–Shetland Escarpment and continental margin north of the Faeroes. *In:* BOTT, M. H. P., SAXOV, S., TALWANI, M. & THIEDE, J. (eds), *Structure and Development of the Greenland–Scotland Ridge.* Plenum Press, New York, pp. 109–119.

VAIL, P. R., MITCHUM, R. M., TODD, R. G., WIDMIER, J. M., THOMSON, S., SANGREE, J. B., BUBB, J. W.

& HATLELID, W. G. 1977. Seismic stratigraphy and global changes of sea level. *In*: PAYTON, C. E. (ed) *Seismic Stratigraphy—Applications to Hydrocarbon Exploration.* Memoir of the American Association of Petroleum Geologists, **26**, pp. 49–212.

WHITE, R. S. 1988. A hot-spot model for early Tertiary volcanism in the N Atlantic. *In:* MORTON, A. C. & PARSON, L. M. (eds) *Early Tertiary Volcanism and the Opening of the NE Atlantic.* Geological Society of London, Special Publication, **39**, pp. 3–13.

H. C. LARSEN & S. JAKOBSDOTTIR, The Geological Survey of Greenland, Øster Voldgade 10, DK-1350, Copenhagen K, Denmark.

Late Cretaceous and early Tertiary extension and volcanism around the Faeroe Islands

M. S. Andersen

SUMMARY: In an area around the Faeroe Islands, seismic reflection profiles have been used to map the depth to the surface of the early Tertiary basalts and the distribution of three different types of reflector configurations within the basalts. Using this information, the early Tertiary physiography of the area has been deduced.

The evolution of the seaward-dipping reflector sequence N and NW of the Faeroe Islands is discussed. In models of simple rifting of a normal lithosphere, extension is accompanied by significant subsidence (probably more than 2 km) before a large quantity of magma is produced. It is suggested that in the study area, subsidence prior to magma generation was reduced or avoided due to an additional increase in the geothermal gradient above that predicted by rifting models.

Finally, the relation between extension, volcanism and subsidence before and after the early Tertiary opening of the NE Atlantic Ocean is discussed.

The Faeroe Islands are situated to the NE of the Hatton–Rockall Bank (Fig. 1). The morphology of the Rockall–Faeroe Plateau area, in contrast to most continental shelf areas, is characterized by high amplitude topographic features with a wavelength of 100–200 km (Roberts *et al.* 1979). This morphology is apparently caused by the very small denudation areas available to supply clastic material to fill the structural lows. In 1985 the Geological Survey of Denmark started a regional interpretation of geophysical data in an area around the Faeroe Islands. The database for this study consists of magnetic, gravimetric and seismic reflection data. Figure 2 shows the distribution of seismic reflection data within the study area.

Due to the lack of well control (only one well, DSDP Hole 336, has penetrated the sediments within the study area) (Talwani *et al.* 1976), stratigraphic correlation has to rely on the principles of seismic stratigraphy (Vail *et al.* 1977) and tentative correlations to wells on the eastern side of the Faeroe–Shetland Channel. The use of seismic stratigraphy in this area is somewhat hampered by interference between eustatic sea-level changes and subsidence.

Internal reflector configuration of the basaltic basement

In an attempt to get additional information on the subsidence, the topography of the surface of the early Tertiary basalts and their internal reflector configuration were studied in some detail. Three distinct reflector configurations were identified in this study:

(1) Parallel-bedded reflector sequences.
(2) Prograding reflector sequences.
(3) Seaward-dipping reflector sequences.

These reflector configurations were originally identified in this area by Smythe (1983).

The parallel-bedded reflector sequences were interpreted as subaerially-erupted plateau basalts. This is in agreement with the interpretation of the basaltic rock content of dredges made in the area (Waagstein 1977, 1988). The prograding reflector sequences were interpreted as subaerial basalt flows, which were cooled at the contact with the sea, became more viscous and therefore created steep escarpments. Smythe (1983) originally identified this reflector configuration along the northern and central parts of the Faeroe–Shetland Escarpment. In this work, it is also identified along the southern part of the Faeroe–Shetland Escarpment, within the volcanic basement (ca. 200 ms below the reflector which defines the surface of the volcanic basement) in the northern part of the Rockall Trough. In addition it is recorded within a buried seamount (Figs 3 & 4), originally termed plutonic centre B by Roberts *et al.* (1983, see Fig. 9) and interpreted by them as an intrusive centre covered by a strato volcano. The author suggests that the less ambiguous name, 'Sigmundur Seamount' is used for this feature.

From MORTON, A. C. & PARSON, L. M. (eds), 1988, *Early Tertiary Volcanism and the Opening of the NE Atlantic,* Geological Society Special Publication No. 39, pp. 115–122.

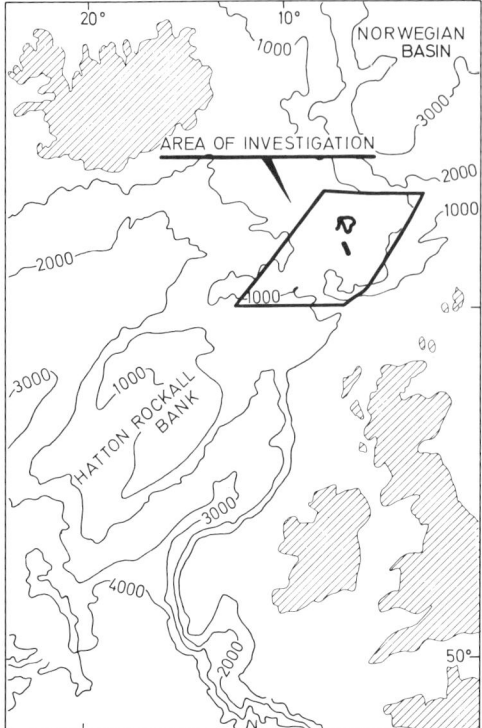

FIG. 1. Location of study area on bathymetric base map. Contour interval is 1000 m.

The origin of basalt escarpments E of the Faeroe Islands and S of the Wyville–Thomson Ridge

If the Faeroe–Shetland Escarpment is not a tectonic feature, is it then possible that other escarpments in the Faeroe Islands region could also be interpreted as primary volcanological features? Within the Faeroese shelf, W and SW

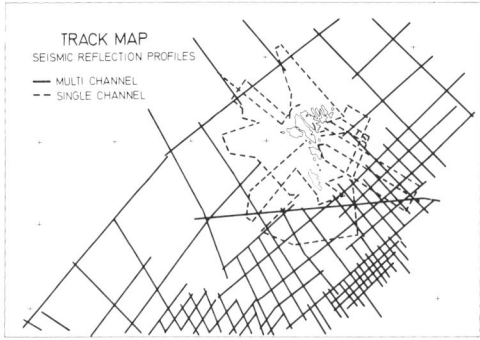

FIG. 2. Seismic reflection profiles used in this study. Full lines are multichannel profiles, stippled lines are single channel profiles.

of the Faeroe–Shetland Escarpment, reflectors below the surface of the basalts are parallel-bedded. On the surface of the basalts, however, a few small escarpments (ca. 50 300 m) are seen. These small escarpments are not associated with prograding reflectors. Below most of these small escarpments the underlying reflectors are easily traced uninterrupted. In this area, there is no other evidence that could be interpreted as the results of erosion of the basalt surface. For this reason, these small escarpments are interpreted as lava fronts. These lava fronts were submerged in a transgressing sea before a prograding reflector sequence was developed, and new small escarpments were formed at a new shoreline nearer to the Faeroe Islands, each small escarpment reflecting the approximate location of the shoreline at the time it was created. S of the Wyville–Thomson Ridge, several large, steep escarpments with approximately linear trends (ca. WNW–ESE) are identified. Some of these can be traced for a considerable distance (see Fig. 3). These escarpments are not associated with a prograding reflector sequence. One of these escarpments ends abruptly at the Sigmundur Seamount. Roberts *et al.* (1983) interpreted these escarpments as faults.

Unless the basalts are considerably older on the Wyville–Thomson Ridge than on the Faeroe Islands, any fault activity on the Wyville–Thomson Ridge must be of late Palaeocene–Eocene age, and should probably be correlated with the Laramide inversion phase in the North Sea rift basins. However, it is also possible that some or all of these escarpments may represent erosional remnants of primary volcanic escarpments.

Some evidence in favour of this hypothesis may be found in the apparent continuation of reflectors below two small escarpments on the Wyville–Thomson Ridge (see Fig. 4).

Discussion of the origin of the seaward-dipping reflectors

A continuous band of seaward-dipping reflectors can be traced from Jan Mayen Ridge in the N to the northern flank of the Iceland–Faeroe Ridge in the S (Smythe 1983). This band is part of a NE-trending zone of seaward-dipping reflectors which can be traced along most of the eastern margin of the NE Atlantic Ocean (Hinz 1981; Talwani *et al.* 1983). Seaward-dipping reflectors have not yet been observed immediately S of the Iceland–Faeroe Ridge, but this may be due to the very poor coverage of multi-channel seismic reflection profiles.

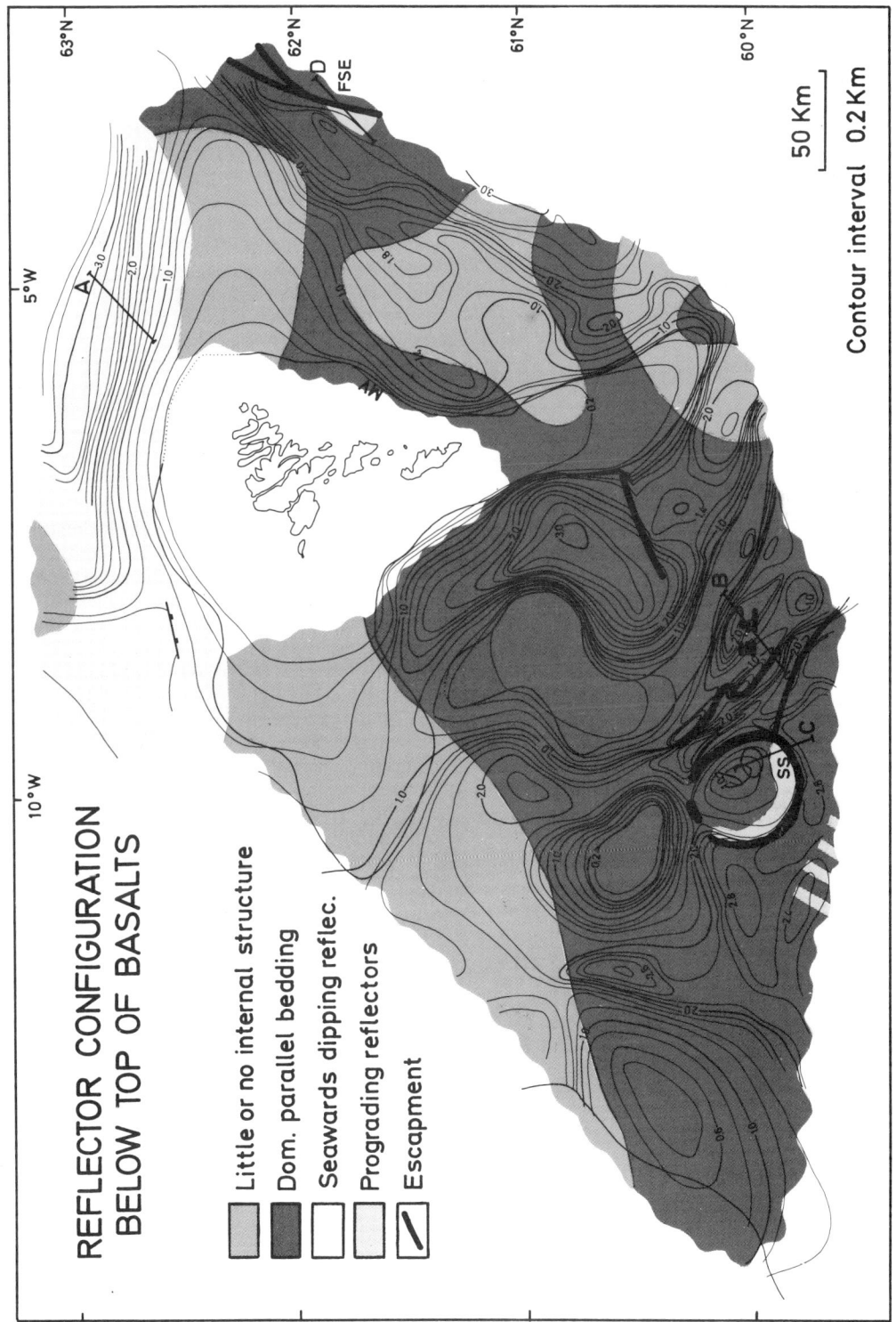

FIG. 3. Depth to the basaltic 'basement' and distribution of reflector configuration within the study area on a base map showing depth to the basaltic basement. Contour interval is 200 m. FSE: Faeroe–Shetland Escarpment; MV: Marginal Valley E of Faeroe Islands; SS: Sigmundur Seamount; and WTR: Wyville–Thomson Ridge.

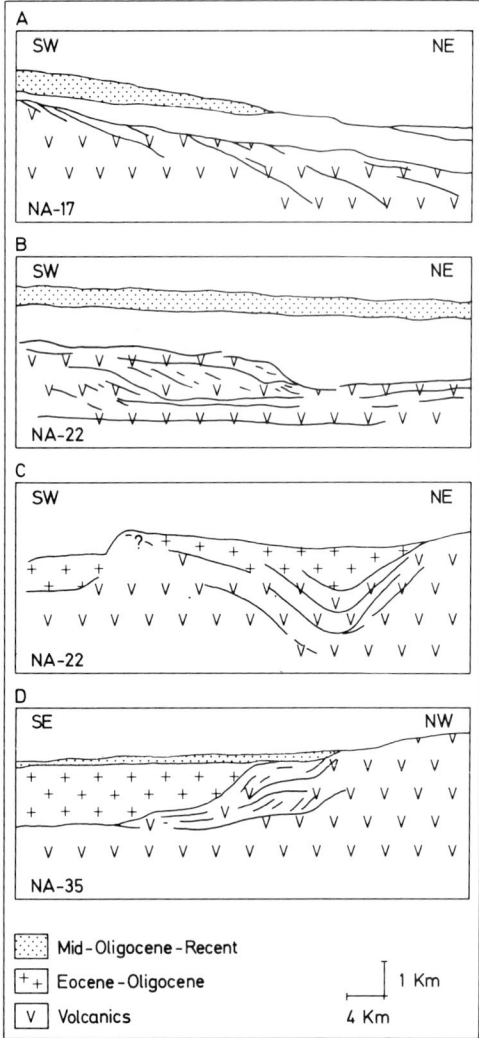

FIG. 4. Four profiles constructed from interpretation of seismic sections. Locations are shown in Fig. 3. (A) Section through the seaward-dipping reflector sequence N of Faeroe Islands. (B) Section across a small basin on the Wyville–Thomson Ridge. Two small escarpments near the centre of the basin are interpreted as primary volcanic features, possibly slightly modified by erosion. (C) Section showing prograded bedding in the basalts at the southern part of Faeroe–Shetland Escarpment. (D) Section showing prograded bedding in Sigmundur Seamount.

N of the Faeroe Islands, the trend of this zone of seaward-dipping reflectors is E–W (see Fig. 3), deviating ca. 45° from the general NW–SE trend of the dipping reflectors. The dip is 2–18° to the N. We can correct the dip for the tilt of the surface of the basalts below the inner part of the slope; the range of dip is then reduced. Just below the surface of the basalts this fossil dip is 1–2°, increasing to ca. 12–15° at 2 km depth. An interpretation of a typical section through the seaward-dipping reflectors is shown in Fig. 4.

N and NW of the Faeroe Islands the seaward-dipping reflectors coincide with a narrow gravity high (Fleischer 1971, see Fig. 7). A similar relation between the seaward-dipping reflectors and gravity is not seen further to the N; this indicates that there is a significant difference in the density distribution along the eastern margin of the NE Atlantic Ocean. The origin of the seaward-dipping reflectors has been much debated. However, it is generally agreed that the seaward-dipping reflectors mainly consist of subaerial or shallow submarine tholeiitic basalts (Hinz 1981; Talwani *et al.* 1983; Roberts *et al.* 1984; Shipboard scientific party 1986).

Hinz (1981) and Roberts *et al.* (1984) suggested that the basalts were erupted during a continental extension phase just before the seafloor spreading. Smythe (1983) suggested that the continuum model of Pálmason (1981) for the generation of oceanic crust along mid-oceanic ridges is also valid in the early stages of seafloor spreading. The seaward-dipping reflectors consist of basalts erupted during the early stage of seafloor spreading. It should be emphasized that Smythe's model predicts an excellent balance between subsidence and volume of erupted volcanic material, in contrast to crustal extension models for the generation of seaward-dipping reflectors. Smythe's model also accounts for seaward-dipping reflectors below the Greenland–Iceland–Faeroe Ridge. At the Greenland end of the ridge, seaward-dipping reflectors can be traced as far out as magnetic anomaly 13 (Larsen & Jakobsdottir 1988), and at the Faeroese end of the ridge, seaward-dipping reflectors can be traced at least 50 km NW of the ocean–continent transition, as located by Bott *et al.* (1976).

Neither of the above mentioned models takes into consideration the increase in geothermal gradient necessary to melt the huge amount of magma that has formed the seaward-dipping reflector sequences. Without additional increase in the geothermal gradient it is not feasible that magma production would start before the stretching of the lithosphere reached a certain threshold value. The first magma would be produced when the stretching factor was approximately 2, and large amounts of tholeiitic magma would be produced when the stretching factor was approximately 4 (Dewey 1982). According to McKenzie (1978), however, stretching of normal continental crust by a factor of 2 would produce a 1.7 km basin subsidence, and stretching by a factor of 4 would produce approximately 2.5 km of subsidence (assuming the basin floor originally was at or below sea-level). This agrees with observations along 'Biscay' type continental margins (Montadert *et al.* 1979). It is, however, very difficult to accept subsidence of this magnitude prior to the main volcanic activity along continental margins that is characterized by seaward-dipping reflector sequences.

The origin of seaward-dipping reflector sequences is discussed in more detail elsewhere in this volume (Mutter & Zehnder 1988; Skogseid & Eldholm 1988; White 1988).

The Faeroe–Shetland Escarpment and the marginal valley E of the Faeroe Islands

The Faeroe–Shetland Escarpment can be traced as a lobate escarpment from near the Jan Mayen fracture zone in the N to the northern part of the Faeroe–Shetland Channel in the S (Talwani & Eldholm 1972; Smythe 1983). The northern part of the escarpment is roughly parallel to the zone of seaward-dipping reflectors below the Møre Plateau, but S of approximately 62.5°N, where the zone of seaward-dipping reflectors turns E–W, the Faeroe–Shetland Escarpment continues with a SW–NE trend into the Faeroe–Shetland Channel. Along this southern section the height of the escarpment diminishes rapidly, and S of 61°N the escarpment vanishes. As mentioned earlier, the escarpment is characterized by a sequence of prograding reflectors (see Fig. 4).

Smythe (1983) suggested that the Faeroe–Shetland Escarpment was formed by basaltic lavas stratigraphically equivalent to the middle and upper basalt formations on the Faeroe Islands, which flowed into the sea above underlying parallel-bedded reflectors. These lower reflectors were interpreted as subaerially erupted basalts stratigraphically equivalent to the Lower Basalt Formation.

In the time between the eruption of the lower and middle basalt formations, the Møre Basin and the northern Faeroe–Shetland Channel subsided sufficiently to allow transgression from the N, creating the environment in which the Faeroe–Shetland Escarpment was formed. Talwani *et al.* (1983, Fig. 1.b) suggested that the Faeroe–Shetland Escarpment continued S of 62°N on a trend which coincides with that of the marginal valley on the shelf E of the Faeroe Islands (Nielsen *et al.* 1979). This feature is, however, distinct from the Faeroe–Shetland Escarpment.

The dip of the surface of the basalts in the marginal valley is never more than 14° and generally less than 5°, whereas the dip of the Faeroe–Shetland Escarpment exceeds 20°. The Tertiary sediments above the basalts are parallel with the basalts in the marginal valley and the basalts on the western side of the marginal valley are truncated by Pliocene–Pleistocene erosion (Nielsen *et al.* 1979, see Fig. 4).

Except for the Faeroe–Shetland Escarpment as defined by Smythe (1983) and the previously mentioned small escarpments, the dip of the surface of the Tertiary plateau basalts E of the Faeroe Islands is considered to be a consequence of the Tertiary subsidence of the Faeroe–Shetland Channel.

Discussion of the relation between extension, volcanism and subsidence

Some important events in the period around the early Tertiary opening of the NE Atlantic Ocean are shown in Fig. 5. Both the amount of pre-spreading extension and the width of the extension zone are poorly constrained, especially on the northern and northwestern margin of the Faeroe Islands:

(1) In the late Cretaceous two *en echelon* extension systems developed around the Faeroe Islands, the Møre extension zone and the Kap Gustav Holm extension zone. Due to the thick basaltic cover and interference with older extension systems, the exact time of the beginning of the extension in these zones is difficult to establish. However, Maastrichtian igneous activity (72.5–68.4 My) and a Campanian uplift is reported in the Faeroe–Shetland Channel (Hitchen & Ritchie 1987). This indicates that the Møre extension zone was active in the Campanian.

Mitchell (1978) has reported Campanian basaltic dykes in the Angmagssalik district, E Greenland. This may indicate that the Kap Gustav Holm extension zone was also active in the Campanian. Marine Cenomanian–Danian sediments in the Kangerdlugssuaq area are considered additional evidence for Upper Cretaceous activity along the Kap Gustav Holm extension zone.

(2) At the end of the Campanian uplift, a connection between the two extension zones was formed N of the Faeroe Islands. The southern part of the Møre extension zone and the northern part of the Kap Gustav Holm extension zone became extinct. This connection is on the same trend as a transfer zone N of the Viking Graben (Nelson & Lamy 1987).

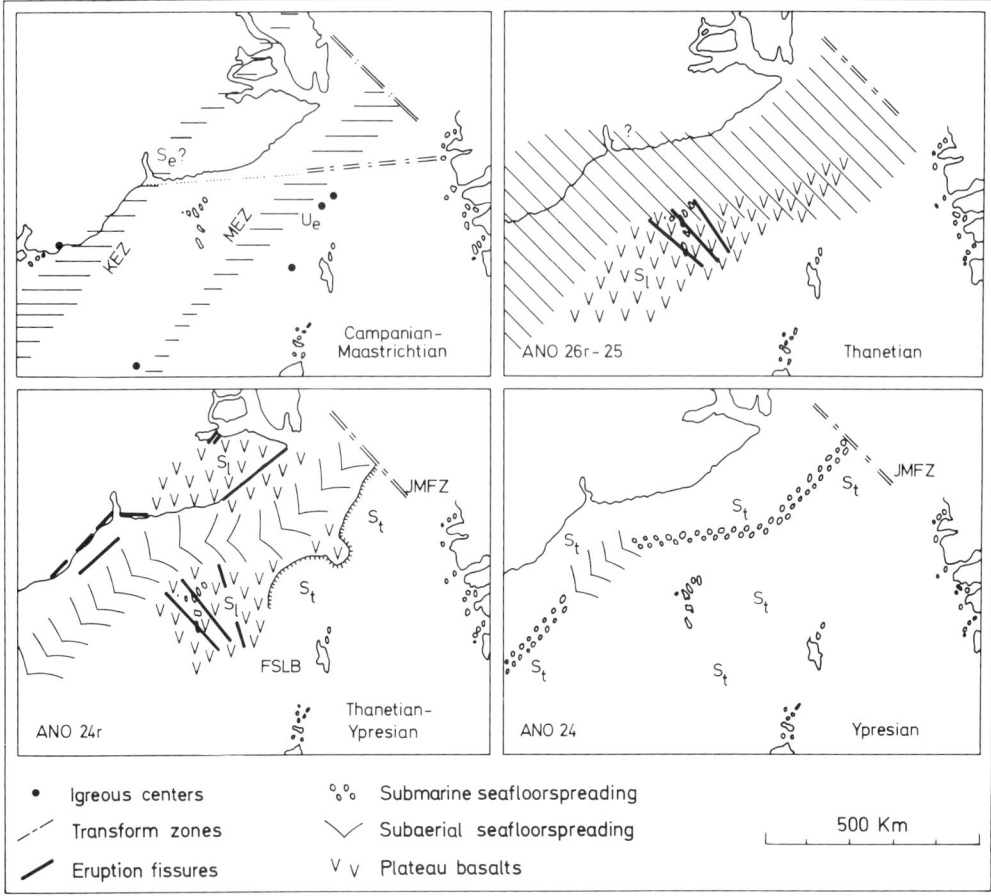

FIG. 5. Cartoon showing four stages during the opening of the NE Atlantic Ocean. Areas affected by crustal extension in Campanian–Maastrichtian and in Thanetian are shown with hatching. JMFZ: Jan Mayen Fracture Zone; KEZ: Kap Gustav Holm Extension Zone; MEZ: Møre Extension Zone. S_e, S_l, S_t and U_e indicate areas under structural, load or thermal subsidence and structural uplift. See also text for further discussion of this figure.

Extension continued along the active extension zone, and a large quantity of basaltic melt was produced in the 'rising' asthenosphere. Due to the load of the overlying lithosphere, magma was forced into the central parts of the extension zone.

(3) According to Waagstein (1988) eruption of the basaltic lavas that formed the lower basalt formation on the Faeroe Islands started in magnetic anomaly chron C26R (late Danian–early Thanetian). On the Faeroe Islands there is evidence that eruptions occurred along NNW–SSE to NW–SE trending fissures (Noe-Nygaard & Rasmussen 1970). However, the main magma chambers were probably located N and NW of the islands. During eruptions, lava migrated to

the SE through the fissures before they emerged at the surface at great distances from the feeding magma chamber. A similar eruptive mechanism has been proposed for volcanoes in the SE Icelandic volcanic zone.

(4) Volcanic activity was interrupted for a short period, evidenced by the 10 m thick coal-bearing formation on the Faeroe Islands, which according to Waagstein (1988) was deposited during the early part of C24R.

It is likely that this interruption in the igneous activity was contemporary with the onset of actual seafloor spreading N and W of the Faeroe Islands. During the early stages of seafloor spreading, eruptions along the spreading axis were subaerial or shallow submarine, and a seaward-dipping sequence was formed as described above. Soon after the start of seafloor spreading magma production along the oblique spreading segment N of the Faeroe Islands increased to levels that allowed renewed volcanic activity on the islands (middle and upper basalt formation). Due to thermal subsidence, the Faeroe–Shetland Basin was flooded from the N, and basalt flows formed the Faeroe–Shetland Escarpment when entering the sea. It is possible that a short-lived land bridge across the southern part of the Faeroe–Shetland Channel did exist at the end of Palaeocene and in the beginning of Eocene (Roberts *et al.* 1983).

(5) At the end of C24R the spreading axis submerged below sea-level N of the Faeroe Islands. This gave rise to a change in eruption type, and the erupted volcanics were concentrated near the spreading axis. NW of the Faeroe Islands subaerial or shallow submarine spreading continued. In the Faeroe–Shetland Channel subsidence gave rise to the present tilt to the E of the basalts, which is seen on the Faeroe Islands and the eastern shelf.

References

BOTT, M. H. P., NIELSEN, P. H. & SUNDERLAND, J., 1976. Converted P-waves originating at the continental margin between the Iceland–Faeroe Ridge and the Faeroe Block. *Geophysical Journal of the Royal Astronomical Society*, **44**, 229–238.

DEWEY, J. F., 1982. Plate tectonics and the evolution of the British Isles. *Journal of the Geological Society of London*, **139**, 371–412.

FLEISCHER, U., 1971. Gravity surveys over the Reykjanes Ridge and between Iceland and the Faeroe Islands. *Marine Geophysical Researches*, **1**, 314–327.

HITCHEN, K. & RITCHIE, J. D., 1987. Geological review of the West Shetland area. *In*: BROOKS, J. & GLENNIE, K. W. (eds) *Petroleum Geology of NW Europe*. Graham & Trotman, London, pp. 737–749.

HINZ, K., 1981. A hypothesis on terrestrial catastrophes. Wedges of very thick oceanward-dipping layers beneath passive continental margins. Their origin and palaeo-environmental significance. *Geologisches Jahrbuch*, **E22**, 3–28.

LARSEN, H. C. & JAKOBSDOTTIR, S., 1988. Distribution, crustal properties and significance of seaward-dipping sub-basement reflectors off E Greenland. *In*: MORTON, A. C. & PARSON, L. M. (eds) *Early Tertiary Volcanism and the Opening of the NE Atlantic*. Geological Society of London, Special Publication, **39**, pp. 95–114.

McKENZIE, D. P., 1978. Some remarks on the development of sedimentary basins. *Earth and Planetary Science Letters*, **40**, 25–32.

MITCHELL, J. G., 1978. Potassium-argon ages from Phanerozoic basic dykes in South-East Greenland. *Geological Survey of Greenland Report*, **90**, 141–146.

MONTADERT, L., ROBERTS, D. G., DE CHARPAL, O. & GUENNOC, P., 1979. Rifting and subsidence of the northern continental margin of the Bay of Biscay. *In*: MONTADERT, L., ROBERTS, D. G. *et al. Initial Reports of the Deep Sea Drilling Project*. US Government Printing Office, Washington, **48**, pp. 1025–1060.

MUTTER, J. C. & ZEHNDER, C. M., 1988. Deep crustal structure and magmatic processes: the inception of seafloor spreading in the Norwegian Greenland Sea. *In*: MORTON, A. C. & PARSON, L. M. (eds) *Early Tertiary Volcanism and the Opening of the NE Atlantic*. Geological Society of London, Special Publication, **39**, pp. 35–48.

NELSON, P. & LAMY, J-M., 1987. The Møre/West Shetland area: a review. *In*: BROOKS, J. & GLENNIE, K. W. (eds) *Petroleum Geology of NW Europe*. Graham & Trotman, London, pp. 775–784.

NIELSEN, P. H., WAAGSTEIN, R., RASMUSSEN, J. & LARSEN, B., 1979. Marine seismic investigations of the shelf around the Faeroe Islands. *Fróðskaparrit*, **27**, 102–112.

NOE-NYGAARD, A. & RASMUSSEN, J., 1970. Geology of the Faeroe Islands (Pre-Quaternary). *Geological Survey of Denmark (I series)*, **25**, 143 pp.

PÁLMASON, G., 1981. A continuum model of crustal generation in Iceland; kinematic aspects. *Journal of Geophysics*, **47**, 7–18.

ROBERTS, D. G., HUNTER, P. M. & LAUGHTON, A. S., 1979. Bathymetry of the northeast Atlantic: continental margin around the British Isles. *Deep Sea Researches*, **26A**, 417–428.

——, BOTT, M. H. P. & URUSKI, C., 1983. Structure and origin of the Wyville-Thomson Ridge. *In*: BOTT, M. H. P., SAXOV, S., TALWANI, M. & THIEDE, J. (eds) *Structure and Development of the Greenland-Scotland Ridge*. Plenum Press, New York, pp. 133–158.

——, BACKMAN, J., MORTON, A. C., MURRAY, J. W. & KEENE, J. B., 1984. Evolution of volcanic rifted margins: synthesis of Leg 81 results on the west margin of Rockall Plateau. *In*: ROBERTS, D. G., SCHNITKER, D. *et al. Initial Reports of the Deep Sea Drilling Project*. US Government Printing Office, Washington, **81**, pp. 883–911.

SHIPBOARD SCIENTIFIC PARTY (ODP LEG 104), 1986. Formation of the Norwegian Sea. *Nature*, **319**, 360–361.

SKOGSEID, J., & ELDHOLM, O., 1988. Early Cainozoic evolution of the Norwegian volcanic passive margin and the formation of marginal highs. *In*: MORTON, A. C. & PARSON, L. M. (eds) *Early Tertiary Volcanism and the Opening of the NE Atlantic*. Geological Society of London, Special Publication, **39**, pp. 49–56.

SMYTHE, D. K., 1983. Faeroe–Shetland Escarpment and continental margin north of the Faeroes. *In*:

BOTT, M. H. P., SAXOV, S., TALWANI, M. & THIEDE, J. (eds) *Structure and Development of the Greenland–Scotland Ridge*. Plenum Press, New York, pp. 77–90.

TALWANI, M. & ELDHOLM, O., 1972. Continental margin of Norway: a geophysical study. *Bulletin of the Geological Society of America*, **83**, 3575–3608.

——, UDINTSEV, G., *et al.*, 1976. *Initial Reports of the Deep Sea Drilling Project*. US Government Printing Office, Washington, **38**, 1256 pp.

——, MUTTER, J. & HINZ, K., 1983. Ocean continent transition under the Norwegian Sea. *In*: BOTT, M. H. P., SAXOV, S., TALWANI, M. & THIEDE, J. (eds) *Structure and Development of the Greenland–Scotland Ridge*. Plenum Press, New York, pp. 121–131.

VAIL, P. R., MITCHUM, R. M., TODD, R. G., WIDIMIER, J. M., THOMPSON, S., SANGREE, J., BUBB, J. N. &

HATELID, W. G., 1977. Seismic stratigraphy and global changes in sea level. *In*: PLAYTON, C. E. (ed) *Seismic Stratigraphy—Application to Hydrocarbon Exploration*. Memoir of the American Association of Petroleum Geologists, **26**, 49–212.

WAAGSTEIN, R., 1977. *The Geology of the Faeroese Plateau*. Thesis (unpublished). Geologisk Central-institut, University of Copenhagen.

—— 1988. Structure, composition and age of the Faeroe basalt plateau. *In*: MORTON, A. C. & PARSON, L. M. (eds) *Early Tertiary Volcanism and the Opening of the NE Atlantic*. Geological Society of London, Special Publication, **39**, pp. 225–238.

WHITE, R. S., 1988. A hot-spot model for early Tertiary volcanism in the N Atlantic. *In*: MORTON, A. C. & PARSON, L. M. (eds) *Early Tertiary Volcanism and the Opening of the NE Atlantic*. Geological Society of London, Special Publication, **39**, pp. 3–13.

M. S. ANDERSEN, Geological Survey of Denmark, Faeroe Division, Debesartrød, FR-100 Tórshavn, Faeroe Islands.

Petrochemistry and isotope geochemistry of early Palaeogene basalts forming the dipping reflector sequence SW of Rockall Plateau, NE Atlantic

R. J. Merriman, P. N. Taylor & A. C. Morton

SUMMARY: A transect across the passive margin SW of Rockall Plateau was drilled during DSDP Leg 81. Site 555 was located on continental crust of the Rockall Plateau. Sites 552 and 553 were located over the wedge of dipping reflectors, and Site 554 drilled the outer high at the transition from the thick wedge of dipping reflectors to normal oceanic crust. New trace-element and Pb-isotope data are presented which characterize three basalt types: (1) Dipping reflector basalts resemble depleted N-type MORBs, strongly depleted in light rare earth elements (LREEs) and other incompatible elements; they have been contaminated by U-depleted continental crust, or alternatively derived from depleted subcontinental lithosphere. (2) Outer high basalts are less strongly depleted N-type MORB and show no evidence of crustal contamination. (3) Site 552 basalts postdate the onset of subsidence of SW Rockall margin; they are relatively evolved N-type MORB derived from a source similar to that of the outer high basalts but show evidence of crustal contamination. Genesis of the dipping reflector basalts is best explained by a stretching/rifting model, not the 'subaerial seafloor spreading' model.

In broad terms, there are two alternative hypotheses that provide plausible explanations for the dipping reflector sequences that characterize Rockall-type passive margins. One regards the sequences as having formed by subaerial seafloor spreading, that is, by extrusion of basalt from a mid-ocean ridge lying above sea-level, analogous to the present situation in Iceland. Talwani *et al.* (1981), Mutter *et al.* (1982) and Smythe (1983) have all invoked such a model for dipping reflector sequences in the NE Atlantic. In contrast, Hinz (1981) proposed that the dipping reflectors consist of basalts and volcaniclastics erupted during the rift stage from centres along the eventual line of separation, the fundamental difference from the 'subaerial seafloor spreading' hypothesis being that, in this case, the dipping sequence is underlain by attenuated continental crust. Roberts *et al.* (1984) proposed a variant of this hypothesis in which extrusion occurred not from a single source but from a series of dykes whose density of intrusion increased toward the eventual line of separation.

During DSDP Leg 81, drilling took place at four sites comprising a transect across the dipping reflector sequence SW of Rockall Plateau (Roberts, Schnitker *et al.* 1984). Sites 552 and 553 were located in Edoras Basin, over the wedge of oceanward-dipping reflectors, with Site 552 in the more landward position (Fig. 1). Sites 552 and 553 are located close to two sites, 404 and

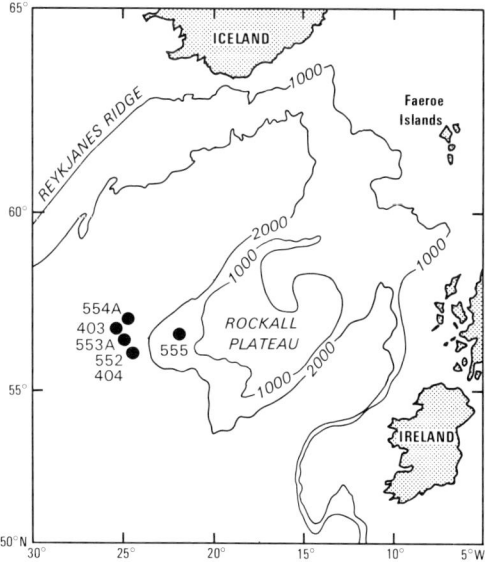

FIG. 1. Location of DSDP Leg 81 and Leg 48 sites on the SW margin of Rockall Plateau. Isobaths in metres.

403 respectively, also drilled in Edoras Basin during DSDP Leg 48 (Montadert, Roberts *et al.* 1979). Site 554 is located on a structure termed the 'outer high', which separates normal oceanic crust from the enigmatic dipping reflector se-

From MORTON, A. C. & PARSON, L. M. (eds), 1988, *Early Tertiary Volcanism and the Opening of the NE Atlantic,* Geological Society Special Publication No. 39, pp. 123–134.

quence of Edoras Basin to the E, whereas Site 555 is situated on the col between Edoras Bank and Hatton Bank, an area underlain by thick continental crust. The principal aim of DSDP Leg 81 was to investigate the nature of the dipping reflector sequence in an attempt to elucidate the processes involved in their formation and, in particular, their significance in the development of the passive margin W of Rockall. Drilling at all four sites terminated in basaltic igneous rocks of early Palaeogene age.

The stratigraphic distribution of early Tertiary basalts and volcaniclastics recovered during DSDP Legs 48 and 81 in the SW Rockall Plateau has been described by Harrison *et al.* (1979) and Morton & Keene (1984), with more detailed geochemical data presented by Harrison & Merriman (1984), Joron *et al.* (1984) and Richardson *et al.* (1984). Sr and Nd isotopic data have been presented by Macintyre & Hamilton (1984), and Pb isotopic data from Hole 553A have been discussed by Morton & Taylor (1987). This paper synthesizes, reviews and discusses the significance of the existing major and trace-element geochemical data on the Leg 81 basaltic rocks, supplemented by further rare-earth element (REE) data resulting from neutron activation analyses of 20 samples, and new Pb-isotope evidence from Sites 552, 554 and 555, augmenting the data from Site 553 presented by Morton & Taylor (1987).

At four sites (403, 404, 553 and 555) a correlatable marker horizon (GM) comprising highly glauconitic sediment has been recognized within the Palaeogene volcanic/volcaniclastic sequence, near its top. It corresponds approximately to the boundary between nannoplankton zones NP10 and NP11, and marks both a major transgression and a change in sediment source (Morton 1984; Morton *et al.* 1984). The horizon was not recognized at Site 552 because drilling was terminated above this level. The GM horizon appears to be coincident with seismic Reflector 1 in Edoras Basin and Reflector 2 in the vicinity of Site 555 (Roberts, Schnitker *et al.* 1984). As discussed by Morton & Keene (1984), the transgressive event that generated the GM horizon can be interpreted either as resulting from margin subsidence due to crustal separation postdating extrusion of the dipping reflector sequence (in the models of Hinz and Roberts), or as the result of the submergence of the active spreading ridge crest below sea-level (in the subaerial seafloor-spreading model). As there appears to be a significant change in basalt geochemistry across the glauconitic marker, the following sections differentiate between basaltic rocks erupted pre- and post-GM.

Petrochemistry of basaltic rocks erupted pre-glauconite marker

Petrography

Basalts and basaltic tuffs occurring below the GM horizon were sampled at Sites 403 and 404 (Leg 48), and at Sites 553 and 555 (Leg 81). The petrography of Leg 48 basaltic tuffs is described in detail by Harrison *et al.* (1979), and Leg 81 basalts are detailed by Harrison & Merriman (1984). Further details of the pyroclastic rocks, including glass chemistry, are provided by Morton & Keene (1984). The following is a summary of these three published accounts.

In Hole 553A recognizable flow units permitted sampling of scoriaceous and agglomeratic tops, highly vesiculated bases and midflow material. Tops and bases of flows are generally aphyric with $< 10\%$ plagioclase microphenocrysts, whereas midflow basalts are plagioclase-phyric or glomerophyric; groundmass textures are pilotaxitic or hyalopilitic. An averaged modal composition (by volume) is plagioclase 36%, pyroxene 48%, Ti-magnetite 8% and accessories plus secondary saponite 8%. Groundmass labradorite (An_{50-70}) coexists with bytownite (An_{76-86}) microphenocrysts and glomerocrysts. Pyroxenes are predominantly augites with rare Mg-pigeonite ($En_{52}Fs_{34}Wo_{14}$). Saponite pseudomorphs after olivine are also rare. Vitroclastic tuffs near the top of Subunit Vb and the base of Subunit Va are hyaloclastites (Morton & Keene 1984).

Massive olivine dolerite forms the lowest Subunit (IVc) sampled in Hole 555, and typically consists of an ophitic intergrowth of labradorite, olivine (Fo_{76-82}), augite and Ti-magnetite. Sparsely vesicular basalt flows and hyaloclastites form the overlying Subunits IVa and IVb. Textures range from aphyric or sparsely plagioclase microphyric at flow margins, to subophitic within the thicker flows. Plagioclase phenocrysts form 7–8% of the phyric basalts, and compositions range from andesine (An_{45}) to bytownite (An_{76}). In contrast, groundmass plagioclase is andesine (An_{45}) to labradorite (An_{57}). The hyaloclastites consist predominantly of argillized vitric fragments and basaltic clasts.

At Site 554, the relationship between the basalts forming the outer high and the GM is not clear, since the latter is not present and biostratigraphic and magnetostratigraphic controls are poor. However, geochemically there are strong similarities (see Fig. 2) between 554 basalts and pre-GM basalts from other sites. The basalts sampled in Hole 554A are microphyric with intergranular to intersertal textures, and consist of approximately 35% plagioclase, 55% pyroxene and 10% opaque oxide and secondary minerals. Plagioclase microphenocrysts and groundmass laths are labradorite (An_{67-68}), and pyroxenes are subcalcic ferroaugite ($En_{33}Fs_{46}Wo_{24}$).

The basalt flows and hyaloclastites are overlain by detrital sandstones and mudstones interbedded with tuff and lapilli tuff, most of which show upward fining consistent with an air-fall origin. Vitric material predominates in these tuffs and much of it appears to be derived from tholeiitic magmas. However, lapilli tuffs at Site 555 contain rare fresh glass, which shows definite alkali basalt affinities (Morton & Keene 1984).

All of the rocks described above are altered in varying degrees. Basalts from Sites 553 and 554 contain saponite and celadonite as the dominant secondary minerals and indicate low temperature (< 150°C) interaction with seawater (Desprairies *et al.* 1984; Harrison & Merriman 1984). Tuffaceous rocks from all four sites contain saponite, celadonite, and zeolites (Morton & Keene 1984), consistent with low temperature alteration. In Hole 555, however, chlorite and possibly actinolite are present in the lowest units, indicating alteration temperatures of up to 250°C.

Geochemistry

Previous geochemical studies of basalts below the GM show that they are low-K, quartz and hypersthene-normative tholeiites, strongly depleted in LREEs, and other incompatible elements (Harrison & Merriman 1984; Joron *et al.* 1984; Richardson *et al.* 1984). Nd- and Sr-isotope data are similar to those of Atlantic MORB, and do not show any clear evidence of continental crustal contamination (Macintyre & Hamilton 1984).

New trace-element data for the Leg 81 basalts are shown in Table 1 and plotted in Figure 2.

TABLE 1. *Instrumental neutron activation analyses (INAA) of DSDP Leg 81 basalts*

(ppm)	552 21 cc 6–9	552 22–1 13–17	553A 38–2 74–76	553A 43–1 99–101	553A 46–2 115–117	553A 50–2 107–109	553A 52–2 18–20	553A 53–1 123–125	553A 54–1 94–96	553A 55–1 99–101
La	5.24	5.81	0.93	0.58	1.37	0.98	1.16	1.11	0.61	0.97
Ce	20.0	18.1	5.96	2.49	2.27	2.85	2.34	2.53	3.76	3.51
Nd	21.1	22.5	6.82	4.12	4.72	4.45	5.18	5.21	6.80	4.89
Sm	7.08	7.35	2.63	1.81	2.20	2.15	1.91	2.04	2.92	2.05
Eu	3.08	2.75	1.17	0.86	1.12	0.92	0.97	0.97	1.25	0.71
Tb	1.87	1.76	0.84	0.72	0.90	0.80	0.67	0.82	0.94	0.69
Yb	7.36	5.79	3.70	3.45	3.36	3.70	2.14	3.04	4.18	1.79
Lu	1.00	0.80	0.47	0.48	0.48	0.46	0.39	0.40	0.63	0.19
Ta	0.29	0.26	0.06	0.07	0.07	0.06	0.12	0.12	0.05	0.02
Hf	6.22	5.37	2.01	2.32	1.71	1.54	1.26	1.68	1.99	1.58
Th	0.14	0.23	0.26	0.43	0.26	0.29	0.44	0.22	0.26	0.26
U	0.24	0.17	0.19	0.14	0.33	0.20	0.32	0.10	0.15	0.19
W	37.2	22.7	33.9	16.5	31.5	42.0	56.8	119	33.8	22.6
Cr	94.5	109	136	47.4	67.5	198	127	110	63.4	95.4

(ppm)	553A 56–2 64–66	553A 57–1 106–108	553A 59–4 64–66	554A 8–1 85–87	554A 10–1 67–71	554A 12–1 7–9	555 70–1 109–111	555 76–4 82–84	555 82–2 140–142	555 90–3 125–127
La	0.86	1.63	1.29	1.39	0.58	0.88	0.94	1.34	2.45	4.68
Ce	3.15	2.04	1.96	4.96	6.30	6.99	5.16	2.13	7.76	11.4
Nd	5.58	3.98	4.87	7.84	7.99	6.56	4.81	4.12	8.95	11.4
Sm	2.03	1.63	1.98	2.48	2.54	2.75	1.93	1.94	2.77	5.05
Eu	0.85	0.72	0.84	0.98	1.12	1.27	1.00	1.07	1.13	2.05
Tb	0.75	0.54	0.73	0.82	0.95	0.93	0.70	0.82	1.01	1.46
Yb	3.70	1.62	2.34	2.86	2.87	3.40	2.84	2.83	3.09	5.99
Lu	0.53	0.30	0.38	0.46	0.41	0.52	0.38	0.41	0.41	0.82
Ta	0.08	0.07	0.05	0.38	0.21	0.13	0.06	0.10	0.11	0.17
Hf	1.31	0.94	1.45	1.92	1.82	1.93	1.58	1.39	2.01	3.86
Th	1.16	0.46	1.26	0.45	0.25	0.30	0.25	0.49	0.42	0.22
U	0.32	0.36	0.31	0.32	0.31	0.45	0.18	0.32	0.13	0.16
W	45.5	34.3	29.7	186	77.4	56.1	30.7	28.3	24.9	85.1
Cr	101	215	326	282	276	288	202	335	72.8	56.4

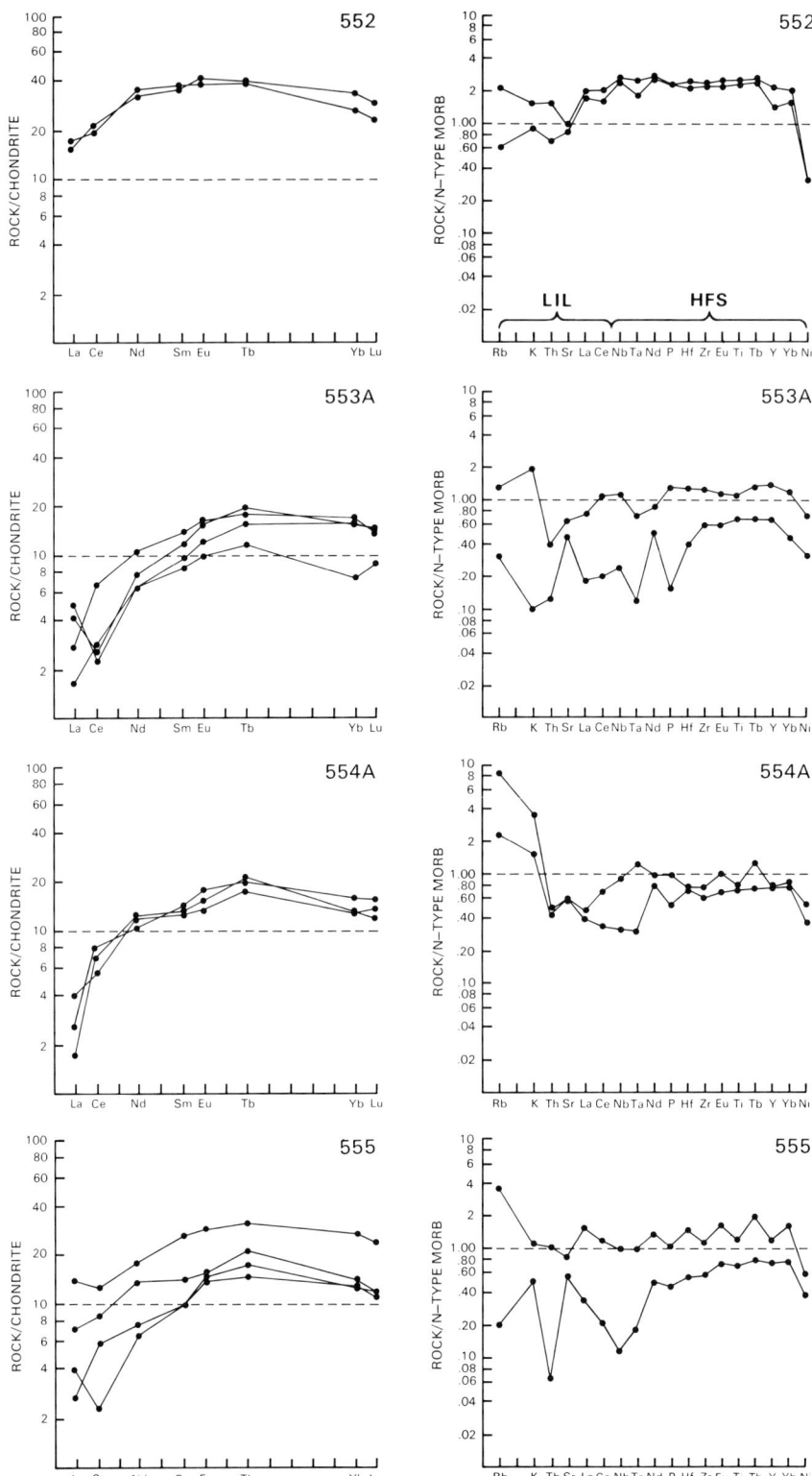

FIG. 2. Chondrite normalized (Wakita *et al.* 1971) REE plots and N-type MORB normalized (Saunders & Tarney 1984) multi-element plots of DSDP Leg 81 basalts.

Concentrations of La, Th and U are well below detection limits and values given must be used with caution. Additional data from Joron *et al.* (1984), have been used in the MORB-normalized plots (Fig. 2). Because of the difficulties of assembling a full set of data for any one sample, maximum and minimum MORB-normalized values are shown as upper and lower 'envelopes' for the range in each hole.

Chondrite-normalized REE plots show that the basalts are strongly depleted in LREEs with very low $(Ce/Yb)_{CN}$ ratios in samples from holes 553A, 554A and 555 (0.15–0.61) and low ratios (0.66–0.76) in hole 552. The MORB-normalized plots are essentially the flat patterns of N-type MORB, with marked depletion in HFS elements Hf to Nb. The LIL elements are also strongly depleted but show a more erratic pattern reflecting variable interaction with seawater. Thus the mobile LIL elements Rb, K and Sr show considerable variation, whereas the immobile LIL elements Ce, La and Th generally show a downwards trend of strong depletion. This trend of increasing depletion with an increase in the incompatibility of elements is consistent with removal of liquids during one or more previous partial melting events in the source region of the pre-GM basalts.

The extremely depleted character of the basalts is also illustrated by Figure 3 (after Saunders 1984, Fig. 6.10), where most samples from hole

553A plot below the field of N-type MORB. If a depleted source with $(Ce/Yb)_{CN} \sim 0.2$ and $Ce_{CN} \sim 0.6$ is assumed, the position of the batch melting curve in Figure 3 suggests 20% partial melting of the source has produced the most depleted 553 basalts. The slightly more evolved character of 554 and some 555 basalts suggests either a different source or smaller degrees of partial melting, compared with 553 basalts. Since the ratio Ce/Yb can be changed by crystal fractionation processes, the ratio La/Ta is a better indication of source character.

Data from Joron *et al.* (1984) have been plotted in Figure 4 together with an N-type MORB trend of La/Ta 18.5, from Saunders (1984). Mean La/Ta ratios for 553 and 555 basalts are significantly greater than N-type MORB, suggesting a high La/Ta end-member source for the dipping reflector basalts (Saunders 1985). Ratios for the 554 basalts are close to typical N-type MORB ratios, suggesting that the outer-high basalts are derived from a less strongly depleted source, where $(Ce/Yb)_{CN} \sim 0.5$. If at least 15% partial melting is assumed to have given rise to the outer-high basalts, the batch melting curve in Figure 3 should originate at $(Ce/Yb)_{CN} \sim 0.5$ and $(Ce)_{CN} \sim 1$.

The relationship between basalt samples in the same hole was explored with a plot of incompatible elements Th and Zr (Fig. 5), using data from Joron *et al.* (1984). While Th shows considerable variation (possibly reflecting the problems of analysing for very low concentrations), Zr variation is notably restricted suggesting, at best, very limited relationships through low-pressure frac-

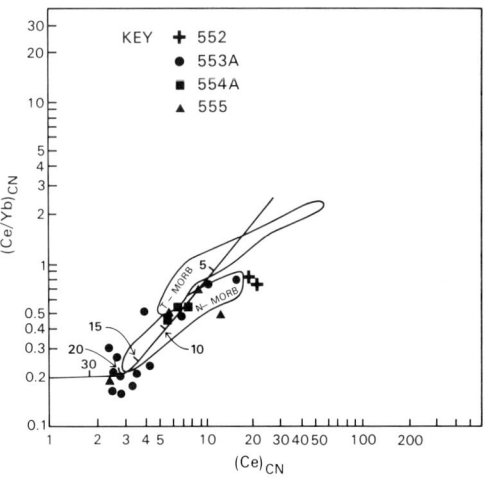

FIG. 3. Log–log plot of chondrite normalized Ce/Yb vs Ce values for DSDP Leg 81 basalts. The fields of N-type MORB and transitional T-type MORB from Saunders (1984, Fig. 6.10). Numbers on batch melting curve indicate percentage melt; they are not absolute values and should be regarded as relative for a given suite of rocks. Source of dipping reflector basalts plots at $(Ce/Yb)_{CN} \sim 0.2$ and $(Ce)_{CN} \sim 0.6$.

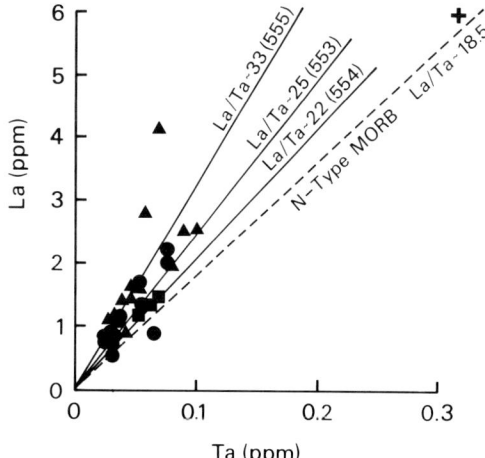

FIG. 4. La vs Ta plot of DSDP Leg 81 basalts. Data from Joron *et al* (1984) Key to symbols shown in Figure 3.

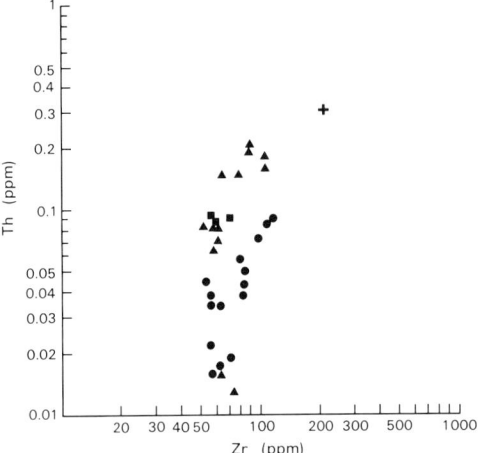

FIG. 5. Log–log plot of Th vs Zr for DSDP Leg 81 basalts. Data from Joron *et al.* (1984). Key to symbols in Figure 3.

tionation processes in holes 553A and 555. The general lack of significant fractionation trends is also seen in Figure 3, where data points tend to lie subparallel to the batch melting curve, rather than display a low-angle trend, typical of low-pressure crystal fractionation.

Petrochemistry of basaltic rocks erupted post-glauconite marker

Petrography

Some 20 tuff beds occur above the GM and are generally thinner and finer grained than those beneath the GM. The tuffs are largely composed of lava fragments and fresh vitric material, which is exclusively of tholeiitic composition (Morton & Keene 1984). Lava fragments are vesicular, aphyric or sparsely microphyric, with textures ranging from intergranular to intersertal. Principal crystalline phases are labradorite (An_{57-63}), microphenocrysts and groundmass laths, augite ($En_{39}Fs_{21}Wo_{40}$) and Ti-magnetite (Harrison & Merriman 1984).

Geochemistry

Basalts from Hole 552 show the LREE-depleted patterns of typical N-type MORB, but neither the total REE abundance nor the $(Ce/Yb)_{CN}$ ratios are comparable with the more depleted basalts in Holes 553A and 554 (Fig. 2). Moreover, the MORB-normalized plots for 552 data do not display the marked depletion in HFS and LIL elements which is characteristic of Holes 553A,

554 and 555. However, it is notable that the general downward trend from Nb to Th in Figure 2 (ignoring Sr) is similar to that displayed by Hole 553, though less steeply inclined.

The less depleted character of the 552 basalts is emphasized in Figure 3, where data points plot upslope of the N-MORB field, and are thus more enriched in LREE than 'normal' ridge segment basalts. However, in Figure 4 it is clear that the 552 basalts have evolved from a source similar to that of typical N-type MORB and the outer-high basalts (554), and one distinctly different from the source of the dipping reflector basalts. If a source with REE characteristics similar to that of the 554 basalt is assumed, i.e., $(CeYb)_{CN} \sim 0.5$ and $(Ce)_{CN} \sim 1$, repositioning of the batch melting curve in Figure 3 indicates that 5–10% partial melting of this source could give rise to the 552 basalts. Such an origin is consistent with the greater abundance of REEs and other incompatible elements compared with the outer high basalts.

Lead isotope geochemistry

Previously reported Pb isotopic data from basalts of Site 553 suggests contamination of mantle-derived magmas by lower continental crust or subcontinental lithosphere of mid-Proterozoic age (Morton & Taylor 1987). The Pb isotope data from Site 553 basalts, plus data from the other Leg 81 sites are shown in Table 2 and Figure 6. The analytical techniques followed in acquisition of the new data are identical to those adopted for the 553 study (Morton & Taylor 1987).

Sites 552 and 553

The data from Sites 552 and 553 form a single coherent trend, extending from the less radiogenic part of the N Atlantic mid-ocean ridge basalt (NAMORB) field toward the geochron, and, in a single case, crossing it. As discussed by Morton & Taylor (1987), this trend is best explained in terms of contamination of mantle-derived magmas by ancient U-depleted continental crust or subcontinental lithosphere. The near-horizontal slope of the data from Sites 552 and 553 does not closely match the contamination arrays of the Tertiary volcanics on Skye (Dickin 1981) or in the Faeroes (Gariépy *et al.* 1983), both of which have steeper slopes, and a somewhat complex model for Pb isotopic evolution is required. To generate $^{207}Pb/^{204}Pb$ ratios comparable to modern MORB-source mantle in association with markedly lower $^{206}Pb/^{204}Pb$ requires a source with a long history of separation from convecting

TABLE 2. *Pb-isotope data from DSDP Leg 81 basalts*

Location	Sample	$^{206}Pb/^{204}Pb$	$^{207}Pb/^{204}Pb$	$^{208}Pb/^{204}Pb$
Hole 552	22–2, 66–68	18.199	15.524	37.817
	23–1, 133–135	18.090	15.526	37.765
Hole 553A	38–1, 39–41	18.035	15.567	37.684
	40–2, 79–81	18.111	15.538	37.833
	45–3, 79–81	17.675	15.503	37.349
	47–3, 30–32	17.968	15.493	37.589
	49–3, 75–77	18.180	15.553	37.914
	51–2, 80–82	17.910	15.518	37.584
	53–3, 116–118	18.092	15.511	37.705
	54–4, 60–62	17.190	15.527	37.002
	58–1, 79–81	17.799	15.534	37.526
Hole 554A	7–1, 40–42	18.871	15.567	37.619
	7–4, 27–29	18.527	15.570	37.741
	14–1, 2–4	18.727	16.562	37.795
Hole 555	69–3, 34–36	18.179	15.503	37.655
	82–2, 2–4	19.300	15.641	38.274
	86–1, 126–128	18.324	15.552	37.785
	97–4, 75–77	18.695	15.621	38.279

mantle material, with an early stage of U/Pb enrichment relative to the main mantle reservoir, followed by a later stage with markedly lower U/Pb ratios. A possible scenario is late Archaean crust formation followed by mid-Proterozoic U-depletion during high-grade metamorphism, although this is not a unique solution. In this context, however, it may be significant that both Laxfordian (mid-Proterozoic) and Grenvillian (late-Proterozoic) granulites have been recovered from the adjacent Rockall Plateau (Miller *et al.* 1973).

Sites 554 and 555

The data from Site 554 basalts plot in the centre of the typical NAMORB field, without any evidence for contamination. The lack of contamination is predictable, as the basalts here are regarded as true oceanic crust in both the rifting and subaerial seafloor-spreading hypotheses. The basalts from Site 555 have slightly higher $^{207}Pb/^{204}Pb$ ratios than typical NAMORB and consequently plot above the NAMORB field in Figure 6, toward the hypothetical orogene and upper

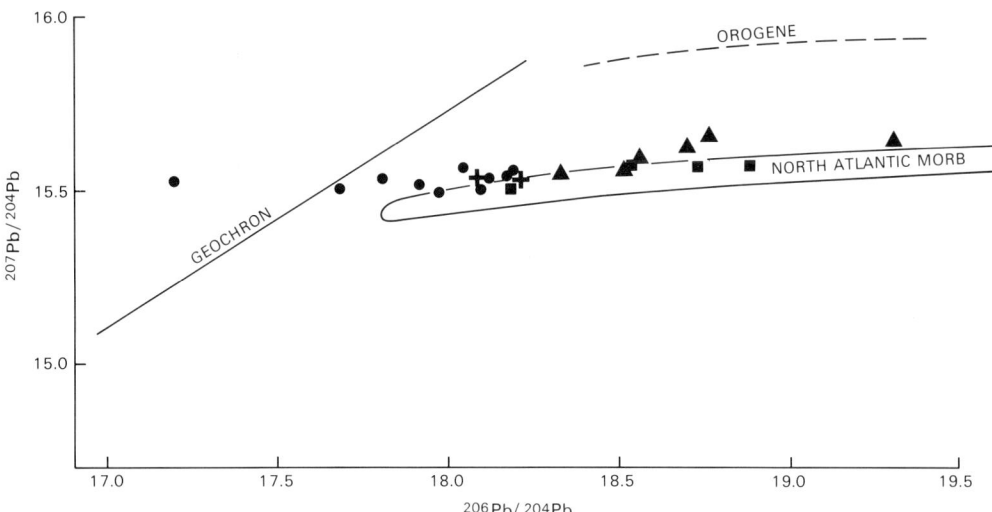

FIG. 6. $^{207}Pb^{204}Pb$ vs $^{206}Pb/^{204}Pb$ plot of DSDP Leg 81 basalts. The compositional field for NAMORB is based on Pb isotopic analyses published by Cohen *et al.* (1980) and Dupré & Allègre (1980) for samples collected along the ridge between latitudes 9°N and 64°N. Key to symbols shown in Figure 3.

crust lines modelled by Zartman & Doe (1981). This pattern is typical of mantle-derived magmas contaminated by material of upper crustal origin, such as sediments or metasediments.

Discussion

Geochemical and Pb isotope data suggest that the Leg 81 basalts SW of Rockall Plateau fall into three groups:

(1) The early Palaeogene basalts comprising the dipping reflector sequence are unusual oceanic tholeiites characterized by depletion in LREE, LIL elements and certain HFS elements with respect to 'normal' ridge segment basalts (N-type MORB). These characteristics suggest that the dipping reflector basalts were derived from a source already depleted in incompatible elements, most likely as a result of previous partial melting. At least 20% partial melting of the very depleted mantle source probably occurred in order to evolve the depleted dipping reflector basalts. Such a source would not normally melt under ridge crests but higher than usual asthenospheric heating appears to have accompanied the stretching and break-up of the Rockall lithosphere (White 1988). Pb isotope data suggests contamination of the Site 555 basalts by upper crustal material, consistent with their location on the Rockall Plateau. In contrast, Site 553 basalts have been contaminated by either long-lived subcontinental lithospheric mantle or with ancient U-depleted continental crust, a situation inconsistent with a simple seafloor-spreading origin for the dipping reflectors.

(2) Basalts from Site 554, forming the outer high, are 'normal' ridge segment basalts (N-type MORB) derived from a mantle source less strongly depleted than that responsible for the dipping reflector basalts. The outer high basalts show no evidence of crustal contamination and are therefore the only Leg 81 basalts that can be regarded as representative of normal seafloor-spreading processes.

(3) The post-GM basalts of Site 552 are clearly the most evolved rocks sampled by Leg 81. They resemble the outer high basalts in possessing similar La/Ta ratios, but show evidence of contamination by either ancient U-depleted continental crust or subcontinental lithosphere. The 552 basalts appear to have been derived from a mantle source similar to that of the outer-high basalts, but represent lower degrees of partial melting.

Any model that seeks to explain the formation of the W Rockall passive margin must embrace the petrochemical differences between the basalt types, summarized above. The models proposed by Talwani et al. (1981), Mutter et al. (1982) and Smythe (1983), are inconsistent with the contaminated character of the dipping reflector basalts. Essentially, these models imply that the dipping reflector basalts are purely the result of seafloor-spreading processes, without any interaction with continental lithosphere. The need to involve continental lithosphere in the genesis of the dipping reflector basalts favours the model proposed by Hinz (1981), and particularly the modification of this model by Roberts et al. (1984). The latter has been further modified in Figure 7, in an attempt to integrate basalt genesis with the formation of the passive margin of SW Rockall.

The model illustrated in Figure 7 proposes that the lithosphere of western Rockall Plateau, prior to stretching and seafloor spreading, was formed of sialic upper crust of Archean to Upper Proterozoic age underplated by basic and ultrabasic lower crust/upper mantle. There is evidence that the continental crust of Rockall Plateau is an extension of the Laxfordian terrain of the Outer Hebrides, with some overprinting by Grenvillian granulite facies metamorphism (Miller et al. 1973; Roberts et al. 1973a). It is likely therefore that the lithosphere W of Rockall has suffered several high-grade metamorphic events with consequent loss of lithophile elements. As a result, the composition of the sialic crust may have approached a Rb, U and Th-depleted condition similar to that of the Scourie gneisses of NW Scotland (Holland & Lambert 1973). Extensive partial melting and hence depletion of the lower crust/upper mantle component of the Rockall lithosphere may have occurred during intrusion of the Scourie dyke swarm, which extends from NW Scotland to Greenland. Further melting and depletion of the subcontinental lithosphere may have occurred during rifting-related basic volcanism in Torridonian times (Lawson 1972; Stewart 1975), subsequently during postulated Permo–Carboniferous rifting of the Rockall Trough (Russell & Smythe 1978), and yet again during the mid- to late-Cretaceous to provide basic magma associated with the development of the Rockall Trough (Roberts et al. 1973b; Harrison et al. 1975).

In the schematic series of events depicted in Figure 7, stretching and thermal disturbance of the lithosphere marked the onset of mantle diapirism prior to rifting and seafloor spreading. As a result of a relatively small increase ($\sim 100°C$) in asthenospheric temperature beneath the Rock-

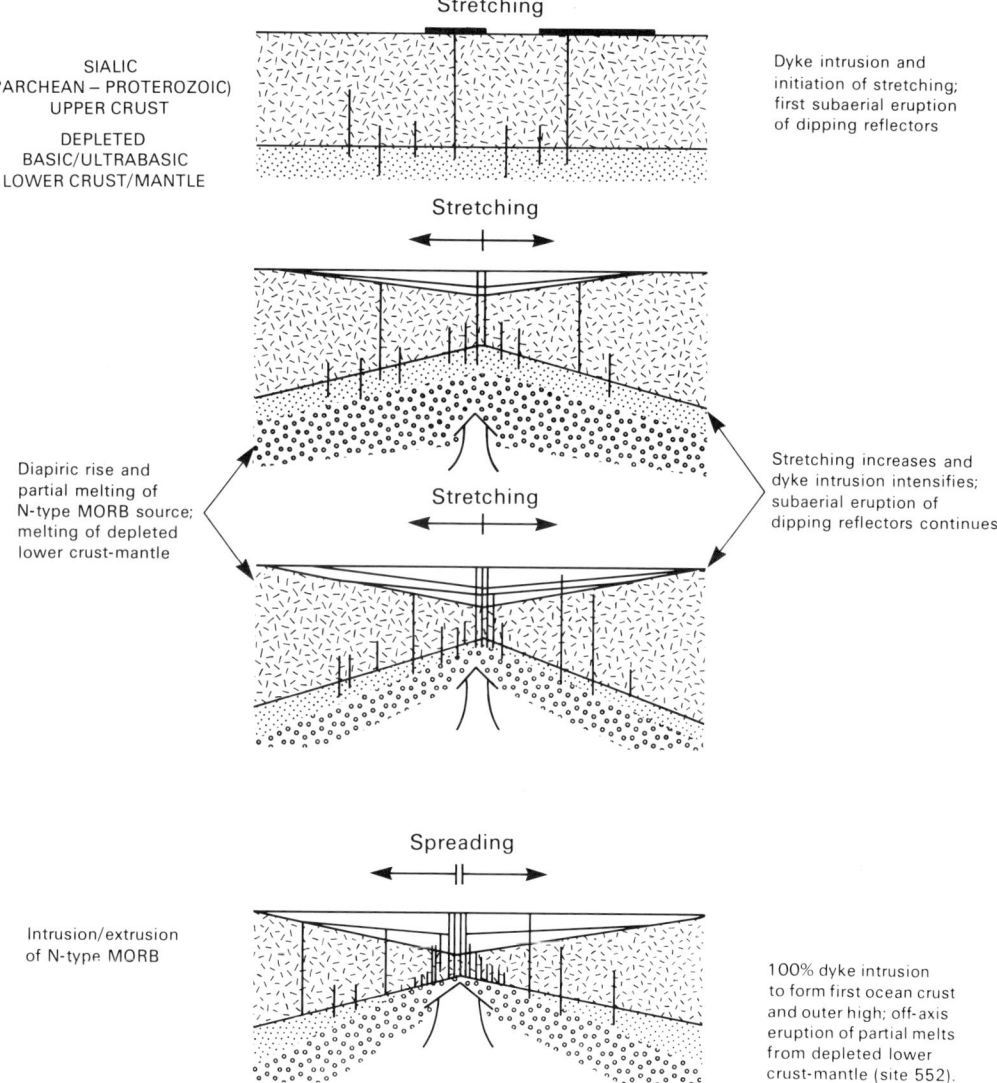

FIG. 7. Schematic model of basalt genesis and structural development of the 'passive' margin of SW Rockall Plateau (modified after Roberts *et al.* 1984).

all lithosphere (White 1988), the first basalt magmas were developed by partial melting of very depleted, semi-refractory upper mantle peridotite forming the base of the lithosphere. Earliest melts were mostly intruded as dykes but some reached the surface to form the earliest subaerial dipping reflector lava flows, in the early part of magnetic anomaly chron C24R (Roberts *et al.* 1984). Continued heating caused extensive melting of the base of the lithosphere, and the increased volume of partial melts accelerated

diapiric doming of the lithosphere. The resulting crustal thinning and stretching permitted melts to form a dyke swarm feeding voluminous subaerial and shallow subaqueous lava flows. These in turn loaded the thin lithosphere, adding to the subsidence induced by evacuation of large volumes of magma. The lava pile then acquired a regional 'inward' dip.

The end of the stretching phase probably coincided with the arrival of a mantle diapir in the base of the thinned Rockall lithosphere, which produced partial melts of less depleted N-type MORB, the typical outer-high basalts. At the onset of spreading the first new ocean crust was formed by dyke swarms feeding submarine lava flows during C24N. At this time subsidence accelerated and permitted influx of fully marine waters over existing nearshore or lagoonal environments of western Rockall, resulting in the extensive development of glauconitic sediments.

Volcanism continued in western Rockall, after the GM, with magmas derived from mantle-source N-type MORB by relatively low degrees of partial melting. These slightly more 'evolved' basalts (552) may have been contaminated by U-depleted crust during the ascent of magma through the subsiding feather edge of the Rockall lithosphere.

Conclusions

The Leg 81 dipping reflector basalts were derived from strongly depleted subcontinental lithosphere by high degrees of partial melting. Basalts forming the outer high were derived from a mantle source uncontaminated by continental lithosphere. These fundamental differences are best explained by the rifting hypothesis put forward by Hinz (1981) and later modified by Roberts *et al.* (1984). The latter proposes that the dipping reflector sequence was formed by mainly subaerial eruption of basalt during rifting of the lithosphere W of Rockall Plateau, and the outer high basalts represent the earliest true oceanic crust.

It remains to be seen whether all dipping reflector sequences that occur at passive margins were formed during rifting, rather than by subaerial seafloor spreading. The dipping reflector sequence SW of Rockall Plateau is somewhat atypical in that it has a faulted contact with the Plateau itself. Moreover, not all dipping reflector basalt sequences in the NE Atlantic show the strong depletion that characterizes the Rockall margin. Those of the Vøring Plateau appear to be N-type MORB-source basalts (Viereck *et al.* 1988), whereas early basalts erupted on the E Greenland margin are LREE-enriched tholeiites (Gill *et al.* 1988). Questions posed by this and other studies reported in this volume, would perhaps be best resolved by a major geochemical sampling programme under the auspices of the ODP, to attempt to detect crustal or subcontinental lithospheric contamination of such sequences.

ACKNOWLEDGEMENTS: Critical comments by Godfrey Fitton and Andy Saunders much improved the original typescript. This contribution is published by permission of the Director, British Geological Survey (NERC).

References

COHEN, R. S., EVENSEN, N. M., HAMILTON, P. J. & O'NIONS, R. K. 1980. U-Pb, Sm-Nd and Rb-Sr systematics of mid-ocean ridge basalt glasses. *Nature*, **283**, 149–153.

DESPRAIRIES, A., BONNOT-COURTOIS, C., JEHANNO, C., VERNHET, S. & JORON, J. L. 1984. Mineralogy and geochemistry of alteration products in Leg 81 basalts. *In*: ROBERTS, D. G., SCHNITKER, D., *et al Initial Reports of the Deep Sea Drilling Project*. US Government Printing Office, Washington, **81**, pp. 733–741.

DICKIN, A. P. 1981. Isotope geochemistry of Tertiary igneous rocks from the Isle of Skye, N.W. Scotland. *Journal of Petrology*, **22**, 155–189.

DUPRÉ, B. & ALLÈGRE, C. J. 1980. Pb-Sr-Nd isotopic correlation and the chemistry of the North Atlantic mantle. *Nature*, **286**, 17–22.

GARIÉPY, C., LUDDEN, J. & BROOKS, C. 1983. Isotopic and trace element constraints on the genesis of the Faeroe lava pile. *Earth and Planetary Science Letters*, **63**, 257–272.

GILL, R. C. O., NIELSEN, T. F. D., BROOKS, C. K. & INGRAM, G. A. 1988. Tertiary volcanism in the Kangerdlugssuaq region, E. Greenland: trace element geochemistry of the Lower Basalts and tholeiitic dyke swarms. *In*: MORTON, A. C. & PARSON, L. M. (eds) *Early Tertiary Volcanism and the Opening of the NE Atlantic*. Geological Society of London, Special Publication, **39**, pp. 161–179.

HARRISON, R. K., & MERRIMAN, R. J. 1984. Petrology, mineralogy, and chemistry of basaltic rocks: Leg 81. *In*: ROBERTS, D. G., SCHNITKER, D., *et al Initial Reports of Deep Sea Drilling Project*. US Government Printing Office, Washington, **81**, pp. 743–774.

——, TRESHAM, A. E., SNELLING, N. J. & RUNDLE, C. C. 1975. Helen's Reef: Petrography, chemistry and K-Ar age determination. *In*: HARRISON, R. K. (ed) *Expedition to Rockall 1971–72. Institute of Geological Sciences Report*, **75/1**, pp. 61–72.

——, KNOX, R. W. O'B. & MORTON, A. C. 1979. Petrography and mineralogy of volcanogenic sediments from DSDP Leg 48, southwest Rockall plateau, Sites 403 and 404. *In*: MONTADERT, L., ROBERTS, D. G., *et al Initial Reports of Deep Sea Drilling Project*. US Government Printing Office, Washington, **48**, pp. 771–785.

HINZ, K. 1981. A hypothesis on terrestrial catastrophes.

Wedges of very thick oceanward dipping layers beneath passive continental margins—their origin and palaeoenvironmental significance. *Geologische Jahrbuch*, **E22**, 3–28.

HOLLAND, J. G. & LAMBERT, R. ST J. 1973. Comparative major element geochemistry of the Lewisian of the mainland of Scotland. *In*: PARK, R. G. & TARNEY, J. (eds) *The Early Precambrian of Scotland and Related Rocks of Greenland*, University of Keele, pp. 51–62.

JORON, J. L., BOUGAULT, H., MAURY, R. C., BOHN, M. & DESPRAIRIES, A. 1984. Strongly depleted tholeiites from the Rockall plateau margin, North Atlantic: Geochemistry and mineralogy. *In*: ROBERTS, D. G., SCHNITKER, D., *et al. Initial Reports of Deep Sea Drilling Project*. US Government Printing Office, Washington, **81**, pp. 783–794.

LAWSON, D. E. 1972. Torridonian volcanic sediments. *Scottish Journal of Geology*, **8**, 345–362.

MACINTYRE, R. M. & HAMILTON, P. J. 1984. Isotopic geochemistry of lavas from Sites 553 and 555. *In*: ROBERTS, D. G., SCHNITKER, D., *et al. Initial Reports of Deep Sea Drilling Project*. US Government Printing Office, Washington, **81**, pp. 775–781.

MILLER, J. A., ROBERTS, D. G. & MATTHEWS, D. H. 1973. Rocks of Grenville age from Rockall Bank. *Nature*, **246**, 61.

MONTADERT, L., ROBERTS, D. G., *et al.* 1979. *Initial Reports of Deep Sea Drilling Project*. US Government Printing Office, Washington, **48**.

MORTON, A. C. 1984. Heavy minerals from Paleogene sediments, Deep Sea Drilling Project Leg 81: their bearing on stratigraphy, sediment provenance, and the evolution of the North Atlantic. *In*: ROBERTS, D. G., SCHNITKER, D., *et al. Initial Reports of Deep Sea Drilling Project*. US Government Printing Office, Washington, **81**, pp. 653–661.

—— & KEENE, J. B. 1984. Paleogene pyroclastic volcanism in the southwest Rockall Plateau. *In*: ROBERTS, D. G., SCHNITKER, D., *et al. Initial Reports of Deep Sea Drilling Project*. US Government Printing Office, Washington, **81**, pp. 633–643.

—— & TAYLOR, P. N. 1987. Lead isotope evidence for the structure of the Rockall dipping-reflector passive margin. *Nature*, **326**, 381–383.

——, MERRIMAN, R. J. & MITCHELL, J. G. 1984. Genesis and significance of glauconitic sediments of the southwest Rockall Plateau. *In*: ROBERTS, D. G., SCHNITKER, D., *et al. Initial Reports of Deep Sea Drilling Project*. US Government Printing Office, Washington, **81**, pp. 645–652.

MUTTER, J., TALWANI, M. & STOFFA, P. A. 1982. Origin of seaward dipping reflectors in oceanic crust off the Norwegian margin by 'subaerial sea-floor spreading'. *Geology*, **10**, 353–357.

RICHARDSON, C., OAKLEY, P. J. & CANN, J. R. 1984. Trace and major element geochemistry of basalts from Leg 81. *In*: ROBERTS, D. G., SCHNITKER, D., *et al. Initial Reports of Deep Sea Drilling Project*. US Government Printing Office, Washington, **81**, pp. 795–806.

ROBERTS, D. G., ARDUS, D. A. & DEARNLEY, R. 1973a. Precambrian rocks drilled from the Rockall Bank. *Nature*, **244**, 21–33.

——, FLEMMING, N. C., HARRISON, R. K. & BINNS, P. 1973b. Helen's reef: A Cretaceous microgabbroic intrusion in the Rockall intrusive centre. *Marine Geology*, **16**, 21–30.

——, BACKMAN, J., MORTON, A. C., MURRAY, J. W. & KEENE, J. B. 1984. Evolution of volcanic rifted margins: Synthesis of Leg 81 results on the west margin of Rockall Plateau. *In*: ROBERTS, D. G., SCHNITKER, D., *et al. Initial Reports of Deep Sea Drilling Project*. US Government Printing Office, Washington, **81**, pp. 883–911.

——, SCHNITKER, D., *et al.* 1984. *Initial Reports of Deep Sea Drilling Project*. US Government Printing Office, Washington, **81**.

RUSSELL, M. J. & SMYTHE, D. K. 1978. Evidence for an early Permian oceanic rift in the northern North Atlantic. *In*: NEUMANN, E-R. & RAMBERG, I. B. (eds) *Petrology and Geochemistry of Continental Rifts*. Reidel, Dordrecht, pp. 173–179.

SAUNDERS, A. D. 1984. The rare earth element characteristics of igneous rocks from the ocean basins. *In*: HENDERSON, P. (ed.) *Rare Earth Element Geochemistry*. Developments in Geochemistry, **2**, pp. 205–236.

—— 1985. Geochemistry of basalts from the Nauru Basin, Deep Sea Drilling Project Legs 61 and 89: Implications for the origin of oceanic flood basalts. *In*: MOBERLY, R., SCHLANGER, S. O., *et al. Initial Reports of the Deep Sea Drilling Project*. US Government Printing Office, Washington, **89**, pp. 499–517.

—— & TARNEY, J. 1984. Geochemical characteristics of basaltic volcanism within back-arc basins. *In*: KOKELAAR, B. P. & HOWELLS, M. F. (eds) *Marginal Basin Geology: Volcanic and Associated Sedimentary and Tectonic Processes in Modern and Ancient Marginal Basins*. Geological Society of London, Special Publication, **16**, pp. 59–76.

SMYTHE, D. K. 1983. Faeroe–Shetland escarpment and continental margin north of the Faeroes. *In*: BOTT, M. H. P., SAXOV, S., TALWANI, M. & THIEDE, J. (eds) *Structure and Development of the Greenland–Scotland Ridge—New Methods and Concepts*. Plenum Press, London, pp. 109–119.

STEWART, A. D. 1975. 'Torridonian' rocks of western Scotland. *In*: HARRIS, A. L., *et al.* (eds) *A Correlation of the Precambrian Rocks in the British Isles*. Geological Society of London, Special Report, **6**, pp. 43–51.

TALWANI, M., MUTTER, J. & ELDHOLM, O. 1981. Initiation of the opening of the Norwegian Sea. *Oceanologica Acta*, **No. SP**, 23–30.

VIERECK, L., TAYLOR, P. N., PARSON, L. M., MORTON, A. C., HERTOGEN, J., GIBSON, I. & the ODP LEG 104 SCIENTIFIC PARTY 1988. Petrogenesis of the Vøring Plateau volcanic sequence drilled on Leg 104: interaction of mantle and crustal derived melts during formation of a constructive plate margin *In*: MORTON, A. C. & PARSON, L. M. (eds)

Early Tertiary Volcanism and the Opening of the NE Atlantic. Geological Society of London, Special Publication, **39**, pp. 69–83.

WAKITA, H., REY, P. & SCHMITT, R. A. 1971. Abundances of the 14 rare earth elements and 12 other trace elements in Apollo 12 samples: 5 igneous and 1 breccia rocks and 4 soils. *Proceedings of the Second Lunar Science Conference*, 1319–1329.

WHITE, R. S. 1988. A hot-spot model for early Tertiary volcanism in the N. Atlantic. *In*: MORTON, A. C. & PARSON, L. M. (eds) *Early Tertiary Volcanism and the Opening of the NE Atlantic*. Geological Society of London, Special Publication, **39**, pp. 3–13.

ZARTMAN, R. E. & DOE, B. R. 1981. Plumbotectonics—the model. *Tectonophysics*, **75**, 135–162.

R. J. MERRIMAN & A. C. MORTON, British Geological Survey, Keyworth, Nottingham, NG12 5GG, UK.

P. N. TAYLOR, Department of Earth Sciences, University of Oxford, Oxford, OX1 3PR, UK.

Early Tertiary volcanism at the western Barents Sea margin

J. I. Faleide, A. M. Myhre & O. Eldholm

SUMMARY: The western Barents Sea margin is composed of two regional shear segments linked by a NE-trending rifted margin SW of Bjørnøya. The primary marginal structures developed in response to the Cainozoic opening of the Norwegian–Greenland Sea. A marginal high characterizes the rifted margin segment. It is suggested that the high forms an edifice of subaerial extrusives emplaced during the initial opening of the southern Greenland Sea in the early Eocene. The extrusives also cover the adjacent continental crust and the volcanic event may have caused the deposition of tuffs in the basinal province of the southwestern Barents Sea. Renewed local volcanic activity in the early Oligocene is related to the major reorganization in plate motion at this time.

Recent multichannel seismic data, including deep seismic reflection and expanded spread profiles, have provided a first-order structural framework for the western Barents Sea margin. The margin is composed of several rifted and sheared segments that developed in response to the Cainozoic opening of the Norwegian–Greenland Sea.

A tectonic regime of regional shear between NE Greenland and Svalbard during the first phase of opening has been established onshore (Harland 1969; Lowell 1972; Kellogg 1975; Håkansson & Stemmerik 1984; Steel et al. 1985) as well as from marine geophysical data and plate reconstructions (Talwani & Eldholm 1977; Myhre et al. 1982). Recently, detailed investigations of the continental margin based on new data of better quality and coverage have yielded important constraints on how the marginal structures formed and developed within this regional tectonic setting. Furthermore, evidence has emerged that a marginal high with associated volcanism is present at the central part of the margin (Eldholm et al., in press; Myhre & Eldholm, in press).

We note that areas of basement elevated with respect to the adjacent oceanic crust exist along many parts of the continental margin in the Norwegian–Greenland Sea (Mutter & Zehnder 1988; Skogseid & Eldholm 1988). These marginal highs, which often are underlain locally by seaward-dipping reflector sequences (Mutter et al. 1982; Hinz et al. 1984), are believed to be formed during the initial opening of the ocean. In particular, Ocean Drilling Programme (ODP) drilling at the Vøring Plateau marginal high has proved that it represents a volcanic structure of early Tertiary age (Eldholm et al. 1987).

Here, the main objective is to describe the early Tertiary volcanism at the western Barents Sea margin in relation to the plate tectonic setting and the Tertiary sediment distribution. Although the main volcanic event in many ways appears similar to that along rifted margin segments further S in the Norwegian–Greenland Sea, we also observe renewed volcanic activity of mid-Tertiary age.

Margin geology

In a regional sense the western Barents Sea margin consists of three main segments (Fig. 1):

(1) A northern, initially sheared and later rifted, margin along the Hornsund Fault Zone.
(2) A central rifted margin complex at the Bjørnøya marginal high.
(3) A southern sheared margin along the Senja Fracture Zone.

The change in crustal type, from continental to oceanic, occurs over a narrow zone and is related to these primary rift and shear structures, which reflect the progressive northward opening of the Greenland Sea. Prominent elongate positive gravity anomalies lie just seaward of the sheared margin segments and a distinct, but smaller amplitude, anomaly is associated with the Bjørnøya marginal high (see Fig. 1) (Eldholm et al., in press).

South of about 73°30'N, the margin forms the western boundary of a regional basinal province (Faleide et al. 1984; Rønnevik & Jacobsen 1984) composed of the deep Cretaceous Tromsø and Bjørnøy Basins and the inverted Senja Ridge (Riis et al. 1986; Brekke & Riis, in press). To the N, the margin limits the elevated Svalbard Platform and its southwestern protrusion, the Stappen High. Off Svalbard, the inner shelf E of the Hornsund Fault Zone is seismically similar to the Svalbard platform, but is located within a region affected by early Tertiary orogenic movements (Steel et al. 1985). The entire margin is characterized by a very thick Cainozoic sedimen-

From MORTON, A. C. & PARSON, L. M. (eds), 1988, *Early Tertiary Volcanism and the Opening of the NE Atlantic*, Geological Society Special Publication No. 39, pp. 135–146.

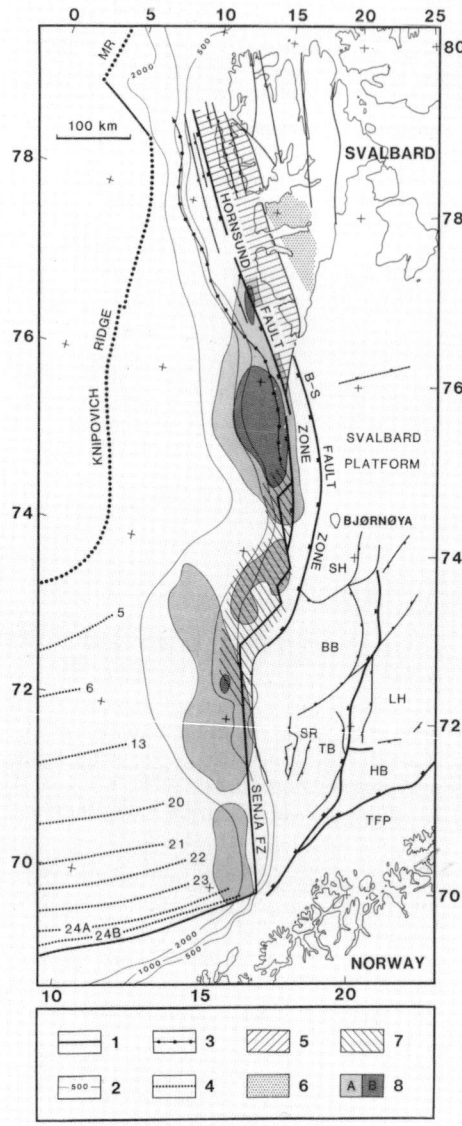

FIG. 1. Main structural features and geological
provinces based on Eldholm *et al.* (in press): **1**
Continent–ocean boundary (COB) and main structural
elements; **2** Bathymetry (m); **3** Limit of identified
oceanic crust (in the seismic record); **4** Magnetic
lineations; **5** Spitsbergen Fold and Thrust Belt; **6**
Tertiary central basin; **7** Bjørnøya marginal high (area
of early Eocene volcanism); **8** Marginal free-air
gravity anomalies (A: 50–100 mGal, B: > 100 mGal).
BB: Bjørnøy Basin; HB: Hammerfest Basin; LH:
Loppa High; SH: Stappen High; SR: Senja Ridge;
TB: Tromsø Basin; TFP: Troms–Finnmark platform;
B–S: Bjørnøya–Sørkapp.

tary wedge (Myhre 1984; Spencer *et al.* 1984). A
large proportion of the wedge consists of Neogene
material and huge deltas have been constructed
in front of major drainage systems. In general,
the maximum sediment thickness overlies the
transition between continental and oceanic crust.

The Cainozoic opening history of the Norwegian–Greenland Sea is characterized by two main
stages yielding different relative plate motion
(Talwani & Eldholm 1977):

(1) Between anomaly 25/24 and anomaly 13
    times (earliest Eocene to earliest Oligocene),
    Greenland moved in a N–NW direction
    relative to Eurasia. Regional shear between
    the Norwegian Sea and Eurasia Basin initi-
    ated the formation of the continental margins
    off the Barents Sea and Svalbard and the
    opening of the southern Greenland Sea.
(2) The relative direction of plate movement
    changed to a W–NW trend at anomaly 13
    time, first causing crustal stretching and later
    seafloor spreading also in the northern Green-
    land Sea.

The direction of early opening makes small but
distinct angles with the sheared margin segments
at the Senja Fracture Zone and the Hornsund
Fault Zone (Fig. 2). This caused non-ideal
transform movements with transtensional and
transpressional components. The Senja Fracture
Zone acted as a leaky transform due to transtension, creating a rhomboid-shaped depression in
which one might visualize a highly oblique
spreading system. This allowed material from
depths greater than normal levels in the mantle
to be intruded, forming an initial high-density
crust in the transtensional zone. Such crust might
explain the prominent positive gravity anomalies
and velocity distribution along parts of the
sheared margin segments (Eldholm *et al.*, in
press). At the same time, transpression caused
folding and thrusting in western Spitsbergen and
a foreland basin was filled from the uplifted
orogenic belt to the W (see Fig. 1) (Steel *et al.*
1985).

The central margin segment, however, was
subject to rifting, volcanism and formation of a
marginal high in early Tertiary times. This
volcanic province was later downfaulted in a
pull-apart tectonic setting.

The Tertiary sediment distribution and the
main structural elements in the western Barents
Sea and Svalbard also reflect the break-up and
early Tertiary opening of the Norwegian–Greenland Sea. In particular, the Senja Ridge, the
Stappen High and the Tertiary Basin in Spitsbergen (see Fig. 1) were affected by events associated
with the opening.

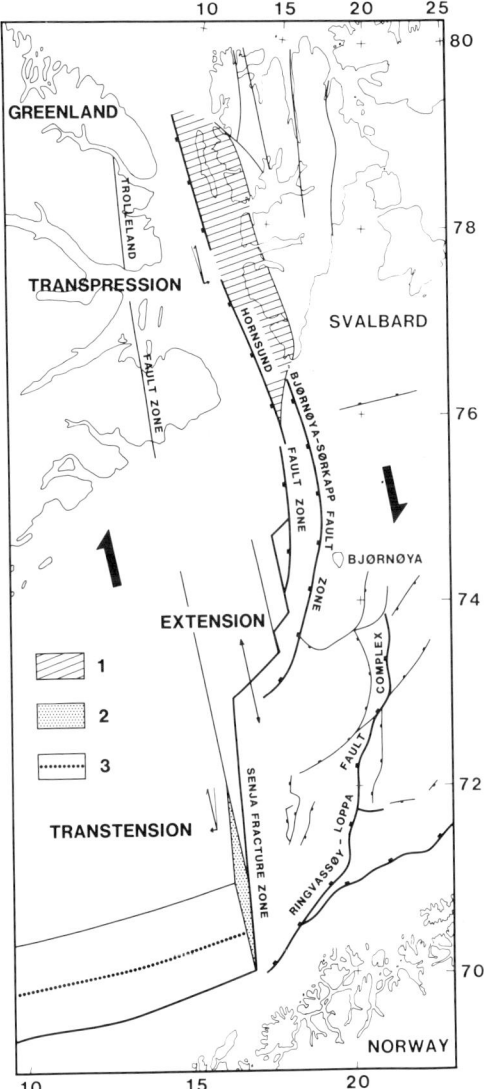

FIG. 2. Plate tectonic setting during early opening of the Norwegian–Greenland Sea and formation of the Bjørnøya marginal high. Reconstruction to anomaly 23 time (Eldholm *et al.*, in press) using the rotation poles of Talwani & Eldholm (1977). **1** Spitsbergen Fold and Thrust Belt; **2** Oblique shear crust; **3** Active spreading axis.

Volcanism at the Bjørnøya marginal high

In the seismic record we have interpreted several features reflecting Cainozoic volcanism:

(1) A characteristic band of high-amplitude reflectors cover the Bjørnøya marginal high.

These reflectors, which form an apparent acoustic basement, have been interpreted as early Eocene volcanic flows (see Figs 4, 5 & 7). Locally, the flows terminate against volcanic mounds (see Fig. 4).

(2) Volcanic peaks that penetrate the flow units at the high indicate later intrusive activity (see Fig. 5).

(3) A prominent seismic marker in the basinal province has been proved to originate from early Tertiary tuff deposits (see Figs 6 & 7).

The Bjørnøya marginal high, which is elevated with respect to the adjacent oceanic crust, is fault-bounded towards the Stappen High and Bjørnøy Basin by several rift and shear structures (see Figs 1 & 3). Both the Senja Fracture Zone and Hornsund Fault Zone form structurally complex intersections with the marginal high. No structuring is observed below the flows in most of the seismic data, but new deep seismic reflection profiles (Gudlaugsson *et al.* 1987) have revealed deeper reflectors at the innermost high (IKU–84–B, Fig. 4). Here, nearly flat-lying reflectors have been downfaulted towards the rifted and sheared segments of the main continental boundary fault (CBF) in Fig. 3. Locally, the volcanics are observed a short distance landward of the fault (Fig. 5). Initially, the flows may have covered a larger area but later vertical movements and erosion have determined the present distribution. The different levels of these volcanics demonstrates large-scale vertical movements in Eocene times, postdating the main volcanic event. Locally, downfaulting in a pull-apart setting exceeds 2000 m. We note that the maximum postvolcanic vertical movement took place along the northern rifted segment W of the Stappen High (BFB–14–74, see Fig. 5) diminishing towards the SW where the marginal high bounds the Bjørnøy Basin.

In terms of regional seismic and gravity signature this province appears similar to other marginal highs in the Norwegian–Greenland Sea. Although we do observe some sub-basement reflector segments (see Figs 4 & 7), no clear seaward-dipping sequences have been mapped. On the other hand, the high is characterized by mounds and constructional features of a volcanic origin. These highs were formed during the earliest opening of the ocean and the crustal transition occurs beneath the inner parts of the highs.

At the northwestern flank of the marginal high (see Figs 3 & 5) we have identified peaks that penetrate the 'acoustic basement' and the oldest Tertiary sediments. Associated local extrusives within the sedimentary section are also recog-

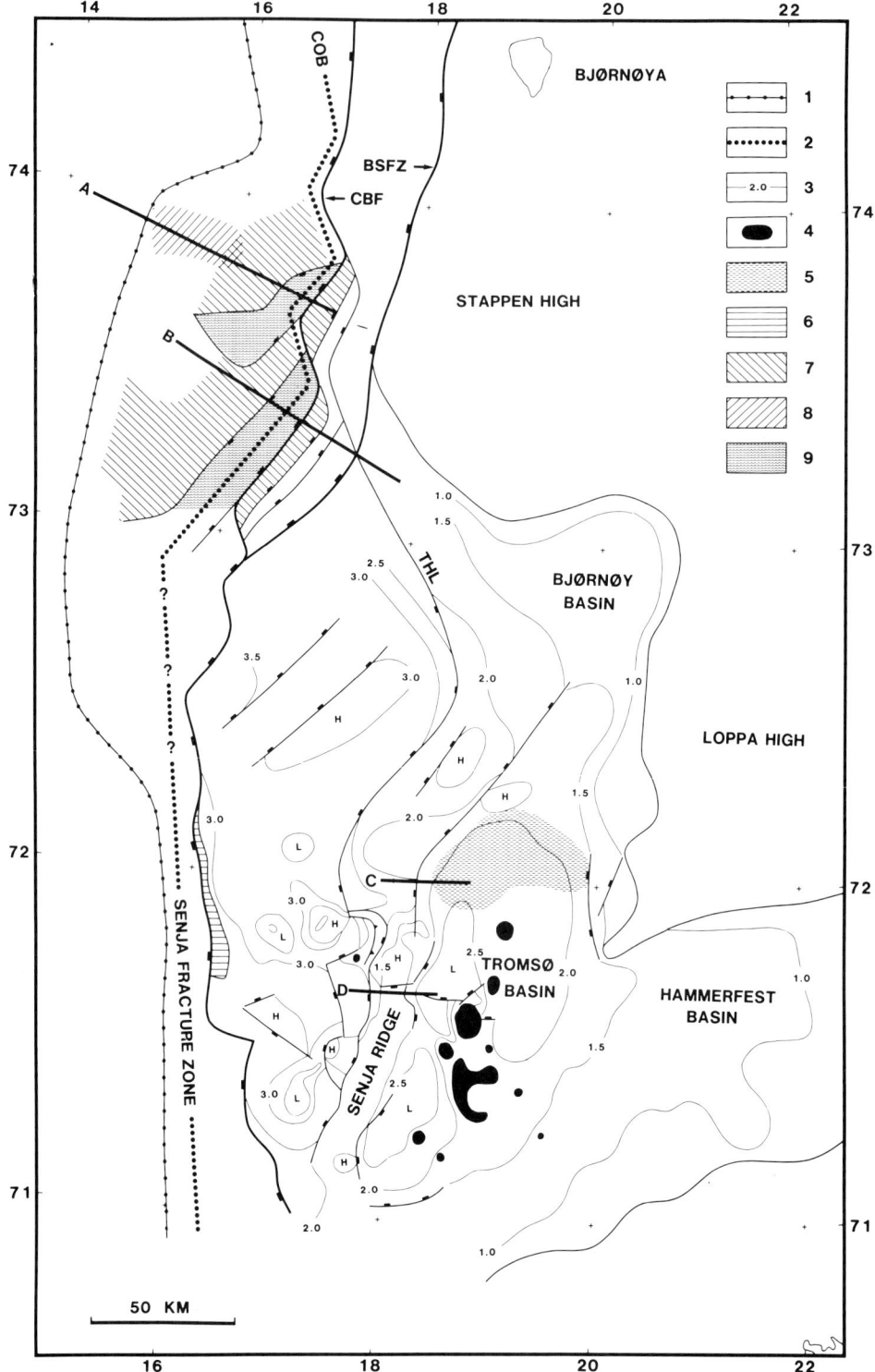

FIG. 3. Main structural elements at the rifted volcanic margin and base Tertiary time structure map (contour interval 0.5 s TWT) in the southwestern Barents Sea. Lines A–D refer to the seismic sections in Figs. 4, 5 & 6. **1** Limit of identified oceanic crust (in the seismic sections); **2** Continent–ocean boundary; **3** Base Tertiary contours (s TWT); **4** Salt diapirs; **5** Prominent tuff marker in the seismic sections; **6** Early Tertiary succession missing; **7** Flat-lying early Eocene flows; **8** Oligocene volcanic overprint; **9** Pull-apart basins. CBF: main continental boundary fault; BSFZ: Bjørnøya-Sørkapp Fault Zone; THL: Tertiary hinge line.

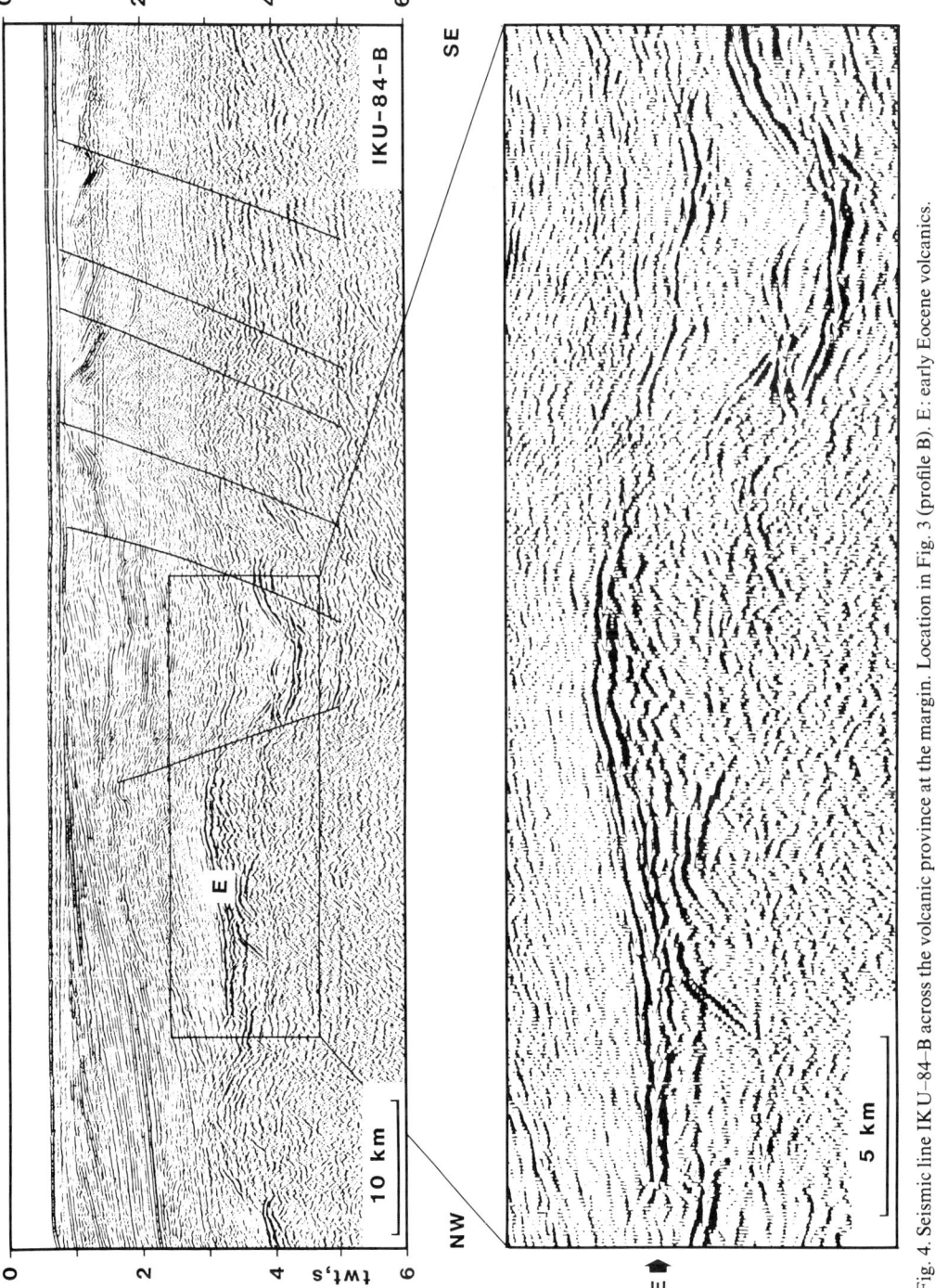

Fig. 4. Seismic line IKU–84–B across the volcanic province at the margin. Location in Fig. 3 (profile B). E: early Eocene volcanics.

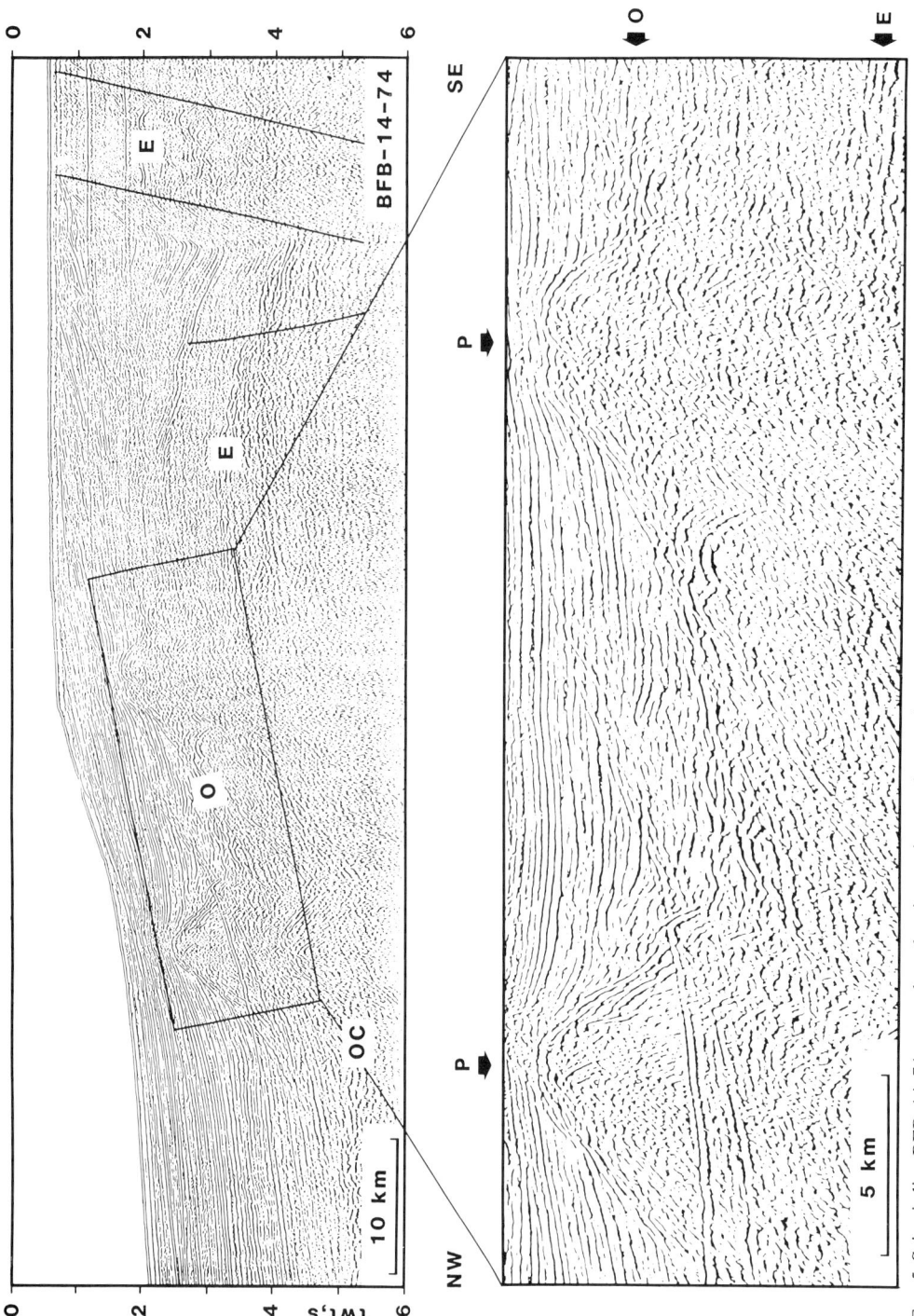

FIG. 5. Seismic line BFB-14-74 across the volcanic province at the margin. Location in Fig. 3 (profile A). E: early Eocene volcanics; O: Oligocene volcanics; OC: oceanic crust; P: basement peaks.

nized, suggesting that this event postdates the early Eocene volcanism. Each of these volcanic peaks had a relief of more than 1000 m and was later buried by the Neogene sediment wedge. We tentatively date these features as of mid-Tertiary age and relate them to a rejuvenation of volcanism associated with the major change in relative plate motion in Oligocene times.

Early Tertiary volcanism is also documented from wells in the southwestern Barents Sea, the Tertiary sequence in Spitsbergen and a high-amplitude reflector in the seismic data interpreted as representing a tuff layer. Volcanic material, tuffaceous claystone, has been reported from wells in the western Hammerfest Basin and dated as of early Eocene/late Palaeocene age (Westre 1984). A number of ash layers in the Palaeocene Firkanten Formation in Spitsbergen (Major & Nagy 1972) indicate adjacent volcanic activity. The seismic tuff marker is especially prominent in the northern Tromsø Basin (see Figs 3, 6 & 7) where it lies close to the Palaeocene–Eocene boundary, according to the Tertiary seismic stratigraphic framework of the southwestern Barents Sea (Spencer *et al.* 1984; Mathisen 1987). The marker horizon terminates against the Senja Ridge and Loppa High. Originally, the tuff may have covered a much larger area, but subsequent uplift and erosion presently restricts it to the basinal provinces.

Tertiary sedimentation and deformation in the southwestern Barents Sea and Svalbard

The Tertiary structural and stratigraphic framework of the southwestern Barents Sea and Svalbard become important when discussing the plate tectonics and timing of the volcanic events at the margin. The base Tertiary time structure map in Fig. 3 is based on a dense grid of seismic lines (5×5 km) S of 72°10′N. In the Bjørnøy Basin to the N, however, the coverage is less dense and the data of lower quality.

The base Tertiary sequence boundary is an unconformity of variable seismic signature. Nevertheless, it can confidently be recognized over large parts of the southwestern Barents Sea. In general, it is only weakly faulted except in the vicinity of the Senja Ridge and along a prominent hinge line crossing the western part of the Bjørnøy Basin (see Fig. 3).

The lowermost Tertiary unit of late Palaeocene age is relatively uniform in thickness. Originally, it probably covered most of the western Barents Sea in a quiet tectonic setting. The Eocene

sequence shows a marked degree of thickening in the Tromsø Basin where it exhibits a westerly progradational pattern. Thin lenses of sediment along the eastern flank of the Senja Ridge reflect two phases of uplift and erosion over the crest of the ridge (see Fig. 6). Minor uplift occurred at the Palaeocene–Eocene transition, whereas a major phase took place in the Eocene. The Eocene sequence is therefore thin, and locally missing, over the ridge crest. The entire early Tertiary sedimentary sequence thins, and locally pinches out, towards a western outer high forming an eastern rift flank along the Senja Fracture Zone (see Fig. 3).

The Neogene and Pleistocene sediments form a westerly thickening and dipping prograding wedge comprised of shelf and slope clastic facies overlying a major Oligocene angular unconformity. The wedge is almost without deformation, although local disturbances occur above the main early Tertiary faults and domal features. The sediments within the wedge are predominantly of Pliocene/Pleistocene age and the various depositional sequences are often separated by major unconformities.

Several positive features at the base Tertiary level in the Senja Ridge area (see Fig. 3) have a complex internal structure indicating that they are related to compressional tectonics. A thick early Cretaceous shale sequence may, however, have enhanced the structuring through mobilization. The central inverted block of the Senja Ridge (NPD-7140-73, see Fig. 6) cannot be followed as far N as 72°N (NPD-7200-73, see Fig. 6), where we only observe blocks gently downfaulted towards the W. Moving off the structure into the Tromsø Basin, we start to recognize the prominent reflector, interpreted as an ash marker associated with the early Tertiary volcanism at the margin.

Mapping of the Tertiary sediment distribution in the Bjørnøy Basin has revealed structural lineaments that are subparallel to the marginal rifted and sheared segments. The most prominent is a NE-trending fault along the northern flank of the high separating the Tromsø and Bjørnøy Basins, which continues N as a NW-trending Tertiary hinge-line crossing the Bjørnøy Basin. The Tertiary basin W of the hinge-line formed in a pull-apart setting similar to that forming the rifted margin. The clinoforms above the top Palaeocene sequence boundary indicate a transport component from the N (Rønnevik & Jacobsen 1984). We suggest that the northern provenance region is a result of inversion of the Stappen High during earliest Eocene times.

The fold and thrust belt of western Spitsbergen (see Figs 1 & 2) is a distinct regional zone of

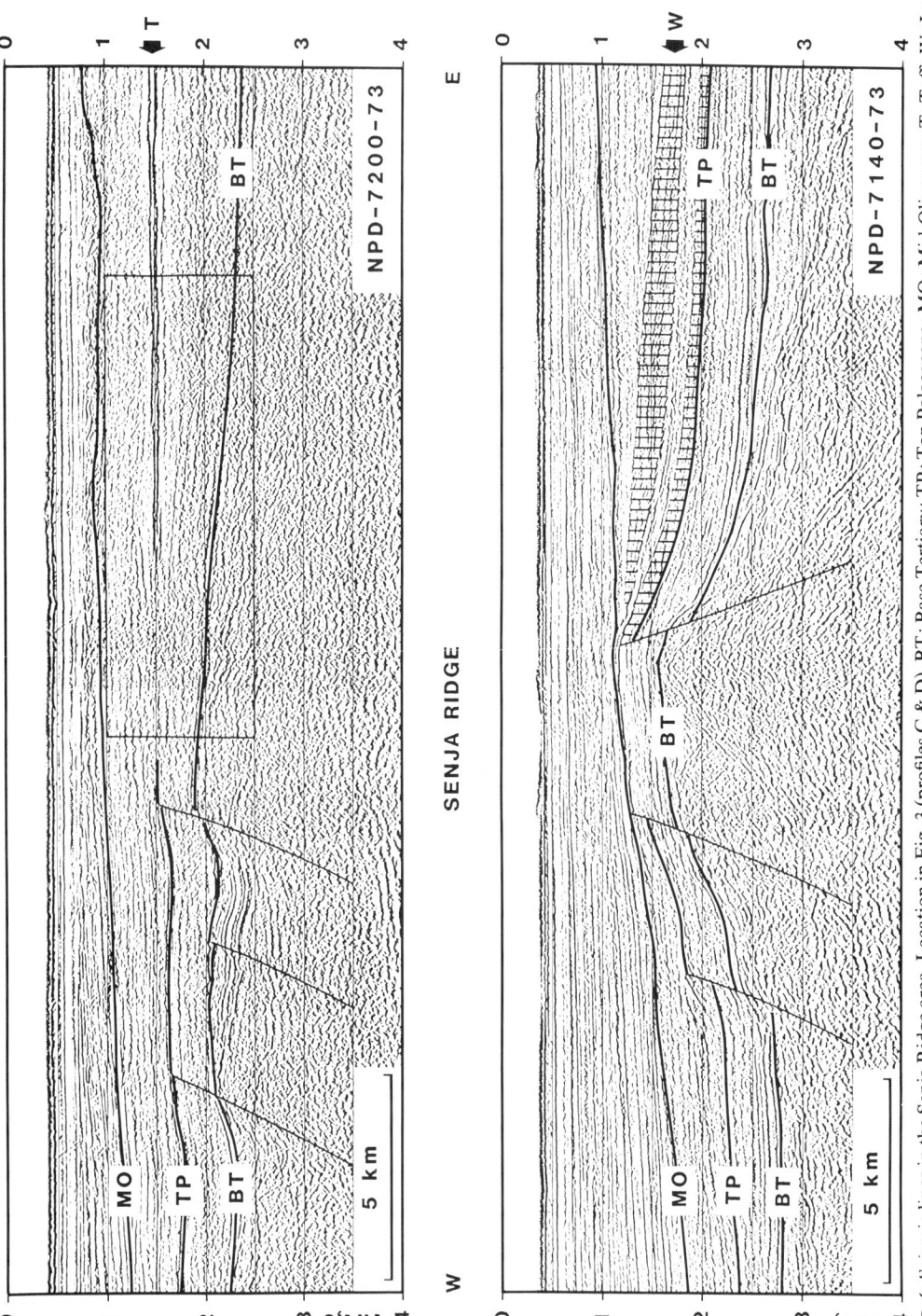

FIG. 6. Seismic lines in the Senja Ridge area. Location in Fig. 3 (profiles C & D). BT: Base Tertiary; TP: Top Palaeocene; MO: Mid-Oligocene; T: Tuff; W: Locally derived sedimentary wedges along the eastern flank of the Senja Ridge. Boxed section on line NPD-7200-77 shown enlarged in Fig. 7.

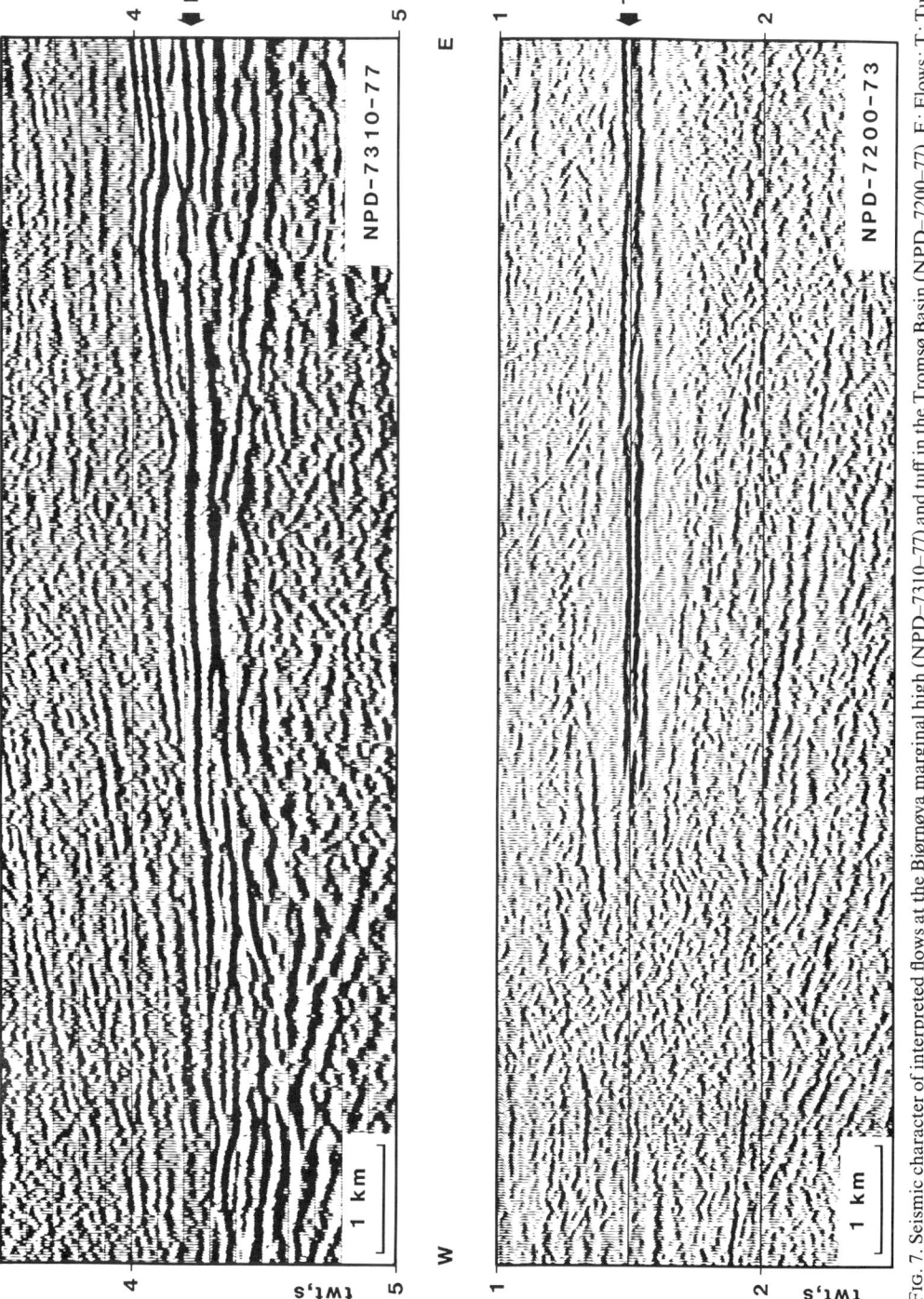

FIG. 7. Seismic character of interpreted flows at the Bjørnøya marginal high (NPD–7310–77) and tuff in the Tromsø Basin (NPD–7200–77). F: Flows; T: Tuff.

elevated basement (pre-Devonian) rocks along the western margin of the island (Steel et al. 1985). It is mainly a product of late Palaeocene–Eocene transpression and marks the approximate position of the early Palaeocene plate boundary. Thus, it should be considered part of the pre-Oligocene strike-slip system of regional shear between Svalbard and NE Greenland (see Fig. 2). Transpressional deformation has caused crustal shortening by about 10 km in the N and 15 km in the S of the island (Birkenmajer 1981), and the fold belt terminates N of Bjørnøya, where only extensional post-Palaeozoic tectonism has been recorded.

The timing of the earliest tectonic events, or at least the significant uplift along the fold and thrust belt, can be constrained by studying the filling of the flanking foreland basin. The central basin on Spitsbergen (see Fig. 1) presently contains up to 2500 m of Cainozoic clastic deposits. An additional overburden of more than 1000 m has been eroded in post-Oligocene times (Manum & Throndsen 1978). The basin development was directly related to, and largely controlled by, the marginal strike-slip movements. It was formed by pronounced subsidence adjacent to the western fold belt with rapid sedimentation taking place just prior to and during the orogenic activity. Major drainage reversal in the central basin, from an eastern to a western source area, and accompanying influx of metamorphic rock fragments caused by the emerging orogenic belt has been dated to the late Palaeocene (Steel et al. 1985). Thus, there is evidence of major tectonic activity, along the palaeo-Hornsund Fault Zone and adjacent areas, somewhat earlier than creation of the oldest dated ocean floor in the Norwegian–Greenland Sea in the earliest Eocene. The climax of orogenic activity along western Spitsbergen is likely to have been about mid-Eocene time according to stratigraphic analysis in the Central Basin (Steel et al. 1985).

Evolution of the rifted margin

Prior to the opening of the Norwegian–Greenland Sea and the eastern Arctic Ocean in earliest Eocene times, a regime of wrench tectonics existed between the Svalbard–Barents Shelf and NE Greenland. This setting was established in Cretaceous times, related to the late Mesozoic crustal extension between Norway and Greenland. Regional shear took place along a system of N–NW-trending faults bounded by the Ring-vassøy–Loppa, Bjørnøya–Sørkapp and Hornsund Fault Zones in the E and the Senja Fracture Zone

and Trolleland Fault Zone trend in the W (see Figs 1 & 2). Complex structuring through Cretaceous times gave rise to the deep Tromsø and Bjørnøy Basins in the southwestern Barents Sea, the first phase of inversion of the Senja Ridge (Riis et al. 1986) and pull-apart basins in NE Greenland (Håkansson & Stemmerik 1984).

The early Tertiary opening occurred within the same tectonic setting (see Fig. 2) and Eldholm et al. (in press) have suggested that the marginal structures reflect a stepwise northeastward propagation of the accreting plate boundary. Non-ideal transform motion caused local transpressional and transtensional deformation. Varying curvature of the plate boundary might explain local variations in structural style. Long shear, or obliquely sheared (transtension) margin segments formed at the Senja Fracture Zone and W of the southern Hornsund Fault Zone linked by a predominantly rifted segment SW of Bjørnøya. At this part of the margin the Bjørnøya marginal high probably represents a volcanic edifice which is characteristic of the volcanic type of rifted passive margins (Skogseid & Eldholm 1988).

Transpression N of the evolving ocean basin created the Spitsbergen Fold and Thrust Belt. The early Tertiary wrench regime was also accompanied by local transpression and transtension forming the present structural framework of the western Barents Sea, in particular the inversion of the Stappen High and the Senja Ridge.

The change in the direction of relative plate movement in the earliest Oligocene time marked a major change to an extensional tectonic regime at the margin N of 74°N. This caused initial crustal stretching and downfaulting along older zones of weakness and subsequent initiation of seafloor spreading and opening of the northern Greenland Sea.

The major phase of volcanism at the margin is related to the early Eocene rifted margin being formed between two regional shear, or transform, fault systems. Here, the initial oceanic crust might have been emplaced above or near sea-level in a similar manner to other marginal highs in the Norwegian–Greenland Sea. Eldholm et al. (in press) indicate that seafloor spreading in the southern Greenland Sea started at about anomaly 23 time, that is 2–4 My later than the opening of the Norwegian Sea. They recognize, however, that an anomaly at 25–24B time of opening of the southern Greenland Sea cannot be ruled out. We also infer that the tuffs mapped by seismic profiles and also drilled in the western Hammerfest Basin are related to this main episode of marginal volcanism.

It is particularly interesting to note that the structural setting of the volcanic rifted margin changes along its strike. The major expressions of extrusion are observed where rifting occurred adjacent to the Stappen High and Svalbard Platform, areas that have experienced less subsidence and have thicker crust than the basin to the S. In contrast, the part of the margin developed at the flank of the Bjørnøy Basin having thin crust shows much less intensive tectonism. The changing tectonic style is of course related to the inversion of the Stappen High. However, we point out the apparent analogy with the changing character of the Vøring and Lofoten–Vesterålen margins off Norway. This might further support the contention of Eldholm *et al.* (1979) and Skogseid & Eldholm (1987) that the pre-rifting

geological setting determines the structural style of the rifted passive margin.

Finally, we have observed mid-Tertiary volcanism and tectonic readjustments which are related to the changing stress pattern at the time of plate reorganization.

ACKNOWLEDGEMENTS: This study has greatly benefited from seismic data collected by the Norwegian Petroleum Directorate, IKU/Esso Exploration and Production Norway and Bundesanstalt für Geowissenschaften und Rohstoffe. Norsk Hydro, Saga Petroleum and Statoil kindly made reprocessed versions of the NPD-73 lines. The work has, in part, been funded by the Norwegian Petroleum Directorate. Norwegian ILP Contribution No. 34.

References

BIRKENMAJER, K. 1981. The geology of Svalbard, the western part of the Barents Sea and the continental margin of Scandinavia. *In*: NAIRN, A. E. M., CHURKIN, M. & STEHLI, F. G. (eds) *The Arctic Ocean.* Plenum Press, New York, pp. 265–329.

BREKKE, H. & RIIS, F. (in press). Tectonics and basin evolution of the Norwegian Shelf between 62° and 72°N. *Proceedings of the 4th annual meeting of the Tectonics and Structural Geology Studies Group (1986).* Special volume of the Journal of the Geological Society of Norway.

ELDHOLM, O., SUNDVOR, E. & MYHRE, A. M. 1979. Continental margin off Lofoten-Vesterålen, Northern Norway. *Marine Geophysical Researches,* **4**, 3–35.

——, THIEDE, J. & TAYLOR, E. *et al.* 1987. *Proceedings of the Ocean Drilling Program,* US Government Printing Office, Washington, **104A**.

——, FALEIDE, J. I. & MYHRE, A. M. (in press). Continent-ocean transition at the western Barents Sea-Svalbard margin. *Geology.*

FALEIDE, J. I., GUDLAUGSSON, S. T. & JACQUART, G. 1984. Evolution of the western Barents Sea. *Marine and Petroleum Geology,* **1**, 123–150.

GUDLAUGSSON, S. T., FALEIDE, J. I., FANAVOLL, S. & JOHANSEN, B. 1987. Deep seismic reflection profiles across the western Barents Sea. *Geophysical Journal of the Royal Astronomical Society,* **89**, 273–278.

HARLAND, W. B. 1969. Contribution of Spitsbergen to understanding of tectonic evolution of the North Atlantic Region. *In*: KAY, M. (ed) *North Atlantic Geology and Continental Drift.* Memoir of the American Association of Petroleum Geologists, **12**, pp. 817–851.

HINZ, K., DOSTMAN, H. J. & HANISCH, J. 1984. Structural elements of the Norwegian continental margin. *Geologische Jahrbuch,* **A75**, 193–221.

HÅKANSSON, E. & STEMMERIK, L. 1984. The Wandel Sea Basin—the Northeast Greenland equivalent to the Svalbard and Barents Shelf. *In*: SPENCER, A. M. *et al.* (eds) *Petroleum Geology of the North European Margin.* Graham & Trotman, London, pp. 97–108.

KELLOGG, H. E. 1975. Tertiary stratigraphy and tectonism in Svalbard and continental drift. *Bulletin of the American Association of Petroleum Geologists,* **59**, 465–485.

LOWELL, J. D. 1972. Spitsbergen Tertiary orogenic belt and the Spitsbergen Fracture Zone. *Bulletin of the Geological Society of America,* **83**, 3091–3102.

MAJOR, H. & NAGY, J. 1972. Geology of the Adventdalen map area. *Norsk Polarinstitutt Skrifter,* **138**, 58 pp.

MANUM, S. B. & THRONDSEN, T. 1978. Rank of coal and dispersed organic matter and its bearing on the Spitsbergen Tertiary. *Norsk Polarinstitutt Årbok 1977,* 159–177.

MATHISEN, K. 1987. *The Senja Ridge—Tertiary structuration in the western Barents Sea.* Cand. scient. thesis, University of Oslo, 78 pp. (in Norwegian).

MUTTER, J. C., TALWANI, M. & STOFFA, P. L. 1982. Origin of seaward dipping reflectors in oceanic crust off the Norwegian margin by 'subaerial seafloor spreading'. *Geology,* **10**, 353–357.

—— & ZEHNDER, C. M. (1988). Deep crustal structure and magmatic processes; the inception of seafloor spreading in the Norwegian–Greenland Sea. *In*: MORTON, A. C. & PARSON, L. M. (eds) *Early Tertiary Volcanism and the Opening of the NE Atlantic.* Geological Society of London, Special Publication, **39**, pp. 35–48.

MYHRE, A. M. 1984. *Marine geophysical studies in the Norwegian–Greenland Sea and adjacent margins.* Doctorate thesis, University of Oslo, 201 pp.

——, ELDHOLM, O. & SUNDVOR, E. 1982. The margin between Senja and Spitsbergen Fracture Zones: implications from plate tectonics. *Tectonophysics,* **89**, 33–50.

—— & —— (in press). The western Svalbard Margin (74–80°N). *Marine and Petroleum Geology.*

RIIS, F., VOLLSET, J. & SAND, M. 1986. Tectonic development of the western margin of the Barents Sea and adjacent areas. *In*: HALBOUTY, M. T. (ed) *Future Petroleum Provinces of the World.* Memoir of the American Association of Petroleum Geologists, **40**, pp. 661–676.

RØNNEVIK, H. & JACOBSEN, H. P. 1984. Structural highs and basins in the western Barents Sea. *In*: SPENCER, A. M. *et al.* (eds) *Petroleum Geology of the North European Margin.* Graham & Trotman, London, pp. 19–32.

SKOGSEID, J. & ELDHOLM, O. 1987. Early Cenozoic crust at the Norwegian continental margin and the conjugate Jan Mayen Ridge. *Journal of Geophysical Research*, **92**, 11471–11491.

—— & —— 1988. Early Cainozoic evolution of the Norwegian volcanic passive margin and the formation of marginal highs. *In*: MORTON, A. C. & PARSON, L. M. (eds) *Early Tertiary Volcanism and the Opening of the NE Atlantic.* Geological Society of London, Special Publication, **39**, pp. 49–56.

SPENCER, A. M., HOME, P. C. & BERGLUND, L. T. 1984. Tertiary structural development of the western Barents Shelf, Troms to Svalbard. *In*: SPENCER, A. M. *et al.* (eds) *Petroleum Geology of the North European Margin.* Graham & Trotman, London, pp. 199–209.

STEEL, R., GJELDBERG, J., NØTTVEDT, A., HELLAND–HANSEN, W., KLEINSPEHN, K. & RYE–LARSEN, M. 1985. The Tertiary strike-slip basins and orogenic belt of Spitsbergen. *Society of Economical Paleontologists and Mineralogists, Special Publication*, **37**, 339–360.

TALWANI, M. & ELDHOLM, O. 1977. Evolution of the Norwegian–Greenland Sea. *Bulletin of the Geological Society of America*, **88**, 969–999.

WESTRE, S. 1984. The Askeladden gas find. *In*: SPENCER, A. M. *et al.* (eds) *Petroleum Geology of the North European Margin.* Graham & Trotman, London, pp. 33–39.

J. I. FALEIDE, A. M. MYHRE & O. ELDHOLM, Department of Geology, University of Oslo, Norway.

Some properties of basalt lava sequences and volcanic centres in a plate-boundary environment

L. Kristjansson & J. Helgason

SUMMARY: This paper discusses the implications that two current studies in Iceland may have in the interpretation of results from marine deep drilling and geophysical surveys in marginal volcanic areas of the N Atlantic. One study, concerning the magnetic remanence of lava flows, shows that the rate of geomagnetic reversals in the Tertiary is higher than is generally expected, and that caution must be exercised in tracing magnetic lineations between coeval submarine and subaerial formations. The other study, which deals with the distribution of major volcanic complexes within the Icelandic lava pile, demonstrates that this distribution is dependent upon various tectonic factors, especially any lateral shifts or direction changes of the active plate boundary that may have occurred.

During recent years, much evidence of marginal volcanism dating from the initial stages of opening of the N Atlantic has been found in the submarine areas off Europe and Greenland. This evidence is mostly derived from geophysical surveys and deep drilling, including Legs 81 and 104 of the Deep Sea Drilling Project/Ocean Drilling Program (DSDP/ODP). Either method yields valuable but rather incomplete information on the geological processes taking place during the initial rifting.

Observations of the accessible early Tertiary areas in Britain, the Faeroes and Greenland are likely to provide important constraints on the interpretation of the offshore results. Further constraints and analogies have emerged from research in Iceland, where a fairly continuous history of events at a divergent plate boundary during the last 15 My can be examined in surface exposures. Early models of crustal rifting in Iceland (Bödvarsson & Walker 1964; Pálmason 1980) were applied by Mutter et al. (1982) and by the Leg 81 Scientific Party (1982) to the generation of subaerial volcanics at the Norwegian and Rockall Plateau margins, respectively. A subsequent model (Helgason 1984) is being used by Skogseid & Eldholm (1987) in their description of volcanism and subsidence occurring during the formation of the Vøring Plateau.

For those interested in comparisons of various geological and geophysical observations between Iceland and marginal volcanic areas, we refer to symposium volumes and review articles available on Icelandic geology (Fridleifsson et al. 1982; Björnsson 1983; Steinthorsson & Jacoby 1985; Saemundsson 1986). It should be kept in mind, however, that the geological history, as observed in Iceland, results from an interplay of hot-spot and rift-zone influences, and is only to a limited extent analogous to the marginal volcanism. The geological environment in Iceland has also been influenced by a glacial climate during the past 3 My.

In this paper we shall concentrate on the description of two related features of the geology of Iceland which are currently being studied and evaluated in the light of the crustal generation model of Helgason (1984). These features are: (1) the palaeomagnetism of lava series and their associated aeromagnetic anomalies; and (2) the distribution of major volcanic centres. The study of these features in Iceland has far-reaching implications for the interpretation of oceanic magnetic survey results, and it may also help in establishing correct stratigraphic and structural relationships in marginal ocean areas.

Magnetic observations in Iceland

The linear magnetic anomalies observed over the submarine ridges SW and N of Iceland are well developed, allowing in many cases definite identification of individual lineations out to an inferred seafloor age of 10 Ma and a more tentative correlation to 20 Ma (Nunns et al. 1983; Vogt 1986). The sources of these anomalies are produced by the combined effects of seafloor-spreading processes and geomagnetic polarity reversals. The reversals will also be recorded as polarity zones in a lava pile emplaced in a subaerial rift-zone environment of the same age range, the resulting remanence contrasts giving rise to magnetic anomalies above these lavas.

At the present stage of research on aeromagnetic anomalies in Iceland, difficulties are being encountered in correlating individual polarity zones or anomaly lineations older than 5–6 Ma between the ocean floor and Iceland. This is due

From MORTON, A. C. & PARSON, L. M. (eds), 1988, *Early Tertiary Volcanism and the Opening of the NE Atlantic*, Geological Society Special Publication No. 39, pp. 147–155.

to the following reasons, some of which will also apply to the Atlantic marginal volcanism:

(1) Inadequacy of the 'geomagnetic polarity time scale'. Palaeomagnetic results have been obtained in several long composite K-Ar dated lava profiles in Iceland, involving magnetic measurements on samples from some 4000 lava flows. From a statistical study of reliably determined palaeofield directions in these surveys, Kristjansson & McDougall (1982) concluded that the rate of geomagnetic reversals in the last 15 My was at least eight per million years on average. In contrast, most versions of the polarity time scale (*e.g.* Harland *et al.* 1982; Lowrie & Kent 1983) indicate that the average reversal rate in this period was only four to five per million years. The discrepancy is most likely to originate from short ($< 50\,000$ years) geomagnetic subchrons whose magnetic field signal is attenuated by the distance between the ocean floor and the survey magnetometer. This fact will account for the presence of 'unexpected' polarity zones in cored deep-ocean sequences (Hall & Robinson 1979; Leg 81 Scientific Party 1982).

(2) The low tilt angle of the lava pile. The average thickness of a polarity zone in the lava pile of Iceland is about 200 metres, and the pile has generally been tilted by less than $10°$. Immediately above a level surface of the lava pile, magnetic anomalies may be expected to alternate from positive to negative at the boundaries of these zones. However, at an altitude of even only a few hundred metres, the magnetic signal has been smoothed to reflect long-wavelength variations in the magnetization of the lavas rather than individual polarity zones. In that context it is significant that the thickness of magnetized crust in Iceland is likely to exceed 3 km (Kristjansson & Watkins 1977).

A good example of such interpretation problems is seen in the 8–15 Ma lava pile of the northwestern peninsula of Iceland (Fig. 1) whose remanence polarity structure has been studied extensively by McDougall *et al.* (1984). Their sampling was carried out in locations indicated with closed circles, giving rise to the polarity columns shown in Fig. 1. Comparison with aeromagnetic profiles flown at 900 m above sea-level and 3 km spacing by Sigurgeirsson (1984) over the area reveals only a faint correspondence with the expected anomaly pattern. For example, a relatively thick sequence of normally magnet-ized lava flows outcropping in the eastern part of the peninsula has been assigned by McDougall *et al.* (1984) to the interval of generally normal geomagnetic polarity at about 10 Ma which causes the marine lineation known as 'Anomaly

5'. Although this anomaly is very prominent in all ocean-ridge magnetic surveys, the corresponding lava series does not give rise to a recognizable signature in the aeromagnetics. Conversely, a prominent positive anomaly occurs over the enclosed area marked with a plus sign in Fig. 1, without having an apparent relation to the polarity zone pattern in the nearby sampled profiles.

(3) Effects of topography. When the tilt of the lava pile is only a few degrees, topographic features may influence the anomaly field, especially if these are elongated erosion features or scarps. Thus, N of point C in Fig. 1 a fjord carved out of mostly normally magnetized lava series is found to cause a small negative aeromagnetic anomaly aligned along a direction which is unrelated to the local tectonic strike. The relative effect of topography will increase with the age and progressive alteration of the lavas, as the alteration of basalts causes a decay of their primary remanence (Wood & Gibson 1976) along with an increase in their induced and viscous magnetization (Kristjansson & Watkins 1977).

(4) The lenticular structure of the lava pile. There is evidence that during the lifetime (around 0.5 My) of each major volcanic centre (see the section on volcanic complexes below), and its associated dyke swarm, relatively rapid accumulation of lava flows will occur in an area having dimensions of the order of 10–20 km by 50 km. Inferred thicknesses of polarity zones in these areas, if crossed by a composite sampling section or an aeromagnetic profile, will then not be in proportion to the corresponding chron lengths in the magnetic polarity time-scale. In agreement with this, Johannesson & Kristjansson (in prep.) have found that similarities between the polarity zone pattern of section A'B and that of A'C in Fig. 1 become progressively smaller with increasing distance between coeval parts of these sections. The problem of stratigraphic correlation between such sections, either by direct mapping or by anomaly tracing, is further compounded by the presence of hiatuses and unconformities (see Fig. 1) caused by frequent migration of the volcanic zone (Helgason 1984). Normal faulting, on the other hand, appears from experience in Iceland to be a less serious problem in such correlations.

(5) Localized magnetic anomalies occur at some of the volcanic centres in Iceland and offshore (Sigurgeirsson 1970; Kristjansson 1976a), particularly in the central regions of western Iceland. Their main characteristics are quite similar to those described for British Lower Tertiary vol-

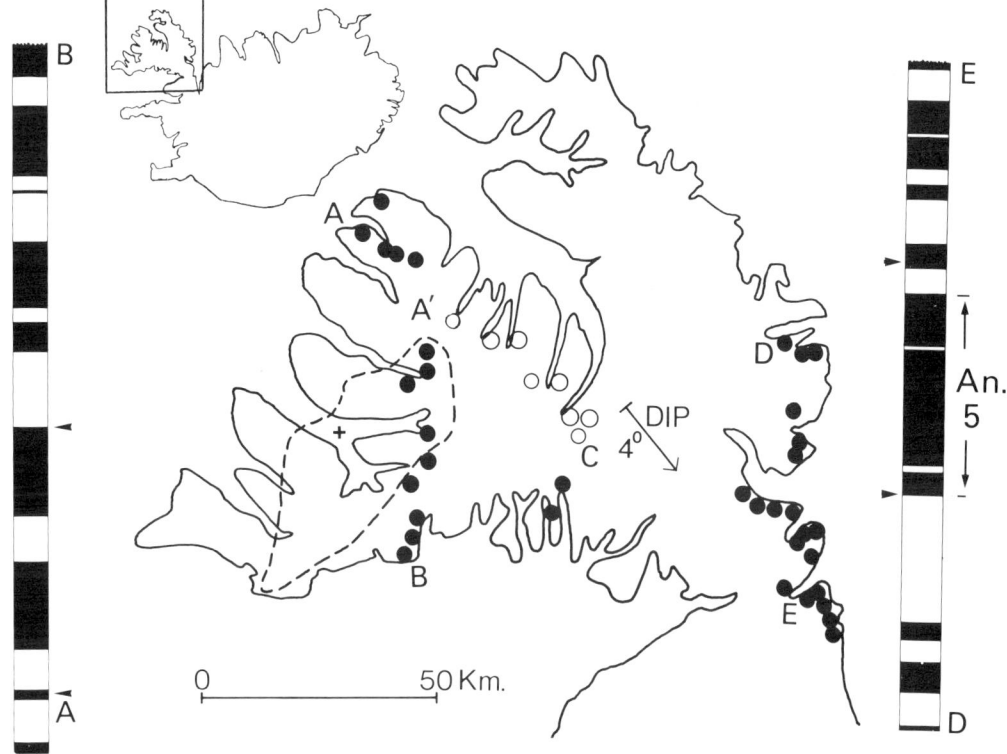

FIG. 1. The northwestern peninsula of Iceland. Palaeomagnetic sampling localities of McDougall *et al.* (1984) and Johannesson & Kristjansson (in prep.) are shown as closed and open circles respectively. Mean dip and strike of the lava pile are indicated. Primary remanence polarity column for the section A–B is on the left, and the corresponding column for D–E is on the right. Dark corresponds to normal magnetization. The cumulative thickness of each column where sampled is 3–4 km. Arrows show positions of unconformities. The ages inferred for column end-points from K-Ar measurements are A: 14.0 Ma, B: 11.9 Ma, D: 12.7 Ma, E: 8.0 Ma. The enclosed region marked + has the most prominent magnetic anomaly of the peninsula.

canic centres (Hall & Dagley 1970; Bott & Tantrigoda 1987). These anomalies may be due to cone sheets, caldera-filling materials (Fridleifsson & Kristjansson 1972), coarse-grained mafic intrusions, or even andesitic rocks (Kristjansson *et al.* 1977); they are generally positive and less than 10 km in size.

(6) Possible lateral offset between magnetic anomaly lineations from typical oceanic crust and magnetic anomalies from coeval lava series. A proportion of outcropping Tertiary lava flows in Iceland must have flowed from feeder dykes now covered by younger lavas a considerable distance down dip (Helgason 1984). It is evident that magnetic anomalies observed over these lavas will be offset, possibly by several km, from a direct onshore continuation of anomalies caused

by contemporaneous submarine volcanism on the same spreading axis. In Iceland, it has been found that dykes are not a major contributing cause of aeromagnetic anomalies (Kristjansson 1985).

In summary, it does not seem to be a straightforward matter to trace magnetic anomalies in detail from areas of typical ocean crust through a shelf or transitional region in order to connect them with anomaly lineations originating in subaerial volcanics. In southwestern Iceland, however, anomaly lineations younger than the latest major reorganization of volcanic zones at 5–6 Ma (Johannesson 1980) are reasonably well developed. Survey work, ground mapping, and modelling studies now in progress may make it possible to follow some of these anomalies, as well as anomaly 5, across the shelf and even across the entire island.

LEGEND

TERTIARY STRATA (3 - 16 M. Y.)

LOWER QUATERNARY STRATA
(0.7 - 3 M. Y.)

UPPER QUATERNARY STRATA
OF BRUHNES AGE (0 - 0.7 M. Y.)

MAINLY SANDUR PLAINS

ACTIVE VOLCANIC SYSTEM

CRUSTAL COMPLEX

GLACIER / ICE CAP

The distribution of volcanic complexes

Volcanic complexes in the Iceland hot-spot frame

The main topographic high of the Iceland region has an inferred lower age boundary coinciding roughly with the onset of the Miocene (25 Ma) and/or seafloor anomaly 6 (Vogt *et al.* 1980). Within the subaerial Iceland segment there are numerous volcanic complexes, which are in many respects similar to the Tertiary complexes of Britain and E Greenland (*e.g.* Walker 1963; Roberts *et al.* 1983) but smaller in size (< 10 km); this may be a reflection of the crustal thickness being less in Iceland than in the continental areas. Their main characteristics include the presence of differentiated rocks, a caldera at the centre of a dyke swarm, high-temperature alteration, intrusive activity, and various geophysical anomalies (Kristjansson 1976a; Saemundsson 1986). Few if any complexes of this type occur on the ocean ridges to the N and SW, i.e. on the Kolbeinsey and Reykjanes Ridges.

Several authors have used volcano spacing in Iceland to model long-term crustal construction processes or to infer hot-spot migration (Vogt 1974; Walker 1974). However, available compilations of volcano complexes (*e.g.* Sigurdsson 1967) are by now somewhat outdated. We present a new map (Fig. 2) of the sites of over 150 volcanic complexes in Iceland and its vicinity, based on various published and unpublished sources. Our main aims are to:

(1) Analyse to what extent these complexes conform to a regional and temporal pattern.
(2) Attempt to use the present distribution pattern to infer large-scale evolution of crustal properties within the Iceland hot-spot frame over the last 25 My, including shifts of the plate boundary, crustal thickness, and linear magnetic anomalies.
(3) Apply the temporal variation in the distribution of volcanic complexes to infer variations in the production of acidic tuff layers in the N Atlantic.

Classification and compilation of volcanic complexes in the Iceland area

The systematic work of G. P. L. Walker and his colleagues in eastern Iceland proved fundamental to an understanding of the interrelationships between plutonic, hypabyssal, and extrusive rocks both of mafic, intermediate and felsic compositions (*e.g.* Walker 1963, 1966). The local occurrence of rhyolitic intrusions and sheets is a first order criterion to detect a partly buried volcanic chamber in an unmapped area. Felsic bodies normally indicate that a magma chamber was present at a shallow crustal level. Exposures of rhyolitic rocks in Iceland similarly indicate that a late-stage volcanic centre had developed, but notable exceptions include the monogenetic Sandfell laccolith in Faskrudsfjordur, eastern Iceland (Hawkes & Hawkes 1933). In other regions mafic intrusions are more conspicuous (Fridleifsson 1977), and some centres in the volcanic zones are recognized by the presence of fissure swarms and geothermal activity.

It is evident that various different aspects may be used in the classification of the volcanic centres, including their tectonic environment, structure, and geochemistry. Only the first of these will be dealt with below, as it is probably both the simplest and best documented aspect.

Interpretation

We studied the distribution or density of volcanic complexes in eight regions (see Fig. 2). In our approach we considered both the tectonic environment, the number of crustal complexes, their average area available for each complex, their distribution within each region and the degree of erosion as well as the available data on age boundaries for the regions. We have not taken into consideration the actual size of each complex: it is in most cases poorly known, and it is also not clear what measure of volcano size should be chosen for comparison between these, e.g. the area of a caldera or a dyke swarm, or a volume based on geophysical observations. Some of our results are presented in Table 1.

Regions 1, 2 and 4 are located at the accreting plate boundary. The mean area available for each complex is only 300–400 km^2. For all the other regions the density of complexes was smaller than this, in particular when it is taken into account that regions 1, 2 and 4 were formed during a relatively short period.

Region 3 is here regarded as a zone dividing northern and southern Iceland (N–S boundary zone) where a change in main tectonic trend from

FIG. 2. A compilation map for known volcanic complexes in the Iceland area (solid dots). The map is divided into eight regions of distinct tectonic character—regions 1, 2 and 4: active volcanism and crustal accretion; region 3: 'N–S boundary zone' with active volcanism and some crustal accretion; region 5: southern Iceland propagating volcanism; regions 6 and 7: Tertiary lava terrains of eastern and northwestern Iceland; region 8: early Miocene terrain off northwestern Iceland.

TABLE 1. *Compilation of crustal complexes in the Iceland region*

Region	Complexes	Area km^2	Area/ Compl. km^2	Approximate duration (Ma)
1	5	2036	407	0–0.7
2	4	1690	422	0–0.7
3	70	33670	481	0–6
4	14	4218	301	0–0.7
5	7	5518	790	0–0.7
6	12	11345	945	6–14
7	9	10876	1208	6.7–14
8	4	22675	5669	14–24

For locations of regions 1–8 see Figure 2

a northeasterly to a northerly direction is taking place. It is therefore transverse in comparison to the main active tectonic belts of Iceland, and includes a large time transgression (from the Miocene to the present) with regard to its volcanic centres. Seventy out of the 150 complexes of Figure 2 fall within this boundary zone. Broadly speaking, this region includes much of the highest topography and the thickest crust in Iceland. We also note that this zone differs from the other regions of Figure 2 in that the number and density of its crustal complexes has a tendency to increase with age. We attribute this increase in part to greater erosion and thus exposure to deeper crustal levels, but in part also to the passive nature of this tectonic environment. Thus, due to lack of crustal accretion in the boundary zone, its E–W orientation, and its extreme width (650 km) volcanism and formation of complexes has persistently been concentrated in this region. When compared to other regions in Iceland it would indeed seem that crustal accretion will lead to burial and 'disappearance' of volcanic complexes. On the other hand we attribute the large amount of time-transgression in region 3 to the long-term existence of separate tectonic trends in the northern and southern volcanic belts.

Region 5 in southern Iceland is highly active volcanically although its crustal accretion is small so far. Here the apparent density of volcanic complexes is low, with the average area per complex being about 800 km^2, but it should be kept in mind that the complexes of this region are generally quite large. We also note that there is a wide area between regions 5 and 3 where no volcanic complexes are present. Although volcanic fissures extend into this intermediate area they are all connected with complexes in either region 3 or 5.

Regions 6 and 7 are Upper Tertiary basalt lava terrains of eastern and western Iceland that formed roughly between 6 and 14 Ma. Here the area available for each complex is around 1000 km^2, i.e. twice as much as at the accreting plate boundary. A few of the complexes in Figure 2 were not included with any of the eight regions; they lie within crust of Tertiary age and have a similar distribution pattern as complexes within regions 6 and 7.

Region 8 is the submarine extension of northwestern Iceland. Only four localized complexes were found in a fairly detailed geophysical survey in this region (Kristjansson 1976b), resulting in an estimated area of 5550 km^2 per volcanic complex. However, this estimate is an upper bound as in the adjacent northwestern peninsula (region 7) most of the exposed volcanic complexes (see Fig. 2) are without prominent localized gravity or aeromagnetic anomalies (Einarsson 1954; Sigurgeirsson 1984). This may be due to less intrusive activity being associated with the complexes here than in other parts of Iceland, or it may be due to geochemical reasons.

Discussion

Except for region 3 (see Fig. 2), crustal complexes clearly tend to decrease in number with age. This raises the question of whether the generation of volcanic centres has increased in the Iceland area between the Upper Miocene and the present day, or whether there is a process that obliterates the centres formed in the volcanic zone when crustal rifting brings these away from the plate boundary. To answer this question it is necessary to bring into focus available models for the long-term crustal construction process in Iceland. Several such models have been proposed, assuming either: (1) that the plate boundary has been stationary for the past 15 My (Pálmason 1980); or (2) that the plate boundary has repeatedly shifted and that these shifts strongly affect the long-term crustal construction process, including the subsidence history and thus the temporal and spatial distribution of volcanic complexes in the Iceland crust (Helgason 1984, 1985).

We note in Figure 2 that no volcanic complexes have been detected in the 0.7–3.1 Ma areas flanking the neovolcanic zones (region 2) in northeastern Iceland. These are the distal areas of the plate boundary to which lavas have flowed, i.e. a depositional environment outside the areas of lava eruption. If the plate boundary had remained stationary for the entire period since 10 or 15 Ma, deep burial of volcanic complexes by distal-type lavas would probably have resulted. It has, however, been suggested (Helgason 1984) that the numerous complexes which have been exposed by the erosion of only approximately 500 m of lava from eastern Iceland, represent palaeo-rift zones. This circumstance is explained by frequent shifting of the plate boundary since the mid-Miocene, so that the rifting and extrusion process does not persist in one location long enough for deep burial of volcanic complexes formed at the rift axis. As an example, the eroded Thingmuli, Breiddalur and Alftafjördur volcanic centres of eastern Iceland were penecontemporaneously active on a common rift zone during the late Miocene.

We regard the low complex density in region 8 as a special feature which may be accounted for in several ways apart from the above explanations based on lack of geophysical signatures. Shifting of the plate boundary may have been infrequent during crustal construction in this region leading to deep burial of volcanic complexes. Another possibility is that it took many millions of years for the crust to develop sufficient thickness or other properties favouring the growth of volcanic complexes. In this context it is significant that Sigurdsson & Loebner (1981), on the basis of acidic tephra layers in marine sediments, show some evidence for a gradual increase in explosive volcanism in the Iceland area from the onset of the Miocene.

Conclusions

In all geological work it is appropriate to extrapolate not only from the present to the past, but also from the exposed to the unexposed. Iceland is certainly quite different from the Vøring Plateau and other marginal volcanic areas of the N Atlantic in key parameters such as age and crustal thickness. However, in Iceland good exposures provide direct observational control on geological properties such as lava accumulation rates, distribution of volcanic complexes, and tectonic patterns. The interrelations of these properties provide clues to processes having general applicability to a variety of geological, geophysical and geochemical studies of subaerial volcanism in marginal areas.

One aspect of the lava pile in Iceland which we have dealt with in the present paper is its remanent magnetization, which is the basis of a very useful means of stratigraphic correlation over short distances (up to tens of kilometres) in the field. Along with K-Ar dating, this property also furnishes an empirical measure of build-up rates in the lava pile, and it provides a way of estimating certain characteristics of the time-scale for world-wide geomagnetic polarity reversals more reliably than by indirect methods such as anomaly lineation inversions. Ideally, the magnetization of an inaccessible tilted lava pile should also allow tectonic directions, ages and other important features to be deduced through an analysis of magnetic survey results. We demonstrate, however, that several sources of error may occur in magnetic anomaly interpretation over lava series, affecting its usefulness in correlation with ocean-floor anomalies. One of these is due to volcanic complexes which cause enhanced production of lava flows during randomly selected periods, and which also give rise to localized (< 10 km) geomagnetic anomalies.

The second aspect concerns our compilation of crustal complexes of volcanic origin within Iceland and the submarine shelf. It shows that their distribution is dependent on crustal age and tectonic setting. In central Iceland an E-W-trending zone separates the volcanic belts of northern and southern Iceland. Time-transgression is continuous from Recent to Miocene strata in this N-S boundary zone, with a high density of crustal complexes, 480 km^2 per complex. In the Tertiary regions of eastern and northwestern Iceland (6-14 Ma age) the complex density is much lower, with approximately one complex per 1000 km^2. Within the neovolcanic zone (0–0.7 Ma) the density of complexes is highest, about 350 km^2 per complex. A very low density is recorded off the NW coast of Iceland. Similarly, large areas of Quaternary age that flank the neovolcanic zone have a very low density of complexes due to severe blanketing effects by more recent distal-type lavas. By analogy, it may be anticipated that the older volcanic segments of the N Atlantic are heterogeneous with regard to the distribution of crustal complexes and that the local tectonic environment during the first 10 My of ocean formation is mainly responsible for this heterogeneity.

References

BJÖRNSSON, S. 1983. Crust and upper mantle beneath Iceland. *In*: BOTT, M. H. P., SAXOV, S., TALWANI, M. & THIEDE, J. (eds) *Structure and Development of the Greenland–Scotland Ridge*. Plenum Press, New York, pp. 31–61.

BÖDVARSSON, G. & WALKER, G. P. L. 1964. Crustal drift in Iceland. *Geophysical Journal of the Royal Astronomical Society*, **9**, 285–300.

BOTT, M. H. P. & TANTRIGODA, D. A. 1987. Interpretation of the gravity and magnetic anomalies over the Mull Tertiary intrusive complex, NW Scotland. *Journal of the Geological Society of London*, **144**, 17–28.

EINARSSON, T. 1954. *A Survey of Gravity in Iceland.* Societas Scientiarum Islandica Publication, **30**.

FRIDLEIFSSON, I. B. 1977. Distribution of large basaltic intrusions in the Iceland crust and the nature of the layer 2/layer 3 boundary. *Bulletin of the Geological Society of America*, **88**, 1689–1693.

—— & KRISTJANSSON, L. 1972. The Stardalur magnetic anomaly, SW-Iceland. *Jökull*, **22**, 69–78.

——, GIBSON, I. L., HALL, J. M., JOHNSON, H. P., CHRISTENSEN, N. I., SCHMINCKE, H.-U- & SCHÖNHARTING, G. 1982. The Iceland Research Drilling Project. *Journal of Geophysical Research*, **87**, 6359–6361.

HALL, D. H. & DAGLEY, P. 1970. Regional magnetic anomalies. *Institute of Geological Sciences Report*, **70/10**.

HALL, J. M. & ROBINSON, P. T. 1979. Deep crustal drilling in the North Atlantic Ocean. *Science*, **204**, 573–586.

HARLAND, W. B., COX, A. V., LLEWELLYN, P. G., PICKTON, C. A. G., SMITH, A. G. & WALTERS, R. 1982. *A Geologic Time Scale.* Cambridge University Press, Cambridge.

HAWKES, L. & HAWKES, H. K. 1933. The Sandfell laccolith and 'dome of elevation'. *Quarterly Journal of the Geological Society of London*, **89**, 378–398.

HELGASON, J. 1984. Frequent shifts of the volcanic zone in Iceland. *Geology*, **12**, 212–216.

—— 1985. Shifts of the plate boundary in Iceland: some aspects of Tertiary volcanism. *Journal of Geophysical Research*, **90**, 10084–10092.

JOHANNESSON, H. 1980. Evolution of rift zones in western Iceland (in Icelandic with an English summary). *Natturufraedingurinn*, **50**, 13–31.

KRISTJANSSON, L. 1976a. Marine magnetic surveys off the west coast of Iceland. *Societas Scientiarum Islandica, Misc. Papers*, **5**, 23–42.

—— 1976b. Central volcanoes on the western Icelandic shelf. *Marine Geophysical Researches*, **2**, 285–289.

—— 1985. Magnetic and thermal effects of dike intrusions in Iceland. *Journal of Geophysical Research*, **90**, 10129–10135.

—— & WATKINS, N. D. 1977. Magnetic studies of basalt fragments recovered by deep drilling in Iceland, and the 'magnetic layer' concept. *Earth and Planetary Science Letters*, **34**, 365–374.

—— & MCDOUGALL, I. 1982. Some aspects of the late Tertiary geomagnetic field in Iceland. *Geophysical Journal of the Royal Astronomical Society*, **68**, 273–294.

——, THORS, K. & KARLSSON, H. 1977. Confirmation of central volcanoes off the Icelandic coast. *Nature*, **268**, 325–326.

LEG 81 SCIENTIFIC PARTY 1982. Leg 81 drills west margin, Rockall Plateau. *Geotimes*, **27**, 21–23.

LOWRIE, W. & KENT, D. V. 1983. Geomagnetic reversal frequency since the late Cretaceous. *Earth and Planetary Science Letters*, **62**, 305–313.

MCDOUGALL, I., KRISTJANSSON, L. & SAEMUNDSSON, K. 1984. Magnetostratigraphy and geochronology

of North-west Iceland. *Journal of Geophysical Research*, **89**, 7029–7060.

MUTTER, J. C., TALWANI, M. & STOFFA, P. L. 1982. Origin of seaward-dipping reflectors in oceanic crust off the Norwegian margin by 'subaerial sea-floor spreading'. *Geology*, **10**, 353–357.

NUNNS, A. C., TALWANI, M., LORENTZEN, G. R., VOGT, P. R., SIGURGEIRSSON, T., KRISTJANSSON, L., LARSEN, H. C. & VOPPEL, D. 1983. Magnetic anomalies over Iceland and surrounding seas. *In*: BOTT, M. H. P., SAXOV, S., TALWANI, M. & THIEDE, J. (eds) *Structure and Development of the Greenland–Scotland Ridge.* Plenum Press, New York, pp. 661–678.

PÁLMASON, G. 1980. A continuum model of crustal generation in Iceland; kinematic aspects. *Journal of Geophysics*, **47**, 7–18.

ROBERTS, D. G., BOTT, M. H. P. & URUSKI, C. 1983. Structure and origin of the Wyville-Thompson ridge. *In*: BOTT, M. H. P., SAXOV, S., TALWANI, M. & THIEDE, J. (eds) *Structure and Development of the Greenland–Scotland Ridge.* Plenum Press, New York, pp. 133–158.

SAEMUNDSSON, K. 1978. Fissure swarms and central volcanoes of the neovolcanic zones of Iceland. *Geological Journal, Special Issue*, **10**, 415–432.

—— 1986. Subaerial volcanism in the western North Atlantic. *In*: VOGT, P. R. & TUCHOLKE, B. E. (eds) *The Geology of North America (Vol. M). The Western North Atlantic Region.* The Geological Society of America, Boulder, 69–86.

SIGURDSSON, H. 1967. The Icelandic basalt plateau and the question of sial. *In*: BJÖRNSSON, S. (ed) *Iceland and Mid-Ocean Ridges.* Societas Scientiarum Islandica Publication, **38**, 32–46.

—— & LOEBNER, B. 1981. Deep-sea record of Cenozoic explosive volcanism in the North Atlantic. *In*: SELF, S. & SPARKS, R. S. J. (eds) *Tephra Studies.* D. Reidel, Dordrecht, pp. 289–316.

SIGURGEIRSSON, T. 1970. *Aeromagnetic profile and contour map of Southwest Iceland (map sheet 3) in scale 1:250,000.* Science Institute, University of Iceland, Reykjavik.

—— 1984. *Aeromagnetic profile map of Northwest Iceland (map sheet 1) in scale 1:250,000.* Science Institute, University of Iceland, Reykjavik.

SKOGSEID, J. & ELDHOLM, O. 1987. Early Cenozoic crust at the Norwegian continental margin and the conjugate Jan Mayen Ridge. *Journal of Geophysical Research*, **92**, 11471–11491.

STEINTHORSSON, S. & JACOBY, W. (eds) 1985. Special Section: Crustal accretion in and around Iceland. *Journal of Geophysical Research*, **90**, 9951–10192.

VOGT, P. R. 1974. Volcano spacing, fractures, and thickness of the lithosphere. *Earth and Planetary Science Letters*, **21**, 235–252.

—— 1986. Geophysical and geochemical signatures and plate tectonics. *In*: HURDLE, B. G. (ed) *The Nordic Seas.* Springer, Berlin, pp. 413–662.

——, JOHNSON, G. L. & KRISTJANSSON, L. 1980. Morphology and magnetic anomalies north of Iceland. *Journal of Geophysics*, **47**, 67–80.

WALKER, G. P. L. 1963. The Breiddalur central volcano, Eastern Iceland. *Quarterly Journal of the Geological Society of London*, **119**, 29–63.

—— 1966. Acid volcanic rocks in Iceland. *Bulletin Volcanologique*, **29**, 375–406.

—— 1974. The structure of Eastern Iceland. *In*: KRISTJANSSON, L. (ed) *Geodynamics of Iceland and the North Atlantic Area*. D. Reidel, Dordrecht, pp. 177–188.

WOOD, D. A. & GIBSON, I. L. 1976. The relationship between depth of burial and mean intensity of magnetization for basalt from eastern Iceland. *Geophysical Journal of the Royal Astronomical Society*, **46**, 497–498.

L. KRISTJANSSON, Science Institute, University of Iceland, Dunhaga 3, 107 Reykjavik, Iceland.

J. HELGASON, National Energy Authority, Grensasvegur 9, 108 Reykjavik, Iceland.

A multiple and propagating rift model for the NE Atlantic

H. C. Larsen

Two stages of propagating rift behaviour can be observed within the seafloor-spreading history of the NE Atlantic. Spreading started along one virtually continuous spreading ridge (early axis in Fig. 1) from S of Greenland and northward to the Greenland–Senja Fracture Zone. This ridge showed a locally sinuous trend with linear segments to the S and to the N (later Reykjanes and Mohns Ridge respectively), and a curved segment in the middle showing partly oblique spreading (later Aegir Ridge). Spreading was initiated above sea-level and apparently progressed from the S towards the N between anomaly 25 and 24 with an apparent propagation rate of around 1 m per year. The suggestion of a propagating behaviour of the early axis is based on the northward narrowing zone of pre-anomaly 24 oceanic crust (Fig. 1) and is not confirmed stratigraphically.

Initial but unsuccessful attempts at spreading along one straight axis may be correlated with a line of intraplate volcanism in E Greenland (see

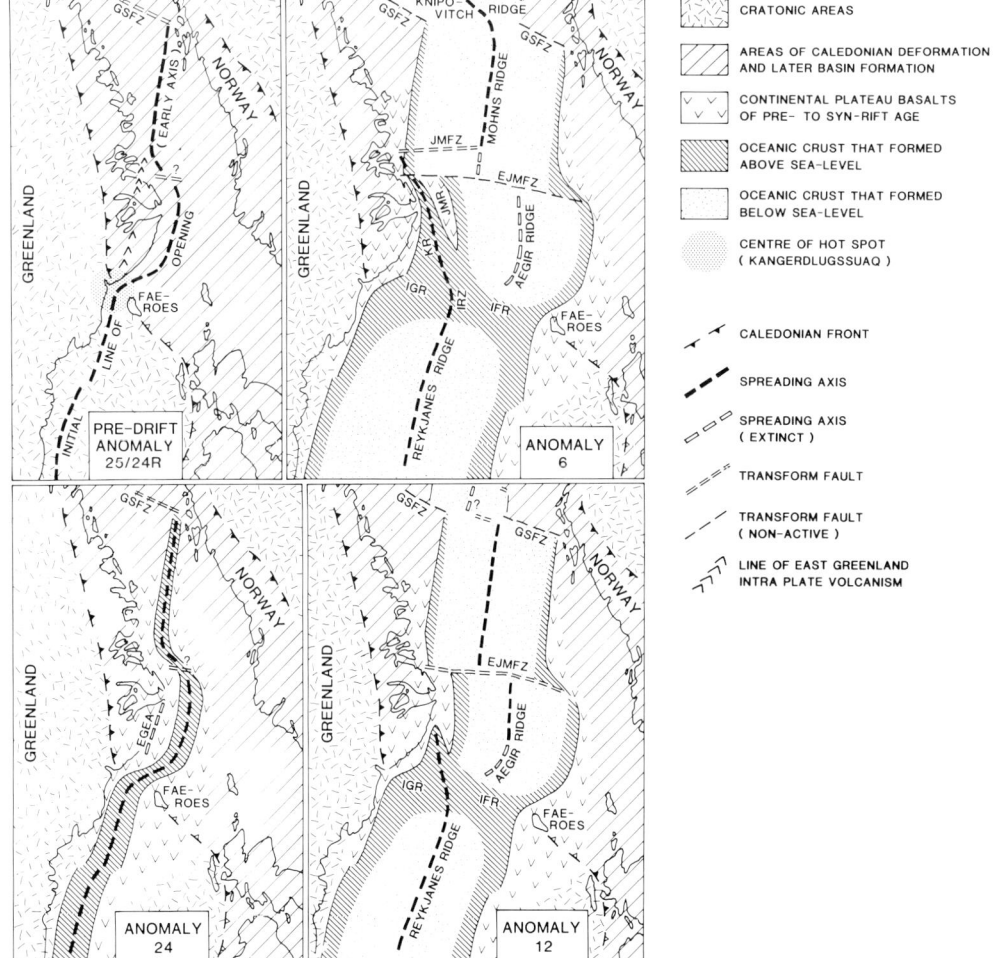

FIG. 1. Development of the NE Atlantic involving the concept of paired propagating/retreating rifts. GSFZ: Greenland–Senja Fracture Zone; EGEA: E Greenland Extinct Axis; IGR: Iceland–Greenland Ridge; IFR: Iceland–Faeroe Ridge; EJMFZ: Early Jan Mayen Fracture Zone; JMFZ: Jan Mayen Fracture Zone; IRZ: Icelandic Rift Zone; JMR: Jan Mayen Ridge; KR: Kolbeinsey Ridge.

From MORTON, A. C. & PARSON, L. M. (eds), 1988, *Early Tertiary Volcanism and the Opening of the NE Atlantic,* Geological Society Special Publication No. 39, pp. 157–158.

Fig. 1) including the source area for the lower part of the E Greenland plateau basalts. An almost successful attempt at straightening the early axis took place around anomaly 24 time and is recorded along part of the E Greenland coast by dyke intrusion, rotation of fault blocks and linear magnetic anomaly formation (E Greenland extinct axis in Fig. 1). Spreading, however, continued along the sinuous axis. By the time of anomaly 22 to 20, the southern straight segment of the axis started to propagate above sea-level, from the position of the hot-spot centre below the Iceland–Greenland Ridge and northward through the middle and curved part of the axis (see Fig. 1). During the early spreading phase a sinistral transform fault was formed between the middle segment and the northern straight segment (early Jan Mayen Fracture Zone). These events led to the formation of the Reykjanes Ridge, the Aegir Ridge (former middle and curved part of the early axis) and the Mohns Ridge. The Reykjanes Ridge continued to propagate northward until about anomaly 6 time at an average rate of 1–2 cm per year. During this period, the Reykjanes Ridge and the Aegir Ridge formed a paired propagating/retreating rift system connected through a pseudotransform fault. An apparently fan-shaped anomaly pattern formed along the Aegir Ridge because of the gradual cessation of spreading along this axis, and a highly diachronous ocean to continent transition formed along the E Greenland margin. A continental sliver (Jan Mayen Ridge) was torn off the Greenland continent by the propagating Reykjanes Ridge, which eventually overstepped the early sinistral Jan Mayen Fracture Zone around anomaly 7–6 time. The present-day large dextral fracture zone was formed in a more northerly position by this event, while spreading ceased completely along the Aegir Ridge. The

Reykjanes Ridge was now the continuous spreading axis between the Charlie Gibbs Fracture Zone and the new dextral Jan Mayen Fracture Zone for about 10 million years. Repeated and increasing eastward displacement of the Icelandic rift zone, however, split up the Miocene Reykjanes Ridge into the present-day Reykjanes Ridge, the Icelandic spreading centre and the Kolbeinsey Ridge N of Iceland.

The propagating rift model complies with the concept of rigid plate behaviour. Deformation and small-scale spreading of anomaly 20–6 age on the Jan Mayen Ridge can, however, be expected as a consequence of the position of the ridge within the pseudotransform-fault zone between the two paired rifts.

Flexure of the southern early axis into a sinuous middle part coincides both with the position of the Kangerdlugssuaq hot-spot centre (see Fig. 1) and with the transition from a southern cratonic area, with a predominant structural grain perpendicular to the initial line of opening, into a northern area affected by Caledonian deformation and later basin formation, with a predominant structural grain oblique to the general trend of the initial line of opening. It is suggested that the change in megatectonic setting and fabric was a main factor in the rift flexure, and that the hot-spot centre possibly aided further in this process by initiating a triple-junction geometry (failed rift in fjord, see Fig. 1). Continued activity of the Kangerdlugssuaq hot-spot centre caused the formation of the elevated Iceland–Greenland Ridge above sea-level, and was probably also instrumental in the propagation of the Reykjanes Ridge from the Iceland–Greenland Ridge and northward to the Jan Mayen Fracture Zone.

ACKNOWLEDGEMENT: The paper is published with the permission of the Greenland Geological Survey.

H. C. LARSEN, The Geological Survey of Greenland, Øster Voldgade 10, DK-1350 Copenhagen K, Denmark.

E Greenland and the
Faeroe Islands

Tertiary volcanism in the Kangerdlugssuaq region, E Greenland: trace-element geochemistry of the Lower Basalts and tholeiitic dyke swarms

R. C. O. Gill, T. F. D. Nielsen, C. K. Brooks & G. A. Ingram

SUMMARY: Tholeiitic dyke swarms in the Kangerdlugssuaq area can be divided, like the Lower Basalts, into two broad categories: (1) an early tholeiitic 'Picrite–ankaramite Series' (PAS) with relatively high levels of incompatible elements; and (2) a less incompatible-element enriched 'Tholeiitic Series' (TS) to which the majority of the dykes belong. The latest of the tholeiitic dykes tend to have the lowest incompatible element contents. The PAS dyke samples have steep light-enriched rare earth patterns and strongly fractionated spidergrams ($[Nb/Yb]_N > 20$), with no conclusive elemental evidence of significant crustal contamination. TS dykes have less steep rare earth element (REE) patterns and spidergrams, the latter having negative Sr anomalies. Corresponding PAS and TS lavas have similar patterns, although the Nb/La, Ta/La and P/Nd ratios are significantly lower.

The PAS dyke samples are more enriched in highly incompatible elements than typical ocean island basalt (OIB) tholeiites. Derivation of these magmas from a sub-lithospheric OIB source, such as that presently contributing to Icelandic volcanism, is tenable only if the picritic magmas underwent subsequent enrichment by incorporation of incompatible elements from sub-continental lithosphere or from continental crust. It is difficult to derive the observed enrichment entirely from the continental crust. The alternative is to envisage an OIB-source jet (the precursor to that currently situated under Iceland) that penetrates upward into the sub-continental lithosphere, and supplies parental picritic magmas which, in passing to the surface, scavenge incompatible elements from enriched lithospheric mantle.

Later tholeiitic lavas and dykes tend to be characterized by less steep incompatible-element profiles, resulting in two late dykes with spidergrams identical to sub-aerial Icelandic tholeiites. This trend may be interpreted in terms of a decreasing contribution of highly incompatible elements from the progressively thinner lithosphere.

Recent geophysical reconstructions of the E Greenland continental margin have indicated that the original continental separation took place as close as 15 km from the present coastline, between Kap Gustav Holm (66°N) and the mouth of Kangerdlugssuaq (68° 10'N) (Larsen 1984 & pers. comm.). Along this stretch of coast and to the E of Kangerdlugssuaq runs one of the most impressive continental-margin dyke swarms exposed on land anywhere in the world (Wager & Deer 1938; Deer 1976; Nielsen 1978; Myers 1980; Nielsen & Brooks 1981; Brooks & Nielsen 1982a). The proximity to oceanic crust and the zone of initial continental rupture makes this part of Greenland an unparalleled geological laboratory for studying the processes accompanying the initiation of a new ocean basin during intracontinental rifting. The petrology and chemistry of the coast-parallel dyke swarm has been studied previously by Nielsen (1978), Brooks & Nielsen (1978) and Rucklidge et al. (1979). This paper is the first part of a detailed chemical study of the dyke swarm and equivalent lavas

Regional outline

The basalt succession in the Kangerdlugssuaq area can be divided into two broad sections (Table 1; Wager 1947; Nielsen et al. 1981). The Lower Basalts (including hyaloclastites and pyroclastics) consist predominantly of tholeiitic basalts but include up to 15% by volume of picrites, MgO > 20% (Nielsen et al. 1981). Above the Lower Basalts lies a succession of water-lain tuffs followed by more uniform, sub-aerial basalts known collectively as the Plateau Basalts. They have been investigated in detail in the Scoresby Sund area (Larsen & Watt 1985; Larsen et al., in press), and are therefore not included in the present study.

The basalts and the underlying basement are cut by an intense coast-parallel swarm dominated by tholeiitic basic dykes. Most of these dykes (> 70%) were intruded prior to the development of the coastal flexure. They were followed by less prominent swarms of younger, post-flexure dykes of both transitional-tholeiitic and alkaline com-

From MORTON, A. C. & PARSON, L. M. (eds), 1988, *Early Tertiary Volcanism and the Opening of the NE Atlantic,* Geological Society Special Publication No. 39, pp. 161–179.

positions. The chronology of these phases of dyke injection, as described by Nielsen (1978), is summarized in Table 1.

The dykes described in this paper are from three key areas near to the mouth of Kangerdlugssuaq (Fig. 1):

(1) Hængefjeldet (Nielsen 1978), directly S of the Skaergaard intrusion, where the most intense part of the coast-parallel dyke swarm cuts the lower part of the Plateau Basalts, the Irminger Formation (Soper *et al.* 1976).
(2) The NW part of Kraemer Ø (Kraemer Island), where a complex set of dyke swarms cuts basement gneisses and earlier Tertiary intrusions (Nielsen 1978; Brooks & Nielsen

1978). This area lies inland of the coast-parallel dyke swarm.
(3) Fladø (Flat Island), an island some 50 km to the SSW, straddling the densest part of the coast-parallel dyke swarm where it is exposed cutting Archaean basement (Bridgwater *et al.* 1978).

Sampling areas and petrography

Lavas

The geology of the Lower Basalts has been described by Brooks *et al.* (1976), Soper *et al.*

TABLE. 1. *Correlation between Tertiary events in the Kangerdlugssuaq–Scoresby Sund region, including results of 1986 fieldwork by W. S. Watt (pers. comm.)*

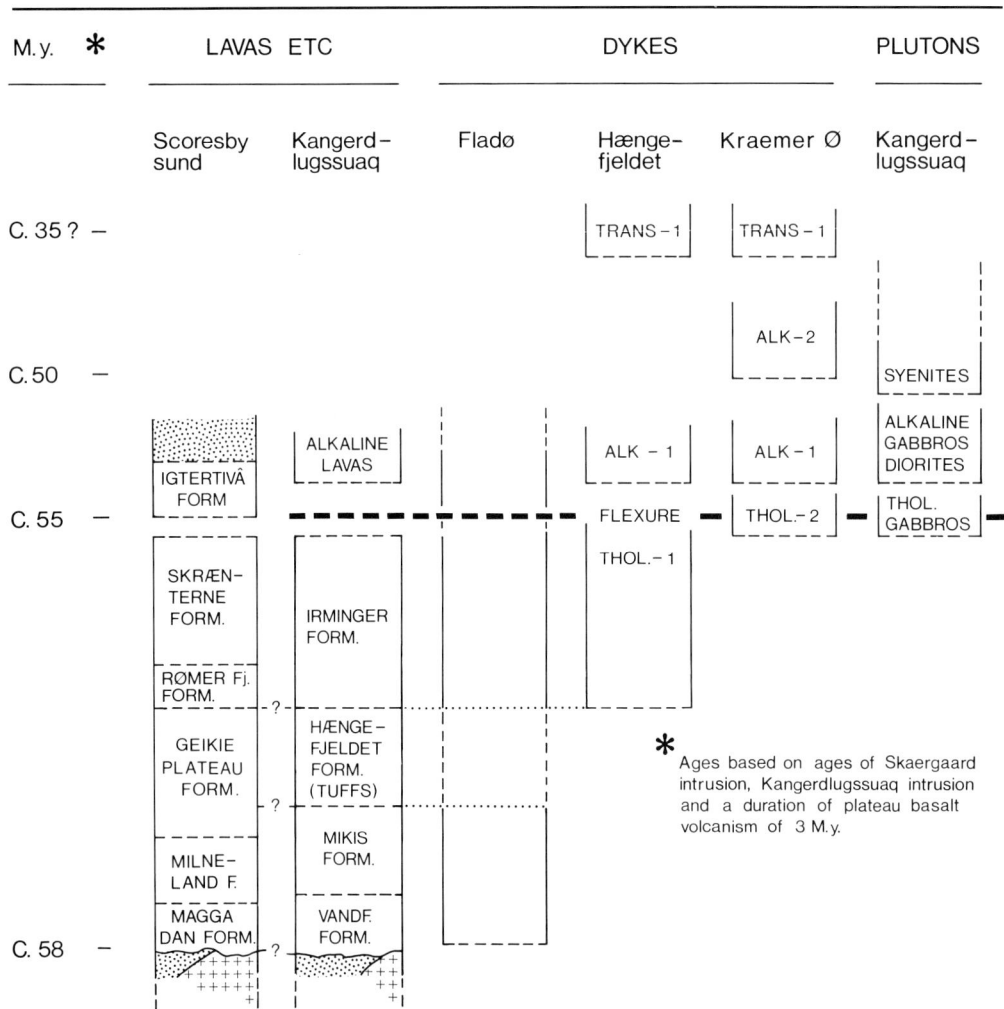

* Ages based on ages of Skaergaard intrusion, Kangerdlugssuaq intrusion and a duration of plateau basalt volcanism of 3 M.y.

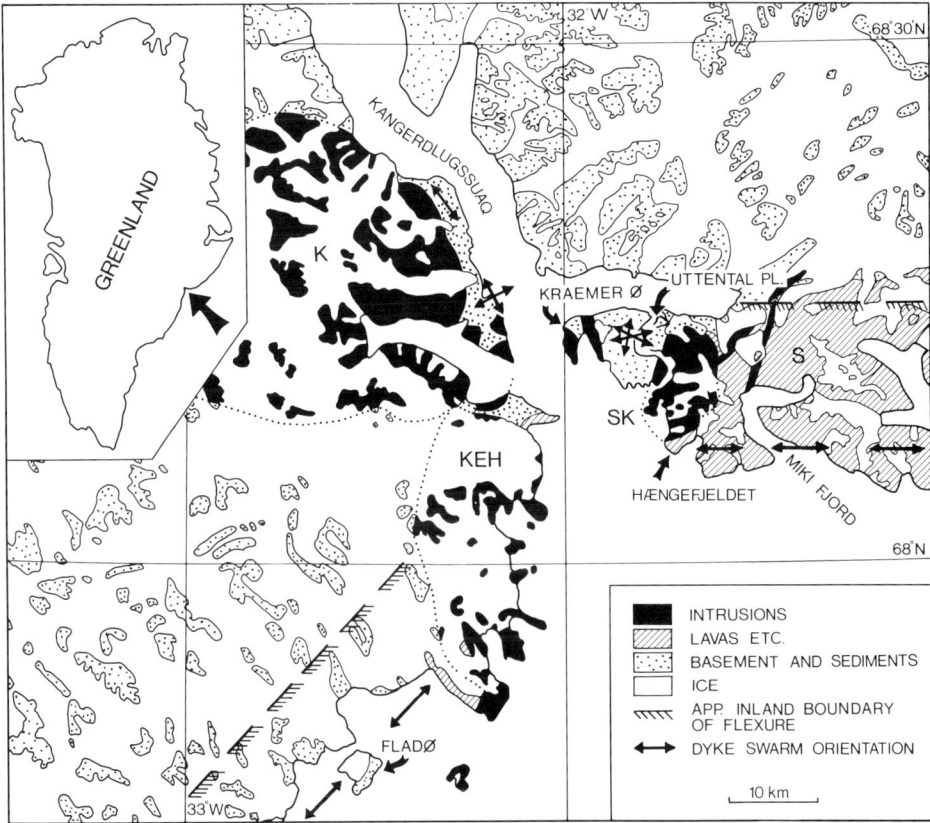

FIG. 1. Sampling areas (rectangles) for Tertiary dykes in the Kangerdlugssuaq area, E Greenland. S: Sødalen; K: Kangerdlugssuaq intrusion; SK: Skaergaard intrusion; KEH: Kap Edvard Holm complex.

(1976) and Nielsen *et al.* (1981), who divide them into the Vandfaldsdalen Formation (about 550 m at the base of the basalt succession) and the Mikis Formation (750–900 m thick). Their geochemistry has been discussed briefly by Nielsen *et al.* (1981) and Brooks & Nielsen (1982a, b). In the present work we have analyzed in more detail samples collected for these earlier studies from Mikis Fjord (and the adjoining valleys of Vandfaldsdalen and Sødalen) and Jakobsen Fjord E of the mouth of Kangerdlugssuaq (see Fig. 1). The Lower Basalts range from fresh olivine-phyric picritic and olivine tholeiite compositions (with almost universally pseudomorphed olivines) through olivine-free tholeiites to occasional tholeiitic andesites. Tholeiitic ankaramite variants are rich in olivine and green chrome-endiopside phenocrysts, with rarer chromite phenocrysts (the term 'ankaramite' is used for want of a better name). Vesicles are very common in olivine tholeiites and not uncommon in all other types. In general the basalts are partly altered up to greenschist facies assemblages, with prehnite,

quartz and albitic plagioclase, epidote, calcite and zeolites, depending on the distance from later dykes and intrusions (Bird *et al.* 1986).

Sample numbers of the lavas (Table 2) refer to collections in the Geological Museum of the University of Copenhagen.

Dykes in the Kangerdlugssuaq area

The dykes from Hængefjeldet and Kraemer Ø were described by Nielsen (1978), who distinguished three generations of tholeiite dykes:

(1) THOL-1 (Hængefjeldet): early northerly dipping (40–45°) dykes of the main coast-parallel swarm (58–55 Ma), predating the coastal flexure. They are mostly aphyric or sparsely porphyritic dolerites. Augite and plagioclase are unaltered in most samples, but olivine has invariably been replaced by chlorite ± opaques, and interstitial material is often altered to chlorite-rich assemblages. Both

TABLE 2. *Representative chemical analyses*

	Lower Basalts			Flado dykes								Hængefjeldet dykes		KraemerØdyke
	PAS	TS	MT	PAS	PAS			TS			depleted	THOL-1	THOL-1	THOL-2
	GM 20351[1]	GM 40031[1]	GM 20332	GGU 26791B[2]	GGU 26791C[3]	GGU 267902B	GGU 267904B	GGU 267906	GGU 267907	GGU 267916-4	GGU 267909	GM 20381[4]	GM 20376[4]	GM 27700-2[4,5]
Weight % oxide														
SiO_2	48.1	46.6	47.1	47.5	46.1	51.1	48.3	49.2	48.1	48.4	48.1	48.5	48.5	49.6
TiO_2	2.68	1.65	2.12	3.14	2.92	1.48	2.59	3.25	3.17	2.90	1.29	2.34	3.50	1.39
Al_2O_3	7.94	12.41	10.72	12.98	11.37	11.29	13.04	13.05	12.92	13.22	16.42	13.49	12.61	15.75
Fe_2O_3	4.18	2.22	7.59	4.17	3.24	1.89	6.84	4.90	4.67	2.91	2.63	3.69	2.82	—
FeO	8.90	7.76	5.64	8.47	8.67	8.59	7.72	9.06	9.90	9.70	7.26	8.56	11.61	9.11[5]
MnO	0.16	0.15	0.16	0.16	0.19	0.16	0.23	0.22	0.22	0.19	0.16	0.20	0.24	0.15
MgO	11.01	7.88	11.80	7.07	10.78	11.66	5.46	4.45	5.37	6.22	6.99	6.30	5.06	8.29
CaO	10.77	11.14	9.44	8.85	11.13	9.44	10.91	8.29	9.75	11.00	12.82	11.63	9.59	13.22
Na_2O	1.48	1.48	2.00	3.12	1.21	1.92	2.34	3.43	2.96	2.53	2.03	2.50	2.78	2.13
K_2O	0.38	0.13	0.38	1.18	0.61	0.75	0.51	1.12	0.90	0.37	0.18	0.38	0.77	0.25
P_2O_5	0.27	0.16	0.31	0.42	0.33	0.14	0.26	0.58	0.42	0.31	0.11	0.33	0.45	0.16
LOI	3.20	7.88	3.09	2.46	2.69	1.69	1.92	2.39	1.87	1.77	2.00	1.78	1.70	0.00[5]
Total	99.1	99.5	100.3	99.5	99.2	100.2	100.1	99.9	100.3	99.5	100.0	99.7	100.0	100.0[5]
ppm metal														
Rb	5	3	3	30	18	33	10	29	20	6	2			
Sr	208	407	233	746	460	242	290	360	386	386	205	250	192	249
Y	24	20	27	27	24	22	42	53	44	36	25			
Zr	171	109	127	200	173	121	175	251	219	171	73	162	215	87
Nb	17	12	8	47	39	12	25	47	38	23	9			
Ba	135	65	70	512	250	163	133	345	313	149	42	125	125	44
La	17.3	15.2	10	33	25	8.6	15.1	30	21	12	4.2	15	15	5.2
Ce	49	36	27.7	79	62	31	42	81	57	36	14	38	38	13.8
Nd	27	19	17.3	35	29	15.4	22.2	36	28	24	9.4	20	22	9.7
Sm	7.15	4.69	5.6	8.18	6.9	4.17	6.76	9.8	7.91	7.14	3.21	5.58	5.84	2.87
Eu	2.08	1.63	1.68	2.37	1.89	1.22	2.00	2.65	2.22	1.85	1.03	1.90	2.09	1.04
Tb	0.99	0.65	0.87	0.96	0.79	0.60	1.11	1.42	1.21	0.93	0.59	0.98	1.10	0.51
Yb	1.77	1.62	2.00	1.61	1.28	1.10	2.70	3.32	2.64	2.07	1.71	2.15	2.73	1.23
Lu	0.20	0.21	0.27	0.24	0.19	0.21	0.46	0.63	0.50	0.35	0.28	0.32	0.56	0.19
Hf	4.93	3.25	3.57	5.37	4.35	3.01	4.43	6.05	4.98	3.83	1.85	4.20	5.06	1.88
Ta	0.89	0.59	0.47	2.09	1.72	0.59	1.20	2.17	1.62	0.97	0.42	1.25	1.10	0.40
Th	1.32	0.90	0.61	2.38	1.94	0.94	1.23	2.76	1.82	1.22	0.45	1.55	1.40	0.30
Mg No	64.2	62.5	66.1	54.4	65.7	70.0	44.8	40.5	44.0	51.0	59.9	52.2	41.4	65.2

[1] Major element data from Nielsen et al. (1981), where locality details are given; [2] Margin of dyke; [3] Central zone of dyke; [4] Major element and Sr, Zr and Ba data from Nielsen (1978), where locality details are given; [5] Analysis summed to 100% volatile-free. Total Fe expressed as FeO. GM and GGU refer to the collections of the Geological Museum (University of Copenhagen) and Grønlands Geologiske Undersøgelse (Geological Survey of Greenland) respectively. LOI = loss on ignition.

calcite and zeolites are common vein-filling materials at Hængefjeldet.

(2) THOL-2 (Kraemer Ø and Uttental Plateau— see Fig. 1): a less dense swarm of later dykes which radiate from the mouth of Kangerdlugssuaq, and appear to be contemporaneous with the Skaergaard intrusion, i.e. ~ 55 Ma (Brooks & Nielsen 1978). The dykes are aphyric or plagioclase-phyric dolerites, generally less altered than THOL-1 dykes, although olivine is always altered to chlorite.

(3) TRANS-1: a much younger swarm of porphyritic, northerly-dipping (70–90°), coast-parallel dykes (41–36 Ma), transitional in composition between tholeiitic and alkaline affinity. The swarm postdates the flexure and the major syenitic magmatism in the Kangerdlugssuaq area. The dykes are commonly olivine-phyric dolerites. Those from Kraemer Ø are often unaltered.

Sample numbers (see Table 2) refer to the Geological Museum collections.

Dykes on Fladø

Unlike the localities discussed above, the dykes on Fladø have not previously been described in detail. Much of Fladø is permanently ice-covered, but the E side of the island offers excellent glaciated exposures with dyke densities over 50%, where the relative ages of dykes can be determined. The dyke samples were collected during brief helicopter visits (by Gill in 1979) as part of the regional geological mapping programme of the Geological Survey of Greenland (GGU), and the sample numbers (see Table 2) refer to their collections. There was insufficient time for systematic mapping of the dyke swarm, which is complicated by the numerous intersections and mutual displacements. Five sampling localities were chosen where three to five successive dykes intersected and a relative age control could be established, but isolation from the areas where the dyke chronology is more fully understood prevents any direct age correlation. The majority of dykes dip westward or north-westward at angles between 45 and 60° and presumably predate the development of the coastal flexure. The youngest dykes are generally vertical and apparently postflexure. Local strike measurements suggest a crude tendency for successive dykes to have progressively more northerly strikes, but all dykes observed were broadly coast-parallel. The level of exposure here lies below the first basalts erupted, and the dyke collection may include early members that were feeders to the

Lower Basalts, in contrast to Hængefjeldet where the level of exposure lies above the Lower Basalts, and the dykes are regarded as feeders solely to the overlying Plateau Basalts.

The majority of the dykes collected from Fladø are fine-grained dolerites with ubiquitous plagioclase, augite (both commonly zoned) and opaques. They are generally aphyric or sparsely plagioclase-phyric, with groundmass grain sizes varying from 0.2–1.5 mm. Plagioclase is often slightly sericitized, and chlorite or epidote are common interstitial constituents, presumably replacing glass or fine-grained matrix.

Several of the Fladø dykes merit individual description. Sample numbers 267901A–D come from an early, 2 m-wide composite porphyritic dyke with indistinctly banded margins. The dyke is amygdaloidal and rather altered, particularly in the margins where the plagioclase has been extensively sericitized. Pseudomorphs 1–2 mm in size comprising opaques and epidote are presumed to be relics of olivine phenocrysts. Augite, present both as microphenocrysts and in the groundmass, shows marked normal, sector and oscillatory zoning, and is commonly rimmed by a distinct overgrowth of different composition. Brown hornblende and chlorite are common replacement products, particularly in the patchy, banded marginal zone in which pyroxene has been completely replaced.

Sample numbers 267902A–C are from another early porphyritic dyke collected at the same site, approximately 2.5 m wide but irregular in outcrop and developing several apophyses. It consists of 2 mm-sized augite and pseudomorphed olivine phenocrysts, the latter sometimes surrounded by aggregates of augite crystals, set in an intersertal matrix of fresh plagioclase, augite and opaques with interstitial chlorite. The texture suggests the composition of this dyke may have been slightly modified by accumulation. It was impossible to establish the ages of these dykes in relation to each other, but they are clearly the earliest two dykes at the site of collection. Chemical analyses (Fig. 2, see Table 2) show they are the most magnesian of the dykes studied.

Sample 267909 comes from a dyke which dips 50° to the W but appears not to be cut by other dykes, and is therefore thought to be rather late in the sequence represented on Fladø. The rock has a seriate texture with abundant fresh plagioclase phenocrysts (1–3 mm in size) set in a 0.5–1.0 mm-sized doleritic groundmass. The plagioclase commonly shows oscillatory zoning, and scarce augite microphenocrysts are also strongly zoned. As shown below, its chemical composition suggests it is related to members of the younger THOL-2 dyke generation.

Geochemistry

The major element analyses were carried out by X-ray fluorescence (XRF) fusion procedures (Sørensen 1975) at the Geological Survey of Greenland. Trace-element methods and their accuracy are detailed in the Appendix. Representative analyses are shown in Table 2; a complete compilation is available from TFDN. Mg number $[100 \times Mg/(Mg + Fe^{2+})$ in atomic proportions] and CIPW norms have been calculated assuming a standardized $Fe_2O_3 = 0.15$ FeO after Brooks (1977). In cation terms this is equivalent to setting 'Fe^{2+}' $= 0.865$ total Fe. For plotting and modelling, the analyses have been recalculated volatile-free.

Subdivision

Figure 2 shows TiO_2 contents of the dykes and representative samples of the Lower Basalts

plotted against Mg number. The Lower Lava compositions are scattered across a broad band as noted by Brooks & Nielsen (1982b), and can be subdivided as follows:

(1) Picrite-ankaramite Series: the uppermost part of the 'band' consists of picrites (Mg number < 85; MgO < 30) and tholeiitic ankaramites with elevated TiO_2 contents. The most evolved members of this group (which corresponds to the 'picritic trend' of Brooks & Nielsen 1982b) have Mg numbers of about 50 (MgO ~ 6%) and TiO_2 contents greater than 4%.
(2) Tholeiite Series: the lower margin of the 'band' is defined by a series of relatively evolved, less Ti-rich tholeiites with Mg numbers less than 65. These samples comprise the 'tholeiitic trend' of Brooks & Nielsen (1982b). The comparable Plateau Basalts of the Scoresby Sund region show a wider range of TiO_2 content, which Larsen & Watt (1985)

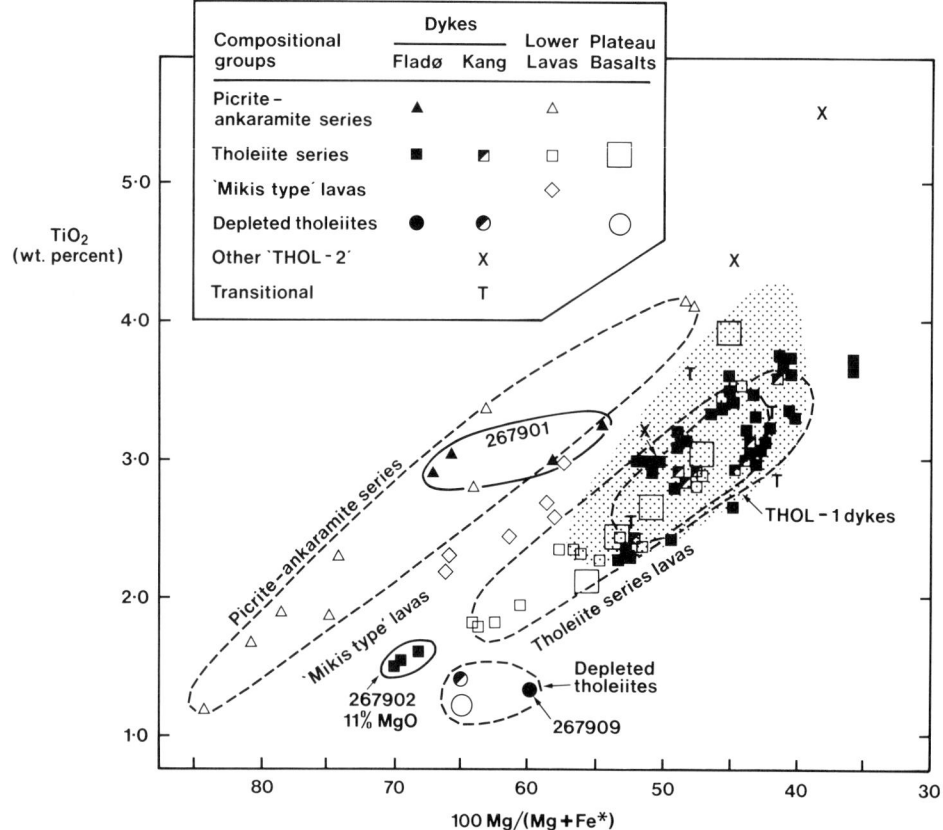

FIG. 2. TiO_2 contents (analyses recalculated volatile-free) plotted against Mg number. The stippled field shows the range of Scoresby Sund main basalts (medium-Ti, high-Ti and 'titan-tholeiite' types) after Larsen & Watt (1985). The large symbols are representative analyses tabulated in their paper.

attribute, at least in part, to the effects of fractionation in open magma chambers (O'Hara 1977; O'Hara & Matthews 1981). This suggests that quite a wide compositional field can be generated by the postgenerational fractionation of the more primitive members of the TS, and the limits of this category are enlarged in Figure 2 (relative to the 'tholeiitic trend' of Brooks & Nielsen 1982b) to take account of this possibility.

(3) Mikis-type lavas: Lower Basalts with compositions lying between the two series defined above were described by Brooks & Nielsen (1982b) as 'probable mixed lavas'. In the present study, in order not to prejudge their origin, we adopt the descriptive term 'Mikis type', after the locality where such lava compositions are particularly prominent.

In broad terms the compositions of the dykes from Fladø and the Kangerdlugssuaq area follow the same pattern as the lavas, but there are important differences in detail. Two of the earliest dykes stand apart in Figure 2. Samples from the composite dyke with banded margins on Fladø (267901, samples A–D) have high Mg numbers, in the range 65–55, and relatively high titanium contents. The more magnesian samples come from the central zone of the dyke, and in their geochemical attributes (see Figs 2, 3, 4, 5 & 6) and petrography (olivine & augite phenocrysts) resemble the picrite–ankaramite suite among the Lower Basalts. The chill specimen and the patchy banded marginal zone, however, plot further to the right (Mg number 58–55); they fall on the periphery of the picrite–ankaramite field, and are closer to Mikis type than the samples from the dyke centre. The dyke fissure seems to have served two successive batches of magma whose compositions do not appear, from their TiO_2 contents, to be related by the same fractionation process that generated the PAS lava compositions. Possibly the banded zone originated as a combination of two incompletely mixed magma components sheared out by laminar flow, the later batch now occupying the centre of the dyke representing the more primitive PAS component. This is the first dyke of such affinity reported from E Greenland.

The other early dyke at the same locality (267902, samples A–C) is also conspicuously magnesian. On the basis of its TiO_2 content (about half that of 267901) it correlates closely with the TS, but is significantly more magnesian than the corresponding members of the Lower Basalts (though similar compositions are found as dykes cutting the Lower Basalts in the Jakobsen Fjord area). It has abundant phenocrysts of olivine and augite but not plagioclase. The possibility of slight olivine-augite accumulation does not diminish the fact that this early dyke represents a significantly more primitive melt than the other TS dykes that succeeded it (see Fig. 2).

Relatively magnesian compositions also occur occasionally among later dykes. Two examples are 267909 from Fladø and a THOL-2 dyke, 27700–2, from Kraemer Ø (see Fig. 2). The latter was regarded by Brooks & Nielsen (1978) as similar to the parental magma of the Skaergaard intrusion. Both dykes have high CaO and Al_2O_3 contents that set them apart from the other dykes analysed (Fig. 3). These characteristics are not the result of plagioclase accumulation, as the rare earth patterns of these two dykes do not have perceptible europium anomalies (see Fig. 5). The low K_2O contents (0.25% or less) indicate a broadly mid-ocean ridge basalt (MORB)-like character, and these dykes closely resemble the compositions of basalts from the central rift zone of Iceland (see Fig. 8). Such compositions are not recorded from the Lower Basalts but minor volumes have been reported in the Plateau Basalts in the Scoresby Sund area (Larsen & Watt 1985, Table 1, analysis 3, Figs 2–4). These samples are sufficiently distinct to be regarded as a fourth compositional category in the geochemical subdivision outlined above.

Most of the other dykes are markedly less magnesian and lie within the bounds of the TS, as defined above. The dykes from Fladø (excluding those already discussed) cover much the same range of Mg number as the preflexure (THOL-1) dykes from Hængefjeldet, in spite of the different level of exposure. This compositional range is similar to that of the corresponding Lower Basalts, although the main density of dyke compositions is displaced to lower Mg values (along the tholeiite trend), coinciding more closely with the Scoresby Sund main basalt field.

In summary, the great majority of the tholeiitic Lower Basalts and dykes in the Kangerdlugssuaq area fall into four geochemical categories:

(1) Picrite-ankaramite Series: Lower Basalts, with one early Fladø dyke.
(2) Tholeiite Series: lavas and dykes, including the majority of Fladø dykes and the THOL-1 swarm.
(3) Mikis-type: lavas only (although the margins of the PAS dyke 267901 may be regarded as belonging to this type).
(4) Relatively 'depleted' tholeiites of Icelandic type: certain relatively late dykes on Fladø and Kraemer Ø. No basalts of this composition are recorded from the Lower Basalts, but

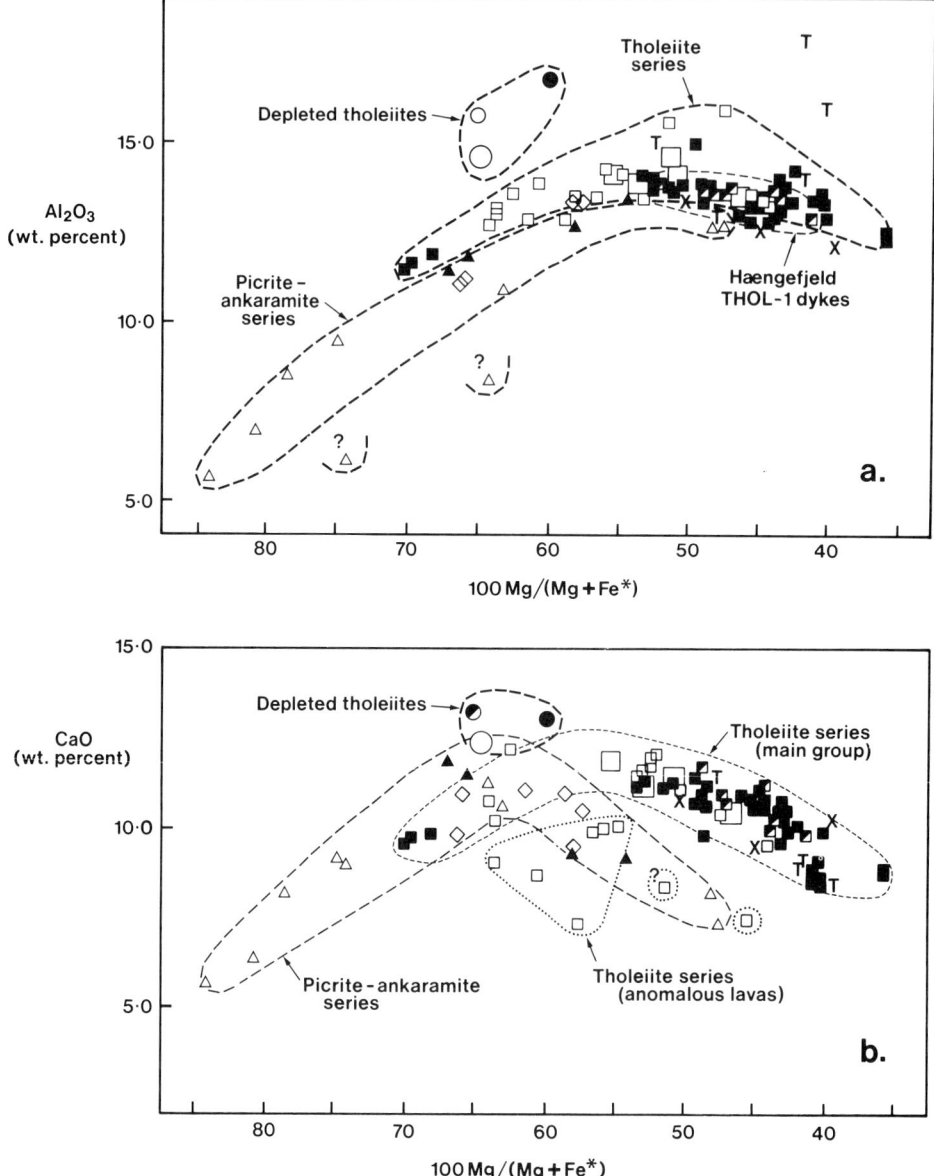

FIG. 3. (a) Al_2O_3 and (b) CaO contents (volatile-free analyses) plotted against Mg number.

they form a minor component of the Plateau Basalts in the Scoresby Sund region.

The THOL-2 dykes, with the exception of depleted members like 27700–2, have variable, often high TiO_2 contents. The compositions of the late THOL-2 and transitional (TRANS-1) dyke swarms are too scattered in Figures 2–4 to allow any general conclusions to be drawn from the few samples so far analysed.

Major element composition

The PAS lavas and dykes form a geochemically coherent group. All are tholeiitic and, except for the most evolved members, strongly olivine normative (up to 25%). They have consistently lower Al_2O_3 contents than the tholeiitic series magmas (see Fig. 3a), but higher concentrations of incompatible elements. Among the more magnesian members SiO_2, TiO_2, Al_2O_3 and CaO

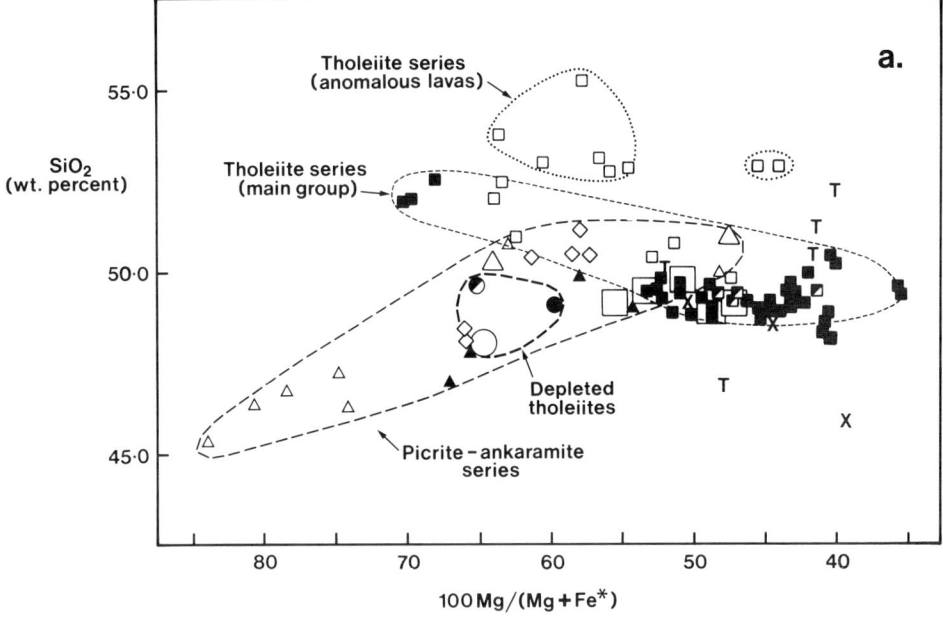

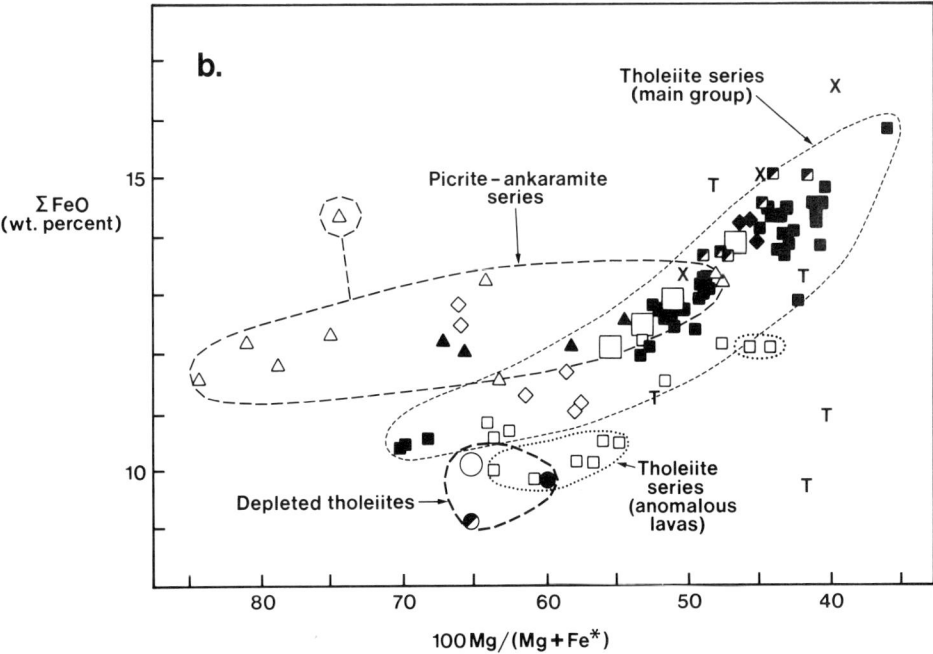

FIG. 4. (a) SiO$_2$ and (b) total iron as FeO (volatile-free analyses) plotted against Mg number. The large triangles represent samples whose incompatible-element abundances suggest contamination by sediment (see Fig. 6).

rise with decreasing Mg number (see Figs. 2–4) in a manner qualitatively consistent with the fractionation of olivine and chrome spinel. Olivine phenocrysts crowded with euhedral chromite crystals are reported in lavas of this composition in the Lower Basalts (Nielsen *et al.* 1981). At Mg numbers about 65, CaO begins an abrupt decline (see Fig. 3b) related to the appearance of chrome-endiopside on the liquid-us; Nielsen *et al.* (1981) record its appearance as a phenocryst phase in ankaramite lavas of similar composition. The plateau in Al_2O_3 at Mg number values of about 55 (see Fig. 3a) presumably represents the onset of plagioclase crystalliz-ation. Phosphorus begins a sharp decline at this point, indicating saturation with apatite as well.

Though a common 'tholeiitic series' can be distinguished among both lavas and dykes in Figures 2, 3a and 4b, in other diagrams it appears as a more dispersed grouping than the PAS. There is a consistently discrepant group of lava samples with high SiO_2 ($> 52.5\%$) and K_2O ($> 0.5\%$) and low CaO ($< 10\%$) and total FeO ($< 10.5\%$). As these samples include the most oxidized of the TS lava samples studied and mostly have water contents in excess of 3%, it seems likely that much of the scatter and inconsistency reflects alteration and/or contami-nation. These samples occur near to the base of the lava succession and may have reacted with basement or with the Cretaceous–Tertiary Kan-gerdlugssuaq sediments underlying the volcanics. The remaining 'tholeiitic trend' lava samples together with the equivalent dykes form a coherent trend of more or less level SiO_2, arching CaO and increasing total iron and K_2O contents, more in keeping with expected magmatic evolu-tion. The THOL-1 dykes from Hængefjeldet form a less scattered group having amongst the lowest Al, Si and K and highest FeO contents of the TS as a whole.

As many of the lavas and dykes of the TS are devoid of phenocrysts, the details of the fraction-ation process by which they are related, if indeed they do belong to a single line of descent, are less readily established than for the PAS.

As a group, the Mikis-type basalts span the region of overlap or the boundary between the PAS and TS fields in Figures 3 and 4, although individuals may show more affinity with one series or the other. Brooks & Nielsen (1982a) proposed that the Mikis-type lavas are the product of mixing between PAS picrite [($\sim 20\%$ MgO) and more evolved TS basalt (MgO $\sim 7\%$)], followed by fractional crystallization. Figures 2–4 are broadly consistent with this model, although the MT compositions are too dispersed to provide a discriminating test.

Incompatible element geochemistry

Plotting P_2O_5, Zr, Nb or Ba contents of the lavas and dykes against Mg number leads to a similar subdivision of the samples as that seen in Figure 2. In all of these plots the PAS is consistently more enriched in incompatible elements than TS rocks of the same MgO content or Mg-number value. The depleted character of dykes 27700–2 and 267909 emerges clearly in each case.

Rare earth elements

Most of the analyzed specimens have straight, light-enriched patterns typical of continental tholeiites (Fig. 5), with no significant europium anomalies. The steepest slopes ($Ce_N/Yb_N > 7$) are shown by the PAS dyke and lavas. Even for samples with MgO $\sim 11\%$, La_N and Ce_N lie in the 50–100 range. The TS as a whole have less steep patterns ($Ce_N/Yb_N < 7$), in spite of being more evolved in terms of their major element chemistry (Mg number, except for 267902, being less than 53). Within this group the THOL-1 dykes tend to have the lowest Ce/Yb ratios (see Fig. 5).

The depleted tholeiite dykes are distinctive in having gently curved, upward-convex patterns (see Fig. 5) more reminiscent of subaerial Icelan-dic basalts than continental tholeiites, with Ce/Yb ratios lower than the other groups (though overlapping with THOL-1).

Incompatible element abundance patterns

Figure 6 shows comprehensive incompatible element patterns for the PAS lavas and dykes, normalized to an idealized chondritic mantle after Thompson *et al.* (1982). The Fladø dyke (267901) shows simple, highly fractionated pat-terns that rise log-linearly from Yb to Nb, interrupted only by small anomalies at Ti and, for one of the samples, Sr. There is no suggestion of the negative Nb–Ta anomaly characteristic of many continental flood basalts (cf. Fig. 9b). With respect to their high $(Nb/La)_N$ and $(Nb/Nd)_N$ ratios, the dyke samples bear closer resemblance to certain ocean island basalts (see Fig. 9a). These ratios exceed those in OIB tholeiites, however, and are more characteristic of ocean island basalts of transitional or alkaline affinity.

The patterns of the analyzed PAS lavas are in overall terms similarly fractionated, but they deviate in several significant respects (see Fig. 6b): (1) the patterns show negative anomalies for P; (2) the abundances of Nb, Ta, La and Ce are lower relative to Nd than in the equivalent dyke (see Fig. 6a); and (3) Rb, Ba and K vary erratically relative to Th (probably as a result of alteration). It is interesting to note that highly fractionated incompatible element profiles are also observed in the early picritic lavas of the Karoo province (see Fig. 9b), and they share with the PAS lavas several of the features outlined

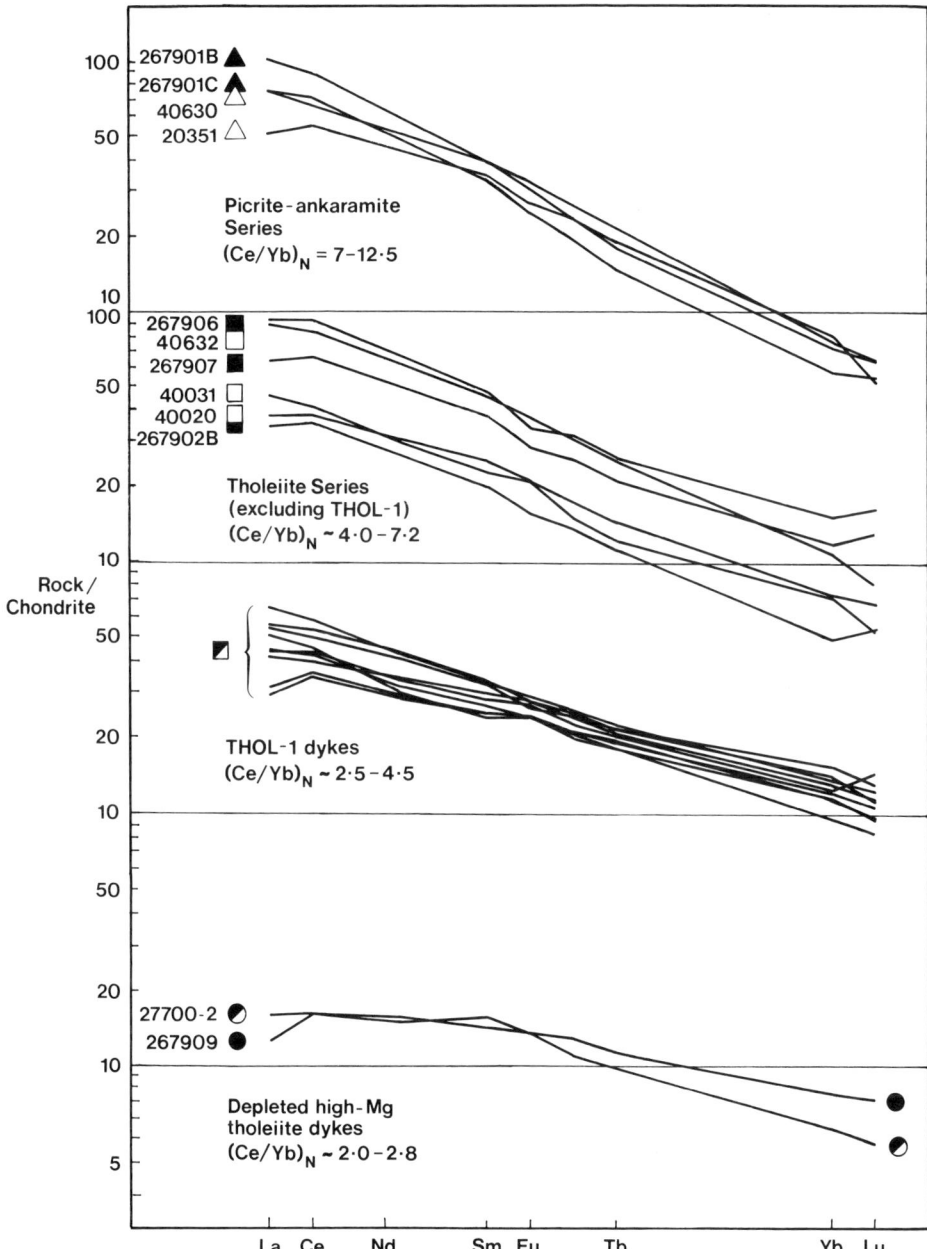

FIG. 5. Rare earth patterns of representative lavas and dykes: (a) PAS lavas and Fladø dykes; (b) TS lavas and Fladø dykes; (c) THOL-1 dykes from Hængefjeldet; (d) depleted Mg-rich tholeiite dykes from Kraemer Ø and Fladø. Chondrite values used are those of Nakamura (1974).

above, notably the negative P anomalies and depressed levels of Nb and Ta. These characteristics have been widely attributed to assimilation of components of the calc-alkaline continental crust (e.g. Thompson *et al.* 1983), although some authors associate them with magma sources in the underlying lithospheric mantle (Cox 1983; Weaver & Tarney 1983).

The cause of the disparity between the PAS dyke compositions and the equivalent lavas is far from clear. One may suppose that both batches of magma had similar opportunity to interact

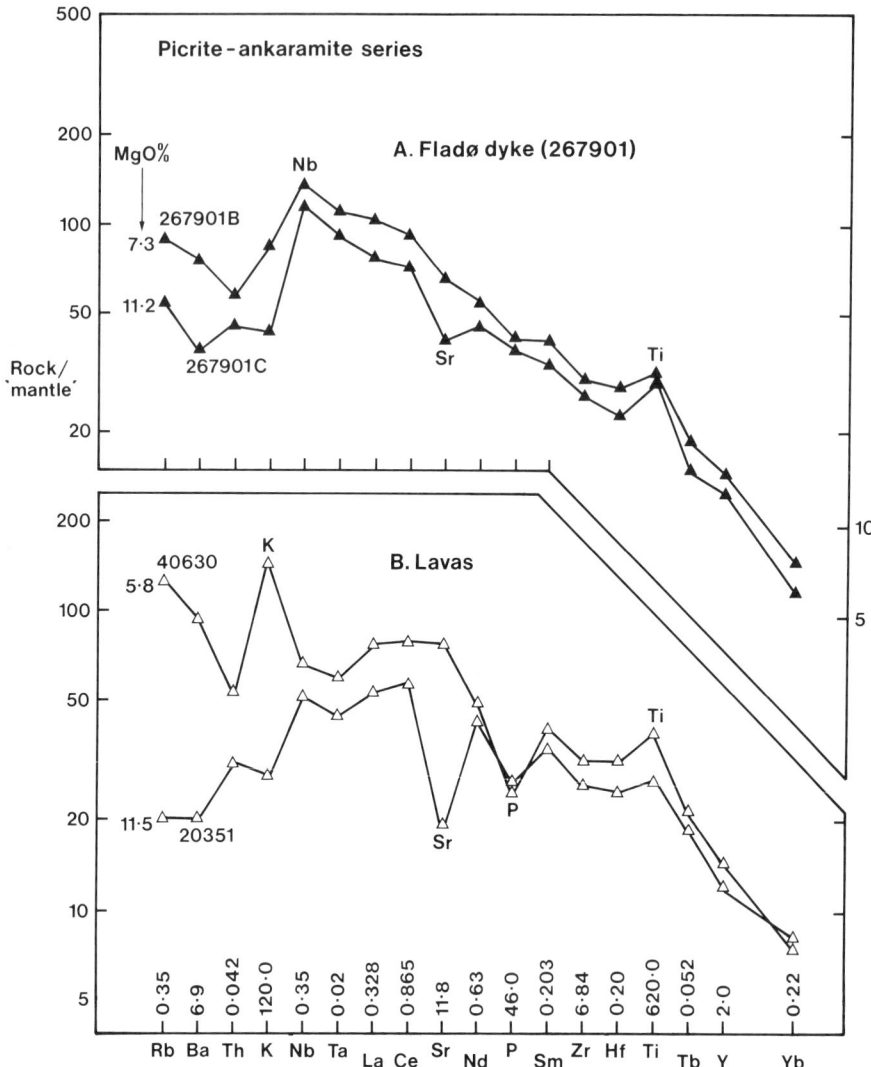

FIG. 6. Normalized incompatible-element profiles of PAS dykes (Fladø) and lavas. Normalizing values (after Thompson *et al*. 1982) are shown at the foot of the diagram.

with sialic basement during transit through the crust. The only observable difference in the crustal rocks encountered is that the Lower Basalts rest on thin Cretaceous–Tertiary sediments (shales, siltstones & sandstones, of which analyses are not yet available). Both samples in Fig. 6b, which come from relatively low levels in the succession, have moderately high SiO_2 contents (see Fig. 4), consistent with contamination by siliceous sediment, which may also account for the relative depletion of the PAS lavas in elements like Nb, Ta and P. However, too few

samples have been analysed for the significance of these differences to be firmly established.

Two TS dykes from Fladø (267906, 267907) have patterns resembling Fig. 6a except that they are less steep and have more marked Sr anomalies (Fig. 7). As with the PAS dyke, the high Nb and Ta values and the lack of a P anomaly are consistent with minimal crustal contamination. The corresponding Lower Lava samples have lower Ba/La, Nb/La, Nb/Nd and Ta/La ratios, however, perhaps reflecting interaction with sialic material. The THOL-1 dykes (shaded) have

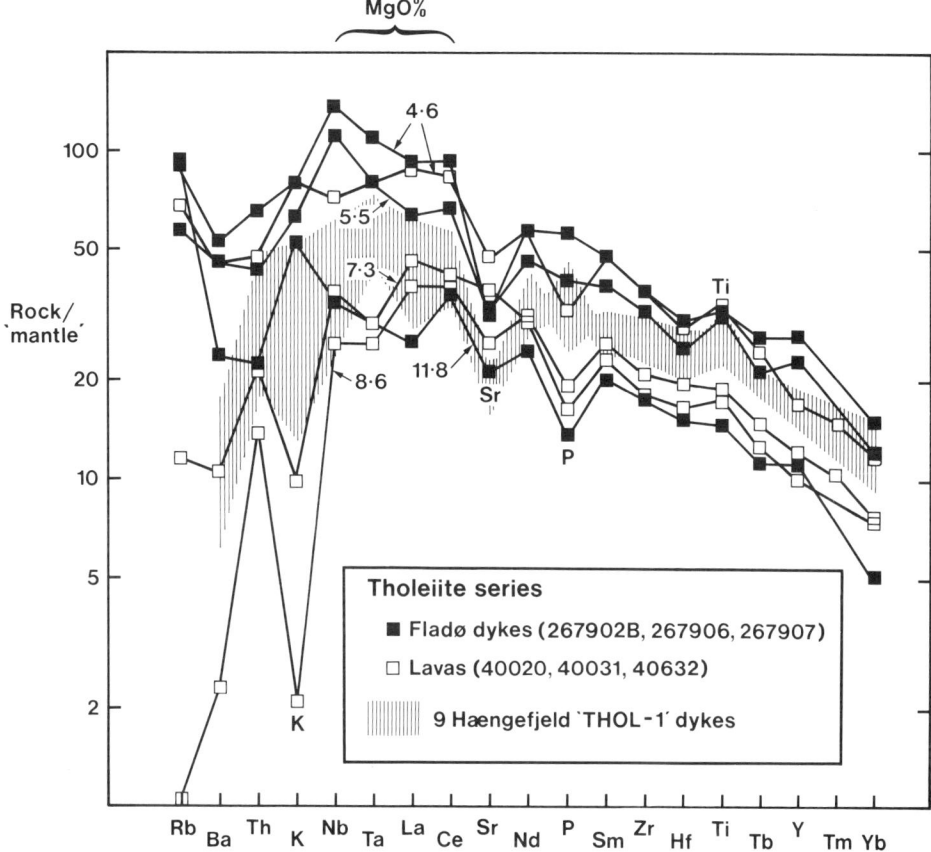

FIG. 7. Normalized incompatible-element profiles of TS rocks. THOL-1 dykes (Hængefjeldet) are represented by the stippled pattern.

similar patterns to the Fladø dykes, but with a slightly lower slope.

The late dykes 27700–2 (THOL-2) and 267909 have the most gently sloping patterns (Fig. 8). Their high Ca and Al contents (see Fig. 3) and relatively low enrichment in highly incompatible elements compare remarkably closely with sub-aerial lavas from Iceland (see shaded band). These dykes, though cutting continental crust, testify to the beginning of the oceanic phase of volcanism in the NE Atlantic.

Discussion

It is clear that in general terms the dykes considered in this paper mirror the range of magma types distinguished in the lavas of the Kangerdlugssuaq area. It is noticeable, however, that dykes and Lower Basalts, particularly among

the TS, tend to lie in different ranges of Mg number (see Fig. 2) and MgO content. The TS dykes correspond more closely with the more evolved Mg number range of the main Plateau Basalts, as indicated by recent work in the Scoresby Sund area (see Fig. 2). This may be an indication that many of the TS dykes at Fladø, like the THOL-1 dykes at Hængefjeldet, were feeders to the Plateau Basalts rather than the Lower Basalts. The cause of these differences is not entirely clear. With increasing thickness of the lava pile, there may have been a tendency for stagnation of magma batches in the feeder system, leading to more advanced differentiation of the main Plateau Basalts compared to analogous compositions among the Lower Basalts.

Several of the analysed dykes have no counter-parts among the lavas in the immediate Kangerd-lugssuaq area. The THOL-2 group have a wide range of incompatible element contents, and their interrelationship cannot be ascertained from the

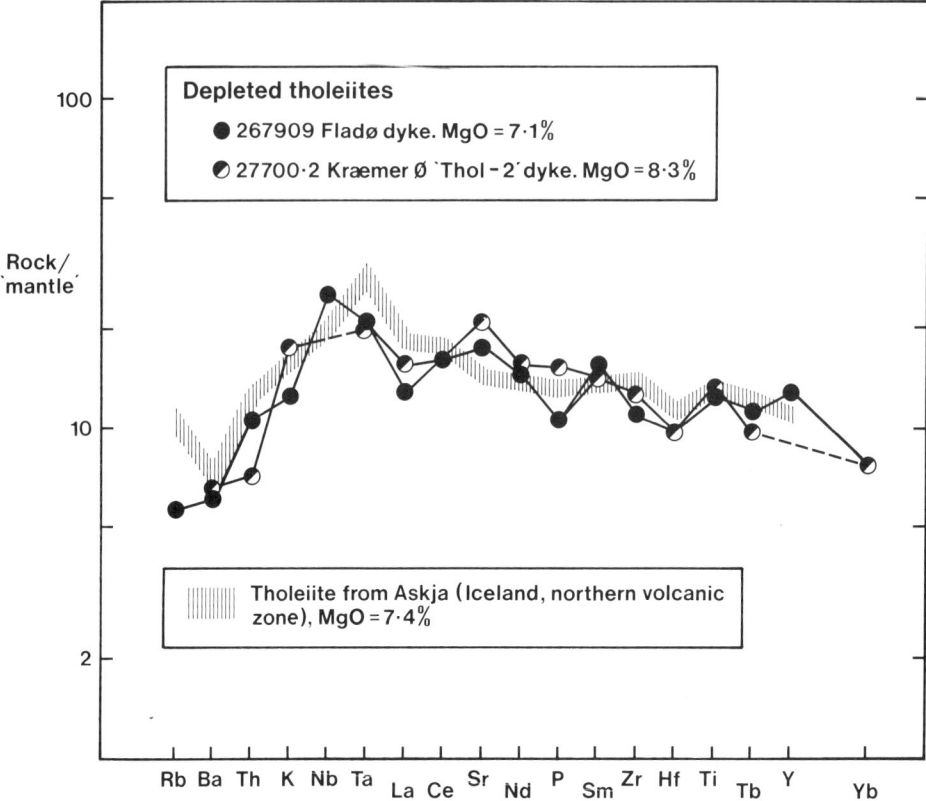

FIG. 8. Incompatible-element patterns for 'depleted' tholeiite dykes. Data for Askja tholeiite from Wood *et al.* (1979).

few samples considered in this study (the same is true of the TRANS dykes). The most distinctive member of the THOL-2 group is dyke 27700–2 from Kraemer Ø which, like the late dyke 267909 from Fladø, has parallels with Icelandic basalts.

Two major petrogenetic problems arise from this work. The first is to explain the origin of the distinctive, steep incompatible-element profiles of the PAS magmas (see Fig. 6), in which the highly incompatible elements are much more fractionated than in continental or oceanic tholeiites generally. The similarity to certain OIB alkali basalts has been noted (Fig. 9). Because the thermal and geochemical anomaly associated with Iceland today was presumably developing in the mantle underneath the Kangerdlugssuaq area in early Tertiary times (Brooks 1973; Brooks & Nielson 1982a, b), it is not unreasonable to seek the ultimate source of the PAS magmas in an OIB source-region below the sub-continental lithosphere. It is important to note, however, that the incompatible-element profiles shown in Fig. 6 are significantly steeper between Nb and Yb

than typical OIB *tholeiites*, and very much steeper than basalts erupted in the central volcanic areas of Iceland today. Postulating an OIB-like source beneath the lithosphere does not therefore account quantitatively for the trace-element signature of the PAS magmas, without a further stage of incompatible-element enrichment. This could occur as a result of the incorporation of crustal material, or at an earlier stage during passage of the magmas through the sub-continental lithosphere.

It is not difficult to accept that certain geochemical characteristics of the PAS magmas, particularly the lava samples examined in this study, have been influenced by interaction with the continental crust. It is a different, much less sustainable argument to suggest that the unusual enrichments in highly incompatible elements are entirely derived from this source. Partial trace-element analyses of typical basement rocks in the area suggest a fairly persistent negative Nb-Ta anomaly, which should be apparent in any magmatic rock assimilating a large proportion of

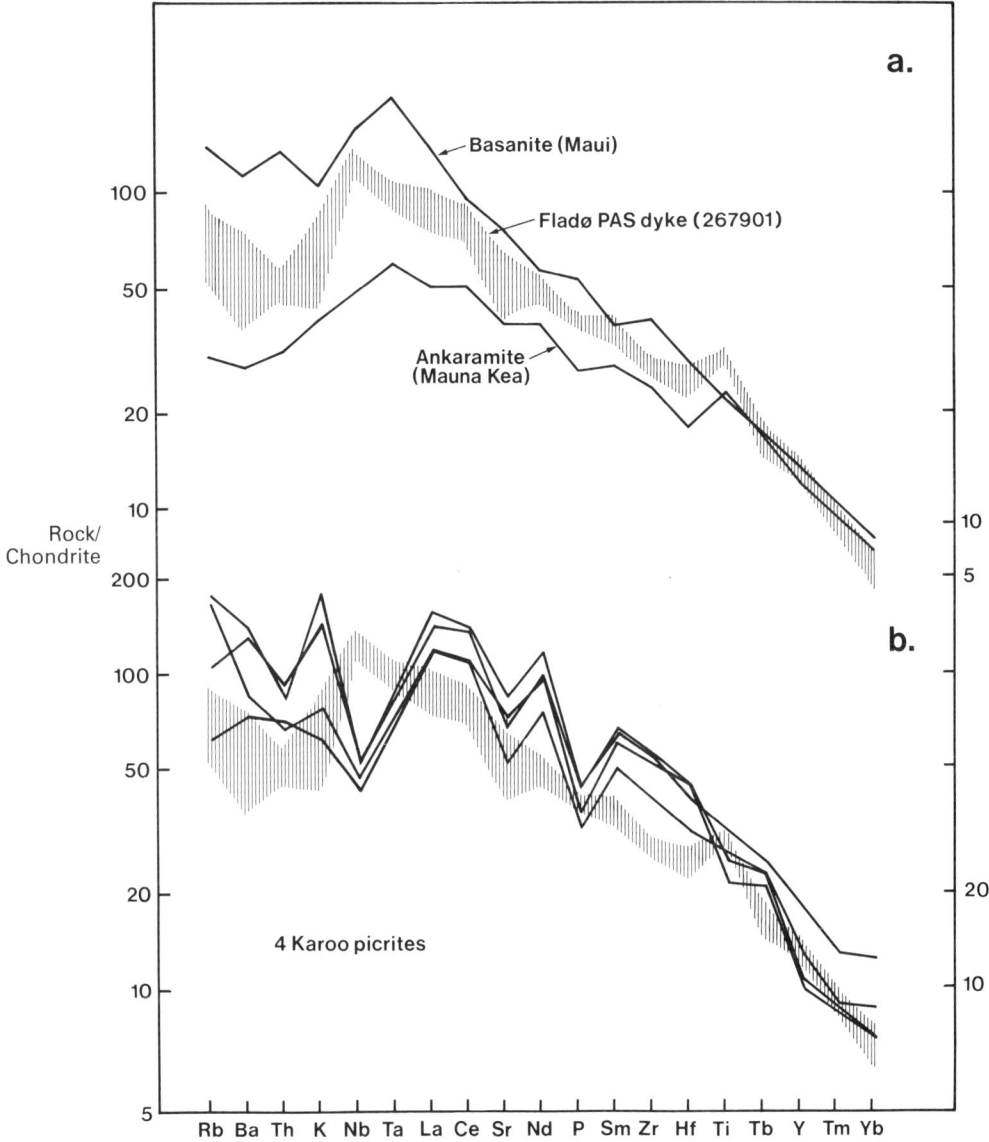

FIG. 9. PAS dyke (shaded field) compared with: (a) ocean island basalts (samples RTH 24 and RTH 31 from Thompson *et al.* 1984); (b) Karoo picrite basalts (data from Cox *et al.* 1984).

its incompatible elements from such materials. No such anomaly is seen in the Fladø PAS dyke, and it is far from prominent in the corresponding lavas. Sr-Nd systematics (Holm 1988) point to a source of enrichment in which Rb/Sr is more enhanced than Sm/Nd. These are not the characteristics of the originally granulite facies (low Rb/Sr) gneisses into which the Fladø dykes have been emplaced.

Menzies *et al.* (1987) have documented evi-

dence of highly enriched mantle peridotites and pyroxenites underneath the Lewisian crust of the Outer Hebrides. Xenoliths of this composition have not been recorded from E Greenland, but in view of the similar basement age it is not unreasonable to infer the existence of similarly enriched domains in the lithospheric mantle under the Kangerdlugssuaq area. One can postulate mechanisms by which parental picrite magmas might scavenge incompatible elements from

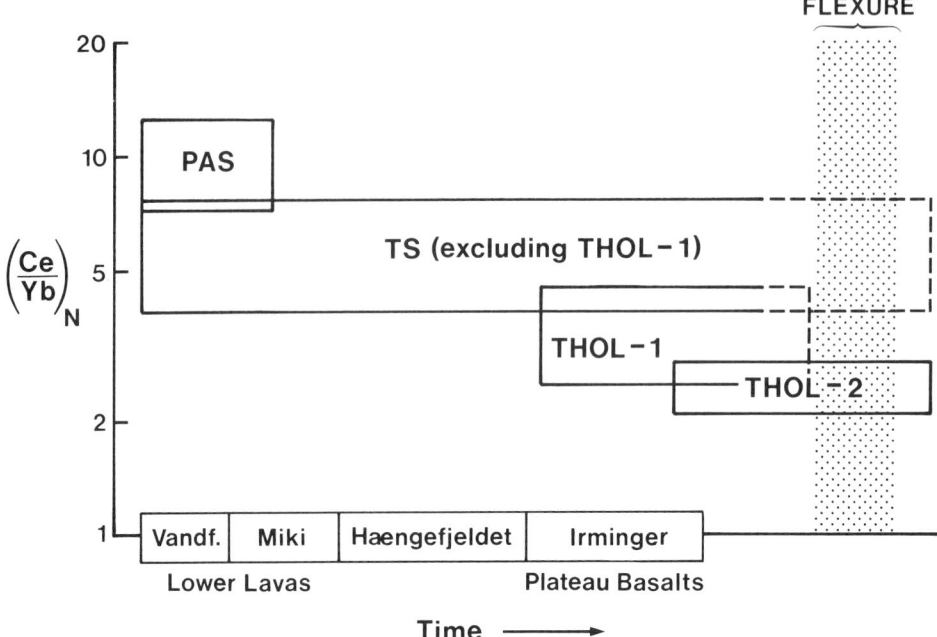

FIG. 10. Evolution of magma type with time. 'THOL-2' here refers solely to the primitive member 27700–2 and the analogous 'depleted' Fladø dyke 267909.

the grain-boundary assemblages of such litho-spheric mantle materials in passing to the surface, and thereby acquire incompatible element signatures with similarly steep profiles. The quantitative influence of enriched lithosphere on the geochemistry of high-Ti picrites is speculative, however, since the volume-extent of such enriched domains in the sub-continental lithosphere is unknown.

The second question concerns the evolution in magma type in time. The enriched, high Ce/Yb PAS lavas are characteristic of the earliest phase of volcanism in the Kangerdlugssuaq area (the Lower Basalts). The less-enriched TS lavas with which they interdigitate continued erupting for a much longer period, as they constitute much of the Plateau Basalt succession. The subsequent THOL-1 dykes seem to represent a further decline in incompatible element content and Ce/Yb ratio (see Figs. 6 & 8). The lowest contents are found in two late dykes similar in composition to Icelandic lavas. The succession of magma types is complex and there would undoubtedly have been more than one magma type available for eruption at any one stage. We nevertheless believe that this broad time progression (Fig. 10) has some petrogenetic significance.

Brooks & Nielsen (1982b) proposed a model in

which the early picrites are the product of relatively deep melting of enriched upper mantle. The upward passage of the hot proto-picrite magmas then provokes melting at higher, more depleted levels, generating the TS magmas. A possible objection to this model is that it assumes a specific style of chemical layering of the upper mantle.

The elemental trends can be explained by mixing of several mantle sources in proportions which vary with time. We envisage an OIB-source 'jet' rising through the asthenosphere and invading the sub-continental lithosphere below the Kangerdlugssauq region. In this model, the picrite melts parental to the PAS are extracted primarily from a zone of melting in the jet near to the base of the lithosphere, and assimilate incompatible elements during their upward passage through the lithosphere. The zone of melting, due to its high temperature, causes 'magmatic erosion' and thinning of the refractory sub-continental lithosphere from beneath. As the process progresses towards eventual parting of the continental block, the contribution of the enriched lithospheric mantle decreases, so that magmas escaping to the surface become progressively less enriched in incompatible elements. The system eventually advances to the present-day steady state in which the jet material rises through the asthenosphere under oceanic lithosphere.

This magmatic thinning and extraction model, by invoking a particularly high degree of melting as a requirement for the advancing invasion of refractory continental lithosphere, offers a possible explanation for the common occurrence of picrite basalts in the earliest stages of the development of continental flood basalt provinces, as in the Karoo and the Deccan. Such a process would have a long time constant, and it is interesting to note that uplift and rifting in the Kangerdlugssuaq area can be traced back as far as the Lower Cretaceous.

ACKNOWLEDGEMENTS: R. C. O. Gill thanks John S. Myers and the Geological Survey of Greenland for the opportunity to collect material from Fladø in 1979. We are indebted to Drs G. F. Marriner, J. C. Bailey, S. J. Parry, N. W. Rogers and I. Sinclair for their help or advice with trace-element analysis. Mary Gill helped in the data preparation, Sarah Viggers typed the manuscript and Christine Flood and Bente Thomas prepared the diagrams.

R. C. O. Gill acknowledges funding from the Royal Society under the European Scientific Exchange Scheme, and the hospitality of the Geological Museum in Copenhagen. C. K. Brooks and T. F. D. Nielsen acknowledge the support of the Danish Natural Science Research Council over the years 1972–7. This paper is published with permission of the Director of the Geological Survey of Greenland.

References

BIRD, D. K., ROGERS, R. D. & MANNING, C. E. 1986. Mineralized fracture systems of the Skaergaard intrusion, East Greenland. *Meddelelser om Grønland, Geoscience*, **16**, 1–68.

BRIDGWATER, D., DAVIES, F. B., GILL, R. C. O., GORMAN, B. E., MYERS, J. S., PEDERSEN, S. & TAYLOR, P. N. 1978. Precambrian and Tertiary geology between Kangerdlugssuaq and Angmagssalik, East Greenland. *Grønlands Geologiske Undersøgelse Rapport*, **83**, 17 pp.

BROOKS, C. K. 1973. Rifting and doming in southern East Greenland. *Nature Physical Science*, **244**, 23–25.

—— 1977. The Fe₂O₃/FeO ratio of basalt analyses: an appeal for a standardized procedure. *Bulletin of the Geological Society of Denmark*, **25**, 117–220.

—— & NIELSEN, T. F. D. 1978. Early stages in the differentiation of the Skaergaard magma as revealed by a closely related suite of dike rocks. *Lithos*, **11**, 1–14.

—— & —— 1982a. The Phanerozoic development of the Kangerdlugssuaq area, East Greenland. *Meddelelser om Grønland, Geoscience*, **9**, 30 pp.

—— & —— 1982b. The E Greenland continental margin: a transition between oceanic and continental magmatism. *Journal of the Geological Society of London*, **139**, 265–275.

——, —— & PETERSEN, T. S. 1976. The Blosseville Coast basalts of East Greenland: composition and temporal variations. *Contributions to Mineralogy and Petrology*, **58**, 279–292.

COX, K. G. 1983. The Karoo Province of southern Africa: origin of trace element enrichment patterns. *In*: HAWKESWORTH, C. J. & NORRY, M. J. (eds) *Continental Basalts and Mantle Xenoliths*. Shiva, pp. 139–157.

——, DUNCAN, A. R., BRISTOW, J. W., TAYLOR, S. R. & ERLANK, A. J. 1984. Petrogenesis of the basic rocks of the Lebombo. *In*: ERLANK, A. J. (ed) *Petrogenesis of the Volcanic Rocks of the Karoo Province*. Special Publication of the Geological Society of South Africa, **13**, 149–169.

DEER, W. A. 1976. Tertiary igneous rocks between Scoresby Sund and Kap Gustav Holm, East Greenland. *In*: ESCHER, A. & WATT, W. S. (eds) *Geology of Greenland*. Geological Survey of Greenland, 405–429.

GOVINDARAJU, K. 1984. 1984 compilation of working values and sample description for 170 international reference samples of mainly silicate rocks and minerals. *Geostandards Newsletter, Special Issue*, **8**, 3–13.

HENDERSON, P. & WILLIAMS, C. T. 1981. Application of intrinsic Ge-detectors to the instrumental neutron activation analysis of rare earth elements in rock and minerals. *Journal of Radioanalytical Chemistry*, **67**, 445–452.

HOLM, P. M. 1988. Nd, Sr and Pb isotope geochemistry of the Lower Lavas, East Greenland Tertiary Igneous Province. *In*: MORTON, A. C. & PARSON, L. M. (eds) *Early Tertiary Volcanism and the Opening of the NE Altantic*. Geological Society of London, Special Publication, **39**, pp. 181–195.

LARSEN, H. C. 1984. Geology of the East Greenland shelf. *In*: SPENCER, A. M., *et al*. (eds) *Petroleum Geology of the N European Margin*. Norwegian Petroleum Society; Graham & Trotman, London, pp. 329–339.

LARSEN, L. M. & WATT, W. S. 1985. Episodic volcanism during the break-up of the North Atlantic: evidence from the E Greenland plateau basalts. *Earth and Planetary Science Letters*, **73**, 105–116.

——, —— & WATT, M. (in press). Geology and petrology of the Lower Tertiary Plateau Basalts of the Scoresby Sund region. *Grønlands Geologiske Undersøgelse Bulletin*.

MENZIES, M. A., HALLIDAY, A. N., PALACZ, Z., HUNTER., R. H., UPTON, B. G. J., ASPEN, P. & HAWKESWORTH, C. J. 1987. Evidence from mantle xenoliths for an enriched lithospheric keel under the Outer Hebrides. *Nature*, **325**, 44–47.

MYERS, J. S. 1980. Structure of the coastal dyke swarm and associated plutonic intrusions of East Greenland. *Earth and Planetary Science Letters*, **46**, 407–418.

NAKAMURA, N. 1974. Determination of REE, Ba, Fe, Mg, Na and K in carbonaceous and ordinary chondrites. *Geochimica et Cosmochimica Acta*, **38**, 757–775.

NIELSEN, T. F. D. 1978. The Tertiary dike swarm of the Kangerdlugssuaq area, East Greenland. An example of magmatic development during continental break-up. *Contributions to Mineralogy and Petrology*, **67**, 63–78.

—— & BROOKS, C. K. 1981. The E Greenland rifted continental margin: an examination of the coastal flexure. *Journal of the Geological Society of London*, **138**, 559–568.

——, SOPER, N. J., BROOKS, C. K., FALLER, A. M., HIGGINS, A. C. & MATTHEWS, D. W. 1981. The pre-basaltic sediments and the Lower Basalts at Kangerdlugssuaq, East Greenland: their stratigraphy, lithology, palaeomagnetism and petrology. *Meddelelser om Grønland, Geoscience*, **6**, 25 pp.

O'HARA, M. J. 1977. Geochemical evolution during fractional crystallisation of a periodically refilled magma chamber. *Nature*, **266**, 503–507.

—— & MATTHEWS, R. E. 1981. Geochemical evolution in an advancing, periodically replenished, periodically tapped, continuously fractionated magma chamber. *Journal of the Geological Society of London*, **138**, 237–277.

RUCKLIDGE, J., BROOKS, C. K. & NIELSEN, T. F. D. 1979. Petrology of the coastal dykes at Tugtilik, southern East Greenland. *Meddelelser om Grønland, Geoscience*, **3**, 17 pp.

SOPER, N. J., HIGGINS, A. C., DOWNIE, C., MATTHEWS, D. W. & BROWN, P. E. 1976. Late Cretaceous–Early Tertiary stratigraphy of the Kangerdlussuaq area, East Greenland and the opening of the northeast Atlantic. *Journal of the Geological Society of London*, **132**, 85–102.

SØRENSEN, I. 1975. X-ray spectrometry at GGU. *Grønlands Geologiske Undersøgelse Rapport*, **75**, 16–18.

TAYLOR, S. R. & GORTON, M. P. 1977. Geochemical application of spark source mass spectrography—III. Element sensitivity, precision and accuracy. *Geochimica et Cosmochimica Acta*, **41**, 1375–1380.

THOMPSON, R. N., DICKIN, A. P., GIBSON, I. L. & MORRISON, M. A. 1982. Elemental fingerprints of isotopic contamination of Hebridean Palaeocene mantle-derived magmas by Archean sial. *Contributions to Mineralogy and Petrology*, **79**, 159–168.

——, MORRISON, M. A., DICKIN, A. P. & HENDRY, G. L. 1983. Continental flood basalts . . . Arachnids rule OK? *In*: HAWKESWORTH, C. J. & NORRY, M. J. (eds) *Continental Basalts and Mantle Xenoliths*. Shiva, pp. 158–185.

——, ——, HENDRY, G. L. & PARRY, S. J. 1984. An assessment of the roles of crust and mantle in magma genesis: an elemental approach. *Philosophical Transactions of the Royal Society*, **A130**, 549–590.

WAGER, L. R. 1947. Geological investigations in East Greenland, Part IV. Stratigraphy and tectonics of Knud Rasmussen Land and the Kangerdlugssuaq region, East Greenland. *Meddelelser om Grønland*, **134**(5), 64 pp.

—— & DEER, W. A. 1938. A dyke swarm and coastal flexure in East Greenland. *Geological Magazine*, **75**, 39–46.

WEAVER, B. L. & TARNEY, J. 1983. Chemistry of the sub-continental mantle: inferences from Archaean and Proterozoic dykes and continental flood basalts. *In*: HAWKESWORTH, C. J. & NORRY, M. J. (eds) *Continental Basalts and Mantle Xenoliths*. Shiva, pp. 209–229.

WOOD, D. A., JORON, J-L., TREUIL, M., NORRY, M. & TARNEY, J. 1979. Elemental and Sr isotope variations in basic lavas from Iceland and the surrounding ocean floor. *Contributions to Mineralogy and Petrology*, **70**, 319–339.

R. C. O. GILL & G. A. INGRAM, Department of Geology, Royal Holloway and Bedford New College, Egham Hill, Egham, Surrey TW20 0EX, UK.

T. F. D. NIELSEN, Geological Survey of Greenland, Østervoldgade 10, 1350 Copenhagen K, Denmark.

C. K. BROOKS, Institut for Petrologi, University of Copenhagen, Østervoldgade 10, 1350 Copenhagen K, Denmark.

APPENDIX

Summary of results obtained on the USGS Columbia River Basalt
BCR-1 during the course of this work.

Element	RHBNC analyses		No of analyses used for mean	Published values
	Mean	Std. dev.		
XRF[1]				
Nb	14	0.8	4	14[3]
Zr	189	8	4	191[3]
Y	42.3	1.0	4	39[3]
Sr	335	18	4	330[3]
Rb	46.5	2.4	4	47[3]
INAA[2]				
La	24.2	2.1	12	24.2[4]
Ce	53.2	4.1	13	53.7[4]
Nd	27.5	1.6	13	28.5[4]
Sm	6.62	0.2	11	6.7[4]
Eu	1.94	0.1	13	1.95[4]
Tb	1.07	0.09	13	1.08[4]
Yb	3.43	0.23	7[5]	3.48[4]
Lu	0.48	0.03	13	0.51[3], 0.54[4]
Ta	0.77	0.05	9	0.79[3]
Hf	5.21	0.21	9	4.90[3], 5.17[4]
Th	5.62	0.25	7	6.04[3]

1. X-ray fluorescence spectrometry (using the Ag $K\alpha$ tube line for matrix correction) on powder pellets, Bedford College, University of London. The standard deviation shows the between-run reproducibility on R. Gill samples during 1984.
2. Instrumental neutron activation analysis by R. C. O. Gill at Imperial College Reactor Centre (University of London), Silwood Park, Ascot, UK. Method, corrections, etc similar to Henderson and Williams (1981). La, (Nd), Sm, Lu counted 4–6 days after end of irradiation. Ce, Nd, Eu, Tb, (Tm), Yb, (Lu), Ta, Hf, Th counted 18–20 days after irradiation. Standard deviation shows the between-batch reproducibility during 1983 and 1984 (Yb for 1984 only[5]).
3. Govindaraju (1984).
4. Taylor and Gorton (1977).
5. Using 177 keV peak which suffers least interference (employed for all samples in the present study). Results from the 63.1 keV line used in earlier work have been excluded from this value.

Nd, Sr and Pb isotope geochemistry of the Lower Lavas, E Greenland Tertiary Igneous Province

P. M. Holm

SUMMARY: New Sr, Nd and Pb isotopic and trace-element data are presented for the Lower Lavas and some Plateau Basalts. Ranges of isotopic compositions are large among the lower and middle Vandfaldsdalen Formation lavas ($^{87}Sr/^{86}Sr = 0.7033-0.7073$; $^{143}Nd/^{144}Nd = 0.5124-0.5130$; $^{206}Pb/^{204}Pb = 15.1-17.5$; $^{207}Pb/^{204}Pb = 14.7-15.6$; $^{208}Pb/^{204}Pb = 35.0-39.1$) and show a decrease in range of values with stratigraphic height. The youngest flows among the Lower Lavas have compositions similar to the Plateau Basalts of the late Skrænterne and Igtertiva Formations as well as of Nansen Fjord and Wiedemann Fjord, all of which are near to Icelandic composition ($^{87}Sr/^{86}Sr = 0.7031-0.7036$; $^{143}Nd/^{144}Nd = 0.5129-0.5131$; $^{206}Pb/^{204}Pb = 17.1-18.4$; $^{207}Pb/^{204}Pb = 15.3-15.4$; $^{208}Pb/^{204}Pb = 37.6-38.0$). The lavas constitute two groups, one of which (group 2) is restricted to the lower parts of the succession. The source for this group can be modelled by mixing between an enriched old lithospheric mantle and an ocean island basalt (OIB)-type mantle with a relatively small component of depleted asthenosphere. The other group (group 1) is the most common and can be modelled by mixing between a typical Icelandic source component and a second, less enriched, lithospheric mantle component. Contamination from continental crust was of minor overall importance.

The geochemical evolution in basic E Greenland Tertiary lavas is believed to reflect the interaction between the asthenospheric mantle, 3 Ga old continental lithospheric mantle and the mantle jet which is presently positioned under Iceland. With time the most enriched lithospheric source quickly decayed while the less enriched lithospheric source was important during most of Vandfaldsdalen Formation times. The OIB-type source prevailed throughout lava eruption with the asthenospheric contribution larger in the Miki Formation and the Plateau Basalts.

The lavas of the East Greenland Tertiary (EGT) Igneous Province are dominantly tholeiitic flood basalts which cover more than 60 000 km² and attain a maximum thickness in excess of 5 km (Wager 1947; Brooks et al. 1976; Nielsen et al. 1981; Brooks & Nielsen 1982a; Upton et al. 1980; Larsen & Watt 1985). The most prominent part of the province is situated to the W of the ca. 400 km coastline of the Blosseville Kyst between Kangerdlugssuaq and Scoresby Sund (Fig. 1). Recent geophysical and geological investigations indicate that the transition from continental to oceanic lithosphere is very close to the present coastline. Parts of a zone of active rifting are preserved at Kap Dalton, where the final Tertiary lava flows of the Igtertiva Formation outcrop in downfaulted blocks (Larsen & Watt 1985). The Kangerdlugssuaq Fjord area was a major focus of igneous activity and has been suggested to be a failed arm of a palaeo-triple junction and the initial site of the present-day Icelandic mantle jet (Brooks 1973; Brooks & Nielsen 1982b).

The Lower Lavas, comprising the earliest activity of the EGT Igneous Province, outcrop immediately to the E of Kangerdlugssuaq Fjord area and represent those volcanics preserved closest to the trace of the proposed mantle jet on the continental crust (Brooks 1973; Vink 1984).

In the Kangerdlugssuaq area the volcanics penetrated high-grade basement gneisses of late Archaean age which had been part of a stable craton for nearly 3 Ga. The Caledonian Orogeny mainly affected areas further to the N and E but locally also to the W (Hamilton 1964; Brooks et al. 1981). A sedimentary cover occurs only sporadically in the Kangerdlugssuaq area reaching a maximum thickness of ca. 200 m near Jacobsens Fjord. It was deposited during subsidence prior to the onset of volcanic activity in the late Cretaceous to Palaeocene (Soper et al. 1976). The continental rupture at the initiation of the opening of the N Atlantic Ocean was thus the first disturbance of the continental lithosphere since late Archaean times. Over this time-span geochemical contrast between reservoirs in the lithospheric mantle and crust would probably have developed large Nd, Sr and Pb isotopic differences. If components from such reservoirs contributed to the genesis of the magmatic rocks they should be more readily recognized than in areas that had been stable for a shorter period. In this paper the isotopic and elemental variation within the stratigraphically well-constrained Lower Lavas will be described, and the possible source components discussed. The field geology, petrography and major element geochemistry of

From MORTON, A. C. & PARSON, L. M. (eds), 1988, *Early Tertiary Volcanism and the Opening of the NE Atlantic,* Geological Society Special Publication No. 39, pp. 181–195.

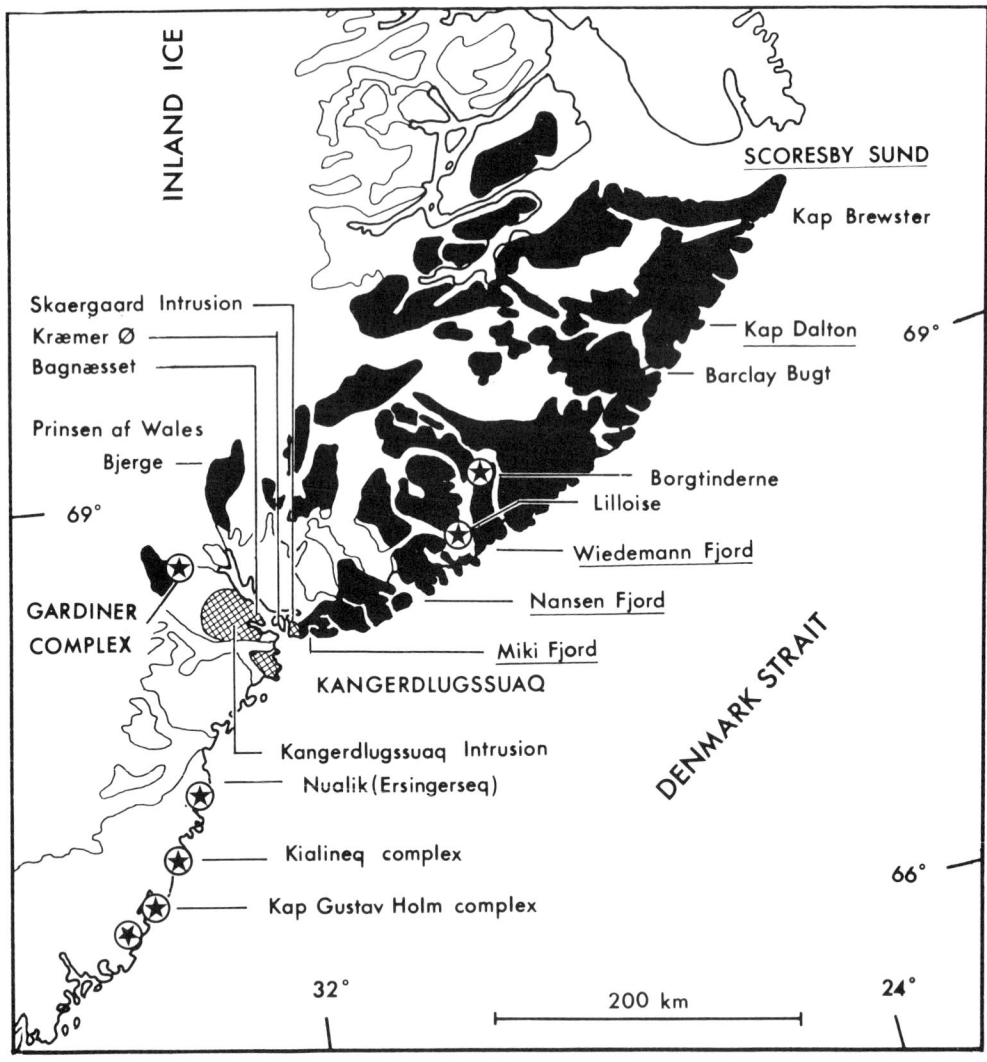

FIG. 1. Tertiary igneous geology of central E Greenland between 66°N and 71°N. Black: basalts; cross-hatched: major intrusions of the Kangerdlugssuaq area; stars: other major central intrusions; unmarked: basement complex. The extent of the ice-cap is also shown. Place names of areas for this study are underlined. After Brooks & Nielsen (1982a, b).

the Lower Lavas and the Scoresby Sund Lavas have recently been discussed (Nielsen *et al.* 1981; Brooks & Nielsen 1982a, b; Larsen & Watt 1985), but published trace-element and isotope geochemistry data are sparse (Carter *et al.* 1979). This paper presents new isotope and trace-element analyses of 23 Lower Lavas and eight flood basalts from localities in Nansen Fjord, Wiedemann Fjord, Kap Dalton and Scoresby Sund. It will be argued that: (1) the geochemical

variation in the lavas mainly reflects different mantle sources; (2) a lithospheric mantle component was prominent only in the early erupted lavas; (3) OIB and mid-ocean-ridge basalt (MORB)-type mantle sources with near Icelandic characteristics were the dominant components for the entire length of the Blosseville Kyst; and (4) the lavas have persistently higher abundances of incompatible elements than tholeiitic rock-types from the active rift zone of Iceland.

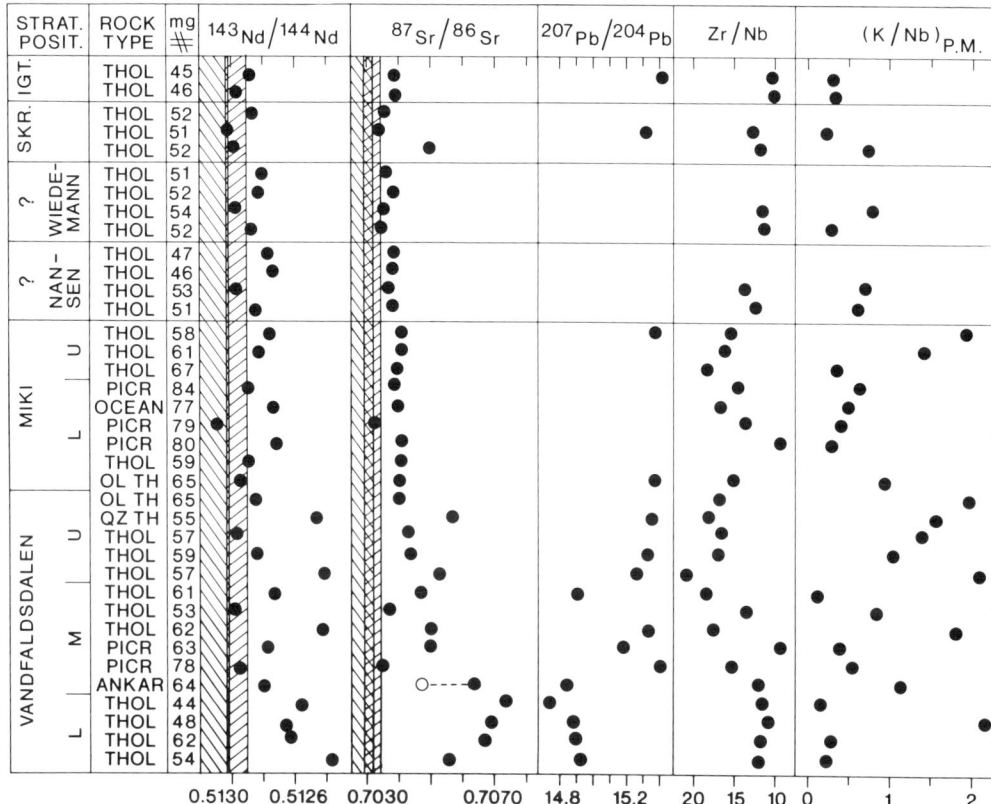

FIG. 2. Summary of the lava stratigraphy and the rock types of the analysed lavas, their isotopic compositions and illustrative elemental ratios. The Lower Lavas: Vandfaldsdalen and Miki formations. Plateau Basalts: Nansen Fjord and Wiedemann Fjord of unknown stratigraphic position, and the Skrænterne and Igtertiva formations. The ruling dipping to the right designates N Atlantic MORB compositions; ruling dipping to the left is the range of basalts from the active volcanic zone of Iceland. Open circle: clinopyroxene separate. See text for explanation.

Presentation of the new analytical data

The widest range of rock-types (Mg number = 78–44) (Fig. 2) is found in the earliest Vandfaldsdalen Formation (ca. 500 m) of the Lower Lavas, which has been subjected to more detailed sampling for this study than the overlying Miki Formation (1400 m) which is characterized by many picritic flows (Nielsen *et al.* 1981). The Plateau Basalts constitute the major upper part of the stratigraphic pile, but is only well studied in the northern part from Kap Dalton to Scoresby Sund (see Fig. 1) (Fawcett *et al.* 1973; Larsen & Watt 1985). Two samples from the youngest Igtertiva formation include the latest lava flow at Kap

Dalton (144805), situated 320 km NE of Miki Fjord. Two samples come from the Skrænterne Formation on the S coast of Scoresby Sund, 400 km NE of Miki Fjord. The stratigraphic position of the lavas of Nansen Fjord (60 km) and Wiedemann Fjord (120 km) is unknown, but they are most probably high in the sequence, judging from the major element chemistry. The samples in this study are thought to represent the full range of variation among the Lower Lavas, and to provide adequate coverage with respect to both age and geographical position of the flood basalts. A total of 32 samples were analysed for Sr isotopes and trace elements, 31 for Nd isotopes and 16, mainly from Vandfaldsdalen Formation, were analysed for Pb isotopes (Tables 1, 2 & 3).

TABLE 1. *Neodymium and strontium isotopic compositions of lavas from the E Greenland Tertiary Igneous Province*

Stratigraphic position	Sample number	Rock type	Mg #	^{147}Sm/^{144}Nd	^{143}Nd/^{144}Nd	^{87}Sr/^{86}Sr	^{87}Sr/^{86}Sr leached	^{87}Rb/^{86}Sr	^{87}Sr/^{86}Sr t = 53 Ma
Lower lavas									
Vandfaldsdalen Formation									
Lower:	CKB-100	tholeiite	54	0.145	0.512373±19	0.70588 ±5[b]	0.706652±22	0.0209	0.70536
	40031	tholeiite	62	0.171	c0.512628±21[a]	0.709397±18	0.706812±38	0.0328	0.70663
	40630	tholeiite	48	0.170	c0.512651±14	0.706901±24	0.70743 ±6[c]	0.168	0.70669
	40632	basaltic andesite	44	0.163	0.512584±21	0.707927±12	0.706383±10[a]	0.1279	0.70733
Middle:	20351	ankaramite	64		0.512805±13	0.704719±58	0.706306±68[a]	0.0766	0.70621
		clinopx separate			0.512846±46				(0.70472)
	40633	picrite	78		0.512961±17	0.705348±10	0.70340 ±4[c]	0.1226	0.70331
	40032	picrite	63		0.512784±15	0.705157±12	0.70497 ±3	0.0835	0.70491
	40634	tholeiite	62	0.155	0.512473±18[a]	0.704910±10	0.704907±10	0.1203	0.70482
	40636	tholeiite	53		0.512985±14	0.703758±12[a]	0.70369 ±4[c]	0.0638	0.70364
	40020	olivine tholeiite	61	0.169	0.512730±28	0.704615±10	0.704500±12	<0.004	0.70450
Upper:	40041	tholeiite	57	0.156	0.512453±20	0.705190±8	0.70523 ±4[c]	0.0785	0.70517
	40625	tholeiite	59		0.512875±17	0.704208±14		0.0789	0.70415
	40626	tholeiite	57		0.512983±16	0.704015±10		0.0707	0.70396
	27116	quartz tholeiite	55	0.199	0.512473±15	0.705618±16	0.705470 ±19	0.1232	0.70538
	27345	olivine tholeiite	65		0.512871±25	0.704221±20		0.169	0.70409
Miki Formation									
Lower:	20332	olivine tholeiite	59		0.512982±14	0.703975±12		0.0369	0.70395
	40657	picrite	80		0.512724±34	0.704484±12		0.258	0.70429
	40646	picrite	79		0.513100±14	0.703122±10		0.0513	0.70308
	40645	oceanite	77		0.512784±12	0.703863±32		0.0676	0.70381
	40651	picrite	84		0.512932±26	0.703777±12		0.1386	0.70362
	40641	tholeiite	67			0.703803±10[a]		0.0475	0.70377
Upper:	40640	tholeiite	61		0.512866±16	0.703885±10[a] 0.703916±6		0.0565	0.70386
	40639	tholeiite	58		0.512801±27	0.703936±14		0.0762	0.70388
Unknown stratigraphic position									
Nansen Fjord	CKB 71-34/16	tholeiite	51		0.512886±23	0.703681±10		0.0565	0.70364
	CKB 71-34/17	tholeiite	53		0.513009±14	0.703489±10		0.0591	0.70344
Wiedemans Fjord	EG7210	tholeiite	52		0.512938±22	0.703261±74		0.0296	0.70324
	EG7211	tholeiite	54		0.513015±17	0.703302±14		0.0565	0.70326
Scoresby Sund lavas									
Skrænterne Formation	EG7108	tholeiite	52		0.513008±16	0.704478±35		0.1305	0.70438
	EG7159	tholeiite	51		0.513040±22	0.703202±70		<0.006	0.70320
Igtertiva Formation	144781	tholeiite	46		0.512983±15[a]	0.703703±10[a]		0.0673	0.70367
	144805	tholeiite	45		0.512889±20	0.703745±10		0.0493	0.70371

Nd and Sr isotopes were measured in Leeds, U.K., except for those marked (c), which were analysed in Copenhagen. (a) denotes duplicate analysis; (b) is from Pankhurst *et al.* (1976). The leached samples were treated with 6N HCl at >100°C in bombs for 12 hours. The Rb/Sr ratios were measured by XRF and have a precision of <1.5%. The detection limit for Rb was 0.5 ppm. The decay constant used was 1.42 10^{-11} a^{-1} for ^{87}Rb.

TABLE 2. *Lead isotopic compositions of lavas from the E Greenland Tertiary Igneous Province*

Stratigraphic position and sample number		Rock type	$^{206}Pb/^{204}Pb$	$^{207}Pb/^{204}Pb$	$^{208}Pb/^{204}Pb$
Lower lavas					
Vandfaldsdalen Formation					
Lower:	CKB-100	tholeiite	15.691	14.988	35.735
	40031	tholeiite	16.162	14.923	35.786
	40630	tholeiite	15.740	14.897	37.793
	40632	basaltic andesite	15.013	14.709	35.688
Middle:	20351	ankaramite	15.645	14.858	36.118
	40633	picrite	17.545	15.415	37.877
	40032	picrite	17.484	15.156	39.136
	41634	tholeiite	16.294	15.344	37.843
	40020	olivine tholeiite	15.456	14.888	34.982
Upper:	40041	tholeiite	17.251	15.269	37.198
	40625	tholeiite	17.277	15.330	37.662
	27116	quartz tholeiite	17.101	15.358	36.833
Mikis Formation					
Lower:	20332	olivine tholeiite	17.322	15.380	37.676
Upper:	40639	tholeiite	17.162	15.397	38.049
Scoresby Sund lavas					
Skrænterne Formation					
	EG7159	tholeiite	18.439	15.291	37.602
Igtertiva Formation					
	144805	tholeiite	18.060	15.404	37.982

Pb isotope ratios were determined in Leeds, UK, using the HBr technique with electrolytic deposition and loading with silica gel. Analytical data were normalized to SRM981 (16.937, 15.491, 36.721).

TABLE 3. *Partial chemical analyses representing the range of variation in the lavas of the E Greenland Tertiary Igneous Province*

	Sample/Formation					
	Group 1			Group 2		
Element	27116 Vandfaldsdalen	20332 Miki	EG7159 Skrænterne	40032 Vandfaldsdalen	40632 Vandfaldsdalen	144805 Igtertiva
Mg#	54	65	51	44	77	45
Ni	98	735	88	39	1099	75
Cr	168	1396	174	44	1748	159
Rb	15	4.2	<0.5	25	9.1	5.5
Ba	270	70	38	334	144	124
K_2O	0.76	0.38	0.09	1.12	0.27	0.27
Nb	9.6	8.5	11	18	14	21
La	14	7	9	22	12	20
Ce	34	21	28	56	28	50
Sr	353	250	231	567	319	287
P_2O_5	0.23	0.31	0.28	0.33	0.22	0.33
Zr	176	130	145	248	130	219
TiO_2	2.22	2.12	2.42	3.39	2.22	3.14
Y	27	25	35	32	17	40

Post-eruptional modification of the isotopic compositions

Exchange at high temperatures of Sr (but not Nd and Pb) isotopes between basaltic lavas and seawater is well documented (e.g. O'Nions *et al.* 1978; O'Nions 1983). Hydrothermal alteration can change the Sr isotopes (but probably not the Nd) towards compositions depending on the origin of Sr in the hydrothermal fluids (Hawkesworth & Morrison 1978). Oxygen isotope studies of the basalts of the Vandfaldsdalen Formation near the Skaergaard intrusion indicate high temperature equilibration of basalt and meteoric water (Taylor & Forester 1979). Recent investigations in the Miki Fjord area have shown ubiquitous low-grade metamorphic alteration of the basalts (Nielsen *et al.* 1981; Bird *et al.* 1986). The alteration is mainly limited to the immediate vicinity of small-scale joints that apparently guided the fluid circulation. Such zones of alteration were deliberately avoided when collecting samples for analysis. However, several samples do contain secondary minerals. These samples were leached with HCl in order to dissolve and remove secondary phases such as calcite and chlorite. The Sr isotope compositions measured subsequent to leaching are thought to represent the magmatic values (see Table 1 for details).

Isotope geochemistry of the lavas

The ranges of Sr, Nd and Pb isotopic compositions of the Tertiary E Greenland Lavas are very large (see Tables 1 & 2):

$$^{143}Nd/^{144}Nd = 0.51237 - 0.513100;$$
$$(^{87}Sr/^{86}Sr)_{53\ Ma} = 0.70308 - 0.70733;$$

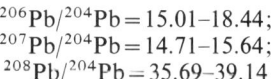

$$^{206}Pb/^{204}Pb = 15.01 - 18.44;$$
$$^{207}Pb/^{204}Pb = 14.71 - 15.64;$$
$$^{208}Pb/^{204}Pb = 35.69 - 39.14$$

The Nd and Pb isotopic ratios have not been corrected for decay of the radioactive parent since extrusion. Nearly all the variation within the data set is present in the Vandfaldsdalen Formation lavas. The Miki Fjord lavas display a relatively much smaller range of isotopic variation, while the flood basalts show relatively little variation. As evident in Fig. 2 there is a clear tendency of decreasing isotopic variation with stratigraphic height.

The Vandfaldsdalen Formation

This formation consists of picrites, basalts and basaltic andesites. In the $^{143}Nd/^{144}Nd$ vs $^{87}Sr/^{86}Sr$ diagram (Fig. 3) early Vandfaldsdalen lavas define a trend towards high $^{87}Sr/^{86}Sr$ ratios. Other Vandfaldsdalen lavas in this diagram trend along the OIB 'mantle array', some plot below the Bulk Earth (BE) composition, and some between BE and the Icelandic field. From these trends three distinct components in the lavas are indicated, of which two can be characterized by time-integrated higher Rb/Sr and Nd/Sm than BE. One of these was relatively very enriched in Rb, which has led to high $^{87}Sr/^{86}Sr$ ratios. The third component has a MORB-type or Icelandic OIB-type characteristic with time-integrated relative depletion in the most incompatible elements (Rb and light rare-earth elements, LREE).

The Pb isotopes are less consistent, but are all very unradiogenic compared to MORB or OIB (including Iceland). A major source component for Pb must, with time, have become relatively depleted in U/Pb and Th/Pb compared to OIB

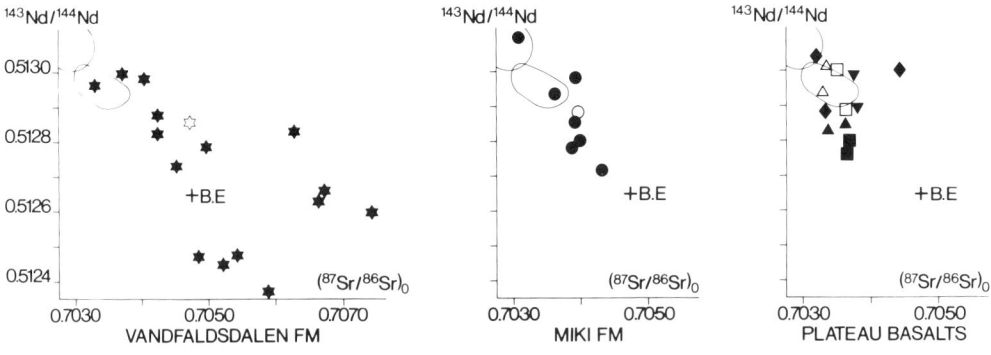

FIG. 3. $^{143}Nd/^{144}Nd$ vs $(^{87}Sr/^{86}Sr)$ t = 53 Ma, for the EGT lavas. Lower Lavas: Vandfaldsdalen Formation (stars). Miki Formation (circles). Plateau Basalts: Wiedemann Fjord (triangles), Nansen Fjord (squares), Skrænterne Formation (inverted triangles), Igtertiva Formation (diamonds). Open star: clinopyroxene separate. Other open symbols are from Carter *et al.* (1979). Also shown are BE (Bulk Earth) and the fields of N Atlantic MORB and the active zone of Iceland (O'Nions *et al.* 1977).

and MORB. From the nature of the $^{207}Pb/^{204}Pb$ vs $^{206}Pb/^{204}Pb$ diagram (Fig. 4) mixing should emerge more clearly than, for example, in the Nd-Sr isotope diagram (see Fig. 3), yet no well-defined trends are seen. However, the samples with very radiogenic Sr also have very unradiogenic Pb, and a trend is present in the $^{87}Sr/^{86}Sr$ vs $^{206}Pb/^{204}Pb$ diagram (Fig. 5).

Clinopyroxene was separated out of ankaramite 20351 and analysed for Nd and Sr isotopes. The Sr isotopic composition is very different from that of the whole rock. Leaching of the whole-rock samples indicates that secondary alteration was of minor importance in this sample.

The Miki Formation

The less variable succession of picritic and tholeiitic lavas of the Miki Formation have Nd and Sr isotope ratios plotting within the depleted part of the OIB mantle array, with one sample (40646) in the MORB field (see Fig. 3). The two samples analysed for Pb isotopes are more unradiogenic than Atlantic MORB and OIB and plot to the left of the geochron.

The Plateau Basalts

All samples analysed from Nansen Fjord, Wiedemann Fjord, Skrænterne Formation and Igtertiva Formation have isotopic Sr and Nd compositions that plot close to the Iceland field in Fig. 3. However, only one sample (EG7203) from Wiedemann Fjord actually plots in the field.

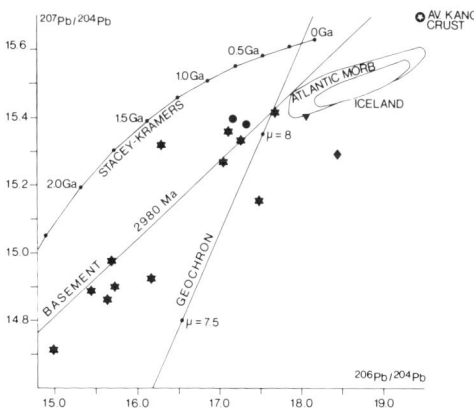

FIG. 4. $^{207}Pb/^{204}Pb$ vs $^{206}Pb/^{204}Pb$. Symbols as in Fig. 3. The field of N Atlantic MORB glasses are from Cohen & O'Nions (1982) and that of Iceland from Sun & Jahn (1975) and Sun *et al.* (1975). The growth curve is from Stacey & Kramers (1975) and the basement isochron from Leeman & Dasch (1976).

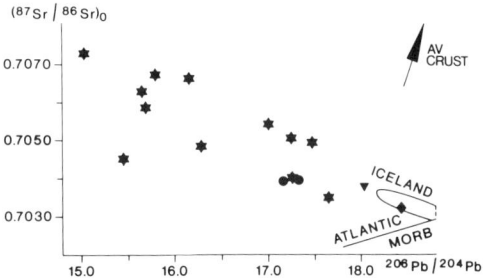

FIG. 5. $^{87}Sr/^{86}Sr$ vs $^{206}Pb/^{204}Pb$. Symbols and sources for MORB and Iceland as in Figs 3 & 4.

Based on Nd/Sr isotope systematics it is not possible to make any distinction between these formations and locations. It may be noted, however, that samples from Nansen Fjord and Wiedemann Fjord have relatively low $^{143}Nd/^{144}Nd$ ratios compared to the $^{87}Sr/^{86}Sr$ ratios. The Pb isotopic compositions of the Skrænterne and the Igtertiva samples are rather different: the Skrænterne samples plot below the Atlantic MORB field, while the Igtertiva samples can be compared to MORB in this diagram. The $^{208}Pb/^{204}Pb$ compositions of both lavas are, however, slightly different from MORB types (not shown).

The depleted component

All analysed lavas have isotopic compositions that conform to the two trends outlined above for the Vandfaldsdalen Formation. The Sr and Nd isotope data suggest that one component has low $^{87}Sr/^{86}Sr$ and high $^{143}Nd/^{144}Nd$ ratios defined by both the end-point of the two trends in the Nd-Sr isotope diagram (see Fig. 3) and in the evolutionary diagram (see Fig. 2) by the development towards a limited isotopic variation in the upper part of the stratigraphic succession. This component could be the Iceland field (see Fig. 3) and the values indicated by the lavas along the entire Blosseville Kyst are: $^{87}Sr/^{86}Sr <$ 0.7030; $^{143}Nd/^{144}Nd > 0.5130$. The Pb isotopes also define near-modern Icelandic compositions for this component. This is seen both in the Pb-Pb diagram (see Fig. 4) and in Sr-Pb diagrams for ^{206}Pb, ^{207}Pb and ^{208}Pb over ^{204}Pb isotopic ratios, e.g. see Fig. 5. The lead isotopic composition thus defined is: $^{206}Pb/^{204}Pb > 18.4$; $^{207}Pb/^{204}Pb > 15.4$ and $^{208}Pb/^{204}Pb > 38.1$. However, all measured Pb isotope analyses are less radiogenic than Icelandic values. Most samples which have Sr and Nd isotopes close to this end-member were not analysed for Pb isotopes. Sample 144805 of the Igtertiva Formation qualifies as an example of this end-member, but the analysed Skrænterne sample has a low $^{207}Pb/^{204}Pb$ ratio. Some samples from all formations have Sr and Nd isotopic compositions much like modern Icelandic lavas.

Abundances of incompatible elements in the lavas with Sr and Nd isotopic compositions close to this end-member (Samples 144805, Igtertiva; EG7159, Skrænterne and 40032, Vandfaldsdalen) are given in Table 3. Compared to typical Icelandic tholeiites, exemplified by ISL28 from the Askja volcano (Wood *et al.* 1979), all the E Greenland samples have higher incompatible element abundances. This is particularly true for sample 144805, which for most elements is enriched 2–3 times relative to the Askja sample. This does not solely reflect differences in degree of fractionation. The combined isotopic and elemental signature of the most depleted E Greenland samples are thus not totally in accord with the composition of the tholeiites presently erupted in the central rift zone of Iceland.

The enriched component

The two trends defined by Sr and Nd isotopes in Fig. 3 require at least two rather distinctive source components with much more radiogenic Sr and much less radiogenic Nd and Pb than the Icelandic type end-member. The OIB mantle array in the Nd-Sr isotope diagram (e.g. Zindler & Hart 1986) encloses even the most enriched members of the one trend, hereafter called group 1. However, the samples in the other trend, group 2, are more radiogenic in their $^{87}Sr/^{86}Sr$ ratio than oceanic lavas. Moreover, Pb isotopes as unradiogenic as those of both groups have never been reported from the oceanic environment. It therefore seems most likely that the two components originate within the continental lithosphere. The separation of the variation inherited from the mantle sources of the basalts from possible effects of subsequent interaction with continental crustal rocks requires a more detailed interpretation of the combined isotope and elemental data.

Inferences from the geochemistry of the two lava groups

The Sm/Nd data of the lavas are presented in Fig. 6. Mainly Vandfaldsdalen lavas have been analysed, but all analysed lavas constitute a trend from near-MORB values towards more enriched compositions with a minimum model age, T_{Nd}, of ca. 700 Ma. However, this trend is neither an isochron nor a two-component mixing line, but

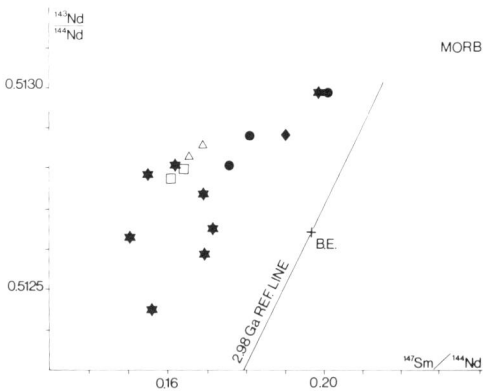

FIG. 6. Sm/Nd isochron plot. Symbols as in Fig. 3.

represents a continuous spectrum of values from $^{147}Sm/^{144}Nd = ca.$ 0.15–0.16 and $^{143}Nd/^{144}Nd = ca.$ 0.5124 towards MORB. Such a trend could result from either variable contributions from different mantle sources or a change in extent of crustal contamination of the magmas during ascent, both involving old reservoirs. The Zr/Nb ratio has proved useful for the discrimination between types of oceanic basalts (e.g. le Roex *et al.* 1983).

In a plot of Zr/Nb vs $^{143}Nd/^{144}Nd$ (Fig. 7a) group 1 defines a trend of relatively high Zr/Nb, clearly separate from the group 2 lavas, thus indicating that the high $^{143}Nd/^{144}Nd$ end-member can be resolved into two and also that some samples with more radiogenic Nd than BE belong to the group 2 lavas. The Plateau Basalts of the entire Blosseville Kyst have Zr/Nb ratios which show continuous variation between the ca. 15 and ca. 10 similar to the depleted end-members of the group 1 lavas and group 2 lavas respectively. Thus, in the Lower Lavas two relatively enriched and two relatively depleted end-members can be distinguished. Compositions intermediate between the two depleted end-members are present in the plateau basalts.

In a plot of Zr/Nb vs Rb/Nb (see Fig. 7b) both groups trend towards the Iceland field at low Rb/Nb ratios, testifying to the relative depletion in the most incompatible elements in the lavas that erupted at relatively late stages of the volcanism. All but one member of group 2 have low Rb/Nb. Except for this, it can be inferred from the consistency in the trend of the data that, for this parameter, a specific end-member composition also exists. The trends are in contrast to the erratic variation in Rb/Nb and Zr/Nb ratios expected from contamination of the lavas by a variety of different crustal rocks or components of crustal rocks. Similar arguments are valid for

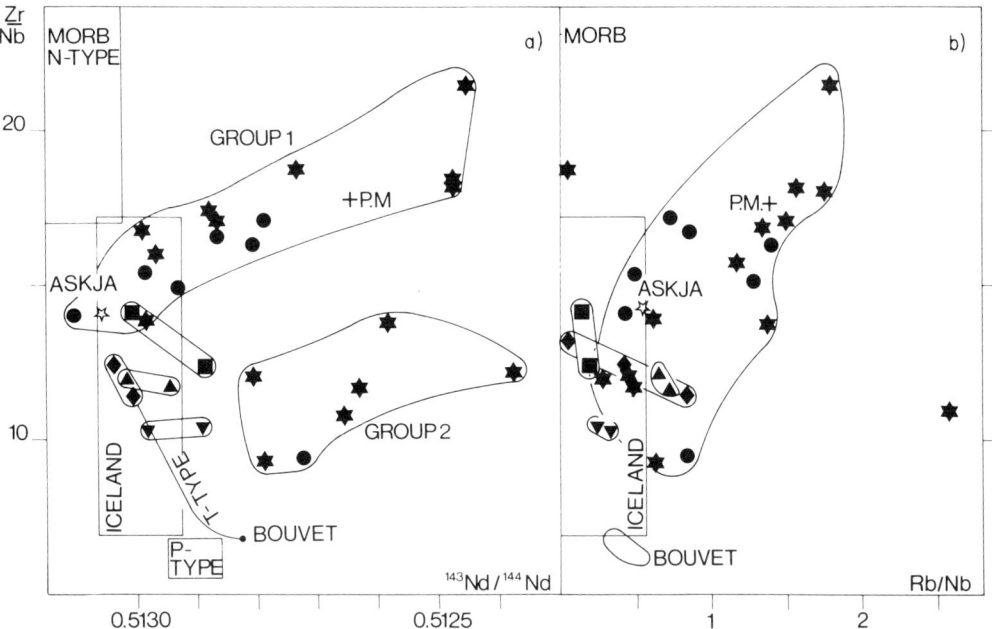

FIG. 7. (a) Zr/Nb vs ^{143}Nd/^{144}Nd; and (b) Zr/Nb vs Rb/Nb. Symbols as in Fig. 3, except for star: Askja lava from Iceland (Wood *et al.* 1979; Condomines *et al.* 1983), and plus: P.M. values from Wood *et al.* (1979) and BE ^{143}Nd/^{144}Nd from Hawkesworth & van Calsteren (1984). The fields shown are N-type, T-type and P-type MORB and Bouvet Island (le Roex *et al.* 1983). See text for discussion.

the trends of the La/Nb ratio vs ^{143}Nd/^{144}Nd (Fig. 8a) and Zr/Nb (Fig. 8b) respectively. In the latter plot it is possible, like in Fig. 7a, to discriminate between four end-members, while in the former plot only one trend is defined.

The overall incompatible element abundance patterns for the two groups contrast both in shape and enrichment level (Fig. 9). The group 1 lavas have less inclined patterns from Y to Nb and less enrichment than group 2 lavas. For example, Nb = ca. 25–30 × P.M. (primordial mantle) in group 1 compared to Nb = ca. 40–50 in group 2. The differences in enrichment cannot be explained only by fractionation effects. The Zr/Nb ratio (see Figs. 7a & 8b) is a measure of the inclination of the middle part of the mantle-normalized abundance patterns (see Fig. 9). Thus, both the level of enrichment and the inclination of the patterns make possible the distinction between the relatively less enriched group 1 and group 2 lavas which plot on the so-called mantle array.

The lavas in group 2 are mainly from the picrite–ankaramite series of Brooks & Nielsen (1982a, b) and Gill *et al.* (1988), but a few other lavas are also included. Further, picrites (see Table 1) are also placed in group 1. Thus, the grouping suggested here differs in some respects from that proposed by Brooks & Nielsen and Gill *et al.* If the combined elemental and isotopic systematics are considered, the late Plateau Basalts partially overlap the geochemical characteristics of the Lower Lavas. The members of the Igtertiva Formation clearly belong to the group 2 lavas as they both have high incompatible element concentrations, comparable to the most enriched lower lavas (e.g. sample 144805; see Table 3 & Fig. 9) and low Zr/Nb ratios (see Figs. 7 & 8). This is also in accordance with their previously noted high P/Ti ratio (Larsen & Watt 1985). The lavas of the Skrænterne Formation and Wiedemann Fjord belong to group 1 (e.g. sample EG7159; see Table 3 & Figs 7, 8 & 9), while the Nansen Fjord samples have La/Nb ratios (see Fig. 8) too high to be assigned undisputedly to group 1, as they would otherwise be from their position in Fig. 7.

A crustal origin?

The pre-Tertiary crust is known from the widespread outcrops of late Archaean gneisses in the Kangerdlugssuaq area and to the S (Bridgwater *et al.* 1978). The basement to the basalts consists

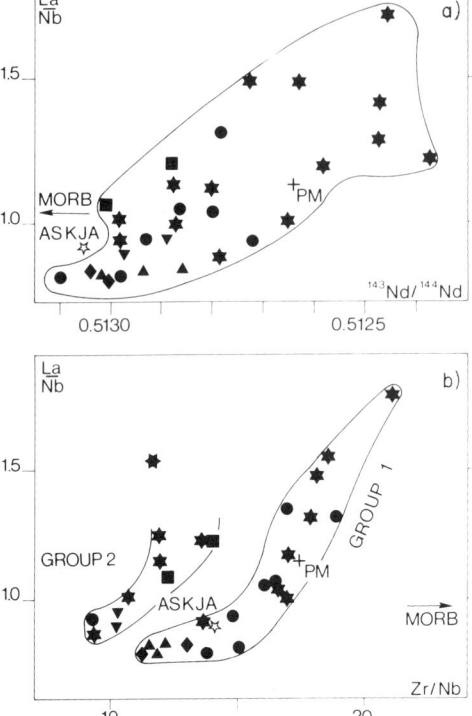

FIG. 8. (a) La/Nb vs $^{143}Nd/^{144}Nd$; and (b) La/Nb vs Zr/Nb. Symbols and fields as in Fig. 7. See text for discussion.

mainly of high amphibolite facies gneisses with scattered inclusions of granulite facies rock types (Leeman & Dasch 1976; Bird *et al.* 1986). Isotopic and elemental compositions of outcropping basement and basement xenoliths in basaltic dykes from the Kangerdlugssuaq area have been analysed (Holm, unpublished data; Hamilton 1970; Leeman & Dasch 1976). Owing to the great age of the crustal rocks the analysed crustal rocks display a very large range of isotopic values, but they do not overlap with the range of the basaltic rocks. Sr and Nd isotopic ranges are $^{87}Sr/^{86}Sr = 0.725–0.931$ and $^{143}Nd/^{144}Nd = 0.5104–0.5111$ (Holm, unpublished data). Weighted means for the lead isotopic compositions are given below.

Generally, trends to high Rb/Nb ratios in basic volcanics can be generated by contamination with crustal rocks in amphibolite facies, which tend to be enriched in Rb and depleted in Nb (e.g. Taylor & McLennan 1981; Thompson *et al.* 1983, 1984), but not by contamination with granulites which have lower Rb/Nb ratios (Weaver & Tarney 1980; Thompson *et al.* 1983).

This would be the case whether whole-rock assimilation, partial melts or a diffusion controlled component was the contaminant. Analysed Kangerdlugssuaq basement rocks have a mean of Rb/Nb ratio of 39. If the enriched endmembers were crustally derived a granulite could explain the Sr and Nd isotopic variation of group 1 lavas and an amphibolite that of group 2. However, contrary to this expectation the group 2 lavas have lower Rb/Nb ratios than group 1 lavas, except for one sample with very high Rb/Nb compared to all other samples (see Fig. 7b). Thus, the combined implication is not of crustal contamination.

As the latest metamorphic crustal pre-Tertiary event was late Archaean, the low Rb/Nb could not have been generated in the crust relatively late to allow a long decay time for Rb. Substantial contamination of tholeiitic basalts generates negative anomalies in the normalized element patterns at Nb, P and Ti (Thompson *et al.* 1983). Such P and Ti anomalies are not seen (see Fig. 9). Thus, the relative behaviour of Nb is probably not an anomaly, but is probably related to a decoupled variation in the most incompatible elements (K, Ba & Rb). Relatively low Rb/Nb

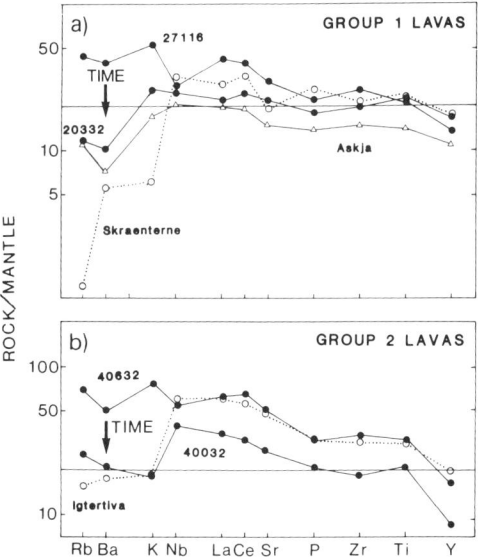

FIG. 9. Time-dependent variation of the mantle normalized elemental abundances of EGT lavas representing the range of variation within a) group 1 lavas; and b) group 2 lavas. Also shown are Plateau Basalts from the Skrænterne and Igtertiva formations and Askja (Iceland). Normalization values are from Thompson *et al.* (1983). See text for discussion.

ratios in the group 2 lavas is a consequence of high Nb and not of low Rb, while high La/Nb in some lavas is caused by the lack of a Nb peak in typical OIB patterns. The relative smoothness of the patterns strongly indicates the insignificance of a felsic crustal component. Finally, there is no clear correlation between degree of fractionation and isotopic composition of the lavas, which might otherwise indicate AFC processes (De Paolo 1981). It thus seems impossible to invoke likely crustal sources that would have had major influences on the displayed trends of trace elements and the isotopes.

The indications from the Pb isotopes have not been used to arrive at the conclusions above. In some instances it has been shown that when basaltic rocks with low Pb concentrations penetrate continental crust, the magmatic Pb isotopes may be totally exchanged with crustal Pb, without the isotope systems of Sr (or Nd) being affected (e.g. Kalsbeek & Taylor 1985). In the case of the EGT lavas there is a good correlation between Sr and Pb isotopes (see Fig. 5) as well as between Pb and Nd isotope ratios. There is thus no indication of a decoupling of Pb isotopes. Moreover, relatively good correlation between Rb and Pb abundances in the lavas seems to make redundant the need to invoke a separate process to explain the Pb geochemistry. Although Pb generally seems to be a more sensitive indicator of crustal contamination than Sr and Nd, examples of uncontaminated basic magmas erupted through old continental crust have been well documented in studies including Pb isotopes (e.g. Kalsbeek & Taylor 1986).

Unradiogenic Pb in the Skye lavas of the British Tertiary Volcanic Province (BTVP) has been ascribed to the effects of assimilation by the lavas of components of predominantly granulite facies rocks (Dickin 1981). Comparison of the EGT and the Skye lavas and their basement rocks reveal differences, although the general trend in the $^{206}Pb/^{204}Pb$ vs $^{207}Pb/^{204}Pb$ diagram (see Fig. 4) is not dissimilar. The EGT lavas have lower Pb concentrations (range < 1–6 ppm, mean 2 ppm) than the Skye lavas (range 1–18 ppm, mean 4.6 ppm) and there is no correlation of $^{206}Pb/^{204}Pb$ and 1/Pb in EGT lavas, as seen in the low-Fe intermediates of Skye. Further, the average of 15 basement samples from the Kangerdlugssuaq area has much more radiogenic Pb [$^{206}Pb/^{204}Pb = 19.43$ (range 14–30)], and the samples have a higher average Pb abundance [19 ppm (range 7–43 ppm)] than the average Lewisian basement ($^{206}Pb/^{204}Pb = 13.95$ & 14.91, Pb concentrations 5 & 12 ppm in amphibolites and granulites respectively). Thus, bulk assimilation of crustal rocks should have a more pronounced effect in the Pb isotopes of the EGT lavas than in the Skye lavas.

Dickin (1981) used these parameters to argue that Pb in the Skye lavas could be modelled by the mixing of mantle-derived Pb with MORB-type Pb isotopic compositions and crustally-derived Pb, which was selectively extracted from the crust in granulite facies by the breakdown of hydrous minerals. The presently available database of crustal lead isotopes from the Kangerdlugssuaq area does not suggest a similar history of contamination of the EGT lavas, although the low number of analyses may be biased towards amphibolite facies rocks. However, the trend of the lavas roughly along the basement isochron with a μ_1 slightly above 8 is very different from the low μ_1 values expected from the Archaean U/Pb depleted granulites. The μ_1 of the lavas is similar to usual mantle values. Moreover, evidence already presented from the trace elements suggests that if the data should be explained by crustal contamination, amphibolites are more likely than granulites. There are thus several indications that measured basement Pb isotopes represent the potential crustal contaminant, but that the Pb isotopic composition of the lavas does not reflect a crustal source.

Two of the assumptions made by Dickin (1981) were that: (1) Pb less radiogenic than the geochron cannot be obtained from the mantle; and (2) mantle heterogeneities cannot persist for a sufficient time to be acknowledged in the isotopes. Both have been shown not to be generally correct (Dupré & Allègre 1983; Hawkesworth *et al.* 1984; Menzies *et al.* 1984; Menzies *et al.* 1987a).

In other work on the BTVP evidence of crustal contamination has been indicated from trace-element patterns and from the relationship between $^{143}Nd/^{144}Nd$ ratios and 1/Nd, Sm/Nd and Fe/Mg ratios (Thompson *et al.* 1982; Thirwall & Jones 1983). Such trends are not seen in the EGT lavas. In the Deccan Traps pronounced crustal contamination was accompanied by the development of several trace-element anomalies (Cox & Hawkesworth 1985). Similar evidence is not found in the EGT lavas, neither is there any clear correlation between SiO_2 content and isotopic composition. Indications from the major elements that several of the magmas had their final evolution before eruption at near atmospheric pressure seem not to be related to the isotopic compositions. Thus, although many of the magmas probably resided for some time in the crust, the overall geochemical impact of this was relatively small. The major contribution to the isotopic and trace-element compositions of the lavas was from the mantle.

The lithospheric component

Enriched domains of the subcontinental mantle have recently been widely acknowledged as lamproitic and other ultrapotassic magmas and a large variety of mantle xenoliths show evidence of old enrichment processes (e.g. Menzies 1983; Hawkesworth *et al.* 1983; Vollmer 1984; Fraser *et al.* 1985; Nelson *et al.* 1986). Ancient enrichment in the lithospheric mantle has been reported from Loch Roag, Isle of Lewis, in mantle xenoliths in a dyke tapping the mantle beneath the Lewisian continental crust (Menzies *et al.* 1987). The basement of NW Scotland is nearly identical in age to the E Greenland basement and in early Tertiary times was located adjacent to it. Old enrichment in the lithospheric mantle has also been noted from W Greenland in lamproites and kimberlites in Precambrian basement (Stecher *et al.* 1987).

Although no direct evidence for the inhomogeneity of the E Greenland subcontinental mantle is at present available it seems highly probable that the trends displayed, particularly in the Lower Lavas, were caused by two distinct types of relatively enriched reservoirs and two depleted oceanic types of reservoirs. This is the conclusion of the systematics of Sr and Nd isotopes and La/Nb, Zr/Nb and Rb/Nb ratios. The end-members thus defined are listed in Table 4. Lithospheric mantle sources for continental flood basalts have been suggested also for the Snake River Plain (Menzies *et al.* 1984) and for other voluminous basalts from the S Western USA (Fitton *et al.* 1987; Ormerod *et al.* 1987).

The contrast in composition of the two enriched components suggests that the one reservoir (of group 2) was more enriched in the most incompatible elements such as Rb. General evidence from metasomatized mantle xenoliths may indicate that this mantle component may have been derived from a potassic fluid. The other enriched component (in group 1) would only need to be slightly enriched compared to BE and thereupon to have sufficient time to decay. If the broad trend among the Pb isotopes in Fig. 4 has an age significance related to events in the mantle, this would have taken place at around 3 Ga. Such a decay period would enable a source with only slightly fractionated Rb/Sr and Nd/Sm relative to BE to evolve to the observed Sr and Nd isotopic values. A two stage model for lead would indicate evolution prior to ca. 3 Ga in an ordinary mantle reservoir with $\mu_1 =$ ca. 8.

The difference in isotopic composition of the clinopyroxene from sample 20351 and that of the whole rock may originate from the mixing of two magmas both from group 2. The clinopyroxene is a chromian endiopside with low Al and probably crystallized at high crustal levels.

There are some unusual requirements for the lithospheric mantle sources in this model. It is very uncommon to get mantle-derived liquids with such unradiogenic Pb. However, unradiogenic Pb from the mantle has been reported from the Smoky Butte lamproite (Fraser *et al.* 1986). How the time-integrated low U/Pb and Th/Pb could be achieved (at the start of stage 2) is unclear, as, unfortunately, U-Th-Pb geochemistry of the mantle is poorly understood (e.g. Sun 1980). Lead isotopic compositions less radiogenic than the geochron have, however, been reported from the mid-Indian Ridge (Dupré & Allègre 1983). They suggested that such compositions represent mantle residuals after extraction of crust, with high U/Pb and Th/Pb plotting to the right of the geochron. The E Greenland subcontinental lithosphere could have preserved more strongly depleted U + Th/Pb because it was stabilized before the ca. 2 Ga event which has caused the apparent Pb-Pb isochron and relatively radiogenic compositions of oceanic Pb.

The lack of relation between Rb/Sr ratios and Sr isotopic composition in most of the group 2 lavas compared to the consistency in the isotopic trends indicates that Rb and Sr were decoupled at a relatively late stage in the mantle source. Rb/Sr ratios are mostly higher than required by the Sr isotopes. Similarly, the Sm/Nd ratios are lower than required by the Nd isotopes.

Group 2 lavas dominate the lower Vandfaldsdalen Formation lavas and do not occur high in the Miki Formation, while group 1 lavas were erupted in the entire region throughout the lava

TABLE 4. *Isotopic compositions and some key trace-element ratios of the end-members of the two groups of lavas*

End member	$^{143}Nd/^{144}Nd$	$^{87}Sr/^{86}Sr$	$^{206}Pb/^{204}Pb$	Zr/Nb	Rb/Nb	La/Nb
Group 1A	0.51245	0.7054	15.5	21	1.8	1.8
Group 1B	0.51295	0.7030	17.5	12	0.5	0.8
Group 2A	0.51258	0.7078	15.0	13	2.5	1.5
Group 2B	0.51278	0.7043	17.5	9	0.5	0.9

sequence. Later, in the Igtertiva Formation, the group 2 lava characteristics reoccur. The time progression already noted within each lava group is thus also evident by the exclusively early occurrence of the group 2 lavas, and may be related to the early mobilization and exhaustion of this component.

A tentative scenario for the Tertiary source evolution of the Lower Lavas is:

(1) In early Tertiary times a mantle jet entered the lithosphere of E Greenland. At the time active rifting was not taking place in the Kangerdlugssuaq area. Dominantly jet-derived magmas interacted with the lithospheric mantle to generate group 2 magmas, which were only erupted at moderate volume over the focus of the jet. Subsequently, group 1 lavas were erupted with a signature of a different and less enriched lithospheric mantle source. Possibly, the enriched end-member of group 2 was more readily incorporated into melts than the group 1 enriched end-member.

(2) As rifting became active, depleted asthenospheric mantle entered the lithosphere to a greater extent and mixed with the jet causing the magmas to have more depleted compositions. At this time the erupted magmas were more voluminous, the lithosphere was hotter, and the enriched component of the group 2 lava was exhausted, while the enriched end-member of group 1 did not become extinct until later.

(3) The mixing of asthenospheric MORB-type magmas and jet-derived OIB-type magmas took place throughout the period of lava extrusion. The Plateau Basalts and depleted end-members of groups 1 and 2 represent mixtures of these components. The group 1 end-member was dominant when active rifting and plate separation took place. In this model, the Igtertiva Formation lavas with group 2 signature would represent initiation of rifting at Kap Dalton.

It seems that one component of the group 2 lavas was derived from the early jet with a relatively small influence of depleted asthenospheric material and, further, that the Icelandic compositions have evolved by a mixing of jet and depleted asthenosphere. The jet composition in

this model has $Zr/Nb \leq 10$ and $^{143}Nd/^{144}Nd \leq 0.5128$ and thus could be like S Atlantic P-type MORB or Bouvet Island lavas (e.g. see Fig. 7).

Summary of the main conclusions

(1) The EGT lavas show a time progression in isotopic and elemental geochemistry from a very wide compositional range in the Vandfaldsdalen Formation lavas towards compositions near to present-day Icelandic which dominate the upper part of the Lower Lavas and the flood basalts along the entire Blosseville Kyst. This is thought to be the result of the initial activity and evolution of a mantle jet.

(2) The lavas can be subdivided into two groups. The group 2 lavas only occur within the early part of the Lower Lavas. End-members of both groups with high La/Nb ratios and time-integrated high Rb/Sr, Nd/Sm, Pb/U and Pb/Th ratios, were erupted early in the volcanic evolution. These end-members were derived from two distinct sources, probably in the lithospheric mantle. Crustal components had less significance.

(3) Of the depleted end-members of the two trends one is near, but not identical, to Icelandic lavas in composition and is characterized by $Zr/Nb =$ ca. 13. The other is thought to be less influenced by an asthenospheric depleted component and to closer approximate to the composition of the presumed mantle jet ($Zr/Nb < 10$).

(4) The flood basalts of the Igtertiva Formation at Kap Dalton belong to the group 2 lavas, indicating a lithospheric mantle component. At the time of eruption of these last plateau basalts, volcanic activity probably had not been centred in this area for long.

ACKNOWLEDGEMENTS: I am grateful to Robin Gill, Kent Brooks and Troels Nielsen for their contribution in discussion to the improvement of the paper. I also thank very much the members of the department in Leeds, UK, who so kindly helped me in their isotope laboratory—a stay which was made possible by the Danish Nature Science Research Council, who also provided the analytical facilities in Copenhagen at the XRF laboratory (J. Bailey) and the isotope laboratory (O. Larsen). G. Lis and J. Petersen made the drawings.

References

BIRD, D. K., ROSING, M. T., MANNING, C. E., & ROSE, N. M., 1986. Geological field studies of the Miki Fjord area, East Greenland. *Bulletin of the Geological Society of Denmark*, **34**, 219–236.

BRIDGWATER, D., DAVIES, F. B., GILL, R. C. O., GORMAN, B. E., MYERS, J. S., PEDERSEN, S., & TAYLOR, P. N. 1978. Precambrian and Tertiary geology between Kangerdlugssuaq and Angmagssalik, East Greenland. *Geological Survey of Greenland Report*, **83**, 17 pp.

BROOKS, C. K. 1973. Rifting and doming in southern East Greenland. *Nature Physical Sciences*, **244**, 23–25.

—— & NIELSEN, T. F. D. 1982a. The Phanerozoic

development of the Kangerdlugssuaq area, East Greenland. *Meddelelser øm Grønland, Geoscience,* **9**, 31 pp.

—— & NIELSEN, T. F. D. 1982b. The E Greenland continental margin: a transition between oceanic and continental magmatism. *Journal of the Geological Society of London,* **139**, 265–275.

——, NIELSEN, T. F. D. & PETERSEN, T. S. 1976. The Blosseville Coast basalts of East Greenland: composition and temporal variation. *Contributions to Mineralogy and Petrology,* **5**, 279–292.

——, FAWCETT, J. J., GITTINS, J. & RUCKLIDGE, J. C. 1981. The Batbjerg complex, East Greenland: a unique ultrapotassic Caledonian intrusion. *Canadian Journal of Earth Sciences,* **18**, 274–285.

CARTER, S. R., EVENSEN, N. M., HAMILTON, P. J. & O'NIONS, R. K. 1979. Basalt magma sources during the opening of the North Atlantic. *Nature,* **281**, 28–30.

COHEN, R. S. & O'NIONS, R. K. 1982. The lead, neodymium and strontium isotopic structure of ocean ridge basalts. *Journal of Petrology,* **23**, 299–324.

CONDOMINES, M., GRÖNVOLD, K., HOOKER, P. J., MUEHLENBACHS, K., O'NIONS, R. K., OSKARSSON, N. & OXBURGH, E. R. 1983. Helium, oxygen, strontium and neodymium isotopic relationships in Icelandic volcanics. *Earth and Planetary Science Letters,* **66**, 125–136.

COX, K. G. & HAWKESWORTH, C. J. 1985. Geochemical stratigraphy of the Deccan Traps at Mahabaleshwar, Western Ghats, India, with implications for open system magmatic processes. *Journal of Petrology,* **26**, 344–377.

DEER, W. A. 1976. Tertiary igneous rocks between Scoresby Sund and Kap Gustav Holm, East Greenland. *In:* ESCHER, A. & WATT, W. S. (eds) *Geology of Greenland.* Geological Survey of Greenland, pp. 405–429.

DE PAOLO, D. J. 1981. Trace element and isotopic effects of combined wall rock assimilation and fractional crystallization. *Earth and Planetary Science Letters,* **53**, 189–202.

DICKIN, A. P. 1981. Isotope geochemistry of Tertiary igneous rocks from the Isle of Skye, N.W. Scotland. *Journal of Petrology,* **22**, 155–190.

DUPRÉ, B. & ALLÈGRE, C. J. 1983. Pb-Sr isotope variation in Indian Ocean basalts and mixing phenomena. *Nature,* **303**, 142–146.

FAWCETT, J. J., BROOKS, C. K. & RUCKLIDGE, J. C. 1973. Chemical petrology of Tertiary flood basalts from the Scoresby Sund area. *Meddelelser om Grønland,* **195/6**, 56 p.

FITTON, J. G., JAMES, D., KEMPTON, P. D. & ORMEROD, D. S. 1987. The role of lithospheric mantle in the generation of Late Cenozoic basic magmas in the Southwestern USA. *Terra Cognita,* **7**, 613.

FRASER, K. J., HAWKESWORTH, C. J., ERLANK, A. J., MITCHELL, R. H. & SCOTT-SMITH, B. H. 1985. Sr, Nd and Pb isotope and minor element geochemistry of lamproites and kimberlites. *Earth and Planetary Science Letters,* **76**, 57–70.

GILL, R. C. O., NIELSEN, T. F. D., BROOKS, C. K. & INGRAM, G. A., 1988. Tertiary volcanism in the

Kangerdlugssuaq region, E Greenland: trace-element geochemistry of the Lower Basalts and tholeiitic dyke swarms. *In:* MORTON, A. C. & PARSON, L. M. (eds) *Early Tertiary Volcanism and the Opening of the NE Atlantic.* Geological Society of London, Special Publication, **39**, 161–179.

HART, S. R. 1984. A large-scale isotope anomaly in the Southern Hemisphere mantle. *Nature,* **309**, 753–757.

HAWKESWORTH, C. J. & MORRISON, M. A. 1978. A reduction in $^{87}Sr/^{86}Sr$ during basalt alteration. *Nature,* **276**, 381–382.

—— & VAN CALSTEREN, P. W. C. 1983. Radiogenic isotopes—some geological applications. *In:* HENDERSON, P. (ed) *Rare Earth Element Geochemistry.* Developments in Geochemistry, **2**, pp. 375–421.

——, ERLANK, A. J., MARSH, J. S., MENZIES, M. A. & VAN CALSTEREN, P. 1983. Evolution of the continental lithosphere: evidence from volcanics and xenoliths in southern Africa. *In:* HAWKESWORTH, C. J. & NORRY, M. J. (eds) *Continental Basalts and Mantle Xenoliths.* Shiva Publishing Company, pp. 111–138.

KALSBEEK, F. & TAYLOR, P. N. 1985. Age and origin of early Proterozoic dolerite dykes in South-West Greenland. *Contributions to Mineralogy and Petrology,* **89**, 307–316.

—— & —— 1986. Chemical and isotopic homogeneity of a 400 km long basic dyke in central West Greenland. *Contributions to Mineralogy and Petrology,* **93**, 439–448.

LARSEN, H. C. 1984. Geology of the East Greenland shelf. *In:* A. M. SPENCER *et al.* (eds) *Petroleum Geology of the North European Margin.* Graham & Trotman, London, pp. 329–339.

LARSEN, L. M. & WATT, W. S. 1985. Episodic volcanism during the break-up of the North Atlantic: evidence from the East Greenland plateau basalts. *Earth and Planetary Science Letters,* **73**, 105–116.

LE ROEX, A. P., DICK, H. J. B., ERLANK, A. J., REID, A. M., FREY, F. A. & HART, S. R. 1983. Geochemistry, mineralogy and petrogenesis of lavas erupted along the Southwest Indian Ridge between the Bouvet triple junction and 11 degrees east. *Journal of Petrology,* **24**, 267–318.

LEEMAN, W. P. & DASCH, E. J. 1976. $^{207}Pb/^{207}Pb$ whole-rock age of gneisses from the Kangerdlugssuaq area, eastern Greenland. *Nature,* **263**, 469–471.

MENZIES, M. A. 1983. Mantle ultramafic xenoliths in alkaline magmas: evidence for mantle heterogeneity modified by magmatic activity. *In:* HAWKESWORTH, C. J. & NORRY, M. J. (eds) *Continental Basalts and Mantle Xenoliths.* Shiva Publishing Company, pp. 92–100.

——, LEEMAN, W. P. & HAWKESWORTH, C. J. 1984. Geochemical and isotopic evidence for the origin of continental flood basalts with particular reference to the Snake River Plain, Idaho, U.S.A. *Philosophical Transactions of the Royal Society of London,* **A310**, 643–660.

——, HALLIDAY, A. N., PALACZ, Z., HUNTER, R. H., UPTON, B. G. J., ASPEN, P. & HAWKESWORTH, C. J. 1987a. Evidence from mantle xenoliths for an enriched lithospheric keel under the Outer Hebrides. *Nature*, **325**, 44–47.

——, ROGERS, N., TINDLE, A. & HAWKESWORTH, C. J. 1987b. Metasomatic and enrichment processes in lithospheric peridotites, an effect of asthenosphere–lithosphere interaction. *In:* MENZIES, M. A. & HAWKESWORTH, C. J. (eds) *Mantle Metasomatism.* Academic Press, London, pp. 313–361.

NELSON, D. R., McCULLOCH, M. T. & SUN, S.-S. 1986. The origins of ultrapotassic rocks as inferred from Sr, Nd and Pb isotopes. *Geochimica et Cosmochimica Acta*, **50**, 231–245.

NIELSEN, T. F. D., SOPER, N. J., BROOKS, C. K., FALLER, A. M., HIGGINS, A. C. & MATTHEWS, D. W. 1981. The prebasaltic sediments and the Lower Basalts at Kangerdlugssuaq, East Greenland: their stratigraphy, lithology, palaeomagnetism and petrology. *Meddelelser øm Grønland, Geoscience*, **6**, 25 pp.

O'NIONS, R. K. 1983. Isotopic abundances relevant to the identification of magma sources. *Philosophical Transactions of the Royal Society of London*, **A310**, 591–603.

——, HAMILTON, P. J. & EVENSEN, N. M. 1977. Variations in ^{143}Nd/^{144}Nd and ^{87}Sr/^{86}Sr in oceanic basalts. *Earth and Planetary Science Letters*, **34**, 13–22.

——, CARTER, S. R., COHEN, R. S., EVENSEN, N. M. & HAMILTON, P. J. 1978. Pb, Nd and Sr isotopes in oceanic ferromanganese deposits and ocean floor basalts. *Nature*, **273**, 435–438.

ORMEROD, D. S., HAWKESWORTH, C. J., LEEMAN, W. P. & MENZIES, M. A. 1987. The identification of subduction-related and within-plate components in basalts from the Western USA. *Terra Cognita*, **7**, 621.

SOPER, N. J., HIGGINS, A. C., DOWNIE, C., MATTHEWS, D. W. & BROWN, P. E. 1976. Late Cretaceous-early Tertiary stratigraphy of the Kangerdlugssuaq area, east Greenland, and the age of opening of the north-east Atlantic. *Journal of the Geological Society of London*, **132**, 85–104.

STACEY, J. S. & KRAMERS, J. D. 1975. Approximation of terrestrial lead isotope evolution by a two-stage model. *Earth and Planetary Science Letters*, **26**, 207–221.

STECHER, O., THY, P. & CARLSON, R. W. 1987. Subcrustal metasomatism below East Greenland: isotopic and geochemical evidence from lamproite and kimberlite dykes. *Terra Cognita*, **7**, 625.

SUN, S.-S. 1980. Lead isotope study of young volcanic rocks from mid-ocean ridges, ocean islands and island arcs. *Philosophical Transactions of the Royal Society of London*, **A297**, 409–445.

—— & JAHN, B. 1975. Lead and strontium isotopes in post-glacial basalts from Iceland. *Nature*, **255**, 527–530.

——, TATSUMOTO, M. & SCHILLING, J.-G. 1975. Mantle plume mixing along the Reykjanes Ridge axis: lead isotopic evidence. *Science*, **190**, 143–147.

TAYLOR (Jr.), H. P. & FORESTER, R. W. 1979. An oxygen and hydrogen isotope study of the Skaergaard intrusion and its country rocks: a description of a 55 m.y. old fossil hydrothermal system. *Journal of Petrology*, **20**, 355–419.

TAYLOR, S. R. & GORTON, M. P. 1977. Geochemical application of spark source mass spectrography—III. Element sensitivity, precision and accuracy. *Geochimica et Cosmochimica Acta*, **41**, 1375–1380.

—— & McLENNAN, S. M. 1981. The composition and evolution of the continental crust: rare earth element evidence from sedimentary rocks. *Philosophical Transactions of the Royal Society of London*, **A301**, 381–399.

THOMPSON, R. N., DICKIN, A. P., GIBSON, I. L. & MORRISON, M. A. 1982. Elemental fingerprints of isotopic contamination of Hebridian Paleocene mantle-derived magmas by Archaean sial. *Contributions to Mineralogy and Petrology*, **79**, 159–168.

——, MORRISON, M. A., DICKIN, A. P. & HENDRY, G. L. 1983. Continental flood basalts... Arachnids rule OK? *In:* HAWKESWORTH, C. J. & NORRY, M. J. (eds) *Continental Basalts and Mantle Xenoliths.* Shiva Publishing Company, 158–185.

——, ——, HENDRY, G. L. & PARRY, S. J. 1984. An assessment of the roles of crust and mantle in magma genesis: an elemental approach. *Philosophical Transactions of the Royal Society*, **A310**, 549–590.

UPTON, B. G. J., EMELEUS, C. H. & HALD, N. 1980. Tertiary volcanism in northern East Greenland: Gauss Halvo and Hold with Hope. *Journal of the Geological Society of London*, **137**, 491–508.

VINK, G. E. 1984. A hotspot model for Iceland and the Vøring Plateau. *Journal of Geophysical Research*, **89**, 9949–9959.

VOLLMER, R., OGDEN, P., SCHILLING, J.-G., KINGSLEY, R. H. & WAGGONER, D. G. 1984. Nd and Sr isotopes in ultrapotassic volcanic rocks from the Leucite Hills, Wyoming. *Contributions to Mineralogy and Petrology*, **87**, 359–368.

WAGER, L. R. 1947. Geological investigations in East Greenland. Part IV. The stratigraphy and tectonics of Knud Rasmussens Land. *Meddeleleser øm Grønland*, **134(5)**, 64 pp.

—— & HAMILTON, E. I. 1964. Some radiometric rock ages and the problem of the southward continuation of the East Greenland Caledonian orogeny. *Nature*, **4963**, 1079–1080.

WEAVER, B. L. & TARNEY, J. 1980. Rare-earth element geochemistry of Lewisian granulite-facies, northwest Scotland: implications for the petrogenesis of the Archaean lower continental crust. *Earth and Planetary Science Letters*, **51**, 279–296.

WOOD, D. A., JORON, J.-L., TREUILL, M., NORRY, M. & TARNEY, J 1979. Elemental and Sr isotope variations in basic lavas from Iceland and the surrounding ocean floor. *Contributions to Mineralogy and Petrology*, **70**, 319–339.

ZINDLER, A. & HART, S. R. 1986. Chemical geodynamics. *Annual Reviews in Earth and Planetary Sciences*, **14**, 493–571.

P. M. HOLM, Institute of Petrology, University of Copenhagen, Øster Voldgade 10, DK-1350 Copenhagen, Denmark.

Cyclical tholeiitic volcanism and associated magma chambers: eruptive mechanisms in E Greenland

A. J. C. Hogg, J. J. Fawcett, J. Gittins & M. P. Gorton

The basaltic Prinsen af Wales Bjerge (PWB) are part of the Tertiary Volcanic Province of E Greenland and lie W of the Blosseville Kyst that extends from Kangerdlugssuaq Fjord to Scoresby Sund. They consist of tholeiitic basalts overlain by alkalic basalts that were erupted 100–150 km W of the original axis of continental rifting and active ocean-floor development during the creation of the N Atlantic ocean. The tholeiites have many features of continental flood basalts but are somewhat enriched in Fe, and in Ti relative to Fe, and have slightly lower Al_2O_3. They have slight enrichment in light rare earth elements (LREEs) (La/Yb = 3–4).

Tholeiitic lavas are aphyric to modestly porphyritic (phenocrysts up to 15%) with plagioclase the most common phenocryst followed by pyroxene and olivine. Phenocryst plagioclase is An_{50}–An_{60} except in strongly zoned crystals where it extends from An_{75}–An_{35}. Groundmass feldspar is An_{35}–An_{45}. Clinopyroxene is sub-calcic augite to occasional pigeonite and shows modest zoning from $Ca_{42}Mg_{43}Fe_{15}$–$Ca_{22}Mg_{50}Fe_{28}$. Olivine is close to Fo_{75}. Groundmass varies from fine-grained and cryptocrystalline to glassy with equant grains of magnetite and laths of ilmenite interstitial to plagioclase and pyroxene; patches of red-brown glass frequently enclose very fine-grained opaque crystals and acicular apatite.

A nunatak within the PWB, visited in 1982, displays four cycles of tholeiitic basalt, each about 50 m thick, which are defined by systematic variations in chemical composition (Sr, Zr, Cr, Ni, TiO_2 & Mg), (Figs 1, 2 & 3). In three of the four cycles the lowermost flows are the most highly differentiated, and successive flows are increasingly primitive.

A plot of Ni vs Zr (Fig. 4) reveals some of the steps in the processes by which three of the cycles formed. The distribution is a distinct V-shaped pattern which, although based on a small number of analyses is duplicated for Ni and Cr in each cycle. This pattern could not be produced by periodic expulsion from a simple magma chamber at successive stages in its differentiation. Such a process would produce smooth curves in which Ni would decline exponentially with advancing fractionation of the magma reservoir. Not only does Ni *not* decline regularly, but in each cycle it declines slightly near the base and then increases

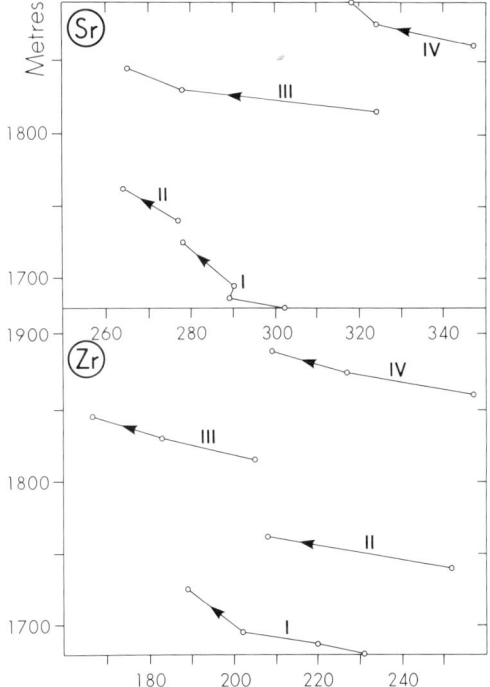

FIG. 1. Variation of Sr and Zr (in ppm) with stratigraphic height (in metres) in the tholeiitic lava sequence of 1982 nunatak.

steadily to the top of the cycle. This is the reverse of what would normally be expected.

The most likely explanation is that of fractionation of magma in discrete chambers accompanied by frequent influxes of new primitive magma and expulsion of some of the already fractionated magma as lava flows. Our view is that the newly injected magma mixes with the already differentiated magma to produce something progressively closer to the primitive parent magma with each successive injection and volcanic expulsion. The last lava is, therefore, much more primitive than the lowest in the pile (first extruded). The process is exemplified by the behaviour of Ni. Crystallization of only very small quantities of olivine will have a profound effect on the Ni content of the magma. The effect is illustrated by Fig. 4 where the dashed line shows the changing composition of the magma

From MORTON, A. C. & PARSON, L. M. (eds), 1988, *Early Tertiary Volcanism and the Opening of the NE Atlantic*, Geological Society Special Publication No. 39, pp. 197–200.

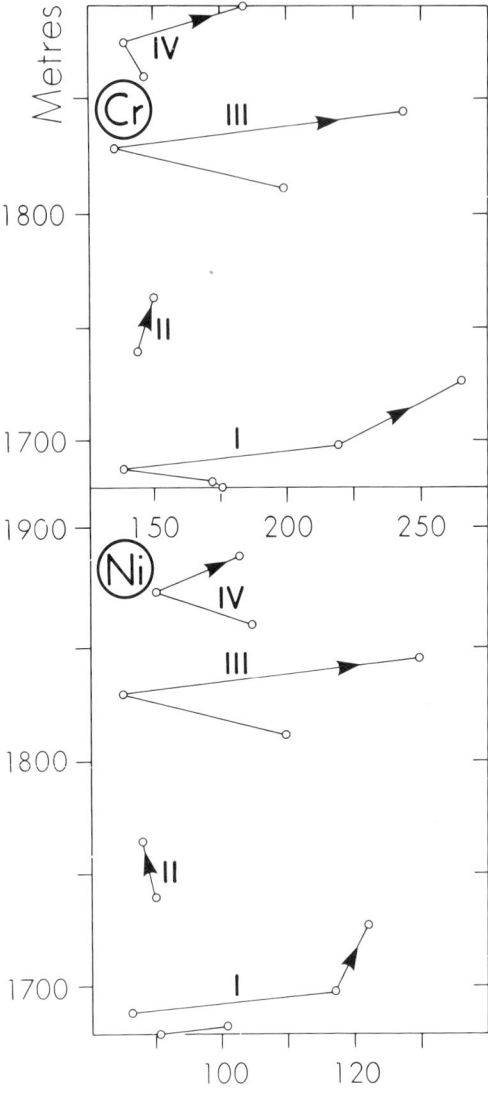

FIG. 2. Variation of Cr and of Ni ppm with stratigraphic height (in metres) in the tholeiitic lavas of 1982 nunatak. Note that arrows indicate the direction of upward progression stratigraphically.

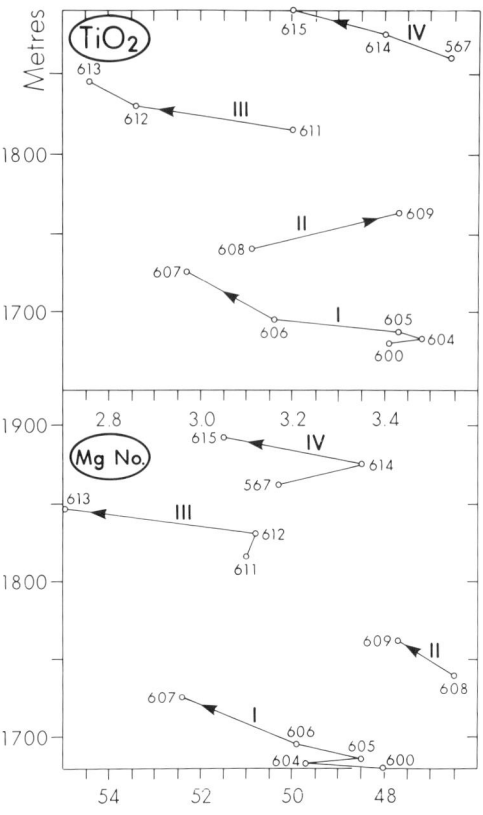

FIG. 3. Variation of TiO₂ (wt.%) and of Mg number with stratigraphic height (in metres) in the tholeiitic lava sequence of 1982 nunatak. Four cycles are labelled.

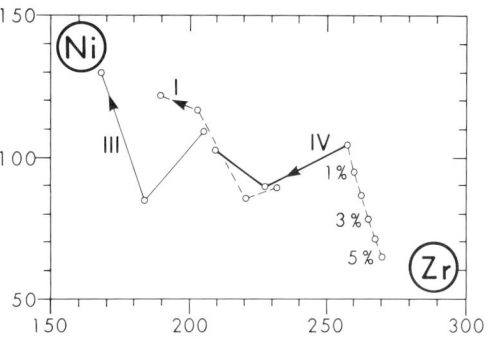

FIG. 4. Variation of Ni against Zr (ppm) in the tholeiitic lavas of 1982 nunatak. The dashed line with data points labelled 1–5% represents the calculated composition of the magma with successive fractionation of 1–5 wt.% olivine (using $K_D = 10$). Note that arrows indicate the direction of upward progression stratigraphically.

with fractionation of 1–5 wt.% olivine (using $K_D = 10$). The theory of the proposed process is illustrated in Fig. 5.

An initial magma, derived by partial melting within the mantle, is intruded into a crustal reservoir and differentiates, initially through the crystallization of olivine. After extensive fractionation has reduced the Ni content substantially and also increased the content of such incompatible elements as Zr, the upper part of the magma reservoir has, for example, changed its composi-

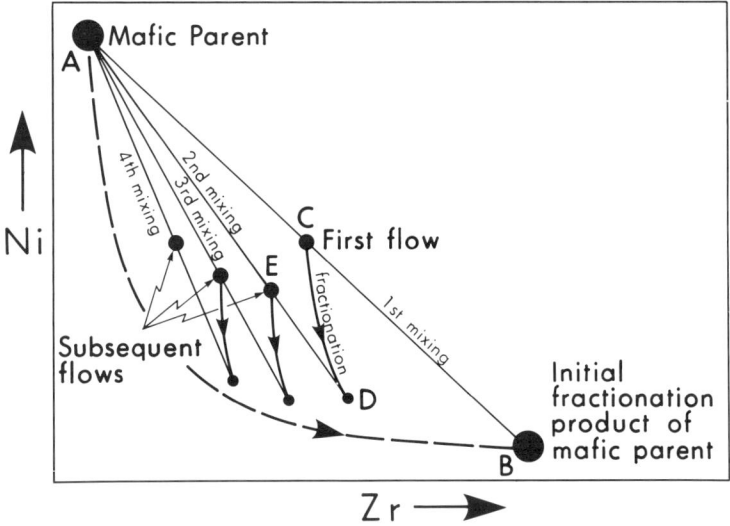

FIG. 5. Illustration of the principle of magma fractionation, injection of more primitive magma that mixes with the already partly fractionated magma, and extrusion of lava flows in a repeated sequence of events. The magma (composition A) that originally filled the chamber evolves to composition B along the hypothetical fractionation curve (dashed line). Injection of more magma of composition A produces a hybrid magma C. Subsequent fractionation causes the magma to evolve to D, whereupon a further injection of primitive magma A generates a hybrid magma of composition E, and so on. With each injection of primitive magma a small proportion of the magma in the chamber is forced out as a flow, and the sequence of flows from the base of a cycle upward changes progressively from B to more primitive compositions.

tion from A to B. After this quiescent interval a series of injections of primitive magma (composition A) commences. The composition of the first hybrid magma lies along a mixing line AB, exactly where is dependent on the relative volumes of the chamber and the injected magma. Point C is chosen by way of illustration. Renewed olivine fractionation changes the magma composition from C to D along a fractionation curve, whereupon a further injection of primitive magma causes renewed mixing, changing the composition of the reservoir to E along a second mixing line DE. Each new influx of primitive magma increases the hydraulic pressure in the reservoir and causes an expulsion of magma, most of which reaches the surface as a lava flow. The composition of the first expelled magma may, of course, be anywhere along a series of mixing lines between A and points along the curve from C to D. If the volume of injected replenishment magma is small the new composition of the reservoir will remain close to the curve CD. How far the composition will evolve from each newly mixed composition will depend on the frequency with which the reservoir is replenished and also on the degree of mixing that occurs prior to eruption. Thus, in the example where mixing changes the reservoir composition from D to E, the composition of the resulting flow may

be anywhere between D and E depending on the degree of mixing that has occurred by the time the eruption takes place. However, the similar trends within the cycles, as revealed by chemical analysis (see Figs 1–4), suggests that mixing prior to eruption is fairly thorough. That, in turn, suggests that the volume of the reservoir is relatively small, for efficient magma mixing is less likely in large reservoirs.

The lavas in the PWB extend for only a few kilometres and are generally less than 10 m thick with volumes less than 0.3 km^3. For illustration, a reservoir of 1 km radius (assuming spherical shape for simplicity) has a volume of 4.2 km^3, and so each flow would require the expulsion of only 0.25–0.7% of the total volume to produce the typical flows of the PWB. It seems likely that such small reservoirs would be widely distributed in the upper crust, especially in a region of active volcanism and crustal disruption, and if many of these are being replenished periodically, after the manner proposed here, there will be numerous sources from which reversely-zoned lava sequences can be supplied.

Such a geological setting is more likely in a region well back from the line of active rifting (100–150 km in the case of the PWB), and we think it unlikely that volcanism in these areas is controlled to any significant extent by very large magma chambers. By contrast, lava extrusion to the E along the line of active rifting was vastly more voluminous and probably did involve the very large magma chambers envisaged by Larsen & Watt (1985). We, therefore, envisage two contrasted styles of volcanism accompanying crustal rifting and the associated ocean-floor development. Volcanism well back from the line of active rifting is controlled by fractionation and repeated magma replenishment in small magma chambers and is characterized by thin flows of small volume. Volcanic cycles of as little as 50 m thickness are probably common. Volcanism along the line of active rifting is characterized by flows of huge volume (> 300 km^3) derived from extremely large magma chambers where high flow rates are common. Cyclical volcanism in this setting is probably more subtle than in the type exposed in PWB.

ACKNOWLEDGMENTS: This study was supported by grants from the Natural Sciences and Engineering Research Council of Canada, and from NATO, and by a University of Toronto Open Fellowship to AJCH. Essential assistance with field logistics came from Dr. T. F. Nielsen of Grønlands Geologiske Undersøgelse, Dr Hauge Anderson of Geodaetisk Institut and Dr C. K. Brooks of the University of Copenhagen. JG is grateful to the Warden and Fellows of Robinson College, Cambridge for a Bye Fellowship in 1983–84, and to Prof. E. R. Oxburgh for use of the electron microprobe and other research facilities in the Department of Earth Sciences of the University of Cambridge on several occasions.

References

LARSEN, L. M. & WATT, W. S. 1985. Episodic volcanism during break-up of the North Atlantic: evidence from the East Greenland plateau basalts. *Earth and Planetary Science Letters*, **73**, 105–116.

A. J. C. HOGG, Department of Geology and Mineralogy, University of Aberdeen, Aberdeen, Scotland AB9 1AS, UK.

J. J. FAWCETT, J. GITTINS & M. P. GORTON, Department of Geology, University of Toronto, Toronto, Ontario M5S 1A1, Canada.

Age constraints on Atlantic evolution: timing of magmatic activity along the E Greenland continental margin

R. H. Noble, R. M. Macintyre & P. E. Brown

SUMMARY: The results of a comprehensive K-Ar investigation of 130 petrographically selected rock and mineral specimens from the E Greenland Tertiary Igneous Province are reported. The new age data refine the chronology of this region. They suggest the following conclusions: (1) the lavas were extruded between 54 and 57 Ma; (2) the pre-flexure dyke swarm was intruded between 51 and 53 Ma; (3) the postflexure dyke swarm was intruded between 49 and 51 Ma; and (4) plutonic activity occurred at 50 Ma (Kangerdlugssuaq, Kraemers Ø, Lilloise, etc.), 42–45 Ma (Borgtinderne, Kap Deichman, etc.), 36 Ma (Kialineq), 38 and 28 Ma (Kong Oscars Fjord, Myggbukta).

The E Greenland Tertiary Igneous Province (Fig. 1), consists of ca. 54 000 km² of basalt which thins inland but attains a thickness of ca. 9000 m in the Kangerdlugssuaq area. The Lower Plateau Basalts are picritic while the Upper Basalts are tholeiitic. Dykes and sills, predominantly of dolerite, are extremely abundant along the coast, comprising 50–100% of the exposure. The sequence has been cut by a large number of plutonic centres of diverse composition—mafic/ultramafic to syenitic, nepheline syenitic and granitic. Deer (1976) listed (for the Kangerdlugssuaq region) the following abbreviated sequence of events: main plateau lavas, dolerite sills, basic intrusions (Skaergaard), coastal flexure and dyke swarm, syenites of Kap Boswell and Kap Deichmann, and the Kangerdlugssuaq alkaline intrusion. Nielsen (1978) recognized a greater degree of complexity in the dyke swarms and postulated six generations of minor intrusions. The main coast-parallel dykes (THOL-I, Nielsen 1978) were believed to have predated the coastal flexure and the Skaergaard intrusion, itself tilted by the flexure. This was questioned by Faller & Soper (1979) as palaeomagnetic results on both THOL-I dykes and their host lavas appeared to indicate that the dykes occupy their attitude of intrusion.

A knowledge of the sequence and timing of events in this important province is obviously a prerequisite to the proper understanding of the opening and evolution of the northern Atlantic. Since the classic studies of Wager & Deer, accounts of the general geology have been presented by Brooks (1973b), Deer (1976), Noe-Nygaard (1976), Soper et al. (1976a, b), Brooks & Nielsen (1982) and Larsen & Watt (1985). The biostratigraphic age of the basalts is clearly defined as late Palaeocene to early Eocene (Soper et al. 1976a, b). Determination of their age is therefore important, as it would provide a calibration point on the geological time-scale,

independent of glauconite dates and their associated uncertainties. Magnetic measurements on the basalts (Hailwood et al. 1973; Faller 1975) are consistent with their eruption in the long reversed-polarity interval immediately preceding anomaly 24 and their age is therefore also important for the calibration of the geomagnetic polarity time-scale.

Models for the geological evolution of the area have been proposed by Larsen (1978), Carter et al. (1979), Brooks (1980), Eldholm & Thiede (1980) and Brooks & Nielsen (1982). Any satisfactory model must, of course, be consistent with the constraints imposed by the established geochronology. However, despite its importance, published age information on the E Greenland province is rather limited. Since the first age measurement (Hamilton 1966) and the work of Beckinsale et al. (1970), the region has received little attention, although the number of publications with geochronological information has been increasing. They include Pankhurst et al. (1976), Brown et al. (1977), Rex et al. (1978a, b), Brooks et al. (1979), Gleadow & Brooks (1979), Odin & Mitchell (1983) and Upton et al. (1984).

This comprehensive K-Ar investigation aimed to establish an improved definition of the geochronology and document the geological history. In all about 230 analyses were made on 150 rocks and mineral separates. The complete listing of results is contained in Tables 1–4. Exact sample locations and petrographic information are obtainable from the authors. The measurements define a complex sequence of igneous events which extend over a time interval exceeding 25 My.

Analytical procedures

These follow Macintyre & Hamilton (1984). Separates of 100–150 μm size were used for most

From MORTON, A. C. & PARSON, L. M. (eds), 1988, *Early Tertiary Volcanism and the Opening of the NE Atlantic*, Geological Society Special Publication No. 39, pp. 201–214.

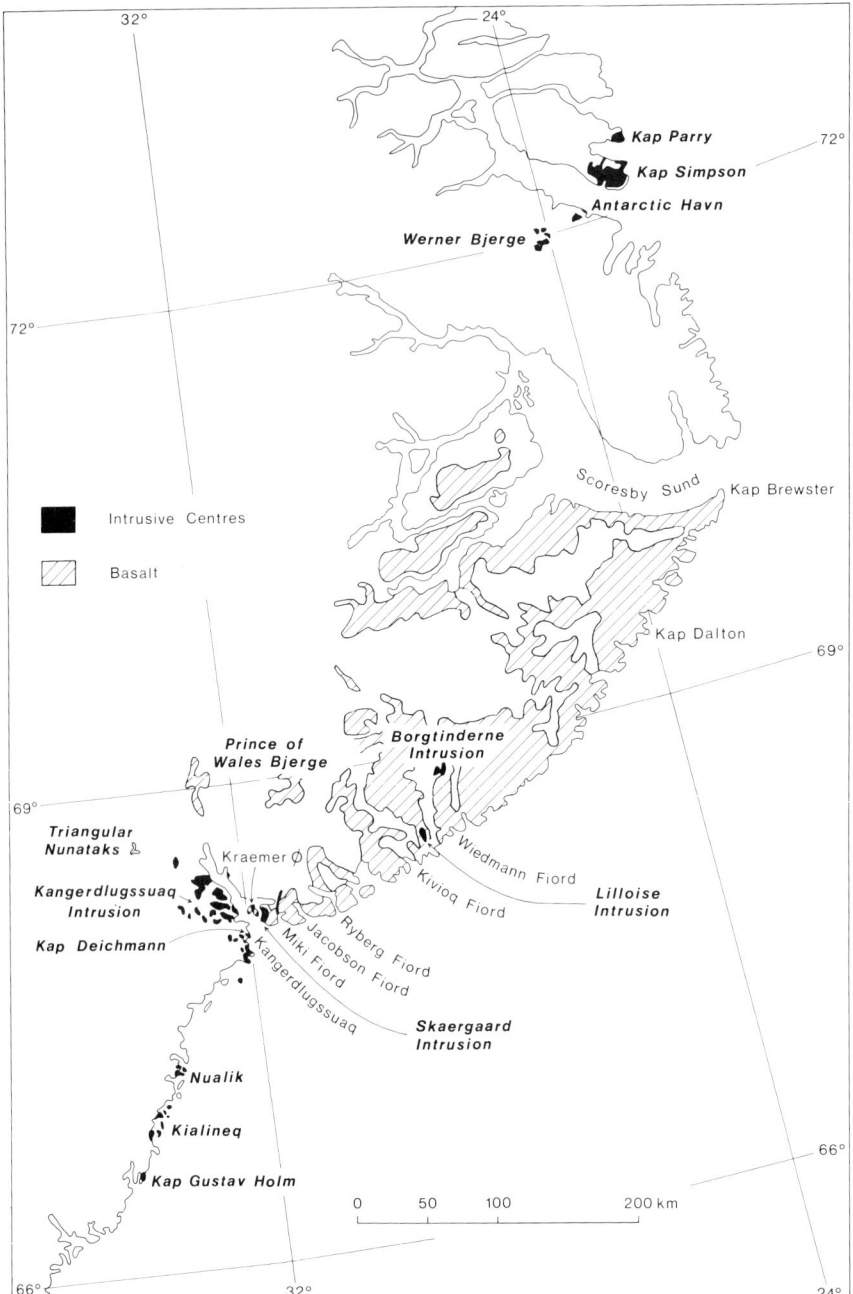

FIG. 1. Generalized geology of E Greenland showing distribution of main intrusive centres along the Blosseville
Kyst, principal sampling areas and localities mentioned in the text.

mineral analyses and powders and chips for
whole rocks. Replicate potassium determinations
were made by flame photometry, the accuracy of
which is estimated at 1.5% at the 2σ level. ^{40}Ar
was analysed in a MS10 instrument with a highly
enriched ^{38}Ar tracer. Errors of 1%, 0.5% and
0.5% have been assigned to the ^{36}Ar, ^{38}Ar and
^{40}Ar peak heights and the errors calculated in a
conventional manner (Baksi 1982). The decay
constants and atomic abundances used in this

work are those recommended by the IUGS Sub-Commission on Geochronology (Steiger & Jäger 1977) ($\lambda = 0.581 \times 10^{-10}$ yr^{-1}, $\lambda_\beta = 4.962 \times 10^{-10}$ yr^{-1}, ^{40}K $= 0.01167$ atom %K) and published ages have been recalculated with these values.

The presence of extraneous argon, that is argon trapped in the system at crystallization and having ^{40}Ar/^{36}Ar significantly different from the atmospheric value, is a severe complicating factor in the application of the K-Ar dating method in this region. Some earlier apparent K-Ar ages seem to have been affected by this problem and as a result now require re-interpretation (Brown *et al.* 1977). Some reference will therefore be made here to ages calculated from K-Ar isochron diagrams, described and fully discussed by Hayatsu (1972) and Hayatsu & Carmichael (1970, 1977). The criteria underlying the proper application of such diagrams are extremely rigorous, so much so that they are rarely, if ever, achieved in practice—a point which cannot be overemphasized. They include the condition that the atmospheric argon contamination in the measurements should be negligible. It is nonetheless possible to demonstrate that, *provided this contamination remains constant among analyses* (and the other criteria are, at the same time, also satisfied), a linear array of points results, which defines an isochron permitting the age to be calculated. However, in such a case, the intercept is no longer the ^{40}Ar/^{36}Ar ratio in the samples at the time of their crystallization. As the *slope* of this line is independent of the *magnitude* of ^{36}Ar, a value for this can be assumed and used to calculate the *theoretical* end-points (corresponding to zero atmospheric argon contamination) of the mixing lines for each sample. These can then be employed in a regression analysis (York 1969) to calculate the age of the samples. The usefulness and effectiveness of this technique, termed ^{36}Ar initialization, has already been examined (Noble 1978).

Basalts and hypabyssal intrusions

The geology of the basalts in the Kangerdlugssuaq area has been comprehensively described by Nielsen *et al.* (1981) and accounts of the dykes given in Nielsen (1975, 1978) and Myers (1980) and references therein.

Basalts

The results obtained by Beckinsale *et al.* (1970) on several flows of the Upper Basalts around Scoresby Sund ranged from ca. 45–60 Ma. The younger ages were attributed to argon loss as a result of alteration and the highest ages were regarded as being the closest approximation to the time of extrusion. It was pointed out by Odin & Mitchell (1983) that these ages were distributed bimodally around 46 and 56 Ma. From the same region they presented their own results for six lavas whose ages ranged from 48–53 Ma. They preferred the mean age of 50 Ma for the time of extrusion, but the older age (ca. 53 Ma) was preferred by Hailwood *et al.* (1979). The age of the basalts has therefore not been well-defined.

Incipient groundmass chloritization is common in the basalts. However, 14 flows, adjudged on petrographic evidence to be the freshest available, were dated from localities along the Blosseville Coast in Miki, Jacobsen, Ryberg and Wiedemann Fjords (see Fig. 1). These results are presented in Table 1.

Most ages are impossibly high and the effects of widespread contamination with extraneous argon of anomalous isotopic composition are evident. Generally, there is a correlation between high ages and increasing degree of alteration of the basalts (although this is not always the case). The Blosseville Group basalts from Ryberg, Jacobsen and Wiedemann Fjords were all from the main tholeiitic basalt pile but their exact stratigraphic position is not certain. The only completely fresh sample from Wiedemann Fjord (440 WR) yielded an age of 53.0 ± 1.0 Ma and two samples from Ryberg Fjord, which showed only very slight incipient ground-mass alteration, gave ages of 53.0 ± 2.9 and 54.9 ± 3.4 Ma (Table 1). The latter could reflect overprinting as the dyke swarm is very intense in this area. From the lowest volcanic formation (Vandsfaldsdalen Formation) only two flows, exposed in Miki Fjord, were considered suitable for dating. Plagioclase was separated from each and both the feldspars and whole-rocks analysed. Although the plagioclase ages are less than the whole-rock ages they are again impossibly high (see Table 1). Mineral whole-rock isochrons for the two flows yield 56 ± 2.1 and 57.8 ± 3.8 Ma. This age concordance might be of significance, although it would be rather surprising if the conditions previously referred to were strictly satisfied in both cases.

Eight fresh samples were analyzed from the Triangular Nunataks and Prince of Wales Bjerge (Anwar 1955; Brooks & Rucklidge 1974; Evans 1983). These lavas occur inland and overlie the Plateau Basalts in the Kangerdlugssuaq area (Deer 1976). They are more alkaline, usually olivine/pyroxene-phyric, and comprise a variety of petrographic types including picrite, ankaramite and hawaiite (Evans 1983). The ages are listed in Table 1 and with one exception (sample

TABLE 1. *Potassium-argon analyses of lavas, E Greenland. Legend—WR: whole rock; BI: biotite; AM: amphibole; PL: plagioclase; NE: nepheline; PX: pyroxene.* $K(wt.\%) = 0.8382\ ^{40}K(ppm)$

Sample No.	^{40}K (ppm)	$^{40}Ar^*$ ($\times 10^{-6}$ sccs/g)	$^{40}Ar^*$ (%)	Age (Ma)
Wiedemann Fjord				
446 WR	0.160	0.457	31.2	87.5 ± 2.7
440 WR	0.259	0.444	62.2	53.0 ± 1.0
430 WR	0.311	1.407	59.1	136.7 ± 2.6
428 WR	0.630	1.385	52.8	67.7 ± 1.4
422 WR	0.272	1.002	43.5	112.1 ± 2.6
10 WR	0.527	2.016	54.8	116.2 ± 2.3
7 WR	0.166	0.354	61.1	65.7 ± 1.2
Miki Fjord				
RP 31 WR	0.262	1.255	44.8	144.4 ± 3.3
		1.223	54.0	140.8 ± 2.8
PL	0.865	2.022	49.6	71.9 ± 1.5
		2.131	72.5	75.7 ± 1.3
RP 30 WR	0.373	1.361	40.0	111.0 ± 2.8
PL	0.412	1.103	57.3	82.1 ± 1.6
Ryberg Fjord				
KI 117 WR	0.165	0.243	20.7	45.6 ± 2.2
KI 114 WR	0.130	0.223	18.1	53.0 ± 2.9
KL 104 WR	0.147	0.261	16.2	54.9 ± 3.4
Jacobsen Fjord				
KN 101 WR	0.149	0.371	19.5	76.5 ± 3.9
Appollo Gletscher				
KG 202 WR	0.199	1.653	24.5	243.5 ± 9.8
Prince of Wales Bjerge				
2223 WR	1.366	2.243	52.0	50.8 ± 1.4
2224 WR	1.696	2.951	72.0	53.8 ± 1.2
2225 WR	1.580	2.733	70.1	53.8 ± 1.2
		2.824	73.8	55.6 ± 1.2
2230 WR	1.208	2.285	67.7	58.4 ± 1.4
2231 WR	1.671	2.873	65.0	53.1 ± 1.2
2240 WR	0.860	1.676	49.7	60.1 ± 1.7
		1.510	61.5	54.1 ± 1.4
2209 WR	1.879	2.568	33.4	41.9 ± 1.6
		2.496	57.2	40.9 ± 1.1
Triangular Nunataks				
3340 WR	1.421	2.687	66.0	58.4 ± 1.4

2209) they lie in the range 51–60 Ma. They are therefore similar to the ages reported by Brooks *et al.* (1979) for alkaline mafic lavas in a nunatak zone much further N (74°). These authors tentatively concluded that the average age of 56 Ma was the best estimate of the time of extrusion, and we draw a similar conclusion from our results (average 55 Ma, omitting 2209). These alkaline lavas are apparently restricted to the inland areas and, as a result, are more distant from the incipient spreading axis (intraplate). It is noteworthy that in both areas they appear to be devoid of the extraneous argon which seems

so characteristic of the Plateau Basalts and makes their precise age determination difficult. Bearing in mind the constraints imposed by later igneous events (such as Skaergaard with an age around 54 Ma (Gleadow & Brooks 1979) it would appear that extrusion of the basalts probably took place *within* the interval 57–53 Ma.

Sills

There is some uncertainty regarding the position of the dolerite sills within the sequence of igneous events. The sill complex is strongly developed E of Kangerdlugssuaq where it apparently predates the dyke swarm and coastal flexure (Wager 1947). However, Nielsen (1978) has suggested that the earliest dykes (THOL-I), when followed inland, gradually turn over to form the sills.

Only three sills proved suitable for dating (Nos RP6, RP7, KB104) and the results obtained are listed in Table 2. Slight alteration is evident in all these samples but the ages of 52.5 ± 1.5, 52.0 ± 1.2 and 52.7 ± 1.3 Ma suggest that the sill complex, and possibly the preflexure coastal dykes, have an age of about 52 Ma.

Dykes

In the coastal area of Kangerdlugssuaq, Nielsen (1978) recognized three generations of coast-parallel dykes. Attempts to date members of these proved unsuccessful. Chloritic alteration was almost universally present to some degree and a spread of whole-rock ages with a preponderance of impossibly high results was obtained (see Table 2). The freshest specimens yielded a few ages between 51 and 49 Ma, but little confidence can be placed in these results without further investigation.

The relationship between the dyke swarms identified by Nielsen and those further along the Blosseville Kyst is not known. In the region of Wiedemann Fjord (see Fig. 1) there are coast-parallel (around 080 degrees) and inland-trending (around 010 degrees) suites. Carbonate veins also trend 010 degrees. Dykes of both swarms are mainly lamprophyres and age determinations were made on amphiboles separated from eight dykes (see Table 2). The ages show a spread from 47–35 Ma. Two samples of one dyke (454 & 455) have average ages of 40.4 and 45.9 Ma, suggesting that argon loss may be responsible for the observed spread of ages. The only inland-trending dyke analyzed (460) yields 35–36 Ma and it seems possible that the intrusion of the inland swarm caused variable loss of argon from the other dykes.

From the proximity of Kivioq Fjord (see Fig. 1) five dykes were dated (see Table 2). These are all members of a swarm trending inland at around 040 degrees, which consists of coarse dolerites cut by similar trending lamprophyres. The dolerites (241, 218, 226) yielded average ages of 46, 49 and 50 Ma. The lamprophyres (21, 234), which are petrographically similar to those dated from Wiedemann Fjord, yielded whole-rock ages of 52.0 ± 1.0 (average of two) and 53.7 ± 1.1 Ma.

Although a considerable spread is shown by all the dyke analyses the freshest dykes can be interpreted as members of the post-flexure swarm with ages of 49–51 Ma.

Intrusions

The high-level, epizonal intrusions have been dated by the analysis of mineral separates, predominantly amphiboles and micas. Post-intrusion interaction with hydrothermal convective systems, if operative, would most probably have overprinted the age of an intrusion. Sheppard *et al.* (1977) showed that for the Lilloise intrusion no interaction had occurred with groundwater, but noted that this was in contrast with most other investigated epizonal intrusions. For the majority of the E Greenland intrusions, which postdate the major regional episodes of dyke injection, it is believed that any overprinting resulting from groundwater interaction would be essentially contemporaneous with the age of intrusion. The ages measured can therefore probably be considered as approximating closely the times of emplacement. These are presented in order of decreasing age.

Gardiner intrusion

This 6 km diameter ultramafic intrusion (Frisch & Keusen 1977) lies in the upper region of Kangerdlugssuaq (see Fig. 1) and cuts both Precambrian gneisses and Tertiary basalts. Phlogopite from a melilite-rich rock yields an age of 53.9 ± 1.1 Ma (Table 3). This is in agreement with the fission-track ages of 48–54 Ma reported by Gleadow & Brooks (1979). The Gardiner intrusion is therefore of comparable age to the pre-flexure Skaergaard intrusion, for which fission-track ages of around 54 Ma have been reported.

Kraemer Ø syenite

This quartz syenite forms the SW tip of Kraemer Ø near the Skaergaard intrusion (see Fig. 1). The syenite contains basalt xenoliths in profusion and cuts dykes of Nielsen's (1978) THOL-2 and ALK-I suites. This constrains its age to less than 52 Ma.

Seven hornblendes and one biotite sample that were analyzed yielded an apparent range in ages from 54–64 Ma (see Table 3). It is apparent that these minerals contain extraneous argon, as documented for amphiboles in the Kialineq region (Brown *et al.* 1977). The samples may be divided into two groups with ages around 54 Ma and between 57 and 63 Ma (Table 3). Interestingly, on an isochron diagram, group 1 yields 49.5 ± 0.6 Ma and group 2 (after initialization) 49.3 ± 2 Ma, ages more in keeping with the field evidence.

TABLE 2. *Potassium-argon analyses of dykes and sills, E Greenland. Legend as for Table 1.*

Sample No.	^{40}K (ppm)	$^{40}Ar*$ ($\times 10^{-6}$ sccs/g)	$^{40}Ar*$ (%)	Age (Ma)
Miki Fjord				
RP 28 PL	0.519	1.502	63.8	88.6 ± 1.6
WR	0.309	1.878	83.3	181.3 ± 3.0
Jacobsen Fjord				
RM 130 WR	1.823	2.373	32.3	40.4 ± 1.2
		2.261	34.7	38.5 ± 1.1
RM 129 WR	0.954	1.654	29.6	53.6 ± 1.8
RM 127 WR	0.317	0.439	25.2	42.9 ± 1.7
RM 126 WR	0.898	2.016	66.6	69.1 ± 1.2
RM 125 WR	0.174	1.880	59.4	310.7 ± 5.9
		0.459	8.8	80.9 ± 9.7
RM 124 WR	0.801	1.827	71.8	70.2 ± 1.2
RM 123 WR	0.915	1.875	90.7	63.2 ± 1.0
RM 122 WR	0.175	1.065	36.3	181.6 ± 4.9
		0.826	14.7	142.4 ± 9.8
RM 121 WR	0.877	1.451	36.8	51.2 ± 1.4
KN 107 WR	0.384	0.633	39.7	51.0 ± 1.3
Ryberg Fjord				
RM 136 WR	0.581	1.472	50.1	77.8 ± 1.6
RM 133 WR	1.088	1.943	58.4	55.2 ± 1.1
RM 132 WR	0.507	0.798	64.7	48.7 ± 0.9
RP 40 WR	1.241	2.029	68.4	50.6 ± 0.9
RP 21 PX	0.084	0.483	27.0	172.0 ± 6.2
PL	2.780	6.517	72.4	72.1 ± 1.3
WR	0.939	3.391	74.0	109.9 ± 1.9
RP 7 WR	0.387	0.651	46.4	52.0 ± 1.2
RP 6 WR	0.317	0.540	42.1	52.7 ± 1.3
RP 2 WR	0.236	7.046	86.0	754.6 ± 12.5
RP 1 WR	0.562	4.122	60.7	216.6 ± 4.1
RM 42 WR	0.451	1.123	48.1	76.5 ± 1.6
RM 41 WR	0.849	2.165	59.9	78.3 ± 1.5
RM 39 WR	0.230	0.768	38.7	101.9 ± 2.6
RM 38 WR	0.754	2.038	60.4	82.9 ± 1.6
RM 77 WR	0.210	2.365	63.2	322.7 ± 5.9
RM 67 WR	0.167	0.830	56.7	149.5 ± 2.9
RM 61 WR	1.631	3.139	78.1	59.4 ± 1.0
RM 60 WR	0.451	0.615	19.8	42.3 ± 2.1
Apollo Gletscher				
KB 105 WR	1.432	39.19	69.7	702.4 ± 12.4
KB 104 WR	0.414	0.703	33.9	52.5 ± 1.5
Kivioq Fjord				
241 WR	0.736	1.049	48.7	44.2 ± 0.9
		1.052	38.8	44.3 ± 1.1
		1.151	30.3	48.4 ± 1.6
234 WR	1.837	3.004	70.7	50.6 ± 0.9
		3.179	69.6	53.5 ± 0.9
227 WR	1.172	2.036	51.0	53.7 ± 1.1
226 WR	0.581	0.931	59.3	49.6 ± 0.9
218 WR	0.311	0.496	30.7	49.4 ± 1.6
PL	0.357	0.562	46.8	48.7 ± 1.1
		0.553	39.5	48.0 ± 1.2
215 WR	2.267	3.871	73.8	52.8 ± 0.9

TABLE 2. *continued*

Sample No.	^{40}K (ppm)	^{40}Ar* ($\times 10^{-6}$ sccs/g)	^{40}Ar* (%)	Age (Ma)
Wiedemann Fjord				
464 AM	1.300	1.939	52.1	46.2 ± 0.9
		1.973	67.0	47.0 ± 0.8
		2.015	47.5	48.0 ± 1.0
		1.680	38.7	40.1 ± 1.0
		2.275	53.9	54.1 ± 1.1
460 AM	0.956	1.102	74.7	35.8 ± 0.6
		1.070	20.0	34.8 ± 1.7
		1.052	17.3	34.2 ± 2.0
457 AM	1.253	1.595	77.6	39.5 ± 0.7
		1.742	43.1	43.1 ± 1.0
455 AM	0.686	0.920	66.8	41.6 ± 0.7
		0.891	11.3	40.3 ± 3.7
		0.866	38.7	39.2 ± 1.0
454 AM	0.625	0.903	67.6	44.8 ± 0.8
		0.946	40.6	46.9 ± 1.2
444 AM	0.781	1.180	70.2	46.8 ± 0.8
		1.172	43.5	46.5 ± 1.1
434 AM	1.396	1.559	73.9	34.7 ± 0.6
		1.559	75.4	34.7 ± 0.6
412 AM	1.152	1.763	71.0	47.4 ± 0.8

Kangerdlugssuaq intrusion

This well-known intrusion (Wager 1965; Kempe *et al.* 1970) was dated by Pankhurst *et al.* (1976) who obtained a Rb-Sr mineral isochron of 49.9 ± 1.0 Ma and a whole-rock isochron of 50.0 ± 1.9 Ma. A similar age had been obtained by Beckinsale *et al.* (1970) on biotite from a pulaskite.

Argon analyses on arfvedsonite from veins in nordmarkite and transitional pulaskite, and nepheline from a vein in pulaskite yielded ages of 50.5 ± 0.9, 52 ± 1.0 and 51.3 ± 0.9 Ma respectively (see Table 3). These separates are from consanguineous veins essentially contemporaneous with the main units of the intrusion (Kempe pers. comm.) and the results confirm the previous age assignments (see also Gleadow & Brooks 1979). Biotites, however, yield younger ages and appear to have experienced some argon loss. Amphiboles from two adjacent minor syenite intrusions (Peak 2005 and Kaerven) are similar to those of Kraemer Ø in that they were found to contain extraneous argon.

Lilloise intrusion

This intrusion (see Fig. 1) is a layered gabbroic body cut by numerous late, cognate, felsic sheets (Brown 1973). Sheppard *et al.* (1977) found that it had not undergone interaction with a late hydrothermal system. Potassium-argon mineral ages, originally attributed by Beckinsale *et al.* (1970) to the Lilloise intrusion, are now believed to have been made on material from the Borgtinderne syenite (Deer 1976; Gleadow & Brooks 1979).

Twelve amphiboles and biotites were separated from 11 samples. Argon determinations on amphibole separates from the layered series, from cognate sheets within the intrusion, and from metabasalt at the immediate contact yielded concordant ages of 49.4 ± 2.0 Ma (Table 4). Biotites yield similar ages, indicating rapid cooling as expected of a high level intrusion and substantiated by the concordance of zircon and apatite fission-track ages (Gleadow & Brooks 1979).

Borgtinderne nepheline syenite

This intrusion occurs inland from the Lilloise Bjerge and some 45 km from the coast (see Fig. 1). It has an outcrop of about 8 × 10 km and is surrounded by flat-lying flood basalts (Brown *et al.* 1978). Numerous cognate aplite syenite sheets and a late suite of lamprophyre dykes cut the intrusion. Pale and dark varieties of syenite are present, the dark being heterogeneous and attributed to assimilation of basalt (Brown *et al.* 1978). Potassium-argon ages between 41 and 50 Ma were reported by Beckinsale *et al.* (1970) and

TABLE 3. *Potassium-argon analyses of syenites from the Kangerdlugssuaq area. Legend as for Table 1.*

Sample No.	^{40}K (ppm)	$^{40}Ar^*$ ($\times 10^{-6}$ sccs/g)	$^{40}Ar^*$ (%)	Age (Ma)
Kangerdlugssuaq syenites				
Main Intrusion				
4789 BI	8.983	12.40	61.3	42.8 ± 0.8
4678 NE	5.166	8.567	66.8	51.3 ± 0.9
4666 AM	1.730	2.823	62.5	50.5 ± 0.9
4582 BI	9.162	13.99	64.3	47.3 ± 0.9
		13.42	35.5	45.4 ± 1.3
2046 AM	1.563	2.628	56.7	52.0 ± 1.0
		2.597	46.6	51.4 ± 1.1
Minor Intrusions				
4636 AM	1.277	2.352	34.4	56.9 ± 1.6
2683 AM	1.078	2.438	39.6	69.6 ± 1.7
Kraemer Ø syenite				
Group 1				
8045 AM	0.803	1.566	31.3	60.2 ± 1.9
		1.635	65.1	62.8 ± 1.1
8041 BI	8.399	17.49	26.1	64.2 ± 2.4
AM	0.754	1.396	68.0	57.2 ± 1.0
		1.441	59.5	59.0 ± 1.1
8026 AM	0.979	1.839	66.1	58.0 ± 1.0
		1.887	70.7	59.5 ± 1.0
8025 AM	0.625	1.279	58.5	63.1 ± 1.2
		1.217	34.7	60.1 ± 1.7
Group 2				
8032 AM	0.707	1.256	66.8	54.9 ± 1.0
8029 AM	0.854	1.500	58.6	54.3 ± 1.0
8027 AM	0.827	1.456	53.0	54.4 ± 1.1
Kap Edvard Holm syenites				
Kap Boswell syenite				
3552 AM	0.387	0.455	17.2	36.5 ± 2.1
Hutchinson Gletscher syenite 1				
3525 BI	8.327	11.52	75.4	42.9 ± 0.7
AM	1.205	1.581	57.3	40.7 ± 0.8
Kap Deichman syenite 1				
3522 BI	8.387	11.14	74.3	41.2 ± 0.7
AM	0.636	0.853	35.5	41.6 ± 1.2
		0.749	36.8	36.6 ± 1.0
3521 BI	7.945	10.65	77.2	41.6 ± 0.7
AM	0.806	0.963	45.9	37.1 ± 0.8
Kap Deichman syenite 2				
3491 BI	7.313	10.17	18.6	43.1 ± 2.3
		9.927	63.7	42.1 ± 0.8
AM	0.996	1.310	49.2	40.8 ± 0.9
3490 AM	1.058	1.329	41.5	39.0 ± 0.9
3486 AM	0.910	1.185	22.7	40.4 ± 1.8
3196 BI	8.947	11.88	73.4	41.2 ± 0.7
Syenite veins				
Sortskaer				
3460 AM	0.802	1.262	30.8	48.7 ± 1.5
Nugalik				
489 BI	7.969	11.36	87.8	44.2 ± 0.7
487 BI	7.886	11.17	81.7	43.9 ± 0.7
		10.99	76.1	43.2 ± 0.7
Igdilitarajik				
457 BI	8.267	11.90	79.9	44.6 ± 0.8

TABLE 4. *Potassium-argon analyses of some other E Greenland intrusions. Legend as for Table 1.*

Sample No.	^{40}K (ppm)	$^{40}Ar*$ ($\times 10^{-6}$ sccs/g)	$^{40}Ar*$ (%)	Age (Ma)
Lilloise intrusion				
Cumulate series				
187 AM	0.876	1.461	44.7	51.6 ± 1.2
136 AM	0.976	1.602	66.9	50.8 ± 0.9
134 AM	0.901	1.468	76.7	50.4 ± 0.9
Meta-basalt and xenoliths				
131 AM	0.952	1.476	64.2	48.0 ± 0.9
		1.420	35.0	46.2 ± 1.3
170 BI	7.683	12.06	73.3	48.5 ± 0.8
		11.78	45.8	47.5 ± 1.1
105 AM	1.169	1.885	82.2	49.9 ± 0.8
		1.824	79.3	48.3 ± 0.8
105 BI	7.874	12.36	77.9	48.6 ± 0.8
		12.10	72.7	47.6 ± 0.8
Dykes, sheets and pegmatites				
175 AM	1.000	1.740	53.5	53.8 ± 1.1
		1.773	49.1	54.8 ± 1.2
148 BI	7.659	11.97	75.7	48.4 ± 0.8
		12.52	77.9	50.6 ± 0.9
AM	0.977	1.483	57.7	47.0 ± 0.9
128 AM	0.988	1.677	82.0	52.5 ± 0.9
19 BI	8.804	13.36	80.1	47.0 ± 0.8
Borgtinderne syenite				
Pale syenite				
T 33 AM	1.980	2.966	77.2	46.4 ± 0.8
T 32 AM	1.706	2.678	62.8	48.6 ± 0.9
		2.650	64.1	48.1 ± 0.9
T 7 BI	6.490	10.19	62.5	48.6 ± 0.9
Dark syenite				
T 23 AM	1.551	2.277	62.0	45.5 ± 0.8
		2.196	41.2	43.9 ± 1.1
T 16 AM	2.314	3.269	63.2	43.8 ± 0.8
T 15 AM	1.658	2.353	70.9	44.0 ± 0.8
T 11 AM	1.010	1.387	52.7	42.6 ± 0.9
Kialineq syenite complex (dykes)				
7565 BI	5.309	4.997	55.7	29.3 ± 0.6
7541 WR	0.947	1.116	38.8	36.6 ± 0.9
7531 WR	1.132	1.238	48.0	34.0 ± 0.7
AM	0.692	0.761	43.7	34.2 ± 0.8
		0.788	41.6	35.4 ± 0.9
7529 BI	4.796	5.028	58.6	32.6 ± 0.6
Kap Parry alkaline complex				
1104 PX	0.173	0.190	15.9	34.1 ± 2.2
AM	1.849	2.016	54.1	33.9 ± 0.7
		2.106	44.4	35.4 ± 0.8
3134 PX	0.132	0.157	14.5	36.9 ± 2.6
Dykes, N of Scoresby sund				
1205 AM	1.384	1.249	40.6	28.1 ± 0.7
1204 BI	6.669	6.191	74.2	28.9 ± 0.5
		6.494	30.9	30.3 ± 1.0
AM	1.420	1.373	42.1	30.1 ± 0.7
1103 WR	3.221	2.210	49.3	21.4 ± 0.5
		2.220	46.4	21.5 ± 0.5

fission-track ages of around 47 Ma were reported by Gleadow & Brooks (1979).

Amphiboles were separated from six samples of syenite and biotite from a pegmatite. Analyses of three amphibole separates from the dark syenite have an isochron age of 45.4 ± 3.0 Ma (see Table 4). Two amphiboles and a biotite from the pale syenite yield an isochron age of 43.4 ± 2.0 Ma. The mean age of 44.6 ± 2.5 Ma can be taken as the best estimate of the age for the intrusion. This is similar to that measured for the Kap Deichmann syenite (see below).

Kap Deichmann syenite

This fayalite–hedenbergite syenite (Deer 1976) is a plug-like mass, 4–5 km in diameter, on the W side of Kangerdlugssuaq. The main intrusion (syenite 1) is cut by an arcuate body of syenite filled with xenoliths of gabbro and basalt (syenite 2).

Coexisting amphibole and biotite were separated from each of two samples of the main syenite. These four results indicate a mean age of 39.8 Ma (see Table 3). The results are rather unusual in that the amphibole ages appear to be less than the biotite ages.

From syenite 2 three amphiboles and two biotites yield a mean age of 40.8 Ma (see Table 3), and so this is essentially contemporaneous with the main intrusion.

Two other syenite bodies occur immediately to the NW of the Kap Deichmann syenite. The Hutchinson Glacier syenite I, which cuts the Deichmann syenite, yields ages of 40.7 ± 0.8 Ma on biotite and 42.9 ± 0.7 Ma on coexisting amphibole (see Table 3). To the S the Kap Boswell syenite gives an amphibole age of 36.5 ± 2.1 Ma.

Syenite veins

Syenite veins are found at a number of localities in and around the entrance to Kangerdlugssuaq Fjord. Their relationship to the major plutons is not clear. Specimens were dated from Amdrups Pynt (Nugalik) within Kangerdlugssuaq and a nearby group of skerries (Sortskaer). Both localities are immediately up-fjord from Kap Deichmann. Two biotites from Amdrups Pynt gave apparent ages close to 44 Ma, similar to the Deichmann syenite, while amphibole from Sortskaer gave 48.7 Ma (see Table 3), more comparable to the Kangerdlugssuaq and Kraemer Ø syenite ages. Biotite from veins at Igdlitarajik, a group of small islands about 50 km SW of Kangerdlugssuaq, gave an apparent age of 44.6 Ma (see Table 3).

Kialineq complex

This granite–quartz syenite complex is the most southerly considered here. A Rb-Sr isochron age of 35 ± 2 Ma and a K-Ar amphibole and mica isochron age of 35.7 ± 2.6 Ma, have already been reported together with a brief geological summary (Brown et al. 1977). Fission-track ages consistent with the original interpretation have since been published (Gleadow & Brooks 1979). Several additional ages, obtained on lamprophyric dykes cutting the complex, are listed in Table 4. A dyke cutting the Qajarsak granite (7529) yielded a biotite age of 32.6 Ma; a dyke cutting Lilleo Island gave whole-rock and amphibole ages around 34 Ma and a dyke cutting Aliuarssik Island, but predating the granite, has a whole-rock age of 36.6 Ma. These measurements are consistent with the earlier conclusions.

Biotite from the Aliuarssik granite (7565) yielded an apparent age of 29.3 ± 0.6 Ma which appears younger than the other members of the complex. Although this granite is, on the evidence of intrusive relationships, one of the latest, this age differs from other reported measurements. (Beckinsale et al. 1970; Gleadow & Brooks 1979).

Kap Parry alkaline complex

Tertiary granite and syenite intrusions occur both N and S of Kong Oscars Fjord, the largest being located at Werner Bjerge, Kap Simpson and Kap Parry (see Fig. 2). Beckinsale et al. (1970) obtained an apparent age of 28.7 Ma on a biotite from Werner Bjerge (Bearth 1959; Brooks et al. 1982), which differed markedly from the ages of other intrusions they dated. However, fission-track ages of around 30 Ma have since been reported by Gleadow & Brooks (1979).

From the Kap Parry complex (Schaub 1942; Engell 1975) coexisting pyroxene and amphibole gave ages of 34.1 ± 2.2 Ma and 33.9 ± 0.7 Ma respectively (see Table 4). Another pyroxene gave 36.9 ± 2.6 Ma. The concordance of these results within experimental errors, and their similarity to those of the Kialineq intrusions, indicates that the average age (35.1 Ma) is probably the best estimate of the age of the Kap Parry complex.

A kaersutite-bearing dyke (1204) at Antarctic Havn (see Fig. 1) yielded ages of between 28 and 30 Ma for coexisting biotite and kaersutite. A second specimen of the same dyke (1205) gave 28.1 ± 0.7 Ma for kaersutite. The position of this dyke in the local igneous sequence is uncertain as it is impossible to differentiate between the regional and local dykes (Kapp 1960). These ages are similar to those reported from the Myggbukta complex in the Hold with Hope Region further

N (Upton *et al.* 1984), and may represent another period of igneous activity around this time (Rex *et al.* 1978b).

Discussion

The K-Ar ages of the Blosseville coast basalts, when considered with those of the Skaergaard and Gardiner Plateau intrusions, suggest that extrusion took place in the interval 57–53 Ma. Their magnetic stratigraphy is now known (Tarling *et al.* 1988) in enough detail to exclude the presence of a normal polarity interval in the pile, as was the case in the Faeroes (Tarling & Gale 1968), and extrusion in the single reversed epoch prior to anomaly 24 appears most probable. The age of this has long been a matter of considerable debate (Ness *et al.* 1980; Backman *et al.* 1984) and a number of different assignments have been made (e.g. Hailwood *et al.* 1979; Lowrie & Alvarez 1981; Berggren *et al.* 1985; Odin & Curry 1985). The results reported here, similar to those reported by Macintyre & Hamilton (1984) for DSDP sites 553 and 555, favour the early assignment of Hailwood *et al.* and render the assignment of Odin & Curry less probable. However, the present imprecision of the measurements does not warrant any readjustments to the time-scales.

Although the problems encountered with extraneous argon have resulted in a large spread of measured ages, it is extremely interesting that the argon systematics of the basalts are similar to those in lavas recovered from the deeper levels at DSDP Sites 553 and 555 (Macintyre & Hamilton 1984) and from Well 163/6–1A in the Rockall Trough (Morton *et al.* 1988). This undoubtedly reflects similarities in the mode of genesis of these rocks.

The E Greenland basalts were extruded from high-level magma chambers onto thinned continental crust, from a linearly elongated source region off-shore from the present outcrop (Brooks *et al.* 1976; Brown & Whitley 1976). The extrusions began in shallow submarine conditions and the sequence closed in marine conditions (Soper *et al.* 1976). The coastal flexure and related coast-parallel dyke swarm were initiated early in the stage of active formation of new oceanic crust off-shore from the present outcrop, probably by a mechanism of shelf-subsidence similar to that invoked by Bott (1973). As post-flexure dykes which are cut by the Kraemer Ø intrusion have been identified (Nielsen 1978) the flexure must have formed around the interval 52–50 Ma.

The plutonic and hypabassal activity which

followed the basalt extrusion was concentrated along the coastal strip and the line of Kangerdlugssuaq. Pre-flexure gabbroic plutons (e.g. Skaergaard, Kap Edvard Holm and in the Kialineq complex) are essentially a continuation of the previous tholeiitic basic magmatism; they contrast sharply with the alkaline penecontemporaneous Gardiner intrusion. This suggests a geographic control of contemporaneous magmatism with the more alkaline activity occurring further inland away from the plate margin.

There also appears to be an overall evolution of the type of magmatism with time which is discernible over the whole province. Thus, the immediate post-flexure intrusions at about 50 Ma differ markedly from the earlier gabbroic plutons. The Lilloise intrusion has strong alkaline affinities, the Kangerdlugssuaq syenite intrusion terminates with foyaiite and the nearby Kraemer Ø intrusion is fayalite syenite. Prolific lamprophyre dyke intrusion also occurred at about 50–42 Ma.

Further activity took place in the interval 45–42 Ma. The Borgtinderne nepheline syenite was intruded inland while in the outer part of the Kangerdlugssuaq area there are quartz syenites, notably at Kap Deichmann. Cessation of spreading in the Labrador Sea took place around 43 Ma (anomaly 19) and it is tempting to relate the magmatism to this major readjustment in plate motions.

The next peak of magmatism occurred around 36 Ma when granite and quartz syenite plutons were emplaced at Kialineq and around Kong Oscars Fjord. Around this time (anomaly 13) there was another major adjustment in plate geometry as the Greenland Sea opened (Talwani & Eldholm 1977). The age measurements further indicate that major plutonic activity continued down to 28 Ma, as previously proposed for the British Tertiary Volcanic Province (Horne & Macintyre 1975, Seeman 1984), and now geophysically confirmed (Tate & Dobson, 1988).

Thus, the generalized evolution of the province in post-basalt times shows a progression from tholeiitic, through basic alkaline magmas, to feldspathoidal syenites, quartz syenites and granites, with a strong suggestion that at any given time the magmatism was more alkaline the further inland it occurred.

A feature of all the later quartz syenite and granite intrusions is their association with large volumes of mixed acid–basic rocks, and with the development of diorites and net-vein complexes (spectacularly displayed at Kialineq). Finally, within the range of ages from 57–56 Ma—28 Ma pulses of magmatic activity seem to be coeval with major adjustments to plate geometry and spreading. There is, however, a need for a greater refinement of the age information and a unifying hypothesis which can link these two phenomena.

ACKNOWLEDGEMENTS: This investigation formed part of a PhD thesis by the first author. Most samples were collected during the Universities' E Greenland Expeditions of 1972 and 1974 and a few were made available through the generosity of various individuals. We would particularly like to thank Professors W. A. Deer and E. A. Vincent, D. R. C. Kempe, N. J. Soper, T. McMenamin, C. K. Brooks, I. B. Evans and Amax Exploration Inc. The work was supported by a NERC studentship to R. H. Noble and carried out at Aberdeen University and in the Isotope Geology Unit, SURRC, E Kilbride which is supported by NERC and the Scottish Universities. Through the Director we wish to thank the Geological Survey of Greenland for discussion.

References

ANWAR, Y. M. 1955. The petrography of The Prince of Wales Bjerge lavas. *Meddelelser øm Grønland*, **135**.

BACKMAN, J., MORTON, A. C., ROBERTS, D. G., BROWN, S., KRUMSIEK, K. & MACINTYRE, R. M. 1984. Geochronology of the Lower Eocene and Upper Paleocene sequences of Leg 81. *In:* ROBERTS, D. G., SCHNITKER, D. *et al. Initial Reports of the Deep Sea Drilling Project*. US Government Printing Office, Washington, **81**, pp. 877–882.

BAKSI, A. K. 1982. A note on the calculation of errors in conventional K-Ar dating. *Chemical Geology*, **35**, 167–172.

BEARTH, P. 1959. On the alkali massif of the Werner Bjerge in East Greenland. *Meddelelser øm Grønland*, **153**.

BECKINSALE, R. D., BROOKS, C. K. & REX, D. C. 1970. K-Ar ages for the Tertiary of East Greenland. *Bulletin of the Geological Society of Denmark*, **20**, 27–37.

BERGGREN, W. A., KENT, D. V., FLYNN, J. J. & VAN COUVERING, J. A. 1985. Cenozoic geochronology. *Bulletin of the Geological Society of America*, **96**, 1407–1418.

BOTT, M. H. P. 1973. Shelf subsidence in relation to the evolution of young continental margins. *In:* TARLING, D. H. & RUNCORN, S. K. (eds) *Implications of Continental Drift to the Earth Sciences*. Academic Press, London, **2**, pp. 675–683.

BROOKS, C. K. 1973a. Rifting and doming in southern East Greenland. *Nature*, **244**, 23–25.

—— 1973b. Tertiary of Greenland—a volcanic and plutonic record of continental break-up. *Arctic Geology*, **19**, 150–160.

—— 1980. Episodic volcanism, epeirogenesis and the formation of the North Atlantic ocean. *Palaeogeography, Palaeoclimatology and Palaeoecology*, **30**, 229–242.

—— & RUCKLIDGE, J. C. 1974. Strongly undersaturated Tertiary volcanic rocks from the Kangerdlugssuaq area, East Greenland. *Lithos*, **7**, 239–248.

—— & NIELSEN, T. F. D. 1982. The East Greenland continental margin: a transition between oceanic and continental magmatism. *Journal of the Geological Society of London*, **139**, 265–275.

——, NIELSEN, T. F. D. & PETERSEN, T. S. 1976. The Blosseville coast basalts of East Greenland: their occurrence, composition and temporal variations. *Contributions to Mineralogy and Petrology*, **58**, 279–292.

——, PEDERSEN, A. K. & REX, D. C. 1979. The petrology and age of alkaline mafic lavas from the nunatak zone of central East Greenland. *Grønlands Geologiske Undersøgelse Bulletin*, **133**.

——, ENGELL, J., LARSEN, L. M. & PEDERSEN, A. K. 1982. Mineralogy of the Werner Bjerge alkaline complex, East Greenland. *Meddelelser øm Grønland, Geoscience*, **7**.

BROWN, P. E. 1973. A layered plutonic complex of alkali basalt parentage: the Lilloise intrusion, East Greenland. *Journal of the Geological Society of London*, **129**, 408–418.

—— & WHITLEY, J. E. 1976. East Greenland basalts and their supposed plume origin. *Nature*, **260**, 232–234.

——, VAN BREEMEN, O., NOBLE, R. H. & MACINTYRE, R. M. 1977. Mid-Tertiary activity in East Greenland—the Kialineq Complex. *Contributions to Mineralogy and Petrology*, **64**, 109–122.

——, BROWN, R. D., CHAMBERS, A. D. & SOPER, N. J. 1978. Fractionation and assimilation in the Borgtinderne syenite, East Greenland. *Contributions to Mineralogy and Petrology*, **67**, 25–34.

CARTER, S. R., EVENSEN, N. M., HAMILTON, P. J. & O'NIONS, R. K. 1979. Basalt magma sources during the opening of the North Atlantic. *Nature*, **281**, 28–30.

DEER, W. A. 1976. Tertiary igneous rocks between Scoresby Sund and Kap Gustav Holm, East Greenland. *In:* ESCHER, A. & WATT, W. S. (eds) *Geology of Greenland*. Grønlands Geologiske Undersøgelse, København, pp. 404–429.

ELDHOLM, O. & THIEDE, J. 1980. Cenozoic continental separation between Europe and Greenland. *Palaeogeography, Palaeoclimatology, Palaeoecology*, **30**, 243–259.

ENGELL, J. E. 1975. The Kap Parry complex, central East Greenland. *Grønlands Geologiske Undersøgelse Rapport*, **75**, 103–106.

EVANS, I. B. 1983. *Geochemical and isotopic studies of the Tertiary volcanic rocks from the Blosseville Coast Region, East Greenland*. PhD thesis, University of Aberdeen.

FALLER, A. M. 1975. Palaeomagnetism of the oldest Tertiary basalts in the Kangerdlugssuaq area of East Greenland. *Bulletin of the Geological Society of Denmark*, **24**, 173–178.

—— & SOPER, N. J. 1979. Palaeomagnetic evidence for the origin of the coastal flexure and dyke swarm in central East Greenland. *Journal of the Geological Society of London*, **136**, 737–744.

FITCH, F. J., HOOKER, P. J., MILLER, J. A. & BRERETON, W. R. 1978. Glauconite dating of the Palaeocene–Eocene rocks of East Kent and the time-scale of Palaeogene volcanism in the north Atlantic region. *Journal of the Geological Society of London*, **135**, 499–512.

FRISCH, W. & KEUSEN, H. 1977. The Gardiner Intrusion, an ultramafic complex at Kangerdlugssuaq, East Greenland. *Grønlands Geologiske Undersøgelse Bulletin*, **122**, 1–62.

GLEADOW, A. J. W. & BROOKS, C. K. 1979. Fission track dating, thermal histories and tectonics of igneous intrusions in East Greenland. *Contributions to Mineralogy and Petrology*, **71**, 45–60.

HAILWOOD, E. A., TARLING, D. H., MITCHELL, J. G. & LØVLIE, R. 1973. Preliminary observations on the palaeomagnetic and radiogenic ages of the Tertiary basalt sequence of Scoresby Sund, East Greenland. *Grønlands Geologiske Undersøgelse Rapport*, **58**, 43–47.

——, BOCK, W., COSTA, L. I., DUPEUBLE, P. A., MULLER, C. & SCHNITKER, D. 1979. Chronology and biostratigraphy of north-east Atlantic sediments, DSDP Leg 48. *In:* MONTADERT, L., ROBERTS, D. G., *et al. Initial Reports of the Deep Sea Drilling Project.* US Government Printing Office, Washington, **48**, 1119–1141.

HAMILTON, E. I. 1966. The isotopic composition of lead in igneous rocks. *Earth and Planetary Science Letters*, **1**, 30–37.

HAYATSU, A. 1972. On the basic assumptions in K-Ar dating methods. *Comments on Earth Sciences; Geophysics*, **3**, 69–75.

—— & CARMICHAEL, C. M. 1970. K-Ar isochron method and initial argon ratios. *Earth and Planetary Science Letters*, **8**, 71–76.

—— & —— 1977. Removal of atmospheric argon contamination and the use and misuse of the K-Ar isochron method. *Canadian Journal of Earth Sciences*, **14**, 337–345.

HORNE, R. R. & MACINTYRE, R. M. 1975. Apparent age and significance of Tertiary dykes in the Dingle Peninsula. S.W. Ireland. *The Scientific Proceedings of the Royal Dublin Society (Series A)*, 5(18), 293–299.

KAPP, H. 1960. Zur Petrologie der subvulkane zwischen Mesters Vig und Antarctic Havn, Ost-Grønland. *Meddelelser øm Grønland*, **153(2)**.

KEMPE, D. R. C., DEER, W. A. & WAGER, L. R. 1970. Geological investigations in East Greenland, Pt. VIII. The petrology of the Kangerdlugssuaq alkaline intrusion. *Meddelelser øm Grønland*, **190(2)**.

KRISTOFFERSEN, Y. & TALWANI, M. 1977. Extinct triple junction south of Greenland and the Tertiary motion of Greenland relative to North America. *Bulletin of the Geological Society of America*, **88**, 1037–1049.

LARSEN, H. C. 1978. Offshore continuation of East Greenland dyke swarm and North Atlantic Ocean formation. *Nature*, **274**, 220–223.

LARSEN, L. M. & WATT, W. S. 1985. Episodic volcanism during break-up of the North Atlantic: evidence from the East Greenland plateau basalts. *Earth and Planetary Science Letters*, **73**, 105–116.

LOWRIE, W. & ALVAREZ, W. 1981. One hundred million years of geomagnetic polarity history. *Geology*, **9**, 392–397.

MACINTYRE, R. M. & HAMILTON, P. J. 1984. Isotopic geochemistry of lavas from sites 553 and 555. *In:* ROBERTS, D. G., SCHNITKER, D. *et al. Initial Reports of the Deep Sea Drilling Project.* US Government Printing Office, Washington, **81**, pp. 775–81.

MORTON, A. C., DIXON, J. E., FITTON, J. G., MACINTYRE, R. M., SMYTHE, D. K. & TAYLOR, P. N. 1988. Early Tertiary volcanic rocks in Well 163/6-1A, Rockall Trough. *In:* MORTON, A. C. & PARSON, L. M. (eds) *Early Tertiary Volcanism and the Opening of the NE Atlantic.* Geological Society of London Special Publication, **39**, pp. 293–308.

MYERS, J. S. 1980. Structure of the coastal dyke swarm and associated plutonic intrusions of East Greenland. *Earth and Planetary Science Letters*, **46**, 407–418.

NESS, G., LEVI, S. & COUCH, R. 1980. Marine magnetic anomaly time-scales for the Cenozoic and Late Cretaceous: a precis, critique and synthesis. *Reviews of Geophysics and Space Physics*, **18**, 753–770.

NIELSEN, T. F. D. 1975. Possible mechanism of continental break up in the North Atlantic. *Nature*, **253**, 182–184.

—— 1978. The Tertiary dyke swarms of the Kangerdlugssuaq area, East Greenland. *Contributions to Mineralogy and Petrology*, **67**, 63–78.

—— & BROOKS, C. K. 1981. The East Greenland rifted continental margin: a reinterpretation of the coastal flexure. *Journal of the Geological Society of London*, **138**, 559–568.

——, SOPER, N. J., BROOKS, C. K., FALLER, A. M., HIGGINS, A. C. & MATTHEWS, D. W. 1981. The pre-basaltic sediments and the Lower Basalts at Kangerdlugssuaq, East Greenland: their stratigraphy, lithology, palaeomagnetism and petrology. *Meddelelser øm Grønland, Geoscience*, **6**.

NOBLE, R. H. 1978. *The Tertiary geochronology of East Greenland.* PhD thesis, University of Aberdeen.

NOE-NYGAARD, A. 1976. Tertiary igneous rocks between Shannon and Scoresby Sund, East Greenland. *In:* ESCHER, A. & WATT, W. S. (eds) *Geology of Greenland.* Grønlands Geologiske Undersøgelse, København, pp. 386–402.

ODIN, G. S. & MITCHELL, J. G. 1983. Dating of the Palaeocene–Eocene Blosseville Group basalts, Scoresby Sund, East Greenland: a review. *Newsletters on Stratigraphy*, **12(2)**, 112–121.

—— & CURRY, D., 1985. The Palaeogene time-scale: radiometric dating versus magnetostratigraphic approach—a review. *Journal of the Geological Society of London*, **142**, 1179–1188.

——, —— & HUNZIKER, J. C. 1978. Radiometric dates from NW European glauconites and the Palaeogene time-scale. *Journal of the Geological Society of London*, **135**, 481–497.

PANKHURST, R. J., BECKINSALE, R. D. & BROOKS, C. K.

1976. Strontium and oxygen isotope evidence relating to the petrogenesis of the Kangerdlugssuaq alkaline intrusion, East Greenland. *Contributions to Mineralogy and Petrology*, **54**, 17–42.

REX, D. C., GLEDHILL, A. R., BRIDGWATER, D. & MYERS, J. S. 1978a. A Rb-Sr whole rock age of 55 ± 7 My from the Nualik plutonic centre, East Greenland. *Grønlands Geologiske Undersøgelse Rapport*, **95**, 102–105.

——, ——, BROOKS, C. K. & STEENFELT, A. 1978b. Radiometric ages of Tertiary salic intrusions near Kong Oscars Fjord, East Greenland. *Grønlands Geologiske Undersøgelse Rapport*, **95**, 106–109.

SCHAUB, H. B. 1942. Zur geologie der Traill Insel (nordost Grønland). *Eclogae Geologicae Helvetiae*, **35**, 1–54.

SCHNITKER, D. 1979. Chronology and biostratigraphy of northeast Atlantic sediments, DSDP Leg 48. *In:* MONTADERT, L., ROBERTS, D. G. *et al. Initial Reports of the Deep Sea Drilling Project.* US Government Printing Office, Washington, **48**, 1119–1141.

SEEMAN, U., 1984. Tertiary intrusives on the Atlantic continental margin off south-west Ireland. *Irish Journal of Earth Sciences*, **6**, 229–236.

SHEPPARD, S. M. F., BROWN, P. E. & CHAMBERS, A. D. 1977. The Lilloise intrusion, East Greenland—hydrogen isotope evidence for the efflux of magmatic water into the contact metamorphic aureole. *Contributions to Mineralogy and Petrology*, **63**, 129–147.

SOPER, N. J., DOWNIE, C., HIGGINS, A. C. & COSTA, L. I. 1976a. Biostratigraphic ages of Tertiary basalts on the East Greenland continental margin and their relationship to plate separation in the north east Atlantic. *Earth and Planetary Science Letters*, **32**, 149–157.

——, HIGGINS, A. C., DOWNIE, C., MATTHEWS, D. W. & BROWN, P. E. 1976b. Late Cretaceous–early Tertiary stratigraphy of the Kangerdlugssuaq area, East Greenland, and the age of opening of the

north-east Atlantic. *Journal of the Geological Society of London*, **132**, 85–104.

STEIGER, R. H. & JÄGER, E. 1977. Subcommission on geochronology: Convention on the use of decay constants in geo and cosmochronology. *Earth and Planetary Science Letters*, **36**, 359–362.

TALWANI, I. & ELDHOLM, O. 1977. Evolution of the Norwegian–Greenland Sea. *Bulletin of the Geological Society of America*, **88**, 969–999.

TARLING, D. H. & GALE, N. H. 1968. Isotopic dating and palaeomagnetic polarity in the Faeroe Islands. *Nature*, **218**, 1043–1044.

——, HAILWOOD, E. A. & LØVLIE, R. A palaeomagnetic study of the lower Tertiary Lavas in E Greenland and comparison with other lower Tertiary observations in the northern Atlantic. *In:* MORTON, A. C. & PARSON, L. M. (eds) *Early Tertiary Volcanism and the Opening of the NE Atlantic.* Geological Society of London Special Publications, **39**, pp. 215–224.

TATE, M. P. & DOBSON, M. R. Syn- and post-rift igneous activity in the Porcupine Seabight Basin and adjacent continental margin W of Ireland. *In:* MORTON, A. C. & PARSON, L. M. (eds) *Early Tertiary Volcanism and the Opening of the NE Atlantic.* Geological Society of London Special Publications, **39**, pp. 309–334.

UPTON, B. G. J., EMELEUS, C. H., BECKINSALE, R. D. & MACINTYRE, R. M. 1984. Myggbukta and Kap Broer Ruys: the most northerly of the East Greenland Tertiary igneous centres (?). *Mineralogical Magazine*, **48**, 323–43.

WAGER, L. R. 1947. Geological investigations in East Greenland. IV. The stratigraphy and tectonics of Knud Rasmussens Land and the Kangerdlugssuaq region. *Meddelelser øm Grønland*, **105(3)**.

—— 1965. The form and internal structure of the alkaline Kangerdlugssuaq intrusion, East Greenland. *Mineralogical Magazine*, **34**, 487–497.

YORK, D. 1969. Least squares fitting of a straight line with correlated errors. *Earth and Planetary Science Letters*, **5**, 320–324.

R. H. NOBLE & P. E. BROWN, Department of Geology and Mineralogy, Marischal College, University of Aberdeen, Broad Street, Aberdeen AB9 1AS, UK.

R. M. MACINTYRE, Scottish Universities Research and Reactor Centre, East Kilbride, Glasgow G75 0QU, UK.

A palaeomagnetic study of lower Tertiary lavas in E Greenland and comparison with other lower Tertiary observations in the northern Atlantic

D. H. Tarling, E. A. Hailwood & R. Løvlie

SUMMARY: Rock samples from five separate profiles, totalling 2.3 km, in the E Greenland lower Tertiary plateau basalts exhibit well-defined magnetic vectors and stability to both alternating magnetic fields and thermal demagnetization. Stability is confirmed by baked-contact evidence and the preservation of a secular variation record. The remanences are all of reversed polarity and the age relationships suggest that these were acquired during chron C24R, immediately prior to the formation of oceanic magnetic anomaly 24. Radiometric age determinations from the N Atlantic Igneous Province suggest that the plateau basalts of E Greenland may be of similar age to those of the Faeroes, but differ in age from those of the British Tertiary Igneous Province. Palaeomagnetic pole positions for each province suggest an age similarity between the Faeroes and British plateau basalts, while their age relationship to the E Greenland basalts is not clear. The pole positions generally confirm the published continental reconstructions for this period. The simplest polarity solution is for the magnetization of the British and Faeroes basalts to be predominantly associated with chrons C24R and C26R, with only chron C24R being represented in E Greenland. However, this model is dependent on the radiometric constraints on the Palaeocene/Eocene boundary.

Lower Tertiary igneous rocks from the N Atlantic region were studied in the very early days of palaeomagnetism (Mercanton 1926), mostly to establish the global existence of reversely polarized rocks. Numerous studies have been undertaken subsequently (see below). The palaeomagnetic results presented here are based on a sample collection obtained during a Geological Survey of Greenland expedition to Scoresby Sund in 1972, and will be compared with other data from the N Atlantic Tertiary Igneous Province. There were three main objectives of this work: (1) to determine an improved early Tertiary palaeomagnetic pole for the E Greenland basalts; (2) to establish the magnetostratigraphic zonation of this formation for chronological and tectonic purposes; and (3) to investigate if such lava sequences could be correlated between outcrops using the geomagnetic secular variation record. The evidence for secular variation records, magnetic stability and the composition of the carriers of the remanence will only be summarized here as they will be discussed more fully elsewhere (Hailwood et al., in prep.). Any tectonic interpretation of such data requires an evaluation of the age of magnetization of the rocks involved; only coeval palaeomagnetic poles can be utilized in the development and evaluation of palaeogeographic reconstructions. The E Greenland palaeomagnetic data will therefore be discussed first, followed by an assessment of the ages of different parts of the N Atlantic Igneous Province. Finally, reconstructions based on the geometric matching of bathymetric features and linear oceanic magnetic anomalies will be tested against the palaeomagnetic data.

Palaeomagnetic properties of the E Greenland plateau basalts

The sampling was undertaken on five partially overlapping plateau basalt sequences S and W of Scoresby Sund (Fig. 1), each of which was 500–800 m thick. The total thickness sampled represents some 2.3 km. Profiles 1–3, located in Gåsefjord (the Rævbræ, Gåseland and NE Sydbræ profiles respectively), were near the base of the plateau basalts, and basement rocks were exposed at the bottom of profiles 2 and 3. Profiles 4 and 5 (SE Sydbræ and Skrænterne) were towards the top of the sequence, with profile 5 thought to include the youngest exposed lavas in this region. In these areas the regional dip is small, about 1° to the SE, being outside the coastal flexure which affects the Tertiary volcanics further S. The rocks sampled were remarkably fresh and unaltered, with very few intrusive dykes. Between three and six sun-compass-oriented drill cores were collected from each of a total of 166 separate lava flows. The thickness of the flows sampled varied between a few centi-

From MORTON, A. C. & PARSON, L. M. (eds), 1988, *Early Tertiary Volcanism and the Opening of the NE Atlantic*, Geological Society Special Publication No. 39, pp. 215–224.

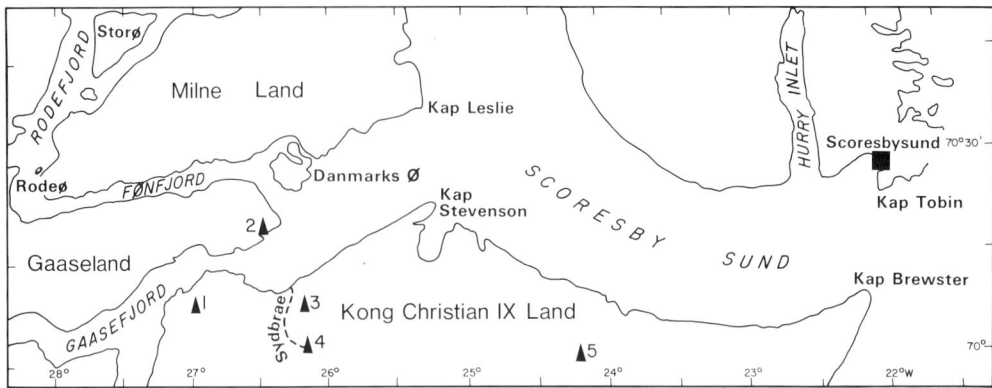

FIG. 1. E Greenland sampling profiles: **1** Rævbræ; **2** Gåseland; **3** NE Sydbræ; **4** SW Sydbræ; **5** Skrænterne.

metres and tens of metres. The thicker flows were sampled near their base and/or near their upper surface. Two specimens were cut from each core, one for thermal demagnetization and one for alternating field demagnetization.

Thermomagnetic analysis of small pieces from the bottom of the cores suggested that high-Ti titanomagnetites are dominant, but low-Ti titanomagnetites are commonly present, especially in the more basal sections. Alternating field demagnetization was undertaken on at least one specimen per site, in fields incremented by 5 mT up to 30 mT, and some one-third of these specimens were further demagnetized in 10 or 20 mT steps up to 90 mT peak field. About 50% of the specimens showed a single reversed component of remanence throughout the full treatment range and a further 40% showed reversed single component remanences in applied fields greater than 5–10 mT. Only a few samples showed lower stability and less well-defined vectors, but even these indicated a consistent reversed component during demagnetization at fields between 10 and 40 mT. Thermal demagnetization identified very similar remanences, with 78% of the specimens showing a well-defined vector during treatment in the temperature range 300–400°C. Instability was largely confined to specimens characterized by Curie points in the range 300–400°C, which probably contain low-Ti titanomagnetite. These specimens frequently showed viscous effects and irregular directional behaviour.

Comparison of the stable reversed vectors isolated by both thermal and alternating magnetic field demagnetization indicated that the components were statistically identical in virtually all cores. The stable magnetic vectors isolated during partial demagnetization are considered to be primary, i.e. to have been acquired at, or shortly

after, emplacement. This conclusion is based mainly on the observation that the upper parts of several flows (the 'entablature' parts) show different directions to those isolated in the centre ('collonade') or base of the same flow, yet identical directions of magnetization could be found in the basal part of the overlying flow. This 'contact' test clearly suggests that the magnetization in these flows can be associated with the emplacement of specific lava flows and therefore any regional magnetic effects, such as those arising from burial metamorphism, must be very small. However, it also indicates that care must be taken in interpreting the magnetization of thin flows as they could have acquired a uniform magnetization throughout, due to baking by an overlying flow. Further evidence that the stable remanent vector is primary is that the within-site scatter is small (mostly <6°), particularly in the case of thin lavas, and if samples from the lower and upper margins of thick lavas are considered as separate spot readings of the past geomagnetic field direction, i.e. to have been magnetized at different times. The between-site scatter was high (15–18°), comparable with that expected from secular variation amplitudes at moderately high latitudes. On this basis, the stable remanences are considered to represent a discontinuous record of secular variation and this is confirmed by the observation that some lavas can be correlated by matching their directions of remanence (Hailwood *et al.*, in prep.). Therefore, the mean directions for each of the five sequences (Table 1) are taken to represent the time-averaged geomagnetic field direction during the time of eruption of the sequences and their consistency does not indicate a later regional remagnetization. Thus, although there are complexities in the magnetic mineralogy and some evidence that low-temper-

TABLE 1. *Mean directions of magnetization and palaeomagnetic pole positions for the Scoresby Sund profiles*

Profile	N	Decl	Incl	α_{95}	South Pole Lat. (N)	South Pole Long. (E)
1 Rævbræ	31	159.0	−66.0	5	68.5	188.5
2 Gåseland	23	161.0	−68.0	7	71.5	189.5
3 NE Sydbræ	36	176.5	−69.0	3	73.5	161.5
4 SE Sydbræ	24	171.2	−64.9	6	68.3	167.9
5 Skrænterne	11	156.8	−62.7	9	64.3	193.8
Mean	5	166.5	−66.3	4	69.7	181.4 ($A_{95} = 5.8°$)

N is the number of flows sampled, α_{95} and A_{95} are, respectively, the radius of a cone of 95% confidence for directions and pole positions.

ature oxidation may have played a significant role in the magnetization of some of the lower lavas, the evidence strongly suggests that the magnetization of these lavas is effectively primary. This is either of thermal origin or associated with deuteric alterations which were completed very shortly after extrusion (normally prior to the emplacement of later flows, but occasionally associated with the emplacement of thick immediately supra-adjacent flows). The mean directions for each of the profiles (see Table 1) can therefore be used as the basis for plate tectonic reconstructions relating to the time of extrusion.

The Lower Tertiary Igneous Province in the N Atlantic

The flat-lying basaltic sequences in W and E Greenland, Iceland, the Faeroes and Britain were once thought to have been emplaced more or less simultaneously and to comprise component parts of a single igneous province (Holmes 1918). Subsequently, a Danian age was demonstrated for the W Greenland basalts (Clarke & Pedersen 1976) and a Miocene age for the oldest Icelandic basalts (Moorbath et al. 1968). Flood basalts and associated rocks in the remaining areas (Britain, the Faeroes and E Greenland) are of early Tertiary age and these areas together comprise the Lower Tertiary Brito–Arctic Igneous Province.

In the British Tertiary Igneous Province, major flood basalts occur at three main localities (Mull, Skye and the Antrim Plateau) but the palaeontological age of sediments associated with these, and with other lower Tertiary igneous formations, is still only poorly constrained as probably Palaeocene or Eocene. Radiometric age and palaeomagnetic polarity data for these regions are fully discussed and referenced elsewhere (Mussett et al., 1988). The conclusions of these authors are that both Mull and Skye show a R/N/R polarity sequence with Mull radiometric ages lying in the range 57–60 Ma and those for Skye in the range 54–59 Ma. Arran also shows a R/N/R polarity sequence and radiometric ages for this area lie in the range 58–60 Ma. The major plateau basalts of Antrim also formed mostly between 58 and 60 Ma, but these, as all other Tertiary lavas in Britain and most intrusives, are entirely of reversed polarity. Both the youngest and oldest accepted radiometric ages in the British Tertiary Igneous Province are for the reversed rocks on Eigg (52–63 Ma). Mussett et al. (1988) also appraise the relationship of the polarity sequences with the magnetostratigraphical time-scales of Berggren et al. (1985) and Harland et al. (1982) (Fig. 2). They conclude that the reversed polarity lavas, if formed primarily during chron C26R, would be more consistent with the Harland et al. (1982) time-scale, while their eruption during chron 24R would be somewhat less consistent with the Berggren et al. (1985) time-scale.

The Faeroe Islands basalts form three stratigraphic sequences of which the Middle and Upper Series are separated from the Lower Series by a coal-bearing sequence which is now considered to be of Late Palaeocene age (Lund 1983; Waagstein 1988; Andersen 1988). Twelve of the 19 radiometric age determinations from the complete sequence gave concordant ages around 55 Ma, but the total range is between 50 and 66 Ma (corrected to new constants), with the older ages occurring in the younger Middle and Upper Series (Tarling & Gale 1968). This suggests that the Lower Series may have suffered argon loss, possibly during burial and metasomatism. How-

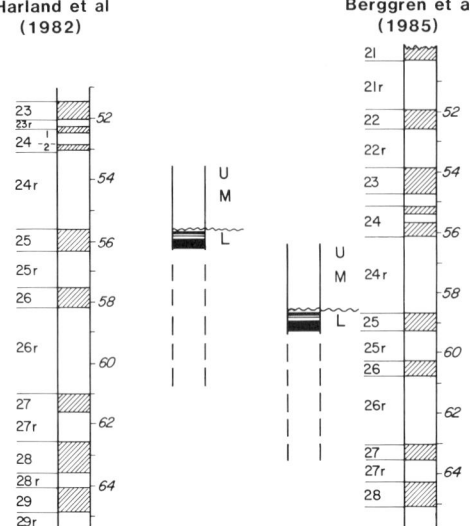

FIG. 2. The reversal sequence in the Faeroe Islands, based on Abrahamsen (1967), Tarling (1970a), Tarling & Gale (1968) and Abrahamsen et al. (1984). Possible correlations with polarity time-scales of Harland et al. (1982) and Berggren et al. (1985) are indicated. Mussett et al. (1988) provide comparable details for the British Tertiary Igneous Province.

ever, the palaeomagnetic record is consistent with that expected from secular variation and the polarity zonations are clear, suggesting that there has been little or no later remagnetization. An alternative explanation is that the Upper Series contains excess radiogenic argon, although this has not been substantiated. The entire sequence of lavas in the Upper and Middle Series is reversely magnetized (Abrahamsen 1967; Tarling 1970a; Tarling & Gale 1968), but the Lower Series contains a thin normal-polarity zone (six flows), shortly below the Coal Horizon, overlying reversed (at least 20 flows) and normal (> 12 flows) zones (Fig. 2). A 2.2 km core drilled through the Lower Series was sampled only at five levels for palaeomagnetic analysis, but all of these were of reversed polarity, suggesting a further very thick reversed zone within the unexposed part of the Lower Series (Schoenharting & Abrahamsen 1984; Abrahamsen et al. 1984).

Even if eruption rates were fast, the Middle and Upper Series must have been formed during a prolonged reversed period, as must the unexposed Lower Series if the five sampled levels are typical of the entire 2 km drilled. Abrahamsen et al. (1984) correlated these two major zones with chrons C22R and C24R on the basis of assumed similarity to the E Greenland basalts. If the mean

radiometric age of 54–55 Ma is taken as correct, then the Upper and Middle Series reversed zone would represent chron C24R on the Harland et al. (1982) scale, and that of the Lower Series chron C26R. If the Upper and Middle Series are regarded as having ages of 58–60 Ma, i.e. assuming argon loss, then their polarity would correspond to chron C26R on the Harland et al. (1982) scale, raising problems of the assignment of the lower polarity-reversed zone. On the Berggren et al. (1985) scale the reversely magnetized parts of the Upper and Middle Series are likely to represent chron C24R, and that of the Lower Series, chron C26R (see Fig. 2). The situation cannot yet be considered to be resolved.

The E Greenland lavas (only those S of Scoresby Sund are considered here) appear to be entirely of reversed polarity (Tarling 1967; Hailwood et al. 1974; Faller 1975; Faller & Soper 1979; Nielsen et al. 1981) and have a well-defined Sparnacian (Palaeocene/Eocene) age based on the occurrence of dinoflagellates correlated to nannoplankton zones NP9 to NP10 in associated sediments (Soper et al. 1976; Nielsen et al. 1981). This age is similar to that of sediments immediately overlying seafloor-spreading anomaly 24 (Maxwell et al. 1970; Sclater et al. 1974; Tarling & Mitchell 1976). In the Kangerdlugssuaq area to the S, radiometric determinations are varied, reflecting difficulties caused by argon diffusion (Noble et al., 1988). Beckinsale et al. (1970) concluded that the extrusives were erupted between 55 and 60 Ma. However the intrusives show a bimodal radiometric age distribution. Remarkably fresh lavas from the five profiles reported in this paper have been radiometrically dated as much younger, between 47 and 52 Ma, with a mean of 50.0 ± 1.4 Ma (Hailwood et al. 1973; Odin & Mitchell 1983). Unlike the structurally complicated area to the S, the palaeomagnetic evidence from the Scoresby Sund area suggests that there has been no remagnetization and hence no resetting of the radiogenic argon content. However, the evidence for some older ages, based on the Rb/Sr method, for intrusives in the S (Gill et al., 1988) suggests a possibly older age for the basalts in this area than that observed for the Scoresby Sund basalts.

The uniform reverse magnetic polarity indicates that the entire lava sequence in this area was emplaced during a single polarity chron, yet the great thickness (ca. 9 km) suggests that a significant interval of time, probably $> 10^6$ years, was required. The longest early Tertiary reverse polarity chrons were C24R and C26R. Since the biostratigraphic age of these basalts is the same as that for the sediments immediately overlying anomaly 24 (Maxwell et al. 1970; Sclater et al.

1974), it is reasonable to suggest that the basalts were erupted immediately prior to the formation of marine anomaly 24, i.e. during chron C24R. The radiometric age of this chron is, however, still in dispute (Tarling & Mitchell 1976; Harland *et al.* 1982; Tarling 1983; Berggren *et al.* 1985; Odin & Curry 1985; Curry & Odin 1982; Snelling 1985; Odin 1986) and this is critical to assessing the relationships within the component parts of the N Atlantic Igneous Province (see below).

In summary, the E Greenland basalts are well dated biostratigraphically and a similar age is indicated for at least the Lower Series basalts in the Faeroes. Age-diagnostic biostratigraphic data are not available for the British Tertiary Igneous Province and there is some uncertainty concerning many of the radiometric age determinations for this province, but values lie predominantly in the range 58–60 Ma. This suggests that these basalts are probably older than those of both the Faeroes (ca. 55–56 Ma) and E Greenland (48–58 Ma). However, within the uncertainties of radiometric age determinations, the possibility still cannot be completely excluded that igneous activity in these three areas was essentially synchronous. It is thus important to utilize other available evidence, such as that from palaeomagnetic pole positions, to help establish age relationships between these areas.

Palaeomagnetic poles and their implications

If palaeomagnetic poles are in agreement for a particular palaeogeographic reconstruction, then this implies that, within the resolution of the apparent polar wander (APW) curves, the poles, and hence the magnetization of the rocks involved, are coeval. However, such a test also requires that: (1) the effects of secular variation of the geomagnetic field have been averaged out; (2) there is a valid model for the average geomagnetic field at that time; and (3) there is a valid palaeogeographic reconstruction. As lavas and shallow dykes generally represent 'spot readings' in the time of the geomagnetic field, usually reflecting the duration of the geomagnetic field at that location over some 1–20 years, a considerable number of observations, spanning a sufficient interval of time, are required in order to properly average out the effects of geomagnetic secular variations. Tarling (1970b) determined that at least 20 palaeomagnetic sites (i.e. discrete sampling levels), evenly and well-distributed through time, are required to result in a mean direction with a precision of 5°. Many published studies involve less than this minimum requirement and the averaging of secular variation is further complicated by the uncertainty in both eruption and secular variation rates. However, deuteric alterations within dykes, and possibly also within thick lavas, may persist for a decade or more after emplacement and thereby assist in averaging out such effects. On this basis, 20 time-independent observations appears to be a reasonable minimum figure for averaging out most secular variation effects, providing that they span a total interval of some 10^3–10^4 years. The adequacy of the average geomagnetic field model is harder to evaluate, but the axial geocentric dipole is generally used in view of the difficulty of defining a more realistic model for specific periods without incorporating further assumptions. In the present analysis, this model will be used, although there is some evidence that better agreement may occur for an offset dipole model (Hailwood 1976). Finally, the validity of any reconstruction depends, fundamentally, on whether it makes geological sense. Assuming, therefore, that these criteria can be satisfied, it is appropriate to evaluate the age differences which should be detectable on such a basis. As the palaeomagnetic pole had an apparent motion relative to the N American plate of some 0.3°/My during much of the Phanerozoic, and as this is similar to that estimated for Europe (Tarling 1983), an age difference of some 10–15 My should correspond to an average polar separation of some 3–5°. On this basis, it can be argued that the potential age-resolution offered by comparison of APW curves is comparable to the total uncertainty in currently available radiometric age determinations for the Greenland and European Igneous Provinces. It is possible that the apparent polar motion was actually faster at this time, allowing better definition of such age differences.

Each of the five E Greenland profiles reported in the present paper spans a sufficient number of individual sites that each mean profile direction should average out secular variation effects. However, the individual mean directions are different from each other, the two Sydbræ profile palaeomagnetic pole positions lying more easterly than the other three (see Table 1). Such a discrepancy could indicate a local clockwise rotation of the Sydbræ area by some 18°, but this is regarded as unlikely. Similarly the discrepancy cannot be due to a trend in the pole positions because of time differences, since the NE Sydbræ, Gåseland and Rævbræ profiles stratigraphically underly the SW Sydbræ and Skrænterne profiles. As there is no other obvious explanation for the differences, it must be assumed that they are due to incomplete averaging of geomagnetic field secular variation so that the mean of all five profiles is a more realistic estimation of the average geomagnetic field direction at this locality during the time of extrusion of these lavas.

The mean palaeomagnetic pole for the Faeroe Islands is based on a total of 303 lavas (1830 cores) spanning the 3 km exposed sequence (Tarling 1970a). Thus, the mean pole primarily relates to the uppermost polarity zone and should have adequately averaged out secular variation effects. This pole was originally considered inconsistent with the then available British Tertiary data (Tarling 1967), but there have been significant improvements in the reliability of the British data since then. The British localities generally show considerable variation in their mean directions (Table 2). However, the pole positions for the Antrim and Skye lavas are almost identical, and not dissimilar from those for Mull. Individual results from other British localities may represent insufficient time to average out secular variation effects, but the

combined mean direction should be comparable to those for the major extrusives of Antrim and Skye. However, the mean directions for these other localities generally show a steeper inclination, so that the corresponding pole positions are closer to the present geographic pole than those of the major extrusives (Fig. 3). As with the palaeomagnetic data for Greenland, neither tectonic rotations nor age differences appear to provide an adequate explanation for the observed scatter. However, the Faeroes pole is consistent with all three mean British poles (see Fig. 3a, b) and this now suggests a similarity in age.

Comparison of the E Greenland poles for anomaly 24 times, and for somewhat earlier, pre-rift times, clearly depends on the validity of the reconstruction. It must be emphasized that presently published reconstructions are compromises as it has not yet proved possible to reconcile completely the bathymetric, magnetic anomaly and fracture zone fits. Apparently consistent reconstructions can be obtained for the N Atlantic N of Iceland, and for S of Iceland, but such

TABLE 2. *British and European Lower Tertiary Igneous Province directions and South Pole locations*

Locality	Decl.	Incl.	α_{95}	N	E	Reference
				South Pole		
The British Tertiary Igneous Province						
Antrim[1]	184.7	−54.3	5	69.6	162.9	Wilson 1970
	198.5	−59.7	10	60.0	175.0	Løvlie *et al.* 1972
Skye[1]						
lavas	183.4	−58.3	2	71.5	165.2	Wilson *et al.* 1972
dykes[2]	186.0	−60.0	5	82.3	162.2	Wilson *et al.* 1982
Mull: lavas and dykes						
normal	4.4	71.1	4	81.9	84.8	Mussett *et al.* 1980
R steep	206.2	−51.7	12	59.9	126.2	Mussett *et al.* 1980
R shallow	182.2	−30.7	9	50.0	170.9	Mussett *et al.* 1980
combined[1]	193.0	−58.8	9	71.0	141.9	Mussett *et al.* 1980
Rhum & Canna[2]	196.5	−57.9	9	68.7	136.0	Dagley & Mussett 1981
Arran dykes[2]	179.0	−65.2	10	81.7	179.8	Dagley *et al.* 1978
Ardnamurchan[2]	180.0	−63.0	3	77.0	175.0	Dagley *et al.* 1984
Muck lavas[3]	176.3	−63.8	3	78.4	186.9	Dagley & Mussett 1986
Eigg lavas[3]	179.7	−59.3	4	73.2	174.5	Dagley & Mussett 1986
Lundy[2]	183.6	−62.7	2	84.0	150.7	Mussett *et al.* 1976
UK dykes[2]	169.5	−61.4	9	75.2	225.9	Dagley 1969
	Mean BTIP pole		5	76.9	166.7	1 localities ([1, 2, 3])
			5	73.2	164.3	5 lava localities ([1, 2])
			7	71.0	156.8	3 main lava localities ([1])
Faeroes[1]	185.0	−66.5	2	77.0	161.0	Tarling 1970a
	176.0	−69.0	6	80.0	159.0	Abrahamsen 1967
	171.9	−72.2	4	84.0	218.0	Løvlie & Kvingedahl 1975
	175.1	−53.9	2	62.0	182.4	Løvlie & Kvingedahl 1975

N is the number of sites sampled, A_{95} is the radius of a cone of 95% confidence. See Fig. 3 for a definition of superscripts 1, 2 and 3.

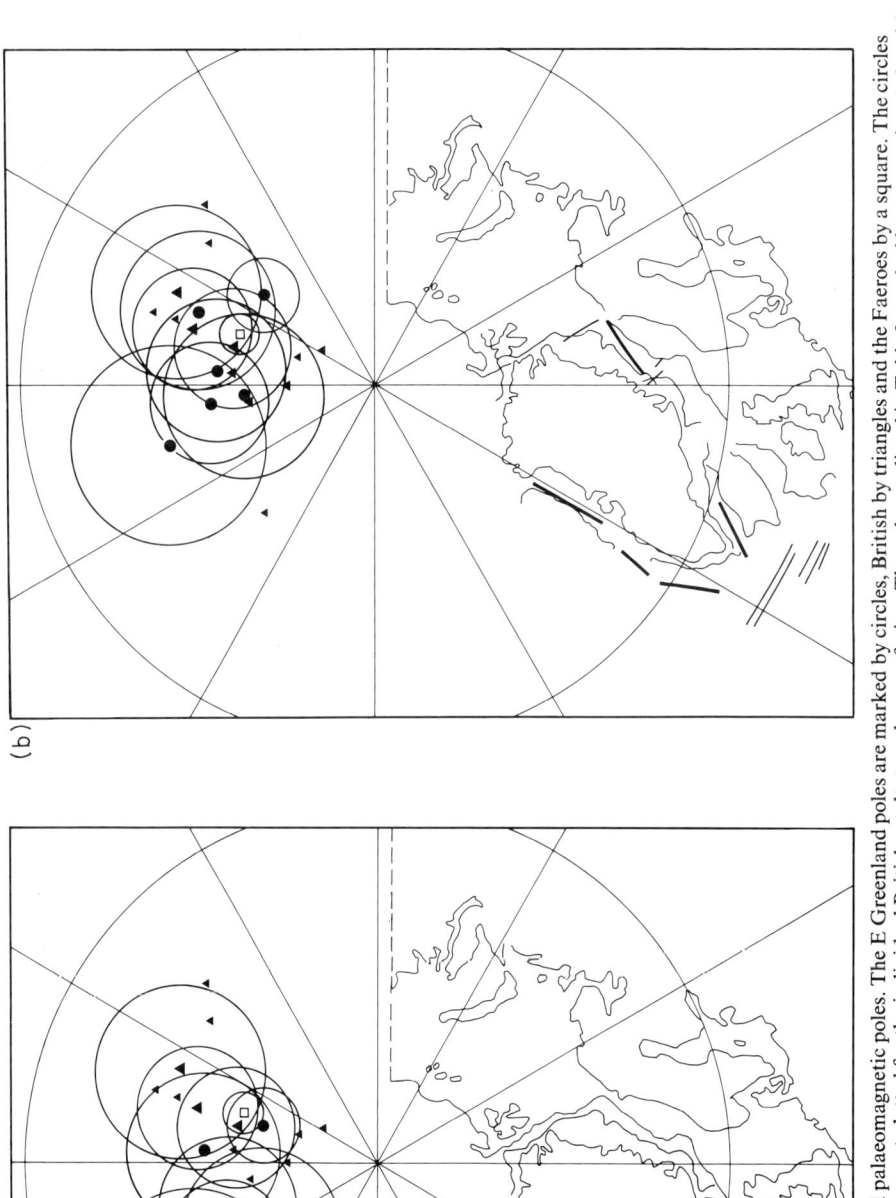

FIG. 3. N Atlantic reconstruction and palaeomagnetic poles. The E Greenland poles are marked by circles, British by triangles and the Faeroes by a square. The circles represent 95% confidence limits, but are not plotted for the individual British results to reduce confusion. The poles are listed in Table 2, marked by superscripts 1, 2 and 3. (a) *Pre-Tertiary configuration.* The reconstruction has been based on a rotation towards Europe of the Rockall–Faeroes area by −4.2° about an Euler pole 65.8°N, 60.3°W, and of Greenland by 16.8° about a pole of 67.7°N, 103.0°E. The Faeroe Islands are indicated for location purposes only. (b) *Anomaly 24 times.* Rockall and the Faeroes are in their present positions relative to Europe, and Greenland has been rotated by 9.9° about a Euler pole at 43.2°N, 121.9°E. The solid lines indicate the location of anomaly 24, and the narrow lines mark major fracture zones.

reconstructions are mutually inconsistent (with a discrepancy of 100–200 km). However, such discrepancies are small in comparison with the uncertainties in the extant palaeomagnetic data. The reconstruction for the period prior to opening (see Fig. 3a) is similar to that of Roberts (1975) and was used by Soper *et al.* (1976). Greenland is closed to Europe using an Euler pole and rotation of 67.7°N, 103.0°E, 16.8°. As a part of this reconstruction, the Rockall–Faeroes area was rotated to Europe by 7.5° about a pole at 62.6°N, 3.2°E. The evidence for compression of the Sverdrup Basin may require some further closure of Greenland against Svalbard, possibly by 10° about a pole at about 80°N, 100°W, but this is essentially a subjective estimate and has not been used in this reconstruction. The best compromise for matching anomaly 24 both N and S of the Iceland Ridge (see Fig. 3b) is based on an Euler pole for rotating Greenland relative to Europe of 43.2°N, 121.9°E, with a rotation of 9.9°. While this rotation matches anomaly 24 within the N Atlantic, it causes a mismatch of the same anomaly within the Labrador Sea and an overlap between Svalbard and northern Greenland. However, any reduction of this overlap would cause unacceptable mismatches along the Charlie Gibbs fracture zone. Similarly, the locations of the Greenland and Senja fracture zones and E and W Jan Mayen fracture zones are not quite reconciled (see Fig. 3b).

The E Greenland and European poles appear to match better for the anomaly 24 reconstruction than for the pre-rift model (see Fig. 3), although the overlap of the circles of confidence in both instances suggests that both reconstructions are broadly consistent with the palaeomagnetic data. Nonetheless, the improvement suggests that the extrusives are more likely to have formed immediately prior to anomaly 24 times, rather than much earlier. Thus, on balance, the palaeomagnetic data are more consistent with an age similarity between the igneous activity in the three areas than with evidence for a significant age difference. However, a significant age difference still cannot be ruled out. It is of interest to note that the improvement in the data base over

the last 17 years is now making more realistic evaluations possible and so future palaeomagnetic studies of the N Atlantic Igneous Province should make further contributions to dating the igneous activity in the now-separated parts of this province.

Conclusions

It must be concluded that the existence, or not, of a significant age difference between the lower Tertiary igneous rocks in the separate parts of the N Atlantic Igneous Province still remains to be established. However, it now seems most probable that the lower Tertiary igneous activity in Britain and the Faeroes was more or less contemporaneous, with the majority of the British reversed-polarity lavas being emplaced synchronously with the reversed-polarity Lower Series of the Faeroe Islands. These were probably erupted during chron C26R. The Upper and Middle Series of the Faeroes, other parts of the British Tertiary Igneous Province and the E Greenland basalts, S of Scoresby Sund, are most likely to have been erupted during chron C24R. The radiometric ages of these chrons must, however, be resolved by further radiometric and palaeomagnetic work on rocks which can be directly related to the biostratigraphic time-scale. For this reason, the less-disturbed volcanic sequences near Scoresby Sund still appear to offer the greatest potential for determining this relationship.

ACKNOWLEDGEMENTS: We are particularly grateful to scientists of the Geological Survey of Greenland for their assistance and advice throughout this work, particularly Dr S. Watt. We also particularly thank Dr J. G. Mitchell, of Newcastle University, for numerous discussions on the radiometric observations from this intriguing region, Dr N. Abrahamsen, of Aarhus University, for his assistance on the Faeroes polarity sequence, and Dr N. J. Soper, of Sheffield University, for comments on the biostratigraphic dates of the E Greenland sequences. This article is published with the kind permission of the Director of the Geological Survey of Greenland.

References

ABRAHAMSEN, N. 1967. Some paleomagnetic investigations in the Faeroe Islands. *Meddeleser Danske Geologiska Forhandling*, **17**, 371–354.
——, SCHOENHARTING, G. & HEINESEN, M. 1984. Palaeomagnetism of the Vestmanna core and magnetic age and evolution of the Faeroe Islands. *In*: BERTHELSEN, O., NOE-NYGAARD, A. & RASMUSSEN, J. (eds) *The Deep Drilling Project 1980–1981 in the Faeroe Islands.* Foroya Frodskapferlag, pp. 93–108.

ANDERSEN, M. S. Late Cretaceous and Early Tertiary extension and volcanism around the Faeroe Islands. *In*: MORTON, A. C. & PARSON, L. M. (eds) *Early Tertiary Volcanism and the Opening of the NE Atlantic.* Geological Society of London, Special Publication, **39**, 115–122.

BECKINSALE, R. D., BROOKS, C. K. & REX, D. C. 1970. K-Ar ages for the Tertiary Basalts of East Greenland. *Bulletin of the Geological Society of Denmark*, **20**, 27–37.

BERGGREN, W. A., KENT, D. V. & FLYNN, J. J. 1985. Palaeogene geochronology and chronostratigraphy. *In*: SNELLING, N. J. (ed) *The Chronology of the Geological Record.* Memoir of the Geological Society of London, **10**, pp. 141–186.

CLARKE, D. B. & PEDERSEN, A. K. 1976. Tertiary volcanic province of West Greenland. *In*: ESCHER, A. & WATT, W. S. (eds) *Geology of Greenland.* Geological Survey of Greenland, Copenhagen, pp. 364–385.

CURRY, D. & ODIN, G. S. 1982. Dating of the Palaeogene. *In*: ODIN, G. S. (ed) *Numerical Dating in Stratigraphy.* John Wiley, Chichester, pp. 607–630.

DAGLEY, P. 1969. Palaeomagnetic results from some British Tertiary dykes. *Earth and Planetary Science Letters*, **6**, 349–354.

—— & MUSSETT, A. E. 1981. Palaeomagnetism of the British Tertiary igneous province; Rhum and Canna. *Geophysical Journal of the Royal Astronomical Society*, **65**, 475–491.

—— & —— 1986. Palaeomagnetic and radiometric dating of the British Tertiary igneous province: Muck and Eigg. *Geophysical Journal of the Royal Astronomical Society*, **85**, 221–242.

——, ——, WILSON, R. L. & HALL, L. M. 1978. The British Tertiary Igneous Province: palaeomagnetism of the Arran dykes. *Geophysical Journal of the Royal Astronomical Society*, **54**, 75–79.

——, —— & SKELHORN, R. R. 1984. The palaeomagnetism of the Tertiary igneous complex of Ardnamurchan. *Geophysical Journal of the Royal Astronomical Society*, **79**, 911–922.

FALLER, A. M. 1975. Palaeomagnetism of the oldest Tertiary basalts in the Kangerdlugssuaq area of East Greenland *Meddeleser Danske Geologiska Forhandling*, **24**, 173–178.

—— & SOPER, N. J. 1979. Palaeomagnetic evidence for the origin of the coastal flexure and dyke swarm in central East Greenland. *Journal of the Geological Society of London*, **136**, 737–744.

GILL, R. C. O., NIELSEN, T. F. D., BROOKS, C. K. & INGRAM, G. A. 1988. Tertiary volcanism in the Kangerdlugssuaq region, E. Greenland: trace-element geochemistry of the Lower Basalts and tholeiitic dyke swarms. *In*: MORTON, A. C. & PARSON, L. M. (eds) *Early Tertiary Volcanism and the Opening of the NE Atlantic.* Geological Society of London, Special Publication, **39**, 161–179.

HAILWOOD, E. A. 1976. Configuration of the geomagnetic field in early Tertiary times. *Journal of the Geological Society of London*, **133**, 23–36.

——, TARLING, D. H., MITCHELL, J. G. & LØVLIE, R. 1973. Preliminary observations on the palaeo-magnetism and radiometric ages of the Tertiary basalt sequence of Scoresby Sund, East Greenland. *Grønlands Geologiske Undersogelse Report*, **58**, 43–47.

HALL, J. M., WILSON, R. L. & DAGLEY, P. 1977. A palaeomagnetic study of the Mull lava succession, *Geophysical Journal of the Royal Astronomical Society*, **49**, 499–514.

HARLAND, W. B., COX, A. V., LLEWELLYN, P. G., PICKTON, C. A. G., SMITH, A. G. & WALTERS, R. 1982. *A Geological Time Scale.* Cambridge University Press, 131 pp.

HOLMES, A. 1918. The basaltic rocks of the Arctic region *Mineralogical Magazine*, **18**, 180–223.

LØVLIE, R. & KVINGEDAHL, M. 1975. A palaeomagnetic discordance between the lava sequence and an associated interbasaltic horizon from the Faeroe Islands, *Geophysical Journal of the Royal Astronomical Society*, **40**, 45–54.

——, GIDSKEHAUG, A. & STORETVEDT, K. M. 1972. On the magnetization history of the Northern Irish Basalts, *Geophysical Journal of the Royal Astronomical Society*, **27**, 487–498.

LUND, J. 1983. Biostratigraphy of intrabasaltic coals from the Faeroe Islands. *In*: BOTT, M. H. P., SAXOV, S., TALWANI, M. & THIEDE, J. (eds) *Structure and Development of the Greenland–Scotland Ridge.* Plenum Press, New York, pp. 417–423.

MAXWELL, A. E., VON HERZEN, R. P., ANDREW, J. E., BOYLE, R. E., MILLOW, E. D., HSÜ, K. J., PERCIVAL, S. F. & SAITO, T. 1970. *Initial Reports of the Deep Sea Drilling Project*, U.S. Government Printing Office, Washington, **3**, 806 pp.

MERCANTON, P. L. 1926. Inversion de l'inclinasion magnétique aux ages géologiques. *Terrestrial Magnetism and Atmospheric Electricity*, **31**, 187–190.

MOORBATH, S., SIGURDSSON, H. & GOODWIN, R. 1968. K-Ar ages of the oldest exposed rocks in Iceland. *Earth and Planetary Science Letters*, **4**, 197–205.

MUSSETT, A. E., BROWN, G. C., ECKFORD, M. & CHARLTON, S. R. 1972. The British Tertiary igneous province: K-Ar ages of some dykes and lavas from Mull, Scotland *Geophysical Journal of the Royal Astronomical Society*, **30**, 405–413.

——, DAGLEY, P. & ECKFORD, M. 1976. The British Tertiary Igneous Province: palaeomagnetism and ages of dykes, Lundy Island, Bristol Channel. *Geophysical Journal of the Royal Astronomical Society*, **46**, 595–603.

——, —— & SKELHORN, R. R. 1980. Magnetostratigraphy of the Tertiary igneous succession of Mull, Scotland. *Journal of the Geological Society of London*, **137**, 349–357.

——, ——, —— 1988. Time and duration of activity in the British Tertiary Igneous Province. *In*: MORTON, A. C. & PARSON, L. M. (eds) *Early Tertiary Volcanism and the Opening of the NE Atlantic.* Geological Society of London, Special Publication, **39**, 337–348.

NIELSEN, T. D. F., SOPER, N. J., BROOKS, C. K., FALLER, A. M., HIGGINS, A. C. & MATTHEWS, D. W. 1981. The pre-basaltic sediments and the Lower Basalts at Kangerdlugssuaq, East Greenland: their stratigraphy, lithology, palaeomagnetism and petrology. *Meddeleser om Grønland, Geoscience*, **6**, 25 pp.

NOBLE, R., MACINTYRE, R. M. & BROWN, P. E. 1988. Age constraints on Atlantic evolution—timing of magmatic activity along the E Greenland continental margin. *In*: MORTON, A. C. & PARSON, L. M. (eds) *Early Tertiary Volcanism and the Opening of the NE Atlantic*. Geological Society of London, Special Publication, **39**, 201–214.

ODIN, G. S. 1986. The Palaeogene time scale, geochronological discussion of an interpolated version. *Bulletin of Liaison and Information IGCP Project 196*, **6**, 11–26.

ODIN, G. S. & MITCHELL, J. G. 1983. Dating of the Blosseville group basalts, Scoresby Sund, East Greenland; a review. *Newsletter of Stratigraphy*, **12**, 112–121.

—— & CURRY, D. 1985. The Palaeogene time-scale: radiometric dating versus magnetostratigraphic approach. *Journal of the Geological Society of London*, **142**, 1179–1188.

ROBERTS, D. G. 1975. Marine geology of the Rockall Plateau and Trough. *Philosophical Transactions of the Royal Society London*, **A278**, 447–509.

SCHOENHARTING, G. & ABRAHAMSEN, N. 1984. Magnetic investigations on cores from the Lopra–1 drillhole, Faeroe Island. *In*: BERTHELSEN, O., NOE-NYGAARD, A. & RASMUSSEN, J. (eds) *The Deep Drilling Project 1980–1981 in the Faeroe Islands*, Foroya Frodskapferlag, pp. 109–114.

SCLATER, J. G., JARRARD, R. D., MCGOWRAN, B. & GARTNER, S. 1974. Comparison of the magnetic and biostratigraphic time scales since the late Cretaceous. *In*: VON DER BORCH, C. C., SCLATER, J. G. *et al. Initial Reports of the Deep Sea Drilling Project*, US Government Printing Office, Washington, **22**, pp. 381–386.

SNELLING, N. J. 1985. An interim time-scale. *In*: SNELLING, N. J. (ed) *The Chronology of the Geological Record*. Memoir of the Geological Society of London, **10**, pp. 261–265.

SOPER, N. J., DOWNIE, C., HIGGINS, A. C. & COSTA, L.

I. 1976. Biostratigraphic ages of Tertiary basalts on the East Greenland continental margin and their relationship to plate separation in the Northeast Atlantic. *Earth and Planetary Science Letters*, **32**, 149–157.

TARLING, D. H. 1967. The palaeomagnetic properties of some Tertiary lavas from East Greenland. *Earth and Planetary Science Letters*, **3**, 81–88.

—— 1970a. Palaeomagnetic results from the Faeroes Islands. *In*: RUNCORN, S. K. (ed) *Palaeogeophysics*. Academic Press, London, pp. 193–208.

—— 1970b. Palaeomagnetism and the origin of the Red Sea and Gulf of Aden. *Philosophical Transactions of the Royal Society of London*, **A267**, 219–226.

—— 1983. *Palaeomagnetism*. Chapman & Hall, London.

—— & GALE, N. H. 1968. Isotopic dating and palaeomagnetic polarity in the Faeroe Islands. *Nature*, **218**, 1043–1044.

—— & MITCHELL, J. G. 1976. Revised Cenozoic polarity time scale. *Geology*, **4**, 133–136.

WAAGSTEIN, R. 1988. Structure, composition and age of the Faeroes basalt plateau. *In*: MORTON, A. C. & PARSON, L. M. (eds) *Early Tertiary Volcanism and the Opening of the NE Atlantic*. Geological Society of London, Special Publication, **39**, 225–238.

WENK, E. 1961. Tertiary of Greenland. *In*: RAASCH, G. O. (ed) *Geology of the Arctic*. Toronto University Press, **1**, pp. 278–284.

WILSON, R. L. 1970. Palaeomagnetic stratigraphy of Tertiary lavas from Northern Ireland. *Geophysical Journal of the Royal Astronomical Society*, **20**, 1–9.

——, DAGLEY, P. & ADE-HALL, J. M. 1972. Palaeomagnetism of the British Tertiary igneous province: the Skye lavas. *Geophysical Journal of the Royal Astronomical Society*, **28**, 285–293.

——, HALL, J. M. & DAGLEY, P. 1982. The British Tertiary Igneous Province: palaeomagnetism of the dyke swarm along the Sleat coast of Skye. *Geophysical Journal of the Royal Astronomical Society*, **68**, 317–323.

D. H. TARLING, Department of Geological Sciences, Plymouth Polytechnic, Drake Circus, Plymouth, Devon PL4 8AA, UK.

E. A. HAILWOOD, Department of Oceanography, Southampton University, Southampton SO9 5NH, UK.

R. LØVLIE, Department of Geophysics, University of Bergen, Bergen, Norway.

Structure, composition and age of the Faeroe basalt plateau

R. Waagstein

SUMMARY: The lower Tertiary Faeroe basalt plateau is centred around the Faeroe Islands in the northern part of the supposed Faeroe–Rockall microcontinent. It consists of subaerial lavas, divided into three formations. The lower formation is probably thickest (> 3 km) in the southern or central part of the islands, whereas the middle and upper formations are probably thickest (> 2 km in total) just N of the Faeroes, close to the line of opening between the Faeroes and Greenland. The lower formation consists of tholeiites relatively low in TiO_2. The lower part of the middle formation shows an upward progression from high-Ti olivine tholeiites to high-Ti tholeiites, whereas the entire upper and rest of the middle formations consists of a contrasting population of high-Ti tholeiites and low-Ti MORB-like olivine tholeiites to tholeiites. The MORB type is confined to the northern Faeroes, increasing in abundance upwards. Similar MORB-like basalts have been recovered from further SW along the microcontinent (Faeroe Bank, Bill Bailey Bank, Lousy Bank and the Rockall Plateau), being associated with high-Ti tholeiites near the Faeroes and away from the Tertiary line of opening (Faeroe and Bill Bailey Banks). In addition, transitional to mildly alkaline basalts have been recovered on the E side of the Faeroe Bank Channel and on Bill Bailey Bank.

Based on field revision of the magnetostratigraphy of the lower formation it is tentatively suggested that the lower formation was extruded during chrons C26R to C25N and the two higher formations during C24R. Accumulation rates seem to have varied systematically and to have been related to the abundance and type of sediments between the lavas, which may indicate that extrusive activity ceased in the Faeroes before the opening of the NE Atlantic in C24R, and explain the presence of conjugate strike-slip faults in the NW Faeroes.

The spatial distribution of the various basalt types fits a model of shifting axes of intraplate volcanicity, shifting from the Faeroes to E Greenland and in part back again.

The Faeroe Rise (Bott & Watts 1971) forms a 1500 km long NE–SW trending belt of shallow banks N and W of the British Isles. The rise may be divided into three sections: (1) the Faeroe Block (the Faeroe Islands and insular shelf); (2) Faeroe Bank, Bill Bailey Bank and Lousy Bank (Fig. 1); and (3) the Rockall Plateau (George Bligh Bank, Hatton Bank and Rockall Bank). The Rockall Plateau is continental (Roberts 1975), as is the Faeroe Block judging from geophysical and isotopic evidence (Bott et al. 1974; Garièpy et al. 1983; Hald & Waagstein 1983). Reconstructions of the N Atlantic area prior to continental splitting suggest that these two sections of the rise are joined by continental crust beneath the intervening section of small banks, forming an unbroken Faeroe Rise microcontinent (Bott & Watts 1971, Roberts et al. 1983). The microcontinent is bounded to the E by oceanic crust of probable Lower Cretaceous age within the Rockall Trough and possibly also within the narrow Faeroe–Shetland Channel farther N (Fig. 1; Hanisch 1984); it is bounded to the S by Upper Cretaceous oceanic crust and to the W and N by lower Tertiary oceanic crust (Laughton 1975).

At about the time of separation of the Faeroe Rise from Greenland in the lower Tertiary, large outpourings of basalt flooded the Faeroe Rise and E Greenland in a wide belt along the line of opening. The thick coherent sheet of lower Tertiary basalts that erupted E of the line of opening is called the Faeroe basalt plateau in this paper. The lavas extend eastwards into the Faeroe–Shetland Channel, the northern part of the Rockall Trough (see Fig. 1) and the Hatton–Rockall Basin of the Rockall Plateau (Smythe 1983; Roberts et al. 1983, 1984).

Although the lower Tertiary volcanism was no doubt connected with the continental break-up, the relationship between the different volcanic and tectonic processes or stages and their exact timing is still a matter of debate. In this paper the stratigraphy and structure of the Faeroe basalt plateau and the chemical compositions of the lavas are reviewed. The geochemical review is based on samples from exposed sections, two deep drill-holes in the Faeroes and dredged samples from the insular shelf and Faeroe Bank, Bill Bailey Bank and Lousy Bank (Figs 1, 2 & 3; Noe-Nygaard & Rasmussen 1968; Waagstein 1977; Hald & Waagstein 1984; Waagstein & Hald 1984). The magnetostratigraphy of Tarling & Gale (1968) has been adjusted by new field work and is used in a discussion of the age and accumulation rates of the basalts. Finally, the

From MORTON, A. C. & PARSON, L. M. (eds), 1988, *Early Tertiary Volcanism and the Opening of the NE Atlantic,* Geological Society Special Publication No. 39, pp. 225–238.

R. Waagstein

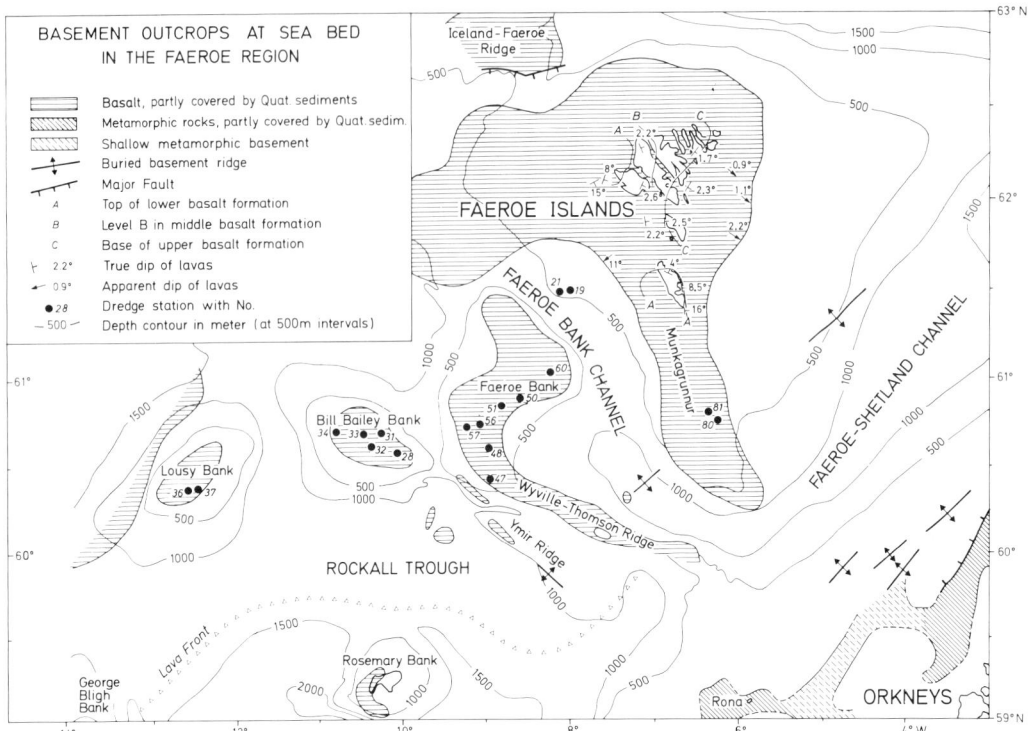

FIG. 1. Basement outcrops on the seabed in the Faeroe region. The map is adapted from Waagstein (1977) with later revisions of the basalt limits after M. S. Andersen (pers. comm.). Orkney shelf data are from Bott & Watts (1971).

above results are used to model the evolution of the plateau in connection with the continental break-up.

Stratigraphy of the Faeroe Islands

The Faeroe Islands are underlain by a strong seismic refractor with V_p about 5.9–6.2 km/s interpreted as continental metamorphic basement (Bott *et al.* 1974; Casten 1974). The basement is presumably Precambrian in age because some of the Faeroese basalts seem to be contaminated with very old continental rocks as judged by their high $^{87}Sr/^{86}Sr$ ratios (>0.71) (Gariépy *et al.* 1983; Hald & Waagstein 1983).

Minor gas and oil shows in the lava sequence penetrated by the 2.2 km deep Lopra drill-hole in the southernmost part of the Faeroes (see Figs 2 & 3) suggest that the basalts are locally underlain by marine sediments (Jacobsen & Laier 1984), perhaps forming part of the Faeroe Basin (Ridd 1983) centred in the Faeroe–Shetland Channel.

The Faeroese lava sequence consists of sub-aerial basalt lavas of tholeiitic composition and minor intercalated sediments (Noe-Nygaard & Rasmussen 1968; Rasmussen & Noe-Nygaard 1969, 1970; Waagstein 1977; Hald & Waagstein 1984; Waagstein *et al.* 1984). The sequence is divided informally into a lower, middle and upper basalt formation with a total stratigraphic thickness of at least $5\frac{1}{2}$ km (roughly $3\frac{1}{2}$ km exposed and 2 km drilled).

The lower basalt formation (>3 km) consists of macroscopically near-aphyric or aphyric flows with an average thickness of about 20 m. Most are of simple type with a several metres thick vesicular and rubbly top zone, often covered by red clay or tuff.

A coal-bearing formation of claystones and shales of lacustrine origin is about 10 m thick and rests on the slightly eroded top of the lower formation. In Suduroy two coal seams with a total thickness of 1 m are usually present in the lower part, while basaltic conglomerates and sandstones are locally present at the top.

The middle basalt formation (1.4 km) is dominated by porphyritic compound flows made up of thin amygdaloidal flow-units. The major part of the formation is plagioclase-phyric. However,

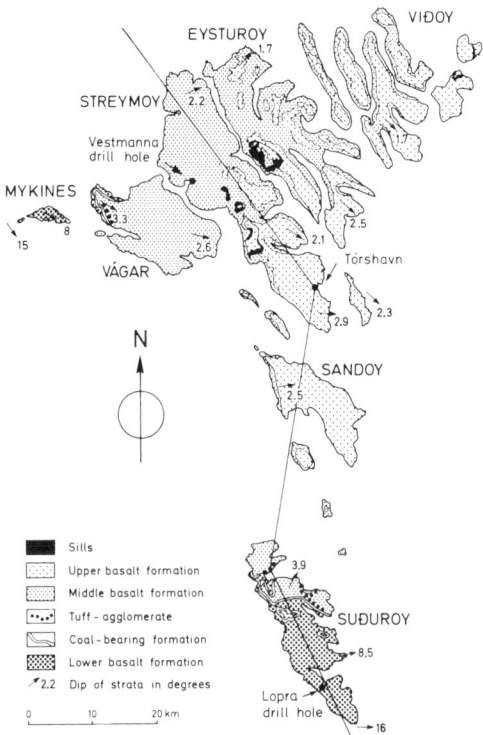

FIG. 2. Geological map of the Faeroe Islands adapted from Rasmussen & Noe-Nygaard (1969, 1970). Dips of strata are from Waagstein (1977). Also shown are the positions of the Vestmanna-1 and Lopra-1 drill-holes and the line of section of Fig. 3.

two olivine-phyric (to near-aphyric) sequences are here recognized as separate members of the formation.

The olivine-phyric Vestmanna member (new name) is a few hundred metres thick regional sequence overlying the coal-bearing formation (see Fig. 3). The greatest thickness (450 m) is recorded in the Vestmanna drill-hole where the member consists of 12 distinct chemical flow groups defining three cycles of upward increasing phenocryst content (Waagstein & Hald 1984). The tuff-agglomerate zone of Rasmussen & Noc-Nygaard (1969, 1970) is here considered a local facies of the Vestmanna member. The tuff-agglomerate zone forms a few kilometres wide NW-trending belt on the islands of Vágar and Suduroy, immediately above the coal-bearing formation. It is rarely more than a few tens of metres thick and consists of tuff, lapilli and bombs, heavily intruded by irregular dykes and sills with similar compositions as the rest of the member, and thus probably partly being feeders

(Waagstein 1977; Waagstein & Hald 1984 and unpublished).

The olivine-phyric Eidi member is taken to include the Eidis series and the B horizon flows of Rasmussen & Noe-Nygaard (1969, 1970) 600 m below the top of the middle basalt formation (see Fig. 3). It is more than 100 m thick in the northernmost part of the Faeroes but decreases in thickness southwards, probably interfingering with the plagioclase-phyric lavas.

The upper basalt formation (>0.9 km) is characterized by larger flow thicknesses than the middle formation due to a greater abundance of simple (i.e. non-compound) flows, usually separated by tuffs. The formation is dominated by near-aphyric and olivine-phyric flows in the northeastern areas, forming the basal Kolla-fjørdur member (formerly known as the C or Kollafjørdur horizon flows) and the major Vidoy member (new name), but dominated by plagioclase-phyric flows in the central Faeroes (Sandoy).

The basalt plateau is cut by numerous basalt dykes in various directions and by a few sills. The sills are intruded around the middle–upper formation boundary and seem to postdate the dykes.

Structure of the Faeroe basalt plateau

The Faeroe Islands

The top of the supposed Precambrian metamorphic basement deepens southwards in the northern Faeroes, perhaps forming an escarpment (Casten 1974). Assuming that the basement is directly overlain by basalt with $V_p = 4.9$ km/s (Pálmason 1965) and using the time terms given by Bott *et al.* (1976), the depth below sea-level is estimated at roughly 3 km around the Vestmanna drill-hole but at almost 5 km around Tórshavn (see Fig. 3). The work by Pálmason and Casten gives similar results. Reliable depth estimates cannot be made farther to the S in the Faeroes because of poor seismic control.

The lower basalt formation is slightly more than 2 km thick at Vestmanna and roughly 3 km thick at Tórshavn according to the above estimates (see Fig. 3; Waagstein 1977). The true thicknesses may be smaller if sediments are present beneath the basalt, but larger if the seismic velocity increases downwards in the basalt. However, even in the latter case it is considered rather unlikely that the thickness at Vestmanna exceeds the stratigraphic thickness of at least 3 km recorded in Suduroy (Hald &

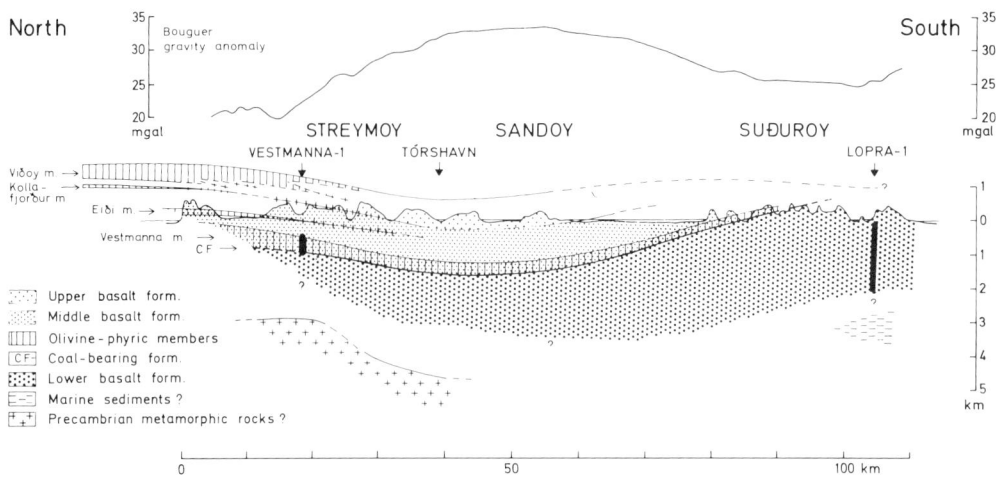

FIG. 3. North–south section through the Faeroe Islands along the line shown in FIG. 2. The basalt stratigraphy is constructed on the basis of exposures and the two drill-holes Vestmanna-1 and Lopra-1 (black bars) (Rasmussen & Noe-Nygaard 1969, 1970; Hald & Waagstein 1984; Waagstein, unpublished). The strata in the northern Faeroes are partly extrapolated from exposures far W and E of the line of section (projecting due E–W). The Precambrian metamorphic basement is inferred from seismic refraction studies (see text). Hydrocarbon shows in the Lopra drill-hole suggest the presence of marine sediments below Suduroy (Jacobsen & Laier 1984). Bouguer gravity data are from Saxov (1969).

Waagstein 1984). The depocentre of the lower basalt formation thus seems to be located in the southern or central part of the Faeroes rather than in the northern part.

The lower formation flows are difficult to map because they are all rather similar. However, interfingering of flows seems to be commonplace judging from the variation of the thickness and number of flows along cliff sections and also from the misfits between overlapping chemical profiles (unpublished analyses). This suggests that the flows were erupted locally rather than from a source region W of the Faeroes as assumed by Rasmussen & Noe-Nygaard (1969, 1970), although no eruption sites have been identified as of yet. The flows tend to form small groups with distinct chemical characters (e.g. Y/Zr ratios) suggesting that they originate from a number of independent volcanic systems (Hald & Waagstein 1984).

The lower formation probably subsided slowly during extrusion, forming a flat low-lying plateau at the time of deposition of the overlying coal-bearing formation. Before the volcanism resumed the plateau was subject to minor faulting (possibly reflecting SW–NE tension) and erosion, and Rasmussen & Noe-Nygaard (1969, 1970) suggested that the plateau was tilted to the E at the same time. This was later questioned by Waagstein (1977), however, and recent drilling at

Vestmanna shows that if any tilting occurred it must have been small (Waagstein & Hald 1984).

The middle basalt formation is about 1400 m thick at Vestmanna (Waagstein & Hald 1984) while the stratigraphic thickness is estimated to be slightly less between Sandoy and Suduroy (see Fig. 3). The upper basalt formation is about 700 m thick in Vidoy in the northeastern Faeroes and about 900 m thick in Sandoy, and in both places the vesicles of the youngest flows are either empty (except for a thin lining of smectite) or contain chabazite or calcite, suggesting that only some few hundred metres of overlying lavas have been removed by erosion. Thus, before erosion, the upper formation probably had a fairly uniform thickness of $1–1\frac{1}{2}$ km over a large area. However, both the middle and upper formations thinned towards the southern part of Suduroy. This is suggested by the fossil geothermal gradient obtained from secondary minerals at Lopra, which places the top of the lava pile 600–1200 m above sea-level, i.e. close to the top of the lower basalt formation (see Fig. 3; Jørgensen 1984). The depocentre of the middle and upper formations therefore seems to be located in the northern part of the Faeroe Block. Detailed mapping of three lava horizons spanning the interval from the Eidi to Kollafjørdur members shows that the horizons mutually diverge by up to $\frac{1}{2}°$ in a northerly direction (see Fig. 3; Waagstein 1977), and this may indicate that during extrusion of the middle formation, the northern submerged margin of the Faeroe Block subsided at a slightly faster rate than the northern Faeroes.

The middle and upper formations show systematic variations in the abundance of olivine-phyric (or near-aphyric) basalt and plagioclase-phyric basalt, respectively. In the southern and central Faeroes, plagioclase-phyric basalt is dominant in all exposures above the basal olivine-phyric Vestmanna member. In the northern and northeastern Faeroes, olivine-phyric to near-aphyric basalt becomes increasingly abundant upwards at the expense of plagioclase-phyric basalt within the same interval. The abundance of the former basalt type seems to increase to the N (see Fig. 3; Waagstein 1977) or E (Rasmussen & Noe-Nygaard 1969, 1970) at all levels in the middle and upper formations. Nevertheless all types of basalt, including the olivine-phyric to near-aphyric type, seem to have been erupted locally. The middle formation lavas were erupted from fissures or oblong vents orientated approximately NW–SE as judged from the distribution of the tuff-agglomerate zone and the silicic Klaksvík flow at the base and near the top of the formation, respectively, and judged from the shape or orientation of several exposed vent agglomerates and supposed feeder dykes (Hald *et al.* 1969; Rasmussen & Noe-Nygaard 1969, 1970; unpublished). The youngest flows mapped belong to the Kollafjørdur member at the base of the upper formation (see Fig. 3) and form flat shields providing no evidence of the orientation of feeder channels (Noe-Nygaard 1968).

The Faeroe Islands form part of two structural domes, a large northern one centred just NW of the islands and a smaller southern one centred around Suduroy (see Fig. 1; Schrøder 1971). The dip of strata is generally 1–4°, but locally increases to 8–16° in the oldest lower formation strata of Suduroy and Mykines near the centres of the domes (see Fig. 2). The main phase of doming seems to have occurred after the extrusion of the upper formation and is probably related to the formation of master joints, minor faults, dykes and sills.

The submerged areas

The Faeroe basalt plateau outcrops on the seabed, or has only a thin cover of Quaternary sediments, in a large continuous area around the Faeroes (see Fig. 1). Most of this area is inside the 500 m depth contour and may be considered part of the Faeroe Block. It forms two domes as mentioned

above. The dome structure is suggested by the curvature of the top of the lower formation mapped magnetically close to land (Schrøder 1971) and by the distribution and/or dip of sequences of thick flows mapped bathymetrically (Waagstein 1977). The dip of the lavas is generally small on the shelf but tends to increase in marginal areas where the basalts become covered by Palaeogene sediments (Nielsen *et al.* 1979). Estimates of the thickness of the upper basalt formation on the eastern and southwestern shelf confirm that only a minor part of the formation is missing on land (Waagstein 1977). Judging by the distribution of basalt types among glacial erratics dredged outside the basalt margin, but inside the former glaciation limit, the upper formation is dominantly plagioclase-phyric E of the Faeroes, as on Sandoy, whereas it has a substantial olivine-phyric component SW of the Faeroes (Waagstein 1977). Some of the olivine-phyric basalts are transitional or mildly alkaline (see below). A small number of erratics recovered consist of carbonate-cemented marine basaltic sandstones and tuffs probably representing the Palaeogene sediments above (Nielsen *et al.* 1979), but there is no evidence of pillow lavas or breccias. Most of the sediments are Eocene in age (Heilmann–Clausen, pers. comm.), i.e. comparable in age to tuffs dredged on the S side of Faeroe Bank (Jones & Ramsay 1982).

South and SW of the Faeroe Block the basalt plateau is exposed locally on the seabed, especially in shoal areas including the Faeroe Bank, Bill Bailey Bank, Lousy Bank and the Wyville–Thomson and Ymir Ridges (see Fig. 1; Waagstein 1977). These high-lying areas were mainly formed by flexuring and show evidence of subsequent erosion (Roberts *et al.* 1983).

Seismic reflection profiles show that the Faeroe basalt plateau extends far eastwards into the Faeroe–Shetland Channel beneath 2–3 km of Cainozoic sediments, terminating in two steps which possibly represent former coast lines (Smythe 1983). The lower and more widespread basalt unit approaches the Shetland Platform in a lobe centred above and probably fed by the buried Erlend volcanic complex (Smythe *et al.* 1983). The top of the lower basalt unit is laterally equivalent to the top Palaeocene ash marker of the North Sea. Smythe *et al.* (1983) equate the lower and upper basalt units with the Faeroes lower and middle formations on the basis of different seismic reflection characteristics, which implies that the Faeroes lower formation predates the ash marker. Their correlation of the top Palaeocene ash marker with the Faeroes tuff-agglomerate zone is more speculative because of the very local distribution and special chemistry of the latter (Waagstein & Hald 1984), precluding a common source for the two volcaniclastic sequences.

South of the Wyville–Thomson Ymir Ridge system the thick Faeroe basalt plateau continues at depth in the Rockall Trough almost to the latitude of Rosemary Bank (see Fig. 1), where the top is dated as uppermost Palaeocene by a long distance seismic tie to IPOD wells in the Bay of Biscay (Roberts *et al.* 1983).

Several volcanic or intrusive centres have been identified or suggested within the submerged areas of the basalt plateau on the basis of seismic, gravity and magnetic data (Roberts *et al.* 1983; Smythe 1983).

Chemical composition

The Faeroe Islands

The basalts from the Faeroe Islands are all tholeiites, i.e. hypersthene normative (assuming $Fe_2O_3/FeO = 0.15$), but show a large compositional variation ranging from picrite to ferrobasalt (MgO 4.5–23%), from low to high titanium types (TiO_2 0.9–4.0%), and from types strongly depleted to types enriched in incompatible elements (Table 1; Noe-Nygaard & Rasmussen 1968; Schilling & Noe-Nygaard 1974; Bollingberg *et al.* 1975; Waagstein 1977; Garièpy *et al.* 1983; Hald & Waagstein 1984; Waagstein & Hald 1984). A few basalt flows have high SiO_2 (52–55%) and $^{87}Sr/^{86}Sr$ (0.710–0.716) suggesting crustal contamination (Hald & Waagstein 1983).

The main compositional variation of the basalts can be shown by diagrams involving MgO, FeO* and TiO_2, e.g. a $TiO_2/FeO*$ vs FeO*/MgO diagram (Fig. 4a) to avoid the effect of plagioclase accumulation (Waagstein 1977). In Figure 4a basalts from different stratigraphic levels are readily separated. The lower basalt formation occupies a field of high FeO*/MgO and relatively low $TiO_2/FeO*$ almost completely separated from the two higher formations. In general, the middle basalt formation shows a progression from low to high FeO*/MgO ratio at a consistently higher level of $TiO_2/FeO*$. This trend is observed up through the olivine-phyric Vestmanna member and the overlying plagioclase-phyric main sequence of the middle formation. In the upper part of the middle formation olivine-phyric basalts reappear in some places within the plagioclase-phyric sequence (e.g. the Eidi member). However, in contrast to the basalts of the Vestmanna member, these are low in both FeO*/MgO and $TiO_2/FeO*$ and MORB-like,

showing depletion in titanium, phosphorus, potassium and other incompatible elements (cf. samples 15506 & 441.01 m in Table 1). The upper basalt formation consists of the same two contrasting basalt populations as the upper part of the middle basalt formation, one low in both ratios and MORB-like, and the other type high in both ratios. The former type is represented by the olivine-phyric Kollafjørdur and Vidoy members, and the latter by the plagioclase-phyric main sequence of the upper formation. The youngest dykes and sills also broadly belong to these two categories (Hald & Waagstein, unpublished data).

Ignoring the few silicic lavas with low FeO*/MgO and 1.0–1.6% TiO_2 the Faroese basalts may thus be divided into four chemical groups on the basis of Fig. 4a: (1) low-Ti olivine tholeiite; (2) low-Ti tholeiite; (3) high-Ti olivine tholeiite; and (4) high-Ti tholeiite (note that 'low' and 'high' are relative terms dependent on the FeO* content). The boundary between olivine tholeiite and tholeiite may be set at a FeO*/MgO ratio of about 1.4 corresponding to the transition from olivine-phyric to plagioclase-phyric basalt in the high-Ti suite (Waagstein 1977; Waagstein & Hald 1984). The basalts of the low- and high-Ti suites differ mineralogically, especially at the tholeiitic end. The most obvious difference is that the low-Ti tholeiites are generally near-aphyric or aphyric (0–5% phenocrysts), whereas the high-Ti tholeiites are generally plagioclase-phyric (10–40% phenocrysts).

The main compositional trends shown by the low- and high-Ti suites of basalt in general cannot be explained by simple low-pressure fractionation, because most of the low-Ti olivine tholeiites have lower $^{87}Sr/^{86}Sr$ initial ratios (<0.7030) than the other chemical groups, and because the ratio between strongly and less strongly incompatible elements (e.g. La/Sm and Zr/Y) tend to increase throughout both suites (Garièpy *et al.* 1983; Hald & Waagstein 1984).

Dredged basalts from the Faeroe Rise

Basalts of local origin have been dredged from 16 sites on Munkagrunnur, Faeroe Bank, Bill Bailey Bank and Lousy Bank (see Fig. 1; Waagstein 1977). Their local origin is deduced from the petrographic characteristics of the individual hauls and from the absence or near-absence of other rock types. Most of the recovered stones (5–50 cm) seem to originate from submerged shore-lines. Glacial erratics dredged near the shelf-edge W of Suduroy (hauls 19, 21) and probably recovered from lodgement till also seem to be of rather local origin judging from the over-

TABLE 1. *Major element composition and CIPW norm of selected basalts from the Faeroe region*

	Low-Ti suite					High-Ti suite				Alkaline suite
	MBF Eidi m. 15506	Faeroe Bank 57–26	UBF Vidoy m. XI, 23	LBF Lopra-1 2177 m	LBF Lopra-1 339 m	MBF VM-1 441.01 m	Bailey Bank 28–1	Faeroe Bank 57–3	MBF Sneis 15569B	Munka-grunnur 81–2
SiO_2	45.13	48.08	47.78	47.43	48.04	44.61	47.92	46.59	46.96	43.38
TiO_2	1.03	1.12	1.34	1.76	3.44	1.52	1.95	3.00	3.99	1.69
Al_2O_3	12.50	14.69	15.21	13.44	12.31	7.82	14.86	14.10	12.89	13.57
Fe_2O_3	1.61	4.40	6.03	4.12	6.63	1.81	4.32	5.88	5.89	13.70
FeO	9.49	7.17	5.67	9.90	10.05	10.46	6.51	7.94	9.35	n.a.
MnO	0.20	0.20	0.19	0.22	0.24	0.20	0.16	0.21	0.21	0.19
MgO	14.91	8.16	7.24	6.85	4.79	23.06	7.93	6.62	6.09	12.12
CaO	10.22	12.43	12.16	11.99	9.42	7.35	11.00	10.62	9.98	9.41
Na_2O	1.80	2.10	2.09	2.05	2.80	1.25	2.73	2.79	2.32	2.53
K_2O	0.15	0.16	0.26	0.17	0.89	0.16	0.28	0.18	0.24	0.45
P_2O_5	0.07	0.09	0.09	0.16	0.42	0.12	0.14	0.27	0.32	0.13
LOI	2.54	1.74	1.79	2.30	1.25	1.26	2.37	1.66	1.15	2.07
Total	99.65	100.34	99.85	100.39	100.28	99.62	100.17	99.86	99.39	99.24
Fe_2O^*/MgO	0.73	1.36	1.53	1.99	3.34	0.52	1.31	2.00	2.41	1.02
TiO_2/FeO^*	0.094	0.101	0.121	0.129	0.215	0.126	0.188	0.227	0.272	0.137
Qz	–	–	–	–	–	–	–	–	0.81	–
Or	0.91	0.96	1.57	1.03	5.33	0.96	1.70	1.09	1.45	2.77
Ab	15.69	18.07	18.12	17.73	24.03	10.76	23.69	24.14	20.06	14.93
An	26.35	30.70	32.12	27.56	18.66	15.51	28.17	25.99	24.58	25.37
Ne	–	–	–	–	–	–	–	–	–	4.00
Di	20.45	25.76	23.91	26.66	21.82	16.58	21.91	21.53	19.79	18.17
Hy	1.77	10.00	11.93	14.80	16.81	9.57	6.83	7.70	21.93	–
Ol	30.48	9.96	7.35	5.78	2.62	41.04	11.53	10.49	–	28.65
Mt	2.16	2.17	2.18	2.66	3.11	2.36	2.04	2.59	2.87	2.46
Il	2.01	2.16	2.61	3.42	6.63	2.94	3.80	5.83	7.74	3.34
Ap	0.17	0.21	0.21	0.38	0.99	0.28	0.33	0.64	0.76	0.31

LBF, MBF & UBF = the Faeroes lower, middle & upper basalt formations. XI, 23 and 81–2 are new analyses; the two Lopra-1 analyses are from Hald & Waagstein 1984, the VM-1 analysis is from Waagstein & Hald 1984, and the remaining analyses are from Waagstein 1977 (with minor revisions). No. succeeded by a hyphen is a dredge station No. CIPW norms calculated assuming $Fe_2O_3/FeO = 0.15$. n.a. = not analyzed.

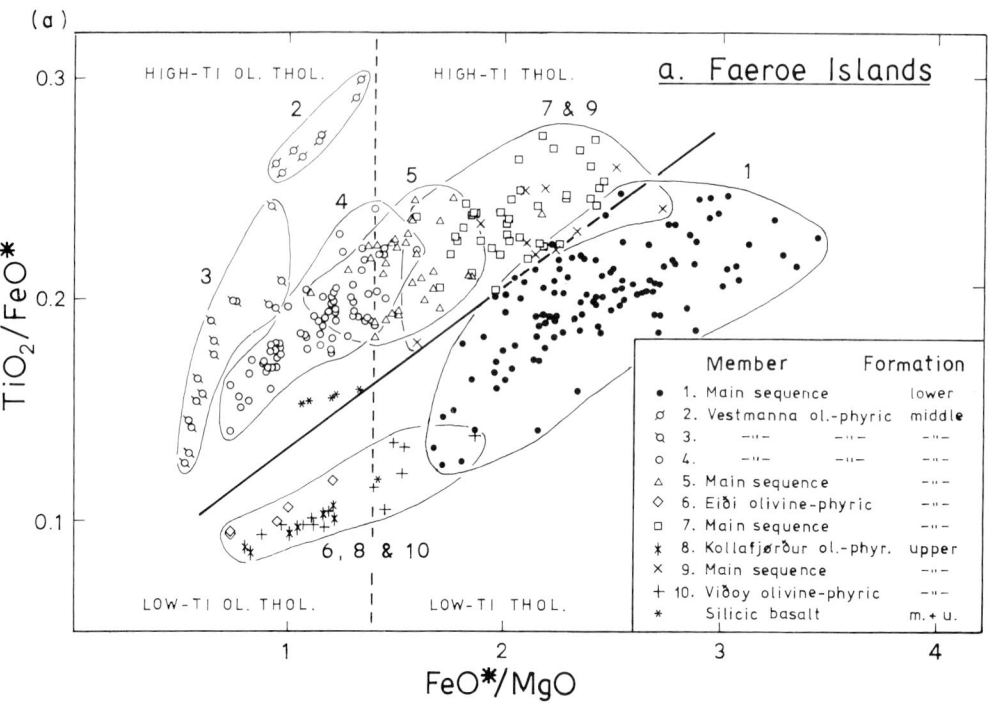

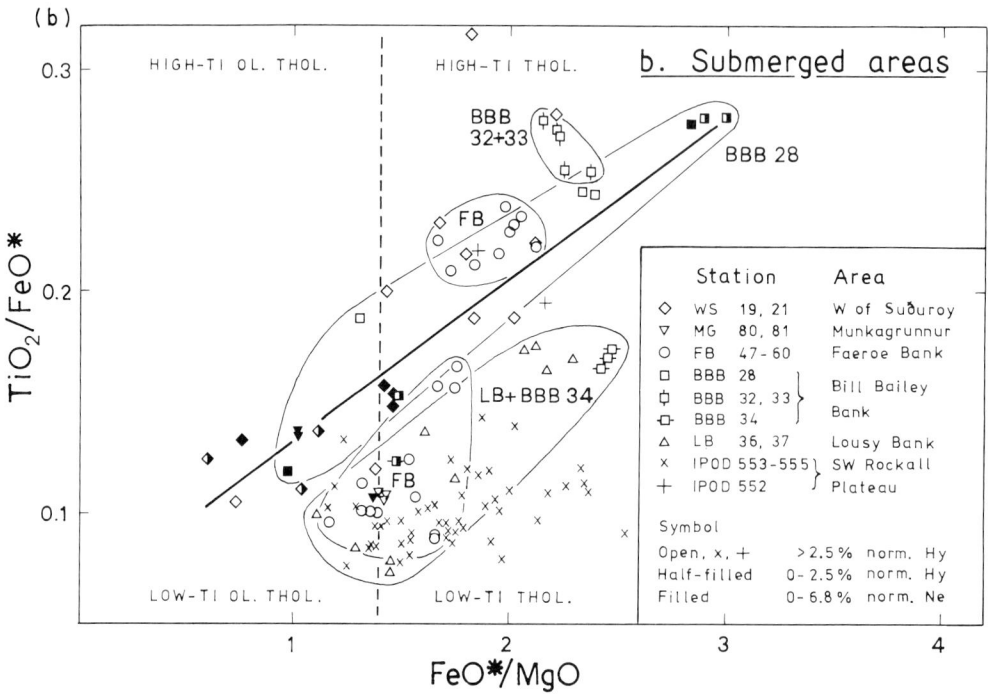

abundance of olivine-phyric basalt in comparison with the adjoining part of the Faeroes.

The dredged basalts fall into the same four main groups as those from land (see Fig. 4b and Table 1) but also include a fifth category of transitional to mildly alkaline basalt (2.5% normative hypersthene to 6.8% normative nepheline when Fe_2O_3/FeO is adjusted to 0.15). No silicic basalt or basalt with MgO higher than 9.4% has been recovered from the banks. The hauls reveal some interesting geographical differences. Three out of four hauls from the eastern margin of the Faeroe Bank Channel (19, 21, 81) contain transitional to mildly alkaline basalts in addition to tholeiitic basalts, probably from a high stratigraphic level judging from the general dome structure of the Faeroe Block. The seven successful hauls on top of the Faeroe Bank sampled a contrasting population of low-Ti olivine tholeiites and tholeiites on the one hand and high-Ti tholeiites on the other (in three cases at the same site), bearing a strong resemblance to the upper succession in the northern Faeroes. The five hauls from Bill Bailey Bank contained a varied population of tholeiitic and transitional (<2.5% normative Hy or Ne) basalts, whereas the two hauls from Lousy Bank only contained olivine tholeiites and tholeiites of the low-Ti suite. Some of the low-Ti basalts from Lousy Bank and the westernmost site on Bill Bailey Bank are relatively evolved and similar in both major and trace-element chemistry to some lower formation tholeiites (Waagstein 1977). However, as only some hundred metres of basalt have been removed by erosion on top of Lousy Bank, judging from commercial seismic data (Andersen, pers. comm.), and as similar basalt types are found among Faeroese dykes and sills, these evolved low-Ti tholeiites are considered relatively late. The basalts from all three banks are therefore correlated with the youngest basalts from the Faeroe Block.

The basalts of the high-Ti suite seem to be concentrated near the Faeroes (Faeroe Bank and Bill Bailey Bank) while the basalts of the low-Ti suite presumably continue via Lousy and Hatton Banks to the SW corner of the Rockall Plateau

where very similar MORB-like basalts were encountered in IPOD wells (see Fig. 4b; Joron *et al.* 1984).

The basalts from the Faeroe Rise are similar to the basalts from Iceland and the Reykjanes Ridge in both major and trace-element composition, except that in comparison to these and other oceanic basalts they are generally low in Nb and Ta relative to other high field-strength elements, e.g. Zr (Fig. 5). This property they share with the basalts from E Greenland and the Hebridean basalt province and also with basalts from continental 'initial rifting' settings elsewhere (Holm 1985).

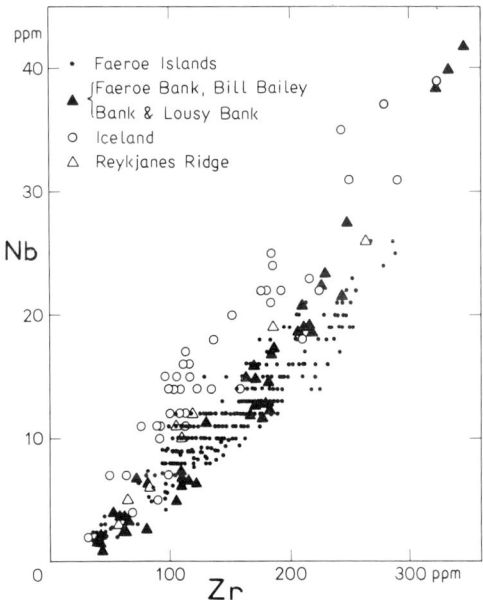

FIG. 5. Nb vs Zr in basalts from the Faeroe Islands, the southwestern banks, Iceland and the Reykjanes Ridge. Sources of data: Hald & Waagstein (1984); Tarney *et al.* (1979); Waagstein (1977); Wood (1978); Wood *et al.* (1979); plus unpublished analyses of the Faeroes middle and upper formations.

FIG. 4. TiO_2/FeO^* vs FeO^*/MgO diagrams for basalts from the Faeroe Rise (FeO^* = total iron recalculated as FeO). The oblique full line and the vertical stippled line mark the proposed boundaries between high-Ti olivine tholeiites, high-Ti tholeiites, low-Ti olivine tholeiites, and low-Ti tholeiites (see text). (a) *Faeroe Islands* Source of data of the numbered sequences: (1) Hald & Waagstein (1984) (max. 1 analysis per flow); (2–4) groups a_1 to a_2, a_3 to b, and c to f_3 in Waagstein & Hald (1984); (5) groups g_1 to g_4 in Waagstein & Hald (1984) and unpublished; (6–10) the Sneis section in Waagstein (1977) plus unpublished analyses of sections XI and 35 of Rasmussen & Noe-Nygaard (1969) and of a few other samples; silicic basalts from Hald & Waagstein (1983). (b) *Submerged areas* Source of data for the areas indicated: (WS and MG) unpublished; (FB, BBB and LB) Waagstein (1977) with minor revisions; (IPOD 552 to 555) Initial Report of the Deep Sea Drilling Project Leg 81. The content of normative hypersthene or nepheline is calculated assuming $Fe_2O_3/FeO = 0.15$.

Age

In order to understand the relationship between the volcanic and plate-tectonic processes that resulted in the opening of the NE Atlantic it is important to accurately date the Faeroese lava sequence. This task is difficult because of the lack of marine fossils for stratigraphic correlation and fresh rocks for K-Ar dating. Long-distance correlation by palaeomagnetic means is promising, but unfortunately there has not been any general agreement as to the proper identification of the observed magnetochrons. Smythe *et al.* (1983) correlate the exposed sequence with chrons C24R and C25N, partly on the basis of seismic work. Roberts *et al.* (1984) correlate the same sequence with C24R alone, assuming some hitherto unrecognized normal events within this magnetochron. Abrahamsen *et al.* (1984) suggest that the exposed *and* drilled sequence correlate with either C25R–24R, C24R–23R or C24R–22R, preferring the latter correlation on the basis of the K-Ar datings of Tarling & Gale (1968) and presumed extrusion rates.

The problem of proper identification of the magnetochrons arises partly from an uncertainty about the number and stratigraphic position of the reversals in the lower basalt formation. In order to remove this uncertainty a tight stratigraphic control has recently been established by the author between the three palaeomagnetic sections used by Tarling (1970) from Suduroy. This was done using air photos, chemical analyses of the basalts and magnetic polarity measurements with a field compass (unpublished work). It shows that only two normal polarity intervals are present and not three as suggested by Tarling. The intervals are thicker than may be presumed from the number of palaeomagnetic sites (lavas) measured by Tarling. On the basis of this revision a new stratigraphic correlation is here proposed. It is suggested that the two relatively thick normal polarity intervals represent C26N and C25N and that the complete drilled and exposed sequence was formed during the interval C26R through to C24R (Fig. 6). The new correlation is in agreement with a palynological dating of the coal-bearing formation as uppermost Palaeocene (Lund 1981), and with the suggested correlation between this stratigraphic level and the top Palaeocene ash marker of the North Sea (Smythe *et al.* 1983). Using the time-scale of Harland *et al.* (1982) (see Fig. 6), the magnetochrons are also within the range of K-Ar ages obtained from the least altered basalts (50–63 Ma) (Tarling & Gale 1968; Rex & Waagstein, unpublished). Unfortunately, the ages of the above magnetochrons are not well established as of yet, the beginning of

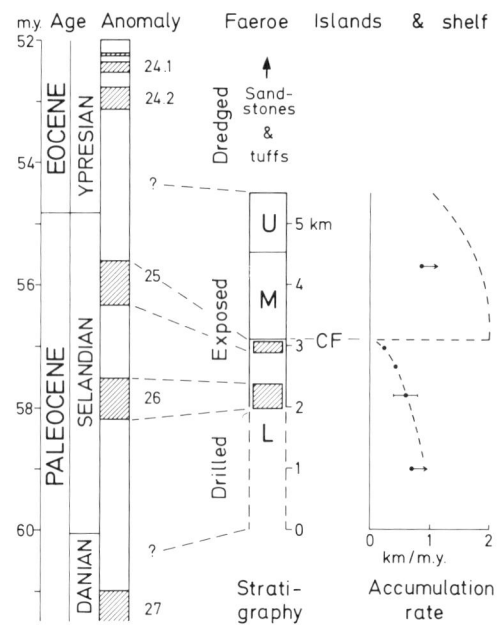

FIG. 6. Magnetic anomaly age of the Faeroes basalt sequence proposed in this paper; see text for discussion. The reversal time-scale is from Harland *et al.* (1982, Table 4.3). The boundaries between the Danian, Selandian and Ypresian are located relative to the magnetic reversals in accordance with Berggren *et al.* (1985).

C24R being dated as 50.2, 53.2 and 56.1 Ma in three recent time-scales proposed by Tarling (1983), Harland *et al.* (1982) and Berggren *et al.* (1985), respectively. The identification of magnetochrons by K-Ar datings is thus still conjectural.

Accumulation rates

Although the absolute ages of the magnetochrons are not yet accurately known, the duration of individual chrons may be estimated with high precision from the width of the magnetic anomalies of the ocean floor (Ness *et al.* 1980). This allows precise estimates of the extrusion rate in some parts of the Faeroese lava pile and sets lower limits to this rate in the remaining part. The thicknesses of the suggested chrons C26N, C25R and C25N in the lower basalt formation on Suduroy are estimated at 400 ± 130 m, 510 ± 30 m and 180 ± 20 m, respectively, the bottom of C26N being constrained by the uppermost cored flow in the Lopra drill-hole, the flow presumably being formed in a reverse magnetic field (Schoenharting & Abrahamsen 1984). Using the time-scale of Harland *et al.* (1982), based on marine magnetic profiles (Ness *et al.* 1980), the duration times are estimated at 0.67, 1.19 and 0.73 Ma, respectively, and the accumulation rates are estimated at 600 ± 200, 430 ± 25 and 250 ± 30 m/ My, respectively (see Fig. 6).

The strong decrease in accumulation rate towards the top of the lower basalt formation is supported by geological observations. The lower formation lava flows are separated by palaeosol (red clay) and tuff (Hald & Waagstein 1984; Parra *et al.* 1987). In the Lopra drill-hole the sediments form about 2% of the lava pile (Hald & Waagstein 1984). The exposed sections of supposed C26N and C25R seem to contain a similar or only slightly higher amount of sediments. However, minor lenses of coal occur sporadically in some beds from near the C26N–25R boundary and upwards, judging from the records of Rasmussen & Noe-Nygaard (1969, 1970), and may be taken as evidence of decreasing volcanic activity. A further decrease is suggested in the uppermost 100 m of the formation where clay and tuff form up to 15% of the sequence, occasionally being associated with minor coal (Hald & Waagstein 1984).

The five widely spaced cores recovered from the Lopra drill-hole in the oldest part of the lower basalt formation supposedly represent a single reverse magnetochron (Schoenharting & Abrahamsen 1984), i.e. C26R. Assuming that the drilled section did not form much faster than the oldest exposed strata, the Faeroes volcanism must have started in the beginning of the Selandian or earlier (see Fig. 6; cf. Morton *et al.*, 1988).

The middle and upper basalt formations were also formed within a single reverse magnetochron (Tarling & Gale 1968), now correlated with C24R. The middle basalt formation accumulated at a very high rate. It is dominated by thick compound flows which often exceed 30 m in thickness in the lower part of the formation. The flows either have no sediments in between or they are separated by thin beds of red tuff rather than soil (Waagstein & Hald 1984; Parra *et al.* 1987). Thus, only 30–40 tuff beds averaging 30 cm in thickness have been recorded through the 1.4 km thick middle formation (unpublished mapping by the author). The lava accumulation rate probably decreased slightly with time because the upper formation flows are more commonly separated by tuff, sometimes containing plant imprints (Rasmussen & Noe-Nygaard 1969, 1970). However, the above observations suggest that the middle and upper formations formed in a much shorter time-span than the lower formation and, assuming a reasonable rate of deposition for the ca. 10 m thick coal-bearing formation in

between, this implies that the extrusion of lavas ceased long before C24N time (see Fig. 6).

Evolution

According to the above interpretations of the magnetochrons in the Faeroes succession, the lower basalt formation predates the supposed onset of seafloor spreading between the Faeroes and Greenland in C24R. The formation extends about 100–250 km SE of the ocean–continent transition line located beneath the Norwegian Sea and the Iceland–Faeroe Ridge (Smythe 1983). Its depocentre seems to be in the southern rather than the northern part of the Faeroes, although no feeders have been identified as yet. The lower formation volcanism in the Faeroes is suggested here to have ended at the very beginning of chron C24R with one or a few reversely magnetized flows, whereas it is generally accepted that the E Greenland basalts were not erupted before C24R. Thus, it seems that the lower formation volcanism, chemically very similar to the plateau basalts of E Greenland (Hald & Waagstein 1984), switched at the beginning of C24R from the Faeroes to the adjoining part of E Greenland.

The following model is proposed to account for this lateral shift of volcanism: In the beginning of C26R, i.e. near the Danian–Selandian boundary, an initial episode of rifting started in the Faeroes resulting in voluminous outpourings of tholeiitic basalt which was relatively low TiO_2. The rift-related volcanism gradually decreased and finally stopped in the Faeroe region. At the beginning of C24R, volcanic activity was initiated along a new continental rift zone created in central E Greenland, then situated immediately N of the Faeroes. According to Larsen & Watt (1985) the first rifting took place far inland in central E Greenland, to form a straight line of opening of the NE Atlantic following the present coast line farther S and N. Later, in C24R, the rift zone shifted to the present shelf of central E Greenland where it was more in line with the Faeroes (Fig. 7). In this period of less than 3 My the 2–7 km thick E Greenland basalt plateau was formed (Nielsen & Brooks 1981).

The NW–SE trend of the feeders of the Faeroes middle basalt formation, almost perpendicular to the axis of opening of the NE Atlantic, suggests that the Faeroes were situated outside the active zone of continental rifting centred in E Greenland. The high-Ti type of volcanism dominating the middle formation may be related to this new tectonic setting. It is suggested that the appearance of MORB-like, low-Ti olivine tholeiites in

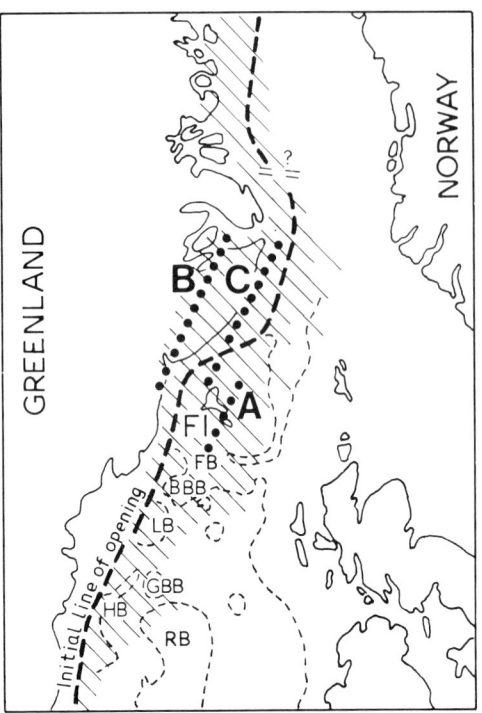

FIG. 7. Reconstruction of the NE Atlantic region before the onset of seafloor spreading in the Lower Tertiary modified after Larsen & Watt (1985) and Larsen (1988). Areas of thick basalt are ruled. The heavy dotted lines A, B and C are suggested successive axes of intraplate volcanic activity characterized by the production of basalt sequences having low to moderate Ti/Fe ratios (see text): (A) the Faeroes lower basalt formation of suggested C26R–C25N age; (B) the lower part of the E Greenland basalt plateau formed in the early part of C24R (1st cycle of Larsen & Watt 1985); (C) the upper part of the E Greenland basalt plateau formed later during C24R (2nd cycle). It is suggested that the gradual incoming of MORB-like basalts from the upper part of the Faeroes middle basalt formation and upwards is associated with a southwestward propagation of axis C into the Faeroe Block. The volcanic activity in the Faeroes perhaps ended before the onset of seafloor spreading in C24R along the initial line of opening (heavy stippled line), whereas MORB-like basalts of C24N age seem to be present in E Greenland (3rd cycle). The light stippled lines are 1000 m depth contours. FI: Faeroe Islands; FB: Faeroe Bank; BBB: Bill Bailey Bank; LB: Lousy Bank; GBB: George Bligh Bank; HB: Hatton Bank; RB: Rockall Bank.

the Eidi member with a northerly source is evidence for the eastward shift of the E Greenland centre of volcanism mentioned above. The increasing abundance of MORB-like basalts in the northern and eastern exposures of the upper formation may then be evidence for a southwestward propagation of this new rift zone into the adjoining Faeroe Block (see Fig. 7).

After the cessation of extrusive activity, the northwestern part of the Faeroes was subject to mild E–W compression at about 80°E to judge from the presence of small conjugate strike-slip faults; cf. the geological map of Rasmussen & Noe-Nygaard (1969, 1970). The faults are often intruded by dykes, and some of these are unsheared, suggesting that the compression occurred before the end of intrusive activity. The compression is tentatively related to plate acceleration during the decoupling of the plates; this implies that the Faeroese lava pile completely predates the onset of seafloor spreading between Greenland and the Faeroes. Such early termination of the Faeroes volcanism is also in accordance with the above suggestions about ages and extrusion rates.

ACKNOWLEDGEMENTS: The writer is grateful to T. F. D. Nielsen and two anonymous reviewers for critical comments on an earlier version of the manuscript, to A. Noe-Nygaard and J. Rasmussen for access to all their samples and field notes, to D. H. Tarling for access to his field maps, and to J. Bailey, G. Hornung and I. Sørensen for chemical analyses. The work on the shelf was made possible with grants from the Danish Natural Science Research Council (SNF). The 'Fiskirannsóknarstovan' (Fishery Research Institute) in Tórshavn kindly provided ship facilities with the 'J. C. Svabo' free of charge.

References

ABRAHAMSEN, N., SCHOENHARTING, G. & HEINESEN, M. 1984. Palaeomagnetism of the Vestmanna core and magnetic age and evolution of the Faeroe Islands. *In*: BERTHELSEN, O., NOE-NYGAARD, A. & RASMUSSEN, J. (eds) *The Deep Drilling Project 1980–1981 in the Faeroe Islands*. Føroya Fróðskaparfelag, Tórshavn, pp. 93–108.
BERGGREN, W. A., KENT, D. V., FLYNN, J. J. & VAN

COUVERING, J. A. 1985. Cenozoic geochronology. *Bulletin of the Geological Society of America*, **96**, 1407–1418.
BOLLINGBERG, H., BROOKS, C. K. & NOE-NYGAARD, A. 1975. Trace element variations in Faeroese basalts and their possible relationships to ocean floor spreading history. *Bulletin of the Geological Society of Denmark*, **24**, 55–60.

BOTT, M. H. P. & WATTS, A. B. 1971. Deep structure of the continental margin adjacent to the British Isles. *In*: DELANY, F. M. (ed) *ICSU/SCOR Symposium on East Atlantic Continental Margins 1970. Report of the Institute of Geological Sciences*, **70/14**, 89–109.

——, SUNDERLAND, J., SMITH, P. J., CASTEN, U. & SAXOV, S. 1974. Evidence for continental crust beneath the Faeroe Islands. *Nature*, **248**, 202–204.

——, NIELSEN, P. H. & SUNDERLAND, J. 1976. Converted P-waves originating at the continental margin between the Iceland–Faeroe Ridge and the Faeroe Block. *Geophysical Journal of the Royal Astronomical Society*, **44**, 229–238.

CASTEN, U. 1974. Eine analyse seismischer Registrierungen von den Färöer Inseln. *Hamburger Geophysikalische Einzelschriften, Geophysikalische Institut der Universität Hamburg*, **21**, 109 pp.

GARIÉPY, C., LUDDEN, J. & BROOKS, C. 1983. Isotopic and trace element constraints on the genesis of the Faeroe lava pile. *Earth and Planetary Science Letters*, **63**, 257–272.

HALD, N. & WAAGSTEIN, R. 1983. Silicic basalts from the Faeroe Islands: Evidence of crustal contamination. *In*: BOTT, M. H. P., SAXOV, S., TALWANI, M. & THIEDE, J. (eds) *Structure and Development of the Greenland–Scotland Ridge*. Plenum Press, New York, pp. 343–349.

—— & —— 1984. Lithology and chemistry of a 2-km sequence of Lower Tertiary tholeiitic lavas drilled on Suduroy, Faeroe Islands (Lopra-1). *In*: BERTHELSEN, O., NOE-NYGAARD, A. & RASMUSSEN, J. (eds) *The Deep Drilling Project 1980–1981 in the Faeroe Islands*. Føroya Fródskaparfelag, Tórshavn, pp. 15–38.

——, NOE-NYGAARD, A. & WAAGSTEIN, R. 1969. On extrusion forms in plateau basalts, pt. 2. The Klakksvík flow, Faeroe Islands. *Bulletin of the Geological Society of Denmark*, **19**, 2–7.

HANISCH, J. 1984. The Cretaceous opening of the northeast Atlantic. *Tectonophysics*, **101**, 1–23.

HARLAND, W. B., COX, A. V., LLEWELLYN, P. G., PICKTON, C. A., SMITH, A. G. & WALTERS, R. 1982. *A Geological Time Scale*. Cambridge University Press, 128 pp.

HOLM, P. E. 1985. The geochemical fingerprints of different tectonomagmatic environments using hygromagmatophile element abundances of tholeiitic basalts and basaltic andesites. *Chemical Geology*, **51**, 303–323.

JACOBSEN, O. S. & LAIER, T. 1984. Analysis of gas and water samples from the Vestmanna-1 and Lopra-1 wells, Faeroe Islands. *In*: BERTHELSEN, O., NOE-NYGAARD, A. & RASMUSSEN, J. (eds) *The Deep Drilling Project 1980–1981 in the Faeroe Islands*. Føroya Fródskaparfelag, Tórshavn, pp. 149–155.

JONES, E. J. W. & RAMSAY, A. T. S. 1982. Volcanic ash deposits of early Eocene age from the Rockall Trough. *Nature*, **299**, 342–344.

JORON, J. L., BOUGAULT, H., MAURY, R. C., BOHN, M. & DESPRAIRIES, A. 1984. Strongly depleted tholeiites from the Rockall Plateau margin, North Atlantic: Geochemistry and mineralogy. *In*: ROBERTS, D. G., SCHNITKER, D., *et al.*; *Initial Reports of the Deep Sea Drilling Project*. US Government Printing Office, Washington, **81**, pp. 783–794.

JØRGENSEN, O. 1984. Zeolite zones in the basaltic lavas of the Faeroe Islands. A quantitative description of the secondary minerals in the deep wells of Vestmanna-1 and Lopra-1. *In*: BERTHELSEN, O., NOE-NYGAARD, A. & RASMUSSEN, J. (eds) *The Deep Drilling Project 1980–1981 in the Faeroe Islands*. Føroya Fródskaparfelag, Tórshavn, pp. 71–91.

LARSEN, H. C. 1988. A multiple and propagating rift model for the NE Atlantic (extended abstract). *In*: MORTON, A. C. & PARSON, L. M. (eds) *Early Tertiary Volcanism and the Opening of the NE Atlantic*. Geological Society of London Special Publication, **39**, pp. 157–158.

LARSEN, L. M. & WATT, W. S. 1985. Episodic volcanism during break-up of the North Atlantic: evidence from the East Greenland plateau basalts. *Earth and Planetary Science Letters*, **73**, 105–116.

LAUGHTON, A. S. 1975. Tectonic evolution of the northeast Atlantic Ocean: a review. *Norges Geologiske Undersøgelse*, **316**, 169–193.

LUND, J. 1981. Eine Ober-Paläozäne Mikroflora von den Färöern, Dänemark. *Courier Forschnung-Institut Senckenberg*, **50**, 41–45.

MORTON, A. C., EVANS, D., HARLAND, R., KING, C. & RITCHIE, D. 1988. Volcanic ash in a cored borehole W of the Shetland Isles: evidence for Selandian (Late Palaeocene) volcanism in the Faeroes region. *In*: MORTON, A. C. & PARSON, L. M. (eds) *Early Tertiary Volcanism and the Opening of the NE Atlantic*. Geological Society of London Special Publication, **39**, pp. 263–269.

NESS, G., LEVI, S. & COUCH, R. 1980. Marine magnetic anomaly time-scales for the Cenozoic and Late Cretaceous. A précis, critique and synthesis. *Reviews of Geophysics and Space Physics*, **18**, 753–770.

NIELSEN, P. H. & BROOKS, C. K. 1981. The E Greenland rifted continental margin: an examination of the coastal flexure. *Journal of the Geological Society of London*, **138**, 559–568.

——, WAAGSTEIN, R., RASMUSSEN, J. & LARSEN, B. 1979. Marine seismic investigation of the shelf around the Faeroe Islands. *Fródskaparrit* (Tórshavn), **27**, 102–112.

NOE-NYGAARD, A. 1968. On extrusion forms in plateau basalts. Shield volcanoes of 'scutulum' type. *Science in Iceland*, **1**, 10–13.

—— & RASMUSSEN, J. 1968. Petrology of a 3000 metre sequence of basaltic lavas in the Faeroe islands. *Lithos*, **1**, 268–304.

PÁLMASON, G. 1965. Seismic refraction measurements of the basalt lavas of the Faeroe Islands. *Tectonophysics*, **2**, 475–482.

PARRA, M., DELMONT, P., DUMON, J. C., FERRAGNE, A. & PONS, J. C. 1987. Mineralogy and origin of Tertiary interbasaltic clays from the Faeroe Islands, northeastern Atlantic. *Clay Minerals*, **22**, 63–82.

RASMUSSEN, J. & NOE-NYGAARD, A. 1969. Beskrivelse til geologisk kort over Færøerne. *Danmarks Geologiske Undersøgelse 1. serie*, **24**, 370 pp.

—— & —— 1970. Geology of the Faeroe Islands. *Danmarks Geologiske Undersøgelse 1. series*, **25**, 142 pp.

RIDD, M. F. 1983. Aspects of the Tertiary geology of the Faeroe–Shetland Channel. *In*: BOTT, M. H. P., SAXOV, S., TALWANI, M. & THIEDE, J. (eds). *Structure and Development of the Greenland–Scotland Ridge*. Plenum Press, New York, pp. 91–108.

ROBERTS, D. G. 1975. Marine geology of the Rockall Plateau and Trough. *Philosophical Transactions of the Royal Society of London*, **A278**, 447–509.

——, BOTT, M. H. P. & URUSKI, C. 1983. Structure and origin of the Wyville-Thomson Ridge. *In*: BOTT, M. H. P., SAXOV, S., TALWANI, M. & THIEDE, J. (eds) *Structure and Development of the Greenland–Scotland Ridge*. Plenum Press, New York, pp. 133–158.

——, MORTON, A. C. & BACKMAN, J. 1984. Late Palaeocene–Eocene volcanic events in the northern North Atlantic Ocean. *In*: ROBERTS, D. G., SCHNITKER, D. *et al. Initial Reports of the Deep Sea Drilling Project*. US Government Printing Office, Washington, **81**, pp. 913–923.

SAXOV, S. 1969. Gravimetry in the Faroe Islands. *Geodætisk Institut, Meddelelse*, No **43**, 24 p.

SCHILLING, J.-G. & NOE-NYGAARD, A. 1974. Faeroe–Iceland plume: rare-earth evidence. *Earth and Planetary Science Letters*, **24**, 1–14.

SCHOENHARTING, G. & ABRAHAMSEN, N. 1984. Magnetic investigations on cores from the Lopra-1 drillhole, Faeroe Islands. *In*: BERTHELSEN, O., NOE-NYGAARD, A. & RASMUSSEN, J. (eds) *The Deep Drilling Project 1980–1981 in the Faeroe Islands*. Føroya Fródskaparfelag, Tórshavn, pp. 109–114.

SCHRØDER, N. F. 1971. Magnetic anomalies around the Faeroe Islands. *Fródskaparrit* (Tórshavn), **19**, 20–29.

SMYTHE, D. K. 1983. Faeroe–Shetland Escarpment and continental margin north of the Faeroes. *In*: BOTT, M. H. P., SAXOV, S., TALWANI, M. & THIEDE, J. (eds) *Structure and Development of the Greenland–Scotland Ridge*. Plenum Press, New York, pp. 109–119.

——, CHALMERS, J. A., SKUCE, A. G., DOBINSON, A. & MOULD, A. S. 1983. Early opening history of the North Atlantic—I. Structure and origin of the Faeroe–Shetland Escarpment. *Geophysical Journal of the Royal Astronomical Society*, **72**, 373–398.

TARLING, D. H. 1970. Palaeomagnetic results from the Faeroe Islands. *In*: RUNCORN, S. K. (ed) *Palaeogeophysics*, Academic Press, London, pp. 193–208.

—— 1983. *Paleomagnetism. Principles and applications in geology, geophysics and archaeology*. Chapman & Hall, London & New York, 379 pp.

—— & GALE, N. H. 1968. Isotopic dating and palaeomagnetic polarity in the Faeroe Islands. *Nature*, **218**, 1043–1044.

TARNEY, J., WOOD, D. A., VARET, J., SAUNDERS, A. D. & CANN, J. R. 1979. Nature of mantle heterogeneity in the North Atlantic: evidence from Leg 49 basalts. *In*: TALWANI, M., HARRISON, C. G. & HAYES, D. E. (eds) *American Geophysical Union, Maurice Ewing Series*, **2**, 285–301.

WAAGSTEIN, R. 1977. *The geology of the Faeroe Plateau*. Licentiatus Scientiarum thesis. University of Copenhagen, 183 pp.

—— & HALD, N. 1984. Structure and petrography of a 660 m lava sequence from the Vestmanna-1 drill hole, lower and middle basalt series, Faeroe Islands. *In*: BERTHELSEN, O., NOE-NYGAARD, A. & RASMUSSEN, J. (eds) *The Deep Drilling Project 1980–1981 in the Faeroe Islands*. Føroya Fródskaparfelag, Tórshavn, pp. 39–65.

——, ——, JØRGENSEN, O., NIELSEN, P. H., NOE-NYGAARD, A., RASMUSSEN, J. & SCHÖNHARTING, G. 1984. Deep drilling on the Faeroe Islands. *Bulletin of the Geological Society of Denmark*, **32**, 133–138.

WOOD, D. A. 1978. Major and trace element variations in the Tertiary lavas of eastern Iceland and their significance with respect to the Iceland geochemical anomaly. *Journal of Petrology*, **19**, 393–436.

——, JORON, J.-L., TREUIL, M., NORRY, M. & TARNEY, J. 1979. Elemental and Sr isotope variations in basic lavas from Iceland and the surrounding ocean floor. *Contributions to Mineralogy and Petrology*, **70**, 319–339.

R. WAAGSTEIN, Geological Survey of Denmark, Føroyadeild, Debesartrød, FR-100 Tórshavn, Faeroe Islands.

Volcanism in Basins to the N and W of the British Isles

The geochemistry and origin of the Faeroe–Shetland sill complex

F. G. F. Gibb & R. Kanaris-Sotiriou

SUMMARY: Geochemical investigations of dolerite cores from four intrusions in the recently discovered Faeroe–Shetland sill complex have established that the sills are of transitional (T) mid-ocean ridge basalt (MORB)-type composition. Some uncertainty surrounds the age of the complex, but there is no doubt that it is, at least in part, of Tertiary age. Comparisons with previously proposed models for the development of sill–sediment complexes during initial stages of seafloor spreading suggests that the Faeroe–Shetland sills may represent an intrusive episode associated with a spreading axis that eventually produced oceanic crust W of the Faeroes.

During the last ten years, hydrocarbon exploration N and W of the Shetland Isles has revealed the existence of horizons of igneous rock within the Mesozoic and Cainozoic sediments of the Faeroe–Shetland and Møre Basins. There is little doubt that many of these igneous rocks are intrusive since they thermally metamorphose and are chilled against the overlying sediments. They are parts of an extensive 'belt' of intrusions which we refer to as the Faeroe–Shetland sill complex. This occurs below, and protrudes beyond, the SE subcrop limit of the Faeroes Lower Series lavas (Fig. 1). The limits of this complex are known from drilling and seismic investigations but, because of the seismic opacity of the lavas, little is known about how far it extends to the NW.

This paper reports on the geochemistry of four tholeiitic sills recovered as core from two wells, 219/20–1 and 208/21–1, and discusses possible models for their origin.

Structure of the sill complex

The exact location of individual sills can only be determined by drilling but the approximate positions of the uppermost members of the complex can be seen on seismic reflection profiles. A NNW–SSE seismic profile across the SE edge of the sill complex in the general vicinity of Well 219/20–1 is shown in Fig. 2b along with a geological interpretation (Fig. 2a). It is evident from Fig. 2 that, whatever the form of the complex below the upper level of sills (which seismically obscure the lower levels), its top is 'transgressive'. We have examined a number of similar seismic profiles and all demonstrate that the top of the sill complex is intruded at stratigraphically higher levels to the NW.

A number of commercial wells have intersected the Faeroe–Shetland sill complex. In Well 208/15–1A (see Fig. 1 for locations of wells) eight intrusions, ranging in thickness from 2 m to more than 340 m, were penetrated in the depth interval 1935–3123 m. The distributions of igneous horizons in Wells 208/21–1 and 219/20–1 are given in Fig. 3 but with arbitrary depth-scale zeros to preserve the confidentiality of the true depths. The lateral dimensions of individual sills are unknown, but it appears from the seismic data that they can extend for several kilometres.

It appears from the seismic and well data available to us that the complex is at least 1500 m thick in places, 'feathers out' to the SE and its upper surface descends to lower stratigraphic levels towards the feather edge.

Age of the sills

There is some uncertainty regarding the age(s) of the sills with suggestions ranging from Campanian to Eocene.

In Well 219/20–1 the host rocks to most of the sills are Campanian, but the uppermost sill occurs in Maastrichtian sediments (see Fig. 3). The intrusions in Well 208/15–1A occur in calcareous mudstones of Lower Palaeocene (Danian) age, while in Well 208/21–1 the lower group of igneous horizons (see Fig. 3), which are unquestionably sills, occurs in uppermost Maastrichtian sediments. The upper group of four horizons in this last well occurs in Upper Palaeocene sediments but we can not be absolutely certain that they are sills since these horizons have not been cored. The stratigraphic age of the host rocks can, of course, only indicate that at least some of the sills are younger than the youngest host rock. Hence, from the above, some of the sills are certainly Danian or younger and, if the top four horizons in Well 208/21–1 are sills, they are Upper Palaeocene or younger.

Hitchen & Ritchie (1987) cite K/Ar whole-rock radiometric ages for members of the Faeroe–

From MORTON, A. C. & PARSON, L. M. (eds), 1988, *Early Tertiary Volcanism and the Opening of the NE Atlantic,* Geological Society Special Publication No. 39, pp. 241–252.

241

F. G. F. Gibb & R. Kanaris-Sotiriou

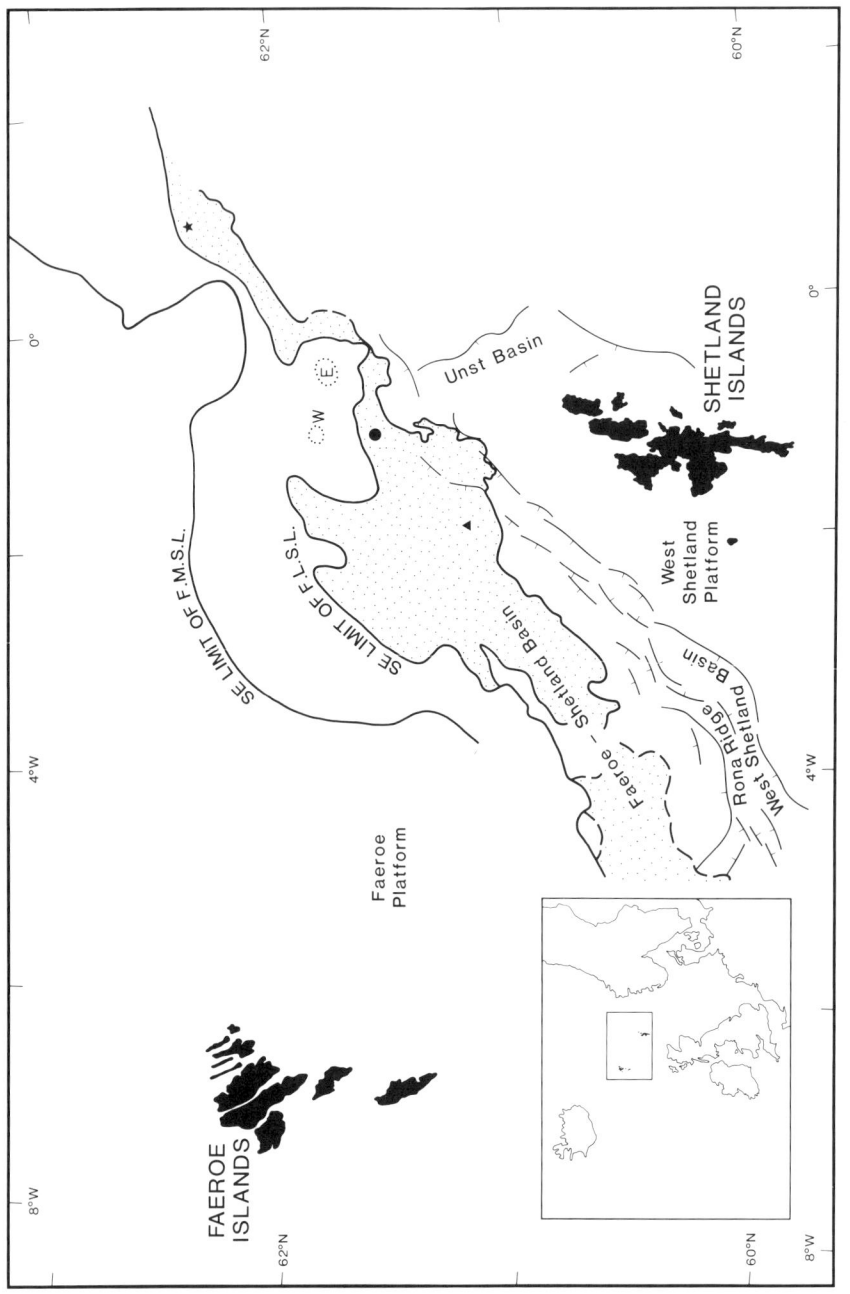

FIG. 1. Map showing the location of the Faeroe–Shetland sill complex (stippled). F.M.S.L. = Faeroes Middle Series Lavas; F.L.S.L. = Faeroes Lower Series Lavas. Geological boundaries are based on data from Ridd (1983), Gatliff *et al.* (1984), Smythe *et al.* (1983), Hitchen & Ritchie (1987) and other sources. Structural elements are after Mudge & Rashid (1987). W, E = W Erlend and Erlend complexes respectively. Well locations are as follows: ∗ = 219/20–1; ● = 208/15–1A; ▲ = 208/21–1.

NW   SE

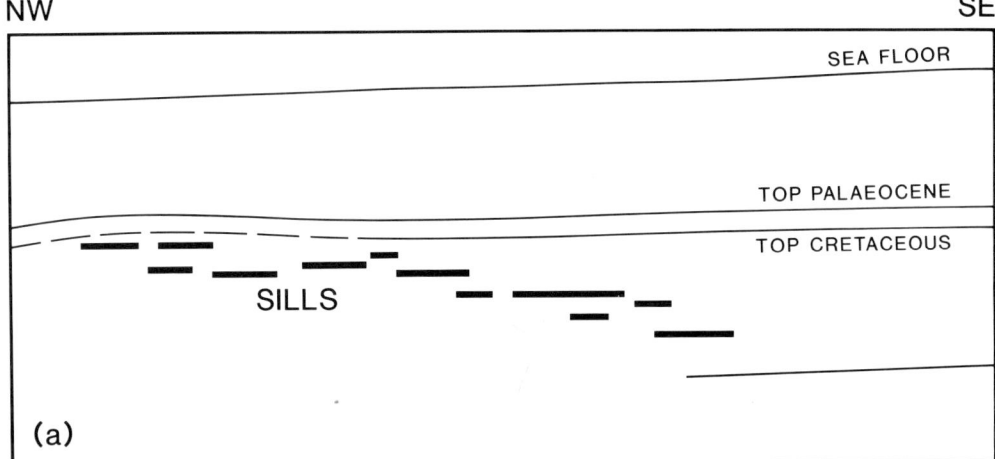

(a)

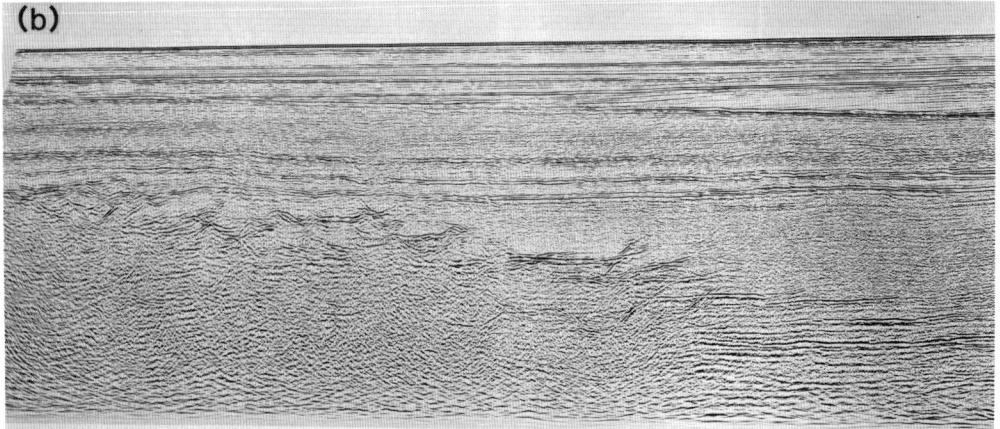

(b)

FIG. 2. (a) Geological interpretation of seismic reflection profile shown in (b); (b) Migrated NNW–SSE seismic profile across the SE edge of the sill complex in the vicinity of Well 219/20–1. The length represented by the profile is 39 km.

Shetland Basin 'intrusive belt' (i.e. the sill complex) from four commercial wells as follows:

Well 208/15–1A   59 Ma (Late Palaeocene)
                   65 Ma (Maastrichtian)
Well 209/6–1X   72 Ma (Campanian)
Well 214/27–1   55 Ma (Early Eocene)
Well 214/28–1   61 Ma (Palaeocene)
                   63 Ma (Palaeocene)

From this they conclude (Hitchen, pers. comm.) that the igneous activity extended from Late Cretaceous to earliest Eocene. Also using radiometric data, Mudge & Rashid (1987) suggested that the sills intruding the Upper Cretaceous sediments in the NE Faeroe Basin (syn.

Faeroe–Shetland Basin) are of Late Cretaceous age, although they recognized that the intrusions in Well 208/15–1A are probably Palaeocene.

Duindam & Van Hoorn (1987) record sills in the Upper Cretaceous and lowermost Palaeocene sequence of the Faeroe Basin but subsequently ascribe the 'dense network of sills' to an intrusive phase 'towards the end of the Cretaceous'. Further, they regard this phase as being 'contemporaneous with the formation of oceanic and transitional crust in the Rockall Trough . . . and Møre Basin'.

In a preliminary account of the Faeroe–Shetland sill complex Gibb *et al.* (1986) suggested that the sills are 'no older than Palaeocene' and, on the basis of chemical similarities to the

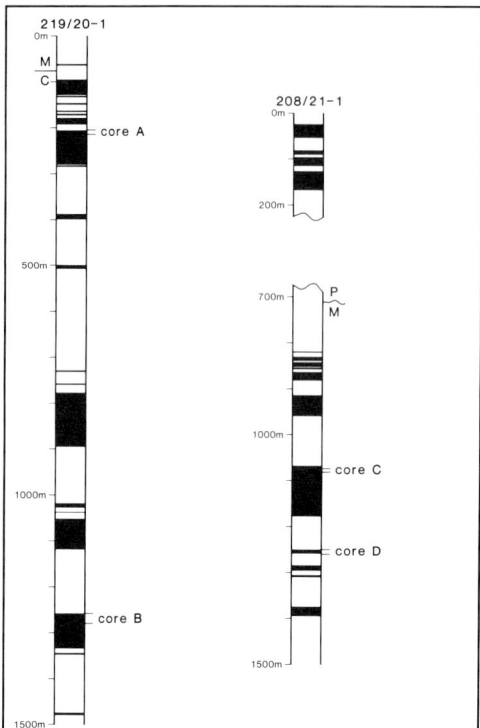

FIG. 3. Distribution of sills (solid) and positions of cored sections in Wells 219/20–1 and 208/21–1. M/C = Maastrichtian/Campanian boundary; P/M = Palaeocene/Maastrichtian boundary. The zero of the vertical scale has been arbitrarily set (see text).

Faeroes Upper Series Lavas, inferred that they are of Eocene age.

The basaltic rocks forming these sills are not ideal materials for whole-rock K/Ar dating, not least because of their low K contents and (?) secondary alteration (see also the discussion on K/Ar dating in Fitch *et al.*, 1988). Attempts made on our behalf to date sill samples from Well 219/20–1 gave results ranging from 44–105 Ma, with samples from the same sill yielding results differing by tens of My. Consequently, radiometric ages from the sills based on whole-rock K/Ar methods must be treated with caution.

A further insight into the minimum age of the sill complex may be gained from migrated seismic reflection profiles such as the one shown in Fig. 2b. Towards the NW, where the topmost sills of the complex are at the highest level, they appear to disturb or disrupt the prominent 'top-Cretaceous' reflector. This could be interpreted in several ways:

(1) The sills could be 'rising through' the top of the Cretaceous sediments so that, at the NW end of the section, they are disrupting and intruding the Palaeocene sediments. This would be consistent with the known occurrence of sills in Danian mudstones (Well 208/15–1A).

(2) The igneous horizons could be transgressing upwards to the *contemporary* surface such that they change from intrusive sills to extrusive lavas at the NW end of the section. In this case the lavas (and sills) would be Palaeocene.

(3) The top-Cretaceous reflector could be 'domed up' over the sill complex as a result of distension due to the emplacement of the sills (as shown in Fig. 2a). In such a case the distension would be greatest where the density of sills is greatest, i.e. to the NW away from the feather edge.

(4) The top-Cretaceous reflector could be 'draped over' the sill complex, due either to deposition on an irregularly domed surface and/or to differential compaction of the underlying sills and sediments. The compaction would be least where the density of the sills is greatest.

In all but the last of these cases the igneous activity would be demonstrably post-Cretaceous. However, the top-Palaeocene reflector (Fig. 2) also appears to be domed up (rather than draped) over the highest part of the sill complex, strongly suggesting a post-Palaeocene emplacement age for at least the youngest of the sills.

In summary, if the sill complex formed over a *relatively restricted period* (e.g. of the order of 1 My) then, on the basis of the host sediments, it must be no older than Lower Palaeocene (Upper Palaeocene if the topmost igneous horizons in Well 208/21–1 are parts of it). Such ages could be consistent with a genetic relationship between the sills and either the late Palaeocene lavas of the Erlend Complex (Gatliff *et al.* 1984) or the Faeroes lavas (Gibb *et al.* 1986). However, these relationships need to be reconciled with the structural inference that the sill complex has a maximum age of Eocene (see also p. 245).

Alternatively, the sills could have been emplaced over a period of several million years, although it may be difficult to envisage an episode of intrusive activity lasting from Campanian to late Palaeocene as the radiometric ages suggest.

Concurrent work (Fitch *et al.* 1988) records Ar/Ar ages of 80 Ma (Campanian) and 50 Ma (mid-Eocene) for two sills in Well 219/28–2. The younger of these two sills is chemically almost identical to the sills described here and is almost certainly part of the same (tholeiitic) sill complex. In contrast, the older of the sills in Well 219/28–2 is of a distinctly alkali olivine basalt affinity and must belong to a different episode of magmatic activity. Hence, while there were at least two episodes of sill emplacement in the area, there are as yet no grounds for believing that the tholeiitic sills of the Faeroe–Shetland Complex were emplaced either individually or in episodes over a period of more than 1–2 My. Further, the 50 Ma Ar/Ar date appears to confirm the suggestion (above and Gibb et al., 1986) that this complex is of Eocene age.

Chemistry of the sills

Two of the sills in each of Wells 208/21–1 and 219/20–1 have been sampled by coring. The relative positions of these cores (A–D) are given in Fig. 3. All the sampled sections, except D, indicate that the sills are internally differentiated. The details of these internal variations will be

published elsewhere and it is sufficient for present purposes to record that all the sampled sections, except for parts of C, are basaltic and that only data for basaltic rocks are presented here.

Petrographically these rocks are dolerites or olivine dolerites exhibiting slight to moderate degrees of alteration. In most of the dolerites there are green/brown pseudomorphs after olivine, and some contain small amounts of slightly altered orthopyroxene. The grain size varies from that of very fine, chilled dolerite to gabbroic or pegmatitic dolerite. The modal mineralogy clearly indicates that the sills are olivine tholeiites.

Forty-five dolerite samples have been analyzed for major and trace elements by X-ray fluorescence spectrometry and for rare-earth elements (REE) by inductively coupled plasma (ICP) spectrometry. Representative analyses for samples from each of the sills are presented in Table 1.

TABLE 1. *Representative analyses and CIPW norms*

wt%	A	B	C	D		ppm	A	B	C	D
SiO_2	45.80	47.64	45.94	42.94		V	185	331	382	287
TiO_2	1.01	1.14	1.48	1.40		Cr	488	270	187	110
Al_2O_3	15.86	14.28	15.72	15.19		Ni	368	102	67	110
Fe_2O_3	3.78	2.90	4.41	2.79		Cu	161	164	168	96
FeO	7.05	8.80	7.72	8.56		Zn	76	80	78	72
MnO	0.16	0.21	0.16	0.18		Rb	6	1	5	2
MgO	10.50	7.96	6.26	7.31		Sr	135	120	182	244
CaO	10.80	13.11	12.37	10.00		Y	18	22	23	27
Na_2O	1.62	1.81	2.15	2.20		Zr	61	70	66	87
K_2O	0.17	0.10	0.28	0.20		Nb	3	2	5	5
P_2O_5	0.07	0.08	0.07	0.12		Ba	56	76	154	114
S	0.16	0.09	0.25	0.28		La	8.0	4.4	2.8	5.4
CO_2	0.14	0.18	1.17	7.03		Ce	12.0	9.8	7.2	12.3
H_2O^+	3.21	1.46	1.86	1.40		Nd	9.2	8.2	6.1	8.5
Total	100.33	99.76	99.84	99.60		Sm	2.8	2.6	2.5	3.1
						Eu	1.0	1.0	1.0	1.1
$Fe_2O_3^t$	11.61	12.68	12.99	12.30		Gd	—	3.7	3.6	4.2
Mg†	72.6	61.7	59.1	60.35		Dy	3.2	4.1	4.1	4.2
						Ho	0.7	0.9	0.8	0.9
*	A	B	C	D		Er	3.5	2.5	2.5	2.6
						Yb	2.0	2.3	2.6	2.5
or	1.06	0.59	1.71	1.30		Lu	0.3	0.4	0.4	0.4
ab	14.22	15.65	18.87	20.48						
an	36.73	31.21	33.69	34.17						
di	14.84	28.49	24.49	16.15						
hy	13.44	11.55	3.75	5.87						
ol	15.49	7.87	12.09	16.46						
mt	2.07	2.23	2.32	2.33						
il	1.99	2.20	2.92	2.92						
ap	0.17	0.19	0.17	0.31						

A—normal olivine dolerite from sill A
B—normal olivine dolerite from sill B
C—coarse ophitic olivine dolerite sill C
D—normal olivine dolerite from sill D

Sills A and B are from Well 219/20–1
Sills C and D are from Well 208/21–1
(see Fig. 3 for positions within wells)

* = CIPW norms (Fe_2O_3/FeO set to 0.15); (analyses rescaled volatile-free)

^{t} = total Fe as Fe_2O_3
† = [MgO/(MgO + FeO)] × 100 (moles)

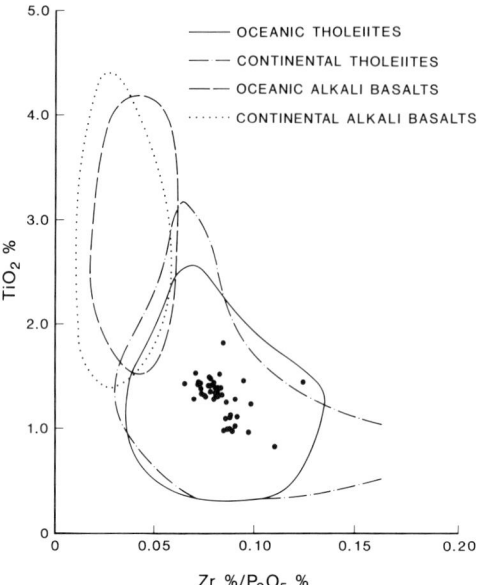

FIG. 4. TiO_2–Zr/P_2O_5 diagram demonstrating the tholeiitic nature of the sill rocks; field boundaries after Floyd & Winchester (1975).

Plotting these analyses on a TiO_2–Zr/P_2O_5 diagram (Fig. 4) indicates that the sills are tholeiitic, and this is confirmed by their normative mineralogy (see Table 1) and relatively high Y:Nb ratios (4–12:1). The TiO_2–K_2O–P_2O_5 diagram (Fig. 5) of Pearce *et al.* (1975) discriminates between fields containing 93% of oceanic and over 80% of continental 'primitive' basalts. The fact that all the analyzed dolerites clearly

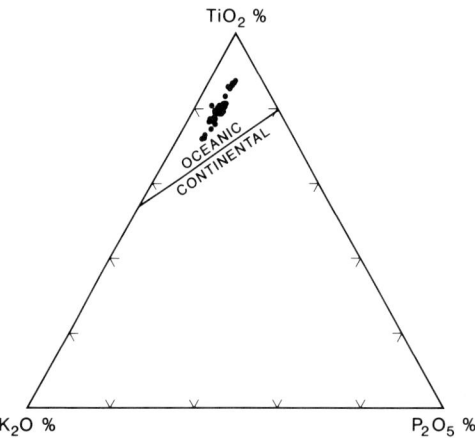

FIG. 5. TiO_2–K_2O–P_2O_5 diagram for the sill rocks; continental/oceanic discriminant after Pearce *et al.* (1975).

plot in the former field is a strong indication that they are oceanic. Using the Ti–Zr–Y diagram (Fig. 6a) in conjunction with a Ti–Zr plot (Fig. 6b) as recommended by Pearce & Cann (1973) it appears that the sill rocks, taken as a single group, are of MORB-type. Unfortunately, as has been pointed out by Meschede (1985, 1986) and others, the Ti–Zr–Y diagram does not distinguish all continental tholeiites from MORB or volcanic arc basalts. However, the sill rocks can be characterized beyond any reasonable doubt using the Nb–Zr–Y diagram of Meschede (1986) (Fig. 7), which is applicable to all tholeiitic and within-plate basalts. From this diagram it is obvious that the sills can only be MORB or volcanic arc low-K tholeiite-type. Quite apart from the geological improbability of a volcanic arc in the Faeroe–Shetland Basin, the latter possibility can be eliminated using the Ti/Cr diagram of Pearce (1975) (Fig. 8) on which the sills plot as MORB.

It appears from Fig. 7 that not only are these sills of MORB-type but they are N (normal) rather than E (enriched or plume)-type MORB. However, Gibb *et al.* (1986) suggested, on the basis of Ti:V ratios, that the Faeroe–Shetland sill complex is T-type MORB (i.e. it is transitional between N- and E-type). Normalized REE patterns for samples from two of the cored sills are presented in Fig. 9a & b. It is evident from these patterns that the sills lack the enrichment in light REE which characterize E-type MORB (Fig. 9c). A notable feature of these sills is that they are relatively low in SiO_2 for ocean-floor basalts and thus invite comparison with other anomalously low SiO_2 basalts in the N Atlantic area, especially those from the Faeroes and Iceland.

Schilling *et al.* (1982) have recorded a temporal variation in the nature of Icelandic basalts. Those of Miocene–Pleistocene age show typical E-type MORB enrichment in light REE whereas younger basalts from the neovolcanic zones have REE patterns ranging from light REE enriched to the light REE depleted type characteristic of normal mid-ocean ridge segments. Normalized REE patterns for some recent basalts from the central rift zone of Iceland are shown in Fig. 9d and the similarities between these T-type MORBs and the sills (Fig. 9a & b) are marked. Probably the most appropriate comparison in view of the geographical relationship of the sill complex to the Faeroe–Iceland Ridge is with the basalts of the mid-Atlantic Ridge N of Iceland (Fig. 9e). These latter basalts are generally regarded as being erupted from a normal ridge segment. While the Faeroe–Shetland sills are clearly different from the more strongly light REE depleted basalts in Fig. 9e, they are remarkably

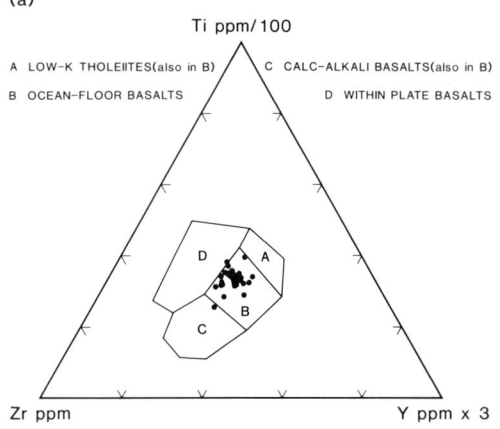

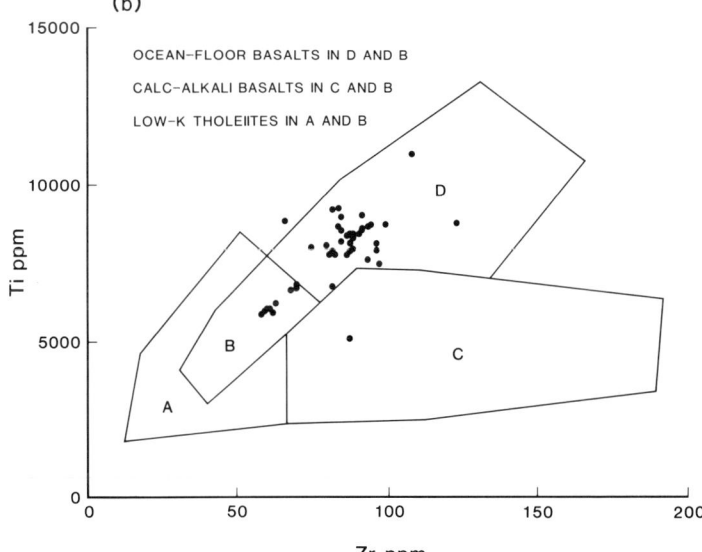

FIG. 6. (a) Ti–Zr–Y diagram for the sill rocks; fields A–D after Pearce & Cann (1973); (b) Ti–Zr diagram for the sill rocks; fields A–D after Pearce & Cann (1973).

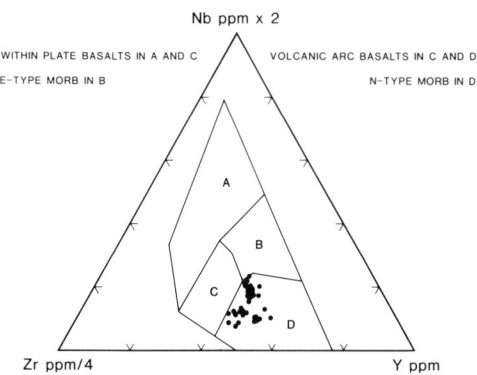

FIG. 7. Nb Zr Y diagram for the sill rocks; fields A–D after Meschede (1986).

similar to the least depleted examples. All the evidence therefore points to the sills being of T-type MORB.

On the basis of major and trace-element chemistry Gibb *et al.* (1986) concluded that the sills are significantly different from the Lower and Middle Series lavas of the Faeroes but that they closely resemble the Upper Series lavas. Comparison of Fig. 9a & b with the REE patterns for Faeroes Upper Series lavas (Fig. 9f) indicates that the REE patterns for the sills fall within the range exhibited by the lavas and, hence, such a correlation is not precluded on chemical grounds.

A more detailed account of the chemistry of the individual sills and their internal variations will be published elsewhere.

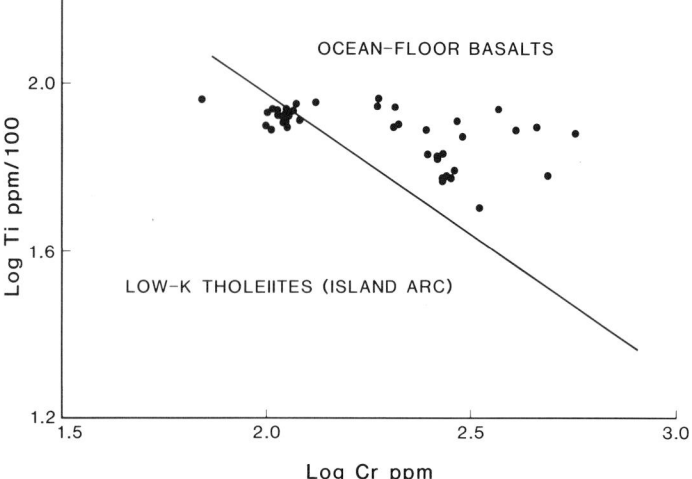

FIG. 8. Ti–Cr diagram for the sill rocks; discriminant line after Pearce (1975).

Discussion

It has been established that the sills of the Faeroe–Shetland Tertiary Complex are of MORB-type and there is little doubt that they were emplaced in an actively subsiding sedimentary basin. There are few documented accounts of sill–sediment complexes of this type. However, we wish to draw attention to certain similarities and 'coincidences' between the Faeroe–Shetland sill complex and the genetic model proposed by Einsele (1985) for sill–sediment complexes associated with active spreading centres such as the Guaymas Basin in the Gulf of California.

Prior to the formation of oceanic crust in a pull-apart basin the continental crust is subjected to tension, listric faulting, rifting and subsidence accompanied by the deposition of synrift sediments. This is not unlike the situation which is generally agreed to have existed in the Faeroe–Shetland Basin, and which produced intermediate-type crust during the Cretaceous and early Tertiary.

When oceanic crust does begin to form at a spreading centre of the Guaymas basin-type, the magma, unlike that at a 'normal' mid-ocean ridge, rarely extrudes onto the contemporary seafloor as lavas but is intruded into the relatively unconsolidated sediments as dykes and sills. By comparing the incidence of sills in such complexes with known spreading rates and the sizes of the intrusions in ophiolitic sheeted dyke complexes, Einsele (1985) suggested that a sill is only added to the sill–sediment complex when an exceptionally large pulse of magma is involved. On such occasions the magma in excess of that required to form the sheeted dyke is intruded laterally at a level where the magma pressure can overcome the lithostatic pressure and tensile strength of the sediments. According to Einsele, such intrusions will occur preferentially in the soft sediments 'some 100 m below the sea floor'. As each sill adds to the lithostatic pressure and increases the tensile strength due to induration of the sediments, successively younger members of the sill complex are intruded at higher and higher levels so that the complex is built up vertically and horizontally as shown in Fig. 10a. It is noteworthy, certainly in the context of the NE part of the Faeroe–Shetland sill complex, that in such a model the sills become stratigraphically lower and buried under increasing thicknesses of younger sediments away from the centre of the basin (Fig. 10a & b). Whether these observations apply equally to the SW part of the complex is uncertain at present, but it does appear to apply to the sills in the Møre and Vøring Basins further to the NE (Skogseid, pers. comm.)

Einsele (1985) has also suggested that if the sedimentation rate at the basin axis remains sufficiently high in relation to the frequency and volume of magmatic activity, the sill–sediment complex will continue to form (Fig. 10b). If, however, the sediment supply decreases, the vertical intervals between the sills will decrease and the magma may reach the sea floor to form pillows and lava flows. Since the rate of sedimentation at the axis of such a basin would normally decrease as the distance of the axis from the source of sediment (usually the margins) in-

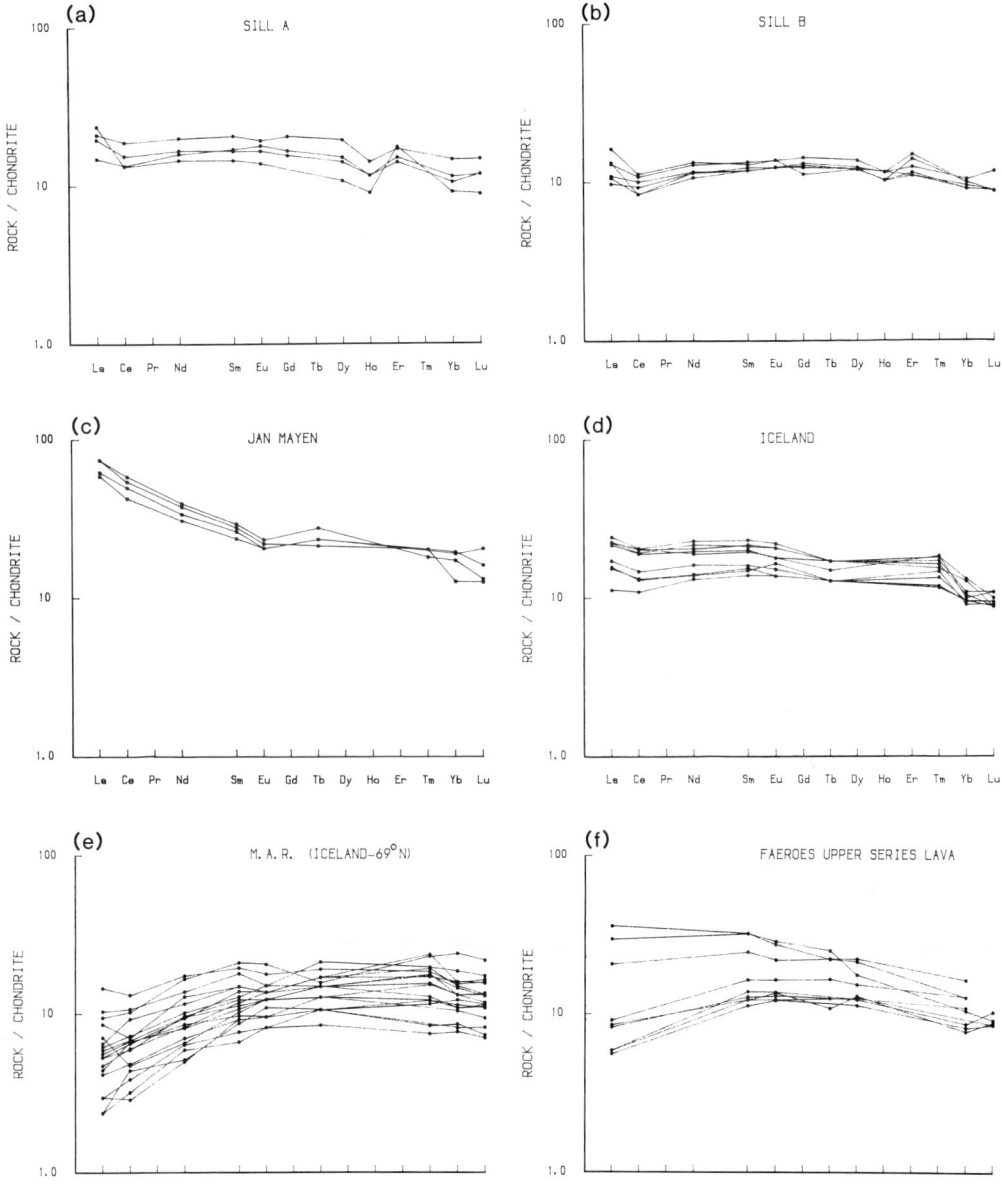

FIG. 9. Chondrite normalized REE diagrams with all REE concentrations normalized to the values given by Wakita *et al.* (1971). (a) Sill A; (b) Sill B; (c) MORBs from Jan Mayen; (d) Recent basalts from central Iceland; (e) Basalts from the mid-Atlantic ridge between Iceland and 69°N; (f) Faeroes Upper Series lavas. Data for (c), (d) and (e) from Schilling *et al.* (1983); data for (f) from Schilling & Noe-Nygaard (1974).

creases, it is to be expected that the sill complex will transgress upwards to the contemporary surface and pass laterally into lavas towards the axis of a mature spreading centre (Fig. 10c). In this context attention is drawn to the spatial relationship between the Faeroe–Shetland sill complex and the Faeroes lavas (Fig. 11), to the transgressive form of the complex in Fig. 2b and to its possible lateral transition from sills to lavas towards the NW (case 2, p. 244)

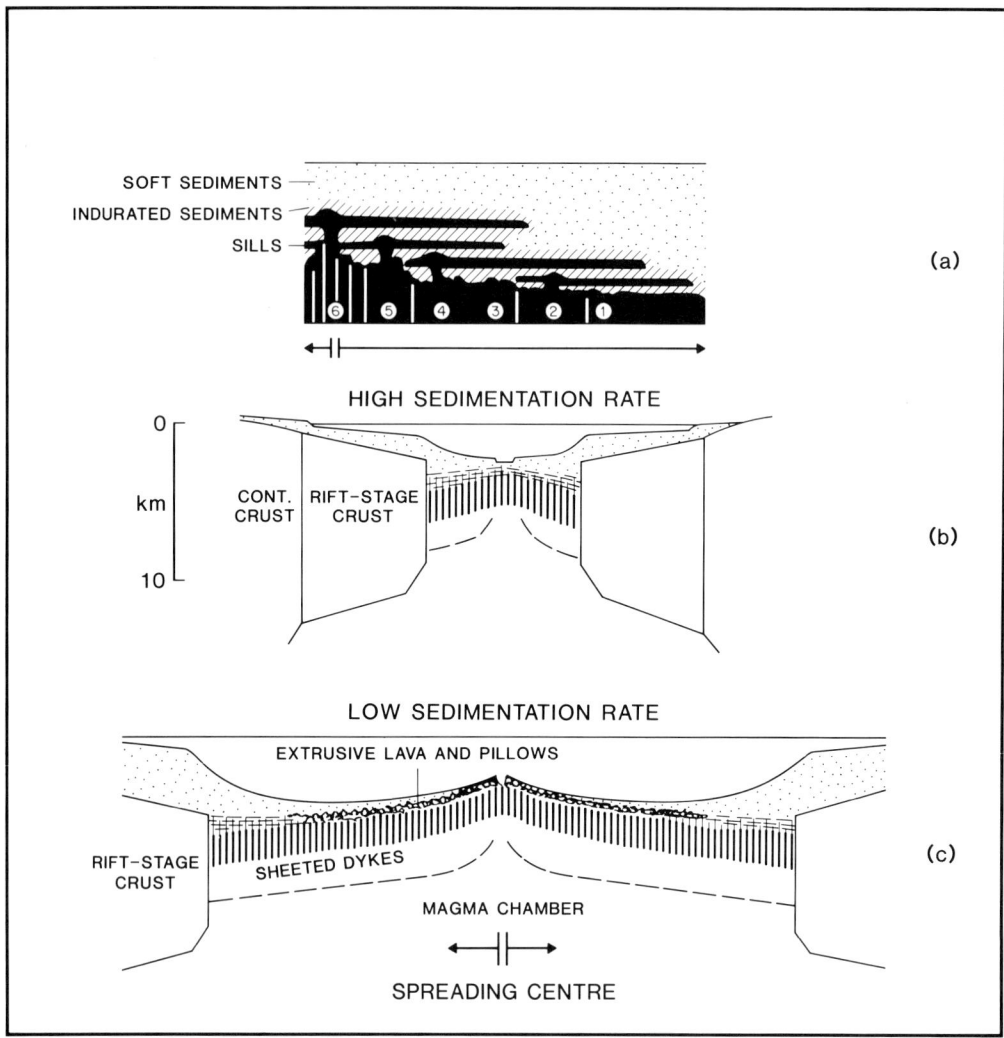

FIG. 10. Aspects of the evolution of sill–sediment complexes after Einsele (1985). (a) Schematic model of vertical and horizontal build-up of a sill sediment complex within a spreading trough, numbers signifying sequence of events; (b) Spreading centre in the early stages of evolution, with high sedimentation rate and continuing growth of sill–sediment complex; (c) More advanced stage of spreading with low sedimentation rate producing extrusion of lavas in central part of the basin. The vertical lines represent sheeted dykes and the horizontal lines sills.

Drawing an analogy between the Faeroe–Shetland sill complex and the model of Einsele is admittedly speculative but it is interesting to consider the inferences of doing so. Firstly, the model would require that oceanic crust in the form of a sheeted dyke complex occurs somewhere beneath the Faeroe–Shetland sill complex. While there is no direct evidence for the occurrence of any such crust, this suggestion could raise questions about the exact nature of the 'Axial Opaque Zone' beneath the SE edge of the

Faeroes Lower Series lavas (Ridd 1983). Secondly, if the model was to apply to the Faeroe–Shetland tholeiitic sill complex, the sills would have to have been emplaced over a protracted period, e.g. at least from Campanian to Eocene. Whilst we have some reservation in accepting this implication, it would be consistent with the radiometric ages and the views of other authors (see p. 243–4).

An earlier model for the genesis of sill–sediment complexes at passive plate margins was

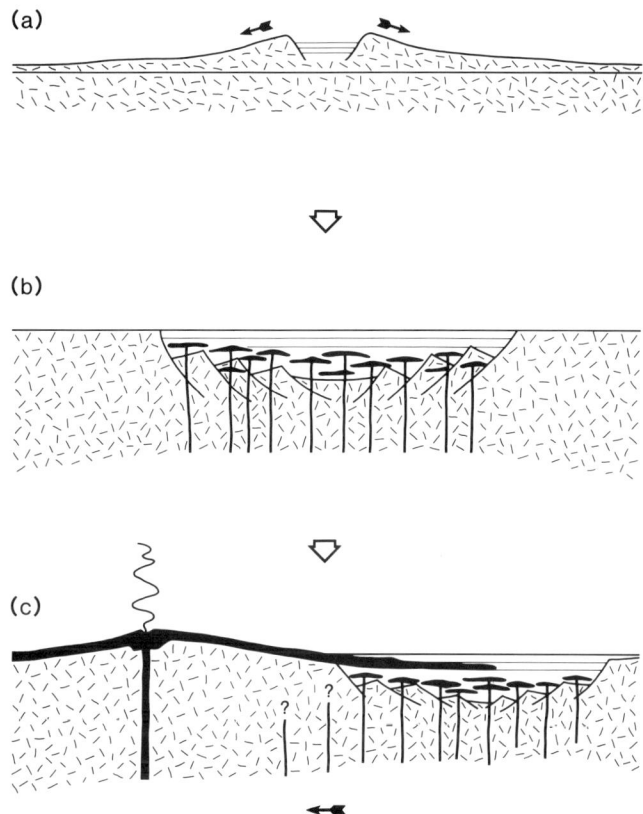

FIG. 11. Model for the formation of sill complexes at passive plate margins, after Sheridan (1981), and its application to the Faeroe–Shetland sill complex. (a) Early stage—uplift, rifting and erosion of continental crust; (b) Intrusion of dykes and sill complex into rift sediments of the Faeroe–Shetland Basin leading to the formation of 'transitional' crust; (c) Migration or 'jump' of active spreading axis to the NW (left) and eruption of Faeroes lavas.

proposed by Sheridan (1981). This model does not invoke the same intensity of dykes feeding the sill complex as the Einsele model (cf. Figs 10 & 11), and hence does not require the generation of oceanic crust contemporaneously with the intrusion of the sill complex. The model thus results in 'a combination of sills and sediments and dykes and continental septa' forming a 'broad zone of heterogeneous crust . . . "transitional crust"' (see Fig. 11). Sheridan also associates this type of crustal configuration with 'ridge-jumps' which would be an appropriate mechanism for the northwesterly relocation of the spreading axis and the ensuing development of oceanic crust W of the Faeroes.

Both the Einsele and Sheridan models have aspects which we consider applicable to the Faeroe–Shetland sill complex. Overall, we think it likely that most of the complex formed in the manner proposed by Sheridan, but to the NW where the sills were emplaced in younger (near contemporaneous?) sediments the mechanism may have merged with that proposed by Einsele.

ACKNOWLEDGEMENTS We are grateful to Conoco (UK) Ltd., Saxon Oil, Lasmo and Elf (UK) for the supply of core material, and to Dr H. Kirchner of Conoco, Dr R. Neves and Prof. C. Downie for invaluable discussions. We thank Drs B. Van Hoorn, K. Hitchen and D. C. Mudge for preprints of papers given at the 3rd Conference on the Petroleum Geology of NW Europe, and Dr F. J. Fitch for a manuscript copy of the paper in this volume.

We are also grateful to Dr J. N. Walsh for the ICP analyses of REE, Mike Cooper for drawing most of the diagrams and Diane Hall for analytical assistance.

Finally, we are particularly grateful to one of the anonymous referees for a detailed and knowledgeable review which led to a considerable improvement of the manuscript.

References

DUINDAM, P. & VAN HOORN, B. 1987. Structural evolution of the West Shetland continental margin. *In*: BROOKS, J. & GLENNIE, K. W. (eds) *Petroleum Geology of NW Europe.* Heyden & Son, London. pp. 765–773.

EINSELE, G. 1985. Basaltic sill–sediment complexes in young spreading centres: genesis and significance. *Geology*, 13, 249–252.

FITCH, F. J., HEARD, G. L. & MILLER, J. A. 1988. Basaltic magmatism of Late Cretaceous and Palaeogene age recorded in wells NNE of the Shetlands. *In*: MORTON, A. C. & PARSON, L. M. (eds) *Early Tertiary Volcanism and the Opening of the NE Atlantic,* Geological Society of London, Special Publication, 39, pp. 253–262.

FLOYD, P. A. & WINCHESTER, J. A. 1975. Magma type and tectonic setting discrimination using immobile elements. *Earth and Planetary Science Letters*, 27, 211–218.

GATLIFF, R. W., HITCHEN, K., RITCHIE, J. D. & SMYTHE, D. K. 1984. Internal structure of the Erlend Tertiary volcanic complex, north of Shetland, revealed by seismic reflection. *Journal of the Geological Society of London*, 141, 555–562.

GIBB, F. G. F., KANARIS-SOTIRIOU, R. & NEVES, R. 1986. A new Tertiary sill complex of mid-ocean ridge basalt type NNE of the Shetland Isles: a preliminary report. *Transactions of the Royal Society of Edinburgh: Earth Sciences*, 77, 223–230.

HITCHEN, K. & RITCHIE, J. D. 1987. Geological review of the West Shetland area. *In*: BROOKS, J. & GLENNIE, K. W. (eds) *Petroleum Geology of NW Europe.* Heyden & Son, London. pp. 737–749.

MESCHEDE, M. 1985. The geochemical character of volcanic rocks of the Basco–Cantabrian basin, northeastern Spain. *Neues Jahrbuch für Geologie Palaeontologie Montasheft*, 2, 115–128.

—— 1986. A method of discriminating between different types of mid-ocean ridge basalts and continental tholeiites with the Nb–Zr–Y diagram. *Chemical Geology*, 56, 207–218.

MUDGE, D. C. & RASHID, B. 1987. The geology of the Faeroe Basin area. *In*: BROOKS, J. & GLENNIE, K. W. (eds) *Petroleum Geology of NW Europe.* Heyden & Son, London. pp. 751–763.

PEARCE, J. A. 1975. Basalt geochemistry used to investigate the past-tectonic environment in Cyprus. *Tectonophysics*, 25, 41–67.

—— & CANN, J. R. 1973. Tectonic setting of basic volcanic rocks determined using trace element analyses. *Earth and Planetary Science Letters*, 19, 290–300.

PEARCE, T. H., GORMAN, B. E. & BIRKETT, T. C. 1975. The TiO_2–K_2O–P_2O_5 diagram: a method of discriminating between oceanic and non-oceanic basalts. *Earth and Planetary Science Letters*, 24, 419–426.

RIDD, M. F. 1983. Aspects of the Tertiary geology of the Faeroe–Shetland Channel. *In*: BOTT, M. H. P., SAXOV, S., TALWANI, M. & THIEDE, J. (eds) *Structure and Development of the Greenland–Scotland ridge.* Plenum Press, New York, 91–108.

SCHILLING, J. G. & NOE-NYGAARD, A. 1974. Faeroe–Iceland Plume: Rare-earth evidence. *Earth and Planetary Science Letters*, 24, 1–14.

——, MEYER, P. S. & KINGSLEY, R. H. 1982. Evolution of the Iceland hotspot. *Nature*, 298, 313–320.

——, ZAJAC, M., EVANS, R., JOHNSTON, T., WHITE, W., DEVINE, J. D. & KINGSLEY, R. 1983. Petrologic and geochemical variations along the mid-Atlantic ridge from 29°N to 73°N. *American Journal of Science*, 283, 510–586.

SHERIDAN, R. E. 1981. Recent research on passive continental margins. *Society of Economic Palaeontologists and Mineralogists Special Publication*, 32, 39–55.

SMYTHE, D. K., CHALMERS, J. A., SKUCE, A. G., DOBINSON, A. & MOULD, A. S. 1983. Early opening history of the North Atlantic: I. Structure and origin of the Faeroe–Shetland escarpment. *Geophysical Journal of the Royal Astronomical Society*, 72, 373–398.

WAKITA, H., REY, P. & SCHMITT, R. A. 1971. Abundances of the 14 rare-earth elements and 12 other trace elements in Apollo 12 samples: five igneous and one breccia rocks and four soils. *Proceedings of 2nd Lunar Science Conference (Suppl. 2: Geochimica et Cosmochimica Acta)*, 1, 1319–1329.

F. G. F. GIBB & R. KANARIS-SOTIRIOU, Department of Geology, University of Sheffield, Sheffield S1 3JD, UK.

Basaltic magmatism of late Cretaceous and Palaeogene age recorded in wells NNE of the Shetlands

F. J. Fitch, G. L. Heard & J. A. Miller

SUMMARY: Two wells drilled by Sovereign in 1984 NNE of the Shetlands (Wells 219/28-1 & 219/28-2) passed through numerous primary basic ashfall tuff horizons within a sequence of abundantly tuffaceous rocks representing the basal early Eocene 'ash-marker'. Small amounts of volcanic debris can be recognized in adjacent strata. In one of the wells, two chemically distinct sills of olivine-dolerite were penetrated below the 'ash-marker' in late Cretaceous rocks. The probable magmatic sequence revealed by petrographic studies, K/Ar and $^{40}Ar/^{39}Ar$ dating is:

(1) The near-sediment/sea water interface intrusion of a dolerite sill into Santonian strata around 80 Ma in early Campanian times (Well 219/28-2).
(2) A major episode of explosive volcanic eruptions depositing many primary basic ashfall tuffs in the basal early Eocene (both wells).
(3) Minor further volcanic activity continuing throughout most of the early Eocene (both wells) with two slightly more significant eruptions in the late early Eocene and early mid-Eocene respectively.
(4) The intrusion of a dolerite sill into Campanian strata close to 50 Ma in mid-Eocene times (Well 219/28-2).

Sovereign Wells 219/28-1 and 219/28-2 were drilled some 170 km NNE of the Shetlands in 1984. They were located on the flank of the Møre Basin (N Shetland Trough) abutting Margareta's Spur. Nearby is the volcanic Erlend Complex and Well 219/20-1 drilled by Conoco (Fig. 1). Both Sovereign wells are SE of the furthest postulated extent of the Faeroes Lower and Middle Series lavas, but Well 219/28-2 is just within the 'intrusion zone' of Ridd (1983). This zone is confirmed from seismic data to be the SE limit of sill injection in Sovereign Blocks 219/27 and 28 (see also seismic evidence quoted by Gibb & Kanaris-Sotiriou 1988).

Evidence of major early Palaeogene Thulean volcanism was encountered in both wells; many horizons of predominantly basic primary ashfall tuff were found to be intercalated within a siltstone, silty claystone and claystone sedimentary sequence of earliest Eocene age, and an abundance of reworked volcanoclastic debris was seen to be present within the sediments themselves. In both wells the ash-rich interval begins abruptly at the base of strata that can be correlated with the Sele and Balder Formations of the Northern North Sea Basin. Stratigraphical palaeontology was undertaken by Palaeoservices Limited of Watford (UK) and their results were reported to Sovereign in 1984 (Project No. 1322) and 1985 (Project No. 1356). Following the usage recommended by Knox (1984), the boundary between the Palaeocene and the Eocene is taken here to be the base of the Sele Formation.

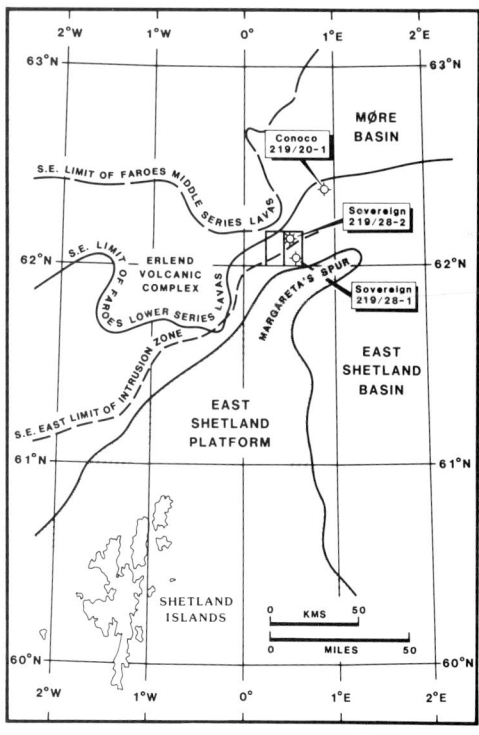

FIG. 1. Location of Sovereign Wells 219/28-1 and 219/28-2 in relation to the Shetlands and other features: (1) Sovereign Well 219/28-1; (2) Sovereign Well 219/28-2; (3) Conoco Well 219/20-1.

From MORTON, A. C. & PARSON, L. M. (eds), 1988, *Early Tertiary Volcanism and the Opening of the NE Atlantic,* Geological Society Special Publication No. 39, pp. 253–262.

F. J. Fitch et al.

In Well 219/28-1, the 'ash-marker' sequence begins with a 3 m thick basic tuff, followed by a further 88 m of sediments rich in tuffaceous debris. This sequence is thought to be the equivalent of the Sele and Balder Formations elsewhere. Above this level, over most of the following 284 m, volcanic debris can still be identified in the sediments of the early Eocene Hordaland Group (Fig. 2). It is probable that rare volcanic debris is present also in the rocks of the Lista Formation below the 'ash-marker', but as the sedimentary rock-types concerned are so closely similar, much of the tuffaceous material seen in ditch cuttings from the level of this formation may have been derived by caving from overlying sequences.

In Well 219/28-2, side-wall cores were available to supplement the ditch cuttings and, thus, it is possible to be certain that many horizons of undoubted primary basic ashfall tuff are intercalated within the particularly ash-rich strata of the interval assigned to the Sele and Balder Formations. At this locality, the 'ash-marker' is some 104 m thick. Above this ash-rich basal layer, ash-bearing sediments of the lower part of the Hordaland Group continue for another 396 m into the early Middle Eocene. Two slightly more significantly ash-rich zones occur a little before and soon after the Early/Middle Eocene boundary (Fig. 3). Again, it is probable that rare volcanic debris is present also in the rocks of the Lista Formation below the 'ash-marker'.

Two thick basic sills are emplaced in the late Cretaceous rocks of Well 219/28-2. An extrusive origin for the rocks was excluded by the recognition of spotted-contact hornfelses at the top and bottom of each of the igneous intervals. Although the principal rock-type of both sills is olivine-dolerite, criteria were established which made it possible to distinguish between them when mixed in junk-basket samples and ditch cuttings. The sills are 34 m (upper) and 46 m (lower) thick respectively with exceptionally extensive contact-metamorphic aureoles; the top 9 m of the upper sill is a sheeted complex of thin basaltic and tachylytic folia intruded into strongly spotted sedimentary hornfelses (see Fig. 4).

Petrographic and chemical studies were made upon ditch cuttings, side-wall cores and junk-basket samples of tuffs, dolerites and hornfelsed sediments to establish the characteristics of the

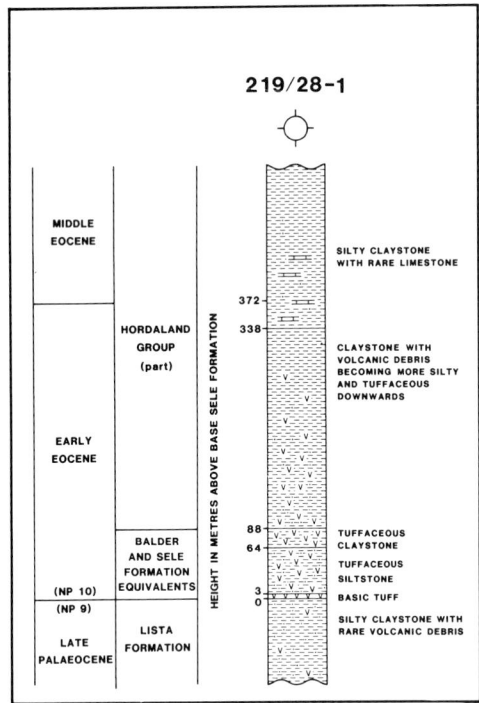

FIG. 2. The early Eocene tuffaceous interval in Sovereign Well 219/28–1. (Following Knox (1984), the Palaeocene/Eocene boundary is taken to be at the base of the Sele Formation, i.e. at the boundary between nannofossil zones NP9 and NP10.)

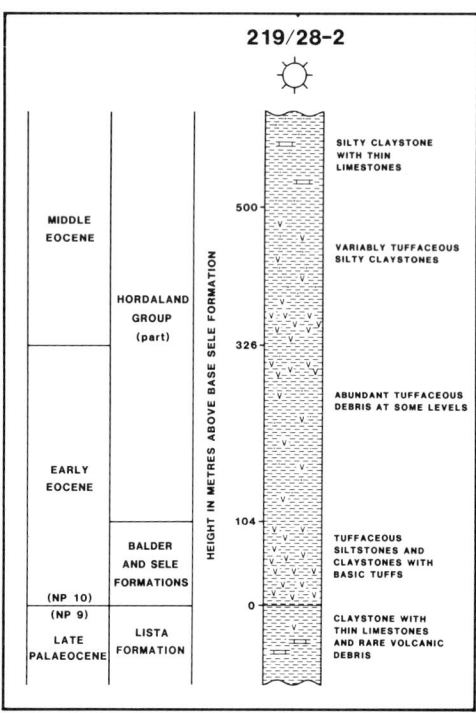

FIG. 3. The early to mid-Eocene tuffaceous interval in Sovereign Well 219/28-2.

igneous rock-types and confirm their intrusive or extrusive nature. Six triplicate K-Ar, five total degassing and two step-heating $^{40}Ar/^{39}Ar$ age determination analyses were undertaken on samples from Well 219/28-2. The two step-heating analyses failed to produce unambiguous results (for a variety of reasons, possibly including ^{39}Ar recoil problems as well as argon migration in and out of the sampled rocks) and are not discussed in the present paper.

Petrography of the basic tuffs

Ditch cuttings were examined at 3 m intervals throughout the major part of the ash-bearing Eocene section of Well 219/28-2. In addition, 15 side-wall cores were taken within the primary tuff-rich basal half of the 104 m of strata assigned to the Sele and Balder Formations. One more side-wall core penetrated typical sediments of the Lista Formation. The amount and thickness of primary ashfall and other volcanic debris that was accumulating in earliest Eocene times in the area sampled by the two Sovereign wells suggests that this was a time of major regional volcanism and some explosive source vents must have been relatively close by. Primary tuffs were not identified in the underlying or overlying sections and the quantity and abrasion level of the detrital grains of volcanic origin is extremely variable. Nevertheless, at some horizons volcanic debris is markedly more abundant and it can be assumed that explosive volcanic activity was occurring at a not too distant locality.

The variably silty, argillaceous sediments of the late Palaeocene Lista Formation contain no more than a minor, rare component of possible volcanic origin. However, the occasional presence of 'floating' quartz grains that appear to be almost completely unworn fragments of that type of partially resorbed, euhedral quartz phenocrysts that are common in rhyolite volcanics may be significant. Above these Palaeocene rocks much of the volcanic debris contained within the predominantly argillaceous Eocene sediments (shales and siltstones, some calcareous, some arenaceous, some pyritiferous, with rare thin limestones) is degraded or partly degraded very fine-grained basalt and basaltic glass. In the ditch cuttings and side-wall cores, in which they are present as a minor component, former fragments of basic glass are completely degraded to form small dark brown argillaceous pellets.

In many examples, it is clear from the admixture and interdigitation of silty or sandy detritus with the volcanic component that much of the primary ashfall tephra was immediately picked up, carried away and re-deposited by currents to become a penecontemporaneous detrital component in what is now a sequence of tuffaceous sediments.

The primary ashfalls, from which the reworked tephra were derived, were mostly basaltic and lithic–vitric in character. Crystal fragments, usually of pyroxene and plagioclase, are present in very small amounts. At least one pure lithic tuff of basaltic composition was seen; good examples of partially palagonitized primary lithic–vitric basic tuffs were sampled in three other side-wall cores. The finest, mostly vitric ash forming the matrix of these rocks is always completely altered to wispy, fibropalagonitic and chloritic material. Alteration of the larger rock fragments, glassy and pumiceous tephra is usually incomplete, with intermediate stages in the sequence, fresh glass—palagonitized glass–fibropalagonite—chlorite/clay mineral aggregate, being present in adjacent grains. Zeolite replacements and a zeolite cement are to be seen in some of the basic tuffs. Acid tuffs, or their reworked products, were not positively identified in the Eocene sequence; the absence of angular shards suggests that if acid volcanics are present, they must be in no more than subordinate amounts.

Petrography of the hornfelses and the dolerite sills

The late Cretaceous sediments into which the two dolerite sills are emplaced (Fig. 4) are a sequence of claystones and shales with thin limestones. Both sills are enclosed by remarkably wide zones of contact metamorphism; this is confirmed by vitrinite reflectance studies.

Above and below the upper sill, contact metamorphism extends over at least 9–12 m. Below the lower sill the aureole is certainly more than 9 m thick, whilst above it contact metamorphism may have been even more extensive. Nevertheless, although ubiquitously spotted, the hornfelses remain very fine-grained (with the exception of those found within the sheeted complex at the top of the upper sill) and do not show any marked high-grade, high-temperature features even quite close to their contact with the sills. The wide zones of contact metamorphism adjacent to basic sills less than 50 m thick are unusual and may indicate either that one or both sills acted as conduits for the flow of considerable volumes of magma or the nearby presence of a much thicker sill complex (see also Gibb & Kanaris-Sotiriou 1988).

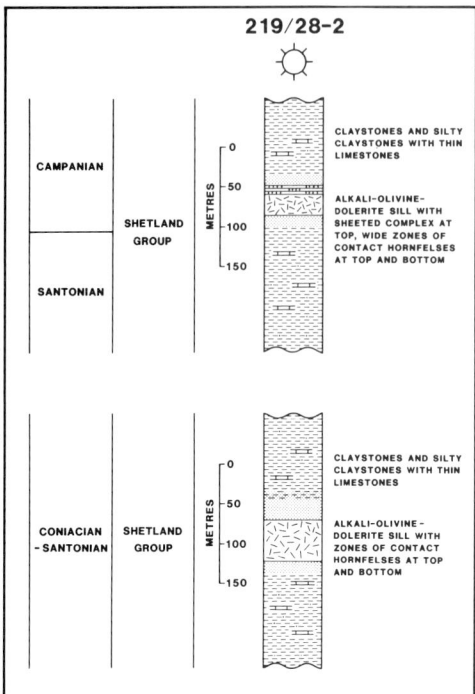

FIG. 4. Two basic sills encountered within late Cretaceous strata in Sovereign Well 219/28–2.

In the more argillaceous strata, contact metamorphism has resulted in progressive recrystallization of the original clay minerals and other, silty, detritus to convert the rocks into fine-grained spotted hornfelses now largely composed of quartz, chlorite, K-Fe micas, plagioclase and carbonate. Initially, spotting resulted from the preservation of residual, only partially recrystalline areas within a more recrystalline and clarified base; later, as thermal metamorphism progressed, the rock became a fine felty aggregate of chlorite and micas with carbonate granules set in a quartzo-feldspathic groundmass. Examination of thin sections suggests that eventually cordierite and/or andalusite grew to form the clearly maculose fabric of the highest grade hornfelses, afterwards becoming degraded to an aggregate of lower temperature minerals when the initial, purely thermal, event was followed by a phase of hydrous alteration. The impure limestones became completely recrystallized, with a form of spotting that results from the growth of siderite/ankerite granules within a calcite-quartz-chlorite-mica-plagioclase background.

The two dolerite sills encountered in Sovereign Well 219/28–2 are both of olivine-dolerite composition but, although superficially alike, they

differ in a number of easily recognizable ways. A specific feature of the upper sill is the presence of a wide upper border zone (some 9 m thick) in which the igneous rock seen in ditch cuttings and side-wall cores was a microporphyritic basaltic glass or a very fine-grained basalt before devitrification/alteration. Glassy marginal chill is probably also present at the lower contact of the upper sill and at both contacts of the lower sill, but at none of these places is it so exceptionally prominent as it is at the top of the upper sill. The central parts of both sills are microgabbroic. The minor, but very noticeable presence of brown amphibole, brown biotite, brassy sulphides and a leucoxene alteration product after ilmenite are particular features of the rock of the lower sill and can be used to distinguish it visually from that of the upper sill. The ubiquitous presence of low temperature hydrous alteration products in both sills tends, however, to blur the distinction.

Initially, samples of the dolerite sills, their constituent minerals and contact hornfelses were subjected to rapid X-ray diffraction (XRD) and microprobe analysis by Dr K Pye of the Department of Earth Sciences, Cambridge. Later, the bulk and trace-element chemistry of typical examples of the two dolerites was determined by X-ray fluorescence (XRF) analysis by the Materials Analysis Service of the University of St. Andrews (Table 1).

An examination of the chemical data obtained from the two dolerite sills intersected by Sovereign Well 219/28-2 shows them to have been derived from typically oceanic basin magmas (according to the TiO_2-K_2O-P_2O_5 discriminant of Pearce et al. 1975). Important differences between the two batches of basaltic magma appear, however, on an alkali–silica plot. It can be seen (Fig. 5) that the analyzed rock from the lower sill plots in the field of the alkali basalts of Cox et al. (1979), whereas the low-potassium rock of the upper sill plots in the field of subalkaline basalts. Nevertheless, the chemistry of the upper sill is undoubtedly tholeiitic: it is reminiscent of that of the Upper Lavas of the Faeroes (Bollingberg et al. 1975) and is very similar to that of a complex of basic sills, also within Campanian strata, encountered in Well 219/20-1 drilled by Conoco some 30 km to the NNE of the Sovereign wells (Gibb et al. 1986; Gibb & Kanaris-Sotiriou 1988). Of particular interest is the presence of hypersthene in the CIPW norm and traces of orthopyroxene in the mode of the upper sill; both features are also shown by the Upper Lavas of the Faeroes and by the sills in the Conoco well. The apparently alkalic nature of the lower sill on the plots of Kuno (1966) and Cox et al. (1979) is distinctive, but otherwise the chemical indica-

TABLE 1. *Bulk major and trace-element chemistry of the basic sills*

	Upper Sill (FM 8741)	Lower Sill (FM 8759)
	wt%	wt%
SiO_2	46.3	46.8
TiO_2	1.10	1.31
Al_2O_3	16.9	17.3
Fe_2O_3	1.78	—
FeO	8.36	8.79
MnO	0.16	0.14
MgO	7.58	6.93
CaO	11.94	12.40
Na_2O	2.10	2.98
K_2O	0.25	1.29
P_2O_5	0.09	0.19
LOI	2.8	1.6
Totals	99.36	99.73
	ppm	ppm
Nb	4	12
Zr	57	92
Y	21	23
Rb	—	17
Sr	186	378
Th	1	2
Pb	11	13
Zn	75	68
Cu	71	88
Ni	138	107
Cr	300	337
V	253	271
Ba	520	849
Hf	1	1
(Ce)	3	15
(La)	—	4
	normative % (CIPW)	normative % (CIPW)
Q	—	—
C	—	—
Or	1.48	7.62
Ab	17.76	10.69
An	35.90	30.00
Lc	—	—
Ne	—	—
Di	18.56	24.89
Wo	—	—
Hy	6.5	—
Ol	11.43	14.11
Mt	2.58	—
Il	2.09	2.49
Ap	0.21	0.45
Cc	—	—

tions are that it is also of tholeiitic derivation. A further unusual feature of the analyzed sample of the lower sill is its exceptionally low Zr/P_2O_5 ratio.

The whole-rock K-Ar dating sample collected from the upper sill (FM 8741) is a coarsely ophitic-textured olivine-dolerite, composed of partially to completely serpentinized/chloritized olivine, unaltered magnetite and large fresh ophitic plates of augite set in a matrix of Ca-rich plagioclase laths with minor accessories including rare orthopyroxene. Alteration is widespread, but of very patchy distribution. The secondary mineralogy includes secondary silica, chlorites, micas, carbonates and zeolites (mainly the calcium-rich zeolite wairakite). The whole-rock dating sample of the lower sill (FM 8759) is a coarse-grained subophitic-textured olivine-dolerite carrying strongly altered olivine with reaction coronas and large augite plates (5–10 mm across) with ragged titaniferous margins. The plagioclase laths in the microgabbroic matrix range from anorthite to labradorite in composition. Amphibole, biotite and pyrite are conspicuously present. The iron ores include titanomagnetite and ilmenite in addition to pyrite. Alteration products include secondary silica, chlorites, micas, carbonates and zeolites (again mainly wairakite).

Biotite was concentrated for K-Ar dating from another virtually identical sample of the lower sill (FM 8760).

Because they are chemically distinct, it is clear that the two sills are the products of different batches of basaltic magma. Nevertheless, when a Campanian age was confirmed for the lower sill by K-Ar dating, the possibility that the glassy basalts at the top of the first sill might be extrusive needed to be considered. The 9–12 m of section at the top of the upper sill, in which fine-grained basalt and tachylyte are the dominant igneous rock-types in the ditch cuttings, is too thick to be just the chilled margin of the underlying dolerite sill, but no vesiculation or other indication of an extrusive origin was observed. Over this interval, abundant chips of spotted hornfelses are associated with fragments of a variety of basaltic rock-types (ranging from aphyric and microporphyritic tachylytes through microlitic and hyalopilitic to fully crystalline fine-grained basalts), most of which are strongly calcified and otherwise altered. The simplest explanation of these observations is that the top of the upper sill is a sheeted complex of thin sills with intervening hornfelsed Campanian sediments over some 9–12 m. This interpretation was confirmed when it became possible to assign a mid-Eocene age to the upper sill.

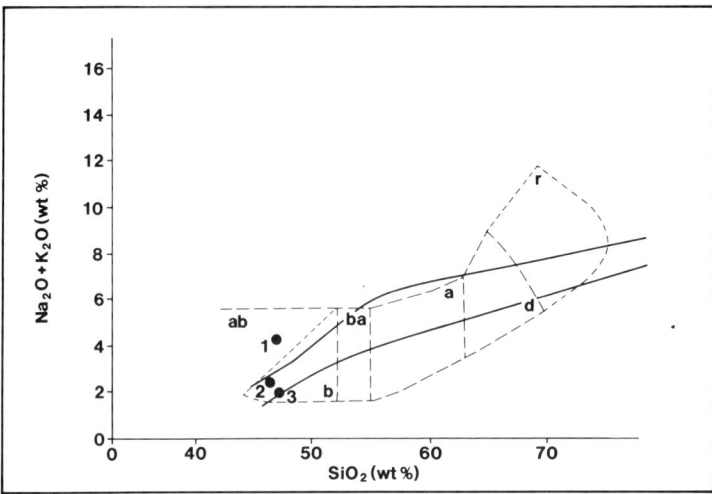

FIG. 5. Alkali–silica diagram. The lower dolerite sill (1) plots in the field of alkali basalt (ab), whilst the upper sill (2) and the average Upper Lavas of the Faeroes (3) plot in the field of subalkaline basalts (b) (broken-line field boundaries after Cox *et al.* (1979); ab = alkali basalt, a = andesite, d = dacite, r = rhyolite). For comparison the fields (reading from top to bottom) of the alkalic, high-alumina and tholeiitic series of Kuno (1966) are also shown (solid lines). Nevertheless (see Table 1 and text), both rocks are most probably derived from tholeiitic magmas.

K-Ar dating of basic igneous rocks

Successful K-Ar dating of basic tuffs, basalts and dolerites is frequently difficult. Geochronologists have been aware of the problems of dating basic igneous rocks for a long time (see, for example, the discussions and references in Miller & Fitch 1964; Dalrymple & Lanphere 1969; Fitch 1972; Evans *et al.* 1973; Fitch *et al.* 1978; Fitch & Miller 1984; Mussett 1986).

In the first instance, difficulties arise from the extremely low potassium content of many basic rocks and minerals. From such substances the volumes of radiogenic argon available for analysis are proportionally small, resulting in high atmospheric correction levels, low precision and the possibility of the introduction of a significant error due to the presence of extraneous initial or introduced argon. Extraction of a dating sample that is a concentrate of the mineral phase (or phases) richest in potassium can sometimes provide a solution to this problem (i.e. extraction of unaltered biotite from an alkalic dolerite), but for many basic rocks this is not feasible and their dating must be attempted from whole-rock samples. Again, many basic igneous rocks are particularly susceptible to low-temperature hydrous alteration; argon loss or introduction and/ or potassium mobility may accompany the alteration process and lead to the K-Ar dates from altered basic rocks being geologically discrepant.

The most successful approach to the dating of basic igneous rocks has been through a combination of careful petrographic selection of unaltered samples and rigorous experimental technique. The introduction of the $^{40}Ar/^{39}Ar$ methods represented a considerable advance in the dating of most holocrystalline basic igneous rocks, but the possibility of argon-recoil problems arising during irradiation prevents the unambiguous application of $^{40}Ar/^{39}Ar$ dating to devitrified glasses, cryptocrystalline rocks and to some rocks rich in very fine-grained alteration products. Confidence in the K-Ar dating of basic igneous rocks is very dependent upon the successful repetition of results from fractional, duplicate and replicate samples of each individual rock and from other samples of the same horizon. This is, of course, a council of perfection, rarely attained in practice. Description of the dating techniques utilized in this study, and an account of how the level of experimental precision was estimated, can be found in Fitch & Miller (1984).

Interpretation of radioisotopic dates

Although none were ideal, three of the side-wall cores from the ash-rich basal early Eocene interval (FM 8789, FM 8791 & FM 8794) were considered to be more suitable for conventional

TABLE 2. *Results of age determinations on rock and mineral samples from Sovereign Well 219/28-2*

Sample	Method	K_2O (%)	Atmos. contam. (%)	V/M	Apparent age & error (Ma)	Average age (Ma)	Age
FM 8789	K/Ar	1.38+	65.7	1.94×10^{-3}	43.1 ± 0.7		Eocence
Tuff	w.r.	0.005	66.1	1.95×10^{-3}	43.3 ± 0.7	43.4 ± 0.7	(Bartonian)
	60/85		64.3	1.97×10^{-3}	43.7 ± 0.7		
FM 8791	K/Ar	1.06+	89.1	4.60×10^{-4}	13.4 ± 0.4		Middle Miocene
Tuff	w.r.	0.01	90.0	4.63×10^{-4}	13.5 ± 0.4	13.8 ± 0.5	
	60/85		90.7	4.74×10^{-4}	13.8 ± 0.5		
FM 8794	K/Ar	1.92+	47.8	3.25×10^{-3}	51.7 ± 0.8		Eocene
Tuff	w.r.	0.01	46.8	3.18×10^{-3}	50.6 ± 0.8	51.7 ± 0.8	(Ypresian)
	60/85		45.1	3.32×10^{-3}	52.8 ± 0.8		
FM 8741	K/Ar	0.21+	59.7	1.258×10^{-3}	177 ± 10		mid-Jurassic
U/sill	w.r.	0.007	59.4	1.242×10^{-3}	175 ± 10	177 ± 10	
	70/100		54.4	1.265×10^{-3}	178 ± 10		
FM 8759	K/Ar	1.23+	41.5	3.299×10^{-3}	81.3 ± 4.1		late Cretaceous
L/sill	w.r.	0.02	41.0	3.332×10^{-3}	82.1 ± 4.1	81.0 ± 4.1	(Campanian)
	70/100		42.8	3.229×10^{-3}	79.6 ± 4.0		
FM 8760	K/Ar	7.26+	20.5	1.888×10^{-2}	78.9 ± 2.4		late Cretaceous
L/sill	biotite	0.04	19.9	1.893×10^{-2}	79.1 ± 2.4	79.1 ± 2.4	(Campanian)
	60/120		20.4	1.901×10^{-2}	79.4 ± 2.4		

w.r. = whole rock; V/M = volume of radiogenic ^{40}Ar (mm)^3 NTP/g of sample; K/Ar ages are calculated using the IUGS constants (Steiger & Jäger 1977).

Sample	J	Atmospheric correction	Age & error (Ma)	Average age (Ma)	Age
FM 8798	0.030970	47.60	49.3 ± 2.0		Eocene
hornfels	0.031100	55.59	51.2 ± 2.4	49.6 ± 2.3	(Lutetian)
SWC (w.r.)	0.031100	56.59	48.4 ± 2.4		
FM 8799	0.031180	46.70	55.3 ± 1.9		Palaeocene
hornfels	0.031180	40.59	61.6 ± 2.6	58.4 ± 2.2	(Thane-
SWC (w.r.)					tian)

Method: $^{40}Ar/^{39}Ar$ total degassing; factor J as defined by Fitch *et al.* (1974, p. 1455).

K-Ar whole-rock dating than the remainder (Table 2). Originally, the selected samples were primary lithic-vitric ashfall basic tuffs in which small fragments (0.5–0.05 mm) of basaltic glass, pumice, glassy microlithic basalt and rare fragments of pyroxene and plagioclase were carried in a variably porous matrix of finer vitric ash and dust. Devitrification and other subsequent alteration processes have caused partial to complete clouding, palagonitization, chloritization and argillification of the ash fragments. The fine-grained matrix was even more susceptible to alteration and to the introduction of a zeolite cement. Visual inspection suggested that the level of alteration in the three tuff samples is roughly as follows: FM 8794 (10%), FM 8789 (25%) and FM 8791 (75%). Thus, sample FM 8794 was the most suitable of the three tuffs for K-Ar dating.

As the tuffs were deposited in rapid succession within the earliest early Eocene, their actual ages must be virtually identical; the average whole-rock K-Ar apparent ages, however (51.7 ± 0.8 Ma (FM 8794); 43.4 ± 0.7 Ma (FM 8789) & 13.6 ± 0.4 Ma (FM 8791)), show a rapid decrease with increase in alteration, indicating progressive argon-loss discrepancy. Thus, the oldest apparent age obtained (52 Ma) must be regarded as a minimum estimate for their true age. At present, there is no acceptable way by which the extent of the argon-loss discrepancy suffered by this basic tuff can be reliably estimated. In our opinion, the error in the age may be as much as 10%; a correction of this magnitude would allow agreement with some recent timescales (e.g., Berggren *et al.* 1985 and Aubry *et al.* 1986) that assign an age of around 57 Ma to the base of the Eocene.

K-Ar whole-rock ages that were manifestly discrepant were also obtained from the rock of the upper of the two sills: an average value of 177 ± 10 Ma for sample FM 8741 could not be accepted for a sill emplaced into Campanian strata not older than 83 Ma! The dolerite of the upper sill is particularly poor in potassium, however, and from a highly zeolitized, altered example of such a rock, widely discrepant whole-rock ages are not unusual. In this case it is probable that the error results largely from the unwanted presence of quite small amounts of excess radiogenic ^{40}Ar. Better K-Ar dating samples were not available from the upper sill; thus a more satisfactory estimate of its age of emplacement was sought from the dating of samples of the contact hornfelses associated with it.

Side-wall core sample FM 8798 is a completely recrystallized spotted hornfels from the contact metamorphosed upper aureole of the upper dolerite sill, fine-grained but not cryptocrystalline, therefore unlikely to be subject to ^{39}Ar recoil problems. In fact, the total degassing $^{40}Ar/^{39}Ar$ ages (49.3 + 2.0 Ma, 51.2 + 2.4 Ma & 48.4 + 2.4 Ma) obtained from three independent whole-rock samples of this hornfels are concordant within experimental error. Side-wall core sample FM 8799 is a less strongly recrystallized spotted hornfels from the contact aureole below the upper sill. Two total degassing $^{40}Ar/^{39}Ar$ age determinations were obtained from whole-rock samples. They were not concordant, however, being 61.6 + 2.6 Ma and 55.5 + 1.9 Ma respectively, and are to be regarded as discrepant. It is possible that complete outgassing of previously accumulated radiogenic argon did not take place during the metamorphism of the Campanian sediment at this depth, causing the apparent ages obtained to lie between the age of the unaltered sediment (approximately 78 Ma) and the age of metamorphism (approximately 50 Ma). Alternatively, because of the much finer grain size of this rock, ^{39}Ar recoil may be causing excess age discrepancy during the analysis of these two samples. Thus, the average apparent age derived from the upper contact aureole is preferred; it would appear that the upper sill was intruded into Campanian strata around 50 Ma in mid-Eocene times.

The rock of the lower dolerite sill promised to provide more satisfactory material for K-Ar dating than either the upper dolerite, the basic tuffs or the contact hornfelses, for the whole-rock sample of the lower intrusion had a potassium content of 1.23%, and it also proved possible to extract a concentrate of fresh primary biotite from the sill as an additional, confirmatory dating-sample. The analysis of these two samples (whole-rock FM 8759 & biotite FM 8760) produced average conventional K-Ar ages (81.0 + 4.1 Ma & 79.1 + 2.4 Ma, respectively) that are identical within experimental error, even though the atmospheric correction levels differ considerably (41.8% & 20.3% respectively). This concordance makes it likely that 80 Ma is the true age of the sill encountered in Santonian rocks.

Regional context of activity

Rocks produced during several episodes of basic magmatism related to the Mesozoic and Cainozoic development of the northern N Atlantic Ocean can be identified in Sovereign Wells 219/28-1 and 219/28-2. The first episode occurred around 80 Ma in late Cretaceous (Campanian) times when (seen in Well 219/28-2 only) a sill of primitive alkali-rich oceanic olivine-dolerite was intruded into Santonian sediments that themselves could not have been deposited more than 8 My previously. It is probable that this batch of basaltic magma reached to within 670 m of the sediment/sea water interface. Volcanic and intrusive basic igneous rocks of the same age have been reported from a number of localities in the northern N Atlantic. The age of the pre-basaltic rhyolites at the base of the nearby Erlend volcano are uncertain, but a Campanian age has been proposed (Ridd 1983; Gatliff et al. 1984).

The volcanic events producing tuffs at the level of the base of the Lista Formation in the North Sea (Knox et al. 1981) and correlated by these authors with the 'basal Thanetian' volcanic episode of Jacqué & Thouvenin (1975) was not identified clearly in the Sovereign wells. It is possible, however, that some of the 'floating' quartz crystals seen in the upper part of the Lista Formation in Well 219/28-2 may be related to nearby acid magmatism similar in character to acid activity known to have occurred elsewhere in the region at this time, e.g. at Tardree (59–58 Ma according to unpublished work in numerous laboratories world-wide) and in the Western Red Hills of Skye (Dickin 1980). The best available age for the Thanet Base Bed of Kent is 60.9 + 0.9 Ma (glauconite recalculated by International Union of Geological Sciences (IUGS) standard constants, Fitch et al. 1978), which is in good agreement with the age of the early volcanic plateau lavas of Mull (60 + 0.5 Ma, Mussett 1986) and with other estimates of the age of the base of the Lista Formation.

The next magmatic climax positively identified in the Sovereign wells is a major episode of explosive volcanism (dated as being older than a minimum age of 52 Ma, but from circumstantial evidence thought to be possibly as old as 57 Ma), producing voluminous basaltic ashfall deposits within the Sele and Balder Formations. This volcanic event correlates with the very widespread 'ash-marker' of the North Sea (Jacqué & Thouvenin 1975; Knox & Morton 1983; Knox 1983, 1984) and is thought to be the lateral equivalent of the top basalts of the Erlend Volcano (Gatliff et al. 1984). Indeed, it is suggested by Gatliff et al. that the Erlend Volcano is one of the principal sources of the tuffs of the North Sea 'ash-marker' horizon; certainly the Erlend Volcano, lying only 50 km to the SW, is ideally situated to have been the source of the ashfalls in the earliest early Eocene at the locality of the Sovereign wells. Of course, other, as yet undescribed, hidden volcanoes may have contributed to the widespread 'ash-marker' horizon. Accepting these correlations, the latest, explosive, basaltic eruptions of the Erlend Volcano, the North Sea 'ash-marker' and the earliest early Eocene sediments of the Sele and Balder Formations can be dated with the Sovereign tuffs as being undoubtedly older than 52 Ma and possibly as old as 57 Ma. If the latter date is correct, then the climax of explosive activity at the Erlend Volcano occurred at the same time as the explosive climax of the Mull Volcano (57 ± 1 Ma, according to Mussett 1986).

In Sovereign Well 219/28–2, two slightly more tuffaceous horizons, one a little before and one a little after the Early/Middle Eocene boundary, are thought to represent specifically enhanced volcanic activity at a not too distant locality. The most likely source area would be the Faeroes, where volcanism was undoubtedly occurring over this period. None of the younger tuffaceous sediments in the Sovereign wells have been dated radiometrically, but the intrusion of a sill of olivine-dolerite in underlying late Cretaceous rocks can be shown to have occurred around 50 Ma. The rock-type of the sill is very similar to that of the Upper basalts of the Faeroes, thus it is possible that the upper (early Middle Eocene) tuff concentration in Well 219/28–2, the dated sill and the Upper basalts of the Faeroes are related and can be tentatively assigned dates close to 50 Ma. The petrologically very similar sills in Conoco Well 219/20–1 30 km to the NNE are most probably also of this age.

Whilst none of the above discussion is incompatible with the main conclusions of Aubry *et al.* (1986) or Berggren *et al.* (1985) regarding Palaeogene correlation and stratigraphy, some unresolved difficulties arise with the conclusions of Odin & Curry (1985) and with attempts to apply the magnetostratigraphic timescale of Harland *et al.* (1982), e.g. as by Mussett (1986). However, as these matters receive attention from several other contributors to this volume, they are not discussed further here.

ACKNOWLEDGEMENTS: The authors wish to thank Sovereign Oil and Gas plc, Dow Chemical Company Limited, Trafalgar House Oil & Gas Limited and North Sea & General Oil Operations plc for permission to publish a report on this work and the many colleagues who have given them assistance and encouragement during its progress.

References

AUBRY, M-P., HAILWOOD, E. A. & TOWNSEND, H. A. 1986. Magnetic and calcareous-nannofossil stratigraphy of the lower Palaeogene formations of the Hampshire and London basins. *Journal of the Geological Society of London*, **143**, 729–735.

BERGGREN, W. A., KENT, D. V. & FLYNN, J. J. 1985. Paleogene geochronology and chronostratigraphy. *In*: SNELLING, N. J. (ed) *Geochronology of the Geological Record*. Memoir of the Geological Society of London, **10**, 141–195.

BOLLINGBERG, H., BROOKS, C. K. & NOE-NYGAARD, A. 1975. Trace element variations in Faeroese basalts and their possible relationships to ocean floor spreading history. *Bulletin of the Geological Society of Denmark*, **24**, 55–60.

COX, K. G., BELL, J. D. & PANKHURST, R. J. 1979. *The Interpretation of Igneous Rocks*. George Allen & Unwin, London, 450 pp.

DALRYMPLE, G. B. & LANPHERE, M. A. 1969. *Potassium-argon dating*. Freeman, San Francisco, 258 pp.

DICKIN, A. P. 1980. Isotope geochemistry of Tertiary igneous rocks from the Isle of Skye N.W. Scotland. *Journal of Petrology*, **22**, 155–189.

EVANS, A. LL., FITCH, F. J. & MILLER, J. A. 1973. Potassium-argon age determinations on some British Tertiary volcanic rocks. *Journal of the Geological Society of London*, **129**, 419–443.

FITCH, F. J. 1972. Selection of suitable material for dating and the assessment of geological error in potassium-argon age determination. *In*: BISHOP, W. W. & MILLER, J. A. (eds) *Calibration of Hominoid Evolution*. Wenner–Gren Foundation of Anthropological Research, New York, pp. 77–91.

—— & MILLER, J. A. 1984. Dating Karoo igneous rocks by the conventional K-Ar and $^{40}Ar/^{39}Ar$ age spectrum methods. *In*: ERLANK, A. J. (ed) *Petrogenesis of the Volcanic Rocks of the Karoo Province*. Geological Society of South Africa, Special Publication, pp. 247–266.

——, FORSTER, S. C. & MILLER, J. A. 1974. Geological Time Scale. *Reports on Progress in Physics*, **37**, 1433–1496.

——, HOOKER, P. J., MILLER, J. A. & BRERETON, N. R. 1978. Glauconite dating of Palaeocene–Eocene rocks from East Kent and the time-scale of Palaeogene volcanism in the North Atlantic region. *Journal of the Geological Society of London*, **135**, 499–512.

GATLIFF, R. W., HITCHEN, K., RITCHIE, J. D. & SMYTHE, D. K. 1984. Internal structure of the

Ereland Tertiary volcanic complex, north of Scotland, revealed by seismic reflection. *Journal of the Geological Society of London,* **141**, 555–562.

GIBB, F. G. F. & KANARIS-SOTIRIOU, R. 1988. The geochemistry and origin of the MORB-type Faeroes–Shetlands Tertiary sill complex. *In*: MORTON, A. C. & PARSON, L. M. (eds) *Early Tertiary Volcanism and the Opening of the NE Atlantic.* Geological Society of London, Special Publication, **39**, pp. 241–252.

——, —— & NEVES, R. 1986. A new Tertiary sill complex of mid-ocean ridge basalt type NNE of the Shetland Isles: a preliminary report. *Transactions of the Royal Society of Edinburgh, Earth Sciences,* **77**, 223–230.

HARLAND, W. B., COX, A. V., LLEWELLYN, P. G., PICKTON, C. A. G., SMITH, A. G. & WALTERS, R. 1982. *A Geologic time scale.* Cambridge University Press, Cambridge, 128 pp.

JACQUÉ, M. & THOUVENIN, J. 1975. Lower Tertiary tuffs and volcanic activity in the North Sea. *In*: WOODLAND, A. W. (ed) *Petroleum and the Continental Shelf of North West Europe (Vol. 1: Geology).* Applied Science Publishers, London, pp. 455–465.

KNOX, R. W. O'B. 1983. Volcanic ash in the Oldhaven Beds of southeast England and its stratigraphical significance. *Proceedings of the Geologists' Association,* **94**, 245–250.

—— 1984. Nannoplankton zonation and the Palaeocene/Eocene boundary beds of NW Europe: an indirect correlation by means of volcanic ash layers. *Journal of the Geological Society of London,* **141**, 993–999.

—— & MORTON, A. C. 1983. Stratigraphical distribution of Early Palaeogene pyroclastic deposits in the North Sea Basin. *Proceedings of the Yorkshire Geological Society,* **44**, 355–363.

——, —— & HARLAND, R. 1981. Stratigraphical relationships of Palaeocene sands in the UK sector of the central North Sea. *In*: ILLING, L. V. & HOBSON, G. D. (eds) *Petroleum Geology of the Continental Shelf of North-West Europe.* Heyden & Son, London, pp. 267–281.

KUNO, H. 1966. Lateral variation of basalt magma type across continental margins and island arcs. *Bulletin Volcanologique,* **29**, 195–222.

MILLER, J. A. & FITCH, F. J. 1964. K-Ar methods with special reference to basic igneous rocks. *In*: HARLAND, W. B., SMITH, A. G. & WILCOCK, B. (eds) *The Phanerozoic Time-Scale.* Geological Society, London, pp. 101–117.

MUSSETT, A. E. 1986. ^{40}Ar-^{39}Ar step-heating ages of the Tertiary igneous rocks of Mull, Scotland. *Journal of the Geological Society of London,* **143**, 887–896.

ODIN, G. S. & CURRY, D. 1985. The Palaeogene timescale: radiometric dating versus magnetostratigraphic approach. *Journal of the Geological Society of London,* **142**, 1179–1188.

PEARCE, T. H., GORMAN, B. E. & BIRKETT, T. C. 1975. The TiO_2-K_2O-P_2O_5 diagram: a method of discriminating between oceanic and non-oceanic basalts. *Earth and Planetary Science Letters,* **24**, 419–426.

RIDD, M. F. 1983. Aspects of the Tertiary geology of the Faeroe–Shetland Channel. *In*: BOTT, M. H. P., SAXOV, S., TALWANI, M. & THIEDE, J. (eds) *Structure and Development of the Greenland–Scotland Ridge.* Plenum Press, New York, pp. 91–108.

STEIGER, R. H. & JAGER, E. 1977. Subcommission on Geochronology: convention on the use of decay constants in geo- and cosmochronology. *Earth and Planetary Science Letters,* **36**, 359–362.

F. J. FITCH, FM Consultants Ltd, 21 Harcourt Drive, Herne Bay, Kent CT6 8DJ, UK.

G. L. HEARD, Sovereign Oil & Gas plc, Portland House, Stag Place, London SW1E 5BH, UK.

J. A. MILLER, Bullard Laboratories, Department of Earth Sciences, University of Cambridge, Madingley Road, Cambridge CB3 0EZ, UK.

Volcanic ash in a cored borehole W of the Shetland Islands: evidence for Selandian (late Palaeocene) volcanism in the Faeroes region

A. C. Morton, D. Evans, R. Harland, C. King & D. K. Ritchie

SUMMARY: A volcanic ash layer, containing basic glass shards compositionally comparable to the basalts of the Faeroe–East Greenland Province, has been discovered in early Selandian sediments from a borehole W of the Shetland Isles. The available magnetostratigraphical, biostratigraphical and tephrachronological information indicates that basaltic volcanism in the province began during magnetic anomaly chron C26R, earlier than the currently accepted C24R interval, but correlating with the main period of basaltic volcanism in the British Tertiary Igneous Province.

In 1982, the Marine Geology Unit of the Institute of Geological Sciences (now the British Geological Survey) drilled Borehole 82/12 at 59°59.86′N, 3°17.75′W (Fig. 1), 113 km WSW of Scalloway (Shetland Islands) as part of the mapping programme of the UK continental shelf. The borehole passed through 16.5 m of Quaternary sediments before encountering a sequence of predominantly hard, non-calcareous, very dark grey claystones, 7.7 m of which were recovered before the hole was terminated at 28 m below seabed.

Volcanic ash

A thin (1 cm maximum) tuff occurs within the claystone sequence at 22.70 m depth. It is composed mainly of pale yellow, angular, vitric shards which are isotropic and show little visible alteration. The shards are generally non-vesicular but, where present, the vesicles vary from spheroidal to highly elongate. Lithic particles are rare, but plagioclase and clinopyroxene are conspicuous minor components.

Major and trace element analyses by X-ray fluorescence (XRF) were not undertaken because of the limited amount of material and the degree of reworking and contamination from the surrounding sediment. However, reasonable estimates of the composition of the volcanic glass were made by electron microprobe analysis. Eight shards were analyzed with a Link Systems energy-dispersive X-ray analyzer attached to a Geoscan electron microprobe (Table 1), confirming both the fresh nature of the glass and its basaltic composition. Analytical totals range from 96.68–98.38%, averaging 97.66%, indicating a lack of significant hydration. The glass is tholeiitic, and is olivine- and hypersthene-normative. A characteristic feature is the consistently

high TiO_2 content (mean 2.76%, range 2.65–2.95%). Plagioclases are mainly labradorite–bytownite (An_{65-78}), and the clinopyroxenes plot close to the diopside–salite–augite–endiopside join on the $MgSiO_3$-$FeSiO_3$–$CaSiO_3$ diagram (Fig. 2).

Age

The claystones are rich in palynomorphs and siliceous microfossils, but poor in calcareous microfossils. Only one calcareous foraminiferid was found, this being *Bulimina midwayensis* Cushman & Parker, from a sample at 18.2 m depth. This is a characteristic late Palaeocene species, restricted to benthic subzone NSB1b in the North Sea (King 1983). The siliceous microfossils include diverse diatoms, radiolaria and silicoflagellates. The diatoms are of general late Palaeocene to early Eocene aspect, but some of the radiolarian specimens are large and are closely comparable to the species used as the index for the early late Palaeocene planktic zone NSP2 of King.

The palynomorph assemblages of three samples (16.70 m, 22.70 m & 26.95 m) are rich, diverse and well-preserved, and include *Alisocysta circumtabulata* (Drugg) Stover & Evitt, *Apteodinium fallax* (Morgenroth) Stover & Evitt, *Cassidium* cf. *fragile* (Harris) Drugg, *Cerodinium striata* (Drugg) Lentin & Williams, *Danea californica* (Drugg) Lentin & Williams, *Hystrichosphaeridium tubiferum* (Ehrenberg) Deflandre, *Membranosphaera maastrichtica* Samoilovich ex Norris and Sarjeant, *Palaeocystodinium australinum* (Cookson) Lentin & Williams, *Palaeoperidinium pyrophorum* (Ehrenberg) Sarjeant, and *Xenicodinium rugulatum* Hansen. Jurassic and Cretaceous forms are also common, and the terrestrial component increases downhole. The

From MORTON, A. C. & PARSON, L. M. (eds), 1988, *Early Tertiary Volcanism and the Opening of the NE Atlantic,* Geological Society Special Publication No. 39, pp. 263–269.

A. C. Morton et al.

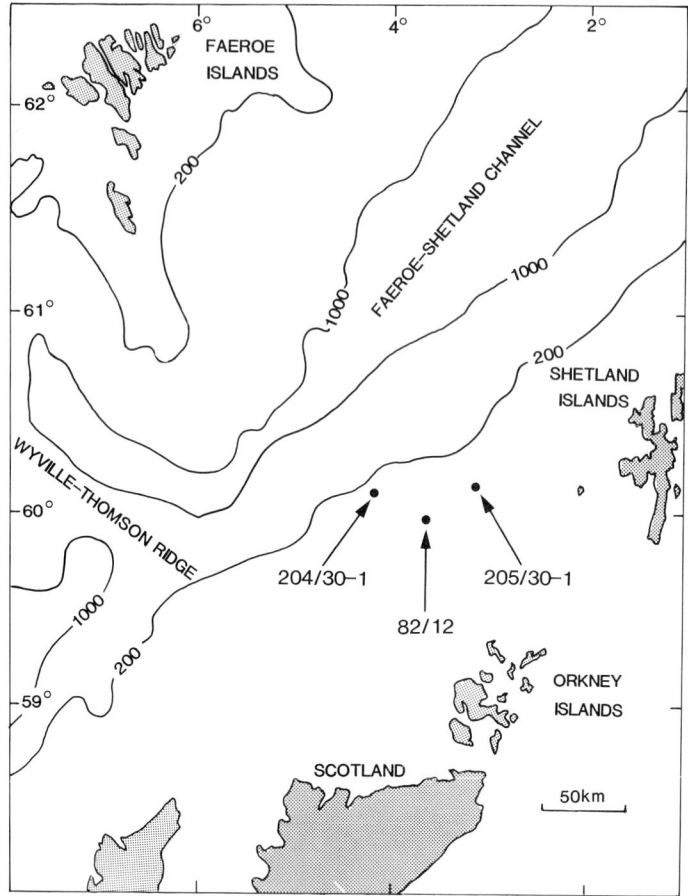

FIG. 1. Location of BGS Borehole 82/12 and commercial wells 204/30–1 and 205/30–1.

TABLE 1. *Major element composition of glass shards from ash layer at 22.70 m depth, BGS Borehole 82/12, determined by electron microprobe. CIPW norms calculated assuming $Fe_2O_3/FeO = 0.15$.*

SiO$_2$	47.53	47.99	47.90	48.10	47.91	47.81	48.43	48.11
Al$_2$O$_3$	12.86	12.79	12.98	13.18	13.08	12.67	13.58	13.12
TiO$_2$	2.67	2.65	2.72	2.74	2.82	2.65	2.84	2.95
FeO	11.62	11.04	11.36	10.99	11.10	11.54	11.43	11.06
MgO	8.28	7.97	8.21	7.39	6.90	8.28	7.02	7.28
CaO	11.70	11.71	11.21	12.12	11.56	11.57	11.73	12.02
MnO	0.11	0.22	0.19	0.19	0.10	0.29	0.17	0.14
Na$_2$O	2.60	2.49	2.59	2.46	2.58	2.40	2.54	2.58
K$_2$O	0.53	0.48	0.46	0.45	0.47	0.39	0.45	0.46
P$_2$O$_5$	0.03	0.00	0.01	0.02	0.01	0.00	0.04	0.02
Total	97.93	97.34	97.63	97.64	96.53	97.60	98.23	97.74
AB	22.40	21.59	22.39	21.26	22.55	20.75	21.82	22.28
OR	3.19	2.91	2.78	2.72	2.87	2.35	2.70	2.77
AN	22.29	22.89	22.95	24.13	23.51	23.18	24.73	23.36
DI	29.91	29.91	27.58	30.38	29.22	28.99	27.95	30.53
HY	0.26	6.09	5.94	6.62	8.27	7.75	8.98	5.36
OL	14.40	9.26	10.80	7.32	5.76	9.54	5.97	7.73
MT	2.26	2.17	2.22	2.15	2.21	2.26	2.23	2.16
IL	5.17	5.16	5.28	5.32	5.54	5.15	5.48	5.73
AP	0.07	0.00	0.02	0.05	0.02	0.00	0.09	0.05

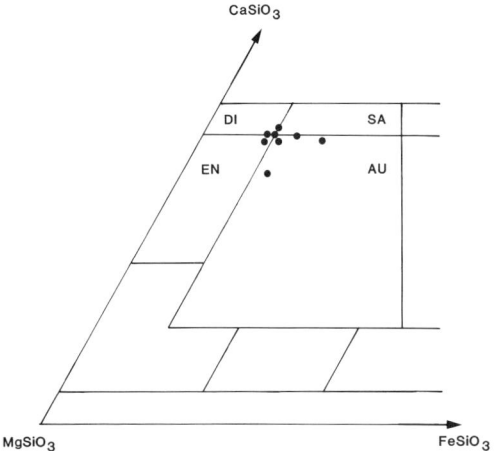

FIG. 2. Clinopyroxene compositions from the volcanic ash layer at 22.70 m depth in BGS Borehole 82/12. DI: Diopside, EN: Endiopside, SA: Salite, AU: Augite.

assemblages appear to be of Danian age and can be confidently placed within the *Palaeoperidinium pyrophorum* informal zone of Knox *et al.* (1981) and possibly within the *Danea mutabilis* zone of Hansen (1977).

However, on regional grounds, it appears unlikely that the sequence is Danian in age. In this part of the W Shetland Basin, the Danian is a calcareous marl, and is overlain by a sand rich in bioclastic material, particularly bryozoa, earning it the informal name 'Bryozoan Sand' (Hitchen & Ritchie 1987). The Bryozoan Sand is overlain in turn by dark carbonaceous mudstones, lithologically comparable to the 82/12 sequence. In the nearby commercial Well 205/30–1 (Figs 1 & 3), the mudstone sequence can be divided into a lower unit (308–390 m) and an upper unit (280–308 m). The lower unit is characterized by a rich siliceous microfossil assemblage with frequent spherical radiolaria, associated with arenaceous foraminifera. Calcareous planktonic foraminifera are rare and possibly caved, but calcareous benthic and agglutinating foraminiferids are present. In contrast, the upper unit contains common planktonic foraminifera, including *Globorotalia chapmani* Parr, and *Globigerina* gr. *triloculinoides* Plummer. *G. chapmani* ranges from the upper part of late Palaeocene Zone P3 to the basal early Eocene Zone P6 (zones of Berggren 1969). The absence of species indicative of Zones P5 and P6 probably indicates that the upper unit can be assigned to the early part of the late Palaeocene (Zones P3/P4 of Berggren 1969).

The abundance of siliceous and scarcity of calcareous microfossils in the lower interval suggests a correlation with the 82/12 claystones.

This is corroborated by the presence of tuffaceous claystones at 344 m in 205/30–1, and is compatible with the seismic evidence which indicates that the cored interval lies above the top of the Bryozoan Sand. The tuff in 82/12 therefore appears to be post-Danian (ie. post Zone P2), and being below an interval belonging to Zones P3/P4 can be assigned to Zone P3 with some confidence. The anomalous Danian age suggested by dinoflagellate cyst assemblages could be the result of reworking, in part or in entirety, from the Danian; the presence of Cretaceous and Jurassic species indicates that significant reworking has occurred. Alternatively, it may simply demonstrate the lack of detailed dinoflagellate investigations across the Danian–Selandian boundary, despite the excellent work of Heilmann–Clausen (1985). Heilmann–Clausen's informal zonation clearly shows that the early Selandian pre-dates the *Alisocysta margarita* informal zone of Knox *et al.* (1981) but can be distinguished from the *Palaeoperidinium pyrophorum* informal zone below. The assemblage in the 82/12 mudstones could undoubtedly contain some reworked Danian elements, pre-date the *Alisocysta margarita* informal zone and be consistent with an early Selandian age. This situation is further complicated by the sparse dinoflagellate floras often recovered from late Palaeocene sediments. In the North Sea and W of Shetland areas, the early part of the late Palaeocene characteristically has low-diversity assemblages, often lacking diagnostic species. Rich assemblages only reappear in the later part of the late Palaeocene (*Apectodinium hyperacanthum* informal zone of Knox *et al.* 1981). The lack of forms diagnostic of this zone is significant and clearly indicates that the sediments predate this zone, if not the *Alisocysta margarita* informal zone alluded to above.

The main period of pyroclastic activity in the North Sea and W of Shetland areas is recorded by the Balder Formation, a sequence of abundant tuffs interbedded with mudstones (phase 2b of Knox & Morton 1983). Well 204/30–1, another commercial well in the vicinity of 82/12 (Figs 1 & 3), contains the Balder Formation equivalent above the *G. chapmani* interval, confirming the broad equivalence of the 82/12 tuff with phase 1 of Knox & Morton.

Discussion

The relatively high TiO_2 content and the tholeiitic nature of the 82/12 tuff are features which suggest that it has a closer affinity with the Faeroe–Greenland Igneous Province than with the British

A. C. Morton et al.

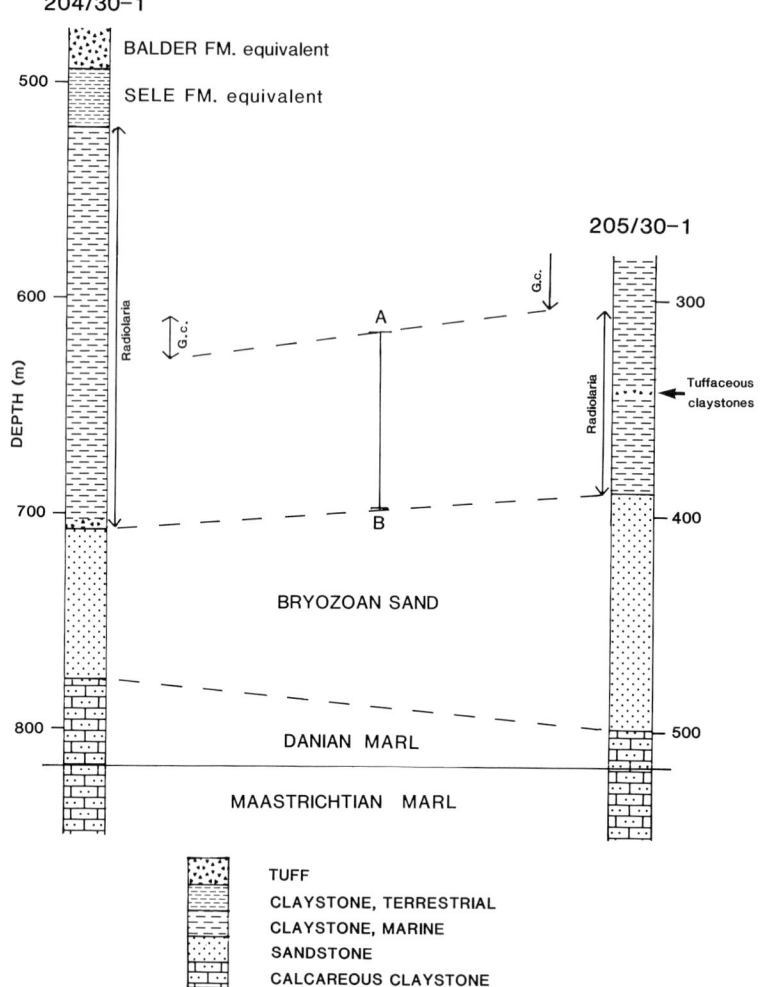

FIG. 3. Correlation of the early Tertiary sequences of 204/30–1 and 205/30–1, and possible stratigraphic limits (A–B) of the tuff-bearing claystones of 82/12. G.c. = *Globorotalia chapmani*.

Tertiary Igneous Province, in which there is a scarcity of high-Ti tholeiites. Moreover, during the Palaeocene, the Faeroe–Greenland area was closer to the site. Although the E Greenland basalts and the Faeroes Lower and Middle Series are generally more iron-rich than the single tuff layer in Borehole 82/12, basalts with similar FeO/ (FeO + MgO) ratios are not uncommon (Larsen & Watt 1985). However, it is generally regarded that volcanism in E Greenland began somewhat later, in the latest Palaeocene *Wetzeliella hypera-cantha* Zone of Costa & Downie (1976), based upon the presence of *Apectodinium homomorphum* (Deflandre & Cookson) Lentin & Williams in a low diversity assemblage from the Vandfaldsda-len Formation, near the base of the E Greenland

basalt pile in the Kangerdlugssuaq area (Soper *et al.* 1976).

The Faeroes sequence is subdivided into three, the Lower, Middle and Upper Series (Noe-Nygaard & Rasmussen 1968). Of these, the Lower and Middle Series are dominated by Ti-rich tholeiites. Below the Middle Series there is an unconformity, with the Lower Series basalts being capped by a coal-bearing sequence typically some 10 m in thickness (Noe-Nygaard & Ras-mussen 1968; Waagstein & Hald 1984). Between the Lower and Middle Series on the Faeroes there is, therefore, a break in basalt extrusion of unknown duration. At the base of the Middle Series, there is a tuff–agglomerate unit up to 100 m thick which has been correlated on seismic reflection profiles and on a lithological basis with the North Sea Balder Formation (Smythe 1983; Smythe *et al.* 1983). This is believed to be of early Eocene (nannoplankton Zone NP10) age (Knox 1984). Geochemical studies of Balder Formation tuffs from commercial boreholes in the area W of Shetland, currently in progress at the BGS, confirm this correlation.

Magnetostratigraphical investigations have proved the presence of two intervals of normal polarity in the uppermost part of the Lower Series (Abrahamsen 1967; Tarling & Gale 1968; Abrahamsen *et al.* 1984). The early Eocene (nannoplankton Zone NP10) age for the basal part of the Faeroes Middle Series indicates that these intervals of normal polarity are not magnetic anomaly chrons C23N and C24N,

as suggested by Abrahamsen *et al.*, but must be C25N or an older normal interval. This is substantiated by the presence of a late Palaeocene pollen assemblage in the coal sequence which separates the Lower and Middle Series (Lund 1983). Smythe (1983) favoured an assignment of both normal polarity intervals, plus the intervening reversed polarity interval, to C25N, but there are two arguments which favour their assignment to C26N. Firstly, a short period of reversed magnetization has not been noted within anomaly 25, whereas C26N in the Gubbio section of northern Italy has been shown to be composite (Lowrie *et al.* 1982). Secondly, C26R is considerably longer than C25R (Harland *et al.* 1982; Berggren *et al.* 1985). If the recorded normal interval is assigned to C25N and a constant

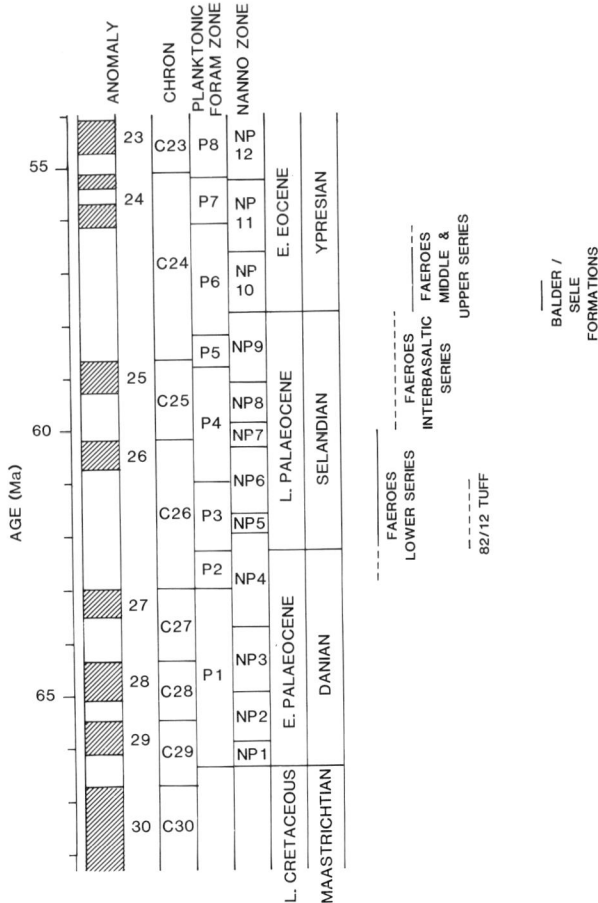

FIG. 4. Proposed time-scale of early Tertiary volcanic events in the Faeroes area, based on tephrachronological, magnetostratigraphical and biostratigraphical information. Magnetobiochronological time-scale is that of Berggren *et al.* (1985): calibration with radiometric time-scales is a matter of controversy and for this reason little weight should be attached to absolute ages at this stage.

extrusion rate is assumed, another normally-magnetized zone corresponding to C26N should be recorded in the lower part of the Lower Series. Abrahamsen *et al.* (1984) have shown that this is not the case. However, if the recorded normal interval is C26N and a constant extrusion rate is assumed, there would be no need to predict a normal interval lower in the Lower Series. This is compatible with the biostratigraphical assignment of the tuff in Borehole 82/12 to Zone P3, which falls in C26R (Berggren *et al.* 1985). The assignment of most of the Faeroes Lower Series to C26R indicates a direct correlation with the main phase of basaltic volcanism in the British Tertiary Volcanic Province, which also falls within C26R, according to Mussett (1984). Waagstein (1988) has reached a similar conclusion, although referring only the lower normal polarity zone to anomaly 26, the upper being referred to anomaly 25.

Conclusions

Combining the magnetostratigraphical, biostratigraphical and tephrachronological evidence, it appears that extrusion of the Faeroes Lower Basalts began early in C26R (about 63 Ma according to the timescale of Berggren *et al.* 1985), and continued until just after the end of C26N (about 60 Ma on the Berggren *et al.* timescale), as shown in Fig. 4. In view of the

major discrepancies in the postulated ages of the early Tertiary epoch boundaries in the various timescales currently in circulation (cf. Curry & Odin 1982; Harland *et al.* 1982; Berggren *et al.* 1985), these absolute dates must be treated with caution.

Although the Faeroes Lower Series basalts are generally correlated with the E Greenland Plateau Basalts, the assignment of most of the Faeroes Lower Series to C26 argues against this. However, it should be recognized that the biostratigraphic assessment of the timing of onset of volcanism in E Greenland is based solely on palynological evidence from a single, extremely low diversity sample. A more thorough investigation of the palynology of the Vandfaldsdalen Formation would now seem to be warranted. If, however, one accepts the late Palaeocene age for the Vandfaldsdalen Formation, the E Greenland succession may be correlated with the Faeroes Middle Series. In this context, it is interesting to note the presence of relatively primitive Ti-rich olivine tholeiites at the base of the Middle Series (Waagstein & Hald 1984), as these are closely comparable to some of the Lower Basalts of the Kangerdlugssuaq area described by Nielsen *et al.* (1981).

ACKNOWLEDGEMENTS: This paper is published with the approval of the Director of the British Geological Survey (NERC).

References

ABRAHAMSEN, N. 1967. Some palaeomagnetic investigations in the Faeroe Islands. *Bulletin of the Geological Society of Denmark*, **17**, 371–384.

——, SCHOENHARTING, G. & HEINESEN, M. 1984. Palaeomagnetism of the Vestmanna core and magnetic age and evolution of the Faeroe Islands. *In:* BERTHELSEN, O., NOE-NYGAARD, A. & RASMUSSEN, J. (eds) *The Deep Drilling Project 1980–1981 in the Faeroe Islands.* Foroya Frodskaparfelag, pp. 93–108.

BERGGREN, W. A. 1969. Cenozoic chronostratigraphy, planktonic foraminiferal zonation and the radiometric time scale. *Nature*, **224**, 1072–1075.

——, KENT, D. V., FLYNN, J. J. & VAN COUVERING, J. A. 1985. Cenozoic geochronology. *Bulletin of the Geological Society of America*, **96**, 1407–1418.

COSTA, L. & DOWNIE, C. 1976. The distribution of the dinoflagellate *Wetzeliella* in the Palaeogene of north-western Europe. *Palaeontology*, **19**, 591–614.

CURRY, D. & ODIN, G. S. 1982. Dating of the Palaeogene. *In:* ODIN, G. S. (ed) *Numerical Dating in Stratigraphy, Part 1.* Wiley, London, pp. 607–630.

HANSEN, J. M. 1977. Dinoflagellate stratigraphy and echinoid distribution in Upper Maastrichtian and Danian deposits from Denmark. *Bulletin of the Geological Society of Denmark*, **26**, 1–26.

HARLAND, W. B., COX, A. V., LLEWELLYN, P. G., PICKTON, C. A. G., SMITH, A. G. & WALTERS, R. 1982. *A Geologic Time Scale.* Cambridge University Press, Cambridge, 128 pp.

HEILMANN-CLAUSEN, C. 1985. Dinoflagellate stratigraphy of the uppermost Danian to Ypresian in the Viborg 1 Borehole, central Jylland, Denmark. *Danmarks Geologiske Undersogelse, Raekke A,* **7**, 1–69.

HITCHEN, K. & RITCHIE, J. D. 1987. Geological review of the West Shetland area. *In:* BROOKS, J. & GLENNIE, K. W. (eds) *Petroleum Geology of Northwest Europe.* Graham & Trotman, London, pp. 737–749.

KING, C. 1983. Cainozoic micropalaeontological biostratigraphy of the North Sea. *Institute of Geological Sciences Report,* **82/7**, 11–40.

KNOX, R. W. O'B. 1984. Nannoplankton zonation and the Palaeocene/Eocene boundary beds of NW

Europe: an indirect correlation by means of volcanic ash layers. *Journal of the Geological Society of London*, **141**, 993–999.

—— & MORTON, A. C. 1983. Stratigraphical distribution of early Palaeogene pyroclastic deposits in the North Sea Basin. *Proceedings of the Yorkshire Geological Society*, **44**, 355–363.

——, —— & HARLAND, R. 1981. Stratigraphical relationships of Palaeocene sands in the UK sector of the central North Sea. *In:* ILLING, L. V. & HOBSON, G. D. (eds) *Petroleum Geology of the Continental Shelf of North-west Europe.* Heyden, London, pp. 267–281.

LARSEN, L. M. & WATT, W. S. 1985. Episodic volcanism during the break-up of the North Atlantic: evidence from the East Greenland plateau basalts. *Earth and Planetary Science Letters*, **73**, 105–116.

LOWRIE, W., ALVAREZ, W., NAPOLEONE, G., PERCH-NIELSEN, K., PREMOLI SILVA, I. & TOUMARKINE, M. 1982. Paleogene magnetic stratigraphy in Umbrian pelagic carbonate rocks: The Contessa sections, Gubbio. *Bulletin of the Geological Society of America*, **93**, 414–432.

LUND, J. 1983. Biostratigraphy of interbasaltic coals from the Faeroe Islands. *In:* BOTT, M. H. P., SAXOV, S., TALWANI, M. & THIEDE, J. (eds) *Structure and Development of the Greenland-Scotland Ridge.* Plenum Press, New York, pp. 417–423.

MUSSETT, A. E. 1984. Time and duration of Tertiary igneous activity of Rhum and adjacent areas. *Scottish Journal of Geology*, **20**, 273–279.

NIELSEN, T. F. D., SOPER, N. J., BROOKS, C. K., FALLER, A. M., HIGGINS, A. C. & MATTHEWS, D. W. 1981. The pre-basaltic sediments and the Lower Basalts at Kangerdlugssuaq, East Greenland: their stratigraphy, lithology, palaeomagne-tism and petrology. *Meddelelser om Grønland, Geoscience*, **6**, 1–25.

NOE-NYGAARD, A. & RASMUSSEN, J. 1968. Petrology of a 3000 metre sequence of basaltic lavas in the Faeroe Islands. *Lithos*, **1**, 286–304.

SMYTHE, D. K. 1983. Faeroe–Shetland Escarpment and continental margin north of the Faeroes. *In:* BOTT, M. H. P., SAXOV, S., TALWANI, M. & THIEDE, J. (eds) *Structure and development of the Greenland-Scotland Ridge.* Plenum Press, New York, pp. 109–119.

——, CHALMERS, J. A., SKUCE, A. G., DOBINSON, A. & MOULD, A. S. 1983. Early opening history of the North Atlantic—1. Structure and origin of the Faeroe–Shetland Escarpment. *Geophysical Journal of the Royal Astronomical Society*, **72**, 373–398.

SOPER, N. J., HIGGINS, A. C., DOWNIE, C., MATTHEWS, D. W. & BROWN, P. E. 1976. Late Cretaceous-early Tertiary stratigraphy of the Kangerdlugssuaq area, east Greenland, and the age of the opening of the north-east Atlantic. *Journal of the Geological Society of London*, **132**, 85–104.

TARLING, D. H. & GALE, N. H. 1968. Isotopic dating and palaeomagnetic polarity in the Faeroe Islands. *Nature*, **218**, 1043–1044.

WAAGSTEIN, R. 1988. Geochemical stratigraphy and structure of the Faeroes basalt plateau. *In:* MORTON, A. C. & PARSON, L. M. (eds) *Early Tertiary Volcanism and the Opening of the NE Atlantic.* Geological Society of London Special Publication, **39**, pp. 225–238.

—— & HALD, N. 1984. Structure and petrography of a 660 m lava sequence from the Vestmanna-1 drill hole, lower and middle basalt series, Faeroe Islands. *In:* BERTHELSEN, O., NOE-NYGAARD, A. & RASMUSSEN, J. (eds) *The Deep Drilling Project 1980–1981 in the Faeroe Islands.* Foroya Frodskaparfelag, pp. 39–70.

A. C. MORTON, R. HARLAND, British Geological Survey, Keyworth, Notts NG12 5GG, UK.

D. EVANS, British Geological Survey, Murchison House, West Mains Rd., Edinburgh EH9 3LA, UK.

C. KING, Paleoservices Ltd., Unit 15, Paramount Industrial Estate, Sandown Rd., Watford WD2 4XA, UK.

D. RITCHIE, British Geological Survey, 19 Grange Terrace, Edinburgh EH9 2LF, UK.

Early Tertiary basalts and tuffaceous sandstones from the Hebrides Shelf and Wyville–Thomson Ridge, NE Atlantic

M. S. Stoker, A. C. Morton, D. Evans, M. J. Hughes, R. Harland & D. K. Graham

SUMMARY: A sequence of Palaeocene basalts and late Palaeocene–early Eocene tuffaceous sandstones has been recovered from three boreholes on the northern Hebrides Shelf and Wyville–Thomson Ridge. The basalts, cored in boreholes 85/5B and 85/7 on the northern Hebrides Shelf, can be seismically correlated with the extensive N Atlantic lava sequence. The basalt in 85/5B displays major element chemistry typical of alkali-basalts but trace and rare-earth element (REE) patterns are typical of olivine-tholeiites. In contrast, the basalt in 85/7 displays geochemical characteristics typical of a depleted tholeiite. Despite their different compositions, the basalts have similar ratios of the highly incompatible elements La and Ta which may reflect different degrees of partial melting of similar source material. The tuffaceous sandstones, cored in borehole 85/2B on the Wyville–Thomson Ridge, were deposited in a warm, shallow marine environment contemporaneous with the reworking of nearby basaltic terrains. *Nummulites rockallensis* Hinte & Wong occurs abundantly throughout the sandstones; its only previous known occurrence is from latest Palaeocene–earliest Eocene sediments at DSDP Site 117A on Rockall Bank. Regional mapping indicates that the tuffaceous sandstones overlie the basalts, thus the latter can be assigned to the Palaeocene (or older). This is consistent with K-Ar dates of 57.3–63.1 Ma obtained from the basalts.

.

On the Rockall Plateau and in the southern Rockall Trough, boreholes drilled by the Deep Sea Drilling Project have enabled an early Tertiary stratigraphic framework to be established (Laughton *et al.* 1972; Roberts *et al.* 1984; Ruddiman *et al.* 1987). In contrast, the early Tertiary sequence in the northern Rockall Trough and on bordering highs is less well known. This paper describes early Tertiary sedimentary and igneous rocks recovered in three boreholes drilled on the north-eastern flank of the trough, on the Wyville–Thomson Ridge and northern Hebrides Shelf, as part of a reconnaissance mapping programme carried out by the British Geological Survey (BGS) (Fig. 1).

The Wyville–Thomson Ridge forms a NW–SE trending bathymetric high which extends from Faeroe Bank to the W Shetland Shelf (see Fig. 1). It has been interpreted from gravity modelling as a thick pile of Palaeocene basaltic lavas extruded along NW–SE trending fissures and contemporaneous with the formation of the Faeroe Shelf and Lousy, Bill Bailey and Faeroe Banks (Roberts *et al.* 1983). On seismic reflection profiles the basalts appear to be overlain by sediments (Himsworth 1973), a view supported by Jones & Ramsay (1982) who dredged early Eocene pyroclastic tuffs from the north-western end of the ridge (see Fig. 1).

Gravity and magnetic studies have suggested the presence of early Tertiary igneous and sedimentary rocks on the Hebrides Shelf (Bullerwell 1972; Jones 1978; Brewer & Smythe 1986). Basaltic rocks, yielding a mean age of 43.4 ± 1.0 Ma by K-Ar dating, have been recovered in dredge hauls between the known early Tertiary centre of St. Kilda and the Precambrian Flannan Isles (Jones *et al.* 1986a), while an early Oligocene chalk has been collected from the Geikic escarpment, NW of St. Kilda (Jones *et al.* 1986b).

The BGS borehole data have been tied to a network of shallow seismic profiles throughout the study area, thus extending these earlier observations and allowing a fuller understanding to be gained of the early Tertiary succession bordering the northern Rockall Trough.

Geological setting of the boreholes

Borehole 85/2B is located on the southern flank of the Wyville–Thomson Ridge (59°34.9′N, 6°32.5′W), at the junction with the Hebrides/W Shetland Shelf (Figs 2 & 3). The airgun profile in Fig. 3 clearly shows a sedimentary cover on the upper part of the ridge structure, although it is known that basalts occur at or near to seabed on the northern margin of the ridge (Roberts *et al.* 1983; Stoker & Hitchen, in press). Along the northern margin of the Rockall Trough, the

From MORTON, A. C. & PARSON, L. M. (eds), 1988, *Early Tertiary Volcanism and the Opening of the NE Atlantic,* Geological Society Special Publication No. 39, pp. 271–282.

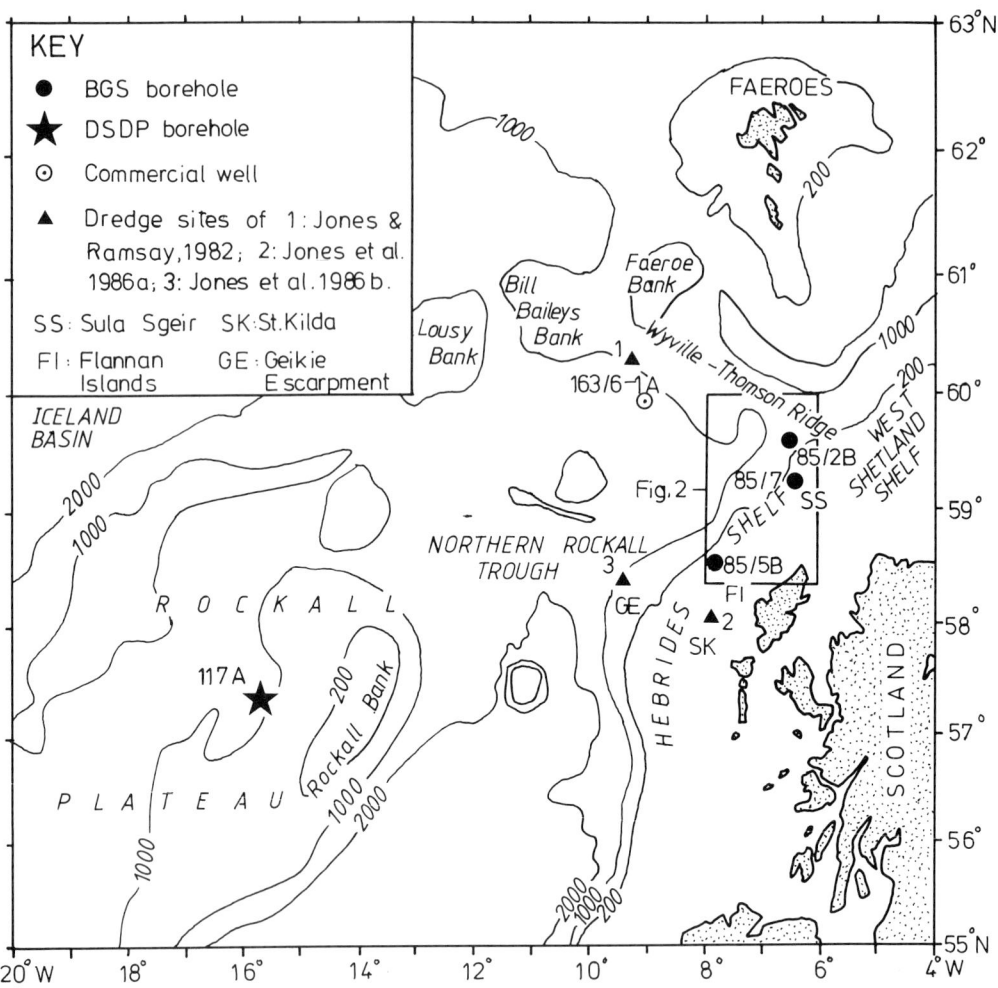

FIG. 1. Map showing BGS and other relevant sample sites and geographic locations referred to in text.
Bathymetric contours in metres. Rectangle shows location of study area expanded in Fig. 2.

sedimentary cover can be divided into two
distinct units, an upper unit characterized by
mounded, contourite sediments, and a lower
undulatory and locally faulted unit (see Fig. 3).
The contact between the two units is marked by
a distinct basin margin angular unconformity, a
relationship first noted by Himsworth (1973).
This unconformity is a composite feature, the
result of repeated basin margin erosion during
the late Palaeogene and early Neogene. The two
units have been assigned preliminary ages of
early Palaeogene and Neogene–Quaternary (Stoker & Hitchen, in press). The mapped outcrop
limits of these two units are shown in Fig. 2.

On the Wyville–Thomson Ridge, the early
Palaeogene sediments lie close to seabed, in
contrast to the basinal areas where they are

overlain by up to 600 m of younger sediments. A
prominent reflector (P) terminates abruptly at an
erosion surface close to the seabed giving rise to
a strong diffraction, D (Fig. 3). This is clearly not
a continuation of reflector P as it cuts the
underlying reflectors. Borehole 85/2B was sited
to investigate the nature of the prominent reflector
in the light of the regional seismic interpretation.
It was drilled in 342 m of water from the drilling
vessel *MV Pholas* in July 1985. The borehole
drilled 4.95 m of Tertiary sediment beneath a
thin (23.40 m) cover of Quaternary diamict. The
hole was terminated within the Tertiary at a
depth of 28.35 m below seabed.

Boreholes 85/5B and 85/7 are located on the
outer northern Hebrides Shelf W of Lewis
(58°29.0′N, 7°53.1′W) and NW of Sula Sgeir

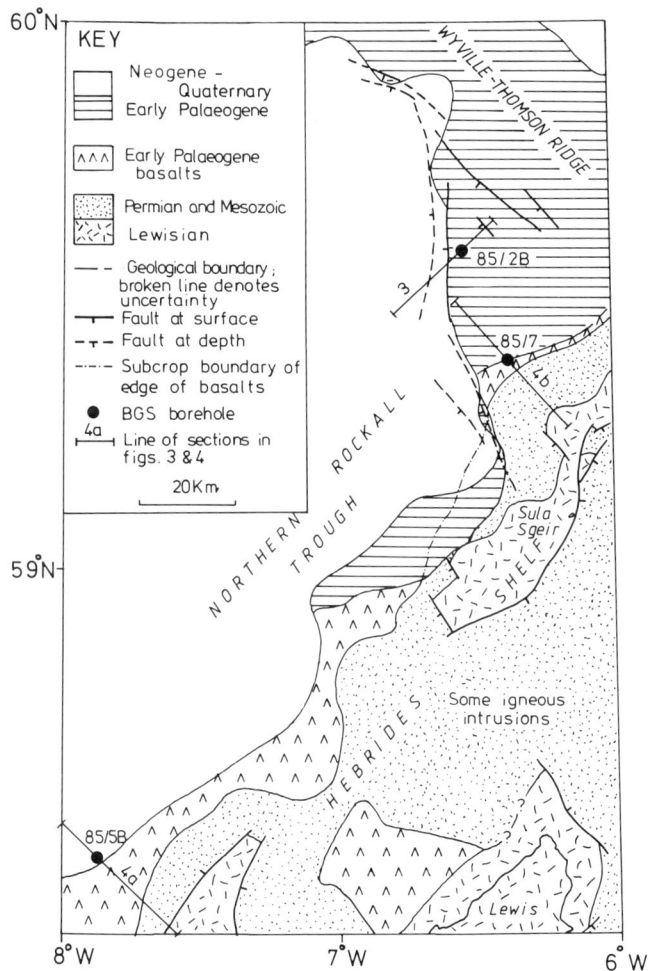

FIG. 2. Schematic geological framework of study area based on BGS regional mapping data. Quaternary cover on Hebrides Shelf and Wyville–Thomson Ridge omitted.

(59°23.2′N, 6°23.6′W), respectively (Fig. 2). The outer shelf in this region consists of a westerly-thickening wedge of Tertiary and Quaternary sediments (Jones *et al.* 1986b; Stoker & Hitchen, in press) underlain by an irregular reflector identified on shallow seismic data and coincident with strong magnetic anomalies (Fig. 4). The mapped outcrop limits of this unit are shown in Fig. 2. The basin margin unconformity is less well defined in this area, although the northern Rockall Trough seismic sequence can be adopted in the region NW of Sula Sgeir. Study of commercial seismic data has enabled us to map the irregular reflector into the northern Rockall Trough, where it correlates with the top of the

Palaeocene basalts (Roberts *et al.* 1983; Stoker & Hitchen, in press) cored in Well 163/6–1A (Morton *et al.* 1988b).

Borehole 85/5B was located near the feather-edge of the Tertiary/Quaternary sediment prism where the irregular reflector could also be sampled (Fig. 4a). The borehole drilled 41.10 m of Pliocene and Quaternary sediment above 2.90 m of Tertiary basalt. The hole was drilled in 140 m of water and was terminated in basalt at a depth of 44.00 m below seabed. Borehole 85/7, similarly located to core the irregular reflector (Fig. 4b), was drilled in 168 m of water. Thin Quaternary (24.15 m) and undifferentiated Quaternary/Tertiary (1.85 m) sediment overlaid 5.00 m of Ter-

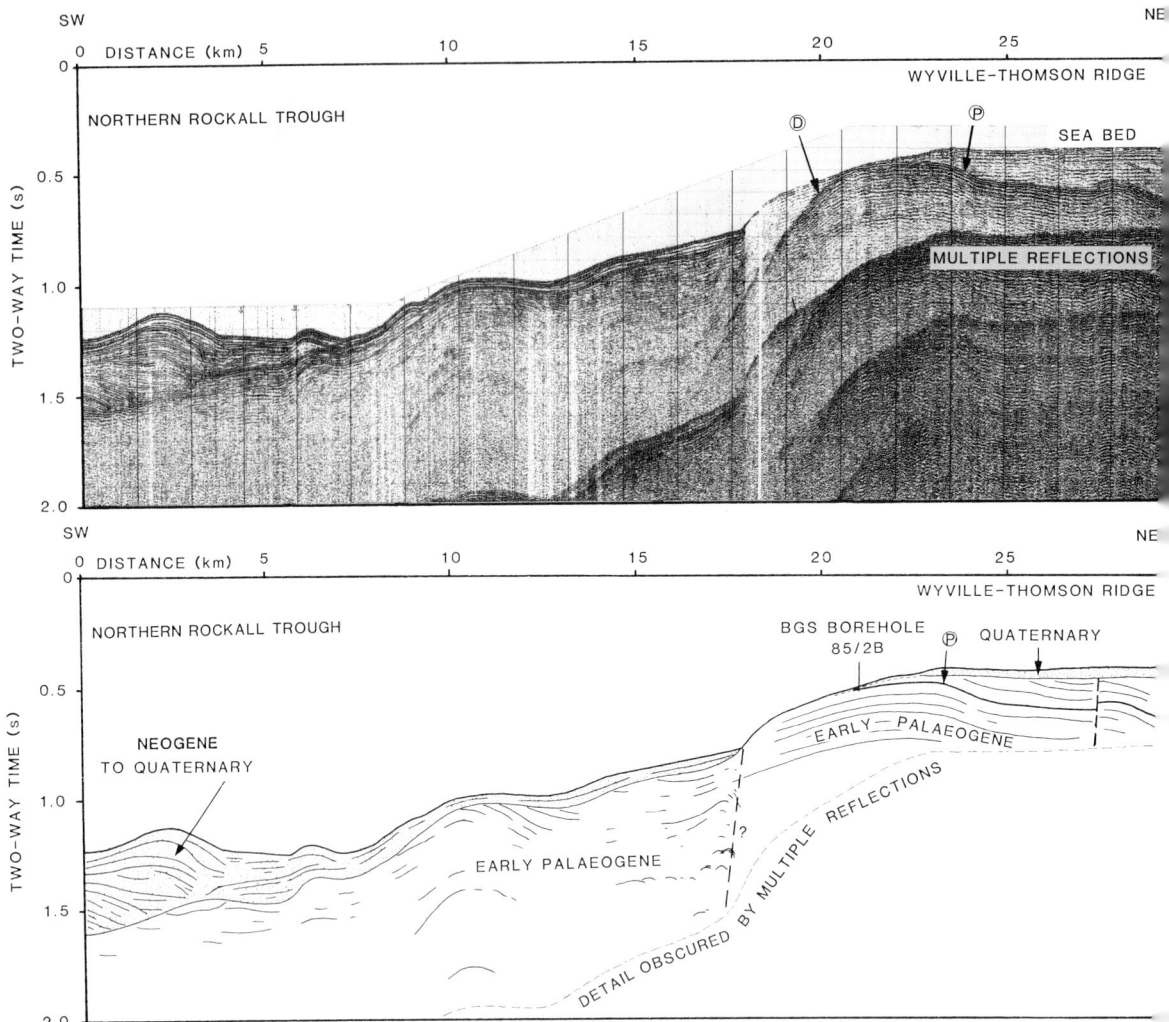

FIG. 3. Single channel airgun profile and interpreted line drawing showing geological setting of BGS Borehole 85/2B. Profile located in Fig. 2. See text for details. P: Prominent reflector; D: Diffraction caused by intersection of P with erosion surface close to seabed.

tiary basalt. This hole was terminated in basalt at a depth of 31.00 m below seabed. Both boreholes were drilled from the drilling vessel *Bucentaur* in October 1985.

Tertiary basalts

Borehole 85/5B

Petrology

The basalt recovered in this borehole is generally coarse-grained and porphyritic but with a glassy groundmass. It is rich in elongate plagioclase

phenocrysts which form up to 52% of the rock, are up to 1 cm in length and show marked flow alignment. The phenocrysts are labradorite (An_{51-59}). Most are unzoned, although margin compositions of An_{38} were recorded in some cases by electron microprobe. Olivine phenocrysts originally formed up to 7% of the rock, but are now wholly replaced by clays.

The phenocrysts are set in a groundmass of plagioclase laths and subhedral titanomagnetite sub-ophitically enclosed by brown clinopyroxene, with minor interstitial feldspar and extensive devitrified glass. The plagioclase laths are of oligoclase–labradorite composition (An_{25-56})

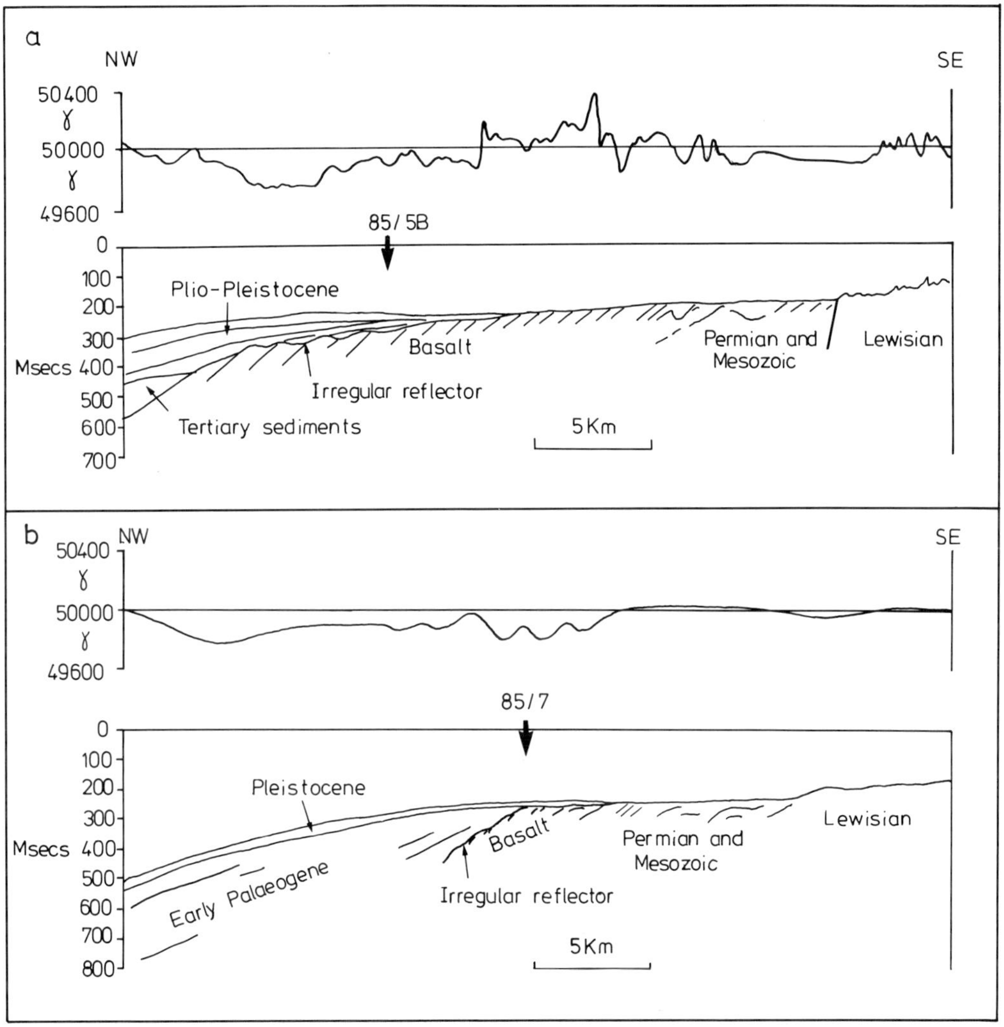

FIG 4. Uncorrected marine magnetic profiles and interpreted line drawing of sparker profiles illustrating the geological setting of BGS Boreholes 85/5B(a) and 85/7(b). Profiles located in Fig. 2. For explanation see text.

and form 6–14% of the rock. The interstitial feldspar is more sodic (An_{14}). Ti-rich salitic pyroxene (Fig. 5) ranges from 10–11% by volume; zoning is uncommon. Devitrified glass forms 19–27% of the rock, with four stages of alteration recognizable: (1) a thin rim of greenish brown smectite with a relatively high K_2O content; (2) a blue-green smectite lower in alkalis; (3) a less dense framework of more coarsely crystalline smectite similar to (2) in composition; and (4) oxidation of secondary clays to iddingsite. Iddingsite development is minor at 41.45 m below seabed, but becomes increasingly significant

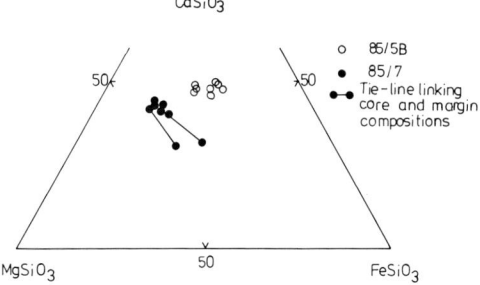

FIG. 5. Variation in the pyroxene compositions in basaltic rocks of Boreholes 85/5B and 85/7.

below this level, and is widely developed at 43.08 m.

Geochemistry

The geochemistry of the 85/5B rocks is comparable with the Plateau Lavas of the British Tertiary Igneous Province (BTIP), such as the Skye Main Lava Series and the Mull Plateau Group, in that they possess major-element chemistry typical of alkali basalts but trace and rare-earth element patterns typical of olivine tholeiites (Table 1),

TABLE 1. *Geochemistry of basaltic rocks, BGS Boreholes 85/5B and 85/7. Major and trace element analyses by X-ray fluorescence carried out by Midland Earth Science Associates. REE analyses by neutron activation at Imperial College Reactor Centre.*

(1) Major element data expressed as %

| | BH 85/5B | | BH 85/7 | |
	41.45 m	42.65 m	28.20 m	30.85 m
SiO_2	44.19	44.12	47.19	46.96
Al_2O_3	15.29	15.90	14.76	13.14
TiO_2	2.50	2.22	1.49	1.74
Fe_2O_3	17.40	16.96	15.29	16.21
MgO	5.85	6.01	6.94	6.67
CaO	7.55	6.95	9.57	9.96
Na_2O	3.85	3.67	2.80	2.83
K_2O	0.25	0.28	0.25	0.15
MnO	0.21	0.23	0.15	0.19
P_2O_5	0.23	0.20	0.11	0.14
LOI	2.62	3.11	1.92	1.58
Total	99.94	99.65	100.47	99.57

LOI = Loss on ignition

(2) Trace and rare earth elements expressed in ppm

| | BH 85/5B | | BH 85/7 | |
	41.45 m	42.65 m	28.20 m	30.85 m
Ba	71	77	25	13
Cr	50	41	146	120
Ni	71	82	85	76
Nb	7	7	5	4
Rb	9	9	3	5
Sr	229	224	119	103
V	129	142	324	301
Y	37	32	36	39
Zn	65	68	105	108
Zr	149	132	90	97
La	6.78	6.41	3.81	3.42
Ce	22.08	18.79	5.83	11.89
Nd	16.56	17.38	9.24	8.41
Sm	5.08	4.57	3.84	3.55
Eu	1.82	1.66	1.34	1.25
Tb	0.98	0.89	0.97	0.88
Yb	2.68	2.48	3.16	3.06
Lu	0.48	0.52	0.64	0.59
Ta	0.44	0.40	0.23	0.24
Hf	3.47	2.77	2.33	2.17

(Morrison *et al.* 1980). An alkalic nature is suggested by the presence of titaniferous pyroxenes, Ti/V ratios of 94–116 (Shervais 1982), and by their position on the alkalis–silica plot (Fig. 6) relative to the Hawaiian Division Line of MacDonald & Katsura (1964). In contrast, the Y/Nb ratio is high (4.6–5.3), characteristic of tholeiites (Pearce & Cann 1973), and REE plots and mid-ocean ridge basalt (MORB)-normalized spidergram patterns are typically tholeiitic. The REE pattern is relatively flat normalized to chondrite (Fig. 7), with the light REE showing a slightly convex-upward profile and slight depletion of the heavy REE: $(La/Yb)_N > 1$. This pattern is comparable with other rocks of the BTIP, and in particular is very similar to that shown by the Mull Plateau Group (Morrison *et al.* 1980) and the Lower Basalt Formation of Antrim (Lyle 1985). It is also comparable to the upper basalts of Well 163/6–1A in the basinal parts of the northern Rockall Trough (Morton *et al.* 1988b). Figure 8 is a spidergram plot of the 85/5B rocks normalized to normal MORB (N-type MORB), showing the 1.5–2 times enrichment of the more hygromagmatophile (HYG) elements and the tendency toward depletion of the less HYG elements.

Paragenesis

An intrusive origin was originally assumed on account of the coarse-grained nature of the rocks in hand specimen. However, further work has revealed an extensive glassy groundmass and the presence of vesicles in the core that may have resulted from a rapidly cooled, possibly extrusive

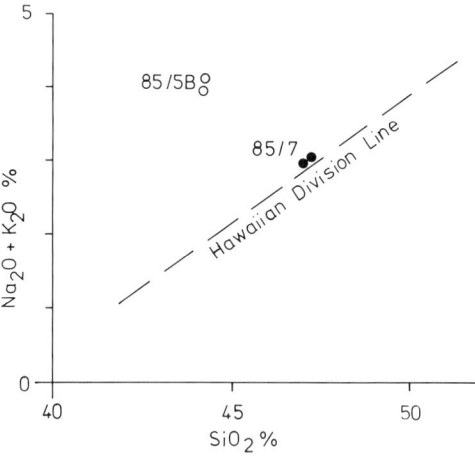

FIG. 6. Alkalis–silica plot demonstrating the more alkalic nature of the 85/5B rocks compared to those from 85/7.

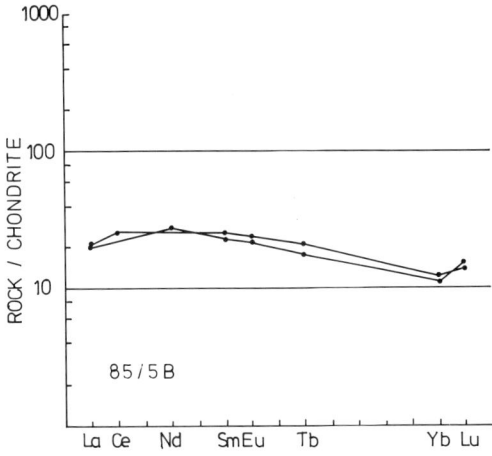

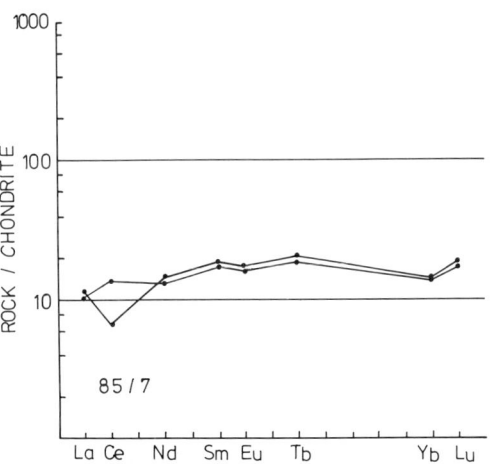

FIG. 7. Chondrite-normalized (REE) plots of 85/5B and 85/7 basaltic rocks.

the older date, and a late Danian to Selandian age is inferred. This compares well with the dates yielded by the basalts of the BTIP (Mussett *et al.* 1988) and with the age of the Lower Basalts of the Faeroes inferred by Waagstein (1988).

Borehole 85/7

Petrology

The igneous section consists of dark grey to black amygdaloidal plagioclase-phyric basalts. The upper sample (28.20 m) is fine-grained: the lower (30.85 m) is slightly coarser. Plagioclase phenocrysts are a minor component of the rock, forming only 1–2%. They are bytownitic, and show zoning from cores of An_{79-85} to margins of An_{66-84}.

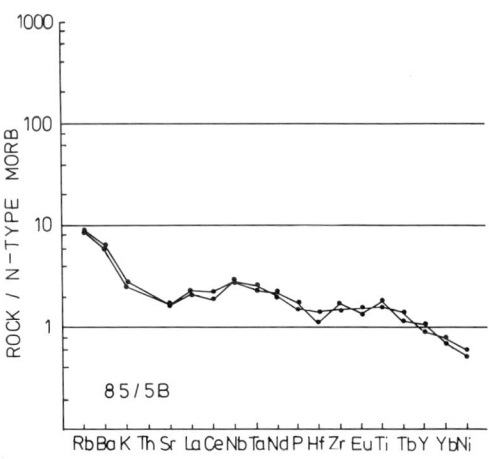

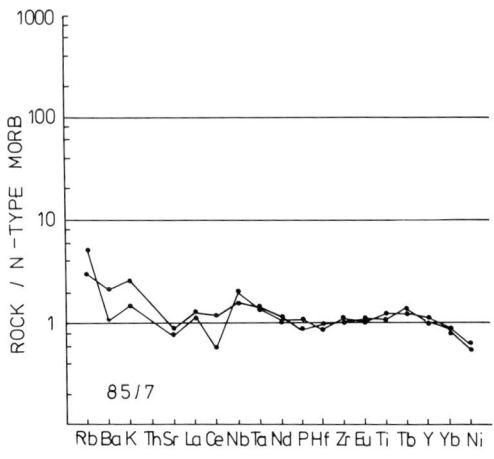

FIG. 8. MORB-normalized spidergrams of 85/5B and 85/7 basaltic rocks

origin. Lavas similar in appearance to these are known from the onshore Tertiary province, the so-called 'big feldspar' lavas of northern Skye (Thomson *et al.* 1980). The continuous nature of the reflectors on the seismic records further supports an extrusive origin for the basalt.

Age of the basalt

Two samples were dated by K-Ar, one (41.45 m) giving an apparent age of 58.0 ± 6.2–63.1 ± 6.4 My, the other (42.65 m) giving a younger age of 38.4 ± 7.5–44.9 ± 8.5 My. Although the latter is similar to age data obtained nearby by Jones *et al.* (1986a) from dredged material, regional stratigraphic evidence (see Discussion) favours

Plagioclase is more common in the groundmass, with labradorite (An_{51-73}) laths forming 35–37% of the basalt. Subhedral titanomagnetite is a minor groundmass phase, forming 3% of the rock. Colourless clinopyroxene is a major phase comprising about 22–23% of the rock, and ophitically encloses plagioclase and opaques. In contrast to its counterpart in 85/5B, it is Ti-poor and plots in the augite field of the pyroxene triangle (Fig. 5). In the coarser-grained, more slowly cooled sample (30.85 m), it shows marked zoning, with typical Fe-enrichment at grain margins. Devitrified glass forms between 32 and 36% of the rock, occurring interstitially. The alteration product is smectite, which also occurs as a vesicle and vein fill, together with analcime.

Geochemistry

In contrast to 85/5B, the basalt cored in 85/7 is a typical depleted tholeiite, comparable in composition to N-type MORB (Table 1).

They have low Ti/V (27–35) and high Y/Nb (7.2–9.8) ratios, and plot close to the Hawaiian division line of MacDonald & Katsura (1964) on an alkalis–silica plot (Fig. 6). They show light REE depletion (Fig. 7), with $(La/Yb)_N < 1$, and have a flat pattern on the MORB-normalized spidergram (Fig. 8). Although differing from the main Plateau Lavas of the BTIP, analogues to the 85/7 basalts may be found in the Preshal Mhor lavas of Skye and Mull (Thompson *et al.* 1980) and the Upper Basalts of Antrim (Lyle 1985). They differ from the basalts of 163/6–1A by being more light REE depleted.

Paragenesis

As with borehole 85/5B the glassy nature of the groundmass and the presence of vesicles suggest an extrusive origin for the basalt in borehole 85/7. On seismic profiles, the basalt can be correlated with the extensive sequence of lavas identified in the northern Rockall Trough (Stoker & Hitchen, in press).

Age of the basalt

Two samples have been dated by K-Ar; the lower sample (30.85 m) gave a very similar date to the 85/5B basalts of 57.3 ± 5.2–62.3 ± 4.7 Ma, while the upper sample (28.20 m) gave an anomalously young date of 11.2 ± 0.9–11.8 ± 2.3 Ma. Regional stratigraphic evidence (see Discussion) favours the former date and a late Danian to Selandian age is inferred.

Tertiary sediments

Borehole 85/2B

Lithology

The sediments consist of dark grey to dark greenish grey, massive to poorly bedded, shelly, muddy, fine tuffaceous sandstones becoming more shaly down-core. Bedding is subhorizontal and is depicted by concentrations of shell debris which vary from single laminae with fragments less than 1 cm in length, to 10 cm thick indurated bands with individual shell fragments up to 4 cm in length. The shells display a random orientation.

Smear slide examination confirms their highly tuffaceous nature with abundant basaltic fragments, quartz, feldspar, heavy minerals (hornblende, pyroxene, zircon) and a clay mineralogy composed solely of smectite. Elongate zeolites are also common. The absence of graded bedding and the high degree of intermixing with terrigenous clastic material indicates that the basaltic detritus is epiclastic rather than pyroclastic.

Palaeontology

The sediments contain a poor dinoflagellate cyst flora including such species as *Cordosphaeridium inodes* (Klumpp) Eisenack, *Cribroperidinium tenuitabulatum* (Gerlach) Helenes and ?*Systematophora* sp. In addition a foraminiferal fauna of *Cibicidoides alleni* (Plummer), *Cibicides westi* (Howe), *Gyroidinoides naranjoensis* (White), *Lenticulina midwayensis* (Plummer), *Melonis affine* (Reuss), abundant *Nummulites rockallensis* Hinte & Wong, *Quinqueloculina plummerae* Cushmann & Todd and *Vaginulinopsis* cf. *longiforma* (Plummer) was recovered.

The macrofauna recovered includes a relatively abundant, but poorly preserved, molluscan assemblage. The gastropods consist principally of small high-spired forms, probably *Turritella* sp., with *Conomitra parva*? (J. de C. Sowerby) and *Streptolathyrus*? also present. The bivalve fauna is dominated by cardiacea, of which the more identifiable forms include *Cardita* cf. *acuticostata* Lamarck, *Cardita crebricostata*? Edwards, and *Venericor planicosta*? Lamarck. Specimens of *Semimodiola elegans* (J. Sowerby) were also found.

Depositional environment

The palaeontological data suggest a marine depositional environment. *N. rockallensis* is indicative of a warm, shallow sea (van Hinte & Wong 1975), and the presence of distinct beds of disorganized macrofossil accumulations may be the result of concentration and reworking by high-energy events in a nearshore setting (Bourgeois & Leithold 1984) and would account for the paucity of dinoflagellate cysts. Weathering of a nearby basalt terrain is regarded as the most likely source of the basaltic detritus and smectite.

Age of the sediments

Smectite-dominated clay mineral assemblages are typical of Palaeocene–Eocene boundary strata in the North Sea and Faeroe–Shetland Channel. This age range is supported by the calcareous microfauna which is generally diagnostic of the late Palaeocene–early Eocene. Although the poor preservation of the molluscs precludes precise identification in most instances, the assemblage is unquestionably Tertiary and suggests an early-to-middle Eocene age. The few dinoflagellate cyst forms present only indicate an early Palaeogene age.

The only previous record of *N. rockallensis* is from DSDP Hole 117A on Rockall Bank on the Rockall Plateau (van Hinte & Wong 1975), where it occurs in core 9. Core 9 was initially ascribed by Laughton *et al.* (1972) to the late Palaeocene, nannoplankton Zone NP9, but subsequent biostratigraphic work has indicated an early Eocene NP10 age (Morton *et al.* 1983). The presence of *N. rockallensis* at BGS Borehole 85/2 therefore indicates an age of latest Palaeocene or earliest Eocene.

Discussion

Regional seismic mapping indicates that the basalts in 85/5B and 85/7 represent a continuation of the widespread northern N Atlantic lava sequence that is extensively developed to the NW of these sites (Stoker & Hitchen, in press; unpublished BGS data). The edge of this sequence can be accurately mapped on the northern Hebrides Shelf (Fig. 2), although this is most probably an erosional limit. The basalts are overlain by volcanogenic sandstones, cored at 85/2B, which consist largely of reworked basaltic detritus. These sandstones were deposited during a marine transgression in the latest Palaeocene or earliest Eocene. Thus, the basalts can be assigned to the Palaeocene (or older), a conclusion wholly in accord with the radiometric age range of 57.3–63.1 Ma. They may therefore be correlated with the Faeroes Lower Series basalts, which are now known to have been extruded in

the magnetochron interval C26R–C24R (Waagstein 1988), confirming seismostratigraphic correlations made by Roberts *et al.* (1983).

Despite their geographical separation and distinctly different compositions, the basalts of 85/5B and 85/7 have similar ratios of the highly incompatible elements La and Ta (La/Ta = 15.4–16.0 for 85/5B and 14.3–16.6 for 85/7), consistent with a homogeneous mantle source in this area (Saunders 1984). The geochemical differences between the two sites can be explained by different degrees of partial melting. The $(Ce)_N$–$(Ce/Yb)_N$ plot (Fig. 9) shows that the basalts of 85/7 could have been generated by 10–20% dynamic partial melting of a source with $(Ce)_N = 1.2$ and $(Ce/Yb)_N = 0.4$, with those of 85/5B generated by about 5% partial melting of the same source. The basalts recovered in Well 163/6–1A, in the northern Rockall Trough (Morton *et al.* 1988b), appear to have had a slightly different mantle source, as La/Ta ratios are slightly higher (16.8–17.5) and they fall onto a partial melt curve with a different origin, at $(Ce)_N = 1.5$ and $(Ce/Yb)_N = 0.8$ (Fig. 9). The basalts of the northern Hebrides Shelf (85/5B, 85/7) and northern Rockall Trough (Well 163/6–1A, Morton *et al.* 1988b) differ markedly in

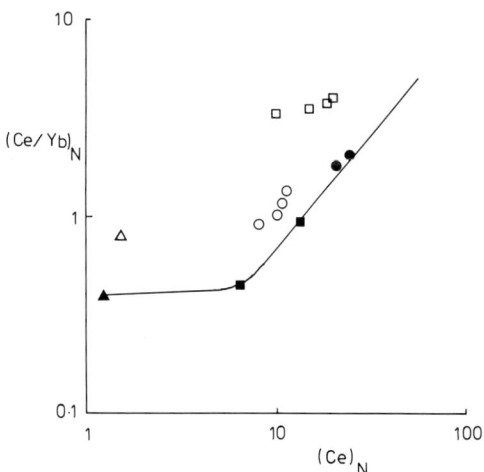

FIG. 9. Basalts from 85/5B and 85/7 plotted on a log–log plot of $(Ce)_N - (Ce/Yb)_N$. Also shown is the dynamic melting curve of a garnet-lherzolite source (after Saunders 1984). Data from 163/6–1A shown for comparison. Key: open triangle: source composition for 163/6–1A lavas; open circle: basalts, cores 5 and 6, 163/6–1A; open square: basalts, core 4, 163/6–1A; filled triangle: source composition for 85/5B and 85/7 basalts; filled circle: basalts, 85/5B; filled squares: basalts, 85/7.

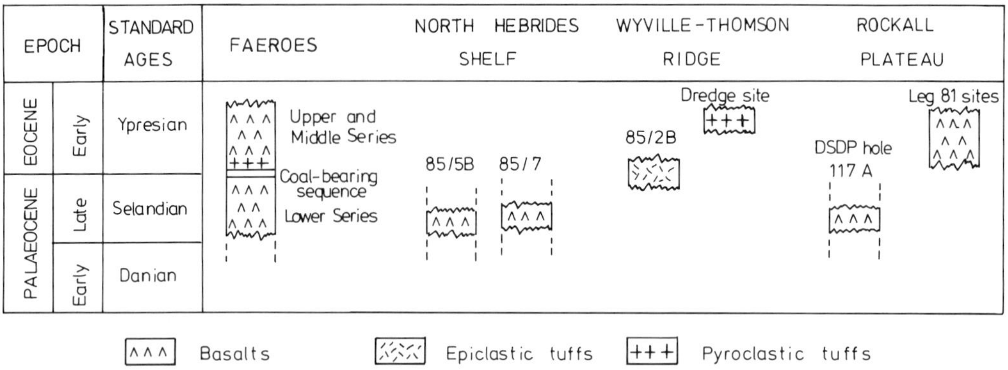

EPOCH		STANDARD AGES	FAEROES	NORTH HEBRIDES SHELF		WYVILLE–THOMSON RIDGE	ROCKALL PLATEAU	
EOCENE	Early	Ypresian	Upper and Middle Series	85/5B	85/7	Dredge site +++ 85/2B	DSDP hole 117 A	Leg 81 sites
PALAEOCENE	Late	Selandian	Coal-bearing sequence Lower Series					
	Early	Danian						

Basalts   Epiclastic tuffs   +++ Pyroclastic tuffs

FIG. 10. Tentative correlation of early Tertiary volcanic and sedimentary events in the NE Atlantic. Information for DSDP Leg 81 sites on SW Rockall Plateau (55–57°N, 20–24°W) taken from Roberts *et al.* 1984.

geochemistry from the Faeroes Lower Series basalts with which they correlate on seismostratigraphic, geochronological and biostratigraphic grounds; the Faeroes Lower Series basalts are typically high Fe-Ti tholeiites with a more enriched mantle source, with a mean La/Ta of 12.9 (data of Gariepy *et al.* 1983). This indicates considerable regional mantle source heterogeneity during the late Palaeocene.

Taken collectively, these data enable us to establish a preliminary early Tertiary stratigraphic framework for the northern Hebrides Shelf and Wyville–Thomson Ridge (Fig. 10). In general terms, a phase of Palaeocene basalt extrusion was followed by the deposition of late Palaeocene–early Eocene volcanogenic shallow-water sandstones. The volcanogenic detritus is regarded as epiclastic in origin, in contrast to the pyroclastic tuffs recovered by Jones & Ramsay (1982), although these latter sediments were dated exclusively as Eocene (late Ypresian NP11–12) and thus may be slightly younger in age. In this scheme, the basalts of the northern Hebrides Shelf and the Wyville–Thomson Ridge are correlated with the Faeroes Lower Series and the basalts of DSDP Hole 117A. They also appear to correlate on geochronological grounds with the basalts of the British Tertiary Igneous Province. Toward the end of the Palaeocene, basalt extrusion terminated and was succeeded by the deposition of the late Palaeocene coal-bearing sequence on the Faeroes and a latest Palaeocene or earliest Eocene marine transgression over the Wyville–Thomson Ridge. Renewed basaltic volcanism then ensued, beginning with the extrusion of the Faeroes Middle Series basalts, contempor-

aneously with the tuffs of the North Sea Balder Formation (Morton *et al.* 1988a) and the development of oceanward-dipping basaltic wedges in the NE Atlantic (Knox & Morton 1988). The pyroclastics recorded by Jones & Ramsey (1982) from the northwestern part of the Wyville–Thomson Ridge therefore belong to this second phase of volcanism.

The occurrence of the N Atlantic lava sequence on the Hebrides Shelf at water depths of less than 200 m has major implications for the subsidence history of the adjacent N Rockall Trough, where the depth to the top of the lavas locally exceeds 2800 m. Clearly, the development of the N Rockall Trough in its present form must post-date this Palaeocene–early Eocene sequence.

ACKNOWLEDGEMENTS: The authors would like to thank Mr K. Hitchen for his comments on an early draft of the paper. We would also wish to thank Drs E. J. W. Jones and N. Fortey whose refereeing of the paper was detailed and helpful. Mrs E. Bates is thanked for typing the manuscript. Dr J. G. Mitchell carried out the whole rock K-Ar analyses of the basalts. We gratefully acknowledge the financial assistance provided by the Department of Energy and the following oil companies which enabled the borehole programme to proceed: Amerada Hess Exploration Ltd, Amoco (UK) Exploration Co., BP Petroleum Development Ltd., Britoil PLC, Elf UK PLC, Enterprise Oil PLC, ICI Petroleum Ltd., Mobil North Sea Ltd., Occidental Petroleum (UK) Ltd., Shell UK Exploration and Production, and Sovereign Oil and Gas PLC. The paper is published with the permission of the Director, British Geological Survey, (NERC).

References

BOURGEOIS, J. & LEITHOLD, E. L. 1984. Wave-worked conglomerates—depositional processes and criteria for recognition. *In:* KOSTER, E. H. & STEEL, R. J. (eds) *Sedimentology of Gravels and Conglomerates. Canadian Society of Petroleum Geologists Memoir*, **10**, pp. 331–343.

BREWER, J. A. & SMYTHE, D. K. 1986. Deep structure of the foreland to the Caledonian orogen, NW Scotland: Results of the BIRPS WINCH profile. *Tectonics*, **5**, 171–194.

BULLERWELL, W. 1972. Geophysical studies relating to the Tertiary volcanic structure of the British Isles. *Philosophical Transactions of the Royal Society of London*, Series A, **271**, 209–215.

GARIEPY, C., LUDDEN, J. & BROOKS, C. 1983. Isotopic and trace element constraints on the genesis of the Faeroe lava pile. *Earth and Planetary Science Letters*, **63**, 257–272.

HIMSWORTH, E. M. 1973. The Wyville–Thomson Ridge. *Quarterly Journal of the Geological Society of London*, **129**, 332–333 (Abstract).

JONES, E. J. W. 1978. Seismic evidence for sedimentary troughs of Mesozoic age on the Hebridean continental margin. *Nature*, **272**, 789–792.

—— & RAMSAY, A. T. S. 1982. Volcanic ash deposits of early Eocene age from the Rockall Trough. *Nature*, **299**, 342–344.

——, MITCHELL, J. G. & PERRY, R. G. 1986a. Early Tertiary igneous activity west of the Outer Hebrides, Scotland—evidence from magnetic anomalies and dredged basaltic rocks. *Marine Geology*, **73**, 47–59.

——, PERRY, R. G. & WILD, J. L. 1986b. Geology of the Hebridean margin of the Rockall Trough. *Proceedings of the Royal Society of Edinburgh*, **88B**, 27–51.

KNOX, R. W. O'B. & MORTON, A. C. 1988. The record of early Tertiary North Atlantic volcanism in sediments of the North Sea Basin. *In:* MORTON, A. C. & PARSON, L. M. (eds) *Early Tertiary Volcanism and the Opening of the NE Atlantic,* Geological Society of London, Special Publication, **39**, pp. 407–419.

LAUGHTON, A. S., BERGGREN, W. A. *et al.*, 1972. Sites 116 and 117. *In:* LAUGHTON, A. S., BERGGREN, W. A. *et al. Initial Reports of the Deep Sea Drilling Project.* US Government Printing Office, Washington, **12**, 395–671.

LYLE, P. 1985. The petrogenesis of the Tertiary basaltic and intermediate lavas in northeast Ireland. *Scottish Journal of Geology*, **21**, 71–84.

MACDONALD, G. A. & KATSURA, T. 1964. Chemical composition of Hawaiian lavas. *Journal of Petrology*, **5**, 82–133.

MORRISON, M. A., THOMPSON, R. N., GIBSON, I. L. & MARRINER, G. F. 1980. Lateral chemical heterogeneity beneath the Scottish Hebrides. *Philosophical Transactions of the Royal Society of London*, Series A, **297**, 229–244.

MORTON, A. C., BACKMAN, J. & HARLAND, R. 1983. A reassessment of the stratigraphy of DSDP Hole 117A, Rockall Plateau: implications for the Palaeocene–Eocene boundary in NW Europe. *Newsletters on Stratigraphy*, **12**, 104–111.

——, EVANS, D., HARLAND, R., KING, C. & RITCHIE, J. D. 1988a. Selandian volcanism in the Faeroes region: evidence from a cored borehole W of the Shetland Islands. *In:* MORTON, A. C. & PARSON, L. M. (eds) *Early Tertiary Volcanism and the Opening of the NE Atlantic,* Geological Society of London, Special Publication, **39**, pp. 263–269.

——, DIXON, J. E., FITTON, J. G., MACINTYRE, R. M., SMYTHE, D. K. & TAYLOR, P. N. 1988b. Early Tertiary volcanic rocks in Well 163/6–1A, Rockall Trough. *In:* MORTON, A. C. & PARSON, L. M. (eds) *Early Tertiary Volcanism and the Opening of the NE Atlantic,* Geological Society of London, Special Publication, **39**, pp. 293–308.

MUSSETT, A. E., DAGLEY, P. & SKELHORN, R. R. 1988. Time and duration of activity in the British Tertiary Igneous Province. *In:* MORTON, A. C. & PARSON, L. M. (eds) *Early Tertiary Volcanism and the Opening of the NE Atlantic,* Geological Society of London, Special Publication, **39**, pp. 337–348.

PEARCE, J. A. & CANN, J. R. 1973. Tectonic setting of basic volcanic rocks determined using trace-element analyses. *Earth and Planetary Science Letters*, **19**, 290–300.

ROBERTS, D. G., BOTT, M. H. P. & URUSKI, C. 1983. Structure and origin of the Wyville–Thomson Ridge. *In:* BOTT, M. H. P., SAXOV, S., TALWANI, M. & THIEDE, J. (eds) *Structure and Development of the Greenland–Scotland Ridge: New methods and concepts.* Plenum Press, New York, pp. 133–158.

——, SCHNITKER, D. *et al.* 1984. *Initial Reports of the Deep Sea Drilling Project.* US Government Printing Office, Washington, **81**, 923 p.

RUDDIMAN, W. F., KIDD, R. B. *et al.* 1987. *Initial Reports of the Deep Sea Drilling Project.* US Government Printing Office, Washington, **94**, 1261 p.

SAUNDERS, A. D. 1984. The rare earth element characteristics of igneous rocks from the ocean basins. *In:* HENDERSON, P. (ed) *Rare Earth Element Geochemistry.* Developments in Geochemistry, **2**, 205–236.

SHERVAIS, J. W. 1982. Ti/V plots and the petrogenesis of modern and ophiolitic lavas. *Earth and Planetary Science Letters*, **59**, 101–118.

STOKER, M. S. & HITCHEN, K. (in press). *Sula Sgeir, Solid and Quaternary Geology.* British Geological Survey 1:250 000 Offshore Map Series.

THOMPSON, R. N., GIBSON, I. L., MARRINER, G. F., MATTEY, D. P. & MORRISON, M. A. 1980. Trace-element evidence of multistage mantle fusion and polybaric fractional crystallisation in the Palaeocene lavas of Skye, NW Scotland. *Journal of Petrology*, **21**, 265–293.

VAN HINTE, J. E. & WONG, Th. E. 1975. *Nummulites rockallensis* n.sp. from the Upper Palaeocene of

Rockall Plateau (North Atlantic). *Journal of Foraminiferal Research*, **5**, 90–101.

WAAGSTEIN, R. 1988. Structure, composition and age of the Faeroe basalt plateau. *In:* MORTON, A. C. & PARSON, L. M. (eds) *Early Tertiary Volcanism and the Opening of the NE Atlantic,* Geological Society of London, Special Publication, **3℃,** pp. 225–238.

M. S. STOKER, D. EVANS & D. K. GRAHAM, British Geological Survey, Murchison House, West Mains Road, Edinburgh EH9 3LA, UK.

A. C. MORTON, M. J. HUGHES & R. HARLAND, British Geological Survey, Keyworth, Nottingham NG12 5GG, UK.

Distribution of early Tertiary lavas in the NE Rockall Trough

M. V. Wood, J. Hall & J. J. Doody

SUMMARY: A study of lavas of early Tertiary age in NE Rockall Trough has revealed that they can be classified on the basis of their seismic reflection character and their relationship with the overlying Tertiary sediments. The lavas have been mapped using commercial seismic reflection data. Our interpretation of the areal distribution of the subaerial flows has major implications for the subsidence history of the area. Three seismic facies are recognized:

(1) High amplitude parallel reflections in laterally continuous sheets with steep edges.
(2) Basinward-thinning wedges of high-amplitude discontinuous reflections.
(3) A single high-amplitude reflection underlain by a zone of small diffractions.

These are interpreted respectively as subaerial flows, hyaloclastite deposits derived from the basalts, and tuffs.

Throughout the sedimentary sequence below the lavas, high amplitude, subhorizontal reflection packages occur. These are interpreted as sills of early Tertiary age intruding Mesozoic sediments. In places, the lavas themselves are intruded by seismically translucent bodies of up to a few kilometres across. These may represent small centres of lava extrusion.

Work has been concentrated on NE Rockall Trough between 59°N and 61°N (Fig. 1). The study area includes Rosemary Bank, Ymir Ridge and the southern margin of the Wyville–Thomson Ridge, the intervening area of deep water, and the NW continental margin of Scotland adjacent to this part of the Trough. Early Tertiary lavas present in the area have been mapped using a grid of multichannel seismic reflection data recorded, mostly, to 6 s two way time (TWT) and unmigrated. Variations in the seismic character of reflectors associated with the lavas have been interpreted as corresponding to lateral facies changes. Interpretation and mapping of the distinctive reflection events reveals a variation in environment of deposition for the lavas throughout the area.

The following account is: (1) a description of the criteria used to classify the reflectors; (2) an interpretation of their distribution and geometry; and (3) a discussion of the implications of this interpretation.

Seismic character and interpretation

Three major seismic facies are recognized and are now described:

Seismic facies A

The lava piles that form the thickest sequences (up to 2.5 km, assuming a velocity of 4.4 km/s) have a very distinctive seismic signature (Fig. 2).

Invariably, their tops are defined by a strong subhorizontal reflector that is laterally continuous and coherent, and hence can be mapped easily over very large areas. To the N similar high amplitude reflectors have been identified by Smythe *et al.* (1983) and Jenyon (1987) as marking the top of piles of basalt flows. Below this reflector occurs a series of subhorizontal parallel reflectors that tend to follow the lava tops. A second high amplitude reflector is often seen, which, in many cases, can be traced unbroken into areas where we believe the lavas are absent or very thin. We interpret this reflector as the base of the lava piles. Figure 3a shows a typical section through a thick lava pile.

These laterally extensive lava sheets terminate in steep scarps. Three explanations may be considered for the formation of these scarps:

(1) They are edges of fault blocks.
(2) Erosion of the lava piles (i.e. valley walls).
(3) They mark the position of a palaeoshoreline.

The first possibility is discounted since continuous reflectors can be traced beneath the scarps, and there is no evidence of faulting in the overlying sediments. If the minimum throw suggested by the size of the scarp is restored, then seismic character on either side of the fault zone is seen to be incompatible. Similarly, the second model of erosion is unattractive since the internal reflections of the flows are not truncated by the scarps but appear to curve downwards in a prograding pattern as they near the scarps (Fig. 3a).

From MORTON, A. C. & PARSON, L. M. (eds), 1988, *Early Tertiary Volcanism and the Opening of the NE Atlantic,* Geological Society Special Publication No. 39, pp. 283–292.

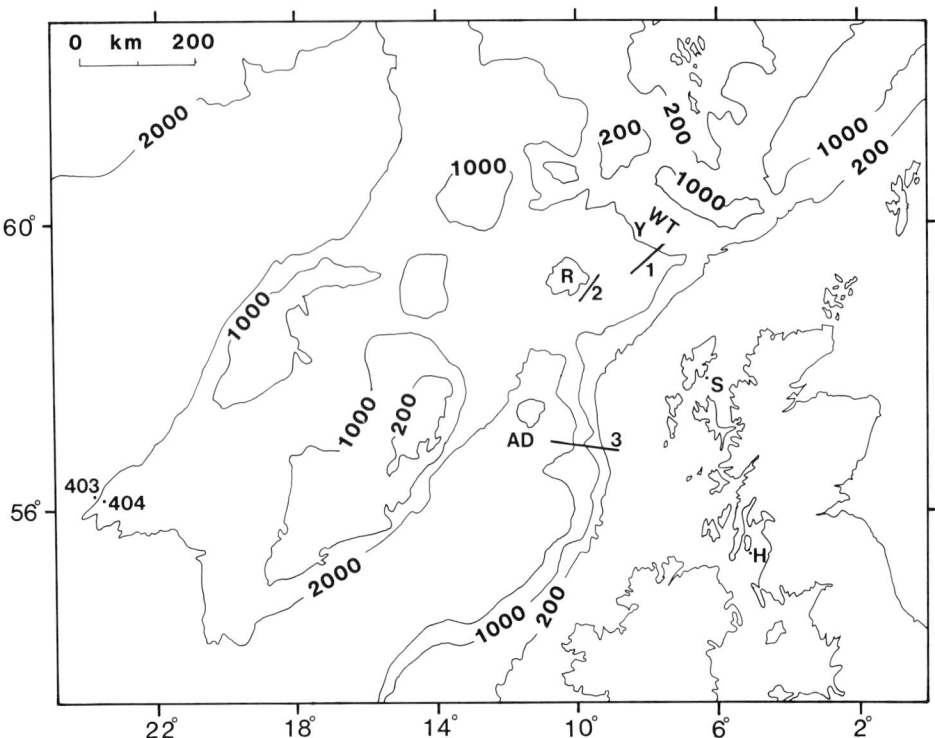

FIG. 1. Regional bathymetry map and location of the study area. R: Rosemary Bank; AD: Anton Dohrn Seamount; WT: Wyville–Thomson Ridge; Y: Ymir Ridge; S: Shiant Isles; H: Holy Island; DSDP Sites 403 & 404. Seismic sections 1–3 are illustrated in Figs 2, 3, 6 & 7.

The preferred interpretation of the scarps is that they mark the position of a palaeoshoreline, similar to the shoreline postulated by Smythe *et al.* (1983) in the Faeroe–Shetland Channel. That is, subaerial flows forming the thick lava sheets came into contact with water that cooled them rapidly and halted their progress. Termination of repeated flow units at, or near, the same position would produce the type of lava edges observed. The relatively static coastline implied by the sudden termination of the lavas may have been due to some underlying fault control at greater depth.

Detailed analysis of the structure of the lava edges is hampered by poor resolution of the seismic data. Simple models for the build-up of such scarps are presented in Figure 4 and demonstrate how edges may develop when lava formation exceeds, keeps pace with, or is exceeded by relative sea-level rise. Bearing in mind that depth conversion of the short seismic section shown in Figure 3a reveals that the true dip of the edge is about 25° (Fig. 3b) the most likely model may be determined. Model a. (see Fig. 4)

may produce the required slope and could also result in the progradation that is apparently seen in Figure 3a; however, this type of progradation is unlikely to occur unless effusion rates are very high. Model b. would produce a near vertical scarp and so is discounted. Model c. could also produce the required slope and would be consistent with the general pattern of subsidence, but with a relative rise of sea-level, some interfingering of lavas and sediments may be expected. No evidence for this can be seen on the seismic data but this may be due to lack of resolution. Thus, we suggest that model c. is the most likely explanation for the scarp development.

Seismic facies B

On many sections, gradually-thinning wedges occur basinward of the lava scarps. Where observed, these occur adjacent to the scarps and thus represent a total change in facies. Their tops are defined by a high amplitude reflector that is disrupted in comparison with the tops of the lava piles. The reflector dips basinward and the entire

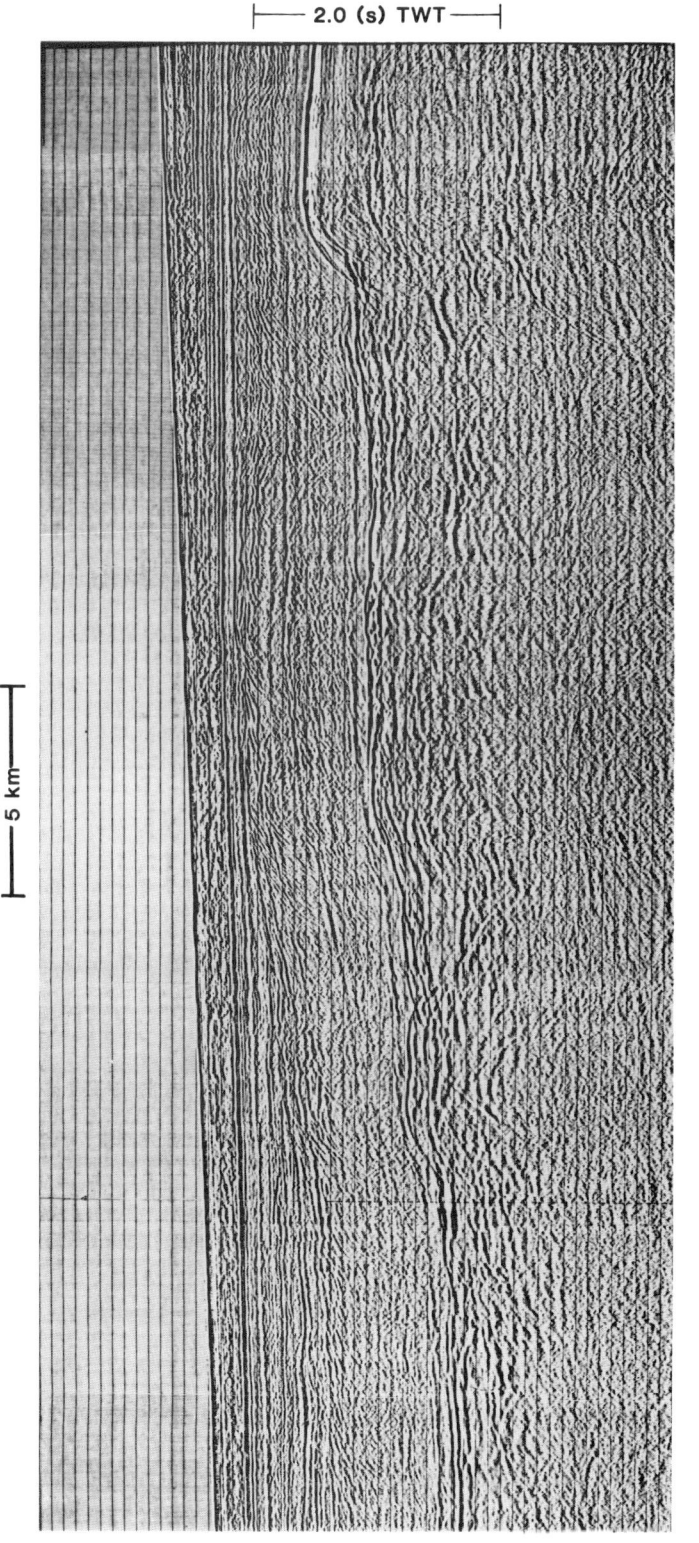

FIG. 2. Seismic section of lava pile, illustrating the seismic character of the lava tops and showing basinward-thinning wedge of reflectors; delta? Location shown as 1 in Fig. 1.

3a)

|←——— 5 km ———→|

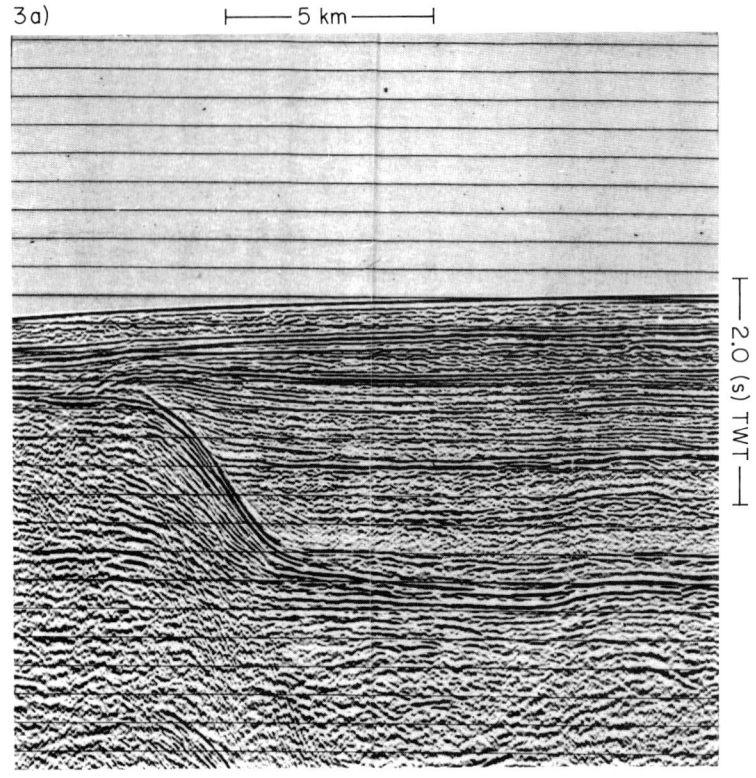

3b   DEPTH SECTION OF LAVA EDGE NE ROSEMARY BANK

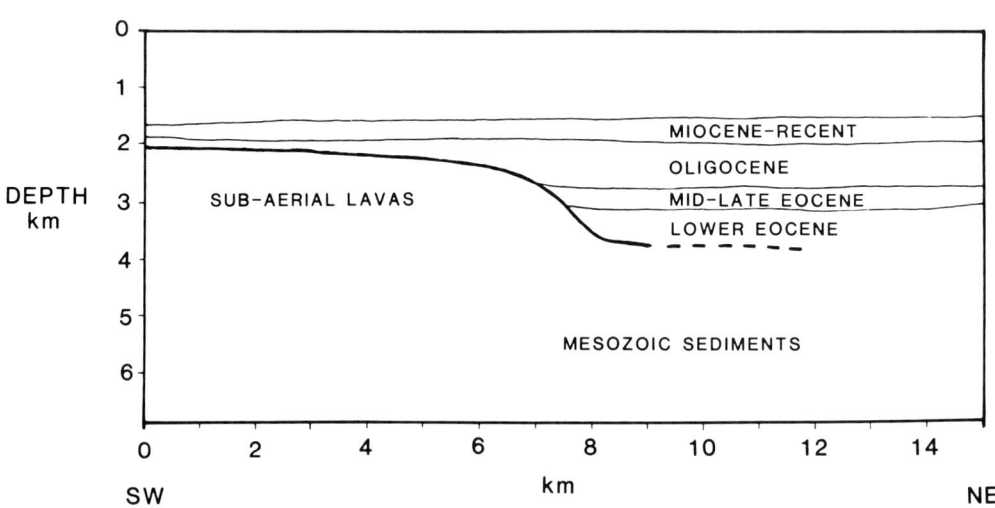

FIG. 3(a) Seismic section through scarp NE of Rosemary Bank, i.e. a thick lava pile: top and base both evident and apparent progradation is visible. Location shown as **2** on Fig. 1. (b) Depth conversion along section 3a, showing average apparent dip of approximately 25°.

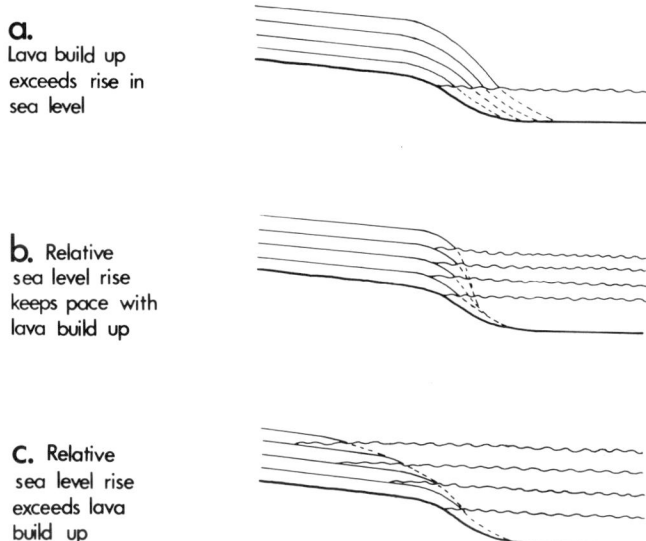

a.
Lava build up
exceeds rise in
sea level

b. Relative
sea level rise
keeps pace with
lava build up

c. Relative
sea level rise
exceeds lava
build up

FIG. 4. Models for build up of lava edges: (a) Lava build-up exceeds relative sea-level rise, requires very high effusion rates. (b) Sea-level rise keeps pace with lava build-up, scarp produced would be too steep to satisfy seismic data. (c) Relative sea-level rise exceeds lava build-up.

pile forms a basinward-thinning wedge of discontinuous reflectors (see Fig. 2).

Because of their relationship with the limits of subaerial flows, disrupted nature and wedge shape, we interpret these as fans of sediment derived from the plateau basalts; some may be deposited as deltas. In areas where the wedges do not have a fan shape, but are developed rather as ribbons parallel to the scarps, facies B may represent hyaloclastite sediment derived from the basalt when it came into contact with the water. Hyaloclastite slope-foot breccias are a characteristic product of the flow of basalt lava from air into water (Jones & Nelson 1970). Large deposits of this kind occur in Iceland (Jones 1968), and on James Ross Island (Nelson 1966). The wedges are not thought to represent subaqueous flows, as their persistence (up to 20 km) seems incompatible with subaqueous extrusion unless effusion rates were very high.

Seismic facies C

The third seismic facies recognized in the lavas is that of a single high amplitude reflection underlain by a zone of small diffractions. These areas occur adjacent to both of the facies described above. This reflection configuration is therefore thought to represent tuffs or other volcaniclastic sediments.

Areal distribution of facies

The areal distributions of the above seismic facies units have been mapped (Fig. 5). The map helps to demonstrate the relationship between present-day features in the Trough, and the distribution of the lavas.

Facies (A) is an area of plateau basalts terminating in a lava scarp that marks the position of a palaeoshoreline. Mapping has revealed that the plateau basalts were prograding into a basin flanked to the E and W by the continental margins of Scotland and Rockall Bank. The area to the N is presently occupied by Wyville–Thomson Ridge. The height of the scarps varies greatly along this shoreline. Scarps of up to 2 km height have been recognized. In other areas, the edge of the plateau basalts is marked by less spectacular features (Fig. 6).

Facies (B) is interpreted as delta fans comprising sediment eroded from the plateau basalts and hyaloclastite deposits derived from the explosive contact of the molten basalts with the sea. It invariably occurs basinward of the subaerial plateau basalts. It would be expected that the positions of the deltas were related to outlets of river systems, which in turn were controlled by land surface topography.

Facies (C) occurs basinward of the plateau basalts and deltaic deposits. It therefore marks

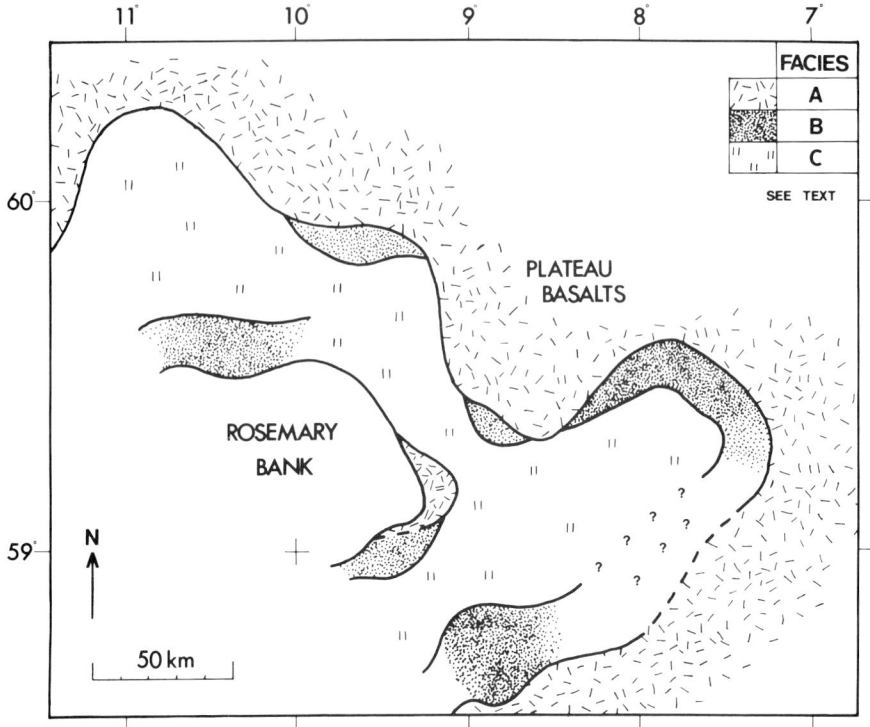

FIG. 5. Map of seismic facies distribution. Produced with an average line spacing of 15 km.

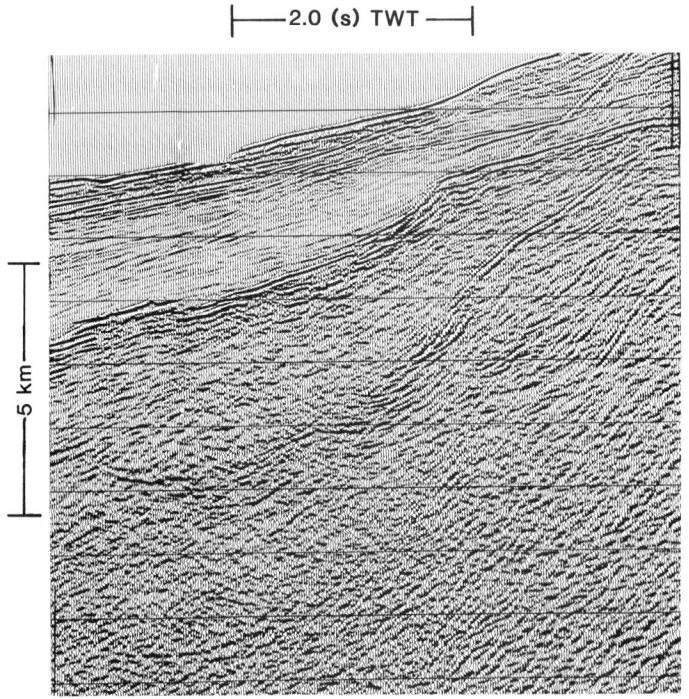

FIG. 6. Seismic section through small scarp on the continental margin. Location shown as **3** in Fig. 1.

the deepest parts of the basin into which the lava material was being dispersed.

Present-day depth distribution of facies

Facies C invariably occurs in the deepest water and is overlain by the greatest thickness of sediments (up to 2.5 km in the area covered by Fig. 5), the overlying Tertiary sequence being complete. This illustrates that the deeper, and more axial, parts of the Trough have remained as such from the early Tertiary to the present day. Facies B tends to occupy a transition zone between this facies and facies A.

The plateau basalts (facies A) occur at shallower depths than the other facies and are very variable in their depth distribution. On the NW Scottish continental margin these occur under a very thin covering of Quaternary sediments and in water depths of about 75 m. This passive shelf margin has been very stable throughout the Tertiary, showing remarkably little subsidence compared to adjacent areas. In the northern part of the Trough the plateau basalts occur below an incomplete Tertiary sequence up to 1.5 km thick, and in water depths of up to 1.5 km. On the Ymir and Wyville–Thomson Ridges the lavas occur again at, or near, the seabed in only a few 100 m of water.

If the plateau lavas were all extruded subaerially and over a relatively short time period, then adjacent areas in this part of the Trough have undergone radically different histories of subsequent subsidence.

Evidence from overlying sediments

The Tertiary sedimentary sequence has been described by Roberts (1975), Roberts *et al.* (1983) and Wood *et al.* (1987). Immediately post-lava sediments infill the topographic lows of the lava flows, and onlap the lava scarps. The lower horizons may have been deposited during or shortly after the later stages of lava extrusion. However, lava extrusion and major sedimentary deposition may not have overlapped significantly, since no evidence of the interfingering of sediments and lava flows has been identified. Sedimentation continued after the cessation of lava extrusion, to fill the basin and cover the lava pile. During transgression over the main area of lava sheets, some other areas remained subaerial and suffered consequent erosion (e.g. Rosemary Bank).

Sub-lava intrusions

Coherent, unambiguous reflections are rare from below the thicker lava sequences. However, where the lavas are thin or absent, numerous high amplitude, discontinuous reflections are observed from the stratigraphic level of the lavas to about 2 s TWT below it, and are laterally continuous for up to 8 km. The reflectors represent real features that can be mapped in three dimensions. These reflectors do not cut those from the lavas.

As these reflectors have a high amplitude, are subhorizontal and laterally discontinuous we interpret them as sills intruding the Mesozoic sedimentary sequence below the lavas. The time of emplacement cannot be determined. However, as they do not appear to intrude the lavas themselves, they are probably also of early Tertiary age. Sill complexes exist to the N in the Faeroe–Shetland Channel (Ridd 1980, 1983), where their positioning may have implications for the positioning of the continental/oceanic crust transition (Gibb *et al.* 1986).

Tertiary sills intruding pre-lava sediments occur within the British Tertiary (Emeleus 1983). Picrite sills are present on the Shiant Isles and similar sills occur in N Skye and Lamlash Bay (Holy Island) (Emeleus 1983). Although the sills are not imaged below the thicker lava piles, it seems likely that they occur throughout this part of Rockall Trough.

On some profiles the reflectors are 'stepped' downward towards the centre of the basin (Fig. 7). This suggests an intrusion mechanism similar to that of the Permo-Carboniferous sills of the Midland Valley of Scotland (Francis 1982). The sills may be intruded along the bedding planes under gravity. Step-down may occur due to loading, and there will be an accumulation of material at the bottom of the basin.

Implications for subsidence

Since the lavas are subaerial and they now occur at depths of up to 3.0 km, they provide us with an estimate of the amount of subsidence that must have taken place during the Tertiary times.

A seismostratigraphic interpretation of the sediments above the lavas has revealed that the subsidence occurs in two distinct phases. Wood *et al.* (1987) describe the Eocene sediments as being of shallow-water type, whereas Oligocene to present-day sediments are deposited in deep water. The change in sediment type during the early Oligocene represents a rapid increase in the subsidence rate of an already subsiding basin.

M. V. Wood et al.

⊢2.0 (s) TWT⊣

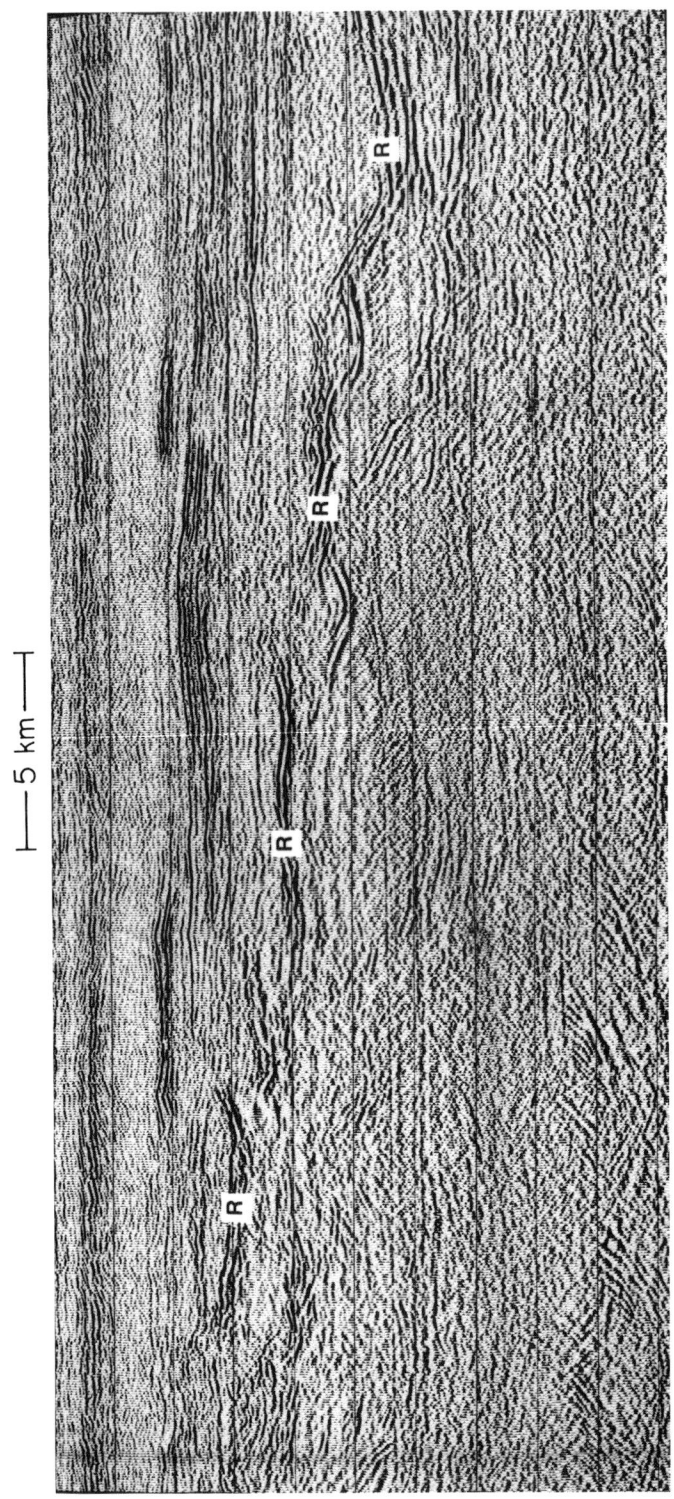

FIG. 7. Seismic section showing stepped nature of sublava reflectors (R) that intrude ?Mesozoic sediments. Location shown as 3 in Fig. 1.

TERTIARY SUBSIDENCE CURVES FOR LAVA TOPS

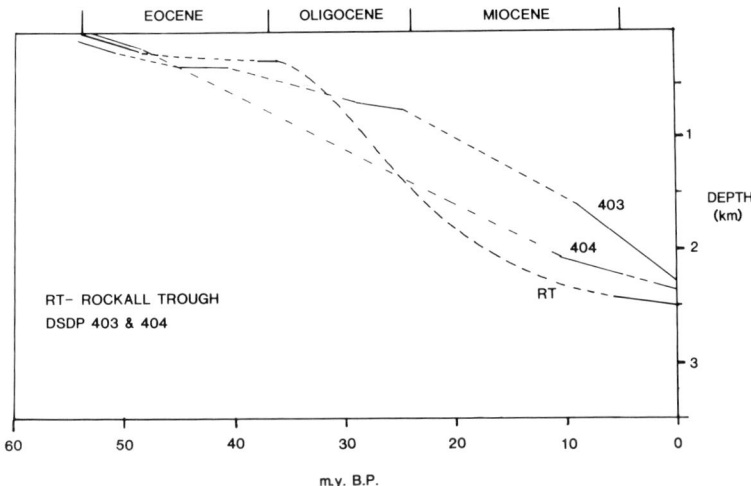

FIG. 8. Compilation of DSDP subsidence curves (for Sites 403 & 404) and NE Rockall Trough curve, showing the similarity in the subsidence pattern between the areas.

Further studies into the possible causes of the subsidence are being undertaken. We must assess the contribution from several factors:

(1) Thermal subsidence which may be related to spreading events to the W. Figure 8 shows the estimated subsidence curve for the Trough and its relation to DSDP data further to the W.
(2) The amplification of subsidence by sediment loading.
(3) The compaction of sub-lava sediments giving a false impression of the amount of subsidence of the Trough.
(4) The possibility of tectonically induced subsidence and its nature.
(5) The amount of crustal flexure that might be expected to occur due to the emplacement of these thick lava piles.

Discussion

We believe that the sinuous nature, lateral consistency and dip of the lava pile scarps can be explained if they mark a palaeoshoreline. This interpretation provides a firm basis for our resulting interpretation of the post-lava subsidence history. Although not well constrained, the subsidence curve, derived from the seismostratigraphic interpretation of the overlying sedimentary pile, provides a good approximation for the subsidence history given the large magnitude of the subsidence that has taken place.

The palaeoenvironmental map (see Fig. 5) is considered a plausible explanation of the seismic data, although the interpretation of the basinward-thinning wedges (facies B) as deltas and hyaloclastite beaches; and of facies C as tuffs, is not well constrained.

Conclusions

(1) Thick, extensive sequences of early Tertiary plateau basalts occur in NE Rockall Trough and can be easily identified and mapped using multichannel seismic reflection data.
(2) The subaerial nature of these lavas (established by the recognition of the lava edges as a palaeoshoreline) has major implications for the Tertiary subsidence history of the trough, as they now occur in water depths of up to 1.5 km and below a sediment cover of up to 1.5 km thickness.
(3) Numerous ?early Tertiary sills intrude the Mesozoic sediments (which we believe have a minimum thickness of 2 km), a commonly observed phenomena throughout the Thulean volcanic province.
(4) Two-phase subsidence of the NE Rockall Trough is related in time, and possibly also causally, to cessation of sea-floor spreading in the Labrador Sea/Baffin Bay area to the W.

ACKNOWLEDGEMENTS: The authors wish to acknowledge the contribution to this work of Shell UK Expro who funded the research and also gave guidance, and also Western Geophysical who kindly allowed us access to the seismic data and granted publication permission. We gratefully acknowledge the assistance of R. T. Cumberland in the drafting of the figures, and D. Maclean for photographic work.

References

EMELEUS, C. H. 1983. Tertiary igneous activity. *In:* CRAIG, G. Y. (ed) *Geology of Scotland.* Scottish Academic Press, Edinburgh, pp. 357–397.

FRANCIS, E. H. 1982. Emplacement mechanism of late Carboniferous tholeiite sills in North Britain. *Journal of the Geological Society of London,* **139**, 1–20.

GIBB, F. G. F., KANARIS-SOTIRIOU, R. & NEVES, R. 1986. A new Tertiary sill complex of mid-ocean ridge basalt type NNE of the Shetland Isles: a preliminary report. *Transactions of the Royal Society of Edinburgh: Earth Sciences,* **77**, 223–230.

JENYON, M. K. 1987. Characteristics of some igneous extrusive and hypabyssal features in seismic data. *Geology,* **15**, 237–240.

JONES, J. G. 1968. Intraglacial volcanoes of the Laugarvatn region, south west Iceland. *Quarterly Journal of the Geological Society of London,* **124**, 197–211.

—— & NELSON, P. H. H. 1970. The flow of basalt lava from air into water—its structural expression and stratigraphic significance. *Geological Magazine,* **107**, 13–21.

NELSON, P. H. H. 1966. The James Ross Island volcanic group of north-east Graham Land. *British Antarctic Survey Report,* **54**, 1–62.

RIDD, M. F. 1980. The petroleum geology west of the Shetlands. *In:* ILLING, L. V. & HOBSON, G. D. (eds) *Petroleum Geology of the Continental Shelf of* *North West Europe.* Heyden & Son, Chichester, 414–425.

—— 1983. Aspects of the Tertiary geology of the Faeroe–Shetland Channel. *In:* BOTT, M. H. P., SAXOV, S., TALWANI, M., & THIEDE, J. (eds) *Structure and Development of the Greenland–Scotland Ridge.* Plenum Press, New York, pp. 91–108.

ROBERTS, D. G. 1975. Tectonic and stratigraphic evolution of the Rockall Plateau and Trough. *In:* WOODLAND, A. W. (ed) *Petroleum Geology and the Continental Shelf of North West Europe.* Applied Science Publishers, London, pp. 77–89.

—— 1983. Structure and origin of the Wyville–Thomson Ridge. *In:* BOTT, M. H. P., SAXOV, S., TALWANI, M. & THIEDE, J. (eds) *Structure and Development of the Greenland–Scotland Ridge.* Plenum Press, New York, pp. 133–158.

SMYTHE, D. K., CHALMERS, J. A., SKUCE, A. G., DOBINSON, A. & MOULD, A. S. 1983. Early opening history of the North Atlantic-I. Structure and origin of the Faeroe–Shetland Escarpment. *Geophysical Journal of the Royal Astronomical Society,* **72**, 373–398.

WOOD, M. V., HALL, J. & VAN HOORN, B. 1987. Post-Mesozoic differential subsidence in the north east Rockall Trough related to volcanicity and sedimentation. *In:* BROOKS, J. & GLENNIE, K. W. (eds) *Petroleum Geology of North West Europe,* Graham & Trotman, London, pp. 677–685.

M. V. WOOD & J. J. DOODY, Department of Geology, University of Glasgow G12 8QQ, UK.

J. HALL, Department of Earth Sciences, Memorial University of Newfoundland, Newfoundland, Canada AB1 3X7.

Early Tertiary volcanic rocks in Well 163/6-1A, Rockall Trough

A. C. Morton, J. E. Dixon, J. G. Fitton, R. M. Macintyre, D. K. Smythe & P. N. Taylor

SUMMARY: Well 163/6-1A, located in the northern part of the Rockall Trough, proved the presence of a thick sequence of extrusive igneous rocks below Upper Palaeocene sediments. K-Ar age dating suggests that the lavas were extruded at approximately 55 Ma. The lava sequence comprises three distinct lithologies. The upper part of the pile consists of olivine tholeiites that show alkalic and picritic tendencies and have a distinct within-plate composition. These are underlain by another group of olivine tholeiites that are much closer in composition to normal mid-ocean ridge basalt (N-type MORB). The Sr and Pb isotopic compositions of the basalts suggests possible contamination by continental crustal material.

The basalts are underlain by a sequence of cordierite-phyric dacites of remarkably homogeneous composition. Their highly aluminous nature, high Ni and Cr contents and their Sr and Pb isotopic compositions indicate that they are not differentiates of a basaltic parent magma, and are considered to have originated by melting of argillaceous and, possibly, arenaceous rocks at depth. Organic-rich black shale lithologies may have been involved.

The Rockall Trough is an area of deep water separating the relatively shallow Rockall Plateau from the main UK landmass (see Fig. 1). It is believed to have formed by a phase of rifting and abortive sea-floor spreading, in either the late Palaeozoic (Russell & Smythe 1977) or the mid-Cretaceous (Roberts et al. 1981). However, there are no mappable magnetic anomalies in the central and northern parts of the Trough, which suggests an alternative hypothesis, that over much of its area it is underlain by attenuated continental crust. It has also been suggested that late Cretaceous–early Tertiary oceanic crust may be present locally, for example around Rosemary Bank (Dietrich & Jones 1980). It has been proposed on the basis of seismic reflection profiling that the basin is filled with Mesozoic sediments and Palaeocene lavas, capped by Eocene and younger sediments (Roberts 1975; Naylor & Shannon 1982).

Well 163/6-1A is at present the only well to have investigated the deep geology of the Trough. It was drilled by the British Natural Oil Corporation (BNOC) on behalf of a consortium of 19 companies and the UK Department of Energy at 59°48.65′N, 8°57.95′W, in the centre of the Trough at its northern limit, just S of the Wyville–Thomson Ridge, in a water depth of 1374 m (Fig. 1). The site lies close to an igneous centre originally termed 'A' by Roberts et al. (1983), and subsequently named 'Darwin'. Igneous centre B of Roberts et al. and Rosemary Bank, known by dredging to consist of late Cretaceous or Tertiary basalt (Dietrich & Jones 1980), also lie close to the site (Fig. 1). A line drawing of a multichannel

seismic profile across the Darwin igneous centre is shown in Fig. 2, and a lithological summary of the 163/6-1A sequence is shown in Fig. 3. The well encountered 689 m of basalt underlying Upper Palaeocene to Recent sediments. Below the basalts, a further 356 m of dacites were drilled before the hole was abandoned. This paper documents the results of investigations into the petrography, geochemistry and radiometric age of the extrusive rocks recovered in the well.

Analytical methods

Following optical petrographical description, determinations of phase chemistry were made by electron microprobe (EM), using a Link Systems energy-dispersive X-ray analyzer attached to a Geoscan electron microprobe (at BGS Keyworth) and a similar ED system on a Microscan V (at Edinburgh University). Vesicle fills were identified optically and by X-ray diffraction (XRD). Samples for geochemistry were selected on a petrographic basis to minimize, as far as possible, the effects of alteration, by choosing those samples without extensive alteration and by avoiding highly amygdaloidal material.

Major and trace element analyses were carried out by X-ray fluorescence (XRF) at Edinburgh University and by MESA (Midland Earth Science Associates). Rare-earth element (REE) abundances were determined by neutron activation analysis (INAA) at the Imperial College Reactor Centre.

From MORTON, A. C. & PARSON, L. M. (eds), 1988, *Early Tertiary Volcanism and the Opening of the NE Atlantic,* Geological Society Special Publication No. 39, pp. 293–308.

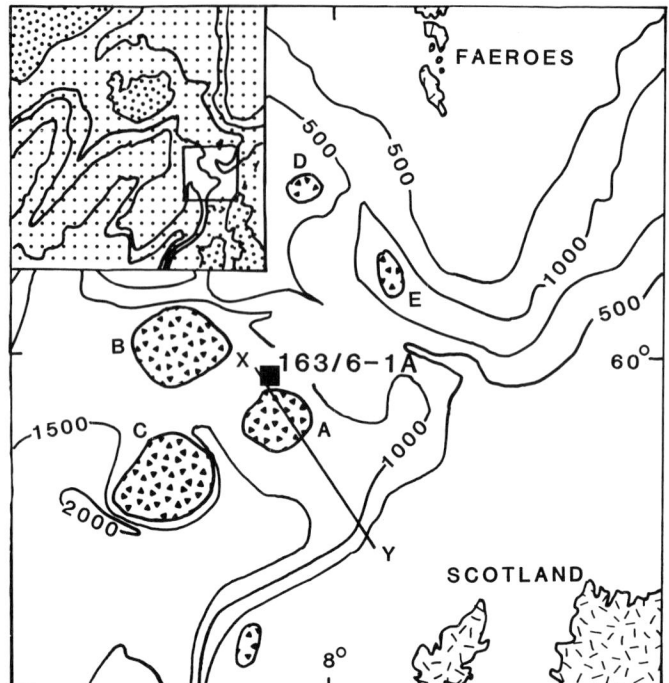

FIG. 1. Location of Well 163/6-1A and distribution of known and inferred igneous centres adjacent to the site. X–Y = line of seismic section shown in Fig. 2. A & B are igneous centres in northern Rockall Trough identified by Roberts *et al.* (1983). C: Rosemary Bank; D: Faeroe Bank Centre; and E: Faeroe Channel Knoll.

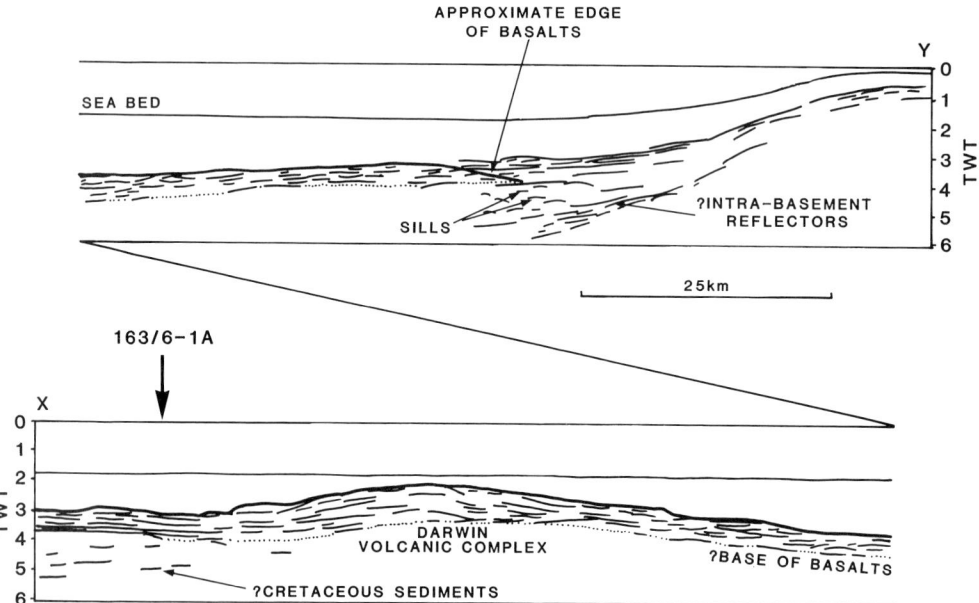

FIG. 2. Line drawing of multichannel line X–Y (Fig. 1) from the Hebridean shelf to 163/6-1A. The buried structural high of the Darwin volcanic complex is the volcanic centre A (Fig. 1) of Roberts *et al.* (1983). The basalts appear to be thin or absent near the edge of the trough, permitting acoustic penetration by a further 2–2.5 s (TWT). To the NW, the lack of reflectors below the presumed base of the basalts is due to insufficient transmission of energy through the basalts.

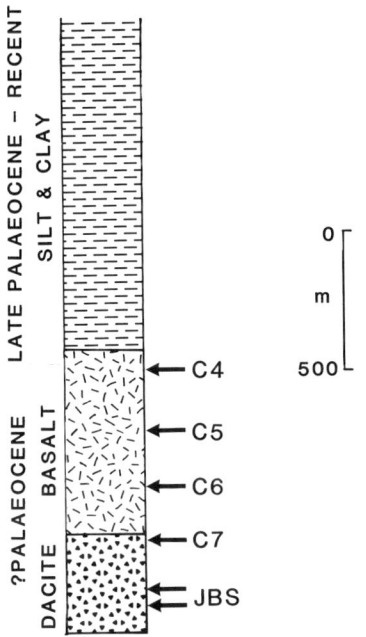

FIG. 3. Skeleton lithological log of sequence drilled in Well 163/6-1A, showing location of cores and junk basket samples.

Sr and Pb isotope geochemistry was carried out in the Department of Earth Sciences, University of Oxford. Sr samples were prepared by eluting Sr from cation exchange resin with 2.5 M HCl. $^{87}Sr/^{86}Sr$ ratios were determined both on whole-rock samples, and on sample residues after leaching in hot 6 M HCl. This procedure removes much of the Sr contained in alteration products, and was followed in an attempt to minimize the effects of any seawater alteration. Samples were loaded on single Ta filaments with H_3PO_4 and analyzed on a VG Micromass 30 thermal ioniza-tion mass spectrometer. Data were corrected for the effects of mass fractionation and for conform-ity with the inter-laboratory Eimer and Amend Sr standard ($^{87}Sr/^{86}Sr = 0.70800$).

Pb samples were prepared by loading the rock sample in 1 M HBr solution onto anion exchange resin, eluting Pb with pure H_2O. They were then loaded on single Re filaments with H_3PO_4 and silica gel activator, and were analyzed on a VG Isomass 54E fully automated thermal ionization mass spectrometer. Analyses were corrected for the effects of mass fractionation.

Whole-rock K-Ar dating was carried out at the Scottish Universities Research and Reactor Centre following procedures described by Macin-tyre & Hamilton (1984).

Basaltic rocks: petrography and chemistry

Three cores 4, 5 and 6, were taken in the basalt sequence.

Core 4

Core 4 (9.2 m long) contains four complete flow units, between 0.91 and 2.87 m thick, and two further partially cored flows, one at the top and one at the base of the core. The basalts are medium to fine-grained, dark grey to black and olivine-phyric. Individual flows have fine-grained, generally aphyric, highly vesicular tops, fine to medium-grained sparsely vesicular centres and fine-grained highly vesicular bases. No pillow structures or glassy rims were observed, and a subaerial origin is inferred.

The basalts of core 4 consist of olivine phenocrysts set in a groundmass of plagioclase laths, granular opaques, and pale to dark brown pyroxenes. The pyroxene is generally sub-ophitic, but in one sample (C4.5) it tends to be intersertal. This sample is from a flow top, unlike the others, which are from flow centres; the textural differ-ence is therefore interpreted as the result of more rapid cooling.

Olivine contents lie between 18 and 40%. Except for one very olivine-rich sample (C4.1), in which the diameters of some olivines exceed 4 mm, the phenocrysts are generally small, around 0.5 mm. The large olivines are zoned from a core composition of Fo_{84-89} to a margin composition of Fo_{77-85}, whereas the smaller grains tend to be unzoned, with a composition of Fo_{71-85}. Several grains contain inclusions of dark brown magne-siochromite. Only two samples (C4.1 & C4.3) retain any fresh olivine; most are altered to smectite. Locally, an intermediate pale brown, low-birefringent alteration phase also occurs, shown by EM to be a serpentine-like mineral. In some cases, red-brown iddingsite is a further alteration product.

Plagioclase laths comprise between 30 and 50% of the rock. They are commonly zoned, from cores of An_{67-72} to rims as sodic as An_{26}. Plagioclase is mainly fresh, but some alteration to smectite and thomsonite has taken place.

Pyroxene contents lie between 13 and 23%. The dark colour is suggestive of a high TiO_2 content, confirmed by EM, which shows the pyroxenes to lie in the salite field (Fig. 4) and to have TiO_2 contents of 2.0–3.3%. This suggests that the basalts have an alkalic nature, although pyroxenes such as these are not uncommon in tholeiites. The pyroxenes plot in a relatively small area on the $TiO_{2(px)}$-$MnO_{(px)}$-$Na_2O_{(px)}$ diagram of Nisbet & Pearce (1976), mainly grouping in the 'within-plate alkalic' field (Fig. 5).

Opaques are a minor component, forming between 3 and 5% of the whole rock. Apart from one ilmenite-bearing sample (C4.1), the opaque phase is titanomagnetite.

The basalts also contain between 7 and 13% smectite as an alteration product of interstitial glass, none of which is preserved in a fresh state.

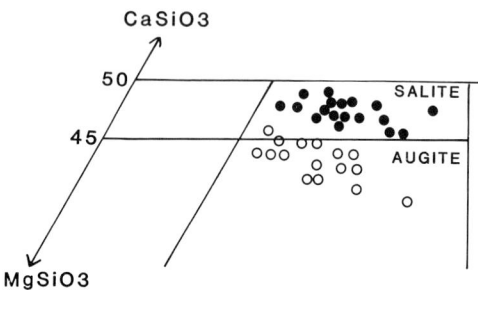

FIG. 4. Composition of clinopyroxenes in the basalts of 163/6-1A. Note the distinctly more calcic nature of the pyroxenes from core 4 (salites) compared to those of cores 5 and 6 (augites).

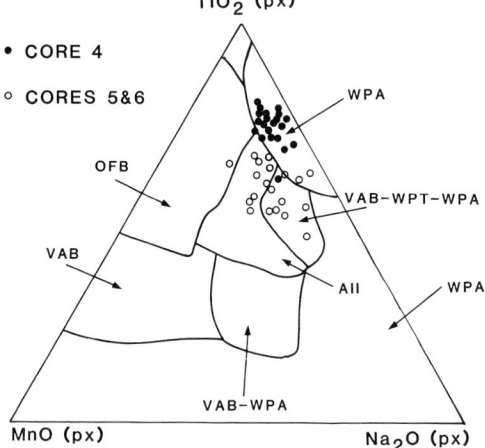

FIG. 5. Minor element geochemistry of clinopyroxenes from the basalts of 163/6-1A, plotted on the discrimination diagram formulated by Nisbet & Pearce (1976). Minor elements were determined by electron microprobe. WPA: within-plate alkali basalt; WPT: within-plate tholeiite; VAB: volcanic arc basalt; OFB: ocean floor basalt.

Areas of interstitial glass appear to have acted as nuclei for secondary clays, which also replace plagioclase and olivine.

Only one sample (C4.5) is markedly vesicular, being from a flow top. The vesicles are filled with spherulitic aggregates of smectite, bladed thomsonite, finely fibrous sheaf-like natrolite and weakly anisotropic analcime.

Apart from one picritic sample (C4.1) with an unusually high olivine content, there is comparatively little chemical variation between samples (Table 1). This is reflected by their CIPW norms, which show them to be olivine tholeiites, with one exception (C4.5). The normative nepheline of this sample is a result of secondary alteration, with an uptake of alkalis causing an excursion into the alkali basalt field.

Although the basalts have a tholeiitic nature, their alkalic tendencies are reflected in the presence of Ti-rich pyroxenes and high whole-rock Ti/V ratios (Shervais 1982), which fall in the narrow range 50–56 (Fig. 6). They show strong enrichment of Rb and Ba and moderate enrichment of K, Th, Sr, La, Ce, Nb, Ta and Nd relative to N-type MORB (Fig. 7). P, Hf, Zr, Eu

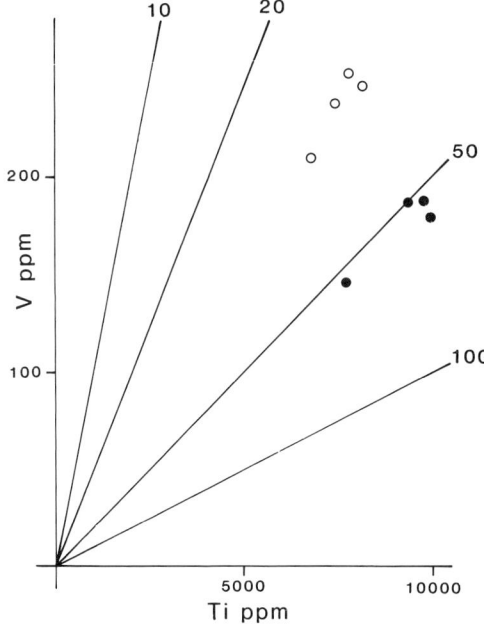

FIG. 6. Ti/V ratios of 163/6-1A basalts, plotted on the diagram devised by Shervais (1982). Increasing Ti/V reflects increasing alkalinity, Ti/V of 50 marking the distinction between mid-ocean ridge and continental flood basalts and alkali basalts.

TABLE 1. *Geochemistry of basalts, core 4, Well 163/6-1A. Major and trace elements by XRF († by MESA, * at Edinburgh). Rare-earth elements by INAA at Imperial College Reactor Centre. Fe_2O_3/ FeO ratios were fixed at 0.15 in normative calculations, following Brooks (1976)*

Sample	C4.1†	C4.2*	C4.3†	C4.4†	C4.5*	C4.6*	C4.7†
SiO_2	41.29	43.70	43.23	44.70	44.05	44.31	44.13
Al_2O_3	9.48	13.04	12.65	14.29	14.81	14.66	14.08
TiO_2	1.26	1.59	1.63	1.66	1.79	1.52	1.57
Fe_2O_3	14.63	13.80	13.81	12.68	12.79	12.63	12.92
MgO	21.32	14.30	14.60	11.40	9.59	10.87	10.98
CaO	5.90	8.18	8.01	8.38	7.57	9.32	9.27
Na_2O	1.24	1.96	1.92	2.50	4.22	2.40	2.19
K_2O	0.07	0.13	0.10	0.16	0.22	0.16	0.15
MnO	0.19	0.17	0.18	0.16	0.16	0.16	0.16
P_2O_5	0.10	0.13	0.13	0.18	0.18	0.15	0.17
LOI	4.72	3.30	3.91	4.10	4.60	3.70	4.41
Total	100.20	100.30	100.17	100.21	99.98	99.89	100.04
CIPW Norm							
Or	0.41	0.80	0.59	0.95	1.35	1.04	0.89
Ab	10.49	17.32	16.25	21.15	24.33	21.33	18.53
An	20.10	27.54	25.60	27.30	22.15	30.23	28.15
Ne					7.31		
Di	6.83	11.31	10.73	10.62	13.44	14.07	13.59
Hy	8.37	6.22	6.14	6.01		1.76	5.80
Ol	42.84	30.87	29.95	23.20	25.03	25.87	21.92
Mt	2.52	2.48	2.38	2.19	2.34	2.29	2.23
Il	2.39	3.15	3.10	3.15	3.61	3.03	2.98
Ap	0.24	0.30	0.31	0.43	0.45	0.39	0.40
Ba	96	59	55	95	69	86	125
Cr	963	812	925	480	451	651	610
Cu	44	42	44	26	60	18	18
Ni	943	492	516	396	249	405	442
Nb	5	6	5	5	8	7	6
Pb		5			7	6	
Rb	4	3	4	4	3	3	4
Sr	158	225	223	257	251	267	251
V	146	233	189	180	274	231	187
Y	14	16	19	21	20	18	21
Zn	82	86	77	81	85	80	72
Zr	77	83	90	111	117	98	114
La	3.7		4.7	6.2			6.2
Ce	9.0		13.5	17.8			17.2
Nd	8.1		11.1	13.2			14.0
Sm	2.29		3.04	3.70			3.69
Eu	0.80		1.11	1.35			1.31
Tb	0.41		0.55	0.62			0.59
Ho	0.31		0.55	0.54			0.52
Yb	0.66		0.91	1.08			1.11
Lu	0.12		0.16	0.20			0.20
Ta	0.20		0.27	0.38			0.35
Th	0.23		0.28	0.38			0.45
Hf	1.59		2.09	2.63			2.50

and Ti values are comparable to N-type MORB, but Tb, Y and Yb are increasingly depleted. According to Pearce (1982), the last feature is characteristic of within-plate basalts, whether they be tholeiitic, transitional or alkalic (fig. 1b, Pearce 1982). They are relatively light rare-earth element (LREE)-enriched, with $(Ce/Yb)_N > 1$ (Fig. 7). A within-plate setting is also suggested by pyroxene chemistry (see Fig. 5), and is further demonstrated by the Pearce & Cann (1973) Ti-Y-

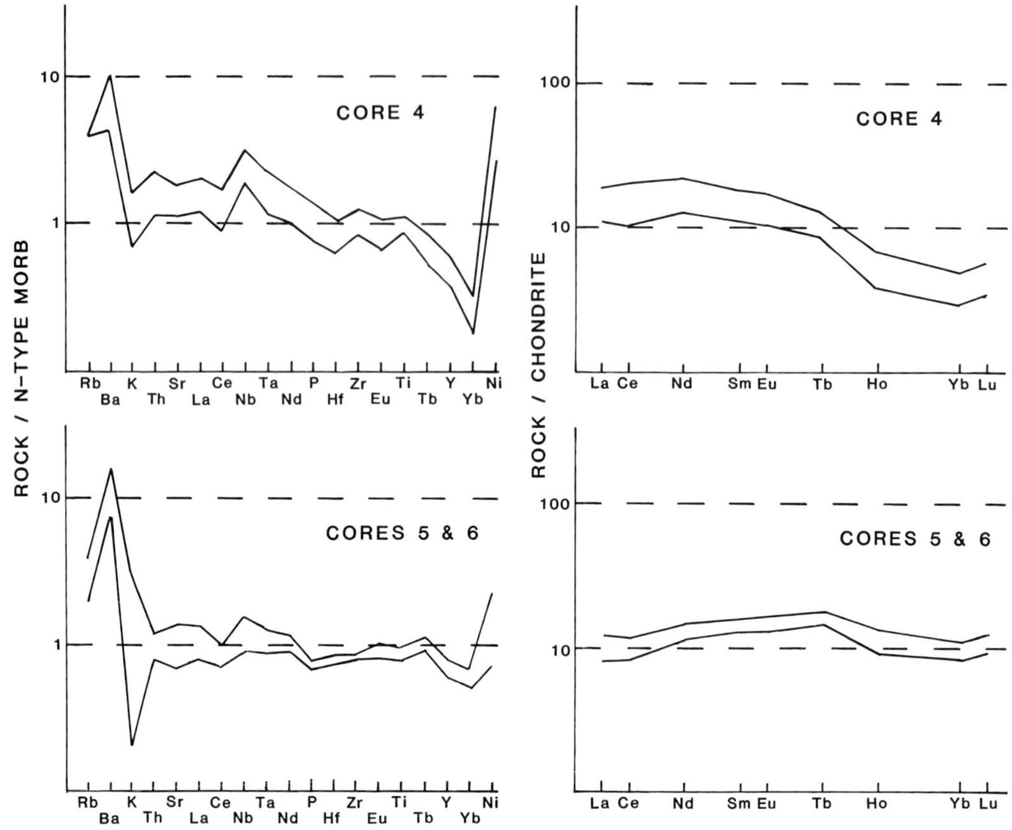

FIG. 7. Multi-element plots, normalized to N-type MORB, and chondrite-normalized rare-earth plots of basalts from 163/6-1A. Data shown are upper and lower limits of elemental abundances.

Zr diagram (Fig. 8). However, it should be emphasized that there are several known instances of MORBs plotting outside the ocean-floor basalt field on this diagram, such as mid-Atlantic Ridge basalts from DSDP Sites 407 and 408 (63°N) and 410 (45°N) (Tarney *et al.* 1979) and many Icelandic basalts (Prestvik 1982). On the Th-Ta-Hf diagram of Wood *et al.* (1979) they fall within the E-type MORB/WPB field, close to the transitional MORB area (Fig. 9).

Core 5

Core 5 (9.2 m long) recovered five flow units, three of which were complete, between 1.69 and 2.67 m thick. The basalts are fine to very fine-grained and sparsely plagioclase- and olivine-phyric. They are paler in colour than core 4 basalts, owing to their greater degree of alteration. The internal subdivisions of the flows are similar to those in core 4, although vesicles are more common in flow centres, and those in flow tops are commonly unfilled. The lack of pillow

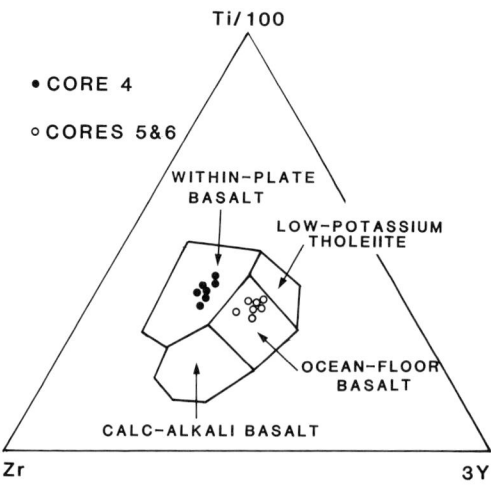

FIG. 8. 163/6-1A basalts plotted on the Ti-Y-Zr diagram of Pearce & Cann (1973).

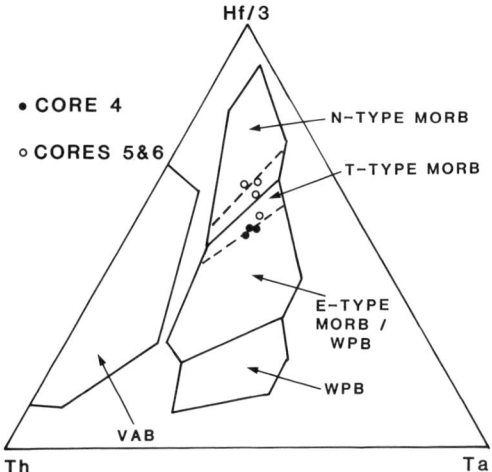

FIG. 9. 163/6-1A basalts plotted on the Th-Ta-Hf diagram of Wood *et al.* (1979).

structures and glassy rims again suggests a subaerial origin.

The basalts consist of plagioclase and olivine phenocrysts, in some cases with a glomerophyric habit, in a groundmass of plagioclase laths and granular pyroxene, with minor opaques tending to occur subpoikilitically.

Highly altered olivine occurs as scattered phenocrysts up to 2 mm in diameter and as microphenocrysts up to 0.1 mm long, and forms 8–9% of the rock. It is wholly replaced by serpentine and/or chlorite and smectite, precluding an assessment of original chemistry.

Plagioclase occurs both as a phenocryst and as a groundmass phase. Phenocryst plagioclase forms 10–12% of the rock, and plagioclase comprises a further 33–37% in the groundmass. In two samples (C5.1 & C5.2), both phenocryst and groundmass plagioclase have been wholly replaced by thomsonite, natrolite and K-feldspar. However, fresh plagioclase (An$_{80-88}$ in the groundmass, phenocrysts zoned from cores of An$_{82-86}$ to margins of about An$_{52}$) is preserved in samples C5.3 and C5.4.

Pyroxene is fresh throughout core 5. It occurs in colourless granules, and forms 20–24% of the rock. Compared to core 4, it is slightly less calcic (see Fig. 4), and is also poorer in TiO$_2$ (0.8–1.6%). Compared to those of core 4, core 5 pyroxenes plot in a different area on the Nisbet & Pearce (1976) TiO$_{2(px)}$-MnO$_{(px)}$-Na$_2$O$_{(px)}$ diagram (see Fig. 5).

Opaques form 3–4% of the rock, and consist of titanomagnetite. Interstitial glass has been wholly replaced by smectite. Smectite, analcime, meso-

lite, natrolite, thomsonite and possible gonnardite occur as vesicle fills.

Although three out of the four samples studied (C5.1, C5.2 & C5.3) have normative nepheline (Table 2), the basalts of core 5 are clearly tholeiitic, the anomalous nepheline-normative character being an effect of alteration. Their tholeiitic nature is shown by their high Y/Nb ratios (5.3–8.7), their low Ti/V ratios (31–32, Fig. 6) and their flat patterns on MORB-normalized spidergrams (see Fig. 7). They therefore bear a strong resemblance to MORB, the only major departures being the high Rb and Ba and a slight tendency for Y and Yb depletion. However, both Rb and Ba are highly mobile elements during alteration and, furthermore, high Ba contents may be due to contamination by drilling mud. They have slightly convex-upward REE patterns (see Fig. 7), with (Ce/Yb)$_N$ close to 1, but relative enrichment of the middle REE Sm, Eu and Tb. They fall into the ocean-floor basalt field on the Pearce & Cann (1973) Ti-Y-Zr diagram (see Fig. 8). However, there are many cases of MORB-like basalts undoubtedly located within-plate, such as the Preshal Mhor and Fairy Bridge basalts of Skye and the low-alkali tholeiites of Mull (Morrison 1978). On the Th-Ta-Hf diagram of Wood *et al.* (1979) they fall in the transitional MORB field (Fig. 9).

Core 6

Core 6 (9.2 m recovered length) contains dark grey medium to fine grained olivine-microphyric basalts, somewhat fresher in appearance than those of core 5. Five flow units were recovered, three of which are complete, ranging in thickness from 0.10–1.01 m. Internal morphology of flows is again similar to that described for core 4, although flow centres are much reduced in thickness because of the relatively thin nature of the flows. Two of the flows have reddened tops, suggesting that extrusion took place subaerially.

The basalts consist of olivine microphenocrysts set in a groundmass of plagioclase laths and opaques, sub-ophitically enclosed by pyroxene. Compaction during crystallization has locally caused sub-ophitic pyroxene to fracture into a mosaic of subgrains which are out of optical continuity.

Olivine is entirely pseudomorphed by smectite and as with core 5 this precludes an assessment of its original composition. Olivine pseudomorphs comprise 11–19% of the rock.

Plagioclase occurs only as a groundmass phase, and is generally unaltered, although in one sample (C6.3) there is a significant proportion of smectite pseudomorphs after plagioclase laths. Plagioclase

TABLE 2. *Geochemistry of basalts, core 5, Well 163/6-1A. Major and trace elements by XRF († by MESA, * at Edinburgh). Rare-earth elements by INAA at Imperial College Reactor Centre. Fe_2O_3/FeO ratios were fixed at 0.15 in normative calculations, following Brooks (1976)*

Sample	C5.1†	C5.2*	C5.3*	C5.4†
SiO_2	45.75	44.46	45.04	45.18
Al_2O_3	14.10	15.14	15.81	14.22
TiO_2	1.30	1.21	1.27	1.13
Fe_2O_3	12.47	11.92	12.20	11.88
MgO	8.79	8.15	8.53	11.90
CaO	7.56	8.81	10.98	10.56
Na_2O	4.76	4.23	2.76	1.59
K_2O	0.30	0.24	0.13	0.02
MnO	0.19	0.17	0.14	0.20
P_2O_5	0.10	0.08	0.08	0.09
LOI	5.01	6.30	3.40	3.58
Total	100.33	100.71	100.34	100.37

CIPW Norm

Or	1.77	1.52	0.76	0.12
Ab	25.49	22.57	19.90	13.45
An	16.22	23.16	31.66	31.61
Ne	7.11	8.54	2.43	
Di	16.87	18.89	20.15	16.35
Hy				12.55
Ol	21.01	20.46	20.20	17.25
Mt	2.15	2.20	2.19	2.04
Il	2.47	2.45	2.51	2.15
Ap	0.24	0.21	0.20	0.21
Ba	146	162	66	103
Cr	268	268	272	550
Cu	145	101	112	105
Ni	106	98	100	321
Nb	3	3	3	4
Pb		7	8	
Rb	4	4	2	2
Sr	95	207	162	119
V	253	326	327	210
Y	26	23	24	21
Zn	63	76	74	71
Zr	72	57	60	71
La	2.7			4.0
Ce	7.3			10.1
Nd	7.3			8.5
Sm	2.64			2.98
Eu	1.01			1.13
Tb	0.71			0.69
Ho	0.82			0.72
Yb	1.94			1.84
Lu	0.32			0.32
Ta	0.16			0.35
Th	0.18			0.16
Hf	1.85			2.16

originally formed 39–47% of the rock, and is zoned from An_{73-83} in grain centres to margins of An_{36}.

Pyroxene forms 21–32% of the basalts and, in compositional terms, it is similar to that of core 5 and distinct from that of core 4 (see Figs 4 & 5).

Opaques are minor constituents, forming 4–5% of the rock, and consist of titanomagnetite, ilmenite and chromite.

The lower degree of alteration in core 6 means that the CIPW norms are a better guide to character than they were in core 5, with only one sample (C6.3) being anomalously nepheline-normative (Table 3). Their tholeiitic character is confirmed by their high Y/Nb and low Ti/V ratios, by their flat patterns on MORB-normalized spidergrams (see Fig. 7) and by their REE patterns (see Fig. 7). The close similarity of basalts from cores 5 and 6 is also shown on Ti/V (see Fig. 6), Ti-Y-Zr (see Fig. 8) and Th-Ta-Hf (see Fig. 9) diagrams.

Dacitic rocks: petrography and chemistry

Black, very fine-grained cordierite-phyric dacites were recovered in core 7. Large, filled vesicles are sparsely distributed throughout the cored interval. No textural trends were noted, implying that the core (3.2 m long) was taken from within a single flow unit. Two junk basket samples from lower in the succession also consist exclusively of dacite, in variable states of alteration. Black, glassy and apparently relatively fresh material occurs in both, as well as more altered, greenish-grey fine-grained pieces, several of which are sparsely vesicular and traversed by abundant fine veins. One junk basket sample also contains light grey, fine-grained, relatively fresh pieces.

Petrographic study shows that, despite variations in degree of crystallinity and alteration, the dacitic sequence is remarkably homogeneous. All samples contain abundant cordierite phenocrysts or pseudomorphs after cordierite, together forming between 11 and 24% of the rock. The groundmass is generally microcrystalline, but is glassy in sample JB2B, indicating rapid quenching. In the core, cordierite is entirely replaced by yellow-green chlorite, but is fresh in some pieces from the junk basket samples. The cordierite is moderately iron-rich and occurs as euhedral short hexagonal prisms up to 0.5 mm long, often in glomerophyric aggregates. Basal sections show the characteristic sector twinning. Many contain inclusions of rounded bytownite (An_{81}). One sample (JB2A) also contains phenocrysts of

TABLE 3. *Geochemistry of basalts, core 6, Well 163/6-1A. Major and trace elements by XRF († by MESA, * at Edinburgh). Rare-earth elements by INAA at Imperial College Reactor Centre. Fe_2O_3/FeO ratios were fixed at 0.15 in normative calculations, following Brooks (1976)*

Sample	C6.1†	C6.2*	C6.3†	C6.4*
SiO_2	46.42	46.16	46.94	45.37
Al_2O_3	15.14	15.34	14.02	14.77
TiO_2	1.24	1.14	1.36	1.13
Fe_2O_3	12.82	12.03	13.20	11.65
MgO	8.14	9.85	7.76	11.49
CaO	12.37	11.74	10.54	10.84
Na_2O	1.98	2.24	3.54	1.79
K_2O	0.07	0.06	0.08	0.05
MnO	0.19	0.20	0.18	0.19
P_2O_5	0.09	0.07	0.10	0.08
LOI	1.57	2.20	2.37	3.20
Total	100.02	101.01	100.08	100.56

CIPW Norm

Or	0.41	0.34	0.47	0.29
Ab	16.75	19.41	24.49	15.63
An	32.22	32.36	22.13	32.36
Ne				2.63
Di	23.43	21.81	24.26	17.70
Hy	6.80	0.06		8.63
Ol	12.93	21.53	17.16	19.82
Mt	2.20	2.12	2.28	2.08
Il	2.35	2.20	2.58	2.24
Ap	0.21	0.17	0.24	0.20
Ba	198	26	142	27
Cr	394	440	233	650
Cu	69	85	130	88
Ni	225	204	112	302
Nb	3	4		3
Pb		8		8
Rb	3	2	2	1
Sr	145	133	188	127
V	238	301	247	282
Y	27	22	28	20
Zn	59	68	65	77
Zr	77	60	71	61
La	3.6		2.4	
Ce	9.7		9.1	
Nd	9.1		9.5	
Sm	3.20		3.23	
Eu	1.20		1.29	
Tb	0.73		0.83	
Ho	1.02		1.04	
Yb	2.02		2.17	
Lu	0.35		0.43	
Ta	0.15		0.16	
Th	0.24		0.24	
Hf	2.13		1.89	

magnesian hypersthene, again forming glomerophyric aggregates.

Groundmass phases include very fine-grained labradorite and sanidine (EM has detected plagioclase of An_{58} and K-feldspar of Or_{56}) and opaques. Vesicles are filled with smectite, several zeolites (including natrolite, heulandite and mordenite), cristobalite and opaques.

Geochemical studies confirm the homogeneous nature of the dacitic sequence (Table 4). The dacites are remarkable for their high Al content, manifested by the 6% normative corundum and by the abundant cordierite phenocrysts. Ni, Cr and, to a lesser extent, V are remarkably high for such siliceous rocks. Relative to chondrite, they show strong LREE enrichment (Fig. 10), and display a marked negative Eu anomaly. Fig. 10 also includes a MORB-normalized spidergram, emphasizing the strong geochemical differences between these dacites and the overlying basalts.

Isotope geochemistry

Sr isotopes

Initial $^{87}Sr/^{86}Sr$ ratios of seven samples (five basalts, two dacites) are presented in Table 5. Three basalt samples (C4.1, C4.2, C5.4) have very low initial $^{87}Sr/^{86}Sr$ ratios (0.70313–0.70329, reduced to 0.70298–0.70317 after acid leaching), typical of mid-Atlantic Ridge basalts. However, the other two samples analyzed (C5.1 & C6.3) have significantly higher initial $^{87}Sr/^{86}Sr$ ratios of 0.70392–0.70435 (0.70398–0.70457 after acid treatment). These values are above the range for typical MORB, although not beyond the range for basalts derived from mantle sources less depleted than MORB-source mantle. However, these samples have MORB-like chemistry, tending to rule out the latter possibility. The implication, therefore, is that samples C5.1 and C6.3 have been slightly contaminated by continental crustal material. Both dacite samples have very high initial $^{87}Sr/^{86}Sr$ ratios (0.71102–0.71112), and it is notable that acid leaching has the effect of increasing $^{87}Sr/^{86}Sr$ in the residue to 0.71211–0.71216. This suggests that leaching has removed Sr held in alteration products, probably Sr substantially modified in isotopic composition by interaction with seawater Sr, which had an $^{87}Sr/^{86}Sr$ ratio of about 0.7077 in the early Tertiary (Burke *et al.* 1982). These values are well outside the range of MORB, clearly indicating that material of continental derivation has played an important role in the genesis of the dacites.

TABLE 4. *Geochemistry of dacites, Well 163/6-1A. Major and trace elements by XRF († by MESA, * at Edinburgh). Rare-earth elements by INAA at Imperial College Reactor Centre. Fe_2O_3/FeO ratios were fixed at 0.15 in normative calculations, following Brooks (1976)*

Sample	C7.1†	C7.2†	JB1A*	JB1B*	JB2A*	JB2B*
SiO_2	64.48	64.52	62.92	67.61	61.63	63.53
Al_2O_3	16.14	16.08	16.28	15.11	17.11	15.79
TiO_2	0.92	0.90	0.91	0.83	0.95	0.87
Fe_2O_3	6.55	6.70	5.41	2.60	6.26	5.54
MgO	1.77	1.53	1.95	0.82	2.03	1.90
CaO	2.51	2.22	2.57	2.04	2.58	2.65
Na_2O	2.06	2.12	1.25	1.08	1.26	2.13
K_2O	2.63	2.57	2.57	2.04	2.58	2.65
MnO	0.11	0.14	0.09	0.05	0.13	0.12
P_2O_5	0.17	0.16	0.17	0.15	0.15	0.15
LOI	2.99	3.23	4.50	3.30	4.50	5.80
Total	100.34	100.16	99.84	100.06	100.29	99.52
CIPW Norm						
Qz	31.45	32.18	32.15	31.44	30.31	38.87
Co	5.75	6.16	6.19	3.06	7.08	7.22
Or	15.54	15.19	23.61	39.61	22.85	6.55
Ab	17.43	17.94	11.14	9.49	11.17	19.31
An	11.34	9.97	12.22	9.50	12.36	13.00
Hy	11.98	11.67	11.44	4.43	12.80	11.86
Mt	1.13	1.16	0.98	0.46	1.13	1.02
Il	1.75	1.71	1.82	1.64	1.91	1.76
Ap	0.40	0.38	0.44	0.36	0.39	0.39
Ba	588	712	297	492	133	272
Cr	135	97	202	129	149	157
Cu	18	20	16	15	15	14
Ni	55	46	77	43	58	63
Nb	18	16	17	16	19	17
Pb			20	18	18	21
Rb	96	96	93	114	82	89
Sr	149	149	118	119	122	257
Th			17	14	16	13
V	123	127	159	144	163	150
Y	34	35	31	27	33	30
Zn	89	87	93	76	82	101
Zr	207	207	213	200	229	206
La	40.8	40.8				
Ce	96	98				
Nd	41.4	43.9				
Sm	7.80	8.02				
Eu	1.46	1.48				
Tb	1.00	1.03				
Ho	1.41	1.49				
Yb	3.06	3.01				
Lu	0.55	0.54				
Ta	1.32	1.21				
Th	12.7	13.1				
U	2.54	2.23				
Hf	6.16	6.61				

Pb isotopes

Pb-isotopic data are presented in Table 6 and shown on a $^{207}Pb/^{204}Pb$–$^{206}Pb/^{204}Pb$ diagram (Fig. 11). Also shown on this diagram is the area in which typical MORB falls (from Gariepy *et al.* 1983), the Pb-isotopic compositions of the Faeroes Upper Series basalts, also from Gariepy *et*

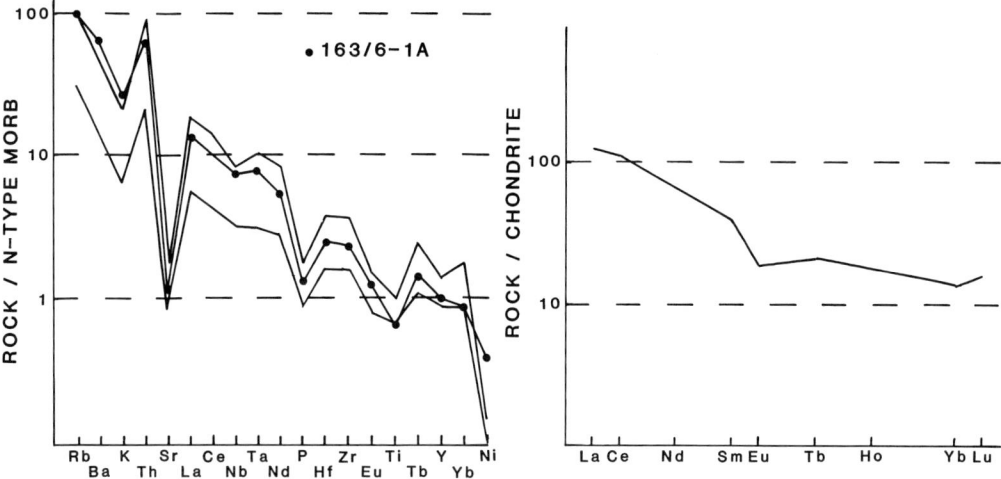

FIG. 10. MORB-normalized spidergram and chondrite-normalized rare-earth plot of 163/6-1A dacites. Also shown on MORB-normalized spidergram is the range of Lower Series peraluminous rocks recovered at Site 642 of the Ocean Drilling Programme on the Vøring Plateau, Norwegian Sea (Eldholm *et al.* 1987).

TABLE 5. *Sr-isotopic characteristics of basalts and dacites, Well 163/6-1A. Initial ratios calculated assuming 55 My in situ Rb decay. Initial ratios for leached samples are approximate, corrected from present compositions using whole-rock Rb/Sr ratios*

	Present		Initial	
Sample	$^{87}Sr/^{86}Sr$ unleached	$^{87}Sr/^{86}Sr$ leached	$^{87}Sr/^{86}Sr$ unleached	$^{87}Sr/^{86}Sr$ leached
C4.1	0.70335 ± 3	0.70316 ± 4	0.70329	0.70310
C4.2	0.70330 ± 6	0.70321 ± 6	0.70326	0.70317
C5.1	0.70445 ± 5	0.70467 ± 5	0.70435	0.70457
C5.4	0.70317 ± 6	0.70302 ± 5	0.70313	0.70298
C6.3	0.70394 ± 5	0.70400 ± 5	0.70392	0.70398
C7.1	0.71258 ± 6	0.71362 ± 5	0.71112	0.71216
C7.2	0.71248 ± 5	0.71356 ± 4	0.71102	0.71211

TABLE 6. *Pb-isotopic characteristics of basalts and dacites, Well 163/6-1A*

Sample	$^{206}Pb/^{204}Pb$	$^{207}Pb/^{204}Pb$	$^{208}Pb/^{204}Pb$
C4.1	18.34	15.55	38.25
C4.2	18.37	15.54	38.25
C5.4	18.18	15.57	38.07
C6.3	17.77	15.56	37.58
C7.1	18.54	15.58	38.60
C7.2	18.57	15.63	38.75

al. (1983), and the lines representing hypothetical 'orogene' and 'upper crust' source compositions, from Zartman & Doe (1981). The Faeroes Upper Series are included because their Pb-isotopic compositions have been modified by contamination with ancient continental basement, because they are closely contemporaneous with the 163/6-1A basalts, and have comparable chemistry, and because 163/6-1A is comparatively close to the Faeroes.

On this diagram, the 163/6-1A basalts fall on the top edge of the MORB field, or slightly above it, and the dacites similarly fall above the MORB field, close to the hypothetical 'orogene' line of Zartman & Doe (1981). This implies that the dacites have initial Pb-isotope characteristics similar to those predicted for magmas derived from mixed crust–mantle sources, and that their isotopic composition is dominated by material from the upper continental crust. The basalts also appear to have been contaminated by upper crustal material. Therefore, the 163/6-1A lavas have interacted with crust of a different nature to that which has contaminated the Faeroes basalts. The effect shown by the Faeroes Upper Series is for deviation from typical MORB toward a single unradiogenic sample with $^{207}Pb/^{204}Pb = 15.184$ and $^{206}Pb/^{204}Pb = 16.100$. Contamination of this type is considered to be due to interaction with ancient U-depleted material typical of the lower crust.

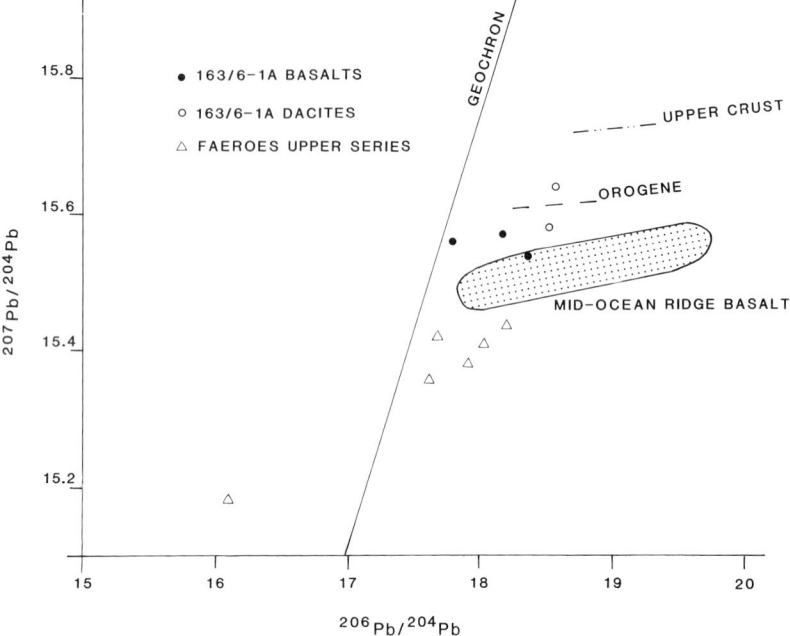

Fig. 11. Pb-isotopic composition of 163/6-1A basalts and dacites, with range of MORB and Faeroes Upper Series basalts (from Gariepy *et al.* 1983) shown for comparison.

K-Ar age measurements

K-Ar ages were determined on six whole-rock samples of basalt and four of dacite. The results of these measurements are presented in Table 7. The ages have been computed in the conventional manner on the assumption that all of the non-radiogenic argon measured in an analysis has an isotopic composition identical to that of modern atmospheric argon. Cores 4 and 5 have low but similar potassium contents, but core 6 is extremely K-deficient. The results of these flame-photometric analyses are in general agreement with those obtained by XRF, demonstrating the efficacy of this technique for K determination even at low levels. However, the results obtained on the junk basket samples are markedly dissimilar, presumably reflecting differences in composition of the subsamples analyzed. As a result of the low potassium contents and relative youthfulness of the basalts, the analytical uncertainty in their age measurements is quite high. Additionally, the extraction and purification of the radiogenic argon from some samples was hampered by the presence of hydrocarbons and, as some of these occur in the same region of the mass spectrum as the argon isotopes, they are potential interferants during mass spectrometric analysis and could lead to computation of misleading results. This

TABLE 7. *K-Ar data, basalts and dacites, Well 163/6-1A*

Sample	K wt %	$^{40}Ar^*$ ($\times 10^{-7}$ scc/g)	$^{40}Ar^*/^{40}Ar_T$	Age (Ma)
Basalts				
C4.2	0.100	2.10	9.02	53 ± 7
		3.42	18.5	(86 ± 6)
C4.6	0.143	3.10	21.5	55 ± 3
C5.2	0.195	4.48	30.2	58 ± 2
C5.3	0.105	2.38	24.8	57 ± 3
C6.2	0.063	1.41	14.9	57 ± 5
		1.93	22.6	(77 ± 4)
Dacites				
C6.3	0.050	1.61	12.1	(81 ± 8)
C7.3	2.19	43.84	88.2	50.9 ± 1.0
C7.4	2.10	46.52	88.0	56.2 ± 1.2
		45.2	69.6	54.6 ± 1.3
JB2A	4.39	133.7	29.9	77 ± 3
		122.1	42.2	70 ± 2
JB2B	0.921	243.6	32.4	580 ± 20
		318.5	42.6	720 ± 20

$^{40}Ar^*$: radiogenic ^{40}Ar; $^{40}Ar_T$: total ^{40}Ar; λ_β: 4.962×10^{-10} yr; λ_e: 0.581×10^{-10} yr; $^{40}K = 1.167 \times 10^{-2}$ atom %; errors computed from $(E_K^2 + (1 + A/R)^2 E_{40}^2 + E_{38}^2 + (A/R)^2 E_{36}^2)^{1/2}$, where $R = (100 - A) = {}^{40}Ar^*/{}^{40}Ar_T$ with error in K, $E_K = 1.5\%$ and errors in peak heights, $E_{40} = E_{38} = 0.5\%$, $E_{36} = 1\%$.

potential problem was recognized at an early stage and careful checks were carried out on all subsequent samples undergoing analysis. No evidence was found for the presence of any hydrocarbons following gas purification and it is not considered that the measured ages are, therefore, uncertain.

For the purposes of discussion, the results in Table 7 are best considered in two groups. The first group consists of all the analyzed basaltic rocks (C4.2, C4.6, C5.2, C5.3, C6.2 & C6.4) and two dacites (C7.3 & C7.4). With three exceptions, most of the apparent ages of this group lie in the restricted range 51–58 Ma, and average 55 Ma. This is probably close to the age of the lavas, as all samples but one (C7.3) fall within analytical error of this. Weathering and alteration can facilitate argon loss, lowering the measured ages. Unfortunately, the least altered samples from core 6 are extremely K-deficient. There does not appear to be any significant difference in age between the dacites and the basalts, a finding which is of significance in their genesis.

The second group comprises the junk basket samples JB2A and JB2B, both of which were analyzed twice. JB2A gave apparent ages of 70 Ma and 77 Ma, and JB2B gave apparent ages of 580 Ma and 720 Ma. Although a late Cretaceous date for the lower part of the dacite sequence is not unreasonable, the apparent age of JB2B is clearly anomalous. Approximately 93% of the measured ^{40}Ar in this sample could not have been derived radiogenically by the *in situ* decay of ^{40}K since the late Cretaceous–early Tertiary. This suggests that JB2A similarly contains extraneous argon.

Both of these samples are distinctive in containing a high proportion of cordierite phenocrysts, many of which are fresh. It has been known for some time that this mineral can host excess argon (Damon & Kulp 1958), located in the wide channels in the crystal structure where it is trapped by cationic 'blocks'. The concentration of extraneous ^{40}Ar in the cordierite of JB2 (ca. 1.3×10^{-4} scc/g) falls in the middle of the range already observed for this mineral (York *et al.* 1969; York & Farquhar 1972, pp. 53–56). JB2B has a glassy groundmass, indicative of rapid chilling. Although it is unlikely that extrusion took place under high hydrostatic pressure, in view of the subaerial nature of the nearby Faeroes lava pile and the presence of reddened tops to basalt flows in core 6, the chilling may have effectively sealed a sample of magmatic gas, packaging and preserving it for posterity. In JB2A, the more microcrystalline nature of the groundmass may have allowed the release of some of the trapped argon; in this instance, the effect on the apparent age is less marked because of the higher potassium content. In the core samples, where the cordierite has been replaced by chlorite, argon release seems to have been complete.

Both the junk basket dacites have high ^{36}Ar contents (up to 3×10^{-7} scc/g). Consequently, sea water must have played a role in the genesis of these rocks, as this is the only reservoir capable of supplying ^{36}Ar in such concentrations (see Allégre *et al.* 1987, table 1). This conclusion is consistent with the increase in ^{87}Sr/^{86}Sr found during acid leaching. This involvement with sea water suggests that the ^{40}Ar excesses found in these rocks could have been derived through the interaction and/or assimilation of older sediments bearing K-rich detritus, which had not been completely degassed on diagenesis. This is similar to the origin proposed for excess argon in the contemporaneous lavas from DSDP Sites 553 and 555 (Macintyre & Hamilton 1984). These lavas have extraneous gas with similar ^{40}Ar/^{36}Ar ratios (ca. 440), but this may be fortuitous as the gas analyzed from Well 163/6-1A may be an admixture of sea water (295) and magmatic (>440) components. The presence of hydrocarbons in the sequence suggests that the sediments that were involved may have been organic in nature.

Discussion

The volcanic sequence in Well 163/6-1A consists of olivine tholeiites overlying cordierite-hypersthene dacites. The basalts may be divided into two groups on geochemical grounds, those of core 4 being significantly different from those of cores 5 and 6. The lower group have distinct MORB-like tendencies, with flat patterns on MORB-normalized spidergrams (see Fig. 7). They plot in the ocean-floor basalt field on the Ti-Y-Zr diagram (see Fig. 8) and in the transitional MORB field on the Ta-Th-Hf diagram (see Fig. 9). However, their pyroxene compositions suggest a within-plate origin (see Fig. 5). The upper basalts, although also olivine tholeiites, are distinctly more alkalic, with higher Ti/V and Nb/Y ratios. They show strong depletion of Y and Yb relative to N-type MORB (see Fig. 7), a feature typical of within-plate basalts (Pearce 1982), and fall into the within-plate basalt field on the Ti-Y-Zr diagram (see Fig. 8). Sr- and Pb-isotope studies suggest that upper continental crustal material has played a minor but significant role in the genesis of both basalt types. Both basalt types have parallels in the British Tertiary Volcanic Province. The basalts of core 4 have

REE patterns comparable to those of the Mull Plateau Group (Morrison *et al.* 1980) and to those of the Lower Basalt Formation of Antrim (Lyle 1985), whereas the basalts of cores 5 and 6 compare well with the Upper Basalt Formation of Antrim (Lyle 1985) and basalts of Preshal Mhor type in Skye (Thompson *et al.* 1980).

The two basalt types have similar highly incompatible element ratios. Mean La/Ta and La/Th values are 17.5 and 15.8 respectively for core 4 basalts, and 16.8 and 16.3 respectively for cores 5 and 6 basalts, although there is greater scatter associated with the ratios from basalts of cores 5 and 6, presumably related to lower analytical precision at the lower elemental abundances associated with this basalt type. Given this, the similarity in La/Ta and La/Th indicates that the same mantle source, under different partial melting conditions, could have been responsible for both magma types. This possibility is explored further in Fig. 12, a chondrite-normalized plot of Ce content against Ce/Yb ratio, as discussed by Saunders (1984). This shows that the basalts of cores 5 and 6 could be produced by high degrees of dynamic partial melting of a source with $(Ce)_{CN} = 1.5$ and $(Ce/Yb)_{CN} = 0.8$, and that those of core 4 could be generated by a lower degree of partial melt of the same source. The displacement of the core 4 basalts to the left of the dynamic melting curve can be interpreted to be the result of the abundant olivine phenocryst population; the sample with the maximum displacement from the curve is C4.1, the most olivine-rich of this group, with 40% modal olivine.

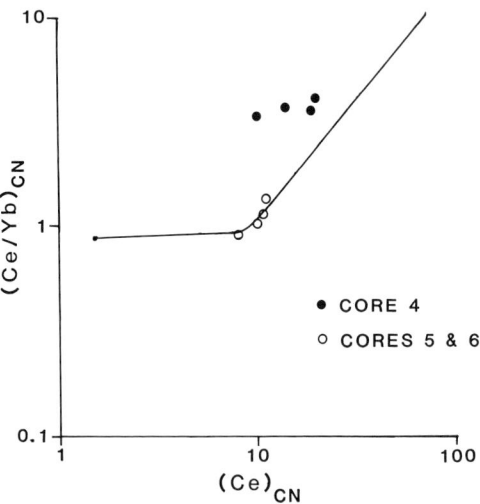

FIG. 12. 163/6-1A basalts plotted on a log-log plot of $(Ce/Yb)_{CN}$ ratio vs $(Ce)_{CN}$ content. Also shown is the dynamic melting curve of a garnet lherzolite source, from Saunders (1984).

Thus, the lower basalts at 163/6-1A appear to represent an early, large-scale melting event, with the succeeding basalts generated by lower degrees of melting as the heat source waned.

The dacites have an unusual chemistry and clearly represent a very specific petrogenetic process. Their corundum-normative character, abundant cordierite phenocrysts and high Ni and Cr contents all argue strongly against an origin by differentiation of an olivine tholeiite parent, which supplied the spatially associated basalts, and suggest that they were either generated by assimilation of argillaceous material or by extensive melting of a quartzo-feldspathic and argillaceous parent.

The dacitic magma has a melting point in excess of 1100° at 1 atm. The total assimilation of aluminous material by a dacitic parent magma would require an extraordinarily high initial magma temperature. Consequently, the melting hypothesis is preferred. In this model, shales, and possibly also sandstones, were in contact with a large body of high-temperature magma for a long period at depth. Their Pb- and Sr-isotopic compositions and high extraneous argon contents are consistent with such a petrogenetic model. Ponding of a large body of high-temperature basaltic magma within a sedimentary pile might be expected to allow the development of anatectic melts. Mixing of such anatectic melts with the basaltic magma may be strongly inhibited by a combination of the density contrast between melts and magma chamber geometry. For example, a pond of anatectic melt lying above basaltic magma could develop in a sheet-like magma chamber, where heat loss takes place almost entirely through the roof, or in a chamber with side walls inclined inwards toward the roof, so that low-density anatectic melts developed on the side walls are constrained to migrate upward along the side walls rather than through the basaltic magma. The development of large volumes of anatectic melt would require a very large heat input which could only be provided by the fractional crystallization of a very large volume of basaltic magma, presumably developing a large pile of cumulate gabbros. Because of the considerable amount of dacitic material produced (in excess of 350 m at the well site, although the lateral extent of the body is unknown), the magma chamber that acted as the heat source must have been large. The most likely location is the large igneous centre detected close by (see Fig. 1).

Some further inferences regarding the nature of the source material may be made from the geochemistry of the dacites. Of particular interest are the high Ni and Cr contents, because these trace elements, along with V, Co, Cu and Zn, are enriched in organic black shale facies (Vine & Toutelot 1970; Cunningham & Gilbert 1984). There is, therefore, a strong possibility that black shales were included in the melt. If this is so, it seems likely that they were of Cretaceous age, as early Cretaceous black shales have been drilled in the Bay of Biscay and Goban Spur areas to the S of the Rockall Trough during DSDP Legs 48 and 80 (Timofeev & Bogolyubova 1979; Graciansky *et al.* 1979; Cunningham & Gilbert 1984). The postulated presence of a thick sedimentary sequence including excellent petroleum source rocks, such as black shales, has significant implications for the petroleum prospects of the Rockall Trough. In this context, it is noteworthy that the basalt sequence on the Faeroes contains migrated hydrocarbons (Waagstein *et al.* 1984), and also that hydrocarbons were detected in the volcanic sequence of the well in question here.

Prior to the drilling of the 163/6-1A dacites, peraluminous rocks were unknown from the NE Atlantic Igneous Province. It is, therefore, remarkable that similar rocks have recently been drilled at Site 642 on the Vøring Plateau (Norwegian Sea) during Leg 104 of the Ocean Drilling Programme. The sequence drilled at Site 642 comprises MORB-like tholeiites that form the dipping reflector sequence, overlying more silicic rocks, including cordierite-, hypersthene- and bytownite-phyric rocks (Eldholm *et al.* 1987).

These have remarkably similar trace element contents to the 163/6-1A dacites (see Fig. 10), and their initial $^{87}Sr/^{86}Sr$ ratios of 0.7115–0.7118 compare closely with the 0.7121–0.7122 values from the 163/6-1A dacites. These two discoveries, although spatially separated, are in broadly comparable tectonic settings and both are located close to the continental margin of the European landmass. Thus, although peraluminous extrusive and intrusive rocks in the NE Atlantic Igneous Province remained undiscovered until 1980, they appear to form an important part of the province, and may prove to be particularly significant in a tectonic context.

ACKNOWLEDGEMENTS: The cooperation of many individuals and groups was required in this study, and thanks are due to all involved. Britoil kindly provided junk basket samples to supplement the core study. Dr B Atkin (MESA) and Mrs D James (Edinburgh) organized the XRF analyses, and Dr I Sinclair and Dr S Parry (Imperial College) carried out the neutron activation analyses. The Isotope Geology Unit at SURRC receives financial support from NERC and the Scottish Universities, which is gratefully acknowledged. We are grateful to Britoil, on behalf of the consortium partners, and the UK Department of Energy for giving permission to publish this paper, which is published with the approval of the Director, British Geological Survey (NERC). The line drawing of the seismic section is published by permission of Merlin Geophysical Ltd.

References

ALLÉGRE C. J., STAUDACHER T. & SARDA P. 1987. Rare gas systematics: formation of the atmosphere, evolution and structure of the Earth's mantle. *Earth and Planetary Science Letters*, **81**, 127–150.

BROOKS C. K. 1976. The Fe_2O_3/FeO ratio of basalt analyses: an appeal for a standardized procedure. *Bulletin of the Geological Society of Denmark*, **25**, 117–120.

BURKE W. H., DENISON R. E., HEATHERINGTON E. A., KOEPNICK R. B., NELSON H. F. & OTTO J. B. 1982. Variation of seawater $^{87}Sr/^{86}Sr$ throughout Phanerozoic time. *Geology*, **10**, 516–519.

CUNNINGHAM R. & GILBERT D. 1984. Organic facies of Cenozoic and Cretaceous sediments from Deep Sea Drilling Project sites 549 and 551, northern North Atlantic. *In*: de GRACIANSKY, P. C. & POAG, C. W. *et al. Initial Reports of the Deep Sea Drilling Project*. US Government Printing Office, Washington, **80**, 1073–1079.

DAMON P. E. & KULP J. L. 1958. Excess helium and argon in beryl and other minerals. *American Mineralogist*, **43**, 443–459.

DIETRICH V. J. & JONES E. J. W. 1980. Volcanic rocks from Rosemary Bank (Rockall Trough, NE Atlantic). *Marine Geology*, **35**, 287–297.

ELDHOLM O., THIEDE J., TAYLOR E. *et al.* 1987.

Proceedings of the Ocean Drilling Program. US Government Printing Office, Washington, **A104**, 783 pp.

GARIEPY C., LUDDEN J. & BROOKS C. 1983. Isotopic and trace element constraints on the genesis of the Faeroe lava pile. *Earth and Planetary Science Letters*, **63**, 257–272.

GRACIANSKY P. C. de, AUFFRET G. A., DUPEUBLE P., MONTADERT L. & MULLER C. 1979. Interpretation of depositional environments of the Aptian/Albian black shales of the north margin of the Bay of Biscay (DSDP sites 400 and 402). *In*: MONTADERT, L., ROBERTS, D. G. *et al. Initial Reports of the Deep Sea Drilling Project*. US Government Printing Office, Washington, **48**, 877–907.

LYLE P. 1985. The petrogenesis of the Tertiary basaltic and intermediate lavas of northeast Ireland. *Scottish Journal of Geology*, **21**, 71–84.

MACINTYRE R. M. & HAMILTON P. J. 1984. Isotope geochemistry of lavas from sites 553 and 555. *In*: ROBERTS, P. G., SCHNITKER, D. *et al. Initial Reports of the Deep Sea Drilling Project*. US Government Printing Office, Washington, **81**, 775–781.

MORRISON M. A. 1978. The use of 'immobile' trace elements to distinguish the palaeotectonic affinities of metabasalts: applications to the Palaeocene basalts of Mull and Skye, northwest Scotland. *Earth and Planetary Science Letters*, **39**, 407–416.

——, THOMPSON R. N., GIBSON I. L. & MARRINER G. F. 1980. Lateral chemical heterogeneity in Palaeocene upper mantle beneath the Scottish Hebrides. *Philosophical Transactions of the Royal Society of London*, **A297**, 229–244.

NAYLOR D. & SHANNON P. M. 1982. *The Geology of Offshore Ireland and West Britain*. Graham & Trotman, London, 161 pp.

NISBET E. G. & PEARCE J. A. 1976. Clinopyroxene composition in mafic lavas from different tectonic settings. *Contributions to Mineralogy and Petrology*, **63**, 149–160.

PEARCE J. A. 1982. Trace element characteristics of lavas from destructive plate boundaries. *In*: THORPE R. S. (ed) *Andesites*. Wiley, New York, pp. 525–548.

—— & CANN J. R. 1973. Tectonic setting of basic volcanic rocks determined using trace element analyses. *Earth and Planetary Science Letters*, **19**, 290–300.

PRESTVIK T. 1982. Basic volcanic rocks and tectonic setting: discussion of the Zr-Ti-Y diagram and its suitability for classification purposes. *Lithos*, **15**, 241–247.

ROBERTS D. G. 1975. Marine geology of the Rockall Plateau and Trough. *Philosophical Transactions of the Royal Society of London*, **A278**, 447–509.

——, MASSON D. G. & MILES P. R. 1981. Age and structure of the southern Rockall Trough. *Earth and Planetary Science Letters*, **52**, 115–128.

——, BOTT M. H. P. & URUSKI C. 1983. Structure and origin of the Wyville–Thomson Ridge. *In*: BOTT M. H. P., SAXOV S., TALWANI M. & THIEDE J. (eds) *Structure and Development of the Greenland–Scotland Ridge*. Plenum Press, New York, pp. 133–158.

RUSSELL M. J. & SMYTHE D. K. 1977. Evidence for an early Permian oceanic rift in the northern North Atlantic. *In*: NEUMANN E. R. & RAMBERG I. B. (eds) *Petrology and Geochemistry of Continental Rifts*. Reidel, Dordrecht, pp. 173–179.

SAUNDERS A. D. 1984. The rare earth element characteristics of igneous rocks from the ocean basins. *In*: HENDERSON P. (ed) *Rare Earth Element Geochemistry*, Developments in Geochemistry, **2**, 205–236.

SHERVAIS J. W. 1982. Ti/V plots and the petrogenesis of modern and ophiolitic lavas. *Earth and Planetary Science Letters*, **52**, 115–128.

TARNEY J., SAUNDERS A. D., WEAVER S. D., DONNELLAN N. C. B. & HENDRY G. L. 1979. Minorelement geochemistry of basalts from Leg 49, North Atlantic Ocean. *In*: LUYENDYK, B. P., CANN, J. R. *et al. Initial Reports of the Deep Sea Drilling Project*. US Government Printing Office, Washington, **49**, 657–691.

THOMPSON R. N., GIBSON I. L., MARRINER G. F., MATTEY D. P. & MORRISON M. A. 1980. Traceelement evidence of multistage mantle fusion and polybaric fractional crystallization in the Palaeocene lavas of Skye, NW Scotland. *Journal of Petrology*, **21**, 265–293.

TIMOFEEV P. P. & BOGOLYUBOVA L. I. 1979. Black shales of the Bay of Biscay and conditions of their formation. *In*: *Initial Reports of the Deep Sea Drilling Project*. US Government Printing Office, Washington, **48**, 831–853.

VINE F. J. & TOUTELOT E. B. 1970. Geochemistry of black shale deposits—a summary. *Economic Geology*, **65**, 253–273.

WAAGSTEIN R., HALD N., JØRGENSEN O., NIELSEN P. H., NOE-NYGAARD A., RASMUSSEN J. & SCHOENHARTING G. 1984. Deep drilling on the Faeroe Islands. *Bulletin of the Geological Society of Denmark*, **32**, 133–138.

WOOD D. A., JORON J-L. & TREUIL M. 1979. A reappraisal of the use of trace elements to classify and discriminate between magma series erupted in different tectonic settings. *Earth and Planetary Science Letters*, **45**, 326–336.

YORK D. & FARQUHAR R. M. 1972. *The Earth's Age and Geochronology*. Pergamon Press, New York, 178 pp.

——, MACINTYRE R. M. & GITTINS J. 1969. Excess radiogenic ^{40}Ar in cancrinite and sodalite. *Earth and Planetary Science Letters*, **7**, 25–28.

ZARTMAN R. E. & DOE B. R. 1981. Plumbotectonics—the model. *Tectonophysics*, **75**, 135–162.

A. C. MORTON, British Geological Survey, Keyworth, Notts NG12 5GG, UK.

J. E. DIXON, J. G. FITTON, Grant Institute of Geology, University of Edinburgh, West Mains Road, Edinburgh EH9 3JW, UK.

R. M. MACINTYRE, Scottish Universities Research and Reactor Centre, East Kilbride, Glasgow G75 0QU, UK.

D. K. SMYTHE, British Geological Survey, Murchison House, West Mains Road, Edinburgh EH9 5LA, UK.

P. N. TAYLOR, Department of Earth Sciences, University of Oxford, Parks Road, Oxford OX1 3PR, UK.

Syn- and post-rift igneous activity in the Porcupine Seabight Basin and adjacent continental margin W of Ireland

M. P. Tate & M. R. Dobson

SUMMARY: An igneous province in the Porcupine Seabight Basin and adjacent shelf region W of Ireland records repeated episodes of intrusive and extrusive igneous activity, from the mid-Jurassic to late Oligocene. The 'Province' maintains a regionally distinctive NE–SW and NNW–SSE curvilinear trend that is inherent to the basement. Mid-Jurassic pyroclastic airfall deposits were probably contemporaneous with mid-Kimmerian uplift. Voluminous early Cretaceous volcanism was coeval with an episode of rifting during the Valanginian to Barremian interval in the Porcupine and Goban Spur Basins. In particular, the 150 km long 'Porcupine Median Volcanic Ridge' (PMVR) is interpreted to be extrusive in the S, but progressively intrusive northwards; from an inferred fissure volcano to a laccolith. Aptian–Albian volcaniclastics are of unknown association, but coincide with the onset of oceanic spreading along the adjacent proto-Atlantic margin. Documented late Cretaceous intrusive and extrusive igneous activity was minor in scale although a period of marked inversion is recorded for the NE Porcupine Basin, which may relate to the emplacement of the areally extensive Brendan Igneous Centre. Intermittent intrusive and extrusive activity throughout the Palaeogene was broadly coincident with the full duration of the 'Thulean' Palaeocene igneous event. Further activity, variously dated from Eocene to late Oligocene, is partly attributed to a series of intrabasinal, parallel-aligned plugs of intrusive and extrusive origin situated in the upper Porcupine Basin and termed the 'Slyne Fissures'. These fissures, whose basement-controlled lineation prolongates onshore as an extended group of sheet intrusions, appear to emanate from the Brendan Centre. Further dykes in County Kerry are also believed to extend towards the same centre.

Within the Province a distinction is made between magmatism that is essentially syn-rift and linked to basin development (notably Valanginian–Barremian), and later post-rift activity (Aptian–Tertiary). The size, emplacement character and location of the PMVR may represent a sudden increase in axial volcanism that occurred during advanced stages of basinal rifting, and could be viewed as a prelude to, or transition from rifting to spreading prior to an abrupt failure. Temporal relationships can be proposed between active faulting associated with initial lithospheric stretching (late Kimmerian), late syn-rift igneous activity (Valanginian–early Barremian) and the initiation of subsequent lithospheric stretching (earliest Barremian) and, finally, the onset of seafloor spreading (Albian) on the continental margin W of Goban Spur. The later post-rift activity is described as 'opportunistic' and triggered by exterior tectonic events that utilized existing tensile trends within the Basin. Emplacement of the Brendan Centre was probably responsible for a phase of inversion with its concomitant influence on sedimentation patterns. Magmatism in the Eocene and Oligocene was associated with minor faulting, an unconformity and pronounced post-Eocene subsidence, together indicative of a final regional thermal episode.

The Porcupine Seabight Basin, located on the continental shelf to the W of Ireland (Fig. 1), is a N–S orientated graben, closed to the N by the Slyne Ridge, which plunges southward to form the bathymetric Seabight embayment. It is considered to be an intracratonic rift basin with thinned continental crust (Masson & Miles 1986a) that initially developed during a rifting event in the late Triassic to early Jurassic (Masson & Miles 1986b). Renewed rifting during the late Jurassic to early Cretaceous interval (Ziegler 1982; Masson & Miles 1986a,b) is identified with the regional late Kimmerian unconformity (Ziegler 1981, 1982).

Numerous geophysical surveys connected with hydrocarbon exploration activity since the mid 1970's have revealed a sedimentary succession at least 10 km thick that developed after the initial rifting event (Croker & Shannon 1987; Mac-Donald et al. 1987; Ziegler 1987). In particular, gravity and magnetic surveys have indicated the existence of several igneous centres, thought to be of Tertiary age, located mainly on the margins of the basin (Riddihough 1968; Young & Bailey

From MORTON, A. C. & PARSON, L. M. (eds), 1988, *Early Tertiary Volcanism and the Opening of the NE Atlantic,* Geological Society Special Publication No. 39, pp. 309–334.

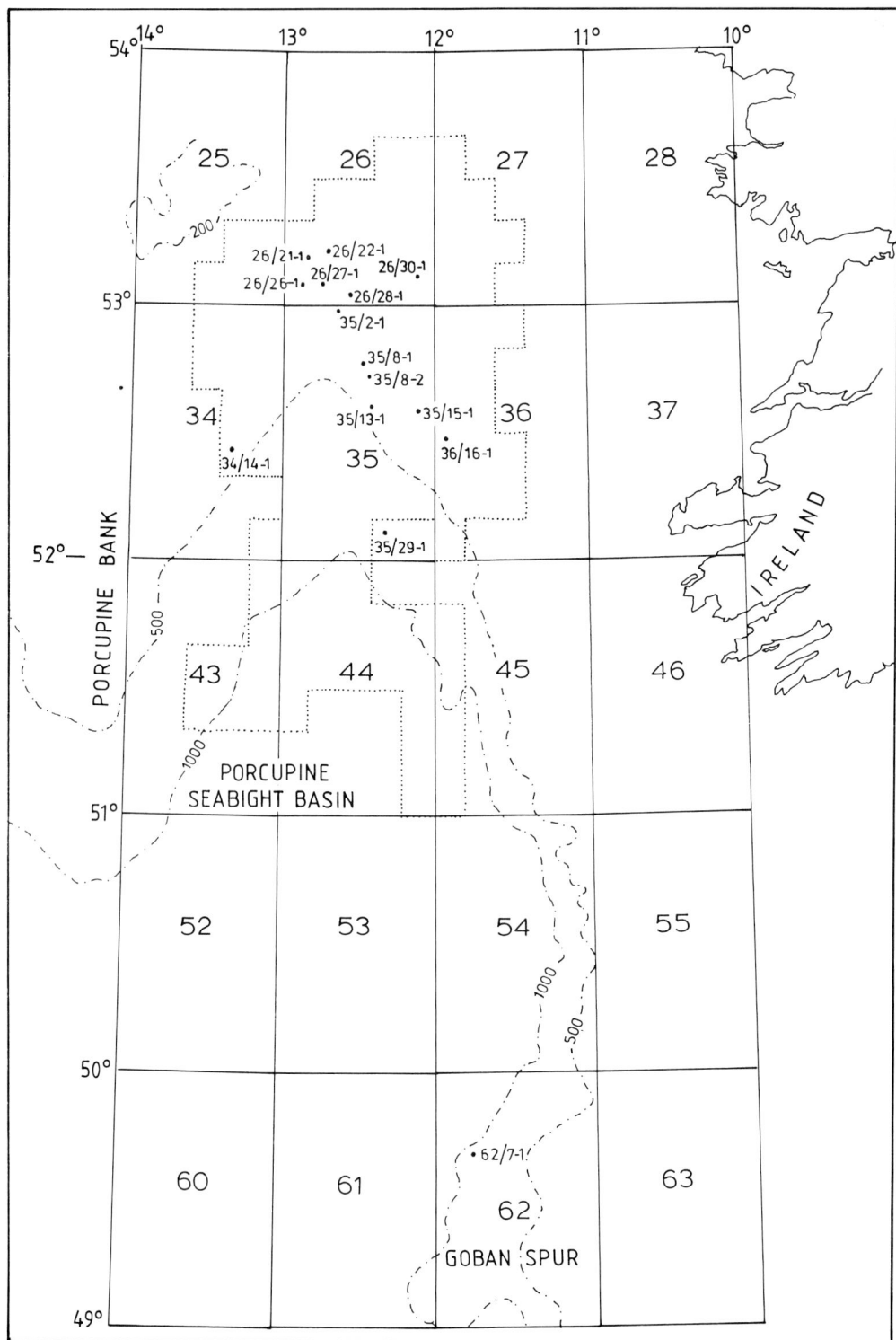

FIG. 1. Map to show location of borehole data and multichannel seismic grid (inside dotted outline). Limited seismic data N of 51°30′N was additionally available.

1974; Bailey 1979; Buckley & Bailey 1979; Riddihough & Max 1976; Max *et al.* 1982; Masson *et al.* 1985). A large median ridge, initially interpreted from magnetic modelling as incipient ocean crust (Young & Bailey 1974) and from gravity modelling as a basic igneous intrusion of dyke-like form (Buckley & Bailey 1975), occurs within the Basin. More recently, Lefort & Max (1984) preferred to interpret the ridge as possible oceanic crust. The application of seismic reflection data revealed its volcanic origin and a Lower Cretaceous age was inferred using magnetostratigraphy (Roberts *et al.* 1981;

Ziegler 1982; Masson & Miles 1986, a, b). Part of a multiple sill complex penetrated in the NE Porcupine Basin has yielded a late Oligocene date (Seemann 1984).

Outcropping onshore in the adjacent peninsulas of County Kerry (Fig. 2), an olivine-dolerite dyke system with a N–S alignment was also dated as late Oligocene and Eocene (Morris 1974; Horne & Macintyre 1975), but has been recently assigned a Palaeocene age by Thompson (1985) and Mitchell & Mohr (1986). Dyke systems in Connacht, of similar composition to the Kerry system, but with a NE–SW trend, together with

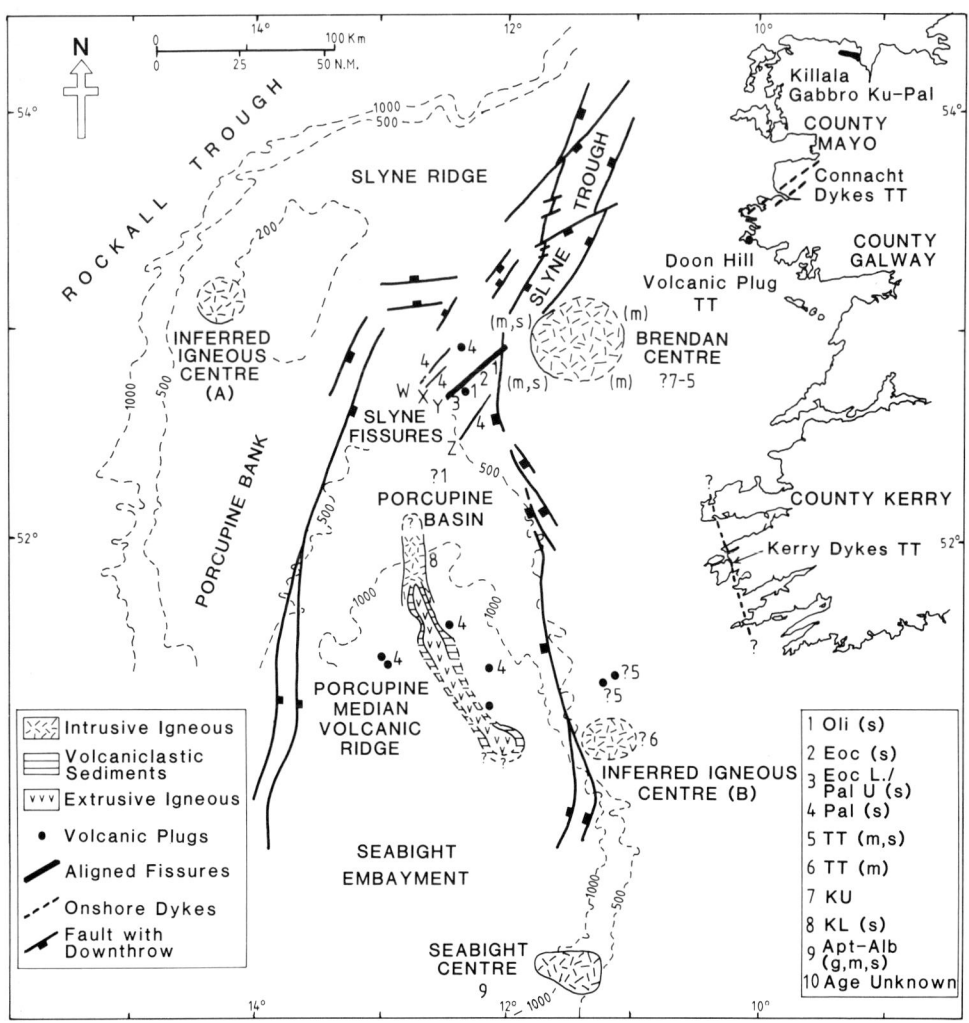

FIG. 2. Summary map of the distribution and age of the components of the Porcupine Igneous Province. Igneous features are mapped from gravity (g), magnetic (m) and multichannel seismic (s) data. Abbreviations: KL: Lower Cretaceous; Apt-Alb: Aptian–Albian; KU: Upper Cretaceous; TT: Tertiary; Pal.: Palaeocene; Eoc.: Eocene; Oli: Oligocene. Basin outline is taken from Naylor & Shannon (1982). Bathymetry in metres.

a volcanic plug at Doon Hill, have been variously dated from Palaeocene to late Oligocene, although a Palaeocene age was preferred due to apparent argon loss (Macintyre *et al.* 1975; Thompson 1985; Mitchell & Mohr 1986). Both groups of dykes are believed to be related to the offshore Brendan Igneous Centre (Fig. 2), but are distinct from the Donegal–Kingscourt trend of the Irish–British Tertiary Province (Macintyre *et al.* 1975; Riddihough & Max 1976; Mohr 1982). A gabbroic intrusion at Killala, County Mayo, gave a regionally anomalous maximum age of 80 Ma, again disputed to be 60 Ma (Thompson 1985; Mohr, pers. comm. 1987).

In the Fastnet Basin, Caston *et al.* (1981), using multichannel seismic profiles and well logs, identified a multiple sill complex and several small volcanic plugs ranging in age from Bathonian to probable Palaeocene (see Fig. 10). To the W in the Goban Spur Basin, a further sequence of lava flows of probable Valanginian age were penetrated by an exploration well sited close to another igneous centre (Cook 1987).

Access to a grid of multichannel seismic reflection data, together with most of the available composite well logs (Fig. 1), has allowed the recognition, dating, subsurface delineation and morphological mapping of the previously recorded and newly identified Mesozoic and Tertiary igneous features. Newly recognized igneous occurrences include: volcaniclastic sediments in the Lower Cretaceous, Palaeocene and Eocene; sill intrusions in the Coniacian and Palaeocene; extrusive activity in the Danian (possibly also the Campanian) and numerous parallel-aligned, small volcanic plugs, together with isolated volcanic plugs variously dated from Palaeocene to late Oligocene (see Figs 2 & 3).

Evidence for Mesozoic and Tertiary igneous activity in the Porcupine Seabight Basin

Distinctions between, for example, lava flows and sills, and fissures and dykes in the subsurface are rarely unambiguous. Consequently, recognized igneous occurrences are initially presented in a descriptive form; subsequent interpretations are based on assessments of their petrographic, petrophysical and seismic character, on the limited dating available and the nature of relevant onshore outcrop. This combined evidence is then amalgamated to make a regional assessment. All chronostratigraphic assignments are based upon

the geological time-scale as defined by Harland *et al.* (1982).

Borehole Data

Inspection of the petrophysical logs for 15 exploration wells drilled in the Porcupine and Goban Spur Basins has revealed a wide stratigraphical range of igneous activity (Fig. 3).

Jurassic

In Block 26/28, cores taken in the Bajocian–Kimmeridgian interval revealed minor amounts of pyroclastic airfall tuff (MacDonald *et al.* 1987). A 4 m thick igneous rock in Well 35/2-1 described as dark grey-green to black weathered dolerite was recovered from Kimmeridgian–Oxfordian shales, interbedded with sandstones and claystones of marginal marine to continental origin, 708 m below the late Kimmerian unconformity.

Lower Cretaceous

Well 35/8-1 contains 314 m of cumulative waterlain ash bands and volcaniclastic sediment, described as scoriacious, with occasional pumice that accumulated with claystones and siltstones in a low energy inner-shelf environment. The volcaniclastics occur within two distinct intervals. The lower section consists of 244 m of Hauterivian and Barremian volcanic-rich sediment unconformably overlying a non-volcanic Valanginian section. These are described as greygreen to black tuffs, angular, blocky, scoriaceous and interlaminated with the claystones. The volcanic content ceases part way up the Barremian interval, but 503 m higher there is a further 70 m of volcanic-rich sediment interbedded with sands, silts and limestone, commencing above the base of an Aptian–Albian interval and terminating close to its top. These are described as black to brown, friable, scoriaceous, waterlain airfall deposits and as light grey to white, reversely graded pumice and black to light grey, blocky, vitric, volcanogenic sediment.

In two further Wells, 26/21-1 and 35/13-1, pyroclastics have been recognized at the Barremian–Aptian boundary and the top Albian interval respectively. Lack of preservation, often due to unfavourable sedimentary facies, may account for the fact that ash bands and volcaniclastic horizons have not been recognized at this level in adjacent wells.

Further S in the Goban Spur Basin (see Fig. 1), Well 62/7-1 penetrated 207 m of basaltic lava

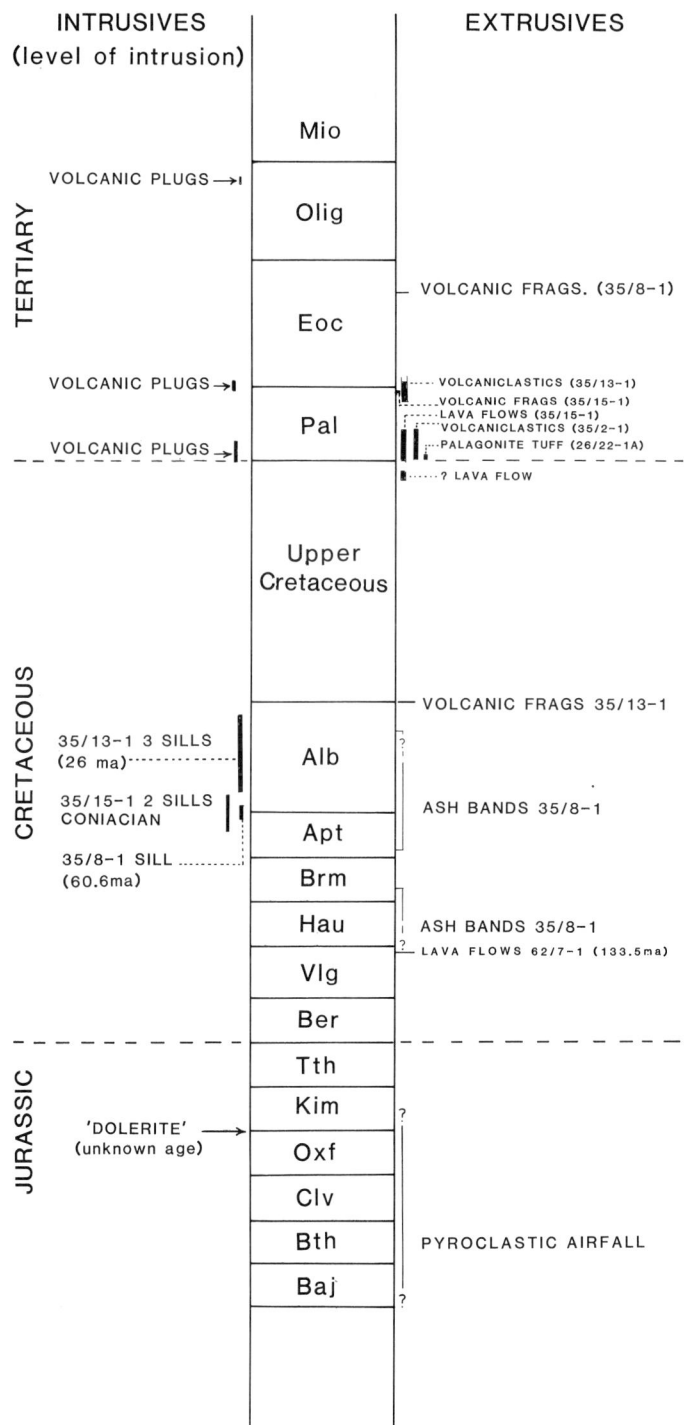

Fig. 3. Stratigraphic summary of all igneous intervals drilled by exploration wells in the region and recorded on composite logs together with additional sources. Upper terminations of Tertiary plugs are inferred from seismic data.

flows. These flows overlie non-marine Bathonian–Callovian sediments and were initially thought to be of the same age (Cook 1987). Subsequent radiometric dating of fresh samples has yielded 85 and 133.5 Ma K-Ar dates (the latter being averaged from 130 ± 2 and 137 ± 2 Ma dates; Esso, pers. comm. 1987). Since the lavas are overlain by a thick Cretaceous section and their extrusive origin is undisputed, the 85 Ma date is invalid. The Valanginian age is consistent with the lavas being overlain by undifferentiated Neocomian sediments. In ac-

cordance with the wireline logs, it is possible to identify 11 major petrophysical log cycles partly displayed on Fig. 4. Sequences typically range between 9 and 30 m. Samples have been variously described as being from black, to grey, to green, fine to medium grained and also as red-brown, light-green, mottled tuff-claystone. In conjunction with the wireline logs, the latter description befits the tops of individual lava flows where followed by periods of dormancy and weathering, whereas the former description probably applies to more basal parts of the flow. Petrographic and

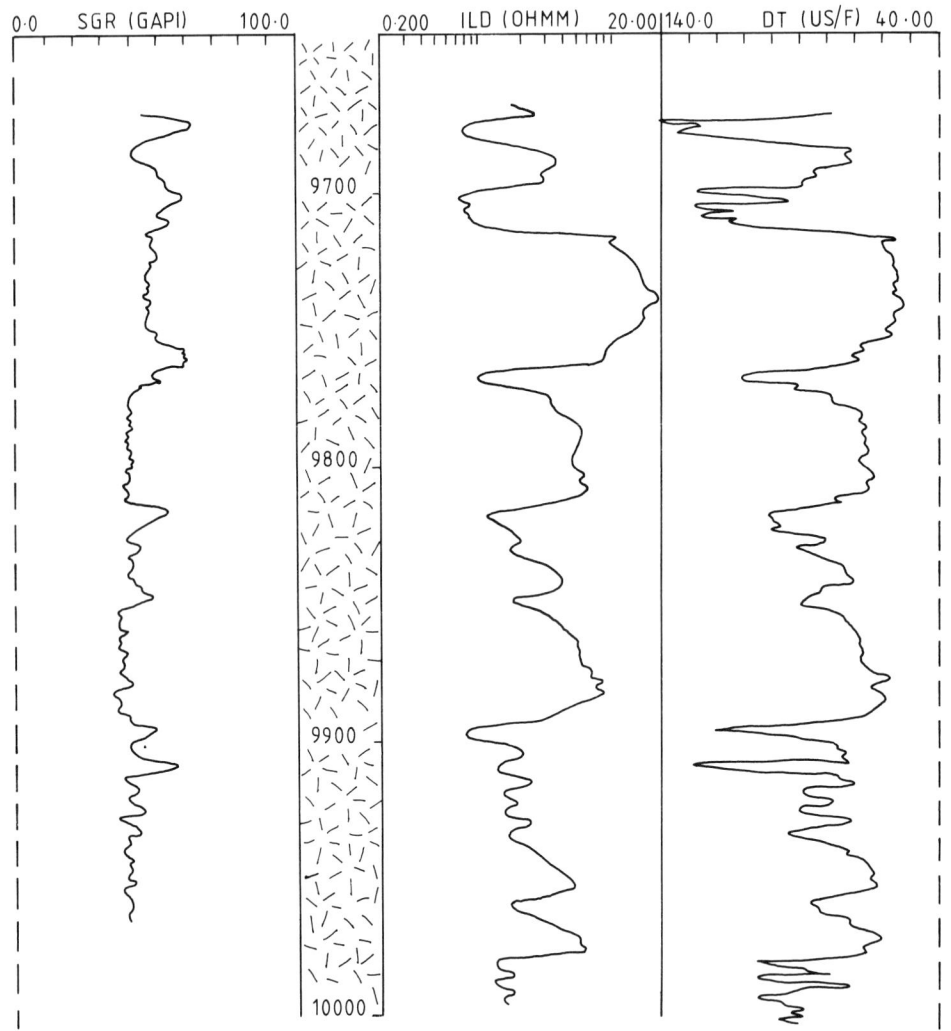

FIG. 4. Cyclical petrophysical log responses measured for basalts penetrated in Goban Spur Well 62/7-1, as displayed on gamma (SGR), deep resistivity (ILD) and density (DT) logs. Lateritic or weathered profiles to lava flows are highlighted by upwardly decreasing resistivity readings (a function of increasing porosity) and decreasing bulk density values, which together commonly culminate with an increase on the gamma log close to the flow top, due to a residual enrichment of heavy minerals and organic material. Depth in feet. (Courtesy of Esso Exploration and Production, Ireland).

geochemical analysis show the basalts to be MgO-rich plagioclase (An_{59-71}) and olivine-phyric tholeiites (Table 1).

Upper Cretaceous

The only record of igneous activity associated directly with Upper Cretaceous sediments is a 2 m undifferentiated igneous interval within a thin sequence of Campanian–Maastrichtian marls in Well 35/15-1. Immediately above, in the Danian, there is an intermittent sequence of lava flows (see next section) of which the 2 m interval appears to be the lowest member. An extrusive origin is preferred here since there are no known sills intruded at this stratigraphic level.

Tertiary

Extrusive igneous activity appears to be concentrated in the Palaeogene, particularly the Danian. Danian igneous intervals (documented in Well 35/15-1) are overlain by a Montian–Thanetian non-volcanogenic succession. Four igneous intervals of varying thickness (5 m, 51.5 m, 36 m, & 15 m) are recorded up the succession. Each of these are interpreted to be extrusive lava flows based on their internal heterogeneity discerned from the logs, together with some evidence for weathering. Near the base they are interbedded with cryptocrystalline, fossiliferous chalk that is occasionally silty. Upwards, the sequence comprises cryptocrystalline to microcrystalline limestone, which is white to pink to purple and occasionally interbedded with volcanics, presumably ash bands or volcaniclastics. Basaltic sequences are described as finely crystalline to glassy, dense, black, speckled with occasional embedded limestone fragments and calcite filled amygdales. The uppermost sequence of basalt is recorded as being purplish-brown to rusty-brown, conglomeratic with limonitic or lateritic developments indicative of weathering. Internal wireline log sequences are poorly defined, perhaps due to reduced weathering in a mostly subaqueous environment. However, the two thickest flows each appear to contain at least four to five flow units. Altogether, this sequence indicates initial emplacement in a submarine environment prior to emergence and subaerial weathering.

Danian volcanics in Well 35/2-1 are described as dark, black, occasionally green, red, purple, and brown, fine to coarse, altered volcanic rock with additional grains of quartz, totalling 17.5 m in thickness. These volcanics are associated with a low gamma reading and interval transit time, both equal to that of the underlying chalk. There is a possible unconformity at the base of the volcanic sequence and a clear unconformity at the upper contact. This volcanic sequence has been interpreted by Elf Aquitaine as a 'submarine volcanic flow'. It is underlain by an outer-shelf chalk facies and overlain by littoral sands of Montian age, followed in turn by deeper water shales. Uniform dips averaging 3° to the S are

TABLE 1. *Whole-rock geochemical analysis by X-ray fluorescence of (a) major and minor constituents, (b) trace elements, and (c) normative calculations for basalts cored in Well 62/7-1 in the Goban Spur Basin (Courtesy of Esso Exploration and Production Ireland).*

(a) Oxide	Wt Percent	(b) Trace element	ppm	(c) Norm. mineral	Wt. percent
SiO_2	45.37	Ba	121	Orthoclase	5.7
Al_2O_3	14.45	Nb	13	Albite	22.0
Fe_2O_3	3.30	Zr	111	Anorthite	26.0
FeO	8.77	Y	15	Diopside	3.9
MgO	12.05	Sr	407	Hypersthene	12.1
CaO	6.97	Rb	22	Olivine	19.5
Na_2O	2.53	Zn	84	Magnetite	4.9
K_2O	0.95	Cu	23	Ilmenite	4.1
H_2O	2.70	Ni	67	Apatite	0.5
CO_2	0.51	Pb	8	Calcite	1.2
TiO_2	2.11	U	<2	Water	2.7
P_2O_5	0.21	Th	2		
S	0.01	Cr	102		
F	0.00	Ga	34		
MnO	0.07	La	21		
		Ce	41		

observed in the underlying Cretaceous interval, but immediately above the volcanics the sands dip at 6° to the N, gradually decreasing upwards over the next 150 m to 3° to the N, implying that this Danian volcanism was associated with uplift.

In the upper part of the basin (26/22-1), a white to yellow bioclastic and glauconitic Danian limestone contains fragments of hydrated basaltic glass or palagonite. These tuffs consist of black, brown to yellow particles with glassy shards in a white matrix. They are recorded over a 10 m interval. In Well 35/15-1 volcanic granules are recorded within a sandstone sequence of late Palaeocene age. A 43 m-thick sequence of volcaniclastics penetrated in Well 35/13-1 are dated as late Palaeocene/lowermost Eocene; detailed examination has revealed that basaltic lava fragments and devitrified glass constitute 47% of this deposit. Other constituents include quartz, chert, microcrystalline calcite, organic pyritic material and coal fragments, in a coarse to fine, poor-to-moderately sorted sandstone (Shell, pers. comm. 1986). These volcaniclastic coastal sandstones overlie open marine sediments, an environment which was re-established in the early Eocene. Again, the volcanics are associated with marked shallowing and a change in dip direction, perhaps related to uplift. The sandstones are barren, but are underlain and overlain by nannoplankton Zones NP9 and NP10 respectively. In the Middle Eocene of Well 35/8-1, medium-grained sandstones of inner-shelf aspect contain reworked sub-rounded to well-rounded grains of basalt.

Sills have been recorded in three wells (excluding 35/2-1) in the NE Porcupine Basin:

(1) Two sills, 6 m and 61 m thick drilled in Well 35/15-1 have intruded Albian limestones and siltstones. The upper of the two is truncated by an unconformity and overlain by Campanian–Maastrichtian marls. The lower interval, described in the log as dolerite to basalt with rare marble, the latter presumably occurring towards the chilled margins, consists of euhedral olivine, pyroxene, feldspar and abundant magnetite. It produces a uniform and broadly parallel-tracking log response characteristic of sills. A Coniacian radiometric date has been obtained (Phillips Petroleum Company, pers. comm. 1987) although further details are not available. The age of the sills is stratigraphically constrained to be post-Albian and pre-late Campanian/Maastrichtian.
(2) A 174 m-thick sill in Well 35/8-1 intrudes Aptian–Albian strata. It is composed of poikilitic diopsidic augite, subhedral olivine and calcitic plagioclase; there is some indi-

cation of a chilled margin. The rock yields a K-Ar radiometric date of 60.6 ± 4.4 Ma (Phillips Petroleum Company, pers. comm. 1987).
(3) In Well 35/13-1, three sills, 31 m, 33 m and 156 m thick, composed of porphyritic olivine-pyroxene dolerite with vein quartz, intruded early and mid-Albian siltstones during the late Oligocene (25.8 ± 2.6 Ma) causing metamorphism of the surrounding limestones.

Geophysical Records

Jurassic

Seismic resolution of the Jurassic sedimentary succession is poor, particularly in the upper Porcupine Basin where it is buried beneath up to 8 km of younger sediment, although occasional high amplitude reflectors of igneous character are recorded within the Jurassic interval. Toward the S, in the shallower western Fastnet Basin (Caston *et al.* 1981), two probable mid-Jurassic volcanic plugs and a multiple sill complex (Bathonian) intrude the Lower Lias at a depth of 1.5–3.0 km.

Lower Cretaceous

A large, elongate volcanic median ridge with a NNW–SSE trend, at least 150 km long and up to 25 km wide, lies subparallel to, and broadly coincident with, the axis of the Porcupine Basin (see Fig. 2). This ridge is termed the 'Porcupine Median Volcanic Ridge' (PMVR). Previously, it was assigned a pre-Aptian age, based on seismic extrapolation by Roberts *et al.* (1981), Berriasian–Barremian by Ziegler (1982 & 1987) and pre-Cenomanian by Masson & Miles (1986a). Its age has been further constrained by using its reversed magnetization (Young & Bailey 1974), which is thought must predate the Cretaceous normal polarity period extending from the Aptian to early Campanian. Thus, it inherited its reversed magnetism during the Mesozoic M-series (Masson & Miles 1986a).

Previous attempts to map the Ridge using limited seismic data relied upon a pronounced axial gravity high (up to 70 mgal) to define its northern extension (Masson & Miles 1986a). The southern part of this gravity high coincides only with the northern termination of the Ridge and the respective trends differ. The southern part of the Ridge is coincident with a SE-trending aeromagnetic anomaly. This indicates that the gravity anomaly is not intimately linked to the seismically mapped subcrop of the Ridge.

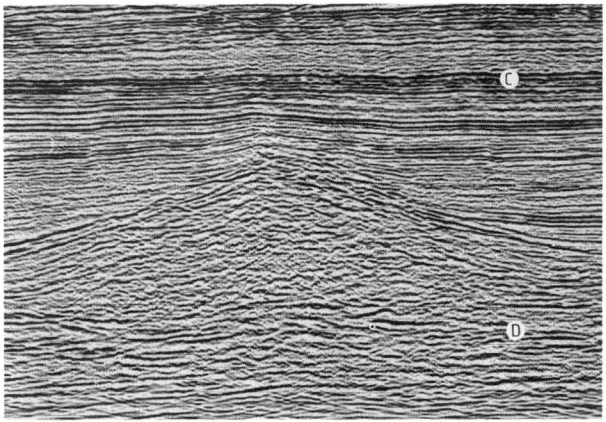

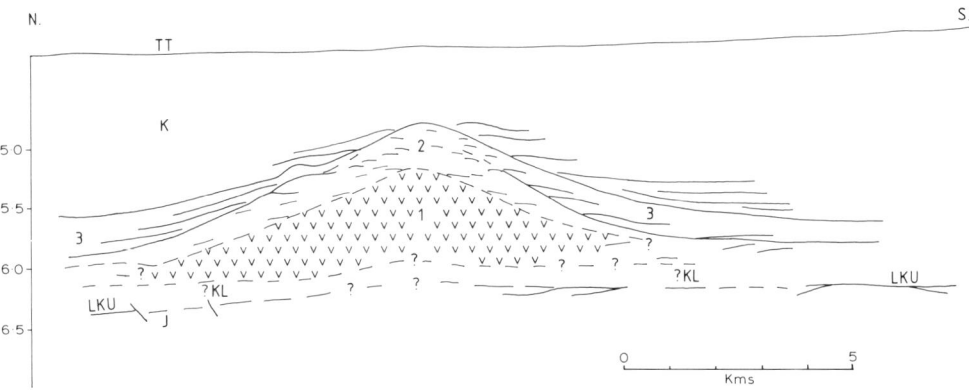

FIG. 5. Seismic reflection profile and interpretation across the southern termination of the Porcupine Median Volcanic Ridge at approximately 12°W. Abbreviations: 'C': Top Chalk; 'D': LKU (late Kimmerian unconformity); TT: Tertiary; K: Cretaceous undifferentiated; KL: Lower Cretaceous; J: Jurassic. Layers 1, 2 and 3 are discussed in the text. Vertical scale is in seconds (TWT)—(Courtesy of Merlin Geophysical Ltd.).

The PMVR is clearly delineated on seismic reflection profiles (see Figs 5 & 6). Towards its southern subcrop the Ridge displays a triangular section with gently curved margins inclined at 25–30° against which reflectors clearly onlap. The basal Cretaceous unconformity, also referred to as the late Kimmerian unconformity (LKU), passes below the base of the Ridge with a possible thin layer of downlapping earliest Cretaceous sediments in between (see Figs 5 & 6d). Progressively further N, the base of the volcanic pile remains at approximately the same depth (6.1–6.2 s TWT) and lies within successively older strata. North of 52°N it is poorly imaged and masked by multiples where the basin deepens abruptly, but probably persists until 52°10′N. Here it underlies the LKU which is noticeably domed (see Fig. 6a). The outline of the Ridge at this point becomes markedly convex and inclined

at a greatly reduced angle. Close to 51°45′N, there appears to be a sector of transition where the morphology is more complex (see Fig. 6b). The flanks are emplaced below the uplifted LKU onto which earliest Cretaceous reflectors onlap. In the centre a protrusion of the Ridge, with an abrupt change to steeper sides, lies within onlapping Cretaceous sediments.

Seismic stratigraphy shows that the PMVR consists of three discrete layers of which only the first pertains to the northern intra-Jurassic extension. Layer 1 is a homogenized core region of chaotic impersistent reflectors. Layer 2 comprises a narrow outer zone surrounding the core; it displays higher amplitude and lower frequency reflectors dipping at low angles down the flanks. Layer 3 is a marginally developed wedge-shaped zone of more parallel and continuous reflectors inclined at higher angles than layers 1 and 2.

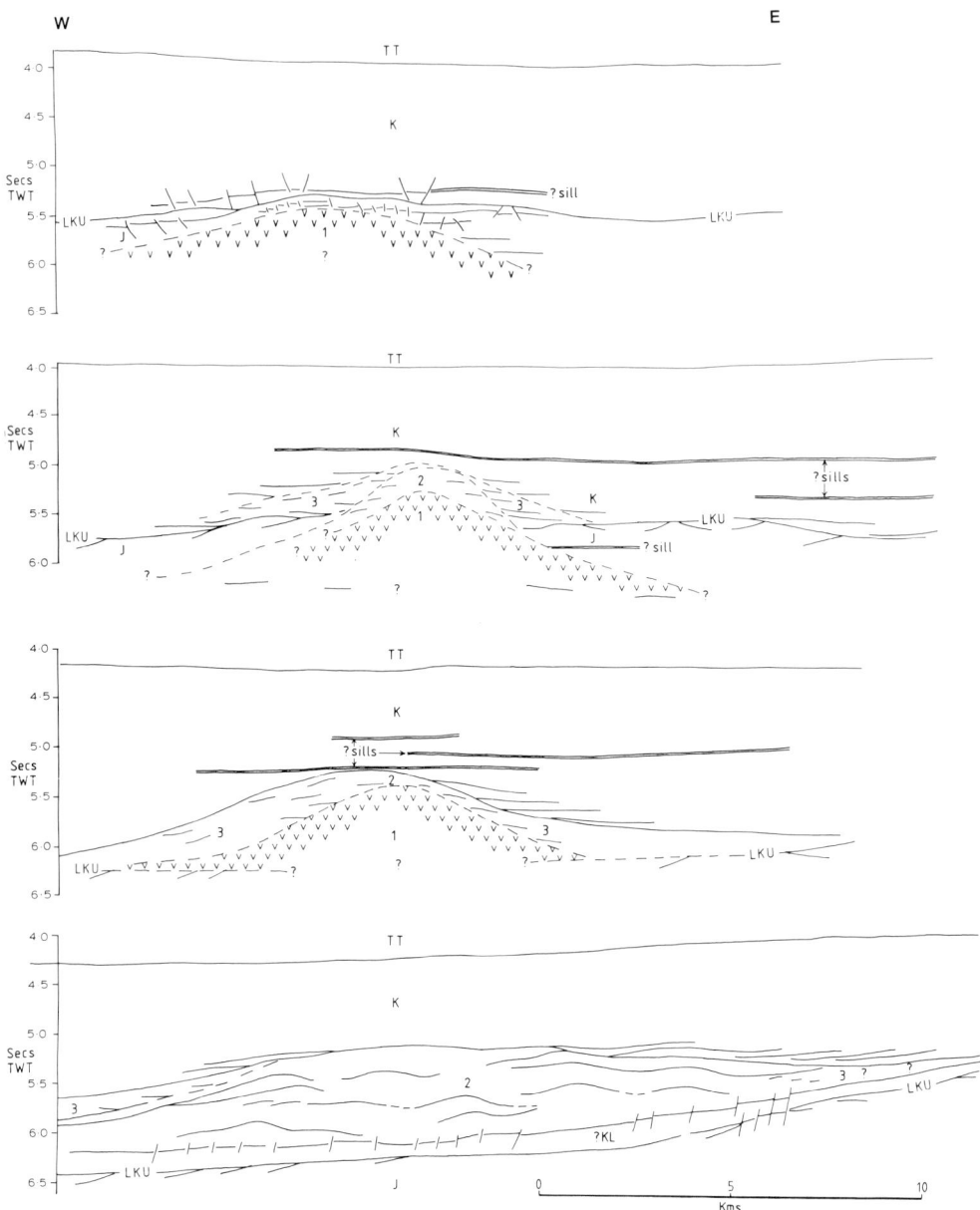

FIG. 6. Line drawings of seismic reflection profiles illustrating the morphology and internal structure of the Porcupine Median Volcanic Ridge from N to S at approximately: 52°N (a), 51°45′N (b), 51°30′N (c), 51°00′N (d). Note the gradual change-over from an intra-Jurassic intrusion to an extrusion with onlapping Lower Cretaceous reflectors. Internal zones 1, 2 and 3 are discussed in the text. Abbreviations as in Fig. 5.

Further S, on the southeastern margin of the Seabight embayment (see Fig. 2), a pronounced magnetic and gravity anomaly at 49°.50′N, 11°.30′W was inferred to have an igneous origin (Roberts *et al.* 1981; Naylor & Shannon 1982). More recent magnetic and gravity modelling

revealed it to be a penetrative igneous body, assigned an Aptian–Albian age from seismic extrapolation (Cook 1987). Since an igneous origin is now evident it is termed the 'Seabight Igneous Centre'. A non-penetrative igneous body further W at 49°30′N, 12°20′W has also been identified (Cook 1987). There is no clear seismic evidence for late Cretaceous igneous activity in the Porcupine Seabight Basin.

Tertiary

Seismic data covering the upper Porcupine Basin reveal four parallel alignments of small, intrabasinal igneous plugs with a NE–SW trend, to be termed the 'Slyne Fissures', W, X, Y and Z (see Fig. 2). Adjacent to the central and broader Fissure Y, three subordinate and narrower alignments can tentatively be discerned. Additional isolated plugs of similar age, together with numerous smaller, vertical disturbances on the seismic data, are also apparent, especially to the N of Fissure W (Jaunich 1983; Kirton & Donato 1985; Sowerbutts 1987).

The most prominent Fissure, Y, is of variable width (less than 2 km) and age. Borehole calibrated seismic data indicate broad diachronism from S to N, from late Palaeocene/lowermost Eocene (see Fig. 8) through a zone of very narrow Palaeocene and Eocene plugs close to the midpoint, prior to emplacement at a high stratigraphic level (see Fig. 9) that approximates to the late Oligocene. Further N, the Fissure Y plug alignment intercepts basement where no confident age inferences can be made. The remaining Fissures (W, X & Z) are narrower, although they intermittently thicken along their trend, whilst the additional volcanic plugs are isolated and do not extend onto adjacent seismic lines. Altogether Fissures W, X and Z are observed to gently dome or disturb the top Chalk (up to and including the Danian) reflectors, inferring at least a Palaeocene age, whilst Y is apparently younger, or has been reactivated.

Fissures W, X and Z are near parallel, strike at 035–040° and prolongate towards the Slyne Trough. However, the younger Fissure Y strikes at 050° and prolongates toward the onshore dyke systems of Connacht. There is no clear intrabasinal magnetic anomaly associated with the thicker Fissure Y, yet where this trend intercepts basement at 52°55′N, a sharp magnetic anomaly commences (Riddihough 1975). Further NW of the Irish coastline, several elongate magnetic anomalies of broadly NNW-trend may also have an igneous origin (Riddihough & Max 1976). Adjacent to the Slyne Fissures there is an increase in density of small normal faults with negligible downthrow, the upper and lower effects of which terminate at the base of the Oligocene and Palaeocene, respectively.

A multiple sill complex is inferred from the seismic data (Badley 1985) and confirmation of these intrusions has been described earlier for the upper Porcupine Basin. Despite an almost ubiquitous distribution of inferred sills throughout the Porcupine Basin, three zones of increased density are noted, with a preference for intrusion within the mid-Cretaceous interval:

(1) In the NW, associated with Slyne Fissures, sills appear as concordant bodies radiating from the margins of volcanic plugs, notably Fissures Y and Z, and occasionally occur in isolation. Less commonly, sills in the Tertiary interval are developed between the Palaeogene fault sets.

(2) Sills occur in close proximity to the PMVR, again mainly within the mid-Cretaceous interval, upwards to 0.1–0.4 s TWT above its apex and extending up to 15 km laterally.

(3) In the SW, at a similar depth interval, sills are developed adjacent to the two Tertiary volcanic plugs W of the PMVR and close to the western basin margin. These sills are gently inclined intrusions concordant with the dipping strata, which terminate close to the rising LKU and within onlapping sediments. Assuming a Palaeocene and/or an Oligocene age for the sills, they would have been emplaced beneath approximately 1.5–2.0 km and 3.0–3.5 km of sediment, respectively.

The existence of three large intrabasement igneous centres has been inferred from geophysical data. The previously termed 'Brendan Centre' (Cole & Crook 1910; Mohr 1982) is located on the NE margin of the Porcupine Basin close to the junction with the Slyne Trough. This is the largest of the intrabasement igneous centres and is associated with a clear subspherical magnetic anomaly 50 km in diameter and coincident with a 70–90 mgal gravity high. The complete absence of seismic reflectors both within and over the centre signifies its presence and proximity to the surface, but lends no further morphological detail. The Centre has not been directly dated, nor is it presently possible to suggest an age from available geophysical data.

Inferred igneous centre A on the Porcupine Bank (see Fig. 2) was initially identified by Bailey (1979) and Young & Bailey (1974) using magnetic data. The magnetic anomaly is 25 km in diameter and is superimposed on a broader 70 mgal gravity high. A crossing seismic line demonstrates a doming of uppermost basement reflectors over the anomaly, but no age inferences can be made. Igneous centre B on the eastern basin margin 100 km SW of Mizen Head is also inferred from a pronounced magnetic anomaly and a 50 mgal gravity high. The former indicates a diameter of 30 km, but again its age is unknown. However two nearby small, high-level plugs identified from seismic and a marine magnetic survey (unpublished) are probably of Tertiary age and may be satellite plugs to centre B.

Outcrop in Western Ireland

Igneous rocks occur at four locations along the western coast of Ireland (excluding County Donegal) and represent the most eastern expression of the Porcupine igneous activity. However, most of the age dates are subject to controversy. Considered sequentially from N to S, their distribution is shown on Fig. 2 and age dates ordered in Table 2.

(1) Killala, County Mayo. A large gabbroic intrusion (the Ros Gabbro) elongated E–W and 450 m wide is now recognized to extend 15 km W of Killala Bay (Mohr, pers. comm. 1987). It is holocrystalline with an unusual composition, comprising quartz associated with alkali ferromagnesian minerals. It has been dated at 80 Ma (Campanian), which is regarded as a maximum age (Macintyre *et al.* 1975). A minimum age of 58 Ma (Thanetian) has been obtained from a cross-cutting dyke system which similarly displays an apparently anomalous E–W trend. Also in the same area are a series of flow-banded multiple dykes thought to have an age range similar to the gabbro.

(2) Counties Galway and Mayo, Connacht. Three distinct dyke systems and associated sills have been recognized. They are composed of analcime-olivine dolerite and display strong geochemical affinities with intrusions of western County Kerry. The dyke systems trend NE–SW, and an inferred zone of dilation more than 50 km wide was calculated for this suite of intrusions. Mitchell & Mohr (1986) obtained a range of K/Ar ages for these dolerites from 57–27 Ma, the ages broadly younging southwestwards. The maximum 57 Ma date was considered to be close to the original emplacement date. Deviations from this date were rejected on the grounds of systematic argon loss. However, no acceptable cause was identified.

(3) Doon Hill, Connacht. This is a 60 m high, 320 m diameter, vertically inclined igneous body composed of tholeiitic olivine dolerite. Its mineral chemistry is consistent with a near-surface intrusive plug (Mohr 1982). An age of 60 Ma (Macintyre *et al.* 1975; Thompson 1985) has been determined.

(4) Western County Kerry. Several dykes, all belonging to one system, make up this most

southerly of the post-Palaeozoic Irish igneous rocks. They are olivine dolerites and trend NNW–SSE. Their age is a matter of controversy. Horne & Macintyre (1975) record dates of 42 and 25.3 Ma (mid-Eocene to late Oligocene), whereas Mitchell & Mohr (1986) and Thompson (1985) have determined Lower Palaeocene (65.6 $\pm$ 1.4 Ma) and mid-Eocene (39 $\pm$ 1.0 Ma) dates.

Analysis of the igneous activity

Jurassic

The rare occurrence of Jurassic volcanism in the Porcupine Seabight is probably a function of the limited database. The Jurassic section has been penetrated by only a few wells because of the burial depth and, consequently much may remain unrecognized. Pyroclastic airfall deposits of mid–late-Jurassic age may correspond to the mid-Kimmerian uplift observed in the Porcupine Seabight and Goban Spur (Croker & Shannon 1987; Cook 1987). In the Fastnet Basin (Caston *et al.* 1981), Bathonian intrusives are contemporaneous with the mid-Kimmerian event, and their occurrence appears to coincide with activity on NW–SE sinistral transform faults (Robinson *et al.* 1981).

Cretaceous

Seismic data, together with evidence of volcaniclastic sediments in Well 35/8-1, indicate that the PMVR was initiated during the late Valanginian to early Hauterivian interval. In the S, volcanics appear to have accumulated over a thin sequence of earliest Cretaceous sediments, probably of outer-shelf aspect (see Fig. 5). In the N, the volcanic pile is progressively contained within Jurassic strata as an intrusion (see Fig. 6a). In between, at approximately 51°45′N, the Ridge is marginally emplaced within the uppermost Jurassic interval. This caused an abrupt doming of the overlying LKU, yet along its axis it still maintained an extrusive core that was later onlapped by Lower Cretaceous sediments (see Fig. 6b). The present maximum vertical height of the Ridge provides an estimation for the palaeohydrostatic head of magma (Fig. 7), albeit a minimum value due to erosion. In accordance with the above estimation, it is proposed that the buoyancy-driven hydrostatic head of magma was fed along a linear vent. Where this palaeohydrostatic head lay above the Cretaceous seabed a fissure-type volcano ensued, but further N it gradually penetrated the seafloor sediments and the underlying Jurassic strata, which were domed

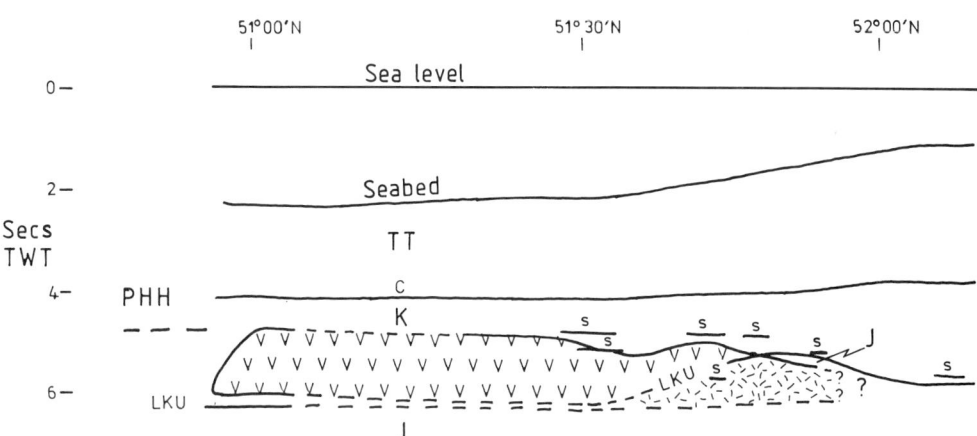

FIG. 7. Cross-section along the axis of the Porcupine Median Volcanic Ridge compiled from seismic reflection data. The remaining vertical extent of the ridge provides a minimum estimate for the palaeo-hydrostatic head of magma (PHH), which may be inferred to diminish northwards where it intrudes Jurassic strata. Abbreviations as in Fig. 5, except s: inferred sill.

upon emplacement. A laccolith is inferred for this intra-Jurassic intrusion, as it is compatible with the emplacement model. Although the convex upper contact can be discerned from seismic data, the lower contact is masked by multiples. Since the base of the Ridge appears constant at 6.1–6.2 s TWT, it is unlikely that the intrusion would abruptly acquire a dyke-like morphology. Moreover, if the base level for the intrusion is assumed to be the same, then a width–height ratio of 1:8 can be calculated. Although a mound-like morphology is evident at its northern termination, it is too speculative to suggest an extrusive or intrusive origin. Adjacent high amplitude reflectors, both concordant and discordant, may represent lava flows or sills respectively.

The palaeohydrostatic head of magma progressively diminished northwards, probably reflecting inhomogeneities in the upper magma column (such as degree of vesiculation) or varying depths to magma generation along the strike of the Ridge. An increase in overburden, prompted by the thickening sedimentary column, as indicated by the LKU, would be insufficient. The peak head of magma occurred at the southern termination, and coincident with the trajectory of an important strike-slip fault believed to have been active during emplacement of the Ridge.

Before this model for emplacement of the Ridge can be developed further, two assumptions are made. First, that the palaeoenvironment in the Hauterivian between 51 and 52° N was an open marine shelf with shallow water depths (ca. 200 m); and second that the Ridge was broadly of basaltic composition. Since the height of the Ridge as calculated from interval velocities is between 2.25 and 3.0 km, diminishing to approximately 1.5 km at 52° N, it follows that volcanism rapidly became emergent and subject to subaerial weathering. A basaltic composition is compatible with the observed morphology, the location of the Ridge in a syn-rift basin, the dark scoriaceous content of the pyroclastics as described in the composite logs and contemporaneous tholeiitic extrusions in the Goban Spur Basin to the S. Analogues appear to be rare, although the 'Volcanic Mound' of Miocene age in the Sea of Japan was similarly extruded during the waning stages of an extensional episode. Despite being circular in plan, it has a broadly similar seismic expression to the Ridge and is at least 6.5 km in diameter. Drilling indicates that it consists of basaltic lavas, tuffs and multiple dolerite intrusions (Suzuki 1983).

The structure of the PMVR is inferred to consist of three zones (see Fig. 6a–d):

(1) A core zone comprising upwardly accreted submarine lava flows associated with pillow lavas and hyaloclastites adjacent to an axial fissure venting system that eventually became subaerially emergent. North of 51°45′N, this passes into an inferred laccolith, comprising a structurally homogeneous, petrographically

layered, gabbroic rock surrounded by a metamorphic aureole.

(2) A narrow, stratified outer zone merging with the core, composed of interbedded lavas, pyroclastics and occasional doleritic intrusions, developed on the flanks. Early subaerial exposure would encourage rapid weathering and the formation of lateritic horizons.

(3) A laterally thickening and well-stratified outermost zone inclined at higher angles than zone 2, which rapidly declines N of 51°45'N. This onlapping sequence almost certainly corresponds to post-volcanic erosional products, essentially comprising fanglomerates and gravity-induced debris flows.

Overall, the topography of the Ridge appears uniform. At 51°55'N it narrows noticeably over a short distance; this is consistent with modern analogues for fissure eruptions where intermittent cones develop at sites of preferential lava extrusion. Fig. 6d illustrates the subdued topography of the southern termination of the Ridge, where it broadens with progressively declining dip angles and only internal zones 2 and 3 are developed. The laterally unconfined lava appears to have flooded radially away from the southern end of the fissure zone, prograding over previously solidified flows and producing a shield morphology. There is insufficient evidence to suggest that the evolving Ridge propagated undirectionally with time. The unconfirmed occurrence of basal Cretaceous sediments underlying the southern part is the only tenuous control currently available.

Further evidence for Cretaceous igneous activity is recorded in the Aptian/Albian in the NE Porcupine Basin. The available biostratigraphic dating does not constrain the timing of this volcanism, although the evidence does indicate that it was episodic. Ash bands, together with occasionally coarse pyroclastics and volcaniclastics, indicate proximal activity, even though their precise association remains uncertain. Late Albian volcanism, recorded at DSDP Sites 550 and 551 off the Goban Spur, may be associated with the Seabight Igneous Centre (see Figs 2 & 10, Table 2). It is also possible that the adjacent PMVR, or one of the other igneous centres, may have contributed volcaniclastics in the NE Porcupine Basin, especially considering the coarse nature of these deposits. The Aptian–Albian date prognosed for the Seabight Igneous centre (Cook 1987) may only constitute a minimum age since there is substantial evidence for reactivation of these igneous centres. Indeed, lava flows of late Valanginian age were drilled only 30 km to the SW.

There is emerging evidence that widely scattered dates for late Cretaceous magmatism may be linked with the emplacement of the Brendan Centre. Well 35/15-1 penetrated sills of probable Coniacian age and a poorly defined igneous interval in the Campanian–Maastrichtian for which an extrusive origin is preferred. The Campanian Killala Gabbro in County Mayo is not believed to be genetically related, but signifies contemporaneous activity. This presumed episode of late Cretaceous igneous activity is coincident with a marked unconformity developed between Albian–Cenomanian strata and the late Campanian–Maastrichtian in all wells drilled in the NE Porcupine Basin. This unconformity cannot be entirely attributed to peneplanation during the Cenomanian transgression as it is believed to be associated with localized inversion. The Brendan Centre (Fig. 2) is a particularly voluminous feature whose magnetic anomaly greatly exceeds those calculated for the largest onland centres such as Mull (Bott & Tantrigoda 1987). Assuming diapiric emplacement of the centre occurred during the Turonian to early Campanian interval (probably nearer the latter), then this event may account for, or be associated with, the observed inversion in the NE part of the basin. The 70 m thick Coniacian sill may also have been emplaced in connection with this inversion event. Whilst the earliest phase of emplacement of the Brendan Centre may date back to the late Cretaceous, there is indirect evidence to suggest that activity extended into the Lower Tertiary. Submarine lava flows and volcaniclastic sediments of Danian age penetrated some 30 km SW and downdip of the Centre, and are more likely to have originated from either it or a satellite centre, and not from the nearby fissure Z. There is no evidence of post-Danian extrusions emanating directly from the centre, but it almost certainly acted as a magma source for adjoining centres such as the Slyne Fissures and adjacent plugs.

Tertiary

The Slyne Fissures are considered to form several discrete, broadly parallel alignments with both extrusive and intrusive character, as confirmed by adjacent borehole data. The available data indicate three periods of activity for this complex: Lower Palaeocene, Upper Palaeocene/lowermost Eocene and late Oligocene, with accessory evidence for activity in the Eocene. Each of these periods is supported by radiometric dates derived from connected sills, volcaniclastics dated by oil company biostratigraphy and seismic extrapolation. The Eocene and late Oligocene dates may

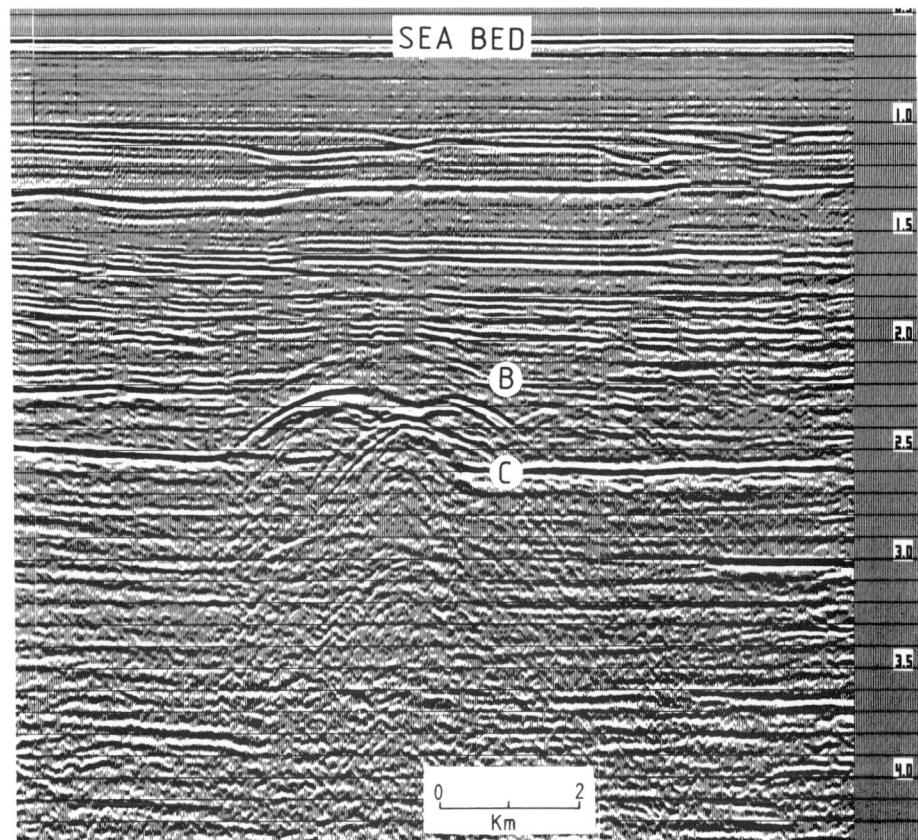

FIG. 8. Seismic reflection profile across a volcanic plug at the southern end of Slyne Fissure Y. Seismic reflectors dated as 'B' late Palaeocene/lowermost Eocene and 'C' top Chalk. Minor compactional faulting may be discerned above the plug. (Courtesy of Western Geophysical). Depth in seconds (TWT).

represent discrete episodes, but earlier activity is likely to have maintained a continuum throughout the Palaeocene and earliest Eocene.

Commencement of activity associated with the Slyne Fissures is believed to have occurred in the lowermost Palaeocene (Danian). The volcanic episode described earlier from Well 35/2-1 is interpreted as consisting of fanglomeratic, gravity-induced debris flows that developed in conjunction with extrusive activity close to Fissure W.

The late Palaeocene/lowermost Eocene date relates to a barren biostratigraphic interval of littoral sandstones, which in at least one well contains a thick sequence of volcaniclastics, underlain and overlain by inner-shelf sediments. Seismic profile correlation of the volcanic plug at the southern end of Slyne Fissure Y (Fig. 8) located within 10 km of these wells indicates that it is approximately the same age. A further nearby well records similar littoral sandstones (presumably not barren) to be within the latest Palaeocene.

In addition, a narrow volcanic plug at 52°46'N and part of Fissure Y are considered to be approximately mid-Eocene using seismic extrapolation. The only recorded volcaniclastics of this age are rounded volcanic grains recovered from a well 5 km to the S of the plug.

The approximate late Oligocene date obtained for isolated plugs (probably of intrusive origin) at the northern end of Fissure Y (Fig. 9) reinforces previously calculated dates of 26 Ma (Seemann 1984), 25–42 Ma (Horne & Macintyre 1975) and 30–36 Ma (Mitchell & Mohr 1986), which can no longer be regarded as anomalous. There is no evidence presently available to suggest that these plug intrusions were associated with extrusive events, although they do appear to have been concurrent with a major phase of widespread

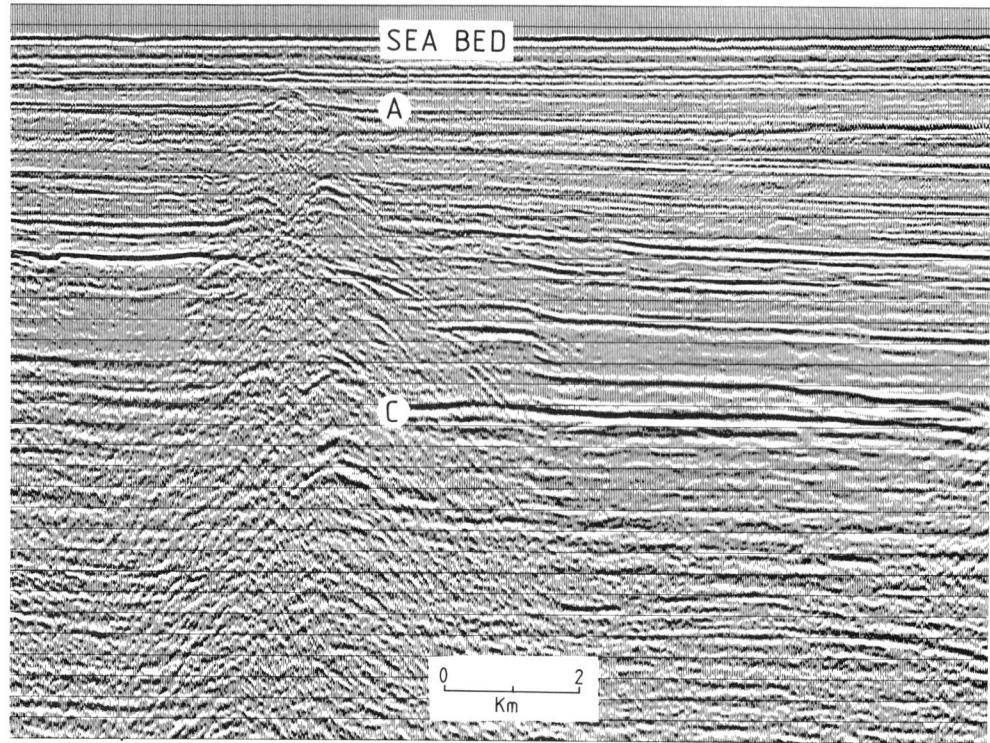

FIG. 9. Seismic reflection profile across a plug located towards the N of Slyne Fissure Y. Seismic reflector 'A' is dated as approximately late Oligocene. (Courtesy of Western Geophysical). 'C' is top Chalk.

dyke and sill emplacement, within and beyond the Slyne Fissures themselves. The top sill penetrated in 35/13-1 and dated by Seemann (1984) is clearly distinguishable on seismic data; it lies between Fissure Y and Z extending over a distance of at least 11 km. This sill is 155 m thick and three internal interfaces detected by abrupt deflections on the various logs may be indicative of multiple injection. Two similar interfaces were also detected in a thicker (174 m) Palaeocene sill in Well 35/8-1, which itself radiates out from an inferred Palaeocene plug associated with Fissure Y.

The alignment of the Slyne Fissures appears to relate to a Caledonoid structural grain in the basement, but their temporal distribution and relationship to this trend is more complex. Fissures W, X and Z prolongate towards the Slyne Trough and N of Ireland. Temporally, they appear to be restricted to the Palaeocene period. However, Fissure Y shows a broad northeastward age progression from Palaeocene to late Oligocene. This fissure system can be extrapolated via the northern margin of the Brendan Centre into the similarly trending dyke systems of Connacht, which themselves have a similar age range (ignoring argon loss), younging southwestwards towards the Brendan Centre (see Table 2). Their combined strike is greater than 250 km.

On the assumption that all the calculated onshore dates are valid then—prior to the late Palaeocene—activity occurred along all four Slyne Fissures, but was later confined to Fissure Y, the Connacht dyke group. Subsequent activity continued intermittently through to the late Oligocene.

Intrusive activity along the County Kerry dyke system may have had a similar history, but perhaps commencing in the Danian (65 $\pm$ 1.4 Ma; Mitchell & Mohr 1986) and linked to the Brendan Centre. A southward extension of this intrusive system is evidenced from seismic data to extend further S into the Celtic Sea Basin (Ziegler, pers. comm. 1986). This reiterates the earlier contention of Riddihough & Max (1976) who postulated a N–S trending zone of Tertiary intrusions on the Irish Shelf, E of the Porcupine Basin, on the basis of magnetic data.

Regional affinities

Punctuated episodes of igneous activity S and W of Ireland were broadly contemporaneous with igneous activity elsewhere in NW Europe, both during the Mesozoic (Harrison *et al.* 1975, 1977, 1979; Woodhall & Knox 1979; Harrison 1982) and in the Tertiary (Sutherland 1982). These events in turn relate to tectonism on a continental scale. However, it should be emphasized that an extended perspective is not presented here, but only relevant contemporaneous activity and occurrences W of Britain since the Jurassic (Fig. 10 & Table 2), excluding the well documented Thulean activity of onshore Scotland and northern Ireland.

Minor pyroclastic airfall within a Bajocian–Kimmeridgian interval in the upper Porcupine Basin (MacDonald *et al.* 1987) may have been broadly concurrent with the emplacement of Bathonian intrusives in the Fastnet Basin (Caston *et al.* 1981). Such activity is believed to have been linked with mid-Kimmerian uplift and erosion witnessed in both these basins and on the Goban Spur (Robinson *et al.* 1981; Cook 1987; Croker & Shannon 1987).

Two episodes of igneous activity are recorded for the Lower Cretaceous W of Britain. Valanginian–Barremian volcanism was essentially syn-rift. It was concurrent with the final phase of extensional tectonism and block faulting in the sedimentary basins W of Britain, that is identified with the late Kimmerian event. The initiation of the PMVR in the late Valanginian to early Hauterivian appears to have been contemporaneous with the extrusion of thick lava flows in the Goban Spur Basin 130 km to the S. On the northern margin of the Western Approaches Basin, emplacement of the Wolf Rock–Epsom Shoal igneous association similarly occurred in the latest Valanginian to basal Hauterivian interval, whilst sills on the SW margin of the same basin have also been dated as Valanginian (Ziegler 1982).

Widespread Aptian–Albian activity was post-rift and coeval with the onset of seafloor spreading at the adjacent Goban Spur–Biscay continental margin. Volcaniclastics of this age in the Porcupine Basin appear to follow a late Barremian quiescence. The Seabight Igneous Centre on the southeastern margin of the Seabight embayment is of inferred Aptian–Albian age and extrusive in character (Cook 1987). Further contemporaneous extrusive activity is known from DSDP Sites 550 and 551 close to the eastern margin of the Porcupine Abyssal Plain, where late Albian pillow lavas and hyaloclastites represent the incipient accretion of oceanic crust (Maury *et al.*

1984). Additional extrusive and intrusive activity in this vicinity, observed from seismic reflection data, is attributed to possible syn-rift activity or eastward excursions of an irregular ocean–continent transition (Masson *et al.* 1984). These events are in turn linked with the widespread Austrian unconformity in the mid-Aptian (Ziegler 1982).

Late Cretaceous igneous activity assumed for the Brendan Centre, and possibly for several other centres further north (see Table 2), may tentatively be connected with ongoing tectonic events in the adjacent Rockall Trough. Some workers have suggested that the late Cretaceous dates recorded can be regarded as a precursor to the main Thulean event (Harrison 1979; Woodhall & Knox 1979; Ziegler 1982). However, the Barra Volcanic Ridge System in the southern Rockall Trough is suggested to be of late Campanian to early Maastrichtian age, based on its normal magnetization relative to anomalies 33 and 32, and is probably earlier (Megson 1987; Scrutton & Bentley 1987). These are thought to have been emplaced during the early stages of the formation of the Charlie Gibbs Fracture Zone. Furthermore, immediately S of the Rockall Trough, earliest oceanic crust is considered to be of Santonian or late Campanian age. Thus, irrespective of the precise dates, igneous activity and tectonism are believed to be associated with the late Cretaceous evolution of this trough, and to be separate from Thulean magmatism and the onset of the seafloor spreading in the Greenland–Norwegian Sea.

Tertiary igneous activity in the Porcupine Basin commenced in the Danian, preceding the Thulean climax in the Thanetian, but contemporaneous with the earliest lava extrusions in Antrim, Mull and Skye (Curry *et al.* 1978). A dyke system of possible Palaeocene age in the Haig Fras Basin has recently been recognized from magnetic data (G. Day, pers. comm. 1987). The dyke system is thought to comprise five reversely magnetized dykes on the N side of the Haig Fras granite, trending NE–SW and parallel to the Haig Fras and Celtic Sea Basin margins. A probable Palaeocene volcanic plug and some sills also occur close by in the Fastnet Basin (Caston *et al.* 1981).

The late Palaeocene/lowermost Eocene volcaniclastics in the NE Porcupine Basin are underlain and overlain by nannoplankton Zones NP 9 and 10, respectively, and occur within a barren sequence. The widespread ash marker horizons of the Balder Formation (NP 10, Lower Eocene) in the North Sea have also been recognized W of the Goban Spur (Knox & Morton 1983; Knox 1984). A further ash horizon in the Upper Palaeocene (NP 9) is also recognized at Site 549

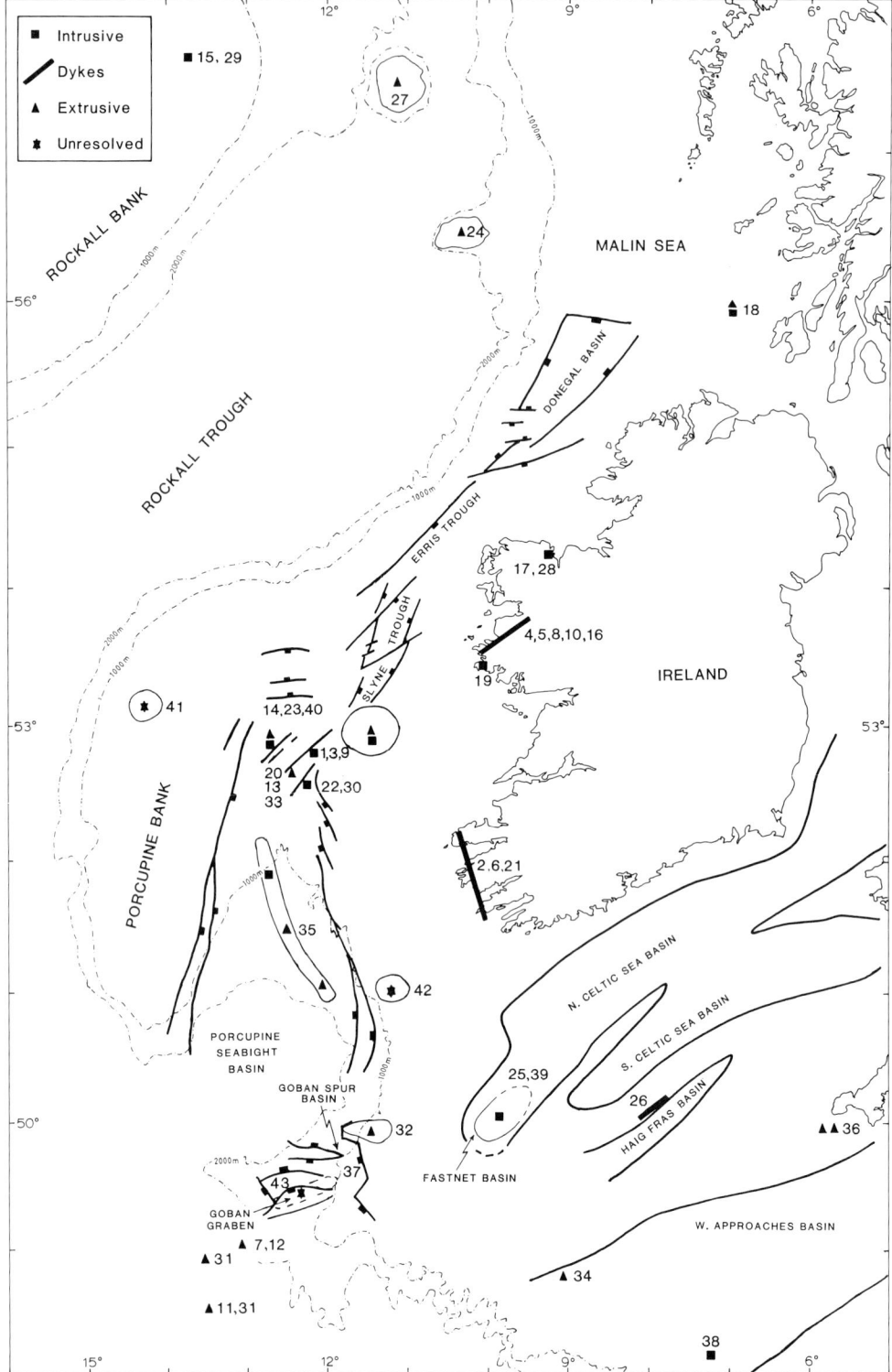

FIG. 10. Regional igneous occurrences on the continental shelf W of the British Isles from Jurassic to Tertiary revealing their relationship to the evolving sedimentary basins. Excludes the established onland British–Irish Thulean Province. Numbers refer to Table 2. Basin outline taken from Naylor & Shannon (1982) and Cook (1987).

TABLE 2. *Timing of Jurassic–Tertiary igneous activity relative to contemporaneous tectonic events for continental shelf W of the British Isles. Numbers refer to Fig. 9.*

Age ,method	Occurrence	Nature	Petrology	Tectonic event	Author
1. Late Oligocene(s)	Slyne Fissure Y, Porcupine Basin	Volcanic plugs	Unknown	Contemporaneous with re-organization of sea-floor spreading in the Arctic Atlantic	This paper
2. 25.3 Ma (K/Ar)	County Kerry	Dyke	Olivine dolerite		Horne & Macintyre 1975 Riddihough & Max 1976
3. 25.8 ± 2.6 Ma (K/Ar)	Slyne Fissure Y	Sill	Olivine dolerite		Seemann 1984
4. 33 ± 3* Ma (K/Ar)	Connacht	Dykes	Olivine dolerite		Mitchell & Mohr 1986
5. 41–42* Ma (K/Ar)	Connacht	Dykes	Olivine dolerite		Mitchell & Mohr 1986
6. 42 Ma* (K/Ar)	County Kerry	Dykes	Olivine dolerite		Horne & Macintyre 1975
7. Upper Eocene (NP 18)	Goban Spur DSDP Site 549	Ash bands	Rhyolitic		Knox 1984
8. 47–48* Ma (K/Ar)	Connacht	Dykes	Olivine dolerite		Mitchell & Mohr 1986
9. Eocene(s)	Slyne Fissure Y	Volcanic plug	Unknown		This paper
10. 52 ± 3* Ma (K/Ar)	Connacht	Dykes	Olivine dolerite	Thulean igneous event: onset of seafloor spreading in the Norwegian-Greenland Sea	Mitchell & Mohr 1986
11. Early Eocene (NP 10; Balder Fm)	Goban Spur DSDP Site 550	Ash bands	Basaltic-andesitic		Knox 1984
12. Late Palaeocene (NP 9)	Goban Spur DSDP Site 549	Ash bands	Rhyolitic		Knox 1984
13. Late Palaeocene/lowermost Eocene (NP 9 or 10)	Slyne Fissure Y and NE Porcupine Basin	Volcanic plug and volcaniclastics	Basaltic		This paper
14. Thanetian (b)	NE Porcupine Basin	Volcaniclastics	Unknown		This paper
15. 55 ± 1 Ma (K/Ar)	Rockall Island	Igneous Centre	Aegirine-granite		Harrison 1982
16. 56 ± 2* Ma (K/Ar)	Connacht	Dykes	Olivine dolerite		Mitchell & Mohr 1986
17. 58.2 ± 1.6 Ma (K/Ar)	Killala (alias Ros Gabbro) County Mayo	Dykes	Dolerite		Macintyre et al. 1975
18. 58.6 ± 1.2 Ma (K/Ar)	Blackstones Bank, Malin Sea	Igneous Centre	Ultramafic with granophyre		Macintyre, pers. comm. 1987
19. 59.85 ± 1.3 Ma (K/Ar) (also 62 Ma)	Doon Hill, Connacht	Volcanic plug	Olivine dolerite		Macintyre et al. 1975 (Thompson 1985) Phillips Petroleum Company, pers. comm. 1987
20. 60.6 ± 4.4 Ma (K/Ar)	Slyne Fissure Y	Sill	Dolerite		Mitchell & Mohr 1986
21. 65.5 ± 1.4 Ma (K/Ar)	County Kerry	Dykes	Olivine dolerite		This paper
22. Danian (b)	NE Porcupine Basin: possible source, Brendan Centre	Lava flows	Basaltic		This paper
23. Danian (Maastrichtian–Montian) (b)	NE Porcupine Basin	Volcaniclastics and Palagonite tuff	Basaltic		This paper

TABLE 2—*cont.*

Age (method)	Occurrence	Nature	Petrology	Tectonic event	Author
24. Palaeocene (K/Ar)	Hebrides Terrace	Seamount	Basalt		Omran & Whittington 1987
25. ?Palaeocene(s)	Fastnet Basin	Volcanic plug and inferred sill	Unknown		Caston et al. 1981
26. Tertiary (m)	Haig Fras Basin	Inferred dyke system	Unknown		G. A. Day, pers. comm. 1987
27. Late Maastrichtian (b)	Anton Dohrn	Seamount	Alkali basalt	Contemporaneous with tectonic activity in the Rockall Trough	Jones et al. 1974
28. 80 Ma (K/Ar) (also 60 Ma; $^{40}Ar/^{39}Ar$)	Killala (alias Ros Gabbro) County Mayo	Large dyke	Gabbro		Macintyre et al. 1975 (Thompson 1985)
29. 83 Ma (K/Ar)	Helen's Reef Rockall	Igneous Centres	Olivine dolerite		Harrison et al. 1975
30. Coniacian	NE margin of Porcupine Basin	Sills	Dolerite		This paper, Phillips Petroleum Company, pers. comm. 1987
31. Late Albian	DSDP Sites 550 & 551 Goban Spur continental margin	Pillow lavas & hyaloclastites	Basalt	Onset of seafloor spreading in the Biscay–Goban Spur rift system	Maury et al. 1984
32. Aptian/Albian (s)	Seabight Igneous Centre	Igneous Centre	Unknown		Cook 1987
33. Aptian/Albian (b)	Porcupine Basin	Volcaniclastics and ashbands	Unknown		This paper
Aptian	Southern England	Volcanigenetic Fuller's Earth	Unknown		Jeans et al. 1977
34. 110 ± 2.8 Ma (K/Ar)	Western Approaches Basin	Lava flows, tuffs and intrusives	Basaltic		Bennet et al. 1985
35. ?Valanginian–Hauterivian to Barremian	Porcupine Median Volcanic Ridge	Fissure volcano and intrusion	?Basaltic	Rifting episode affecting basins S and W of Britain: late Kimmerian event	Masson & Miles 1986a This paper
36. 131 Ma (K/Ar)	Wolf Rock–Epsom Shoal	Volcanic plugs	Phonolite		Mitchell et al. 1975 Harrison et al. 1975, 1977, 1979
37. 133* Ma (K/Ar)	Goban Spur Basin	Lava flows	Tholeiitic basalt		This paper Esso, pers. comm. 1987
38. Valanginian	Western Approaches Basin	Sills	Unknown		Ziegler 1982
Early Cretaceous (pre-Valanginian	Waddenzee Centre (alias Zuidwal plug)	Volcanic plug	Trachyte		Woodhall & Knox 1979
Berriasian-Hauterivian	Offshore E Yorkshire	Ash bands	Unknown		Lott et al. 1986

TABLE 2—*cont.*

Age (method)	Occurrence	Nature	Petrology	Tectonic event	Author
Mid-Cretaceous or earlier(?)	Barra Volcanic Ridge System	Elongate igneous bodies	Unknown	Mid-Kimmerian event	Megson 1987 Scrutton & Bentley 1987
39 Bathonian 170 ± 4* Ma (K/Ar)	Fastnet Basin	Volcanic plugs and sills	Olivine dolerite		Caston *et al.* 1981
40. Bajocian–Kimmeridgian (b)	Porcupine Basin	Pyroclastic airfall	Unknown		MacDonald *et al.* 1987
41. Unknown	Inferred Igneous Centre 'A', on the Porcupine Bank	?Igneous Centre	Unknown	Unknown	Bailey 1979 Young & Bailey 1974 This paper
42. Unknown	Inferred Igneous Centre 'B', SW of Mizen Head	?Igneous Centre	Unknown		This paper
43. Unknown	Goban Graben	Non-penetrative igneous body	Unknown		Cook 1987

*Refers to groups of dates that are averaged.
(b) dated biostratigraphically; (s) dated from seismic data; (m) magnetic data.

volcanism in the Porcupine Basin may have contributed to either top Palaeocene or Lower Eocene ash horizons. However, facies correlation of a non-barren sequence in a nearby well would indicate a late Palaeocene age for this episode.

Connacht and County Kerry dykes show a marked concentration of radiometric dates in the late Eocene interval, the same age as an ash band (NP 18) off the Goban Spur at Site 549 (Knox 1984). Conversely, the Oligocene dates obtained for plugs and sills in the Slyne Fissures have only the onshore dykes in County Kerry and Connacht as a contemporary. However, this regionally anomalous date is believed to be linked to deformations associated with the onshore prolongation of oceanic transform faults, which were active during the contemporaneous reorganization phase of seafloor spreading axes in the Labrador Sea and the Norwegian–Greenland Sea (Ziegler 1987; Faleide *et al.* 1988). In SE Greenland numerous intrusions have recently been dated as Oligocene (Noble *et al.* 1988).

Tectonic and igneous implications

The igneous activity described for the Porcupine Seabight Basin and adjacent shelf region is defined as the 'Porcupine Igneous Province', spatially and temporally distinct from the British–Irish Thulean Province. Activity was episodic and coeval with regional tectonic events in the N Atlantic and the continental shelf of NW Europe. Collectively, the Province encompasses the NNW–SSE PMVR (early Lower Cretaceous); igneous activity in the Goban Spur Basin (early Lower Cretaceous); the NE–SW Slyne Fissures (Palaeogene); the Brendan Centre (?Upper Cretaceous–Palaeogene); two inferred igneous centres, one on the Porcupine Bank and the other SW of Mizen Head (unknown age); the Seabight Igneous Centre (Aptian–Albian), together with the two dyke groups of Connacht and County Kerry (Palaeogene) that are developed as marginal onshore representatives of the Province. Magnetic anomalies of likely igneous origin observed to the N and E of the Porcupine Basin are probably spatially and temporally related and therefore also intrinsic to the Province.

Earlier workers recognized various separate elements of the Province and inferred emplacement dates. Riddihough & Max (1976) and later Mohr (1982) referred to the directional disparity between the onshore dykes in Kerry and Connacht compared to the well established Thulean Donegal–Kingscourt swarm and, on the basis of their strike, related them to the offshore Brendan Centre. However, in the absence of detailed offshore data the regional significance of the dyke systems could not be fully appreciated.

Within the Province a distinction is made between igneous activity directly linked to basin development that is essentially late syn-rift and later post-rift activity. In accordance with the early history of the Province, relationships can be proposed between the period of active faulting associated with lithospheric stretching (late Jurassic to earliest Cretaceous), late syn-rift igneous activity (late Valanginian to early Barremian) and the initiation of subsequent lithospheric stretching (earliest Barremian) and, finally, sea-floor spreading (Albian) on the continental margin W of the Goban Spur. For example, the PMVR, itself established in the ?late Valanginian to early Hauterivian, is a voluminous feature (in excess of 4000 km^3), with an elongate morphology aligned close to the rift axis and, with the exception of similar elongate igneous bodies in the Rockall Trough (Megson 1987; Scrutton & Bentley 1987), appears to be unique. It may have evolved during the sudden increase in axial volcanism that predated an attempted transition from rifting to spreading prior to an abrupt failure in the early Cretaceous.

Rifted basins commonly contain significant quantities of syn-rift volcanics. Ziegler (1982) comments that basinal volcanism is often associated with the uplift of a wide-radius rift dome located over the basin's axis as a consequence of stretching and thermal thinning of the lithosphere. This crustal thinning can be achieved either through asthenospheric upwelling or by the formation of a 'rift pillow', as with the tensile failure model. The syn-rift LKU below the PMVR probably had a similar origin. Its presence is clearly marked by a hiatus, considerable erosion and onlapping sediments. However, the extent to which the PMVR is a function of either magmatism connected to domal uplift or, as its organized axial morphology predicts, associated with a particularly advanced stage of rifting, may be resolved by calculation of β stretching values (Foucher *et al.* 1982). This must await the availability of deeper seismic reflection profiles.

Subsequent post-rift activity is described as 'opportunistic' magmatism triggered by external tectonic events, since prevailing subsidence related processes following thermal decay were insufficient to generate further magmatism. The siting of post-rift igneous activity appears to relate to established tectonic patterns. Inferred late Cretaceous activity in the Porcupine Seabight and that recorded for W of Scotland was concurrent with and probably genetically related to, further crustal extension, igneous activity and transform faulting in the adjacent Rockall

Trough, together with the initiation of the Charlie Gibbs fracture system (Megson 1987; Scrutton & Bentley 1987). Later in the Palaeogene, a continuum of magmatic activity appears to have existed from earliest Palaeocene to early Eocene, an interval that correlates with the full duration of the Thulean event and deposition of the Balder Formation. Post-early Eocene and Oligocene intrusive activity in the Basin appears to be temporally unique to this Province, as there is no record of igneous rocks of this age elsewhere in the British Isles. However, England (1988) demonstrates that NW–SE extension in the Eocene and Oligocene superceded the earlier NW–SE dextral shear (and its associated dilation) that was pervasively exploited by intrusions during the Palaeocene in W Britain. Minor extensional faulting terminating in a basal Oligocene unconformity (Croker & Shannon 1987), combined with accelerated post-Eocene subsidence (MacDonald *et al.* 1987), is indicative of a significant thermal event in the northern part of the Basin.

The spatial distribution of the igneous components has an important bearing on the structural evolution of the Basin and adjacent shelf. The N–S extension axis of the main Porcupine Basin contains two major trends (NE–SW and NNW–SSE), which in turn controlled the distribution of basin magmatism. The PMVR lies oblique to the basin's axis which is at variance with emplacement controlled extensional tectonics, albeit assuming an isotropic basement. Moreover, its NNW–SSE trend parallels the *en echelon* basement controlled eastern boundary faults of the basin. These faults may have governed the orientation of rising magma through the brittle lithosphere.

The Brendan Centre lies close to the intersection of these two basement trends, where increased fracturing of the crust along such deep seated faults may have facilitated less forceful emplacement of magma. The two sets of sheet intrusions recognized, the Slyne Fissure–Connacht group (including the Doon Hill plug) and the County Kerry system, appear to have been emplaced along these two trends with extensions towards the Brendan Centre (with the exception of Slyne Fissures W and X which bypass its margins). The intersections of two sheet intrusions at right angles about an igneous centre is unique and not observed elsewhere in the British–Irish Thulean Province (Mohr 1982). Furthermore, the Kerry and Connacht–Slyne Fissure dyke systems are isolated from the distinctive Irish and British NW–SE dyke swarm and were therefore emplaced into a separate regional stress regime. This is unrelated to a rift-parallel dilation

origin that is commonly recognized adjacent to tensile crustal regimes.

Slyne Fissure Y shows an overall younging towards the Brendan Centre. Similarly, Mitchell & Mohr (1986) recognized an apparent age gradient with younging towards the Brendan Centre (interpreted as a systematic argon loss) together with evidence for lateral injection (Mohr, pers. comm. 1987). Thus, after emplacement of the Brendan Centre, repeated lateral injection of magma into the two intersecting basement lineaments may be inferred throughout the Palaeogene. There is no evidence to indicate that extrusive activity was associated with the intrabasement Connacht dykes, although where these basement trends intercepted the extensional tectonic regime of the Porcupine Seabight Basin, minor extrusive activity also occurred.

Additional structural and igneous modifications may have been imposed by two E–W trending transcurrent faults. The trajectory of the Clare Lineament, believed to be a sinistral transform fault and a precursor to the Charlie Gibbs fracture system, is considered to prolongate ESE from the Porcupine Bank at 51°30′N across the Basin and towards the Fastnet–Celtic Sea Basin boundary (Robinson *et al.* 1981; Naylor & Shannon 1982; Megson 1987). More importantly, this proposed trajectory coincides with the southern termination of the PMVR, which is aligned at 45° to the transform. In the southern Rockall Trough two sets of linear igneous bodies, aligned at 45° to the Charlie Gibbs fracture system, are believed to agree with predictions for the orientation of igneous bodies developed abutting transform faults (Lonsdale & Shor 1979; Megson 1987). However, evidence from Jurassic faulting and intrusions in the Fastnet Basin indicate a sinistral translation for the extended Clare lineament. Assuming this remained unchanged in the Lower Cretaceous and emplacement occurred orthogonal to the direction of least compressive stress, then it is incompatible with the NNW obliquity observed for the PMVR. Again, the extension of a further E–W dextral transform fault at 53°20′N in the northern Porcupine Basin passes through inferred centre A on the Porcupine Bank (Ziegler 1982).

Evidence is available which demonstrates that certain preferred sites of magmatism are susceptible to later igneous reactivation. For example, the inferred multiple sill complex that surrounds the apex of the PMVR is constrained to be post mid-Cretaceous. Assuming a late-Cretaceous origin for the Brendan Centre, then evidence exists for subsequent reactivation in the Palaeogene, associated with the Slyne Fissures and Connacht sheet intrusions. Similar Palaeocene reactivation of an established late Cretaceous igneous centre has been proven for the Erlend Complex (Gatliff *et al.* 1984). Indeed, the existence of a late Cretaceous igneous base, overlain by further Thulean related magmatism may be extended to other igneous centres and explain the variability of derived ages.

Diapirism associated with the emplacement of these igneous centres will have locally enforced radial inversion and controlled sedimentation shortly thereafter. The Great Stone Dome in the Baltimore Canyon Basin E of New Jersey, intruded in the late Aptian–early Albian, is 5–8 km wide and upon emplacement uplifted a dome 1.6 km high and 29 km in diameter. Peneplanation of the dome took 10 Ma, and it then acted as an incompressible buttress to compacting sediments (Crutcher 1983). In the same way, but on a larger scale, the Brendan Centre (50 km in diameter) would, during the late Cretaceous and Palaeocene, have had a commanding influence on the geological evolution of the NE Porcupine Basin. Radial inversion, as already discussed, is believed to have instigated extensive erosion of Cretaceous strata, and during

the Palaeocene regression would have determined the thickness and direction of progradation of the clastic sediments. Certainly, the occurrence of volcaniclastics drilled in wells adjoining the Slyne Fissures is invariably associated with sandstones, perhaps as a consequence of either domal uplift or sedimentary aggradation from a nearby extrusive core. Furthermore, the ?mid-Tertiary uplift in Connacht (Dewey & McKerrow 1963) coincides with the strike of the Connacht dyke systems, now considered to have been intruded approximately at this time.

ACKNOWLEDGEMENTS: Our special thanks are owed to Merlin Geophysical Limited and Western Geophysical for access to an extensive grid of seismic reflection data. We are also indebted to Amerada Hess Limited, British Petroleum Development Limited, Chevron Exploration North Sea, Elf Aquitaine, Esso Exploration and Production Limited, Phillips Petroleum Company and Shell Internationale Petroleum Maatschappij B.V. for providing lithological, radiometric, geochemical data and interpretations. The funding for this project has been provided by the Natural Environment Research Council. Without the help of the above mentioned this paper would not have been possible.

References

BADLEY, M. E. 1985. *Practical Seismic Interpretation.* International Human Resources Development Corporation, Boston.

BAILEY, R. J. 1975. The geology of the Irish continental margin and some comparisons with offshore Eastern Canada. *In:* YORATH, C. J., PARKER, E. R. & GLASS, D. J. (eds) *Canada's Continental Margins and Offshore Petroleum Exploration*, Canadian Society of Petroleum Geologists, Memoir 4, Calgary, pp. 313–340.

BENNET, G., COPESTAKE, P. & HOOKER, N. P. 1985. Stratigraphy of the Britoil 72/10-1A Well, Western Approaches. *Proceedings of the Geological Association*, **96**, 255–261.

BOTT, M. H. P. & TANTRIGODA, D. A. 1987. Interpretation of the gravity and magnetic anomalies over the Mull Tertiary intrusive complex, NW Scotland. *Journal of the Geological Society of London*, **144**, 17–28.

BUCKLEY, J. S & BAILEY, R. J. 1975. A free air gravity anomaly contour map of the continental margin. *Marine and Geophysical Research*, **2**, 185–194.

CASTON, V. N. D., DEARNLEY, R., HARRISON, R. K., RUNDLE, C. C. & STYLES, M. T. 1981. Olivine-dolerite intrusions in the Fastnet Basin. *Journal of the Geological Society of London*, **138**, 31–46.

COLE, G. A. J. & CROOK, T. 1910. *On rock specimens dredged from the floor of the Atlantic off the coast of Ireland and their bearing on submarine geology.* Memoir of the Geological Survey of Ireland, Dublin.

COOK, D. R. 1987. The Goban Spur—Exploration in a deep water frontier basin. *In:* BROOKS, J. &

GLENNIE, K. W. (eds) *Petroleum Geology of NW Europe.* Graham & Trotman, London, pp. 623–632.

CROKER, P. F. & SHANNON, P. M. 1987. The evolution and hydrocarbon prospectivity of the Porcupine Basin, Offshore Ireland. *In:* BROOKS, J. & GLENNIE, K. W. (eds) *Petroleum Geology of NW Europe.* Graham & Trotman, London, pp. 633–642.

CRUTCHER, T. D. 1983. Baltimore Canyon Trough. *In:* BALLY, A. W. (ed) *Seismic Expression of Structural Styles.* American Association of Petroleum Geologists Studies in Geology, 15(2).

CURRY, D., ADAMS, C. G., BOULTER, M. C., DILLEY, F. C., EAMES, F. E., FUNNELL, B. M. & WELLS, M. K. 1978. *A Correlation of Tertiary Rocks in the British Isles.* Geological Society of London Special Report, **12**.

DEWEY, J. F. & MCKERROW, W. S. 1963. An outline of the geomorphology of Murrisk and NW Galway. *Geological Magazine*, **100**, 260–275.

DURANT, G. P., DOBSON, M. R., KOKELAAR, B. P., MACINTYRE, R. M. & REA, W. J. 1976. Preliminary report on the nature and age of the Blackstones Bank igneous complex, western Scotland. *Journal of the Geological Society of London*, **132**, 319–326.

ENGLAND, R. W. 1988. The early Tertiary stress regime in NW Britain: evidence from the patterns of volcanic activity. *In:* MORTON, A. C. & PARSON, L. M. (eds) *Early Tertiary Volcanism and the Opening of the NE Atlantic*, Geological Society of London, Special Publication, **39**, pp. 381–389.

FALEIDE, J. I., MYHRE, A. M. & ELDHOLM, O. 1988.

Early Tertiary volcanism at the western Barents Sea margin. *In:* MORTON, A. C. & PARSON, L. M. (eds) *Early Tertiary Volcanism and the Opening of the NE Atlantic,* Geological Society of London, Special Publication, **39**, pp. 135–146.

FOUCHER, J. P., LE PICHON, X. & SIBUET, J. C. 1982. The ocean–continent transition in the uniform lithospheric stretching model: role of partial melting in the mantle. *Philosophical Transactions of the Royal Society of London,* **A305**, 27–43.

GATLIFF, R. W., KITCHEN, K., RITCHIE, J. D. & SMYTHE, D. K. 1984. Internal structure of the Erlend Tertiary volcanic complex, north of Scotland, revealed by seismic reflection. *Journal of the Geological Society of London,* **141**, 555–562.

HARLAND, W. B., COX, A. V., LLEWELLYN, P. G., PICTON, C. A. G., SMITH, A. G. & WALTERS, R. 1982. *A Geological Time Scale.* Cambridge University Press, Cambridge, 128pp.

HARRISON, R. K. 1982. Mesozoic magmatism in the British Isles and adjacent areas. *In:* SUTHERLAND, D. S. (ed) *Igneous Rocks of the British Isles.* John Wiley and Sons Ltd., Chichester.

——, JEANS, C. V. & MERRIMAN, R. J. 1979. Mesozoic igneous rocks, hydrothermal mineralisation and volcanogenic sediments in Britain and adjacent regions. *Bulletin of the Geological Survey,* **70**, 57–69.

——, SNELLING, N. J., MERRIMAN, R. J., MORGAN, G. E. & GOODE, A. J. J. 1977. The Wolf Rock, Cornwall: new chemical, isotopic age and palaeomagnetic data. *Geological Magazine,* **114**, 249–263.

——, TRESHAM, A. E., SNELLING, N. J. & RUNDLE, C. C. 1975. Helens' reef: petrography, chemistry, and K-Ar age determination. *Report of the Institute of Geological Sciences,* **75/1**, 61–72.

HORNE, R. R. & MACINTYRE, R. M. 1975. Apparent age and significance of Tertiary dykes in the Dingle Peninsula, SW Ireland. *Scientific Proceedings of the Royal Dublin Society,* **5A**, 293–299.

JAUNICH, S. 1983. Tertiary intrusions on the southwest African margin. *In:* BALLY, A. W. (ed) *Seismic expression of structural styles.* American Association of Petroleum Geologists Studies in Geology **15(2)**.

JEANS, C. V., MERRIMAN, R. J. & MITCHELL, J. G. 1977. Origin of Middle Jurassic and Lower Cretaceous Fuller's Earth in England. *Clay Minerals,* **2**, 11–14.

JONES, E. J. W., RAMSAY, A. T. S., PRESTON, N. J. & SMITH, A. C. S. 1974. A Cretaceous guyot in the Rockall Trough. *Nature,* **251**, 129–131.

KIRTON, S. R. & DONATO, J. A. 1985. Some buried Tertiary dykes of Britain and surrounding waters deduced by magnetic modelling and seismic reflection methods. *Journal of the Geological Society of London,* **142**, 1047–1057.

KNOX, R. W. O'B 1984. Stratigraphic significance of volcanic ash in Palaeocene and Eocene sediments at Sites 549 and 550. *In:* DE GRACIANSKY, P. C., POAG, C. W. *et al.* (eds) *Initial Reports of the Deep Sea Drilling Project,* US Government Printing Office, Washington, **80**, pp. 845–850.

—— & MORTON, A. C. 1983. Stratigraphical distribu-

tion of early Palaeogene pyroclastic deposits in the North Sea Basin. *Proceedings of the Yorkshire Geological Society,* **44**, 355–363.

LEFORT, J. P. & MAX, M. D. 1984. Development of the Porcupine Seabight: use of magnetic data to show the direct relationship between early oceanic and continental structures. *Journal of the Geological Society of London,* **141**, 663–674.

LONSDALE, P. & SHOR, A. 1979. The oblique intersection of the Mid-Atlantic Ridge with Charlie-Gibbs transform fault. *Tectonophysics,* **54**, 195–209.

LOTT, G. K., FLETCHER, B. N. & WILKINSON, I. P. 1986. The stratigraphy of the Lower Cretaceous Speeton Clay Formation in a cored borehole off the coast of north-east England. *Proceedings of the Yorkshire Geological Society,* **46**, 39–56.

MACDONALD, H., ALLAN, P. M. & LOVELL, J. P. B. 1987. Geology of oil accumulation in Block 26/28, Porcupine Basin, offshore Ireland. *In:* BROOKS, J. & GLENNIE, K. W. (eds) *Petroleum Geology of NW Europe.* Graham & Trotman, London, 643–651.

MACINTYRE, R. M., MCMENAMIN, T. & PRESTON, J. 1975. K-Ar results from western Ireland and their bearing on the timing and siting of Thulean magnetism. *Scottish Journal of Geology,* **11**, 227–249.

MASSON, D. G. & MILES, P. R. 1986a. Structure and development of Porcupine Seabight Basin, offshore southwest Ireland. *Bulletin of the American Association of Petroleum Geologists,* **70**, 536–548.

—— & —— 1986b. Development and hydrocarbon potential of Mesozoic sedimentary basins around margins of North Atlantic. *Bulletin of the American Association of Petroleum Geologists,* **70**, 721–729.

——, MONTADERT, L., & SCRUTTON, R. A. 1984. Regional geology of the Goban Spur continental margin. *In:* DE GRACIANSKY, P. C., POAG, C. W. *et al* (eds). *Initial Reports of the Deep Sea Drilling Project,* US Government Printing Office, Washington, **80**, pp. 1115–1139.

——, MILES, P. R., MAX, M. D., SCRUTTON, R. A. & INAMDAR, D. D. 1985. A free-air gravity anomaly map of the Irish continental margin and a new gravity model across the southern Porcupine Seabight. *Geological Survey of Ireland, Report Series,* RS **85/4**.

MAURY, R. C., *et al.* 1984. Oceanic tholeiites from Leg 80 sites: geochemistry and mineralogy. *In:* DE GRACIANSKY, P. C., POAG, C. W. *et al* (eds). *Initial Report of the Deep Sea Drilling Project.* US Government Printing Office, Washington, **80**, pp. 939–948.

MAX, M. D., INAMDAR, D. D. & MCINTYRE, T. 1982. Compilation magnetic map: The Irish continental shelf and adjacent areas. *Geological Survey of Ireland, Report Series,* RS **82/2**.

MEGSON, J. B. 1987. The evolution of the Rockall Trough and implications for the Faeroe–Shetland Trough. *In:* BROOKS, J. & GLENNIE, K. W. (eds) *Petroleum Geology of NW Europe.* Graham & Trotman, London. pp. 653–665.

MITCHELL, J. G. & MOHR, P. 1986. K-Ar systematics in Tertiary dolerites from West Connacht Ireland. *Scottish Journal of Geology,* **22**, 225–240.

——, MACINTYRE, R. M. & PRINGLE, I. R. 1975. K-Ar and Rb-Sr isotopic age studies on the Wolf Rock nosean phonolite, Cornwall. *Geological Magazine*, **112**, 55–61.

MOHR, P. 1982. Tertiary dolerite intrusions of West-Central Ireland. *Proceedings of the Royal Irish Academy*, **82B**, 53–82.

MORRIS, P. 1974. A Tertiary dyke system in SW Ireland. *Proceedings of the Royal Irish Academy*, **74B**, 179–184.

NAYLOR, D. & SHANNON, P. M. 1982. *Geology of Offshore Ireland and West Britain*. Graham & Trotman, London.

NOBLE, R. H., MACINTYRE, R. M. & BROWN, P. E. 1988. Age constraints on Atlantic evolution—timing of magmatic activity along the E Greenland continental margin and of some complementary activity in NW Europe. *In:* MORTON, A. C. & PARSON, L. M. (eds) *Early Tertiary Volcanism and the Opening of the NE Atlantic,* Geological Society of London, Special Publication, **39**, pp. 201–214.

OMRAN, M. A. & WHITTINGTON, R. J. 1986. Geophysical studies of the Hebrides Terrace seamount. *Geological Society Newsletter*, **15(6)**, 43.

RIDDIHOUGH, R. P. 1968. Magnetic survey off the North Coast of Ireland. *Proceedings of the Royal Irish Academy*, **66B**, 27.

—— 1975. A magnetic map of the continental margin west of Ireland including part of the Rockall Trough and Faeroe Plateau. *Dublin Institute of Advanced Studies Geophysical Bulletin*, **34**.

—— & MAX, M. D. 1976. A Geological framework for the continental margin to the west of Ireland. *Geological Journal*, **11**, 109–120.

ROBERTS, D. G., MASSON, D. G., MONTADERT, L. & O. DE CHARPAL 1981. Continental margin from the Porcupine Seabight to the Armorician marginal basin. *In:* ILLING, L. V. & HOBSON, G. D. (eds) *Petroleum Geology of the Continental Shelf of NW Europe*. Heyden & Son, London, pp. 455–473.

ROBINSON, R. W., SHANNON, P. M. & YOUNG, D. G. G. 1981. The Fastnet Basin: An integrated analysis. *In:* ILLING, L. V. & HOBSON, G. D. (eds) *Petroleum Geology of the Continental Shelf of NW Europe*, Heyden & Son, London, pp. 444–454.

SCRUTTON, R. A. BENTLEY, P. A. D. 1987. Seismic investigations into the structure of southern Rockall Trough. *In:* BROOKS, J. & GLENNIE, K. W. (eds) *Petroleum Geology of NW Europe.* Graham & Trotman, London, pp. 667–675.

SEEMANN, U. 1984. Tertiary intrusives on the Atlantic continental margin off SW Ireland. *Irish Journal of Earth Sciences*, **6**, 229–236.

SOWERBUTTS, W. T. C. 1987. Magnetic mapping of the Butterton dyke: an example of detailed geophysical surveying. *Journal of the Geological Society of London*, **144**, 29–34.

SUTHERLAND, D. S. (ed) 1982. *Igneous Rocks of the British Isles*. John Wiley & Sons, Chichester.

SUZUKI, U. 1983. The Volcanic Mound. *In:* BALLY, A. W. (ed) *Seismic Expression of Structural Styles*. American Association of Petroleum Geologists Studies in Geology, **15(2)**.

THOMPSON, P. 1985. *Dating of the British–Tertiary Igneous Province by the Ar^{40}/Ar^{39} stepwise degassing method*. PhD thesis (unpublished), University of Liverpool.

WOODHALL, D. & KNOX, R. W. O'B. 1979. Mesozoic volcanism in the northern North Sea and adjacent areas. *Bulletin of the Geological Survey of Great Britain*, **70**, 34–56.

YOUNG, D. G. G. & BAILEY, R. J. 1974. An interpretation of some magnetic data off the west coast of Ireland. *Geological Journal*, **9**, 137–146.

ZIEGLER, P. A. 1981. Evolution of sedimentary basins in NW Europe. *In:* ILLING, L. V. & HOBSON, G. D. (eds) *Petroleum Geology of the Continental Shelf of north-west Europe*, Heyden & Son, London, pp. 3–39.

—— 1982. *Geological Atlas of Western and Central Europe*. Shell Internationale Petroleum, Maatschappij B. V.

—— 1987. Evolution of the Arctic–North Atlantic Rift System. *American Association of Petroleum Geologists Memoir*.

M. P. TATE & M. R. DOBSON, Department of Geology, University College of Wales, Aberystwyth, SY23 3DB UK.

British Tertiary Igneous Province

Time and duration of activity in the British Tertiary Igneous Province

A. E. Mussett, P. Dagley & R. R. Skelhorn

SUMMARY: The time and duration of igneous activity in the separate component areas of the British Tertiary Igneous Province (BTIP) are investigated using radiometric dates that pass internal consistency tests, used in conjunction with palaeomagnetic polarities and the available stratigraphic information.

Reliable results are available only for the currently subaerial parts of the Province. It is found that activity occurred within the approximate interval 63–52 Ma, with most activity at about 59 Ma. Later activity was predominantly of acid magmas though basic rocks preponderate in the province as a whole. The types and span of igneous activity in the separate areas followed no common pattern but areas of more complex geology tend to have had a longer span of activity.

Magnetic polarities are predominantly reversed, with all the lavas having this polarity. This predominance is probably largely, but not entirely, due to reversed polarity intervals being longer than normal ones at this time, and with much activity occurring in a single reversed interval. The sequence of polarities found within the BTIP cannot be fully reconciled with the polarity timescales of either Harland *et al.* (1982) or Berggren *et al.* (1985).

Introduction and methodology

The BTIP consists of many separate areas of igneous activity, each area displaying some or all of: (1) central intrusive complex; (2) lavas; (3) linear dyke swarms; and (4) other minor intrusions (Fig. 1). Though in any particular area the sequence of igneous activity is often well known this is not so for the duration of the igneous activity. Nor, because of the general lack of suitable sedimentary formations, can the age of the rocks be assigned by palaeontological means more precisely than to the Palaeocene (Curry *et al.* 1978). Finally, even the relative ages of the areas are not known, with the exception of a few adjacent areas, notably the Small Isles and Skye.

The obvious solution to the above problems is to use radiometric dating, but it has proved unexpectedly difficult to obtain accurate dates. Most dates have been determined using the K-Ar method and, though in the early days of its application it performed a valuable role in constraining the time of activity, it became apparent that the dates had a considerable scatter, often being incompatible with the known sequence. It used to be thought that this scatter of dates resulted from variable argon loss, and so dates could at least be regarded as minimum ages, but it is now known that dates can be spuriously high as well as low. Therefore, there is clearly a need to be able to recognize reliable dates.

The Rb-Sr isochron method, though not free

problems in its application to the BTIP, has proved to be one way to solve this problem, and has provided a number of valuable dates. However, as it can be applied only to the acid rocks it is not the whole answer, for basic rocks are preponderant in the Province and large areas or sequences of activity, notably the lavas and dyke swarms, are without acid members.

The usefulness, in the BTIP, of the K-Ar method has been retrieved by the development of the ^{40}Ar-^{39}Ar step-heating method. Where a substantial number of successive step ages have the same value (and also satisfy other criteria, see below), then the age is probably geologically meaningful, a conclusion supported by the agreement of dates with the known sequence of formation. However, only about one in three or four samples of BTIP rocks has yielded an acceptable plateau, a result consistent with the scatter of dates mentioned above.

In the following account dates are considered reliable only if they fall into one of the following categories:

(1) Rb-Sr isochron dates, where the isochron satisfies the usual tests of validity (Brooks *et al.* 1972).
(2) Ar-Ar plateau ages that satisfy the criteria of Lanphere & Dalrymple (1978) and Dalrymple *et al.* (1980).
(3) Conventional K-Ar dates where values obtained for different types of rocks or minerals are concordant.

From MORTON, A. C. & PARSON, L. M. (eds), 1988, *Early Tertiary Volcanism and the Opening of the NE Atlantic,* Geological Society Special Publication No. 39, pp. 337–348.

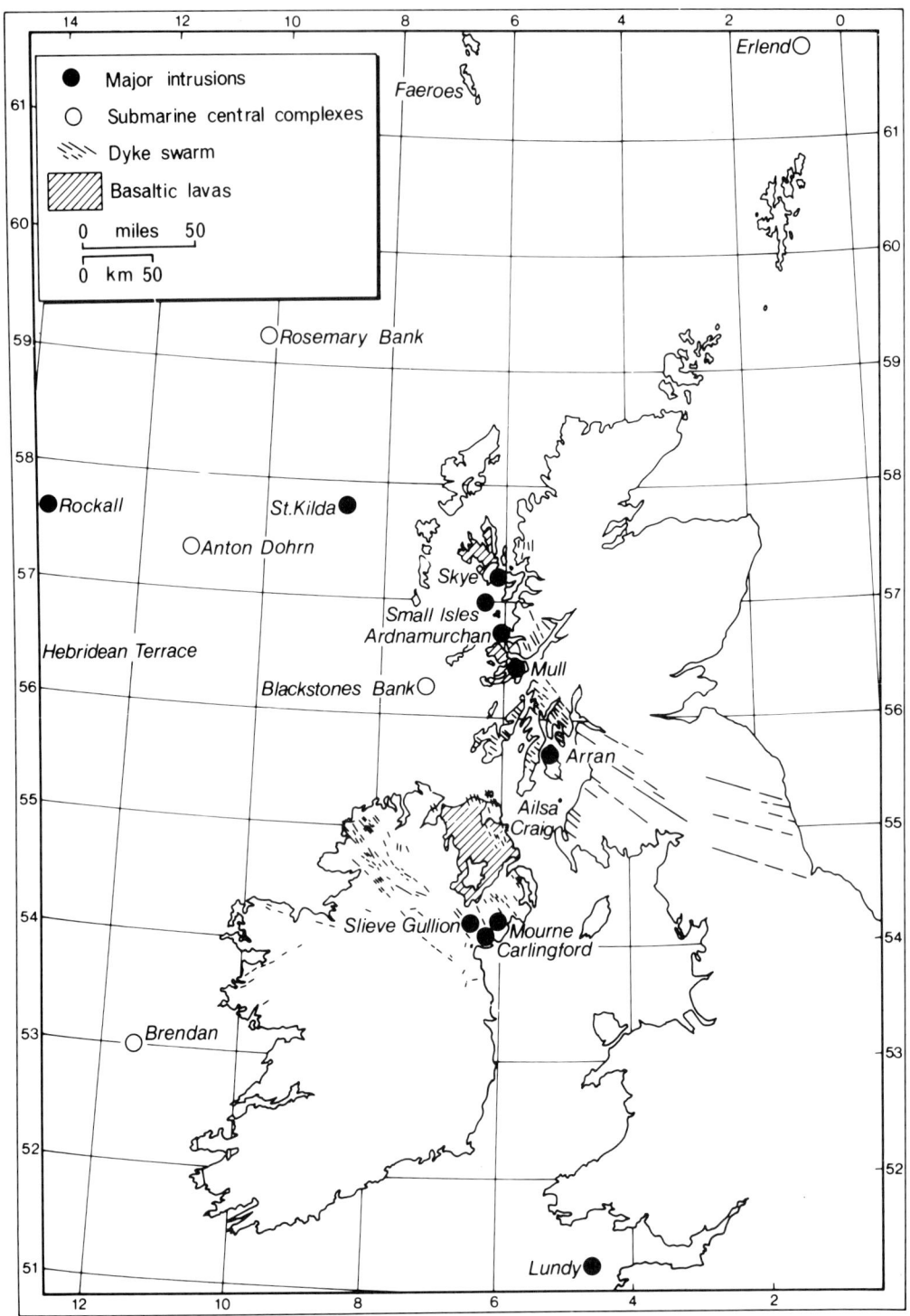

FIG. 1. The British Tertiary Igneous Province (based on Emeleus (1983), Gatliff *et al.* (1984) and Mohr (pers. comm.)).

All dates have been calculated, or recalculated, using the decay constants, etc. recommended in Steiger & Jager (1977). Errors are quoted at the 1σ level.

It is necessary to consider whether dates are for the time of crystallization or have been affected by reheating or the ubiquitous hydrothermal alteration. Results from Mull (Mussett 1986) show that a lava sampled from within the epidote zone of alteration surrounding the central intrusive complex gives the same date as comparatively fresh samples remote from the centre; from this it is concluded that the time of crystallization is being dated.

The role of palaeomagnetism

The errors of radiometric dates for the BTIP are about 1 My and, as it turns out that much of the activity in a given area can occur in this interval, the resolution of events is poor. Palaeomagnetism can help here. The direction of the Earth's magnetic field is not constant but changes (on a geological time-scale) in two ways: (1) secular variation in which the direction wanders within a cone of angle about 20° (Townsend & Hailwood 1986) with a period of a few thousand years; and (2) complete inversions which recur within about $\frac{1}{2}$–3 My for the period of interest (see Fig. 2).

Secular variation can, in principle, be used to test whether two bodies cooled at the same time, and has been employed with some success for, e.g. correlating lava sequences in Northern Ireland (Wilson 1970), but in general has proved to be of limited use because the direction within a single body often varies with position, for reasons which are not understood. In contrast, magnetic polarity has proved very useful. The frequency of inversion is about right, not so slow that a single polarity interval could include all the activity of the BTIP, not so rapid that there would be little chance of recognizing whether two bodies with the same polarity belonged to the same magnetic interval. At present the rocks of the BTIP cannot be assigned unambiguously to a particular polarity chron, because of the frequent revisions of the polarity time-scale. The polarity time-scale of Harland *et al.* (1982) was claimed to be accurate to 1 My, but the time-scale of Berggren *et al.* (1985), which has been accepted by the Committee on Geochronology as the standard time-scale for the Cainozoic, has increased the age of particular chrons by about $2\frac{1}{2}$ Ma for the period of interest, and increased their durations by up to $\frac{1}{2}$ My. Both polarity time-scales are shown in Fig. 2.

In the sections that follow each area will be considered in turn; first a summary of the succession will be given, then the available radiometric and palaeomagnetic data. A summary is given on pp. 345–346 and, particularly, in Fig. 2, but it may be stated here that none of the reliable dates falls outside the approximate interval 62–52 Ma. Our understanding of the age and duration of the BTIP has improved, and changed, considerably since a similar review in 1973 (Macintyre 1973).

A recent general account of the Scottish part of the BTIP is given in Emeleus (1983), and of the Irish part in Holland (1981).

Mull

A great outpouring of lavas covered what are now Mull and adjacent Morvern and their pre-Tertiary rocks, building up to a great thickness, of which up to 1800 m remains but which probably reached 2200 m (Walker 1971). The lavas may be divided into the earlier Plateau Group, mostly of olivine basalts, overlaid in central and SE Mull by the Central Group of tholeiites and basaltic andesites.

Intruded into the lavas is a central intrusive complex, comprising a range of rock types from basic to acid, forming intrusions concentric about three successive centres. The last major intrusion of the central complex is the Loch Ba Felsite, a ring-dyke outlining the third centre.

Basaltic dykes are extremely numerous and most belong to a NW–SE-trending linear swarm whose axis passes through the complex. They were intruded throughout the whole period of activity, some no doubt having fed the lavas, with those cutting the Loch Ba Felsite being the youngest known members of the igneous succession. Details of the succession are given in Skelhorn *et al.* (1969) and the Mull memoir (Bailey *et al.* 1924).

Lavas from near the base of the pile have been dated using the Ar-Ar method, four of which yielded plateaux which were concordant at 60 ± 0.5 Ma (Mussett 1986). Samples from the first and third of the three centres showed a difference in age, with the Loch Ba Felsite giving 56.5 ± 1.0 Ma, a result consistent with an Rb-Sr whole-rock date of 58.2 ± 1.3 Ma, obtained from centre 3 granites (Walsh *et al.* 1979). Thus activity appears to have been continuous over the interval 60–57 Ma.

All the lavas and the earliest members of centre 1 have R polarity, and the rest of the central complex have N polarity, while the dykes cutting the Loch Ba Felsite, and so post-dating the central

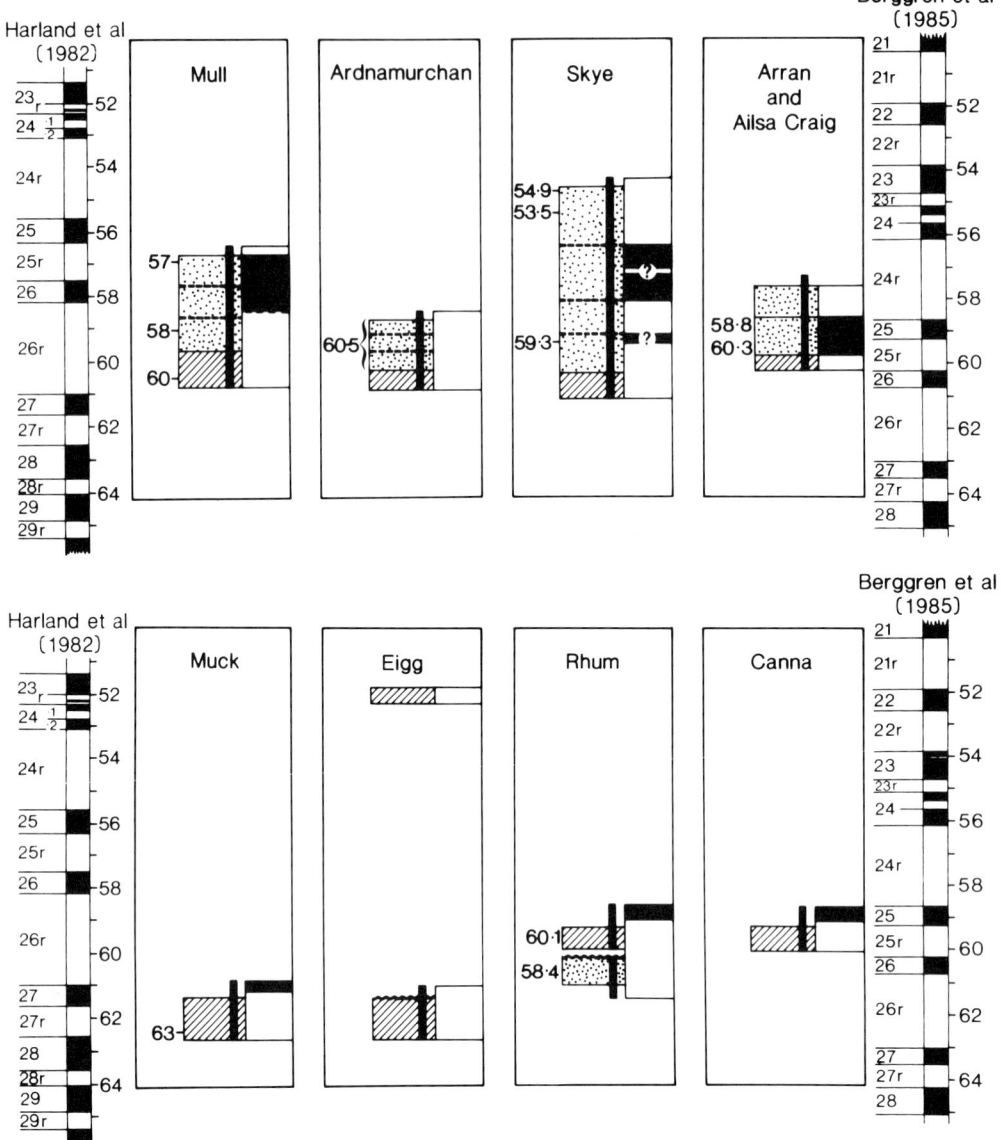

FIG. 2. Summary of results. Each area is assigned a box; within each box the central column summarizes the geological succession, while reliable dates are shown to the left and the magnetic polarities to the right. To either side of the boxes are shown the polarity time-scale according to Harland *et al.* (1982) and Berggren *et al.* (1985) (key is bottom right). See p. 345–346 for further comment.

intrusive complex, are R polarity (Mussett *et al.* 1980). However, there is evidence that the heat of later intrusions remagnetized earlier intrusions and, by observing if the polarity of a given body changes on nearing later intrusions, it has been deduced that the change of polarity occurred early in the development of the second centre (Dagley *et al.* 1987). Thus, activity commenced

within an R chron, continued through an N chron, and extended, perhaps only just, into another R chron. By referring to Fig. 2, it can be seen that the N parts of the central complex formed within about $\frac{1}{2}$ My, and the whole igneous sequence in not more than about $4\frac{1}{2}$ My, a result consistent with the dates given above.

It is believed that the closure of the rocks to

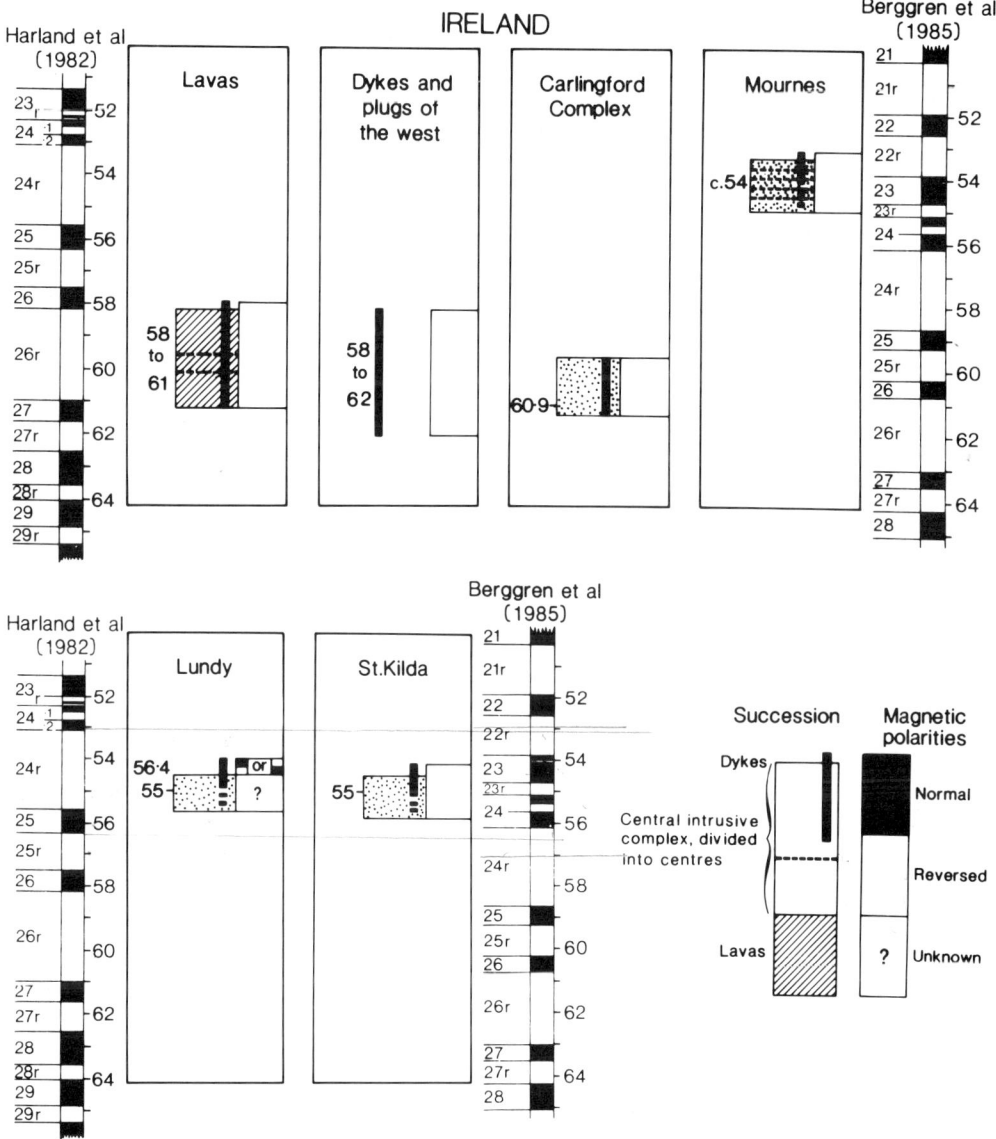

argon, and hence the event dated, did not significantly post-date emplacement of the rocks, for calculations show that the whole central complex, treated as a single intrusion, would have cooled to closure temperatures in about $\frac{1}{2}$ My at the centre and $\frac{1}{3}$ My at the edge (Mussett 1986), and this calculation neglects the great cooling which would have resulted from the large hydrothermal circulation, the existence of which is deduced from oxygen isotope ratios (Taylor & Forester 1971; Forester & Taylor 1976).

Ardnamurchan

Most of this peninsula, which lies only a few km to the N of Mull, consists of Tertiary igneous rocks, emplaced into Moine Schists and a thin sequence of Mesozoic sediments. The oldest of the Tertiary igneous rocks are basalt lavas which probably formed part of the Mull and Morvern succession, but now are seen only in comparatively small areas. The strongly arcuate intrusions of the central complex belong to three centres,

Details of the succession are given in Richey & Thomas (1930) and Gribble *et al.* (1976).

Mitchell & Reen (1973) applied the conventional K-Ar method to samples from all three centres, in several instances dating different specimens from a single body, or acid and basic parts of a composite intrusion, or different minerals from a single sample. Though the dates do not agree completely within the analytical error, their spread is only 6 My, with the mean age for each centre indistinguishable at about 60.5 ± 2.0 Ma, agreeing with an Rb-Sr date of 60.0 ± 1.7 Ma obtained from rocks of centre 3 (Beckinsale & Walsh, pers. comm.).

All the succession, lavas and central complex, appears to be R (Dagley *et al* 1984), so that the radiometric and palaeomagnetic evidence together show that activity probably began at about the same time as in nearby Mull, but was less prolonged.

Skye

This is an area that roughly matches Mull in size and complexity of igneous activity. Much of the island is covered with lavas (basalts, mugearites and trachytes) that can be divided into five series, and reach a thickness of up to 600 m (Anderson & Dunham 1966). As in Mull, a central intrusive complex was emplaced into the lavas. It consists of four centres: the Cuillins, mostly of basic and ultrabasic rocks, are followed in turn by the Strath na Creitheach, Western and Eastern Red Hills centres, in each of which granites predominate. Also, as in Mull, dykes, mostly belonging to a NW–SE trending swarm, were intruded throughout the succession. Detailed accounts of the succession can be found in the Skye memoir (Harker 1904), Bell (1976), Emeleus (1983) and Bell & Harris (1986).

The lavas have not yet been reliably dated. An Rb-Sr feldspar date of 59.3 ± 0.7 Ma has been obtained (Dickin 1981) for the Coire Uaigneich granite, one of the last intrusions of the Cuillins centre, while the age of the Eastern Red Hills centre has been established by an Rb-Sr whole-rock/feldspar date of 53.5 ± 0.4 Ma for the Beinn an Dubhaich epigranite (Dickin 1981) and an Ar-Ar date of 54.9 ± 0.6 Ma (Mussett, unpublished work) for a composite sill, one of the last intrusions of this centre and hence of the central complex as a whole. These dates indicate that activity spanned about 6 Ma but it is not known if activity was continuous. An Rb-Sr whole-rock date of 58.7 ± 0.9 Ma (Dickin 1981), but of poor reliability, for the Loch Ainort epigranite, em-

placed about half-way through the development of the Western Red Hills centre, suggests a gap between the formation of the Western and Eastern Red Hills centres.

The lavas all have R polarity (Wilson *et al.* 1972; unpublished work of the authors). Of the centres, the first one, the Cuillins, is also R except for the Coire Uaigneach granophyre and other late units, while the succeeding Strath na Creitheach centre is R. The Western Red Hills centre is N except for the Loch Ainort epigranite which is R, and the last centre, the Eastern Red Hills, is entirely R (Dagley & Mussett 1980). Thus activity began and ended in a period of R polarity, but it is not clear how many polarity intervals came between, though it is evident that there must have been more than the single N found in Mull and probably Arran.

The Small Isles

These comprise the four islands of Rhum, Canna, Eigg and Muck. In Rhum, Tertiary igneous rocks occupy the southern half of the island and consist of a central intrusive complex, lavas laid down on the eroded surface of the complex and many dykes (Emeleus & Forster 1979; Emeleus 1980). In having lavas that post-date the central complex, Rhum differs from the other central complexes. In the remaining three islands activity mainly consisted of basalt lavas, which still mostly cover them, and dykes, numerous in Eigg and Muck. At the S of Eigg is the prominent Sgurr, a ridge formed of a massive pitchstone lava which filled a valley cut in the basalt lavas.

The relative ages of the igneous activity of the Small Isles and Skye are known. The Eigg lavas are cut by dykes of the Rhum swarm, which in turn are cut and metamorphosed by the Rhum central complex. The Rhum central complex was followed, after uplift and erosion, by lavas. Within the Rhum lavas are found conglomerates which include clasts of rock of the centre (Emeleus 1985), and clasts thought to derive from the Western granophyre of Rhum are found in conglomerates in the Canna lavas (Emeleus 1973) and near the base of the Skye lava pile (Meighan *et al.* 1981). Thus, the basic lavas of Rhum, Canna and Skye are at least partly contemporaneous. This reveals a long sequence: (1) Eigg basic lavas; (2) dykes of the Rhum swarm; (3) Rhum centre; (4) erosion interval; (5) lavas of Rhum, Canna and Skye; and (6) Skye central complex comprising four centres. The basaltic lavas of Muck are probably contemporaneous with those of Eigg. The timing of this sequence is fairly well constrained by dates. Lavas from near the base of the piles of Muck and Eigg have given Ar-Ar dates of 63.0 ± 3.4 Ma and 63.3 ± 1.8 Ma respectively (Dagley & Mussett 1986). These are the oldest reliable dates yet found in the BTIP. The Western granophyre of Rhum yielded 59.8 ± 0.4 Ma and a Rhum lava 61.4 ± 0.4 Ma, indicating that the erosion interval was short (Mussett 1984). The Cuillin Centre of Skye followed at about 59 Ma and activity on Skye ended at about 54 Ma (see previous section).

The Sgurr of Eigg pitchstone has given an Rb-Sr date of 52.1 ± 0.5 Ma (Dickin & Jones 1983), the youngest reliable date yet obtained in the BTIP. Thus, Eigg has both the oldest and youngest rocks of the BTIP, with presumably a long inactive interval between.

All the rocks are magnetized R except for a small proportion of dykes cutting the lavas of all four islands (Dagley & Mussett 1981, 1986), which adds to the evidence that activity did not span a long time (the Sgurr excepted) (see Fig. 2).

quartz porphyry sills and dykes (mostly part of composite intrusions) sampled in the W and S of Arran are N (Mussett *et al.* 1987). If this is true of all the quartz porphyries it is the only example known in the BTIP of a correlation of polarity with petrology, and the quartz porphyries may provide a useful stratigraphic marker.

An Rb-Sr whole-rock isochron date of 60.3 ± 0.8 Ma for the northern granite was obtained by Dickin *et al.* (1981), confirming an Ar-Ar date of 60.3 ± 0.6 Ma obtained by Evans *et al.* (1973) but for which no acceptance criteria were used. An Ar-Ar date of 58.5 ± 0.8 Ma has been determined for one of the quartz porphyry dykes (Mussett *et al.* 1987).

Palaeomagnetic and radiometric data are summarized in Fig. 2. In the absence of firm evidence to the contrary, a polarity sequence of R–N–R is suggested for Arran, with the northern granite and the quartz porphyry intrusions in the same N period.

Arran and Ailsa Craig

The igneous succession of Arran is poorly known because there is a lack of significant cross-cutting relationships between the component bodies, many of which are sills intruded into pre-Tertiary rocks, cut by or cutting only Tertiary dykes of unknown age. The small island of Ailsa Craig, to the S of Arran and consisting of a mass of riebeckite microgranite cut by a few basalt dykes, is usually coupled with Arran. Attempts to deduce a succession using petrographic criteria are hindered by recurrence of magma types. All that is firmly known is that the northern granite preceded the small Central Complex. Lavas are found only as remanié masses within the Central Complex and are presumed to have occurred early in the succession. Further details may be found in MacDonald & Herriott (1983), Tomkieiff (1969) and the Arran memoir (Tyrrell 1928).

As there are few radiometric dates the palaeomagnetic polarities will be discussed first. Both N and R polarities are found, and the cross-cutting relationships within a group of dykes on the S shore show that igneous activity must have spanned at least the sequence R–N–R. Though there is no firm evidence to exclude additional polarity intervals, all the known relationships can be tentatively accommodated within these three polarity intervals (Hodgson 1981). As the northern granite is N, the later and reversely magnetized Central Complex is assigned to the second R interval. An interesting result is that all the

Ireland

Activity in Ireland differed somewhat from that in Scotland, for neither the very extensive Antrim lavas nor the dykes of the NW are associated with a central complex (though dykes of the W and SW may be related to a centre W of Ireland, see Fig. 1). The Tertiary igneous rocks of Ireland are described in Emeleus & Preston (1969) and Holland (1981).

There are very few Rb-Sr dates for Ireland and the Ar-Ar method has been less successful than for Scottish rocks, with few age spectra yielding a fully satisfactory plateau; for these analyses an attempt has been made to identify the mechanism causing the poor spectrum (such as diffusive argon loss or inherited argon) and so assign a minimum or maximum age to the sample. In some areas from which several samples have been analyzed it has been possible to bracket the activity to within a few My; that the maximum and minimum age estimates do not conflict gives one confidence in these interpretations of the spectra.

The Antrim lava succession is divided into three. The oldest rocks are the Lower Basalts, of olivine type, resting on Cretaceous chalk. Then followed the Interbasaltic Formation, mostly representing a long quiescent interval during which the lavas were deeply weathered to laterite, but which also includes the Causeway tholeiitic Basalts (formerly termed Middle Basalts) of the N (to which the Giant's Causeway belongs) and a small occurrence of acid volcanics at Tardree

Youngest are the Upper Basalts, again of olivine type, along with some doleritic plugs.

Ar-Ar dates indicate that the whole lava succession was formed in the approximate interval 61–58 Ma (Thompson 1986). This conflicts with fission track dates of 65.5 ± 3.6 and 64.6 ± 5.0 Ma (Fitch & Hurford 1977) for the Tardree Rhyolite, which belongs to the Interbasaltic Formation. However, a new fission track date for the Tardree Rhyolite and an Ar-Ar date for the nearby Sandy Braes porphyritic obsidian are concordant at 59 Ma (Thompson *et al.* 1984), suggesting an unappreciated error in the earlier fission track date.

Dykes are common in Ireland, particularly in the northern part. In the NW, in parts of Donegal and around Killala Bay, many trend NW–SE. In the mid-west the few but large and persistent dykes trend roughly ENE–WSW, while a N–S trending dyke system (the 'Dingle Dyke') cuts the tips of the peninsulas of the SW; it has been suggested that these dykes relate to a submerged centre off the W of Ireland (Riddihough & Max 1976) (see Fig. 1).

In the Killala Bay area the age of the dykes and of the elongated Killala Gabbro has been constrained to about 58 Ma, while a sample of the Blind Rock Dyke of Donegal yielded a good plateau with age 61.7 ± 0.5 Ma (Thompson 1986). Analysis of samples from two localities of the Dingle Dyke did not give a reliable plateau but indicate an age around 59 Ma, and give no support to the published conventional K-Ar dates of 26 and 43 Ma (Horne & Macintyre 1975) (see Fig. 1).

The Droimchogaidh Sill, Connacht, is composed of gabbro but with picritic and syenitic differentiates. Mohr *et al.* (1984) published an Ar-Ar date of 55 ± 1 Ma for a syenite sample, but work on further samples, some of different composition, indicate that this is a minimum age, possibly because of diffusive argon loss, and that the time of emplacement lies between 58 and 61 Ma.

In north-eastern Ireland are three central complexes, which are the adjacent Slieve Gullion and Carlingford Complexes (which may be one at depth), and to their NE the Mourne Mountains which sweep down to the sea. The sequence of intrusions in the Carlingford Complex is not well known but the succession may be divided into: (1) basalt lavas; (2) intrusions preceding the layered gabbros; (3) layered gabbros; (4) vents and intrusions; (5) Carlingford granophyre; and (6) intrusions post-dating the Carlingford granophyre (Le Bas 1960, 1967; Halsall 1974). A sample of one of the early gabbros, preceding the layered gabbros, has given an Ar-Ar date of 58.7 ±

1.2 Ma, while the granophyre has given 60.9 ± 0.5 Ma (Thompson 1986). The older rock appears to give the younger date. However, 60.9 Ma is preferred because the granophyre sample is fresher than the gabbro and gives a longer plateau, and it is thought the gabbro analysis may have been affected by recoil losses.

The Slieve Gullion central intrusive complex has been dated by the Rb-Sr isochron method as 57.6 ± 0.5 Ma on the sheeted complex (Meighan *et al.* 1988), while O'Connor (1988) has dated porphyritic felsite and granophyre of the Ring Complex, the earliest of the major intrusions, at 58.5 ± 1.2 Ma (MSWD 3.5) and 54.8 ± 1.4 Ma (MSWD 6.0) respectively. Though the MSWDs show the presence of geological scatter these data support an age similar to that of the bulk of the BTIP.

The Mourne Mountains consist principally of five granites, numbered G1-5 in order of emplacement. Ar-Ar dates have been published as 59.6 ± 1.6 Ma for G2 and 59.5 ± 1.6 Ma for G5 (Evans *et al.* 1973). However, new Ar-Ar determinations for G2 and G4 have given substantially lower ages, 54.9 ± 0.6 and 53.3 ± 0.6 Ma respectively (Thompson *et al.* 1987), a value supported by Rb-Sr dates of 56.3–55.7 Ma for granites G1–4 and 51.5 ± 0.9 Ma for G5 (Gibson *et al.* 1987). The reason for the discrepancy between the earlier and later dates is not understood.

The Irish Tertiary igneous rocks have been extensively sampled for palaeomagnetic polarity determinations. The great majority are R, which includes the entire lava sequence, the dykes of the NW and SE, rocks of Carlingford Complex and most of the rocks in the Mournes area; the only known N rocks are an acid vein or dyke in G5 and a dyke of the Mournes swarm.

These results suggest that the bulk of activity occurred within a single R interval, none of which exceeded about 3 My at this time, with only the Mourne Mountain granite being significantly younger.

Lundy

This is the most southerly known area of the BTIP. It consists of a granite, bordered on the SE by Devonian slates, and cut by numerous dykes, mostly basalt.

The granite has been dated by the Ar-Ar method as 54.8 ± 1.4 Ma (Fitch *et al.* 1969), while an Rb-Sr date of 57 ± 2 Ma, of unstated reliability, has been published (I.G.S. 1981). One of the dykes has given an Ar-Ar date of 56.4 ± 0.3 Ma (Mussett, unpublished work), suggesting that at least some of the dykes were intruded soon after the granite.

The granite has been sampled for palaeomagnetic measurement but was not stable enough to yield a polarity. The dykes are both R and N, though predominantly the former (Mussett *et al.* 1976), showing that their intrusion at least straddled a polarity inversion.

St. Kilda

This group of islands far out in the Atlantic is the eroded remains of a central intrusive complex. An Rb-Sr date of 55 ± 0.5 Ma has been obtained for the Conachair granite, the last of the intrusions (Brook 1984). Most of the units on the main island were sampled for palaeomagnetic measurements and all were found to be R (Morgan 1984).

Other areas

There are few dates, and none which meet the reliability criteria of p. 337, available for the remaining areas shown in Fig. 1, most of which are submerged and many of which are assigned to the BTIP by analogy with those already discussed.

The island of Rockall is granitic and rises from the extensive Rockall Plateau which separated from the British Isles in pre-Tertiary times, leaving the Rockall Trough in its place. Rb-Sr whole-rock dates of 57 ± 7, 54 ± 4 and 54 ± 4 Ma (Hawkes *et al.* 1975) show that the granite is Tertiary and suggest it is probably one of the younger parts of the BTIP. Palaeomagnetic measurements were inconclusive but suggest that the polarity is R (Lewis *et al.* 1975).

Helen's Reef is microgabbro and lies 3 km E of Rockall. K-Ar whole-rock dates ranged from 79 ± 3 to 114 ± 3 Ma (Harrison *et al.* 1975), illustrating the scatter of dates often resulting from the conventional K-Ar method. Because of this large scatter, the proximity to Rockall, and evidence that the K-Ar method can give dates that are high as well as low, the apparent pre-Tertiary age must be treated with reservation. Jones *et al.* (1972) reported K-Ar dates from 45–56 Ma for basic rocks dredged 14 km E of Rockall.

Mitchell *et al.* (1976) obtained conventional K-Ar dates for eight basaltic samples dredged from Blackstones Bank; these ranged from 46–58 Ma but Mitchell *et al.* concluded that the best estimate for the age of the rocks is 58.6 ± 0.9 Ma, the mean of the four highest and most concordant dates. A conventional K-Ar date of 72 Ma for a single dyke sample (Durant *et al.* 1976) should be treated with caution. There are no palaeomagnetic polarities for any of these areas.

Summary and conclusions

The igneous successions, together with available reliable dates and palaeomagnetic polarities, are summarized in Fig. 2. Caution is needed before drawing conclusions from the figure, and the reader should bear in mind the following points:

(1) The dates have analytical errors (1σ) ranging roughly from $\frac{1}{2}$–2 My.
(2) The sequence of igneous activity in a given area may not always be perfectly known, and the point in the succession to which a date applies can only be indicated approximately on the figure, because of the limitations of the scale.
(3) Frequent revisions of the polarity time-scale suggest that there may be errors of 1–2 My in the time of a particular chron, though changes in the durations of chrons are likely to be considerably smaller. The time-scales of Harland *et al.* (1982) and Berggren *et al.* (1985) are both shown.

With those reservations in mind the following conclusions may be drawn:

(1) Just as the areas of the BTIP show considerable diversity in their succession, at least at the present level of erosion, so too does the duration and time of activity differ. In Mull, activity was continual for abut 4 My; in Eigg activity was early and late; in Ireland nearly all the extrusive activity was completed within a single reversed polarity interval, so having a maximum duration of about 3 My; in Skye activity spanned about 6 My but whether there was a considerable gap is not yet clear.
(2) All the lavas (including the Sgurr of Eigg pitchstone) were formed when the magnetic field was reversed, most of them probably during the single chron 26r if the polarity time-scale of Harland *et al.* (1982) is accepted, but in chrons 25r and/or 24r according to the time-scale of Berggren *et al.* (1985).
(3) Activity spanned the approximate interval 63–52 Ma, but most activity took place at about 59 Ma.
(4) There is no discernible geographical pattern in the sequence of igneous activity.
(5) The more complex areas took longer to form.

Mull and Arran both span a polarity sequence of at least R–N–R, and Skye probably more, while the available dates show that in Mull and Skye activity spanned about 4 and 6 My respectively. In contrast, all the other areas probably formed in a single or at most two polarity intervals and the duration of activity has not been detected radiometrically.

(6) The younger dates of the BTIP are all associated with acid bodies, e.g. the Western Red Hills of Skye, Lundy, St. Kilda, Sgurr of Eigg, Mourne Mountain granites, Rockall (?), though not all acid bodies belong to this younger phase of activity.

(7) The various bodies cannot be assigned unambiguously to the polarity time-scales of either Harland *et al.* (1982) or Berggren *et al.* (1985) according to the available dates, nor even accommodated without conflict. How-

ever, the conflict is considerably less for the former time-scale.

(8) Reversed magnetic polarity predominates in the BTIP, exemplified by the dyke swarms which generally spanned the activity of an area. The proportion of normally magnetized dykes ranges from 0–25%, depending on the area.

This preponderance of reversed polarity can be partly attributed to reversed periods being longer, on average, than normal ones at the relevant time, and perhaps also because much activity chanced to occur within a single reversed period. Even so, it seems an element of chance has to be invoked to explain why the early dykes of Eigg and Muck, most of the dykes of Lundy and the Sgurr of Eigg are also preponderantly reversely magnetized.

References

ANDERSON, F. W. & DUNHAM, K. C. 1966. *The Geology of Northern Skye*. Memoir of the Geological Survey of Scotland. HMSO, Edinburgh.

BAILEY, E., CLOUGH, C. T., WRIGHT, W. B., RICHEY, J. E. & WILSON, G. V. 1924. *Tertiary and Pre-Tertiary Geology of Mull, Loch Aline and Oban*. Memoir of the Geological Survey of Scotland. HMSO, Edinburgh.

BELL, J. D. 1976. The Tertiary intrusive complex on the Isle of Skye. *Proceedings of the Geologists' Association*, **87**, 247–271.

BELL, B. R. & HARRIS, J. W. 1986. *An Excursion Guide to the Geology of the Isle of Skye*. Geological Society of Glasgow, Glasgow.

BERGGREN, W. A., KENT, D. V., FLYNN, J. J. & VAN COUVERING, J. A. 1985. Cenozoic Geochronology. *Bulletin of the Geological Society of America*, **96**, 1407–1418.

BROOK, M. 1984. The age of the Conochair Granite. *In:* HARDING *et al.* (1984) St. Kilda: an illustrated account of the geology. *Report of the British Geological Survey*, **(16)**7, 40–41.

BROOKS, C., HART, S. R. & WENDT, I. 1972. Realistic use of two-error regression treatments as applied to rubidium-strontium data. *Reviews of Geophysics and Space Science*, **10**, 551–77.

CURRY, D., ADAMS, C. G., BOULTER, M. C., DILLY, F. C., EAMES, F. E., FUNNELL, B. M. & WELLS, M. K. 1978. *A correlation of Tertiary rocks in the British Isles*. Geological Society of London, Special Report, **12**, 72 pp.

DAGLEY, P. & MUSSETT, A. E. 1980. Palaeomagnetism of the Tertiary intrusive complex on the Isle of Skye. *Geophysical Journal of the Royal Astronomical Society*, **61**, 210 (abst.).

—— & —— 1981. Palaeomagnetism of the British Tertiary Igneous Province: Rhum and Canna.

Geophysical Journal of the Royal Astronomical Society, **61**, 475–491.

—— & —— 1986. Palaeomagnetism and radiometric dating of the British Tertiary Igneous Province: Muck and Eigg. *Geophysical Journal of the Royal Astronomical Society*, **85**, 221–242.

——, —— & SKELHORN, R. R. 1984. The palaeomagnetism of the Tertiary igneous complex of Ardnamurchan. *Geophysical Journal of the Royal Astronomical Society*, **79**, 911–922.

——, —— & —— 1987 Polarity stratigraphy and duration of the Mull Tertiary igneous activity. *Journal of the Geological Society of London*, **144**, 985–996.

DALRYMPLE, G. B., LANPHERE, M. A. & CLAGUE, D. A. 1980. Conventional and $^{40}Ar/^{39}Ar$ K–Ar ages of volcanic rocks from Ojin (site 430), Nintuku (site 432), and Suiko (site 433) Seamounts and the chronology of the volcanic propagation along the Hawaiian–Emperor Chain. *In:* JACKSON, E. D., KOISUM, I. I. *et al. Initial Reports of the Deep Sea Drilling Project*. US Government Printing Office, Washington, **55**, 659–676.

DICKIN, A. P. 1981. Isotope geochemistry of Tertiary Igneous Rocks from the Isle of Skye. *Journal of Petrology*, **22**, 155–189.

—— & JONES, N. W. 1983. Isotopic evidence for the age and origin of pitchstones and felsites, Isle of Eigg, NW Scotland. *Journal of the Geological Society of London*, **180**, 691–700.

——, MOORBATH, S. & WELKS, H. J. 1981. Isotope, trace element and major element geochemistry of Tertiary igneous rocks, Isle of Arran, Scotland. *Transactions of the Royal Society of Edinburgh*, **72**, 159–170.

DURANT, G. P., DODSON, M. R., KOKELAAR, B. P., MACINTYRE, R. M. & REA, W. J. 1976. Preliminary report on the nature and age of the Blackstones Bank Igneous Centre, western Scotland. *Journal of the Geological Society of London*, **132**, 319–326.

EMELEUS, C. H. 1973. Granophyre pebbles in Tertiary conglomerate on the Isle of Canna, Inverness-shire. *Scottish Journal of Geology*, **9**, 157–159.

—— 1980. *Rhum: Solid Geology Map, (1:20 000 scale)*. Nature Conservancy Council, Scotland.

—— 1983. Tertiary igneous activity. *In*: CRAIG, G. Y. (ed) *Geology of Scotland*. Scottish Academic Press, Edinburgh.

—— 1985. The Tertiary lavas and sediments of north-west Rhum, Inner Hebrides. *Geological Magazine*, **122**, 418–437.

—— & PRESTON. 1969. *The Tertiary Volcanic Rocks of Ireland; Field excursion guide*. IAVCEI symposium, Oxford.

—— & FORSTER, R. M. 1979. *Tertiary Igneous Rocks of Rhum, Inner Hebrides; Field guide*. Nature Conservancy Council, Geology and Physiography Section, Newbury.

EVANS, A. L., FITCH, F. J. & MILLER, J. A. 1973. Potassium-argon age determinations on some British Tertiary Igneous rocks. *Journal of the Geological Society of London*, **129**, 419–443.

FITCH, F. J. & HURFORD, A. J. 1977. Fission track dating of the Tardree Rhyolite, Co. Antrim. *Proceedings of the Geologists' Association*, **88**, 267–274.

——, MILLER, J. A. & MITCHELL, J. G. 1969. A new approach to radiometric dating in orogenic belts. *In*: KENT, P. E., SATTERTHWAITE, G. E. & SPENCER, A. M. (eds) *Time and Place in Orogeny*. Geological Society of London, Special Publication, **3**, 157–195.

FORESTER, R. W. & TAYLOR, H. P. 1976. ^{18}O-depleted igneous rocks from the Tertiary Complex of the Isle of Mull, Scotland. *Earth and Planetary Science Letters*, **32**, 11–17.

GATLIFF, R. W., HITCHEN, K., RITCHIE, J. D. & SMYTHE, D. K. 1984. Internal structure of the Erlend Tertiary volcanic complex, north of Shetland, revealed by seismic reflection. *Journal of the Geological Society of London*, **141**, 555–562.

GIBSON, D., McCORMICK, A. G., MEIGHAN, I. G. & HALLIDAY, A. N. 1987. The British Tertiary Igneous Province: Rb-Sr ages for the Mourne Mountain granites. *Scottish Journal of Geology*, **23**, 221–225.

GRIBBLE, C. D., DURRANCE, E. M. & WALSH, J. N. 1976. *Ardnamurchan, a guide to geological excursions*. Edinburgh Geological Society, Edinburgh.

HALSALL, T. J. 1974. *The minor intrusions and structure of the Carlingford Complex, Eire*. PhD thesis (unpublished). University of Leicester.

HARDING, R. R., MERRIMAN, R. J. & NANCARROW, P. H. A. 1984. St. Kilda: an illustrated account of the geology. *Report of the British Geological Survey*, **16(7)**, 40–41.

HARKER, A. 1904. *The Tertiary igneous rocks of Skye*. Memoir of the Geological Survey of Great Britain. HMSO, Edinburgh.

HARLAND, W. B., COX, A. V., LLEWELLYN, P. G., PICKTON, C. A. G., SMITH, A. G. & WALTERS, R.

1982. *A Geological Time-scale*. Cambridge University Press. 128 pp.

HARRISON, R. K., TRESHAM, A. E., SNELLING, N. J. & RUNDLE. 1975. Helen's Reef: petrography, chemistry and K-Ar age determination. *In*: HARRISON, R. K. (ed) 1975. Expeditions to Rockall 1971–72. *Report of the Institute of Geological Sciences*, **75/1**, 61–72.

HAWKES, J. R., MERRIMAN, R. J., HARDING, R. R. & DARBYSHIRE, D. P. F. 1975. Rockall Island: new geological, petrological, chemical and Rb-Sr age data. *In*: HARRISON, R. K. (ed) 1975. Expeditions to Rockall 1971–72, *Report of the Institute of Geological Sciences*, **75/1**, 11–51.

HODGSON, B. D. 1981. *Magnetostratigraphy of the Tertiary igneous rocks of Arran, Scotland*. PhD thesis (unpublished). University of Liverpool.

HOLLAND, C. H. 1981. *A Geology of Ireland*. Scottish Academic Press, Edinburgh.

HORNE, R. R. & MACINTYRE, R. M. 1975. Apparent age and significance of Tertiary dykes in the Dingle Peninsula, SW Ireland. *Scientific Proceedings of the Royal Dublin Society*, **A5**, 293–299.

I.G.S. 1981. *Annual Report for 1980 and 1981*; Institute of Geological Sciences. HMSO, London.

JONES, E. J. W., MITCHELL, J. G., SHIDO, F. & PHILLIPS, J. D. 1972. Igneous rocks dredged from the Rockall Plateau. *Nature Physical Science*, **237**, 118–120.

LANPHERE, M. A. & DALRYMPLE, G. B. 1978. *The Use of $^{40}Ar/^{39}Ar$ Data in Evaluation of Disturbed Systems*. US Geological Survey, Open-file report, **78–701**, 241–243.

LE BAS, M. J. 1960. The petrology of the Layered Basic Rocks of the Carlingford Complex, Co. Louth. *Transactions of the Royal Society, Edinburgh*, **64**, 169–200.

—— 1967. On the origin of the Tertiary granophyres of the Carlingford Complex, Ireland. *Proceedings of the Royal Irish Academy*, **65B**, 325–338.

LEWIS, A. G., BECKMAN, G. E. J. & MERRIMAN, R. J. 1975. Notes on geomagnetic work and palaeomagnetic data. *In*: HARRISON, R. K. (ed) 1975. Expeditions to Rockall 1971–72. *Report of the Institute of Geological Sciences*, **75/1**, 50–51.

MACDONALD, J. G. & HERRIOT, A. 1983. *MacGregor's Excursion Guide to the Geology of Arran*. Geological Society of Glasgow, Glasgow.

MACINTYRE, R. M. 1973. Lower Tertiary geochronology of the north Atlantic continental margins. *In*: *Geochronology and Isotope Geology of Scotland*; *Field Guide and Reference*. Third European Colloquium on Geochronology, Oxford.

MEIGHAN, I., HUTCHISON, R., WILLIAMSON, I. & MACINTYRE, R. M. 1981. Geological evidence for the different relative ages of the Rhum and Skye Tertiary central complexes. *Journal of the Geological Society of London*, **139**, 659 (abst.).

——, McCORMICK, A. G., GIBSON, D., GAMBLE, J. A. & GRAHAM, I. J. 1988. Rb-Sr isotopic determinations and the timing of Tertiary central complex magmatism in NE Ireland. *In*: MORTON, A. C. & PARSON, L. M. (eds) *Early Tertiary Volcanism and the Opening of the NE Atlantic*. Geological Society of London, Special Publication, **39**, pp. 349–360.

MITCHELL, J. G. & REEN, K. P. 1973. Potassium-argon ages from the Tertiary ring complexes of Ardnamurchan peninsula, Western Scotland. *Geological Magazine,* **110,** 331–349.

——, JONES, E. J. W. & JONES, G. T. 1976. The composition and age of basalts dredged from the Blackstones igneous centre, western Scotland. *Geological Magazine,* **113,** 525–533.

MOHR, P., MUSSETT, A. E. & KENNAN, P. S. 1984. The Droimchogaidh Sill, Connacht, Ireland. *Geological Journal,* **19,** 1–21.

MORGAN, G. E. 1984. Palaeomagnetism. *In:* HARDING *et al.* (eds) St. Kilda: an illustrated account of the geology. *Report of the British Geological Survey,* **(16)**7, 38–39.

MUSSETT, A. E. 1984. Time and duration of Tertiary igneous activity of Rhum and adjacent areas. *Scottish Journal of Geology,* **20,** 273–279.

—— 1986. ^{40}Ar-^{39}Ar step-heating ages of the Tertiary igneous rocks of Mull, Scotland. *Journal of the Geological Society of London,* **143,** 887–896.

——, DAGLEY, P. & ECKFORD, M. 1976. The British Tertiary Igneous Province: palaeomagnetism and ages of dykes, Lundy Island, Bristol Channel. *Geophysical Journal of the Royal Astronomical Society,* **46,** 595–603.

——, —— & SKELHORN, R. R. 1980. Magnetostratigraphy of the Tertiary igneous succession of Mull, Scotland. *Journal of the Geological Society of London,* **137,** 349–357.

——, ——, HODGSON, B. D. & SKELHORN, R. R. 1987. Palaeomagnetism and age of the quartz-porphyry intrusions, Isle of Arran, Scotland. *Scottish Journal of Geology,* **23,** 9–22.

O'CONNOR, P. J. 1988. Strontium isotope geochemistry of Tertiary igneous rocks, NE Ireland. *In:* MORTON, A. C. & PARSON, L. M. (eds) *Early Tertiary Volcanism and the Opening of the NE Atlantic.* Geological Society of London, Special Publication, **39,** pp. 361–363.

RICHEY, J. E. & THOMAS, H. H. 1930. *The Geology of Ardnamurchan, north-west Mull and Coll.* Memoir of the Geological Survey of Scotland, HMSO, Edinburgh.

RIDDIHOUGH, R. P. & MAX, M. D. 1976. A geological framework for the continental margin to the west of Ireland. *Geological Journal,* **11,** 109–120.

SKELHORN, R. R. 1969. The Tertiary igneous geology of the Isle of Mull. *Geologists' Association of London,* Guide No **20.**

STEIGER, R. H. & JAGER, E. 1977. Subcommission on geochronology: convention on the use of decay constants in geo- and cosmochronology. *Earth and Planetary Sciences Letters,* **36,** 359–362.

TAYLOR, H. P. & FORESTER, R. W. 1971. Low-^{18}O igneous rocks from the intrusive complexes of Skye, Mull and Ardnamurchan, Western Scotland. *Journal of Petrology,* **12,** 465–497.

THOMPSON, P. 1986. *Dating the British Tertiary Igneous Province in Ireland by the ^{40}Ar-^{39}Ar stepwise degassing method.* PhD thesis (unpublished). University of Liverpool.

——, WATT, S. & DURRANI, S. A. 1984. Concordant fission track and ^{40}Ar-^{39}Ar age determination for the Interbasaltic Formation of Northern Ireland. *Geophysical Journal of the Royal Astronomical Society,* **77,** 326 (abst.).

——, MUSSETT, A. E. & DAGLEY, P. 1987. Revised ^{40}Ar-^{39}Ar age for granites of the Mourne Mountains, Ireland. *Scottish Journal of Geology,* **23,** 215–220.

TOMKEIEFF, S. I. 1969. *Isle of Arran.* Geologists' Association of London, Guide No **32.**

TOWNSEND, H. A. & HAILWOOD, E. A. 1986. Magnetostratigraphic correlation of Palaeogene sediments in the Hampshire and London basins, southern UK. *Journal of the Geological Society of London,* **142,** 957–982.

TYRRELL, G. W. 1928. *The Geology of Arran.* Memoir of the Geological Survey of Scotland. HMSO, Edinburgh.

WALKER, G. P. L. 1971. *Distribution of Amygdale Minerals in Mull and Morven (Western Scotland),* West Memorial Volume, 181–194.

WALSH, J. N., BECKINSALE, R. D., SKELHORN, R. R. & THORPE, R. S. 1979. Geochemistry and petrogenesis of Tertiary granitic rocks from the Island of Mull, northwest Scotland. *Contributions to Mineralogy and Petrology,* **71,** 99–116.

WILSON, R. L. 1970. Palaeomagnetic stratigraphy of Tertiary lavas from Northern Ireland. *Geophysical Journal of the Royal Astronomical Society,* **20,** 1–9.

——, DAGLEY, P. & ADE-HALL, J. M. 1972. Palaeomagnetism of the British Tertiary igneous province: the Skye lavas. *Geophysical Journal of the Royal Astronomical Society,* **28,** 285–293.

A. E. MUSSETT & P. DAGLEY, Department of Geological Sciences, University of Liverpool, Liverpool L69 3BX, UK.

R. R. SKELHORN, Department of Geology, City of London Polytechnic, London E1 2NG, UK.

Rb-Sr isotopic determinations and the timing of Tertiary central complex magmatism in NE Ireland

I. G. Meighan, A. G. McCormick, D. Gibson, J. A. Gamble & I. J. Graham

SUMMARY: Full Rb-Sr isotopic data and whole-rock regressions are presented for each of the Mourne Mountains granites (G1–G5 inclusive) and a set of granophyres/microgranites from the Slieve Gullion central complex. The former were intruded at approximately 56 Ma (G1–G4 inclusive) and later (G5), and *may* all be younger than the major intrusions of the Slieve Gullion and Carlingford central complexes. The initial Sr isotope ratios are > 0.7060 for all five Mourne granites, the Slieve Gullion granophyres/microgranites analyzed and the rhyolitic rocks of the Tardree area (Co. Antrim), and crustal involvement is indicated in their petrogenesis. In the E Mourne Centre the first (the least silicic and most Sr-rich) granite has a significantly higher initial ratio (0.7129 ± 2) than G2 (0.7109 ± 3) and G3 (0.7104 ± 6), and the isotopic data provide further support for the recent revision to the mapping of the first and second granites. Preliminary Rb-Sr isotopic data for a comprehensive suite of Tertiary rhyolites from NE Ireland, some separated by > 50 km, define a linear array whose explanation is considered.

The Lower Tertiary major intrusions of NE Ireland were emplaced into Southern Uplands-type crust N of the Iapetus Suture but well to the S of the main outcrop of Tertiary volcanic rocks (the Antrim Plateau basalts) which in places are seen to rest on an eroded land surface developed on Cretaceous chalk (Fig. 1, Table 1). The volcanic suite includes some rhyolites, but these are subordinate to the basic lavas. Traditionally, the major intrusives have been considered in terms of three central complexes, namely Mourne Mountains (with subdivision into an eastern and younger, western centre), Slieve Gullion (separated into an earlier ring complex and a later central or sheeted complex) and Carlingford (see Emeleus 1982, Fig. 29.19). The essential details of each are summarized in Table 1, from which it is apparent that in terms of rock types and volcanic history the Mourne Mountains are distinct from Slieve Gullion–Carlingford.

In marked contrast to other parts of the British Tertiary Igneous Province, such as the Mull and Skye central complexes, there is a general paucity of published radiometric age determinations on Irish intrusive rocks, with none for Slieve Gullion. Furthermore, the relative ages of the Mourne Mountains, Slieve Gullion and Carlingford rocks have not been firmly established. On field evidence Carlingford is believed to postdate Slieve Gullion (see Emeleus & Preston 1969; Emeleus 1982) but although Evans *et al.* (1973) suggested that both of these predate the Mourne centres, others continue to view the Carlingford–Mourne Mountains relative age relationship as uncertain (e.g. Emeleus 1982). Evans *et al.* did present some K-Ar and ^{40}Ar-^{39}Ar biotite ages for certain Mourne Mountains granites, and a

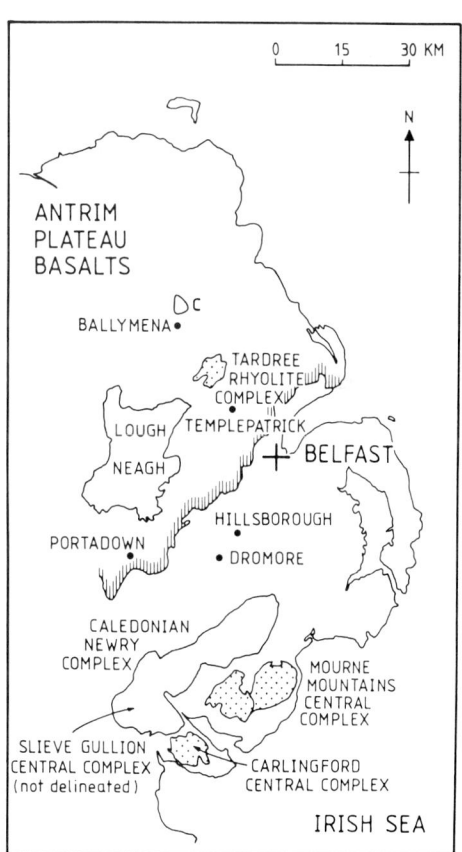

FIG. 1. Map of NE Ireland illustrating localities and some geological features discussed in the text. The hatched line represents the S boundary of the main outcrop of the Antrim Plateau basalts. C: approximate region of the Cloughwater–Quarrytown–Kirkinriola Tertiary rhyolites.

The labels within the figure read: ANTRIM PLATEAU BASALTS; BALLYMENA; C; TARDREE RHYOLITE COMPLEX; TEMPLEPATRICK; LOUGH NEAGH; BELFAST; HILLSBOROUGH; PORTADOWN; DROMORE; CALEDONIAN NEWRY COMPLEX; MOURNE MOUNTAINS CENTRAL COMPLEX; SLIEVE GULLION CENTRAL COMPLEX (not delineated); CARLINGFORD CENTRAL COMPLEX; IRISH SEA; 0 15 30 KM; N

From MORTON, A. C. & PARSON, L. M. (eds), 1988, *Early Tertiary Volcanism and the Opening of the NE Atlantic,* Geological Society Special Publication No. 39, pp. 349–360.

TABLE 1. *Essential details for the Tertiary central complexes of NE Ireland*

	Mourne Mountains	Slieve Gullion	Carlingford
Country rocks	Silurian metasedimentary rocks (turbidites)	Silurian metasedimentary rocks (turbidites) + Caledonian Newry granodiorite	Silurian metasedimentary rocks (turbidites) + Carboniferous sedimentary rocks
Lithology of the major intrusions	Granites (none peralkaline)	Gabbros, dolerites, granophyres/microgranites/felsites (none peralkaline)	Gabbros and granophyres (none peralkaline)
Tertiary volcanic activity	No evidence that Tertiary surface volcanicity occurred, i.e. no lavas or vent agglomerates: Silurian rocks are seen to roof the granites in places	Tertiary lavas and vent agglomerates	Tertiary lavas and vent agglomerates
Principal form of the major intrusions	Ring-dykes	Ring-dykes and sheets	Ring-dykes

whole-rock K-Ar result for one Carlingford sample, but there has been a considerable time interval between their study and the first Rb-Sr age determinations for Irish Tertiary rocks (Gibson *et al.*, 1987). The latter is a preliminary presentation of ages and initial strontium isotope ratios for Mourne granites, G1–G5 inclusive.

The present contribution includes the full Rb-Sr isotopic data and whole-rock regressions for the Mourne Mountains granites, G1–G5 inclusive, and results obtained by the same method for some granophyres/microgranites of the Slieve Gullion sheeted complex. The latter yield the first

radiometric age determination for this central complex. Only certain aspects of the strontium isotope data are discussed, as: (1) this contribution is concerned primarily with ages; and (2) Rb-Sr systematics *alone* may not firmly establish the source(s) of acid magmas.

Rb-Sr data for the Mourne Mountains granites

Essential details for the Mourne granite intrusions, G1–G5 inclusive, are summarized in Table 2. The Eastern Centre intrusions (G1, G2 & G3)

TABLE 2. *Some details for Mourne Mountains granites G1–5 inclusive*

Intrusion	G1	G2	G3	G4	G5
Revised nomenclature following recent remapping[1,2,3]	G1 'Roof'	G2 (Revised) (this includes a subordinate 'mafic facies' and has been subdivided into G2 (Outer) and G2 (Inner), the latter essentially equivalent to the original G2)	—	—	—
No. of magmatic pulses[1,2,3,4]	1	>2	>2	2	2
Summary of mineralogy	Calcic amphibole–biotite ± fayalite granite: Alkali feldspar + plagioclase	Biotite granite (amphibole unusual): Alkali feldspar + plagioclase	Biotite granite: Alkali feldspar + plagioclase	Biotite granite: Alkali feldspar + plagioclase	Calcic amphibole–biotite granite: Alkali feldspar + plagioclase
SiO_2 (wt%)[2,3,4]	67.8–71.7	73.0–80.6	77.6–79.1	75.1–78.9	74.3–78.8

1 = Meighan *et al.* (1984); 2 = Hood (1981); 3 = Meighan (1981); 4 = Gibson (1984).

were distinguished and mapped by J. E. Richey (1928) but recent work has necessitated revision of his classic map, extending the outcrop of the second granite (G2) at the expense of the first (G1) as designated in Fig. 2(b) of Meighan *et al.* (1984). Additionally, these authors proposed the subdivision of G2 (Revised) into G2 (Outer) and G2 (Inner). Their new and revised map also displays a subordinate 'mafic facies' of G2 (Outer), a somewhat enigmatic unit which was delineated by Hood (1981) and can superficially resemble G1 'Roof' (see Table 2), the most mafic granite in the Mourne central complex. Major intrusive activity later shifted to the region of the Western Centre (Richey 1928), where two granite ring-dykes were emplaced (Emeleus 1955; Gibson 1984).

Table 3 presents Rb-Sr data for the Mourne granite intrusions, G1–G5 inclusive, all of which were obtained at the Scottish Universities Research and Reactor Centre (SURRC), E Kilbride. The Rb and Sr concentrations were determined by isotope dilution. Following standard extraction procedures, isotopic ratios were measured mostly on a V.G. Micromass 30B mass spectrometer, but a few of the duplicates were also run on the V.G. Isomass 54E instrument at E Kilbride. The standard analytical procedures of the SURRC laboratory (Halliday *et al.* 1984) were employed throughout the study, including analysis of NBS 987 for $^{87}Sr/^{86}Sr$. The regression analysis followed the methods of York (1969), see Table 4, and Gibson *et al.* (1987), which include information on errors. For all regressions with mean squares of weighted deviates (MSWD) > 1, scatter errors are quoted throughout the paper; these include a measure of the scattering of the points about the best straight line.

The Rb-Sr whole-rock age determinations and the initial strontium isotope ratios are collated in Table 4. For G2 (Revised) it should be noted that separate regressions for G2 (Outer) and G2 (Inner) yield very similar ages and initial ratios (which are not distinguishable from each other or the data of the overall regression). As indicated elsewhere (Gibson *et al.*, 1987), all five Mourne Mountains major intrusions have relatively young crystallization ages in the general context of the British Tertiary igneous activity. G1–G4 inclusive were emplaced at approximately 56 Ma and G5 *apparently* at about 52 Ma. However, the MSWD value for the G5 regression (Table 4) suggests that an attempt should be made to confirm this very young age by obtaining individual isochrons for its two magmatic pulses (Table 2) and perhaps ^{40}Ar-^{39}Ar data also (see Gibson *et al.* 1987).

The high Rb/Sr ratios of the granites make it difficult to interpret initial strontium isotope ratios calculated for individual samples from each intrusion. Nevertheless, the quality of the whole-rock isochrons for the Eastern Centre intrusions (see Figs. 2–4 and Table 4), and especially that for G3, suggests that none of the individual magmatic pulses involved in the emplacement of G2 and G3 (Harry & Richey 1963; Meighan *et al.* 1984; see Table 2) had a highly distinctive strontium isotopic signature. The separate regressions for G2 (Outer) and G2 (Inner) also support this, but further work is necessary here for the Eastern as well as the Western Centre granites.

The strontium isotope data are also very relevant to the recent revision of the first and second granites, see above and Meighan *et al.* (1984). As Table 4 indicates, G1 'Roof' and G2

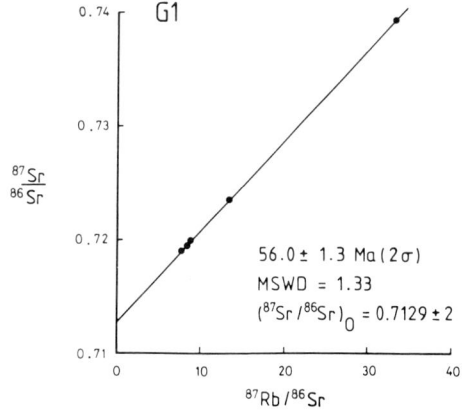

FIG. 2. Five point, Rb-Sr whole-rock isochron for E Mourne Mountains granite G1 'Roof' (see Table 3).

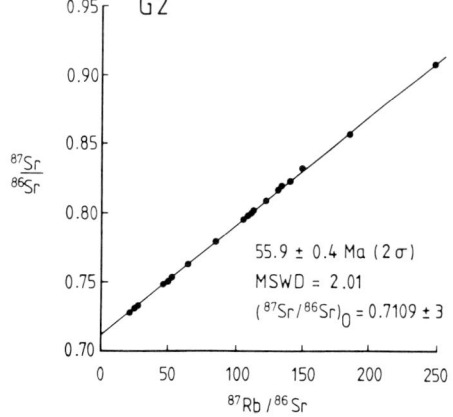

FIG. 3. Twenty-seven point, Rb-Sr whole-rock isochron for E Mourne Mountains granite G2 (Revised) (see Table 3). Because of overlaps not all the data points are plotted.

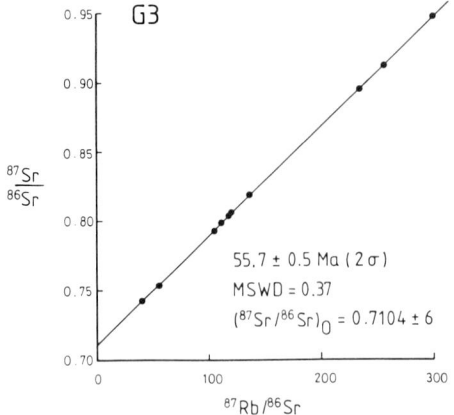

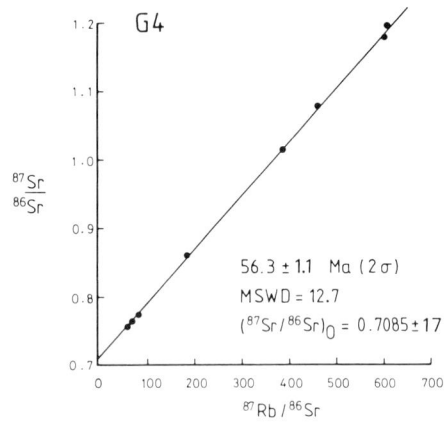

FIG. 4. Ten point, Rb-Sr whole-rock isochron for E Mourne Mountains granite G3 (see Table 3).

FIG. 5. Eight point, Rb-Sr whole-rock regression plot for W Mourne Mountains granite G4 (see Table 3).

TABLE 3. *Rb-Sr measurements for Mourne Mountains granites G1–G5 inclusive*

	Sample number	Irish Grid reference	Rb μg/g	Sr μg/g	$^{87}Rb/^{86}Sr$	$^{87}Sr/^{86}Sr$
G1 'Roof'						
	18	J 359 285	281.5	61.6	13.241	0.72343 + 4
	19	J 365 279	241.1	83.5	8.370	0.71949 + 7
	108	J 335 287	249.5	83.7	8.6301	0.71992 + 16
	137	J 346 290	230.5	87.2	7.661	0.71900 + 9
	146	J 344 284	367.2	32.0	33.304	0.73938 + 10
G2 (Revised)						
M	22	J 350 228	325.2	45.74	20.610	0.72759 + 13
M	29	J 368 268	341.4	40.30	24.567	0.72997 + 9
M	90	J 275 234	310.3	33.03	27.251	0.73226 + 8
M	132	J 302 284	332.5	35.0	27.554	0.73271 + 7
M	134	J 320 302	343.0	36.9	27.030	0.73282 + 7
O	30	J 376 251	415.7	8.80	138.25	0.82191 + 9
O	31	J 375 253	354.7	19.70	52.324	0.75241 + 5
O	32	J 364 257	436.8	14.97	84.985	0.77872 + 16
O	33	J 368 274	452.6	10.08	131.39	0.81540 + 15
O	35	J 382 270	418.2	8.17	149.85	0.83122 + 14
O	36	J 379 269	456.4	12.0	110.90	0.79838 + 9
O	47	J 376 268	418.4	11.62	105.04	0.79433 + 5
O	52	J 313 298	423.6	11.08	111.99	0.79917 + 9
O	129	J 301 286	405.7	11.22	105.54	0.79448 + 7
O	131	J 303 294	419.6	10.87	112.69	0.80104 + 9
I	20	J 342 249	393.1	23.10	49.445	0.74958 + 5
I	21	J 344 260	461.9	10.05	134.37	0.81852 + 18
I	23	J 319 229	342.5	11.73	85.078	0.77782 + 5
I	24	J 327 227	368.7	16.58	64.697	0.76200 + 8
I	25	J 373 277	451.7	7.18	184.62	0.85570 + 8
I	26	J 371 278	458.0	9.58	139.94	0.82182 + 8
I	27	J 352 276	478.9	7.59	185.27	0.85762 + 9
I	44	J 318 288	454.6	12.3	107.89	0.79704 + 8
I	46	J 314 295	421.2	14.54	84.391	0.77818 + 11
I	67	J 374 291	490.8	5.85	247.35	0.90668 + 10
I	89	J 302 262	444.8	10.57	122.99	0.80780 + 9
I	124	J 324 273	371.9	23.44	46.081	0.74722 + 9

TABLE 3. *continued*

Sample number	Irish Grid reference	Rb μg/g	Sr μg/g	^{87}Rb/^{86}Sr	^{87}Sr/^{86}Sr
G3					
92	J 293 240	414.8	10.89	111.15	0.79801 ± 10
93	J 295 255	410.2	8.81	136.16	0.81852 ± 16
94	J 303 261	420.1	10.23	119.93	0.80571 ± 14
95	J 304 260	393.0	20.76	55.00	0.75366 ± 10
96	J 278 237	468.6	5.42	254.92	0.91115 ± 12
97	J 287 244	400.6	11.22	104.12	0.79266 ± 12
98	J 285 204	485.9	6.14	233.18	0.89491 ± 14
99	J 285 204	437.5	4.35	298.13	0.94692 ± 18
100	J 315 253	393.9	9.79	117.56	0.80379 ± 14
147	J 311 214	339.3	24.57	40.087	0.74226 ± 16
G4					
G4/60	J 255 269	498.8	2.50	603.153	1.19535 ± 3
G4/252	J 267 219	460.5	2.33	599.600	1.17749 ± 12
G4/32	J 244 276	446.8	3.08	459.320	1.07636 ± 20
G4/109	J 222 255	441.1	3.41	384.970	1.01348 ± 8
G4/132	J 186 222	443.7	7.09	183.756	0.86005 ± 5
G4/265	J 248 201	391.7	14.00	81.451	0.77269 ± 5
G4/161	J 246 207	368.7	15.68	68.407	0.76361 ± 7
G4/38	J 243 266	355.5	17.81	58.013	0.75478 ± 9
G5					
G5/229	J 239 245	414.7	8.94	135.575	0.81549 ± 6
G5/162	J 251 207	355.9	13.81	74.987	0.76823 ± 9
G5/63	J 229 263	359.6	13.28	78.813	0.77168 ± 4
G5/231	J 238 241	352.6	17.98	57.580	0.75613 ± 3
G5/90	J 230 242	322.4	20.27	46.201	0.74701 ± 5
G5/376	J 225 208	330.2	22.06	43.478	0.74594 ± 4
G5/283	J 227 223	322.7	28.59	32.752	0.73909 ± 5
G5/234	J 215 224	309.0	32.51	27.572	0.73492 ± 6
G5/101	J 222 241	310.3	34.05	26.436	0.73380 ± 16
G5/395	J 208 213	288.2	35.53	23.524	0.73096 ± 4

For G2 (Revised): M: mafic facies; O: G2 (Outer) and I: G2 (Inner) as in Meighan *et al.* 1984. The analytical uncertainties (2σ) quoted for ^{87}Sr/^{86}Sr refer to the least significant digit(s). Responsibilities for data: G1, G2, G3–A.G.M.; G4, G5–D.G.

TABLE 4. *Summary of Rb-Sr ages and initial strontium isotope ratios for Mourne granites G1–G5 inclusive*

Intrusion (or part of intrusion)	Whole-rock isochron age (Ma) with number of samples in brackets	MSWD	(^{87}Sr/^{86}Sr)$_0$
G1 'Roof'	56.0 ± 1.3 (5)	1.33	0.7129 ± 2
G2 (Revised)	55.9 ± 0.4 (27)	2.01	0.7109 ± 3
G2 (Outer)*	56.3 ± 0.8 (10)*	0.92	0.7105 ± 11
G2 (Inner)*	56.0 ± 0.5 (12)*	0.99	0.7105 ± 6
G3	55.7 ± 0.5 (10)	0.37	0.7104 ± 6
G4	56.3 ± 1.1 (8)	12.7	0.7085 ± 17
G5	51.5 ± 1.8 (10)	21.9	0.7142 ± 10

* Separate regression for typical, G2-type granite specimens: all these data are included in the full, 27 point isochron for G2 (Revised).

The Rb-Sr regression analysis followed the method of York (1969): all uncertainties (2σ) are scatter errors, except those for G2 (Outer), G2 (Inner) and G3 which are *a priori* errors (see Gibson *et al.* 1987). For the ^{87}Rb/^{86}Sr ratios an average error of 0.7% was applied to the data for G1, G2 and G3 and one of 0.5% to G4 and G5 (both 1σ): for the ^{87}Sr/^{86}Sr ratios the individual analytical uncertainties were utilized. λ^{87}Rb = 1.42 × 10^{-11} y^{-1}. The uncertainties for (^{87}Sr/^{86}Sr)$_0$ refer to the least significant digit(s).

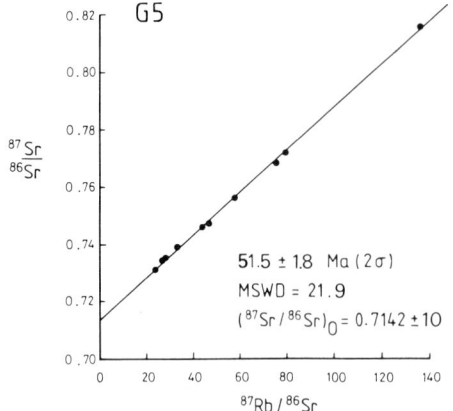

FIG. 6. Ten point, Rb-Sr whole-rock regression plot for W Mourne Mountains granite G5 (see Table 3).

(Revised) have very similar ages but significantly different initial ratios. The 'mafic facies' of G2 (Revised) (see Table 3) falls well off the G1 'Roof' isochron (Fig. 7) and despite some superficial petrographic similarity cannot form an integral part of this granite as the two are quite distinct isotopically (cf. Richey 1928, Plate 53). Other, less mafic, samples of G2 (Outer) (see Table 3), which also were collected from localities originally within Richey's first granite, contribute to

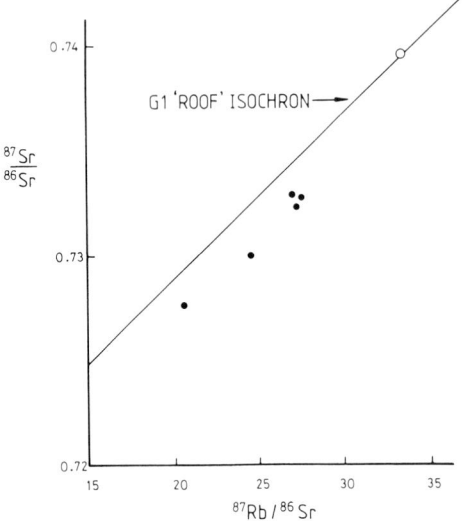

FIG. 7. The upper part of the Mourne Mountains G1 'Roof' isochron of Fig. 2 on an enlarged scale showing one data point (open circle). The five filled circles lying off this isochron are samples of the 'mafic facies' of G2 (Revised) as in Fig. 2 of Meighan et al. (1984) (see Table 3).

the full (27 point) G2 (Revised) isochron (see Fig. 3 and Table 4); and thus they too plot off that for G1 'Roof'. Furthermore, as noted above, separate regressions for G2 (Outer) and G2 (Inner), which is essentially equivalent to Richey's G2 granite, yield very similar ages and initial ratios, the latter quite distinct from G1 'Roof' (see Table 4). Consequently, it can be concluded that the Rb-Sr isotopic data fully support the recent revision of the Eastern Centre granites. Unfortunately, however, they do not allow the second and third granites to be distinguished.

Despite the different (Southern Uplands-type) crustal setting from W and NW Scotland, the initial strontium isotope ratios (Hood et al. 1981; Meighan et al. 1984; Gibson et al. 1987; see Table 4) are comparable to those for many Scottish Tertiary acid bodies (e.g. Walsh et al. 1979; Dickin 1981; Dickin et al. 1981). Furthermore, the data (see Table 4) allow the following conclusions to be drawn for the Mourne Mountains central complex:

(1) There was no simple, systematic change in $^{87}Sr/^{86}Sr$ during the granite emplacement sequence.

(2) The first Mourne granite, the least silicic (see Table 2) and most Sr-rich (see Table 3) of the entire central complex, has the highest initial ratio, and by contrast the more fractionated granites G2, G3 and G4 have lower values of $(^{87}Sr/^{86}Sr)_0$; it is thus evident that G1 cannot be a simple hybrid composition produced by mixing at depth of typical G2 magma and a basaltic one.

(3) For the Mourne 56 Ma intrusive sequence, as for the Mull (Walsh et al. 1979) and other Scottish Tertiary central complexes, the earliest acid magma had the most radiogenic strontium isotopic composition.

(4) All the members of the Mourne acid differentiation series (Meighan et al. 1984) cannot be comagmatic sensu stricto, e.g. the G2 (Revised) and G3 granites cannot be closed system fractional crystallization products of the more radiogenic G1 'Roof' type magma. However, if the granites as a whole are interpreted as being crustally contaminated basaltic differentiates (Meighan et al. 1984), the isotope data for the Eastern Centre do show that the extreme acid magma differentiation indicated by geochemical evidence (Meighan & Gamble 1972; Meighan 1979, 1981; Meighan et al. 1984) occurred after $^{87}Sr/^{86}Sr$ ratios of over 0.7090 had been achieved. Also, there is no evidence for marked isotopic heterogeneity within any of the individual magma batches.

Finally, it is necessary to comment on the ultimate petrogenetic implications of the results. Accepting the limitations imposed by interpreting strontium isotopic data in isolation, for the Mourne Mountains central complex the following are indicated:

(1) None of the granites is an isotopically uncontaminated, closed system differentiate of typical Tertiary basic magma; each contains crustal strontium (Meighan *et al.* 1984) and, as noted above, in the Eastern Centre the *least silicic* granite (see Table 2) has the *highest* initial ratio. Nevertheless, initial ratios within and above the Mourne range do not necessarily preclude the presence of a *substantial* basaltic differentiate component (Meighan 1979; Dickin *et al.* 1981), particularly in the light of this holding even for the Coire Uaigneich granophyre, Isle of Skye (Meighan 1979, 1981) with a very high ($^{87}Sr/^{86}Sr)_0$ value of approximately 0.731 (Dickin & Exley 1981).

(2) Apropos crustal melting models, in the absence of comprehensive strontium isotope data for County Down it is necessary to utilize the available results for the comparable Lower Palaeozoic sedimentary rocks of the Southern Uplands (Halliday *et al.* 1980; Tindle 1982). Considering their 56 Ma ratios in isolation from any other data, it is apparent that the granites *could* have originated by partial melting of certain greywacke compositions with relatively low Rb/Sr ratios, but the shales and many other greywackes must be rejected on any single source model as these had $(^{87}Sr/^{86}Sr)_{56 Ma}$ values well in excess of 0.7150.

(3) Any SE extension at depth of the Caledonian Newry igneous complex constitutes another possible source for the Mourne acid magmas. Typical granodiorites of its NE pluton (Meighan & Neeson 1979; Neeson 1984) had $(^{87}Sr/^{86}Sr)_{56 Ma}$ in the range 0.7073–0.7102, whereas the intermediate rocks adjacent to these had 56 Ma ratios ranging from 0.7058–0.7082 (Meighan & Halliday, unpublished data). Clearly, the G1 'Roof' magma $(^{87}Sr/^{86}Sr = 0.7129 \pm 2)$ cannot have been generated simply by partial melting of Newry complex rocks similar to those exposed at its NE end. However, on the Sr isotope data alone, the more fractionated Eastern Mourne granites G2 and G3 (see Tables 2, 4) *could* have been derived from some of the analyzed Newry granodiorite compositions (i.e. those with relatively high Rb/Sr ratios) but not from the intermediate ones. Similarly, Cale-

donian granodioritic and intermediate rocks constitute *possible* sources for the Western Mourne granite G4.

The origin of the Mourne granites is a problem which will be addressed elsewhere in the light of combined isotopic (Sr + Nd + O) and full geochemical data. However, in view of the present discussion, it is appropriate to mention that such an approach indicates: (1) the Mourne acid magmas were not derived solely from one (or more) crustal sources; and (2) each of the granites contains a very substantial basaltic differentiate component.

The Northern Ireland Lower Tertiary rhyolites

The most extensive outcrop of Tertiary acid volcanic rocks is in the Tardree–Sandy Braes region of mid-Antrim (see Fig. 1). However, there are several other occurrences of Tertiary rhyolite, and it is believed that all the eruptions in question were associated in time with the Interbasaltic Formation (Old 1975), for the duration of which there was basaltic eruptive quiescence (except in the N Antrim region). To date, no comprehensive isotopic dating study has been published on these acid rocks and clearly such an investigation might place important constraints on the time interval represented by the Interbasaltic Formation in NE Ireland.

There is still some uncertainty about the age of the Tardree–Sandy Braes rocks with determinations ranging from 65.2 ± 0.8 Ma (a fission track age published by Fitch & Hurford 1977) to ca. 58 Ma by the K-Ar method (R. M. Macintyre, pers. comm.). However, the former value has been abandoned in favour of an age close to 59 Ma (Fitch *et al.* 1988). Also, an ^{40}Ar-^{39}Ar age of 60.7 ± 0.6 Ma (2σ) has been determined on K feldspar phenocrysts from typical Tardree rhyolite (D. R. Lux & D. Gibson, unpublished data) as part of a programme of age dating work on the full rhyolite suite.

Two aspects of preliminary Rb-Sr results are considered in this paper:

(1) Table 5 contains the first published Rb-Sr results for the Tardree–Sandy Braes rocks. Although a precise age obtained by using this method is not presented here, it is apparent from the 56 and 60 Ma data that, regardless of their exact time of eruption, the porphyritic rhyolite and porphyritic obsidian have initial strontium isotope ratios lying towards the lower end of the spectrum for British Tertiary

TABLE 5. *Rb-Sr data for the acid volcanic rocks of the Tardree area, Co. Antrim*

Sample	Irish Grid reference	Rb μg/g	Sr μg/g	$^{87}Rb/^{86}Sr$	$^{87}Sr/^{86}Sr$	$(^{87}Sr/^{86}Sr)_{56\ Ma}$	$(^{87}Sr/^{86}Sr)_{60\ Ma}$
Porphyritic rhyolite	J 191 948	133.2	35.2	10.966	0.71656 ± 12	0.70784 ± 17	0.70721 ± 18
Porphyritic obsidian	J 207 958	114.7	44.6	7.456	0.71328 ± 5	0.70735 ± 10	0.70692 ± 10

Data obtained by A.G.M. at the SURRC, E Kilbride. Rb and Sr values determined by isotope dilution. Uncertainties (2σ) refer to the least significant digit(s). $\lambda^{87}Rb = 1.42 \times 10^{-11}\ y^{-1}$.

acid rocks. Indeed, their values (see Table 5) are closely comparable to those of the crustally contaminated basaltic differentiates of Centre 2 (Beinn Chaisgidle Centre), Mull (Walsh *et al.* 1979; Walsh & Clarke 1982) and a similar mode of origin has been argued for the Northern Ireland rhyolites (Meighan 1979; Meighan *et al.* 1984). This problem is being investigated further by a combined isotopic approach.

(2) Figs. 8(a) and (b) illustrate an interesting aspect of the Rb-Sr whole-rock data obtained to date for the Northern Ireland rhyolites.

The population comprises ten samples from essentially nine different field localities in five rhyolite regions (Cloughwater–Quarrytown–Kirkinriola, Tardree–Sandy Braes, Templepatrick, Hillsborough/Dromore and Ballydugan; see Figs. 8(a) & (b)). The maximum distance between specimen localities is just over 55 km, involving Cloughwater (N of Ballymena) and Ballydugan (approximately E of Portadown) (see Fig. 1). Although there are several, isolated bodies (some widely separated) and these rocks are not all comagmatic (Meighan 1981), the data define a linear array and yield the following results (the regression analysis was as described in Gibson *et al.* 1987 and Table 4 for the E Mourne granites; 2σ scatter errors are quoted):

(1) 57.2 ± 1.3 Ma 'age', initial ratio of 0.7109 ± 0.0010 and MSWD of 12.3 (nine points; Tardree and Sandy Braes samples excluded).

(2) 60.3 ± 2.8 Ma 'age', initial ratio of 0.7073 ± 0.0009 and MSWD of 75.6 (all 11 points).

One possible interpretation of the nine point regression, based on the Rb-Sr systematics alone, is that the various rhyolite magmas in question

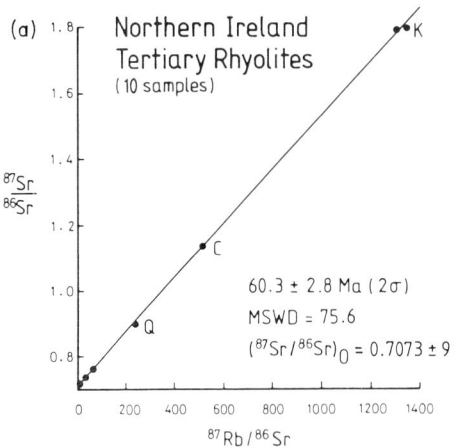

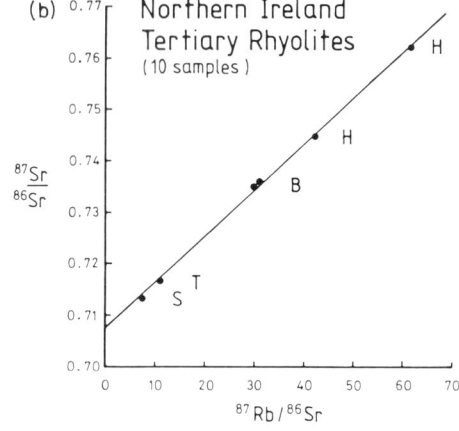

FIG. 8. (a) Eleven point, whole-rock Rb-Sr regression plot for nine Northern Ireland Tertiary rhyolite specimens and one obsidian. Because of overlaps not all the data points are plotted. Sample details (with Irish Grid references): K: Kirkinriola region (D 113 068) (duplicate analyses plotted); C: Cloughwater (D 110 096); Q: Quarrytown region (D 133 082). A rhyolite from the Templepatrick region (J 229 857) lies close to the Quarrytown point and is not plotted. The other data points are explained in the Fig. 8 (b) caption. All the data were determined by A.G.M. at the SURRC, E. Kilbride. (b) The lower part of the regression of Fig. 8(a) on an enlarged scale. Sample details (with Irish Grid references): H: Hillsborough/Dromore region (J 180 570) (two separate localities); B: Ballydugan (Shanes Hill) region (J 065 532) (two samples); T: Tardree Mountain (J 191 948); S: Sandy Braes obsidian (J 207 958). Details of all these Fig. 8 rocks are contained in Meighan (1981).

originated by partial melting of rather similar (as opposed to *extremely* different) crustal sources over a very narrow time interval, and varying degrees of fractional crystallization occurred immediately after magma genesis. However, there is evidence that the rhyolites are contaminated basaltic differentiates (Meighan 1979, 1981; Meighan *et al.* 1984) which display varying degrees of fractionation (Meighan & Gamble 1972). On this model, the nine point linear array, with its slope corresponding to an 'appropriate' (i.e. around 60 Ma) 'age', suggests that individual, essentially coeval acid magma bodies or their precursors achieved $^{87}Sr/^{86}Sr$ ratios which did not differ *vastly* prior to (further) fairly rapid differentiation, the latter accounting for the very high Rb/Sr ratios in some of the rocks. Thus, over an extensive area the magmas in question *may* have undergone (selective) Sr isotopic equilibration with (probably upper) crustal rocks whose Lower Tertiary $^{87}Sr/^{86}Sr$ ratios were not highly dissimilar, cf. Meighan (1979), Meighan *et al.* (1984) and Walsh *et al.* (1979), who also refer to selective contamination.

The ages of the Carlingford and Slieve Gullion central complexes

The evidence for the Carlingford central complex postdating Slieve Gullion originated in the field observations of W. A. Traill in the last century (Emeleus & Preston 1969; Emeleus 1982). In the SE part of the Slieve Gullion area a small stock of aphyric pyroxene granophyre developed a fine-grained margin against, and also veins, the porphyritic granophyre of the ring complex (Richey & Thomas 1932). As well as the ring-dyke, the stock truncates members of the later sheeted complex, but is itself cut, on the S side of Clermont Carn (J 099 157), by basic cone-sheets

belonging to the Carlingford central complex (see Emeleus 1982, Fig. 29.19).

Recently, Thompson (1985) has performed a most valuable ^{40}Ar-^{39}Ar age dating study on the Tertiary igneous rocks of Ireland which provides some important information on general age relationships. For Carlingford, his data suggest that the best age estimate is 60.9 ± 0.5 Ma (1σ), the result of a whole-rock ^{40}Ar-^{39}Ar 'plateau' determination on a granophyre sample 'from the Windy Gap region. This complements the K-Ar minimum age of 60.0 Ma (corrected value following Steiger & Jäger 1977) obtained by Evans *et al.* (1973) for a late basaltic sheet.

For the geologically older Slieve Gullion central complex, Fig. 9 and Table 6 present the first

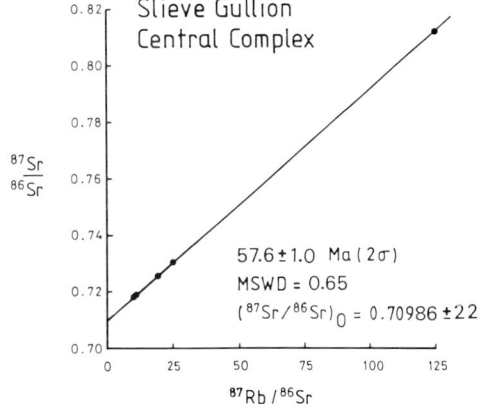

FIG. 9. Five point, Rb-Sr whole-rock isochron for granophyres/microgranites from the Slieve Gullion central complex (see Table 6). See latter and Graham (1985) for analytical details. The regression analysis followed the method of York (1969). For the $^{87}Rb/^{86}Sr$ ratios a blanket error of 1.0% (1σ) was applied to all the data: for the $^{87}Sr/^{86}Sr$ ratios the individual analytical uncertainties were utilized. The uncertainties quoted from the regression are a priori errors (2σ).

TABLE 6. *Rb-Sr measurements for Slieve Gullion central complex granophyres/microgranites*

Sample	Irish Grid reference	Rb μg/g	Sr μg/g	$^{87}Rb/^{86}Sr$	$^{87}Sr/^{86}Sr$
AT 56 T	J 036 234*	212.4	63.3	9.7454	0.71790 ± 6
72/13 T	J 036 234*	212.7	57.5	10.7315	0.71861 ± 11
72/40	J 017 201	245.7	37.3	19.1165	0.72534 ± 11
72/16 T	J 036 234*	246.1	28.6	24.9661	0.73029 ± 11
L5	J 008 240	314.6	7.4	123.8200	0.81176 ± 24

All the samples are from the 'central' or 'sheeted' complex region of Slieve Gullion. T: specimen from access tunnel (Gamble *et al.* 1976), the asterisked grid references are for the tunnel's portal.

All data obtained in Wellington, New Zealand (I.J.G). Rb and Sr concentrations were determined by isotope dilution, excepting L5 for which X.R.F.S. values are reported (see Graham 1985). The analytical uncertainties (2σ) quoted for $^{87}Sr/^{86}Sr$ refer to the least significant digit(s).

published radiometric age determination and strontium isotope data. This work was performed in Wellington, New Zealand using a V.G. Micromass 30 B mass spectrometer (see Table 6, Fig. 9 and Graham (1985) for analytical/regression details). $^{87}Sr/^{86}Sr$ measurements on NBS 987 at the E Kilbride and Lower Hutt (Wellington) laboratories agree within the analytical uncertainties. Ideally, such a whole-rock Rb-Sr regression should be based only on samples which are known to belong definitely to the same intrusion. However, this is not always possible on geological grounds and, indeed, Walsh et al. (1979) have demonstrated for Centre 3 (Loch Bà Centre), Mull that material from different, neighbouring Tertiary granophyre intrusions can yield an apparently reliable Rb-Sr isochron. For the Slieve Gullion data given here it must be noted that all the samples do not pertain to one intrusion. The five point isochron (see Fig. 9) yields an age of 57.6 ± 1.0 Ma (2σ) and an initial ratio of 0.70986 ± 0.00022, and it is interesting to note that the latter is closely comparable to the values for the more fractionated 56 Ma Mourne granites G2–G4 inclusive (see Table 4). Consequently, the discussion already provided is relevant to the Slieve Gullion rocks. The 56–60 Ma ratios for the Newry granodiorite samples mentioned above lie in the range 0.7072–0.7102 and clearly, on these results alone, partial melting of the Caledonian granitoid seems a distinct possibility for the genesis of the Slieve Gullion Tertiary acid magmas. However, such a single crustal source is not supported by other geochemical data (Gamble 1975; see also the discussion in Meighan et al. (1984) for the geochemically comparable Mourne granites). A combined isotopic study (in progress) should cast further light on this problem as, following Meighan (1979, 1981), all the Irish Tertiary acid bodies probably contain a substantial basaltic differentiate component.

The ages of Tertiary major intrusive events in NE Ireland

Doubtless, considerably fuller Rb-Sr and other radiometric age data will eventually be published for some of the Tertiary major intrusions in NE Ireland (e.g. Slieve Gullion) and more tightly constrained timings will emerge for the activity at each central complex. Meanwhile, the present data allow the following conclusions to be made:

(1) Rb-Sr regressions establish definitively that the main episode of major intrusive activity in the Mourne Mountains central complex

occurred at approximately 56 Ma. However, the intrusion of the fifth granite here was a later (? 52 Ma) event. Thus, as discussed by Gibson et al. (1987), some of the igneous activity in question may have been closely associated in time with the main opening of the NE Atlantic Ocean.

(2) Some acid rocks from the sheeted complex of Slieve Gullion have been dated at approximately 57–58 Ma by the Rb-Sr method. However, from his $^{40}Ar-^{39}Ar$ study, Thompson (1985) proposes a 60–61 Ma age for the geologically younger Carlingford central complex.

(3) Clearly, the additional data mentioned above are vital in respect of resolving the dilemma introduced by the Slieve Gullion and Carlingford age determinations. Assuming that Slieve Gullion is geologically older than Carlingford, the following alternatives emerge: (1) major intrusive activity at all three Irish Tertiary central complexes occurred mainly at approximately 56–58 Ma and thus was somewhat later than most of that at some of their Scottish equivalents (see Gibson et al., 1987); (2) the Slieve Gullion and Carlingford central complexes were both emplaced some 4–5 My before the 56 Ma Mourne Mountains granites G1–G4 inclusive. However, if the basic cone-sheets of the Clermont Carn region are significantly younger than the main activity at Carlingford, the currently available radiometric ages and the field evidence could be reconciled.

ACKNOWLEDGEMENTS: This work constitutes part of a major Sr, O and Nd isotopic study of the Mourne central complex involving The Queen's University of Belfast and the Isotope Geology Unit of the Scottish Universities Research and Reactor Centre, E Kilbride. We thank Q.U.B. for Grant W2/ST which financed isotopic determinations at the SURRC where Drs O. van Breemen, A. N. Halliday, A. E. Fallick and Mr J. Hutchinson provided essential assistance such that we are indebted to them in many ways. Dr D. N. Hood's large collection of analyzed E Mourne granitic rock powders was invaluable in selecting samples for the isotopic study. The following gave kind permission to use their unpublished data: Drs P. Thompson, A. E. Mussett, R. M. Macintyre, D. N. Hood and A. G. Tindle. We are most grateful to David Jamison, Elizabeth Lawson, Gail Lyons and Eric McKelvey of the Department of Geology, Queen's University for assistance of various types. We also thank Frank Fitch and Jane Evans for helpful comments. A.G.M. and D.G. acknowledge postgraduate studentships awarded by the Department of Education for Northern Ireland. I.J.G. publishes with the permission of The Director, Institute of Nuclear Sciences (D.S.I.R.).

References

DICKIN, A. P. 1981. Isotope geochemistry of Tertiary igneous rocks from the Isle of Skye, N.W. Scotland. *Journal of Petrology*, **22**, 159–189.

—— & EXLEY, R. A. 1981. Isotopic and geochemical evidence for magma mixing in the petrogenesis of the Coire Uaigneich Granophyre, Isle of Skye, N.W. Scotland. *Contributions to Mineralogy and Petrology*, **76**, 98–108.

——, MOORBATH, S. & WELKE, H. J. 1981. Isotope, trace element and major element geochemistry of Tertiary igneous rocks, Isle of Arran, Scotland. *Transactions of the Royal Society of Edinburgh: Earth Sciences*, **72**, 159–170.

EMELEUS, C. H. 1955. The granites of the Western Mourne Mountains, County Down. *Scientific Proceedings of the Royal Dublin Society*, **27**, 35–50.

—— 1982. The British Tertiary Province: the central complexes. *In*: SUTHERLAND, D. S. (ed) *Igneous Rocks of the British Isles*. John Wiley & Sons, Chichester, pp. 369–414.

—— & PRESTON, J. 1969. *Field excursion guide to the Tertiary volcanic rocks of Ireland*. IAVCEI Symposium, Oxford.

EVANS, A. L., FITCH, F. J. & MILLER, J. A. 1973. Potassium-argon determinations on some British Tertiary igneous rocks. *Journal of the Geological Society of London*, **129**, 419–443.

FITCH, F. J. & HURFORD, A. J. 1977. Fission track dating of the Tardree Rhyolite, Co. Antrim. *Proceedings of the Geologists' Association*, **88**, 267–274.

——, HEARD, G. L. & MILLER, J. A. 1988. Basaltic magmatism of Late Cretaceous and Palaeogene age recorded in wells NNE of the Shetlands. *In*: MORTON, A. C. & PARSON, L. M. (eds) *Early Tertiary Volcanism and the Opening of the NE Atlantic*. Geological Society of London, Special Publication, **39**, pp. 253–262.

GAMBLE, J. A. 1975. *The structure, petrology and geochemistry of the Central Complex of Slieve Gullion*. PhD thesis, The Queen's University of Belfast.

——, OLD, R. A. & PRESTON, J. 1976. Subsurface exploration in the Tertiary Central Complex of gabbro and granophyre at Slieve Gullion, Co. Armagh, Northern Ireland. *Report of the Institute of Geological Sciences*, **76/8**, 17 pp.

GIBSON, D. 1984. *The petrology and geochemistry of the Western Mourne Granites, Co. Down, N. Ireland*. PhD thesis, The Queen's University of Belfast.

——, McCORMICK, A. G., MEIGHAN, I. G. & HALLIDAY, A. N. 1987. The British Tertiary Igneous Province: Young Rb-Sr ages for the Mourne Mountains Granites. *Scottish Journal of Geology*, **23**, 221–225.

GRAHAM, I. J. 1985. Rb-Sr geochronology and geochemistry of Torlesse metasediments from the central North Island, New Zealand. *Chemical Geology (Isotope Geoscience Section)*, **52**, 317–331.

HALLIDAY, A. N., STEPHENS, W. E. & HARMON, R. S. 1980. Rb-Sr and O isotopic relationships in three zoned Caledonian granitic plutons, Southern Uplands, Scotland: evidence for varied sources and hybridization of magmas. *Journal of the Geological Society of London*, **137**, 329–348.

——, FALLICK, A. E., HUTCHINSON, J. & HILDRETH, W. 1984. A Nd, Sr and O isotopic investigation into the causes of chemical and isotopic zonation in the Bishop Tuff, California. *Earth and Planetary Science Letters*, **68**, 379–391.

HARRY, W. T. & RICHEY, J. E. 1963. Magmatic pulses in the emplacement of plutons. *Liverpool and Manchester Geological Journal*, **3**, 254–268.

HOOD, D. N. 1981. *Geochemical, petrological and structural studies on the Tertiary granites and associated rocks of the Eastern Mourne Mountains, Co. Down, Northern Ireland*. PhD thesis, The Queen's University of Belfast.

——, MEIGHAN, I. G., GIBSON, D. & McCORMICK, A. G. 1981. The Tertiary granites of the Eastern and Western Mourne Centres, N. Ireland. *Journal of the Geological Society of London*, **138**, 497.

MEIGHAN, I. G. 1979. The acid igneous rocks of the British Tertiary Province. *Bulletin of the Geological Survey of Great Britain*, **70**, 7–8 & 10–22.

—— 1981. *The petrogenesis of the acid igneous rocks of the British Tertiary Province*. PhD thesis, The Queen's University of Belfast.

—— & GAMBLE, J. A. 1972. Tertiary acid magmatism in NE Ireland. *Nature (Physical Science)*, **240**, 183–184.

—— & NEESON, J. C. 1979. The Newry igneous complex, County Down. *In*: HARRIS, A. L., HOLLAND, C. H. & LEAKE, B. E. (eds) *The Caledonides of the British Isles—reviewed*. Geological Society of London, Special Publication, **8**, 717–722.

——, GIBSON, D. & HOOD, D. N. 1984. Some aspects of Tertiary acid magmatism in NE Ireland. *Mineralogical Magazine*, **48**, 351–363.

NEESON, J. C. 1984. *The geology and geochemistry of the Newry Igneous Complex, Northern Ireland*. PhD thesis, The Queen's University of Belfast.

OLD, R. A. 1975. The age and field relationships of the Tardree Tertiary Rhyolite Complex, County Antrim, Northern Ireland. *Bulletin of the Geological Survey of Great Britain*, **51**, 21–40.

RICHEY, J. E. 1928. The structural relations of the Mourne Granites (Northern Ireland). *Quarterly Journal of the Geological Society of London*, **83**, 653–688.

—— & THOMAS, H. H. 1932. The Tertiary ring complex of Slieve Gullion (Ireland). *Quarterly Journal of the Geological Society of London*, **88**, 776–849.

STEIGER, R. H. & JÄGER, E. 1977. Subcommission on geochronology: convention on the use of decay constants in geo- and cosmochronology. *Earth and Planetary Science Letters*, **36**, 359–362.

THOMPSON, P. 1985. *Dating the British Tertiary Igneous Province in Ireland by the ^{40}Ar-^{39}Ar stepwise degassing method*. PhD thesis, University of Liverpool.

TINDLE, A. G. 1982. *Petrogenesis of the Loch Doon Granite Intrusion, Southern Uplands of Scotland.* PhD thesis, The Open University.

WALSH, J. N. & CLARKE, E. 1982. The role of fractional crystallization in the formation of granitic and intermediate rocks of the Beinn Chaisgidle Centre, Mull, Scotland. *Mineralogical Magazine*, **45**, 247–255.

——, BECKINSALE, R. D., SKELHORN, R. R. & THORPE, R. S. 1979. Geochemistry and petrogenesis of Tertiary granitic rocks from the Island of Mull, Northwest Scotland. *Contributions to Mineralogy and Petrology*, **71**, 99–116.

YORK, D. 1969. Least squares fitting of a straight line with correlated errors. *Earth and Planetary Science Letters*, **5**, 320–324.

I. G. MEIGHAN, A. G. MCCORMICK, D. GIBSON, Department of Geology, The Queen's University of Belfast, Belfast BT7 1NN, N. Ireland.

J. A. GAMBLE, Research School of Earth Sciences, Victoria University of Wellington, Private Bag, Wellington, New Zealand.

I. J. GRAHAM, Institute of Nuclear Sciences (D.S.I.R.), Private Bag, Lower Hutt, New Zealand.

Strontium isotope geochemistry of Tertiary igneous rocks, NE Ireland

P. J. O'Connor

The Tertiary igneous rocks of NE Ireland form an integral part of the British Tertiary Igneous Province (BTIP) and comprise extrusive plateau basalts (the Antrim Basalts), an extensive NW–SE dyke swarm and intrusive central complexes at Slieve Gullion, Carlingford and the Mourne Mountains. The igneous rocks of the central complexes are predominantly either acid (leucogranites, epigranites, granophyres, felsites) or basic (gabbros, dolerites, basalts) with very subordinate intermediate compositions represented.

The strontium isotope compositions of a diverse suite of 50 whole-rock samples from the Irish subprovince were determined to establish the isotopic signatures of the Irish intrusive centres and to ascertain the extent to which these signatures reflect the petrogenesis of contemporaneous acid and basic magmas.

In Table 1, the initial $^{87}Sr/^{86}Sr$ ratios (at 60 Ma) of the basic intrusive and extrusive rocks studied show a progressive increase from 0.7044 ± 0.0001 (Antrim basalts) through 0.7056 ± 0.0002 (Slieve Gullion dolerites) to 0.7062 ± 0.0009 (Carlingford gabbros). Mass balance calculations indicate that the parental mantle-derived basic magmas have been substantially modified by a process of *selective* contamination with crustal ^{87}Sr from a low Sr, high $^{87}Sr/^{86}Sr$ source. At Carlingford, the gabbros define a pseudo-isochron with an apparent age of ca. 910 Ma which is interpreted as a contamination mixing line between magmatic and crustal Sr. The initial Sr isotope ratios of the gabbros show positive correlation with Sr, Rb, Ba, P_2O_5, K_2O

and Na_2O. Similar correlations are apparent in the Cuillins basic pluton, Skye (Moorbath & Thompson 1980) and suggest that relative addition of ^{87}Sr to the basic magmas occurred at high temperature during fractionation, perhaps in more than one cycle, by a process of selective diffusion (volatile transfer of Sr) from the wall rocks of the magma reservoir. The pattern of Sr isotopic variation reflected in the sequence basalt–dolerite–gabbro may be related to the residence time of basic magma in crustal reservoirs; the highest initial ratios are observed in the layered basic–ultrabasic masses at Carlingford and the Cuillins (Skye) which have cooled more slowly and have had the greatest opportunity to interact and equilibrate with crustal Sr.

Closed-system mixing models (Pushkar *et al.* 1972) show that the volumetrically insignificant developments of hybrid rocks found in small scale acid–basic associations (net-veins, sheets, pipes) at Slieve Gullion and Carlingford can be derived by straightforward mixing of acid and basic end-member compositions in definite proportions, e.g. the acid hybrids of the Slieve Gullion central complex (d in Fig. 1) may be derived by mixing peripheral granophyre (b) and dolerite (e) in the proportions 80:20, and a similar relationship prevails at Carlingford. The major and trace-element compositions of the hybrid rocks have been calculated from the empirically derived mixing models and have been found to be in good agreement with the observed compositions.

The Tertiary acid rocks of NE Ireland (Table 1) have initial Sr isotope ratios ranging from

TABLE 1. *Mean compositions of rock suites, NE Ireland*

Rock suite	(n)	Rb(ppm)	Sr(ppm)	$(^{87}Sr/^{86}Sr)i$	Assumed age (Ma)
Antrim basalts	3	2.1 ± 0.9	228 ± 103	0.7044 ± 0.0001	60.0
Carlingford gabbros	4	3.1 ± 4.4	232 ± 104	0.7062 ± 0.0009	60.0
Slieve Gullion dolerites	4	10.3 ± 10	218 ± 23	0.7056 ± 0.0002	60.0
Slieve Gullion acid hybrids	3	166 ± 37	97 ± 40	0.7074 ± 0.0008	58.5
Slieve Gullion peripheral granophyre	3	244 ± 16	65 ± 25	0.7088 ± 0.0008	58.5
Slieve Gullion aplitic granophyres	3	260 ± 25	55 ± 12	0.7083 ± 0.0002	58.5
Slieve Gullion ring complex	6	180 ± 22	36 ± 49	0.7141 ± 0.0005	58.5
Carlingford granophyres (Cooley)	2	212 ± 55	47 ± 23	0.7069 ± 0.0003	58.5
Mourne granite (G2)	1	412	13	0.7107	56.0
Tardree rhyolite	1	134	44	0.7085	56.0

Errors for Rb, Sr and $^{87}Sr/^{86}Sr$ are 2 σ; $\lambda^{87}Rb = 1.42 \times 10^{-11}a^{-1}$

From MORTON, A. C. & PARSON, L. M. (eds), 1988, *Early Tertiary Volcanism and the Opening of the NE Atlantic*, Geological Society Special Publication No. 39, pp. 361–363.

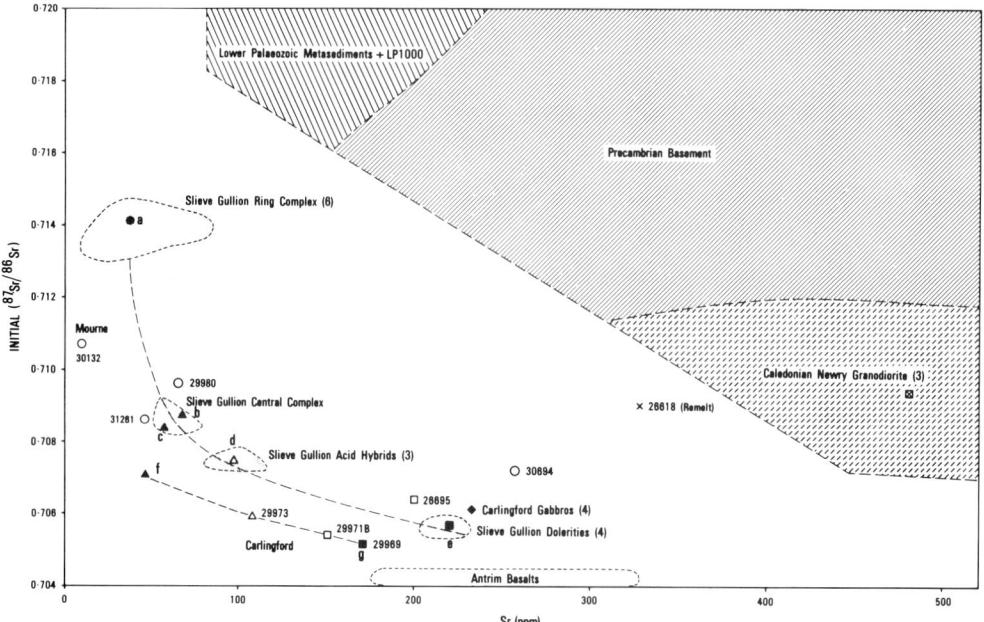

FIG. 1. Mean initial Sr isotope ratios vs Sr (ppm) content in acid–basic rock suites from the Irish Tertiary subprovince (data in Table 1). Curve ae: Slieve Gullion centre; a: Ring complex; b: Central complex peripheral granophyre; c: Central complex aplitic granophyres; d: Central complex acid hybrids; e: Central complex dolerites. Curve fg: Carlingford Centre.

0.707–0.714. When hybrid compositions are excluded, no significant overlap in the initial ratios of the basic and acid rocks is observed, a feature also common to Scottish Tertiary centres. This fact has been used as a fundamental argument against the derivation of the acid rocks of the Tertiary centres solely by fractionation from a parental basic magma (Moorbath & Bell 1965; O'Connor 1976). If the acid magmas were somehow derived from the basic magmas, then the Sr isotope data indicate that the acid magmas were selectively contaminated with crustally derived ^{87}Sr. Inverse hyperbolic mixing trends, such as those depicted in Figure 1 for the Slieve Gullion centre, support the selective contamination hypothesis and are clearly apparent in Sr isotope data reported for Mull (Walsh *et al.* 1979), Arran (Dickin *et al.* 1981) and Skye (Moorbath & Thompson 1980).

At the Slieve Gullion centre there is a definite temporal trend in initial ^{87}Sr/^{86}Sr ratios of the acid rocks to be observed. The earliest acid intrusive activity at 58.5 ± 2.3 Ma is represented by the outer Ring complex suite which is characterized by the highest initial ratios of 0.7141 ± 0.0005 (mean squares of weighted deviates, MSWD $= 3.47$), while the later phases of

activity in the inner Central complex have lower initial ratios (0.708–0.709). Furthermore, the Central complex acid suite of Slieve Gullion has not been derived solely by fractionation from the earlier Ring complex acid suite; successive acid suites display distinctive geochemical and isotopic signatures (O'Connor 1985) and represent successive magma pulses.

Similar temporal trends of decreasing initial ratios are apparent in the Sr isotope data reported for acid suites from other Tertiary centres such as Skye (Moorbath & Bell 1965), Mull (Walsh *et al.* 1979) and the Mournes (Gibson *et al.* 1987). The earliest melts to be emplaced in any particular intrusive cycle have the highest initial ratios, with later melts having lower ratios. These observations suggest that a substantial crustal contribution, perhaps derived by partial melting of reservoir rocks, was involved in the first-formed acid magmas at Slieve Gullion and other centres. Subsequently produced acid magmas have lower initial ratios which reflect a progressive dilution of the crustal component. A combined assimilation–fractionation model (DePaolo 1981) seems to best explain the Sr isotopic relationships observed in the rocks of the Slieve Gullion centre.

References

DePaolo, D. J. 1981. Trace element and isotopic effects of combined wallrock assimilation and fractional crystallization. *Earth & Planetary Science Letters*, **53**, 189–202.

Dickin, A. P., Moorbath, S. & Welke, H. J. 1981. Isotope, trace element and major element geochemistry of Tertiary igneous rocks, Isle of Arran, Scotland. *Transaction of the Royal Society of Edinburgh: Earth Sciences*, **72**, 159–170.

Gibson, D., McCormick, A. G., Meighan, I. G. & Halliday, A. N. 1987. The British Tertiary igneous province: young Rb-Sr ages for the Mourne Mountain granites. *Scottish Journal of Geology*, **23**, 221–225.

Moorbath, S. & Bell, J. D. 1965. Strontium isotope abundance studies and rubidium–strontium age determination on Tertiary igneous rocks from the Isle of Skye, north west Scotland. *Journal of Petrology*, **6**, 37–66.

—— & Thompson, R. N. 1980. Strontium isotope geochemistry and petrogenesis of the early Tertiary lava pile of the Isle of Skye, Scotland, and other basic rocks of the British Tertiary Province: an example of magma–crust interaction. *Journal of Petrology*, **21**, 295–321.

O'Connor, P. J. 1976. Strontium isotope ratios of some acid rocks from Mull and Arran, Scotland. *Geological Magazine*, **113**, 389–391.

—— 1985. Radioelement geochemistry of Irish Tertiary granites. *In: High Heat Production (HHP) Granites, Hydrothermal Circulation and Ore Genesis.* Institution of Mining and Metallurgy, pp. 239–249.

Pushkar, P., McBirney, A. R. & Kudo, A. M. 1972. The isotopic composition of strontium in central American ignimbrites. *Bulletin Volcanology*, **35**, 265–294.

Walsh, J. N., Beckinsale, R. D., Skelhorn, R. R. & Thorpe, R. S. 1979. Geochemistry and petrogenesis of Tertiary granitic rocks from the island of Mull, northwest Scotland. *Contributions to Mineralogy and Petrology*, **71**, 99–116.

P. J. O'Connor, Geological Survey of Ireland, Haddington Road, Dublin 4, Eire.

A review of silicic pyroclastic rocks of the British Tertiary Volcanic Province

B. R. Bell & C. H. Emeleus

SUMMARY: Silicic airfall and ashflow tuffs constitute a significant component of the extrusive products of the British Tertiary Volcanic Province. These deposits typically formed early in the evolution of individual volcanic centres as proximal facies accumulations intercalated with polylithic pyroclastic breccias, and occur throughout the evolution of the Province. Rhyolites and, less commonly, trachytes occur as lavas in close spatial and temporal association and, together with the silicic pyroclastic rocks, are interpreted as the extrusive equivalents of the felsic and felsic–mafic (mixed-magma) subvolcanic intrusions which are important members of the central intrusive complexes of the Province. The field relations of the silicic tuffs suggest that the associated pyroclastic breccias are also often extrusive and together they form crater infills.

The British Tertiary Volcanic Province (BTVP; Fig. 1) consists of a number of intrusive centres, lava fields and associated dyke swarms of Palaeocene age (ca. 62–52 Ma). The Province is located in a tensional tectonic setting within continental crust. A recent analysis of the depth of the Mohorovičić Discontinuity beneath the British Isles (Meissner *et al.* 1986) showed that the majority of the intrusive centres are located within or close to a zone of crustal thinning which was probably initiated in Permian times and which extends in an approximately N–S line down the W coast of Scotland, into the Irish Sea and the Western Approaches (Meissner *et al.* 1986, Fig. 1b).

This paper reviews occurrences of intrusive and extrusive silicic pyroclastic rocks, and lava flows of rhyolite and trachyte, which are important components of the Province and which largely, although not exclusively, occur within the intrusive centres. The paper is, however, not exhaustive in its coverage. We deal only briefly with the occurrences of pyroclastic material associated with the Ardnamurchan, Mull and Arran igneous centres, as these areas are the subject of ongoing studies. In addition, some of the earliest volcanic activity in the BTVP produced ash deposits at the base of the lava fields, but these are not considered further since the deposits are basaltic and clearly relate to the voluminous basaltic fissure eruptions which formed the lava fields and were fed from the regional dyke swarms.

The intrusive centres developed within a variety of basement and cover-sequence rock types and were emplaced at depths as shallow as 1–2 km. The explosive volcanism is usually intimately associated with the earlier stages in the evolution of individual centres. Consequently, interpretation of these rocks is frequently difficult because of the overprinting effects of subsequent intrusive and tectonic events within the centres. Furthermore, pervasive hydrothermal alteration of the rocks of the intrusive centres and their surroundings has, to a certain extent, obscured some of the petrographic features of the fine-grained pyroclastic material.

Many of the terms used to describe pyroclastic deposits are subject to varied usage. In this account the terms employed and their meaning are those of Fisher & Schmincke (1984). In the past the term *explosion breccia* has been used as a variant of *pyroclastic breccia* and a much broader usage, in terms of clast origins, has been attached to *agglomerate* than that now advocated (Fisher & Schmincke 1984, p. 92); the older usage of agglomerate will be implicit when considering work done prior to the 1950's.

Where reference is made to specific areas in Scotland the National Grid position may be given either in the four-figure (e.g. NG 12 34) or six-figure (e.g. NG 123 456) form.

Historical perspective and volcanological interpretation

Since the early observations of Macculloch (1819), von Oeynhausen & von Dechen (1829) and Forbes (1845), there has been sporadic interest in the many occurrences of pyroclastic rocks within the BTVP. Several observations were presented by Judd (1874) and Geikie (1888), with the general concensus that the coarse-grained, heterogeneous deposits represent vent infills, although their time relationships to the lavas, or to the subvolcanic intrusions, were

From MORTON, A. C. & PARSON, L. M. (eds), 1988, *Early Tertiary Volcanism and the Opening of the NE Atlantic,* Geological Society Special Publication No. 39, pp. 365–379.

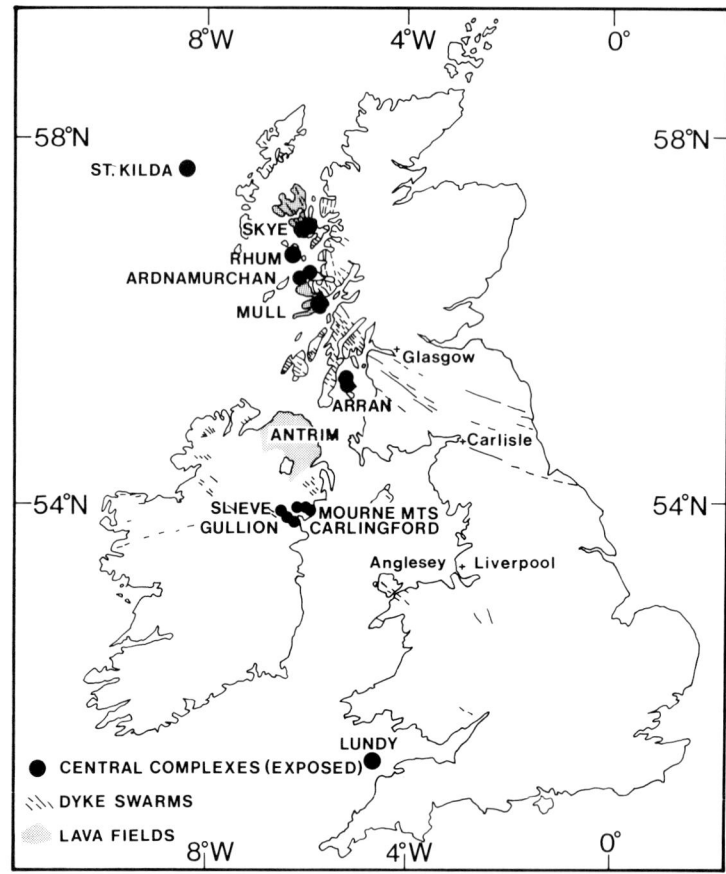

FIG. 1. Sketch map of the British Tertiary Volcanic Province.

deemed obscure. Many of the features described below are based on the writings of the field geologists of the British Geological Survey who present much data in the memoirs produced to accompany the published geological maps.

Skye

Harker (1904) produced a wealth of information on the pyroclastic rocks of Skye for the British Geological Survey (Fig. 2). First, he noted that significant accumulations of pyroclastic material occur: (1) at Kilchrist, and further N at Creagan Dubh [NG 60 21]; (2) in Srath na Creitheach [NG 51 22]; and (3) N of Belig [NG 54 25]. Harker concluded that these pyroclastic rocks are the product of localized episodes of explosive volcanism early in the development of the Skye Centre and are essentially vent infills. Significantly, Harker discussed the polylithic nature of these accumulations and noted that bombs cannot be identified. On the basis of the known local

stratigraphy, it was demonstrated that downward slumping of blocks had occurred. In the Belig occurrence, Harker noted the presence of silicic fragments within a matrix which contains a significant proportion of material of similar composition. Tuffaceous material occurs within all of the accumulations noted above, especially (1) and (2).

N of the Cuillin Complex, at Fionn Choire [NG 45 26], Harker described a thick, although localized, development of intercalated silicic lavas, tuffs and agglomerates. From the field relationships of these volcanic products it was suggested that the point of eruption migrated slowly with time. These evolved extrusive products are intercalated with the plateau lava field, which predates the intrusive centres, and provide good evidence of small, localized chambers that underwent differentiation towards evolved compositions. Thompson (1967) re-interpreted the lavas as silicified trachytes rather than as rhyolites, as Harker believed them to be.

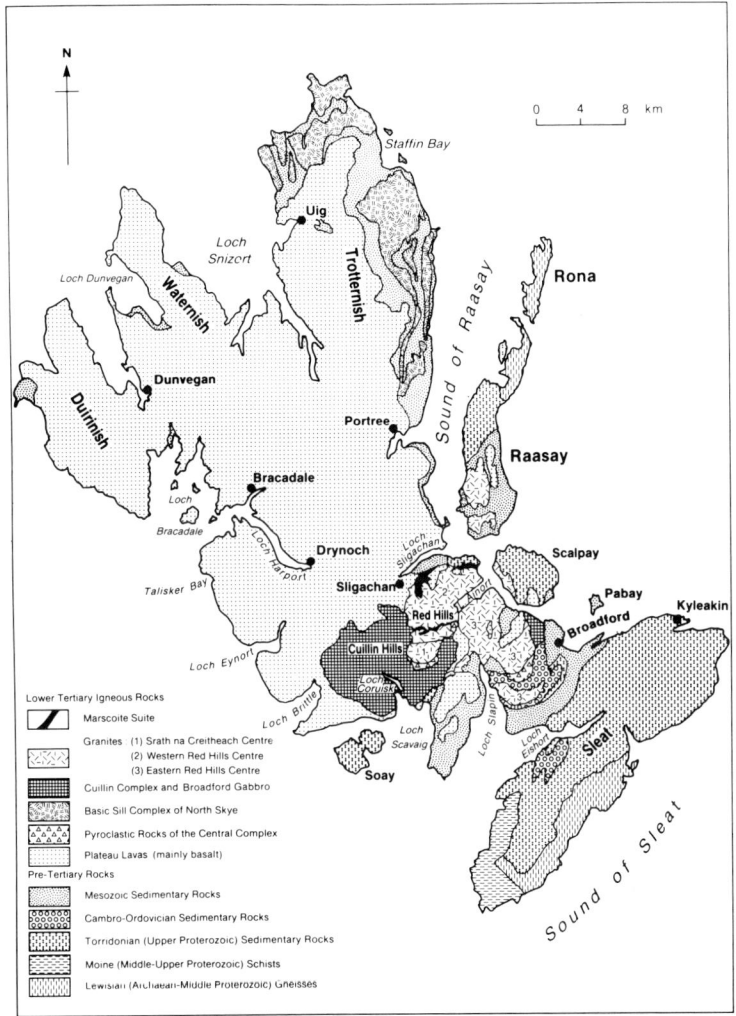

FIG. 2. Geological sketch map of the Isle of Skye.

The deposits at Kilchrist, in the Eastern Red Hills district (Fig. 3), consist of a thick sequence of pyroclastic breccias containing cognate and accidental fragments (Bell 1985). These rocks are intercalated with silicic airfall tuffs and ashflow tuffs which exhibit compaction fabrics defined by fiamme with aspect ratios (length:thickness) of 20:1, or more. The pyroclastic breccias do not appear to be bedded and sorting is minimal. The matrix material is extremely heterogeneous with, in places, juvenile material in a poor state of preservation. Recrystallization effects are, in places, severe, with replacement textures involving chlorite, epidote and clays (kaolinite and montmorillonite).

The upper surface of the largest outcrop of ashflow tuff, within the Allt nan Suidheachan [NG 5962 2078] (see Fig. 3), provides good evidence of rheomorphism (secondary flowage). The overlying pyroclastic breccias occupy erosion hollows within the upper, weathered surface of the tuff.

Within the Allt Coire Forsaidh [NG 6030 2124] (Fig. 3) a pyroclastic breccia with juvenile silicic clasts (rhyolitic and pumiceous) crops out within the main polylithic pyroclastic breccias. This deposit is intensely altered and is dominated by secondary low-temperature minerals.

In summary, we conclude that the Kilchrist deposits represent a thick sequence of polylithic pyroclastic breccias and silicic (airfall and ashflow) tuffs. These accumulations were deposited

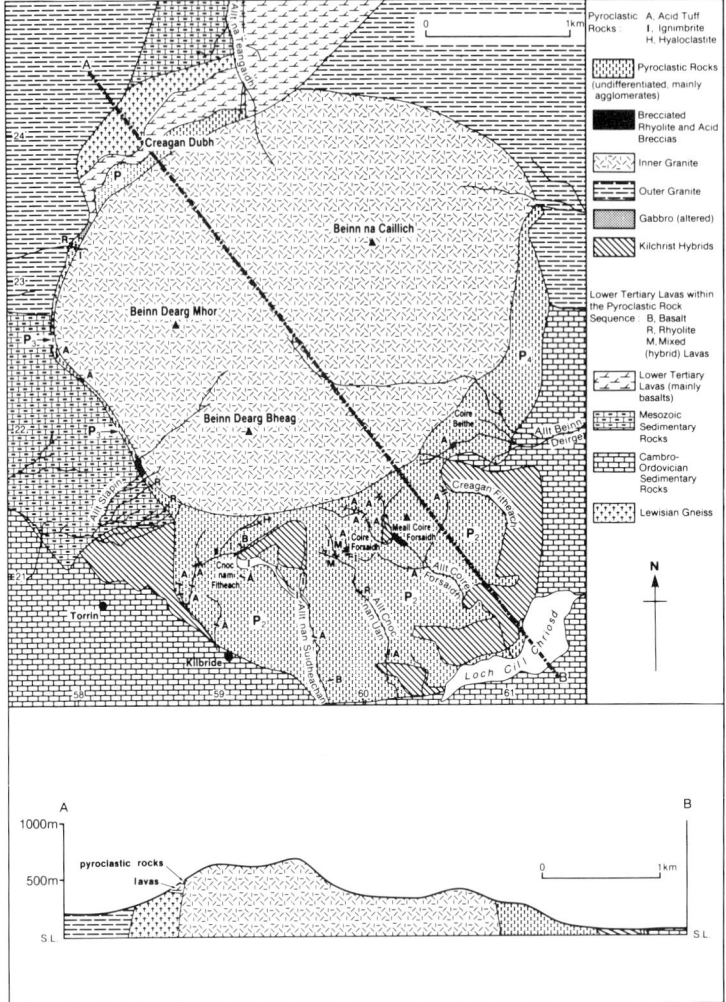

FIG. 3. Geological sketch map of the Kilchrist area, Eastern Red Hills, Isle of Skye.

on an irregular volcanic depression marginal to some form of vent. During, or soon after, the period(s) of eruption subaerial erosion and re-working occurred, resulting in their present field relationships. Subsequently, the volcanic pile was down-faulted as a subsidence block during the emplacement of the Kilchrist ring-dyke (Bell & Harris 1986).

The pyroclastic rocks of Srath na Creitheach [NG 51 22] exhibit many of the features described above. However, within these deposits there is a distinct lack of silicic pyroclastic material. In addition to coarse pyroclastic breccias, Jassim & Gass (1970) recognized a number of fine-grained, well-bedded basaltic tuffs. The breccias are dominated by cognate blocks and slabs of gabbro

and basalt, together with blocks of (juvenile?) trachyte. The whole mass is bounded by a ring-fault, along which the sequence was down-faulted to its present position.

The Belig deposits of the Western Red Hills district [NG 54 23] have been described in detail by Bell (1966). They are dominated by unbedded pyroclastic breccias which contain accidental fragments of the local basement rock types. In addition, a significant component of these depos-its consists of juvenile silicic material in the form of quartz porphyry and microgranite. This close association of silicic rock-types and pyroclastic breccias was interpreted by Bell in terms of gas-streaming from a volatile-rich silicic magma which stoped to a high level within the volcanic centre. The spatially-associated pyroclastic breccias were formed by volatile release during the final stages of emplacement.

The trachytic lavas and associated pyroclastic deposits at Fionn Choire [NG 45 26] have the form of localized outpourings from a vent which was superimposed on the plateau lava-field. Intercalations of locally cross-bedded pyroclastic breccia and tuff with occasional lateritized surfaces suggest that subaerial reworking of volcanic detritus took place within an area of elevated topography.

Rhum

Pyroclastic breccias and porphyritic felsite occur together within the Main Ring Fault (MRF) (Fig. 4) in the Southern Mountains (Hughes 1960) and the Northern Marginal Complex (Dunham 1968). They are cut by numerous basaltic sheets and by gabbros and ultrabasic rocks of the Rhum Layered Complex which have imposed a strong thermal overprint (Emeleus 1987).

The pyroclastic breccias consist essentially of accidental fragments (Lewisian, Archean, gneiss and Torridonian, Upper Proterozoic, clastic sedimentary rocks), together with localized cognate fragments (Tertiary gabbro, dolerite) and very rare juvenile material. These rocks are overlain by or interleaved with porphyritic felsite which occasionally is clearly intrusive, as in Coire Dubh [NM 393 980]. Intrusive tuff (tuffisite, Reynolds 1954) dykes and sheets with a high

proportion of juvenile material were identified cutting pyroclastic breccias by Hughes (1960, plate 13, Fig. 1) and Dunham (1968, plate 25) and along the felsite–breccia contact. Granophyre and microgranite, of compositions identical to the felsites, occur in western Rhum (see Fig. 4).

Judd (1874) identified the silicic rock-types in eastern Rhum as 'felstone lavas and agglomerates'. Subsequently, these felsites were interpreted by Geikie (1888) and Harker (1908) as being intrusive. However, Harker concluded that the associated breccias were related to Lower Palaeozoic thrusts and thus were not of volcanic origin. Bailey (1945) re-interpreted these fragmental rocks as explosion breccias located within the MRF (see Fig. 4) which bounds the Rhum central complex. Further, Bailey recognized a close association between the felsites and explosion breccias and suggested that a genetic relationship existed. He concluded that the felsites and breccias occupied a volcanic conduit associated with the intrusive complex.

A significant observation made by Harker (1908) and alluded to by Bailey (1945) is the development of a 'flow structure' within the felsites. These features within the felsites are further discussed by Hughes (1960) and Dunham (1968), who also stressed the close association of the felsites with polylithic explosion breccias and intrusive tuffs. Essentially, Hughes and Dunham followed the model of Bailey and concluded that these volcanic rocks are intrusive. A re-examination of the Rhum explosion breccias, intrusive tuffs and porphyritic felsite is in progress, following Williams' (1985) discovery that over much of the Cnapan Breaca there is a widespread development of up to several metres of well-bedded lithic lapilli-tuff and lithic and crystal tuffs along the boundary between the breccias and the overlying felsite sheets. Subsequent investigations (Emeleus, unpublished) have shown that towards the E of Cnapan Breaca (at ca. [NM 396 976]) there is a zone, several metres thick, of bedded lapilli-tuffs and tuffs comformable with the base of the felsite but some 50 m below. These bedded deposits occur well within the pyroclastic breccias which, in turn, hereabouts show an ill-defined bedded structure conformable with the other features mentioned. A thin, conformable layer of tuff also occurs within the felsite some metres above the base. The general relationships on the E side of Cnapan Breaca are shown in Fig. 5. A well-defined, discontinuous streaky-banded structure, hitherto interpreted as flow-banding, is conspicuous at and near the base of the Cnapan Breaca and Meall Breac felsites (Fig. 6) and we fully concur

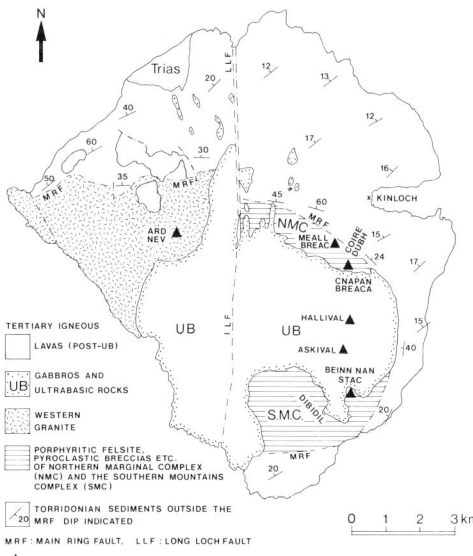

FIG. 4 Geological sketch map of the Isle of Rhum

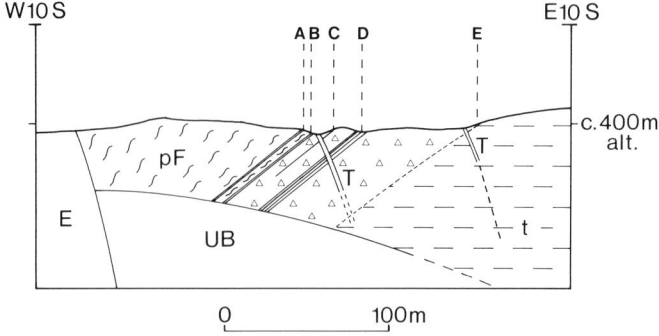

FIG. 5. Cross-section through the Northern Marginal Complex, Rhum, on the NE side of Cnapan Breaca, showing the relationships between: porphyritic felsite (= welded tuff) (pF); bedded tuffs and lapilli-tuffs (close-spaced parallel lines at **A, B & D**); pyroclastic breccias (open triangles); and, Torridonian sedimentary rocks (t) (bedding is horizontal in this section, dips ca. 45° to the S hereabouts) cut by thin tuffisite dykes (T) all underlain and intruded by ultrabasic rocks (UB) and cut by a late gabbro (E). Bedded tuffs within base of felsite at **A**, faint bedded structure in pyroclastic breccias at **C**, ca. 2 m of well-bedded lithic tuffs and crystal tuffs at **D**. Pyroclastic breccia overlies Torridonian rocks at **E**. Section is drawn in approximately the dip direction of the pyroclastic deposits. Horizontal and vertical scales equal.

with Williams' re-interpretation of these as fiammé; surfaces similar to that illustrated by Williams (1985, Fig. 3) occur at many points on these hills and also in the Southern Mountains (R. Greenwood & M. Errington, pers. comm.). All of the pyroclastic rocks of Cnapan Breaca and Meall Breac have been heavily thermally metamorphosed by the later mafic rocks of the Layered Complex (cf. Fig. 4). Despite extensive recrystallization, flattened and partially-flattened relicts of shards are occasionally visible in thin-section where the former contribute to the 'flow-banded' appearance of the rocks (Fig. 7). Thus, there is widespread evidence that the Rhum porphyritic felsites are in fact welded tuffs.

The deposits of Cnapan Breaca, and similar ones on Meall Breac ([NM 39 98]; Figs. 4 & 5) are considered to be a sequence of pyroclastic breccias which pass upwards into interbedded breccias, lapilli-tuffs and lithic and crystal tuffs probably of airfall origin. Upwards, these are interbedded with welded ashflow tuffs which form thick sheets locally capping the succession. These pyroclastic rocks were formed when rhyodacitic magmas rose along the MRF (see Fig. 4) during a period of subsidence and caldera formation (Emeleus *et al.* 1985). They are thus clearly to be regarded as surface deposits, formed on the caldera floor which is represented by unbrecciated Torridonian strata between Cnapan

FIG. 6. Surface of porphyritic felsite showing numerous, aligned, dark fiammé. SW edge of Meall Breac, Rhum (locality I.6 in Emeleus & Forster 1979). Scale: hammer shaft ca. 35 cm.

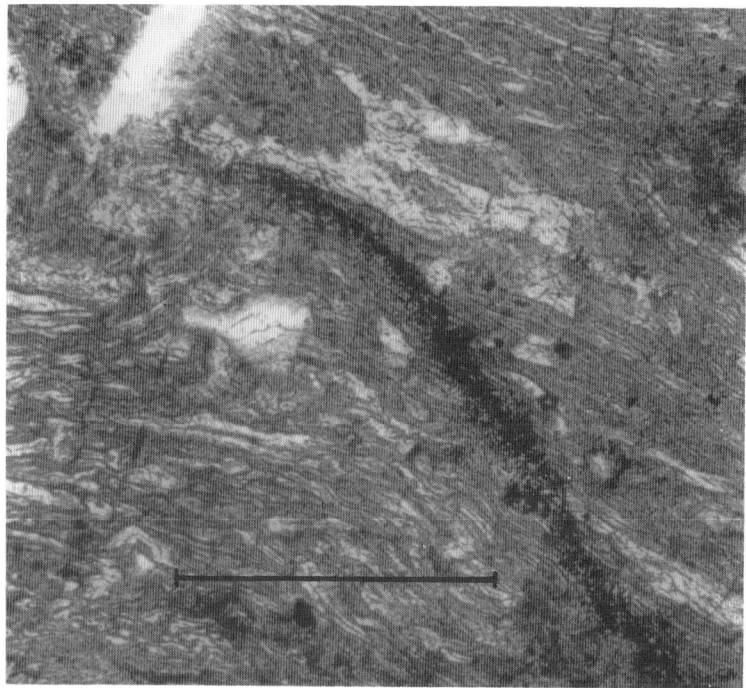

FIG. 7. Photomicrograph of porphyritic felsite matrix showing pumice structure and collapsed shards. Beinn nan Stac, Rhum (No. DU.13876). Scale bar = 0.2 mm. Plane polarized light.

Breaca and the MRF. After the accumulation of the surface pyroclastic deposits, gases continued to stream-off magma still present at depth within the MRF. The gases entrained partially consolidated porphyritic felsite magma and mixed this with small fragments of country-rock dislodged from the sides of cracks and fissures. This gas-entrained mixture was injected into the pyroclastic breccias and underlying Torridonian strata, forming the dykes and irregular areas of tuffisite recognized by Hughes (1960) and Dunham (1968). As noted above, similar features to those found in the Cnapan Breaca–Meall Breac area occur in the Southern Mountains district. However, Hughes (1960) clearly demonstrated that on Sgurr nan Gillean [NM 385 932] undisturbed Torridonian strata occurs between flat-lying sheet-like bodies of felsite and pyroclastic breccia exposed at various levels on the eastern side of the mountain (Hughes 1960, plate 14). Further research on this area by M. Errington may elucidate these apparently contradictory relationships.

Felsite on the N side of the Sgurr nan Gillean and in the small plug-like intrusion N of Cnapan Breaca contains rounded, lobate inclusions of fine-grained mafic rock (Dunham 1968; Forster 1980) indicating that mixed-magmas were present during this early explosive stage in the Rhum intrusive complex.

Ardnamurchan

The composition and field relationships of the pyroclastic rocks associated with the Ardnamurchan igneous complex are documented in detail by Richey & Thomas (1930) and Richey (1938). Of the three centres of igneous activity within the complex it is the oldest, Centre 1, which is dominated by pyroclastic rocks. These volcanic accumulations were considered by Richey to be the product of 'rhythmic eruptions' of an Ardnamurchan volcano. The bedded nature of the intercalated tuffs and pyroclastic breccias is described by Richey and the most complete sequence is superbly exposed in the section at Maclean's Nose [NM 53 61]. We concur with the general model proposed by Richey regarding two distinct vents. Both structures cut deeply into the plateau lava-sequence of Ardnamurchan and into the underlying basement of Upper Proterozoic (Moine) schists. Close to the margins of the outcrops of breccia, significant amounts of accidental material, mainly schist, but with some fragments of the overlying Mesozoic formations, dominate the clast population. Within the main part of the breccias the dominant pyroclasts are juvenile and of trachytic composition. We interpret the 'rhyolite' pyroclasts described by Richey as fragments of ashflow tuff and consider them to be related to the development of the pyroclastic breccias. The ashflow tuff is not found outwith the breccias and consists of partially-devitrified fiammé, with aspect ratios of more than 30:1, in a matrix of fragmented pumice clasts (intensely altered) and welded tuff. Dispersed throughout the ashflow tuff clasts are small fragments of mafic composition and crystals of sodic plagioclase.

Essentially, these deposits have the form of a crater infill on the margin of a shallow volcanic edifice. The pyroclasts of ashflow tuff and trachyte we consider to have been co-magmatic with the main pyroclastic breccias. However, we are not able to substantiate the 'rhythmic eruption' model of Richey, who suggested that cycles of volatile build-up and release within an underlying magma body sequentially and repeatedly gave rise to juvenile tuffs and pyroclastic breccias. It would appear more likely to us that a more continuous production of tuff (ash) was, from time to time, swamped by the slumping of unstable pyroclastic breccias from the margin(s) of a scallop-shaped crater. The nearby biotite trachyte mass at [NM 5550 6265] is dome-shaped and exhibits geochemical affinities with the trachytic component of the pyroclastic accumulations. Significantly, these examples of alkaline magmatism are not recorded elsewhere in the Ardnamurchan igneous complex. However, such compositions abound in the Mull igneous complex, some 25 km to the S. Magmatic 'plumbing' on this scale is a common feature of several modern volcanoes (for example, Hawaii) and, therefore, we suggest that the juvenile pyroclastic material and trachyte plug are more likely related to the Mull volcanic centre.

Mull

Bailey *et al.* (1924) described in detail the pyroclastic rocks associated with the Mull igneous complex and concluded that two groups can be identified: (1) conformable to the lavas, e.g. Coire Mor [NM 68 36] and Loch Spelve [NM 66 25]; and (2) in a transgressive relationship to the pre-Tertiary country-rocks and Lower Tertiary volcanic and subvolcanic rocks as vent infills, e.g. NE and SW of Loch Ba. These vents developed, according to Bailey *et al.*, in close association with the two caldera collapse structures—an early basaltic caldera and a younger silicic caldera,

both of which occur within the central complex. In close spatial association with these agglomerates and tuffs are a number of 'rhyolites' which exhibit 'fluxion' textures.

The pyroclastic breccias intercalated with the lava pile contain accidental and cognate pyroclasts and their overall geometries suggest that they take the form of infills which developed within an irregular plateau lava topography. These accumulations are poorly sorted and it is unlikely that any of the material was transported any great distance.

Bailey *et al.* (1924) suggested that repeated explosions were involved in the vent-forming phase of activity of the Mull igneous complex. Associated with the calderas are fine-grained silicic rocks, such as felsite. One of these, the Loch Ba Felsite, a ring-dyke, has a mixed-magma origin, containing fragments of fine-grained mafic material throughout its mass (Blake *et al.* 1965; Thompson 1980; Marshall & Sparks 1984). Pyroclastic breccias, preserved as a marginal facies to the ring-dyke, have been described by Lewis (1968).

The pyroclastic breccias associated with the Mull calderas are extremely variable in form and magmatic affinities. Early in the evolution of the central complex pyroclastic breccias developed outside the basaltic caldera [NM 61 28]. These deposits show evidence of stratification and in many ways are similar to the interlava deposits noted above. They contain accidental fragments of basement rock-types, together with cognate fragments of early silicic intrusions. We interpret these rocks as having formed exterior to the caldera structure in the form of a proximal talus associated with the early stages of doming within the Mull Volcano (Walker 1975).

Exterior to the calderas are localized developments of rhyolite lava and silicic tuff. The lavas are compound in form, commonly brecciated, and invariably altered.

Pyroclastic breccias are found within the early basaltic caldera, as well as in the younger silicic caldera (Bailey *et al.* 1924). However, in the case of the former, subsequent multiple intrusion of mafic and silicic plutons has destroyed many of the primary features of the pyroclastic rocks. Within both calderas, accidental pyroclasts of pre-Tertiary country-rocks are absent.

Within the younger caldera, centred around Loch Ba [NM 57 38], are substantial accumulations of pyroclastic breccia in close association with compound rhyolite lava flows and tumuli. The pyroclastic breccias contain significant amounts of juvenile silicic material and are interlayered with mafic caldera lavas. We consider that these deposits were formed at about the same time as the exterior Loch Ba ring-dyke which occupies the ring-fault bounding the caldera.

Small, localized accumulations of trachytic tuff and trachyte tumuli, similar to those at Fionn Choire on Skye and Maclean's Nose on Ardnamurchan, are found superimposed on the plateau lava-sequence of N Mull, W of Salen at [NM 5550 4325].

Arran

The model proposed by King (1954) to explain the evolution of the Central Ring Complex of Arran has many features we find agreement with and is similar to the proposed evolution of the younger caldera of the Mull Volcano (see above). We have not studied the complex in detail but offer the following comments based on preliminary observations: (1) the breccias within the marginal portion of the down-dropped block have been derived, predominantly, from erosion of the unstable, topographically-elevated country-rocks outwith the caldera block; (2) we note analogies between the rhyolite lavas and silicic tuffs which crop out in the marginal portions of the complex and the moat rhyolites of more recent calderas such as Long Valley (Bailey *et al.* 1976); and (3) the dacite plugs within the complex most likely represent frozen feeder-channels to the lavas and tuffs.

Ireland

Extrusive rocks

The Antrim lava pile consists of a Lower Basalt Formation, an Interbasaltic Formation and an Upper Basalt Formation (Old 1975; Preston 1981). Silicic pyroclastic rocks and rhyolite lava flows are restricted to the Interbasaltic Formation of mid-Antrim, where they are centred on the Tardree–Sandy Braes area (Fig. 8), while small outcrops occur sporadically for approximately 25 km to the NW, as far as Cloghwater. Deep weathering of these silicic rocks gave rise to workable bauxite deposits (Cole 1896; Cole *et al.* 1912; Eyles 1952). The rocks are frequently 'fluidal and banded' (cf. Cole 1896, plate IV) and at Sandy Braes the presence of agglomerate and tuff involving perlitic obsidian was recognized in addition to rhyolite flows. Subsequently, Cameron & Sabine (1969) identified blocks of welded tuff in the agglomerate of Sandy Braes and suggested that the perlitic obsidian might represent extreme welding. Rhyolitic tuff and a tuffisite dyke were also encountered in boreholes. The

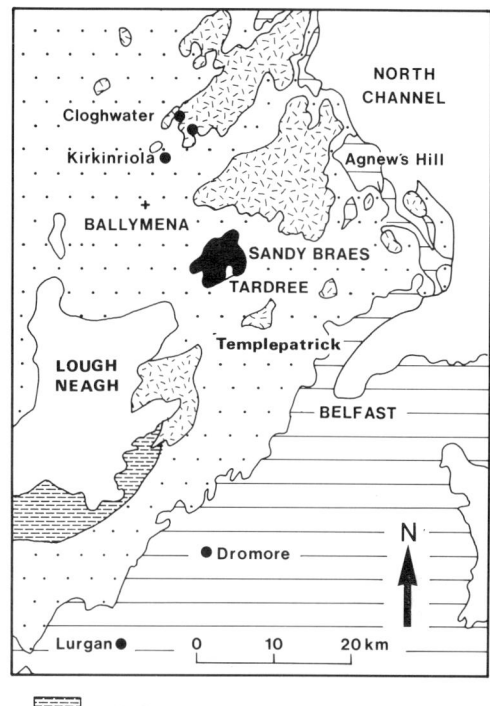

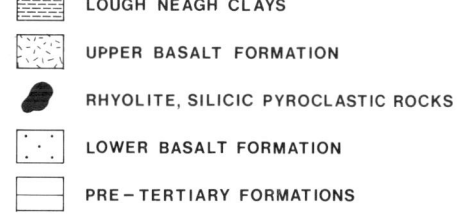

FIG. 8. Geological sketch map of the southern part of the Antrim Tertiary lava field.

sequence of events in the Tardree–Sandy Braes area was, firstly, a period of explosive activity during which pyroclastic breccias with accidental, cognate and probably juvenile fragments were formed. These were followed by a series of rhyolite and obsidian lava flows which built up a dome at Tardree (Old 1975). Nearby, at Sandy Braes, a later vent pierced the flows and was filled with blocks of glassy, welded tuff.

The occurrence of welded tuffs raises the possibility that silicic pyroclastic flows originating at Tardree and Sandy Braes may have extended far afield and been responsible for the scattered rhyolitic bodies encountered to the NW; similarly, ashflows encountered in boreholes at Agnew's Hill (about 10 km ENE of Tardree) may have come from this area (Old 1975). Geochemical investigations have been carried out on the rhyolites and obsidians by

Meighan & Gamble (1972) and Meighan (1979) has suggested that they have the characteristics of extreme fractionates. However, geochemical contrasts between the Tardree rocks and those from Kirkinriola, near Ballymena, argue against a direct connection (Meighan *et al.* 1984).

It has been suggested that the silicic rocks of the Tardree area may overlie a Tertiary central complex (Charlesworth 1963). However, this appears unlikely since the area does not have the pronounced positive Bouguer gravity anomaly which characterizes central complexes of the BTVP (Cook & Murphy 1952; cf. Bott & Tuson 1973). If there is a concealed central complex hereabouts, it was probably short-lived and never developed the characteristic 'root' of dense mafic rocks.

Small quartz porphyry plugs cut Lower Palaeozoic rocks near Dromore and Lurgan (Fig. 8) on the SE edge of the Antrim lavas. However, they are not associated with pyroclastic rocks.

The central complexes: Slieve Gullion

Significant bodies of granite and related silicic rocks are found in all three of the Irish central complexes. However, it is only in Slieve Gullion that there is an extensive development of silicic pyroclastic rocks, although there are small outcrops in the central, ill-exposed part of the Carlingford Complex (LeBas 1960).

The Slieve Gullion Complex consists of early ring-dykes, a central sheeted unit and a late pluton of pyroxene granophyre. For three-quarters of its circumference the ring-dyke portion is formed of porphyritic granophyre, but in the SW quadrant an earlier porphyritic felsite intrudes agglomerate near Forkill (Fig. 9) (Nolan 1877; Richey & Thomas 1932; Emeleus 1962). Nolan originally mapped and described the area, ascribing an aeriform origin to the agglomerates and suggesting that the felsites were a protrusion from the 'elvanite' (that is, the (later) porphyritic granophyre ring-dyke). Richey & Thomas (1932) considered that gases, explosively derived from the silicic magma, which ultimately formed the felsite, were responsible for the formation of the agglomerates. The banded (or flow) structures in the felsite were mapped by Emeleus (1962, Fig. 2, plate 1), who cited evidence for viscous flow and identified several foci of the flow-banding which were thought to mark the conduits followed by the felsite magma. The felsite–agglomerate contacts were found to be sharp and showed evidence of chilling; they are relatively steep except for one locality where the agglomerate, apparently underlain by felsite, is veined by fine-grained, flow-banded felsite to such an extent

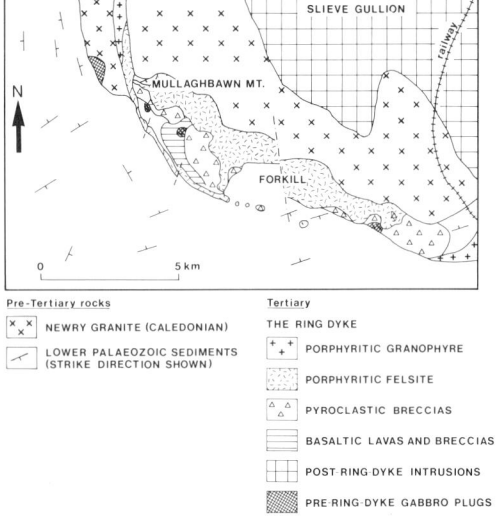

FIG. 9. Geological sketch map of the SW quadrant of the Slieve Gullion Ring-dyke, Co. Armagh, Ireland.

that felsite, rather than tuffaceous material, forms the matrix to the fragments. It was concluded that felsite had risen into agglomerate-filled vents in a viscous, cool condition, moving by laminar flow and occasionally suffering autobrecciation at the margins. The statement by Reynolds (1956) that the silicic rocks of the ring-dyke were an ignimbrite was discounted. Richey & Thomas (1932) also found that breccias (the breccias of Cam Lough) were developed in association with the porphyritic granophyre member of the complex. These are crush breccias, formed in country-rocks outside the ring-dyke and in the outer parts of the porphyritic granophyre ring-dyke.

The polylithic pyroclastic breccias and felsites of the Slieve Gullion ring-dyke have many similarities to the felsites and breccias of the Scottish centres, although there are significant differences. The pyroclastic breccias consist largely of accidental fragments derived from the Caledonian Newry Granite and Lower Palaeozoic strata (indurated shales) of the district, but there are also masses of (altered) olivine basalt, feldspar-phyric basalt and trachytic rocks which Richey & Thomas (1932) considered to have been pieces of a former cover of Tertiary lavas which had subsided into vents. Gabbroic fragments, matching small, local plugs of Tertiary age, are also found throughout the pyroclastic breccias and attest to the thorough mixing experienced by these rocks. Juvenile fragments (porphyritic felsite) are extremely rare.

There are no clear examples of bedding in the breccias. Occasional size-graded structures are either in sparse tuffisite dykes or else are attributable to flow of entrained matrix around large clasts. The relationships between the breccias and the country-rock are generally obscure due to a lack of good exposure, but their apparently irregular contacts, disregarding topography, suggest that they may be steep-sided.

The felsites consist of ca. 30% by volume of small phenocrysts of sanidine, quartz, plagioclase, hedenbergite and fayalite in fine-grained matrices which are finely banded and semi-vitreous in the marginal parts of the felsite bodies, but elsewhere are microcrystalline and of 'stony' appearance. A careful re-examination of relevant thin-sections of the marginal felsite has shown that they exhibit deformed, flattened shards (Fig. 10) and these, together with the fiammé which are conspicuous on weathered surfaces (cf. Emeleus 1962, Fig. 2, plate 4), are conformable with the felsite margins, only fading away where recrystallization of the initially glassy rocks obliterates the finer-scale structures, especially the shards. The wide distribution within the felsite of small (less than 1 cm), highly thermally-metamorphosed fragments of Lower Palaeozoic shale indicate initial high temperatures, mobility and mixing. However, the adjoining country-rocks are free from thermal effects and felsite within a few centimetres of the contacts may develop cataclastic structures indicated by phen-ocrysts and shards that have been fragmented and disrupted by drag against the walls. The felsite behaved as if it had been a cold, viscous mass on emplacement but had previously been through a high temperature, turbulent stage (Emeleus 1962).

Felsite–breccia and felsite–country-rock contacts are frequently steep and the felsite locally intrudes the breccias as thin dykes and small plugs, but only very occasionally is it found as veinlets and apophyses separating individual clasts. Apart from the scattered small inclusions already mentioned, rare xenoliths are recorded.

We consider that the Slieve Gullion felsite–pyroclastic breccia association represents a somewhat deeper level of erosion than that seen on Rhum and in the Kilchrist area on Skye. Volcanic vents are thought to have developed along the SW quadrant of the ring-dyke during central subsidence and surface caldera formation. Steadily degassing, silicic magma rose up the flaring vents, disrupting the country-rocks and subsiding masses of overlying lavas. The magma eventually underwent vesiculation, forming a magma–gas emulsion which swept up the vents, scouring their walls and occasionally sending off-shoots into the breccias, but only mixing with pyroclastic breccias where they were engulfed within the rising magma. As activity waned, cooling, consolidation and further degassing affected the

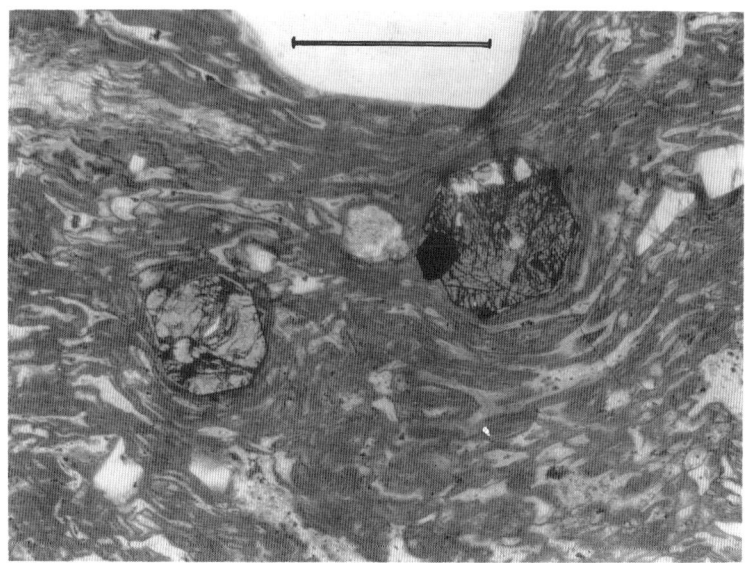

FIG. 10. Photomicrograph of porphyritic felsite with shard structures in matrix. Phenocrysts of quartz, alkali, feldspar, fayalite and hedenbergite also present. Located 0.5 m from pyroclastic breccia, NW side of Mullaghbawn Mountain, Slieve Gullion ring-dyke. (No. G.398b) Scale bar = 0.5 mm. Plane polarized light.

marginal felsite and a glassy selvedge was formed from shards and fiammé pressed against the country-rock by continued upward movement in the more central parts of the vents; sometimes the viscous drag was such that individual phenocrysts shattered and zones of flattened, subparallel shards were disrupted.

We conclude that the features found in the Slieve Gullion ring-dyke developed a short distance below the contemporaneous land surface; they mark a stage intermediate with the wholly intrusive features of some ring-dykes, for example the Slieve Gullion porphyritic granophyre, and materials which accumulated subaerially within a caldera such as on Rhum. The relationships in the SW of the Slieve Gullion ring-dyke may perhaps be most aptly compared with those described by Almond (1977) from the Sabaloka Caldera.

Reynolds (1951) interpreted the central sheeted complex of Slieve Gullion as a series of predominantly silicic agglomerates, tuffs and rhyolite lava-flows with interbedded basaltic lava-flows, cut by dolerite sills. These rocks were considered to have been subsequently transformed into felsites, microgranites, porphyritic granophyres, dolerites and feldspar-phyric gabbros during pneumatolitic and hydrothermal activity within a caldera. Her interpretation was disputed by Bailey & McCallien (1956), who re-interpreted the sheeted centre as forming the upper parts of a series of ring-dykes, with roof and wall relationships, and flat-lying screens of country-rock of Caledonian granodiorite occasionally preserved. Present opinions favour the latter interpretation, although Reynolds' suggestion that a caldera had been present requires reconsideration.

Discussion

Magmatic affinities

The magmatic evolution of the BTVP has been discussed in detail by Thompson (1982). In a general framework, the earliest magmas had alkaline affinities, interpreted in terms of small degrees of partial-melting, giving way, as the thermal and magmatic episode reached its peak, to tholeiitic magmas. Subsequently, as melting waned (and possible high-pressure fractionation of magmas in the lower crust took place), there was a return to alkaline magmatism.

The alkaline affinities of many of the evolved lavas and the juvenile (trachytic) component of the pyroclastic breccias and silicic tuffs suggest either an early or a late position in the magmatic sequence for each volcanic centre. Clearly, field evidence suggests the former for many of the volcanic centres (e.g., Skye, Rhum and Ardnamurchan). However, pyroclastic activity appears to have taken place throughout the evolution of the Mull Centre.

We envisage the explosive volcanism and associated volatile-rich alkaline magmas to be the result of extreme crystal–liquid fractionation within small, high-level chambers superimposed upon the lava fields of the Province.

Regional setting

During the early stages in the development of the Province, magma arrived at the Earth's surface through many fissures as a consequence of the regional stress field. Locally, small magma chambers developed, both at depth and at a high level within the crust (Thompson *et al.* 1972, 1980). The random eruption history of the lavas (no simple fractionation trends with position in the lava stratigraphy) is compelling evidence for many small chambers to have developed. Superimposed on the lava pile, from time-to-time, localized volcanic edifices constructed of evolved lavas and pyroclastic debris developed (e.g. at Fionn Choire on Skye, in Antrim at Tardree–Sandy Braes, and in N Mull). Additionally, at certain structurally favourable sites magma ascent was focused and increased, resulting in the establishment of intrusive centres where mafic magmas were dominant, but at which silicic magmas were generated partly by crystal–liquid fractionation, but also through incorporation of low melting-point constituents from the adjoining crust (Dickin *et al.* 1984).

Many of the deposits we have described are clearly proximal facies of originally widespread deposits. The coarse pyroclastic breccias most likely only developed close to the volcanic structures. However, the associated silicic tuffs must have formed distinct horizons over considerable areas (Jacque & Thouvenin 1975; Knox & Morton 1983). Many of the earliest tuffaceous deposits which have been recorded from boreholes in the North Sea contain aegirine and arfvedsonite and attest to parental alkaline magmas (Knox & Morton 1983). These deposits can be correlated with the ash component of the Lower Tertiary Thanet Beds, as exposed at Pegwell Bay in SE England (Knox 1979). It is likely that these distal pyroclastic accumulations are the products of explosive volcanism of the types described in this contribution. However, further work is required in order to prove correlations between proximal and distal facies of these pyroclastic deposits.

Simplistic calculations, based on known thicknesses of tuffs close to the volcanic centres (approximately 20 cm up to several metres) and in distal areas such as the North Sea (up to 8 cm, Jacque & Touvenin 1975) suggest eruption volumes of the order of tens of cubic kilometres. Historical eruptions of this size include Novarupta (Mt Katmai, Alaska; Hildreth 1983) and Hekla (Iceland; Thorarinsson 1950).

Conclusions

This contribution reports the following observations:

(1) Significant silicic pyroclastic volcanism occurred during early Tertiary times on the eastern margin of the NE Atlantic basin, within a continental setting.
(2) The pyroclastic volcanism occurred throughout the development of the Province.
(3) Much of this pyroclastic material is preserved locally within remnants of dissected calderas.

References

ALMOND, D. C. 1977. The Sabaloka Igneous Complex, Sudan. *Philosophical Transactions of the Royal Society of London*, **A287**, 595–633.

BAILEY, E. B. 1945. Tertiary igneous tectonics of Rhum (Inner Hebrides). *Quarterly Journal of the Geological Society of London*, **100**, 165–188.

—— & McCALLIEN, W. J. 1956. Composite minor intrusions and the Slieve Gullion Complex, Ireland. *Liverpool and Manchester Geological Journal*, **1**, 466–501.

——, CLOUGH, C. T., WRIGHT, W. B., RICHEY, J. E. & WILSON, G. V. 1924. *Tertiary and post-Tertiary geology of Mull, Loch Aline, and Oban*. Memoir of the Geological Survey of Great Britain. HMSO, Edinburgh.

BAILEY, R. A., DALRYMPLE, G. B. & LANPHERE, M. A. 1976. Volcanism, structure and geochronology of Long Valley Caldera, Mono County, California. *Journal of Geophysical Research*, **81**, 725–744.

BELL, B. R. 1985. The pyroclastic rocks and rhyolitic lavas of the Eastern Red Hills district, Isle of Skye. *Scottish Journal of Geology*, **21**, 57–70.

—— & HARRIS, J. W. 1986. *An Excursion Guide to the Geology of the Isle of Skye*. Geological Society of Glasgow, Glasgow.

BELL, J. D. 1966. Granites and associated rocks of the eastern part of the Western Red Hills Complex, Isle of Skye. *Transactions of the Royal Society of Edinburgh*, **66**, 307–343.

BLAKE, D. H., ELWELL, R. W. D., GIBSON, I. L., SKELHORN, R. R. & WALKER, G. P. L. 1965. Some relationships resulting from the intimate association of acid and basic magmas. *Quarterly Journal of the Geological Society of London*, **121**, 31–49.

BOTT, M. H. P. & TUSON, J. 1973. Deep structure beneath the Tertiary volcanic regions of Skye, Mull and Ardnamurchan, north-west Scotland. *Nature, Physical Science*, **242**, 114–116.

CAMERON, I. B. & SABINE, P. A. 1969. The Tertiary welded-tuff vent agglomerate and associated rocks of Sandy Braes, Co. Antrim. *Report of the Institute of Geological Sciences*, **69/6**.

CHARLESWORTH, J. K. 1963. *Historical Geology of Ireland*. Oliver & Boyd, Edinburgh.

COLE, G. A. J. 1896. The rhyolites of the County of Antrim: with a note on bauxite. *Scientific Transactions of the Royal Dublin Society*, **6** (series 2), 77–119.

——, WILKINSON, S. B., McHENRY, A., KILROE, J. R., SEYMOUR, H. J., MOSS, C. E. & HAIGH, W. D. 1912. *The interbasaltic rocks (iron ores and bauxites) of north-east Ireland*. Memoir of the Geological Survey of Ireland. HMSO, Dublin.

COOK, A. H. & MURPHY, T. 1952. Measurements of gravity in Ireland. Gravity survey of Ireland north of the line Sligo–Dundalk. *Dublin Institute of Advanced Studies, Geophysical Memoir*, **2**(4), 36 pp.

DICKIN, A. P., BROWN, J. L., THOMPSON, R. N., HALLIDAY, A. N. & MORRISON, M. A. 1984. Crustal contamination and the granite problem in the British Tertiary Volcanic Province. *Philosophical Transactions of the Royal Society of London*, **A310**, 755–780.

DUNHAM, A. C. 1968. The felsites, granophyres, explosion breccias and tuffisites of the northeastern margin of the Tertiary igneous complex of Rhum, Inverness-shire. *Quarterly Journal of the Geological Society of London*, **123**, 327–350.

EMELEUS, C. H. 1962. The porphyritic felsite of the Tertiary ring complex of Slieve Gullion, Co. Armagh. *Proceedings of the Royal Irish Academy*, **62B**, 55–76.

—— 1987. The Rhum Layered Complex, Inner Hebrides, Scotland. *In*: PARSONS, I. (ed) *Origins of Igneous Layering*. D. Reidel, Dordrecht, pp. 263–286.

—— & FORSTER, R. M. 1979. *Field Guide to the Tertiary Igneous Rocks of Rhum*. Nature Conservancy Council, London.

——, WADSWORTH, W. J. & SMITH, N. J. 1985. The early igneous and tectonic history of the Rhum Tertiary Volcanic Centre. *Geological Magazine*, **122**, 451–457.

EYLES, V. A. 1952. *The composition and origin of the Antrim laterites and bauxites*. Memoir of the Geological Survey of Ireland. HMSO, Belfast.

FISHER, R. V. & SCHMINCKE, H-U. 1984. *Pyroclastic Rocks*. Springer-Verlag, Berlin.

FORBES, J. D. 1845. Notes on the topography and geology of the Cuchullin Hills in Skye and on the traces of ancient glaciers which they present. *Edinburgh New Philosophical Journal*, **40**, 76–99.

FORSTER, R. M. 1980. *A geochemical and petrological study of the Tertiary minor intrusions of Rhum, northwest Scotland.* PhD thesis (unpublished), University of Durham.

GEIKIE, A. 1888. The history of volcanic action during the Tertiary period in the British Isles. *Transactions of the Royal Society of Edinburgh*, **35**, 21–184.

HARKER, A. 1904. *The Tertiary Igneous Rocks of Skye.* Memoir of the Geological Survey of Great Britain. HMSO, Glasgow.

—— 1908. *The Geology of the Small Isles of Inverness-shire.* Memoir of the Geological Survey of Great Britain. HMSO, Glasgow.

HILDRETH, W. 1983. The compositionally zoned eruption of 1912 in the Valley of Ten Thousand Smokes, Katmai National Park, Alaska. *Journal of Volcanology and Geothermal Research*, **18**, 1–56.

HUGHES, C. J. 1960. The Southern Mountains igneous complex, Isle of Rhum. *Quarterly Journal of the Geological Society of London*, **116**, 111–131.

JACQUE, M. & THOUVENIN, J. 1975. Lower Tertiary tuffs and volcanic activity in the North Sea. *In*: WOODLAND, A. W. (ed) *Petroleum and the Continental Shelf of North-West Europe (Vol. 1. Geology).* Applied Science Publishers, Barking, 455–465.

JASSIM, S. Z. & GASS, I. G. 1970. The Loch na Creitheach volcanic vent, Isle of Skye. *Scottish Journal of Geology*, **6**, 285–294.

JUDD, J. W. 1874. The secondary rocks of Scotland. (second paper). On the ancient volcanoes of the Highlands and the relations of their products to the Mesozoic strata. *Quarterly Journal of the Geological Society of London*, **30**, 220–301.

KING, B. C. 1954. The Ard Bheinn area of the central igneous complex of Arran. *Quarterly Journal of the Geological Society of London*, **110**, 323–354.

KNOX, R. W. O'B. 1979. Igneous grains associated with zeolites in the Thanet Beds of Pegwell Bay, north-east Kent. *Proceedings of the Geologists' Association*, **90**, 55–59.

—— & MORTON, A. C. 1983. Stratigraphical distribution of Early Palaeogene pyroclastic deposits in the North Sea Basin. *Proceedings of the Yorkshire Geological Society*, **44**, 355–363.

LE BAS, M. J. 1960. The petrology of the layered basic rocks of the Carlingford Complex, Co. Louth. *Transactions of the Royal Society of Edinburgh*, **64**, 169–200.

LEWIS, J. D. 1968. Form and structure of the Loch Ba ring-dyke, Isle of Mull. *Proceedings of the Geological Society of London*, **1649**, 110–111.

MACCULLOCH, J. 1819. *A Description of the Western Islands of Scotland, Including the Isle of Man: Comprising an Account of their Geological Structure; with Remarks on their Agriculture, Scenery, and Antiquities. Vol. 1: Skye.* Hurst Robinson, London 262–419.

MARSHALL, L. A. & SPARKS, R. S. J. 1984. Origin of some mixed-magma and net-veined ring intrusions. *Journal of the Geological Society of London*, **141**, 171–182.

MEIGHAN, I. G. 1979. The acid igneous rocks of the British Tertiary Province. *Bulletin of the Geological Survey of Great Britain*, **70**, 10–22.

—— & GAMBLE, J. A. 1972. Tertiary acid magmatism in NE Ireland. *Nature, Physical Science*, **240**, 183–184.

——, GIBSON, D. & HOOD, D. N. 1984. Some aspects of Tertiary magmatism in NE Ireland. *Mineralogical Magazine*, **48**, 351–363.

MEISSNER, R., MATTHEWS, D. & WEVER, T. 1986. The "Moho" in and around Great Britain. *Annales Geophysicae*, **4(B6)**, 659–664.

NOLAN, J. 1877. *Explanatory memoir to accompany Sheet 70 of the maps of the Geological Survey of Ireland.* Memoir of the Geological Survey of Ireland. HMSO, Dublin.

OLD, R. A. 1975. The age and field relationships of the Tardree Tertiary Rhyolite Complex, Country Antrim, Northern Ireland. *Bulletin of the Geological Survey of Great Britain*, **51**, 21–40.

PRESTON, J. 1981. Tertiary igneous activity. *In*: HOLLAND, C. H. (ed) *A Geology of Ireland.* Scottish Academic Press, Edinburgh, 213–224.

REYNOLDS, D. L. 1951. The geology of Slieve Gullion, Foughill and Carrickarnan: an actualistic interpretation of a Tertiary gabbro-granite complex. *Transactions of the Royal Society of Edinburgh*, **62**, 85–143.

—— 1954. Fluidization as a geological process, and its bearing on the problem of intrusive granites. *American Journal of Science*, **252**, 577–613.

—— 1956. *Calderas and Ring Complexes.* Gedenboek, H. A., Brouwer, Mouton & Co., 's-Gravenhage, Netherlands, 355–379.

RICHEY, J. E. 1938. The rhythmic eruptions of Ben Hiant, Ardnamurchan, a Tertiary volcano. *Bulletin Volcanologique*, **3** (series 2), 3–19.

—— & THOMAS, H. H. 1930. *The Geology of Ardnamurchan, North-West Mull and Coll.* Memoir of the Geological Survey of Great Britain. HMSO, Edinburgh.

—— & —— 1932. The Tertiary ring complex of Slieve Gullion (Ireland). *Quarterly Journal of the Geological Society of London*, **88**, 776–849.

THOMPSON, R. N. 1967. The 'rhyolite' of Fionn Choire, Isle of Skye. *Proceedings of the Geological Society of London*, **1642**, 212–214.

—— 1980. Askja 1875, Skye 56 Ma: Basalt-triggered, Plinian, mixed-magma eruptions during the emplacement of the Western Red Hills granites, Isle of Skye. *Geologische Rundschau*, **69**, 245–262.

—— 1982. Magmatism in the British Tertiary Volcanic Province. *Scottish Journal of Geology*, **18**, 49–107.

——, ESSON, J. & DUNHAM, A. C. 1972. Major chemical variation in the Eocene lavas of the Isle of Skye, Scotland. *Journal of Petrology*, **13**, 219–253.

——, GIBSON, I. L., MARRINER, G. F. & MORRISON, M. A. 1980. Trace-element evidence of multistage mantle fusion and polybaric fractional crystallisation in the Palaeocene lavas of Skye, NW Scotland. *Journal of Petrology*, **21**, 265–293.

THORARINSSON, S. 1950. The Eruption of Mt. Hekla, 1947–1948. *Bulletin Volcanologique*, **10** (series 2), 157–168.

VON OEYNHAUSEN, C. & VON DECHEN, H. 1829. Die Insel Skye. *Karsten's Archiv fur Mineralogie*, **1**, 56–104.

WALKER, G. P. L. 1975. A new concept of the evolution of the British Tertiary intrusive centres. *Journal of the Geological Society of London*, **131**, 121–141.

WILLIAMS, P. J. 1985. Pyroclastic rocks in the Cnapan Breaca felsite, Rhum. *Geological Magazine*, **122**, 447–450.

B. R. BELL, Department of Applied Geology, University of Strathclyde, Glasgow G1 1XJ, UK.

C. H. EMELEUS, Department of Geological Sciences, University of Durham, Durham DH1 3LE, UK.

The early Tertiary stress regime in NW Britain: evidence from the patterns of volcanic activity

R. W. England

SUMMARY: Major dyke swarms emplaced throughout the Palaeocene volcanic episode in the British Tertiary Volcanic Province indicate NE–SW extension occurred perpendicular to the evolving NE Atlantic continental margin. Sigmoidal N–S orientated secondary swarms cutting Lewisian basement structures indicate a possible dextral shear component within the regional NE–SW extensional stress field, rather than a structural control on dyke emplacement. Both groups of swarms are considered to be the surface expression of linear intrusions emplaced in the lower to middle crust under the influence of the regional extension. Patterns of faulting within the province are consistent with the stresses recorded by the dyke swarms, indicating the stress field persisted into the Eocene during which a change to NW–SE extension occurred.

Much of the modern work in the British Tertiary Volcanic Province (BTVP) has concentrated on the geochemistry and isotope geology of the wide variety of intrusive and extrusive rocks which compose the major centres of activity. However, since the centres were mapped in the first half of this century comparatively little attention has been paid to the influence of regional tectonics on the patterns of magmatism within the Province, with the exception of work by Richey (1939) and Bell (1976). In this contribution the regional stress field is deduced using the extensive dyke swarms as palaeostress indicators. The influence of this stress field is then considered using the patterns of volcanic activity as indicators of the structural evolution of the Province in the early Tertiary.

Dykes as stress indicators

Anderson (1951) demonstrated that dyke emplacement occurred within the plane perpendicular to the minimum principal stress (σ_3). This hypothesis was expanded by Odé (1957), Pollard, (1973), Delaney & Pollard (1981) and Delaney et al. (1986) to account for variations in magma pressure, country-rock inhomogeneity and local variations in stress fields. However, the conclusions of these studies maintain that the overriding factor in controlling dyke emplacement is the σ_3 direction, and that any dyke deviating significantly from the plane containing the maximum and intermediate principal stresses σ_1 and σ_2 will attempt to regain an orientation within that plane (Pollard 1973).

Consequently, the stress trajectories effective during dyke emplacement in the BTVP can be derived from the distribution of swarm axes determined by Speight et al. (1982) (see Fig. 1). These authors emphasized the 'marked en-echelon' component in the orientation of the major swarm axes which clearly indicates dominant NE–SW directed extension. However, the distribution of Tertiary dyke swarms in NW Britain is more complex and widespread. In addition to the major swarms described above there is an additional asymmetric set of sigmoidal secondary swarms, clearly exposed on land, linking the Lewis and Skye, Skye and Mull and Mull and Arran/Kintyre major linear swarms. These secondary swarms have a NNW–SSE to N–S strike, indicating a component of E–W crustal extension. Hence it is evident that while NE–SW extension was the overriding control on dyke emplacement, secondary, less intense, swarms indicate a local modification of the stress field. This modification was probably the result of local structural controls or regional tectonic movements immediately prior to the opening of the N Atlantic. The possible influence of these factors will be discussed below. The existence of further swarms or major dykes has been revealed offshore in the Minches (Ofoegbu & Bott 1985) and in the North Channel and Irish Sea (Caston 1975; Kirton & Donato 1985) by geophysical methods (Fig. 1). Although the relationship between these swarms and their counterparts onshore is unclear it would appear from their more northerly orientation that they form further secondary swarms.

The interpretation of swarm and dilation axes

In any section normal to the strike of a group of dykes the swarm axis for the dykes crossed in

From MORTON, A. C. & PARSON, L. M. (eds), 1988, *Early Tertiary Volcanism and the Opening of the NE Atlantic,* Geological Society Special Publication No. 39, pp. 381–389.

381

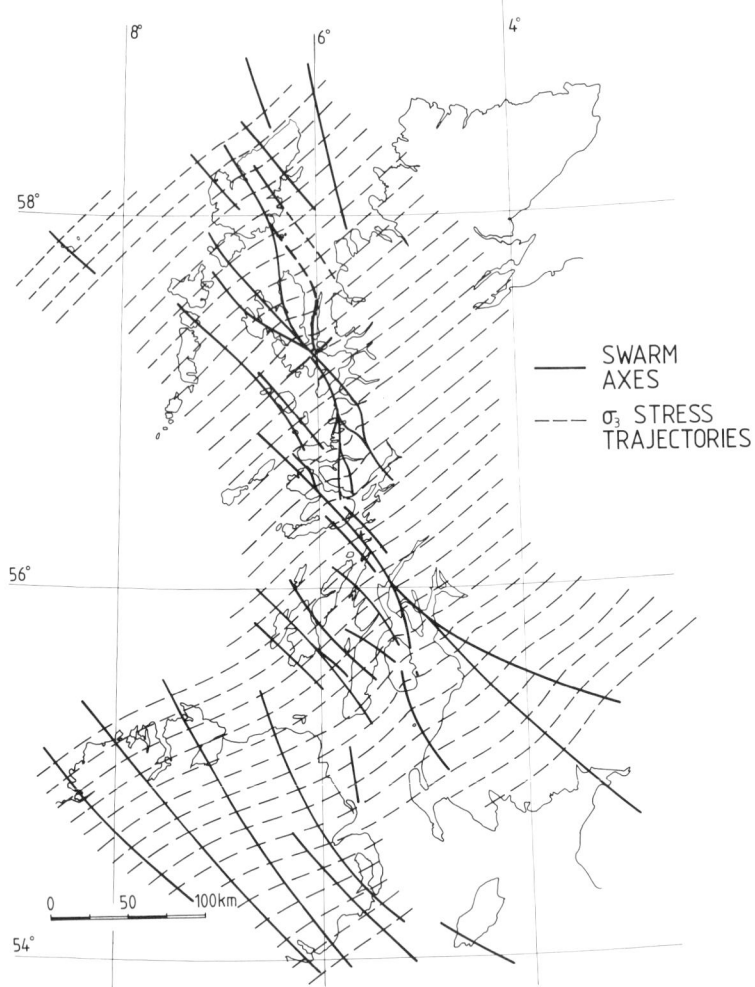

FIG. 1. Map showing the spatial orientation of the minimum principal stress trajectories across the BTVP as derived from local extension directions indicated by the major, secondary and subswarms of dykes. (Pattern of swarm axes taken from Speight *et al.* 1982.)

traversing the section is the mean direction of strike of the dykes. The axis of dilation for the dykes within the same traverse is obtained by plotting dilation as a percentage of the total dilation across the section against the azimuth (strike) of the dykes within the section. This gives a distribution curve with its maxima at the azimuth of the dilation axis (see Fig. 2a). The pattern of swarm and dilation axes are then determined by comparing orientation of swarm and dilation axes for a number of sections, as described above, along the strike of each swarm. It should be noted that there is an intrinsic relationship between an increase in the number of dykes emplaced with a particular strike and an increase in dilation normal to the strike of the

same dykes. This manifests itself as identical and coincident emplacement direction and dilation distribution curves (Fig. 2a). Any deviation from this relationship would suggest an external control on one or both of these parameters.

To form two dykes A and B separated by an angle Θ (see Fig. 2a) under uniform pure shear normal to dyke B, the dilation (z) measured normal to the margin of dyke A will always be less than x, the dilation measured normal to the margin of the dyke B (since y will be equal to x for uniform pure shear) and $z = y.\cos \Theta$. Since dyke emplacement occurs preferentially normal to σ_3, B will lie along the swarm axis and the dilation axis.

If the pure shear stress field is modified to one

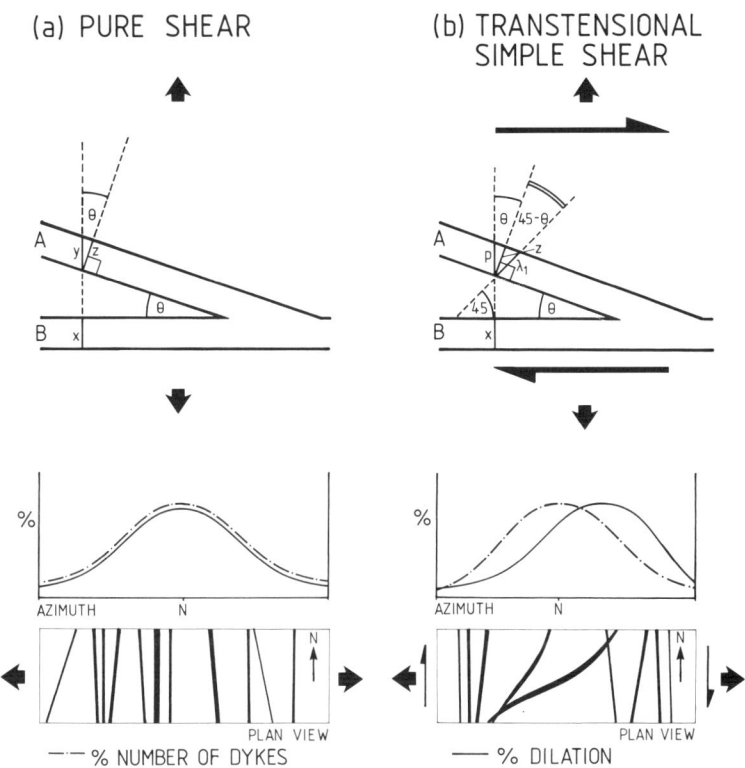

FIG. 2. Geometric models for dyke swarms formed under: (a) pure shear; (b) transtensional simple shear. In each case the upper diagram indicates the mechanism of extension producing the fissure in which the dykes are emplaced and the lower shows a schematic map and distribution curves for a hypothetical swarm resulting from (a) and (b).

of uniform transtensional simple shear (see Fig. 2b) dykes A and B will retain components of dilation x and p (p corresponds to y in the pure shear model) resulting from extension normal to the shear plane (where B lies in the shear plane) but the dilation across A will be increased as a result of shearing. This additional component of extension varies according to the angle the dyke makes with the shear plane and will be at its maximum value when the dyke is orientated at 45° to the shear plane, i.e. perpendicular to the maximum quadratic elongation direction for infinitesimal strain (Ramsay 1967), and zero for a dyke lying in the shear plane. Hence z will be the sum of $p.\cos\Theta + \lambda_1\cos(45° - \Theta)$ (see Fig. 2b). If $p = x$ or $\lambda_1\cos(45° - \Theta) \geq (x - p.\cos\Theta)$, then z will be greater than or equal to x.

The implications of this geometrical consideration are that in the case of a swarm forming under pure shear, swarm and dilation axes will coincide, but this need not be the case for a swarm forming under transtensional simple shear, as indicated in Fig. 2b. If the majority of

the dykes are emplaced in the tensional shear plane, the swarm axis will lie along the strike of the plane. The axis of dilation (which is independent of the number of dykes) may occur at an angle to the shear plane because of increased dilation in dykes which lie out of the shear plane.

In the case of the Tertiary dyke swarms of NW Britain, data from Speight *et al.* (1982) (Fig. 3) indicate that on a regional scale the distribution curves are generally coincident. There is a clockwise separation of 5° between the swarm and dilation axes of the Skye regional swarm and a corresponding 10° shift for the Small Isles Swarm, (Speight 1972). From the hypothesis above, this is indicative of regional NW–SE extension being the dominant control on dyke orientation. However, on a local scale there are notable deviations from this overall pattern (Fig. 4). This figure shows a notably consistent clockwise separation of the swarm and dilation axes within the sigmoidal secondary swarms which increases away from the major NW–SE swarm axis. This separation of swarm and dilation axes

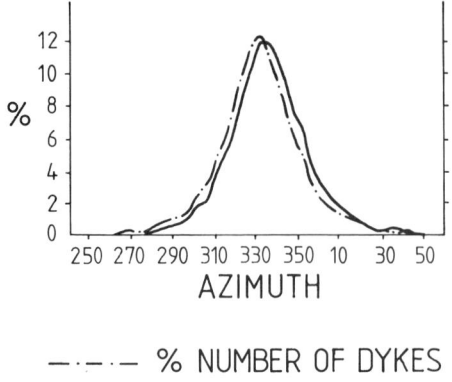

FIG. 3. Distribution curves for number of dykes and dilation against azimuth for the Skye regional swarm (from Speight *et al.* 1982, reprinted by permission of John Wiley & Sons Ltd).

is interpreted, on the basis of the geometric model outlined above, to be the result of dextral transtensional simple shear.

Discussion

The controls on swarm orientation

The general NW–SE to N–S range of orientation of the dykes over a distance of 800 km (Kirton & Donato 1985) indicating NE–SW to E–W extension can only be readily explained by a tectonic control. However, the local variations within the overall pattern suggest either a structural control (Richey 1939; Kirton & Donato 1985) or the existence of a component of dextral shear in the regional stress field.

One of the conclusions of Speight *et al.* (1982) was that the dyke swarms were the surface expression of deeper-seated linear ridges of magma, with the central complexes rising from these ridges. If, as they suggest, both the major and secondary swarms were fed by ridges of magma it must be presumed that these ridges rose from the base of the crust where they were fed with partial melt. Such a lower crustal structure for the swarms is also favoured by Thompson (1982) and Dickin *et al.* (1984) on geochemical grounds, although these authors argue for a lower-crustal dyke swarm (after Patchett 1981) as opposed to a single intrusion.

These conclusions have important implications for the interpretation of both the regional and the secondary swarms of the BTVP. If the swarms are fed by such lower crustal intrusive bodies it

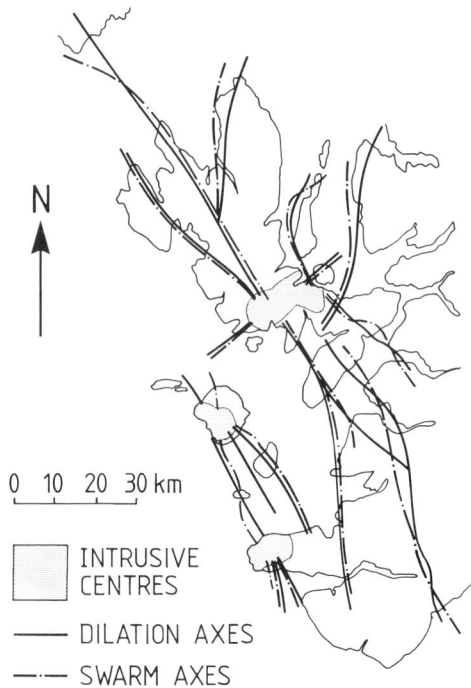

FIG. 4. Pattern of dyke swarm and dilation axes for the Skye and Small Isles regional swarms (details taken from Speight 1972, Fig. 123, with permission).

must be assumed that these intrusions have the same general orientation as the swarms they fed. Hence, the orientation of the dyke swarms of the BTVP would have been controlled at source level (lower crust) by regional stresses or structural grain. Much of the area of the BTVP to the N and W of the Highland Boundary Fault is floored by Lewisian gneiss which has a predominantly NW–SE structural grain illustrated by the orientation of the Precambrian Scourie dykes. While it is not known to what depth this structural grain is effective it is possible that the source of the major NW–SE swarms of the BTVP were influenced by it. However, the secondary N–S swarms cross this structural grain without any real indication of attempting to follow it, or regain an orientation normal to the regional σ_3 defined by the major swarms, as predicted by Pollard (1973) (see above). Hence, it is concluded that the orientation of these swarms could result from local variations in a NE–SW extensional regional stress-field effective throughout the depth of the crust, rather than a structural control. The separation of the swarm and dilation axes within the Skye (see Fig. 4) and Small Isles regional swarms, proposed to indicate a component of dextral shear in the stress field, is consistent with this observation.

Elsewhere in the province there is evidence that the orientation of secondary swarms may have been controlled by pre-existing structures at upper crustal levels. In particular the dykes of the Arran swarm are subparallel to a group of NNW–SSE striking Carboniferous faults within the Clyde estuary (McLean & Deegan 1978), and hence may be interpreted as exploiting a pre-existing structural weakness. However, the main faults of this group lie to the E of, and dip away from, the basic feeder to the dykes, indicated by gravity data to lie beneath Arran and Ailsa Craig. It is probable that the rise of this basic mass resulted in reactivation of the faults, providing an ideal path for dyke emplacement. The control on the orientation of the rising basic mass is unknown. A similar argument could be applied to the dykes of the North Channel, although they cross-cut NW–SE and NE–SW fractures in outcrop observed on the seabed (Caston 1976). In the Morvern area the NNW–SSE dykes of the Morvern swarm may have been deflected toward the N–S structural grain of the Moine rocks (O'Brien 1985) as they passed upward from the Lewisian basement. However, the dykes do not run exactly parallel to the Moine structures and their true orientation may again be the result of a deeper-seated control.

Hence, there is evidence for both control of the secondary swarms by upper crustal structure and variations in the regional stress field. However, the overriding control on their orientation appears to be the direction of the lower crustal intrusive bodies feeding them. The behaviour of the swarms within the Lewisian, which forms the basement to a large part of the BTVP, indicates the control on these intrusions may be stress rather than structure.

The development of the swarms with time

The combined radiometric and palaeomagnetic studies of Dagley, Mussett and others (e.g. Dagley *et al.* 1978, 1984; Dagley & Mussett 1986; Mussett 1986) indicate dyke emplacement occurred throughout the duration of igneous activity, which is strongly supported by field observations. Cross-cutting relationships between dykes and the intrusions of the central complexes indicate dyke emplacement occurred prior to, during and after their formation. Also, the number of dykes seen to cut the central complexes is clearly less than the number comprising the major swarms which are cut by the complexes during their emplacement (Speight *et al.* 1982). This suggests dyke emplacement became less intense as the centres developed. However, a lack of systematic cross-cutting relationships between individual

dykes (Sloan 1971; Speight 1972; Knaap 1973) makes any attempt at interpreting small changes in local or regional stress orientation impossible. Hence, one is left to conclude that on a regional scale the stress orientations controlling dyke emplacement remained approximately constant with time, with intrusion being confined to swarm axes throughout. One possible exception to this may be the Small Isles swarm and associated activity in the Rhum central complex which appears to predate that in Mull, Skye and Ardnamurchan (Meighan *et al.* 1982). This swarm bears no clear relation to the neighbouring Mull and Skye major swarms and the Morvern secondary swarm which are clearly contemporaneous. Therefore, the Small Isles swarm may represent an early failed attempt at forming a major swarm. Initiation of the Mull and Skye major swarms may have been the trigger for the voluminous outpouring of basalt to form the plateau lavas at around 60 Ma, probably as a result of an increase in tensional stresses across the area.

The magnitude of the stress field controlling magmatism

An indication of the magnitude of the inferred stress field can be gained from the following observations. Firstly, there is no strong regional deformation associated with the volcanic activity, with the possible exception of the distortion of an Old Red Sandstone (ORS) dyke swarm and Caledonian faults where they are crossed by the Mull major swarm, as noted by Knill (1960). Secondly, away from the central complexes and swarm axes the amount of dilation decreases over a distance of 5–10 km from approximately 10% to zero (Speight *et al.* 1982), giving a regional extension of below 1% (dilation across major swarms greatly exceeds that across secondary swarms.) Such a small total strain would be compatible with a low magnitude tensional or compressional σ_3. A tensional minimum principal stress is preferred on the basis that such a stress would favour a passive mode of emplacement for the dykes. Field evidence shows that few dykes have brecciated wall rocks consistent with forceful emplacement by magmafracturing. Also, the emplacement of some dykes well away from the swarm axes would suggest that dilation was not purely due to updoming across the intrusive bodies postulated to be feeding the swarms, although this must have been a contributing factor to the density of distribution of the dykes. However, a compressional σ_3 cannot be entirely ruled out. So far discussion of the stress field has been confined to the orientation of the minimum principal stress (σ_3). The dyke swarms can only be used to indicate the orientation of the plane containing the major and intermediate principal stresses. In the absence of significant uplift (Binns *et al.* 1975; Watson 1985), it would appear that the maximum principal stress was orientated in a NW–SE horizontal plane. Such an orientation of the stress field would also be required if a component of dextral shear was present within it.

The central complexes

The regional stress field appears to have no control on the siting of the central complexes. However, the presence of subswarms radial to the centres (Speight *et al.* 1982) indicate that they influenced the stress field within the rocks directly surrounding them. Dilation within the major and secondary swarms increases toward the centres and this has previously been cited as evidence for the swarms originating from, or being controlled by, the centres (Harker 1904; Vann 1978). This hypothesis conflicts with available evidence. Firstly, it has been shown that the major phase of dyke emplacement preceded the central complexes. Secondly, the extent of the radial subswarms associated with the complexes is small, indicating that the extent of their influence on dyke emplacement is small. Thirdly, there is little evidence for lateral injection of the dykes outward from the centres. This increase in dilation can be explained by an increase in the volume of magma available as a source for fissure eruption in the vicinity of the centres as they developed from the magma ridges feeding the dyke swarms. This would result in increased uplift and tension across the swarm axes permitting increased dyke emplacement, which is consistent with a certain amount of overlap between phases of dyke emplacement and development of the central complexes.

One of the features of the central complexes which has received little critical attention is the cause of migration of centres of activity within them. If the secondary swarms do truly represent a component of dextral shear in the regional stress field it would be possible to produce localized zones of tension along the major swarm axes, by analogy to pull-apart basins on major strike-slip faults. Formation of central complexes from rising columns of magma would subsequently cause localization of regional stresses to produce zones of tension through which further batches of magma would preferentially flow. Movement between centres of activity would result from magma rising through migrating zones of tension (Fig. 5). Migration may have been linear, as in the case of the Mull and Carlingford–Slieve Gullion centres or by clockwise rotation as in Skye and Ardnamurchan. Clockwise rotation of the maximum extension direction is again consistent with a component of dextral shear in the regional stress field. However, present evidence indicates that migration in the Mournes centres does not fit within the context of this model.

Late faulting

The lava flows extruded early in the volcanic sequence are crossed by many normal faults commonly striking in NE–SW and NW–SE directions, the former commonly predating the latter (Binns *et al.* 1973). Some of the NW–SE faults are intruded by dykes indicating that they are of Palaeocene age. Binns *et al.* (1975) attribute this faulting to subsidence of the crust beneath the dense lava piles, which is substantiated by the inward dip of Jurassic sediments lying below the Skye plateau basalts—toward the centre of the pile—and the accumulation of Oligocene sediments in a basin floored by lavas to the NW of the Canna ridge. These sediments now lie approximately 250 m below sea-level indicating considerable downwarping during or following their deposition. Watson (1985) noted that the present bases of the lava piles record up to 2.5 km of vertical relief produced by warping and faulting following their extrusion. This was probably the result of movement on the major NE–SW faults bounding the sedimentary basins in the region. In particular the Minch, Raasay and Camasunary–Skerryvore Faults in the Hebrides (Binns *et al.* 1973) and the Tow Valley Fault in NE Ireland (Wilson & Robbie 1966) all cut basalt lava piles of Palaeocene age. Reactivation of these faults indicates a change to dominant NW–SE directed extension across the Province. This change in extension direction could be correlated with the onset of regional uplift, with local downwarping in fault-controlled basins, as described by Watson (1985). The uplift resulted from underplating the crust during the period of active volcanism.

Faulting is common within the central complexes, where its strike is generally parallel to the dyke swarms on which the centres are superimposed. This suggests that the faults are either contemporaneous with dyke emplacement or there is a persistence of the stress-field controlling dyke emplacement after the cessation of activity, depending upon the age relationships between the faults and the intrusions forming the complexes. A number of NW–SE normal faults cut

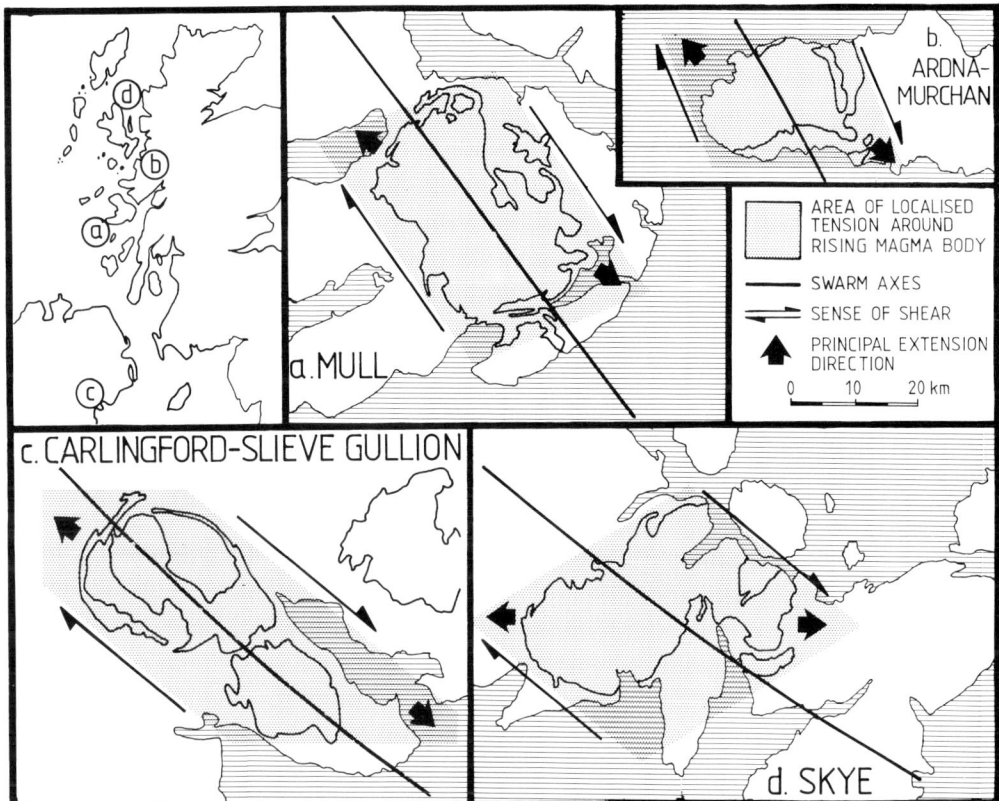

FIG. 5. A possible mechanism for the cause of migration within the central complexes. Dextral transtension across pre-existing dyke swarms may control the emplacement direction of rising magma forming the central complexes. Changes in zones of maximum tension within shaded areas would result in changes in the paths of batches of magma, resulting in changes in centres of activity.

the lavas and parts of the Mull central complex. Mapping indicates that some are contemporaneous with volcanic activity, e.g. the faults within the Loch Ba Centre cut by the felsite ring dyke, whereas those cutting the Loch Don Anticline may, in the absence of cross-cutting relationships with intrusions of the central complex, be later. The Morvern lavas are downthrown against Moine metasediments by a series of N–S faults parallel to the Morvern dyke swarm. The eastern part of the Ardnamurchan Centre is cut by two parallel NNW–SSE faults passing through Faskadale and Loch Mudle. Both these faults are parallel to the axis of the dyke swarm associated with the central complex and a number of faults cutting parts of Centre 3, suggesting a late or post-volcanic origin. The Rhum ultrabasic/basic mass is cut by the N–S-striking Long Loch Fault which displaces the Main Ring Fault dextrally by approximately 700 m. This fault shows evidence of movement both before and after cm

placement of the ultrabasic intrusion and it may have acted as a conduit for rising magma (C. H. Emeleus, pers. comm.). In St. Kilda a set of NW–SE directed faults are cut by late NE–SW orientated faults, dykes and cone sheets (Harding *et al.* 1984). In Northern Ireland, the Western Mourne G4 and G5 granites are cut by NNE–SSW-striking normal faults, downthrowing to the E and W, which clearly postdate intrusion. The Slieve Gullion and Carlingford Centres are cut by a series of dominantly dextral strike-slip faults associated with the Newry Fault, a northwestward extension of which may have influenced deposition of the Oligocene Lough Neagh clays, which thicken toward it (Wilkinson *et al.* 1980). The faults cutting the Mourne granites do not appear to cross Carlingford Lough and, consequently, it is impossible to determine their relationship to the Newry Fault. However, the sense of displacement on these faults shows they must have formed under stress systems of different orientations. The Newry Fault indicates NW–SE directed compression and the faults cutting the Mourne granites NW–SE extension.

The association of the faults, dyke swarms and central complexes described, can be related to the regional stress system. Syn-volcanic to immediately post-volcanic NW–SE to N–S normal or dextral strike-slip faulting is consistent with the stress field proposed to be controlling the dyke swarms and migration of activity within the central complexes. The normal faulting is related to the principal NE–SW directed extension controlling the major swarms and the strike-slip faulting is related to the postulated subsidiary shear component influencing the orientation of the Hebridean secondary swarms. This was followed by a phase of post-volcanic NE–SW normal faulting related to reactivation of major pre-existing structures such as the Minch, Camasunary and Tow Valley faults during Eocene to Oligocene uplift. Such an interpretation indicates a rotation of the regional compression and extension axes during the late Eocene.

Tectonic controls on the stress field

It has been established that the dominant control on the distribution of magmatism in the BTVP is the regional stress field which constrain the orientation of lower crustal intrusions feeding the dyke swarms and the central complexes. This interpretation appears to provide the best explanation for the patterns of activity over such a large area of diverse and complex geology. However, this stress field, with dominant NE–SW extension, does not appear to be consistent with the regional tectonics, namely continental extension on a NE–SW axis. In the absence of good evidence to suggest distortion through 90° of a NW–SE extensional field due to rifting, one is left to conclude that the NE–SW tension recorded by the dyke swarms is the true regional stress field and that the structural development of the BTVP was not directly controlled by the opening of the NE Atlantic, which it immediately predates. The fact that this NE–SW directed extension is synchronous with a similar phase of extension in the subparallel Viking and Central grabens in the North Sea (Dewey 1982; Kirton & Donato 1985) leads to the tentative suggestion that the stress field in NW Britain and Ireland

may have been generated by tectonic events in NE Europe resulting from the collision of the European and African/Iberian plates, as well as the onset of continental rifting in the NE Atlantic. Whatever the cause of the stress field it is apparent from patterns of late Eocene and early Oligocene faulting that its influence decayed during this period, giving way to NW–SE directed extension controlled by continental splitting.

Conclusions

A group of major NW–SE-trending dyke swarms indicate that a stress field with dominant NE–SW extension persisted throughout the period of BTVP volcanism. This stress field is thought to act throughout the depth of the crust and control the orientation of lower crustal intrusions feeding the dyke swarms. An additional set of N–S striking secondary swarms within the Hebrides, cross-cutting structures in the Lewisian basement, would also appear to have been controlled by variations in the same stress field, rather than regional structure. Separation of their swarm and dilation axes suggests they are the result of a minor component of dextral shear within the regional stress field. Further S, in Arran and the Irish Sea, the orientation of secondary swarms appears to have been influenced by upper crustal structure. Whether the orientation of the postulated lower crustal feeders to these swarms is controlled by regional stress or crustal structure is not known. Patterns of faulting within the central complexes and sedimentary basins of the province are consistent with the stress system indicated by the dykes and record subsequent replacement of the Palaeocene stress field by NW–SE extension perpendicular to the evolving NE Atlantic continental margin during the late Eocene.

ACKNOWLEDGEMENTS: I would like to thank Drs C. H. Emeleus and D. H. W. Hutton for help and encouragement in writing this contribution and for reading early drafts, and other members of the Durham department for useful discussions and criticism. Also, Miss K. Gittins and Mr A. Carr for help in preparing the figures and to Dr J. M. Speight for allowing me to use data from his PhD thesis. This work was completed while the author was in receipt of NERC grant GT4/85/GS/27.

References

ANDERSON, E. M. 1951. *The dynamics of faulting and dyke formation, with applications to Britain.* Oliver & Boyd, Edinburgh.

BINNS, P. E., McQUILLIN, R. & KENOLTY, N. 1973. The Geology of the Sea of the Hebrides. *Report of the Institute of Geological Sciences,* **73/14.**

——, —— FANNIN, N. G. T., KENOLTY, N. & ARDUS, D. A. 1975. Structure and stratigraphy of the sea of the Hebrides and Minches. *In:* WOODLAND, A. W. (ed) *Petroleum and the Continental Shelf of NW Europe.* Applied Science Publishers, London, pp. 93–102.

CASTON, G. F. 1975. Igneous dykes and associated scour hollows of the North Channel, Irish Sea. *Marine Geology*, **18**, M77–M85.

—— 1976. The Floor of the North Channel Irish Sea: a Side-scan Sonar Survey. *Report of the Institute of Geological Sciences, 76/7.*

DAGLEY, P. & MUSSETT, A. E. 1986. Palaeomagnetism and radiometric dating of the British Tertiary Igneous Province: Muck and Eigg. *Geophysical Journal of the Royal Astronomical Society*, **85**, 221–242.

——, ——, WILSON, R. L. & HALL, J. M. 1978. The British Tertiary Igneous Province: palaeomagnetism of the Arran Dykes. *Geophysical Journal of the Royal Astronomical Society*, **54**, 75–91.

——, —— & SKELHORN, R. R. 1984. The palaeomagnetism of the Tertiary igneous complex of Ardnamurchan. *Geophysical Journal of the Royal Astronomical Society*, **79**, 911–922.

DELANEY, P. T. & POLLARD, D. D. 1981. Deformation of Host Rocks and Flow of Magma During Growth of Minette Dykes and Breccia-bearing Intrusions near Ship Rock, New Mexico. *US Geological Survey Professional Paper*, **1202**.

——, ——, ZIONY, J. I. & MCKEE, E. H. 1986. Field relations between dikes and joints: emplacement processes and palaeostress. *Journal of Geophysical Research*, **91**, 4920–4939.

DEWEY, J. F. 1982. Plate tectonics and the evolution of the British Isles. *Journal of the Geological Society of London*, **139**, 371–412.

DICKIN, A. P., BROWN, J. L., THOMPSON, R. N., HALLIDAY, A. N. & MORRISON, M. A. 1984. Crustal contamination and the granite problem in the British Tertiary Volcanic Province. *Philosophical Transactions of the Royal Society, London*, **A310**, 755–780.

HARDING, R. R., MERRIMAN, R. J. & NANCARROW, P. H. A. 1984. St. Kilda, an illustrated account of the geology. *Report of the British Geological Survey*, **16(7)**.

HARKER, A. 1904. *The Tertiary Igneous Rocks of Skye.* Memoir of the Geological Survey of Scotland.

KIRTON, S. R. & DONATO, J. A. 1985. Some buried dykes of Britain and surrounding waters deduced by magnetic modelling and seismic reflection methods. *Journal of the Geological Society of London*, **142**, 1047–1058.

KNAAP, R. J. 1973. *The form and structure of the Islay, Jura and Arran Tertiary basic dyke swarms.* PhD thesis, University of London.

KNILL, J. L. 1960. Evidence for Tertiary crustal distortion in mid-Argyll. *Nature*, **185**, 234–235.

MCLEAN, A. C. & DEEGAN, C. E. (eds) 1978. The Solid Geology of the Clyde Sheet 55 N/6 W. *Report of the Institute of Geological Sciences, 78/9.*

MEIGHAN, I. G., HUTCHINSON, R., WILLIAMSON, I. & MCINTYRE, R. M. 1982. Geological evidence for the different relative ages of the Rhum and Skye Tertiary central complexes (abstract). *Journal of the Geological Society of London*, **139**, 659.

MUSSETT, A. E. 1986. ^{40}Ar–^{39}Ar step-heating ages of the Tertiary igneous rocks of Mull, Scotland. *Journal of the Geological Society of London*, **143**, 887–896.

O'BRIEN, B. H. 1985. The geometry of ductile conjugate fold systems in the Ardnamurchan Moine, Scotland. *Geological Journal*, **20**, 91–108.

ODÉ, H. 1957. Mechanical analysis of the dyke pattern of the Spanish Peaks Area, Colorado. *Bulletin of the Geological Society of America*, **68**, 567–578.

OFOEGBU, C. O. & BOTT, M. H. P. 1985. Interpretation of the Minch linear magnetic anomaly and of a similar feature on the shelf North of Lewis by non-linear optimisation. *Journal of the Geological Society of London*, **142**, 1077–1088.

PATCHETT, P. J. 1981. Thermal effects of basalt on continental crust and crustal contamination of magmas. *Nature*, **283**, 555–561.

POLLARD, D. D. 1973. Derivation and evolution of a mechanical model for sheet intrusions. *Tectonophysics*, **19**, 233–269.

RAMSAY, J. G. 1967. *Folding and Fracturing of Rocks.* McGraw-Hill, New York.

RICHEY, J. E. 1939. The Dykes of Scotland. *Transactions of the Geological Society of Edinburgh*, **13**, 393–435.

SLOAN, T. 1971. *The Structure of the Mull Tertiary Dyke Swarm.* PhD thesis, University of London.

SPEIGHT, J. M. 1972. *The Form and Structure of the Tertiary dyke swarms of Skye and Ardnamurchan.* PhD thesis, University of London.

——, SKELHORN, R. R., SLOAN, T. & KNAPP, R. J. 1982. The dyke swarms of Scotland. *In:* SUTHERLAND, D. S. (ed). *Igneous Rocks of the British Isles.* John Wiley & Son, Chichester, 449–459.

THOMPSON, R. N. 1982. Magmatism of the British Tertiary Volcanic Province. *Scottish Journal of Geology*, **18**, 49–107.

VANN, I. R. 1978. The siting of Tertiary volcanicity. *In:* BOWES, D. R. & LEAKE, B. E. (eds) *Crustal Evolution in North Western Britain and Adjacent Regions.* Geological Journal Special Issue, **10**, pp. 393–414.

WATSON, J. 1985. Northern Scotland as an Atlantic–North Sea divide. *Journal of the Geological Society of London*, **142**, 221–243.

WILKINSON, G. S., BAZLEY, R. A. B. & BOULTER, M. C. 1980. The geology and palynology of the Oligocene Lough Neagh clays, Northern Ireland. *Journal of the Geological Society of London*, **137**, 65–75.

WILSON, H. E. & ROBBIE, J. A. 1966. *Geology of the Country around Ballycastle.* Memoir of the Geological Survey of Northern Ireland, HMSO, London.

R. W. ENGLAND, Department of Geological Sciences, University of Durham, Durham DH1 3LE, UK.

A re-evaluation of the origin and nature of layered peridotite, troctolite and gabbro in the Eastern Layered Series of the Rhum ultrabasic complex, Inner Hebrides

J. H. Bédard, R. S. J. Sparks, R. Renner, R. Hunter & M. Cheadle

Mapping of vertical and lateral lithologic variations in the Eastern Layered Series (ELS) on the northern flank of Hallival shows that both peridotite and allivalite (troctolite or gabbro) layers are laterally discontinuous and vary in thickness and lithology. This is particularly evident in some of the allivalites (e.g. Unit 10), where troctolite is replaced by gabbro along strike. Some troctolite layers terminate as isolated, fingered blocks in peridotite. Peridotite generally has sharp contacts against the allivalites, but reaction, dissolution and hybridization effects are developed locally. Peridotitic layers commonly transgress the allivalite layering at a shallow angle, but markedly discordant contacts are also observed. The smaller conformable peridotites cause updoming and downwarping of the host allivalite. Peridotite plugs that are clearly intrusive into allivalite are petrographically and geochemically very similar to the stratiform peridotites and a common origin is proposed. Our preferred interpretation is that most stratiform peridotites in the ELS represent sill-like intrusions of ultramafic magma into a partly solidified layered troctolite complex. However, the peridotites are cumulate rocks and significant fractions of residual basalt must be accounted for. Some of the residual basaltic melt appears to have percolated laterally (up-dip?) through still-porous troctolite and reacted to form gabbro (pyroxenization). The best evidence for pyroxenization of troctolite comes from the top of Unit 9 where the wavy (metre-scale) gabbro–troctolite contact cuts across pre-existing grain size, modal and rhythmic layering, but causes little disturbance to it. Gabbros formed by pyroxenization of troctolite mimic the textures, grain size and rhythmic layering of their troctolitic protoliths. Relict troctolite lamellae and fossil pyroxenization fronts are common. We propose that many of the ELS gabbros formed metasomatically as a result of interaction between the porous host troctolites and low-temperature basaltic melt. The ultimate origin of this basaltic melt is uncertain. It could be: (1) residua released from the peridotite sills; (2) residua from the chamber above that penetrated into high-porosity horizons in the cumulates; (3) related to the Askival Plateau or Atlantic Coire gabbros to the W and SW; (4) related to the basaltic replenishment events documented by Renner & Palacz (1987); or (5) represent residual melt extracted from compacting troctolitic cumulates and concentrated along high-porosity horizons.

References

RENNER, R. & PALACZ, Z. 1987. Basaltic replenishment of the Rhum magma chamber: evidence from unit 14. *Journal of the Geological Society of London*, **144**, 961–970.

BÉDARD, J. H., SPARKS, R. S. J., RENNER, R., HUNTER, R. & CHEADLE, M., Earth Sciences Department, University of Cambridge, Downing Street, Cambridge CB2 3EQ, UK.

From MORTON, A. C. & PARSON, L. M. (eds), 1988, *Early Tertiary Volcanism and the Opening of the NE Atlantic,* Geological Society Special Publication No. 39, p. 391.

391

The North Sea
Sedimentary Record

Palaeogene volcanism: the sedimentary record in Denmark

O. B. Nielsen & C. Heilmann–Clausen

SUMMARY: Approximately 200 volcanic ash layers are known from Palaeocene and Eocene marine clays in Denmark. The oldest layers are present in the upper part of the Upper Palaeocene and are sporadic, thin and acidic in composition. The main volcanic phase took place at the Palaeocene–Eocene transition. During the main phase, a supply from the N to NE is indicated, while a few layers probably had a source to the NW or W. A later volcanic phase is represented by thin and more sporadic ashes in the Lower to Middle Eocene. These are almost totally transformed to smectite.

In the Upper Palaeocene, and partly in the Lower Eocene, smectite is a highly dominant clay mineral in the interbedded clays. Zeolites of the heulandite–clinoptilolite type are also present, while cristobalite is sporadically present in the lower part. A major supply of smectite from the N Atlantic may be indicated. Below the ash-bearing sequence the only evidence for volcanic activity is the presence of very smectite-rich, zeolitic clays with scattered opal-CT and slightly silicified horizons. Calculated chemical compositions of smectites from both ash layers and clay layers between ash layers and below the ash-bearing sequence are almost identical, indicating a volcanic source for the smectite.

The conspicuous black, sandy layers of the Danish Palaeogene Fur Formation were known already by Forchhammer (1835) and their identity as volcanic ash layers was revealed more than a century ago, by Prinz & van Ermengem (1883). Since then, Palaeogene ash layers have been found in several other Danish formations and all over the North Sea Basin. It is the purpose of this study to describe the Danish Palaeogene ashes and associated clays and discuss their age and provenance.

Since their recognition the ash layers, in particular those of the Fur Formation, have been subject to many different analyses. Bøggild (1918), Andersen (1937) and Norin (1940) analyzed their thickness and grain size, while Bøggild (1918), Madirazza & Fregerslev (1969) and Pedersen *et al.* (1975) also focused on their petrographic and chemical composition. Nielsen (1974) and Pedersen (1981) discussed depositional environments and Heilmann–Clausen *et al.* (1985) and Pedersen & Surlyk (1983) established a modern lithostratigraphy for the ash-bearing sequence. The deposits have been dated primarily by means of dinoflagellate cysts (Hansen 1979; Heilmann–Clausen 1982, 1983, 1985; Nielsen & Heilmann–Clausen 1986) and calcareous nannofossils (Perch-Nielsen 1967, 1971; Thiede *et al.* 1980). A biostratigraphic study of silicoflagellates (Perch-Nielsen 1976) from the Fur Formation also gives information on the age.

Localities studied and discussed in the present work are shown in Figure 1. The stratigraphy and chronology of the Danish ash-bearing deposits and their North Sea equivalents are shown in Figure 2.

Methods

Ash and clay layers were subject to: (1) grain-size analysis, after disaggregation by shaking in water for 72 hours and repeated ultrasonic treatment, by combined sieving and settling in Andreassen settling tubes; (2) determination of total organic carbon (TOC), sulphur and carbonate in a LECO induction furnace; (3) chemical analysis of whole rock samples by atomic absorption spectroscopy (AAS); (4) bulk mineralogical composition by means of X-ray diffraction (XRD) and quantified using the principles described by Schultz (1960); (5) clay mineralogy of the <2 μm fraction (XRD).

XRD analyses were performed on a Philips diffractometer, with Cu Kα radiation and an automatic divergence slit. The clay fraction was investigated using oriented smear-slide preparations (Gibbs 1971). Records were made of air-dried, of glycolated (24 hours at 60 °C), and of heated (to 300 °C, and to 550 °C, for two hours) preparations respectively. The minerals were determined according to Brindley & Brown (1980). The semiquantitative estimate of the clay mineral composition is carried out using peak areas corrected with empirically estimated correction factors from the glycolated diffractograms.

Preparations for dinoflagellate biostratigraphy were produced following normal palynological techniques (for details, see Heilmann–Clausen 1985).

The biostratigraphy of Danish Upper Palaeocene to Middle Eocene sediments is mainly based on dinoflagellates, since calcareous microfossils are usually absent. The only exception is

From MORTON, A. C. & PARSON, L. M. (eds), 1988, *Early Tertiary Volcanism and the Opening of the NE Atlantic,* Geological Society Special Publication No. 39, pp. 395–405.

FIG. 1. Map of Denmark and surroundings with localities.

MA	AGE	NP	DINO – ZONES			LITHOSTRATIGRAPHY			
			NW Europe	Den- mark	North – Sea	Ash layers	Denmark		
							NNW (Fur)		SSE (Røjle)
52	MID. EOC.	14				ash numbers of Bøggild (1918)		LILLEBÆLT CLAY FM. (PART)	L3
									L2
53		13							L1
54			K.coleothrypta		HORDA LAND				R6
	EARLY	12*			GROUP			RØSNÆS CLAY FM.	R5
55	EOCENE				(PART)				
56		11*	D.varielongitudum						R4
			D.simile				Knudshoved Mb.		R1–R3
57		10	W.astra + W.meckelfeld.		BALDER FORM.	·series' ·140 ·1	Silstrup Mb.	Vœrum Mb.	
			D.oebisfeld. Acme	7			FUR FM.	ØLST FORM.	
58				6	SELE FORM.	·series' ·33	Knudeklint Mb.	Haslund Mb.	
		9	A.hyperacanthum	5	LISTA FORM.	·39	? GREY CLAY		
59	LATE	8	D. / A.	4	(PART)		HOLMEHUS FORM.		
	PALEOC.		speciosa / margarita						
60		7		3			GREY CLAY (PART)		

FIG. 2. Stratigraphy and age of the Danish ash-bearing deposits and their North Sea equivalents. The lithostratigraphy of Denmark is from Heilmann–Clausen *et al.* (1985) and Pedersen & Surlyk (1983). The North Sea stratigraphy is from Deegan & Scull (1977). Dinoflagellate zones of Denmark and correlations between Denmark and the North Sea are from Heilmann–Clausen (1985). NW European dinoflagellate zones are from Costa *et al.* (1978) and Knox *et al.* (1981). Only calcareous nannoplankton zones (NP Zones) marked with an asterisk are identified in Denmark, other NP Zones are inferred from the dinoflagellate zonation. The calibration between absolute ages and NP Zones is from Berggren *et al.* (1985).

the Lower Eocene Røsnæs Clay Formation, in which the age is evaluated based on both calcareous nannofossils and dinoflagellates.

The age is given in terms of standard calcareous nannoplankton zones (NP Zones) in the Røsnæs Clay Formation. In the non-calcareous formations, the equivalent NP Zones may in some cases be inferred based on dinoflagellate zonation.

The dinoflagellate biostratigraphy of the Upper Palaeocene to Middle Eocene in the North Sea Basin (including Denmark) is developed by, in particular, Costa & Downie (1976), Costa *et al* (1978), Knox & Harland (1979), Knox *et al.* (1981), De Coninck (1975, 1977, 1981) and Heilmann–Clausen (1985). The zonation of Heilmann–Clausen (1985) integrates previous zonations covering the Upper Palaeocene and lowermost Eocene and is used here. For the Lower to Middle Eocene the zonation of Costa *et al.* (1978) is adopted. Additional information is taken from Nielsen & Heilmann–Clausen (1986) regarding identification of the Lower–Middle Eocene boundary.

Results and discussion

The evidence for volcanism

The main evidence for volcanism is the presence of ash layers, and the best proof for a volcanic ash layer is the presence of volcanic glass.

Ash layers containing glass as well as argillized ash layers, in which the glass has been transformed, are known in the Danish Palaeogene. The ash-bearing deposits are shown in Figure 2. The ashes may be separated into three phases:

(1) The layers of the earliest phase occur in the top of the Holmehus Formation and lower part of the Ølst Formation and Fur Formation. Most of the sequence was numbered −1 to −39 by Bøggild (1918), and is known as the 'negative series' (see Fig. 2).

The layers in the 'negative series' generally contain considerable amounts of glass. They have a variety of compositions—basaltic, dacitic, liparitic and peralkaline (Pedersen *et al.* 1975)—and are often partly transformed to clay minerals and zeolites. The composition of the few, thin ashes from the Holmehus Formation is not yet known.

(2) The middle phase, which is the main phase of the Palaeogene volcanism, is present in the upper part of the Fur and Ølst Formations. The ashes of this phase, the 'positive series'

of Bøggild (1918), are numbered +1 to +140 (see Fig. 2).

The layers in the main phase generally contain considerable amounts of glass, they are thick, frequent and appear fresh and unaltered. Their composition is generally basaltic.

(3) The later phase includes approximately 19 thin layers in the Røsnæs Clay Formation and the lower part of the Lillebælt Clay Formation. These layers are argillized and only contain little, if any, glass. They probably had a basaltic composition as they, like the basaltic ashes below, have a relatively high Ti content (Gersner 1980). These altered ash layers still have very sharp boundaries and usually show graded bedding (Fig. 3).

Another line of evidence for volcanism can be found in the interbedded clay and in the clays above and below the ash-bearing sequence. The mineralogy of these clays is well known (Nielsen 1974; Thiede *et al.* 1980; Heilmann–Clausen *et al.* 1985; Nielsen & Heilmann–Clausen 1986). Smectite is the dominant mineral, but zeolites of the heulandite–clinoptilolite type are always present in the Ølst Formation and occasionally present below, besides other minerals. Cristobalite is sporadically present in the Ølst Formation, as sand-sized grains in the clay, and as a diagenetic cement in silicified horizons. Silicified layers are also common in the clay below the Holmehus Formation. Müller (1967) has proposed that smectite, mainly nontronite, zeolites and free silica, which might lead to cristobalite formation, is a typical mineral paragenesis formed by transformation of basaltic material. However, this mineral paragenesis may also develop from non-volcanic precursors.

Figure 4 shows the average smectite content of the clay fraction ($<2\,\mu m$) in the upper part of the

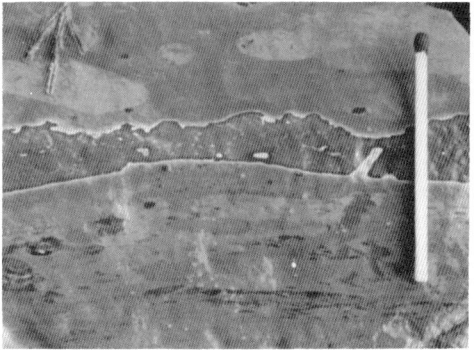

FIG. 3. Argillized ash layer from the Røsnæs Clay Formation. The match is 4.7 cm long.

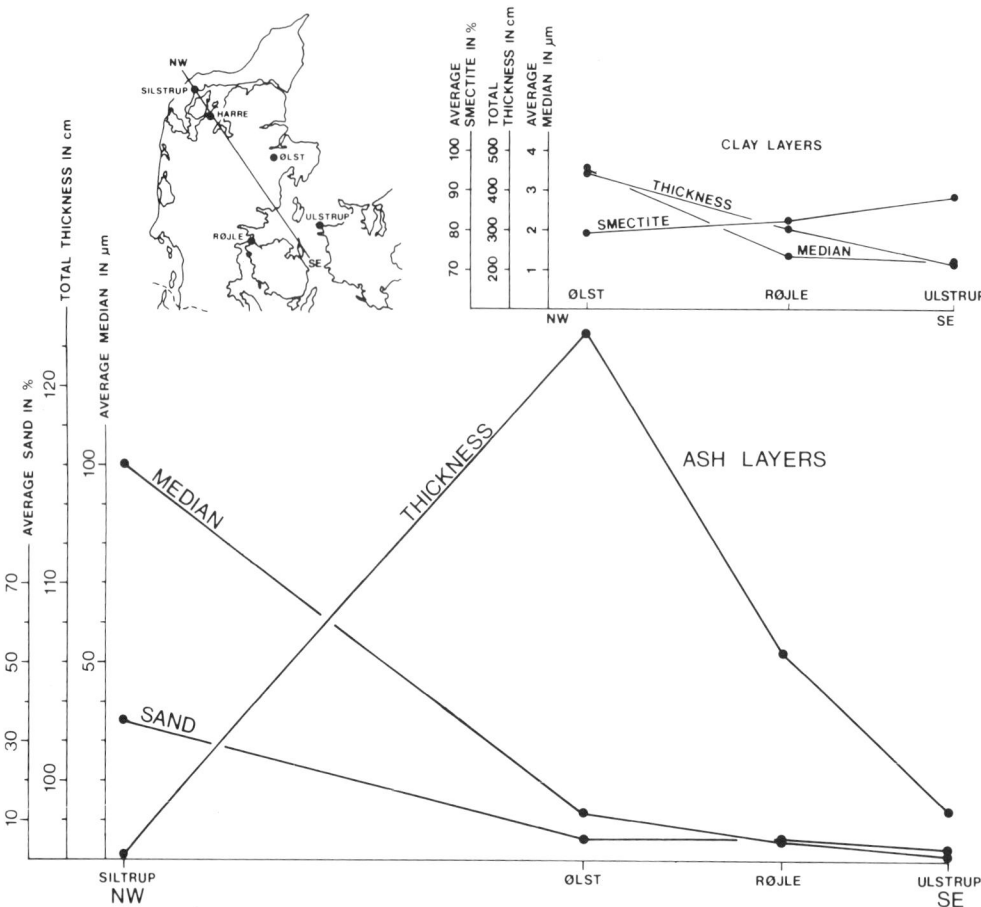

FIG. 4. NW–SE profile showing: Total thickness in cm, average median in μm and average smectite content in % of all identified clay layers. Total thickness in cm, average median in μm and average sand content in % of all ash layers identified with certainty at the four localities (mainly thick basaltic layers in the main phase).

Ølst Formation from three localities. The average mineral composition of the clay from the Holmehus Formation and of transformed ash layers from the Røsnæs Clay Formation is shown in Figure 5. Pyrite and carbonates are not included. It is obvious that the whole rock composition is very similar in the two types of layers, strongly suggesting a volcanic origin of both.

The chemical composition of smectites in argillized ashes from the Røsnæs Clay Formation is compared to the chemical composition of smectites in the Holmehus Formation (Fig. 6).

The whole-rock chemical composition was subject to different calculations in order to determine the chemical composition of the smectites, which make up 85–90% of the whole-rock mineralogy. Carbonates, pyrite, quartz, feldspars, kaolinite and illite were quantified partly by means of chemical analyses, and partly

by X-ray diffraction; a proportional amount of the relevant chemical composition was subtracted from the whole-rock composition. The remaining chemical components were recalculated to 100% (water-free).

It is evident that the composition of the smectites from the two different types of layers is almost identical, and relatively close to the nontronitic composition, mentioned by Müller (1967) as the normal mineral formed by alteration of volcanic material. Thus, the mineralogical and chemical composition of the Holmehus Formation suggests a volcanic origin for the smectites from this unit, though only few ash layers are present. It is most likely that smectites from the clays and marls underlying the Holmehus Formation also have a volcanic origin, since, apart from being carbonate-bearing, they have a very similar mineralogical composition. A nontronitic

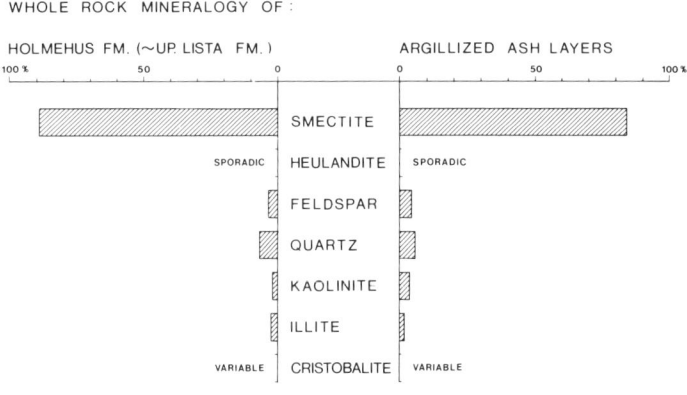

FIG. 5. Average whole-rock mineralogical composition, based on XRD, of the Holmehus Formation and some argillized ash layers in the Røsnæs Clay Formation from Danish outcrops and borings. Recalculated to 100% after subtraction of carbonates and pyrite.

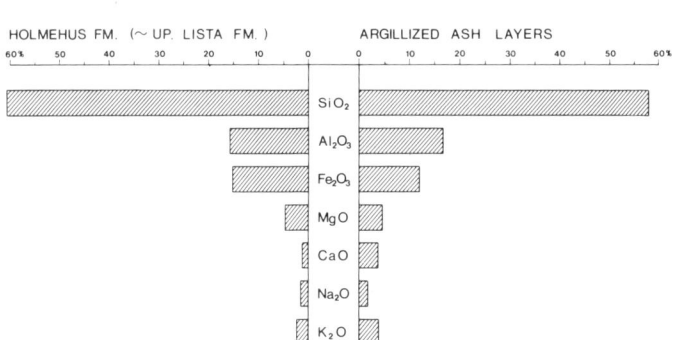

FIG. 6. Average chemical composition of smectites from the Holmehus Formation and some argillized ash layers in the Røsnæs Clay Formation from Danish outcrops and borings. Recalculated to 100% waterfree.

composition of these smectites has not, however, been proved.

Age

The biostratigraphy and the age of the Danish late Palaeocene to Middle Eocene is summarized in Figure 2. The oldest ash layers in the top of the Holmehus Formation occur on the islands of Mors and Fur in northwestern Jutland (see Fig. 1). The ash layers on Fur belong to dinoflagellate Zone 5 of Heilmann–Clausen (1985), probably near the oldest part of the zone. Dinoflagellate Zone 5 may correlate to the upper part of the Lista Formation in the North Sea and to Sables d'Erquelinnes in Belgium (see Heilmann–Clausen 1985). The latter unit contains calcareous nannofossils clearly indicating the NP9 Zone (De

Coninck *et al.* 1981). The correlation with Sables d'Erquelinnes therefore suggests that the oldest ashes in Denmark are of late Palaeocene age, time-equivalent to the NP9 Zone.

The overlying ash sequence in the Ølst and Fur Formations contain the ashes numbered from −39 to +140 by Bøggild (1918) (see Fig. 2). The lowermost of these, Nos −39 to −33, belong to dinoflagellate Zone 6 of Heilmann–Clausen (1985), while the overlying main part, at least above ash layer −19b, belongs to zone 7. Zone 6 is equivalent to the lower part of the *Apectodinium* (formerly *Wetzeliella*) *hyperacanthum* Zone of Costa & Downie (1976), while Zone 7 corresponds to the *Deflandrea oebisfeldensis* acme interval of Knox & Harland (1979) (see Fig. 2). In the North Sea Zone 6 is present in the lower part of the Sele Formation, and Zone 7 in the upper part of the

Sele Formation and in the Balder Formation (see Heilmann–Clausen 1985). In terms of silicoflagellate biostratigraphy the Fur Formation may be referred to as the *Naviculopsis constricta* Zone (Perch–Nielsen 1976).

The exact age of the Ølst and Fur Formations and their North Sea equivalents is uncertain. Most likely, they are of latest Palaeocene to earliest Eocene age, and there is some evidence to suggest that the Palaeocene–Eocene boundary is located close to the base of the Fur Formation. Various ages for these formations have previously been proposed and their chronostratigraphical position will therefore be discussed below. The current use of the Palaeocene–Eocene boundary as equal to the boundary between the calcareous nannofossil Zones NP9 and NP10 (e.g. Berggren *et al.* 1985; Martini & Müller 1986) is adopted here.

A correlation between the ash sequence in the Fur Formation, in the North Sea and the N Atlantic is proposed by Knox (1984) based on variations in the chemical composition of feldspars in the ash layers. According to Knox (1984) the N Atlantic Deep Sea Drilling Project (DSDP) Site 550 contains an ash sequence that may be correlated to the Danish ash series from ash layer No. − 17 to the top of the Fur Formation. The entire ash sequence in Site 550 is included in the lower part of the NP10 Zone, i.e. in the earliest Eocene, and the NP9/NP10 boundary is situated approximately eight metres below the presumed ash layer − 17 in Site 550. In the Fur Formation ash layer − 17 is present in the lower part, and consequently most or all of the Fur Formation may be referred to the NP10 Zone. However, this conclusion is completely dependent on the correct identification of the ash layer − 17 from Site 550. The location of the Palaeocene–Eocene boundary in Denmark is thus based on a long distance correlation, and should therefore be considered with caution.

Heilmann–Clausen (1982) previously used the *W. astra* dinoflagellate Zone as a base Eocene marker, following Costa & Müller (1978). On this basis the Palaeocene–Eocene boundary should be located immediately above the Fur Formation. However, the validity of using the *W. astra* Zone as a basal Eocene marker has been questioned by Morton *et al.* (1983).

The ash layers in the Røsnæs Clay Formation may be confidently dated as Early Eocene, based on calcareous nannoplankton and dinoflagellate assemblages. The calcareous beds of this formation, R4 and R5, are referred to NP Zones 11 and 12 (Thiede *et al.* 1980). The *W. astra* and *W. meckelfeldensis* zones are identified in the basal Knudshoved Member and bed R1 (Heilmann–

Clausen 1982; Nielsen & Heilmann–Clausen 1986) while the *Dracodinium simile, D. varielongitudum* and *Kisselovia coleothrypa* Zones are identified in beds R4 and R5 (Heilmann–Clausen 1983; Nielsen & Heilmann–Clausen 1986).

The youngest ash layers occur in the Lower Lillebælt Clay (beds L1 to L3). Nielsen & Heilmann–Clausen (1986) correlated the dinoflagellate assemblages of these units with assemblages in Belgium, the Netherlands and southern England. They indicated that the Lower–Middle Eocene boundary lies within beds L1 to L3. This is based in particular on the first occurrence of *Areosphaeridium diktyoplokus* in bed R6 and the last occurrence of *Eatonicysta ursulae* in bed L4. Consequently, the latest ash layers, known from Denmark, were deposited approximately at this boundary, i.e. near the NP13/NP14 boundary.

The source of the ash layers

The source, dispersal and deposition of the ashes have been the subject of much discussion. The ashes are all well sorted and well graded. The thickness of ash layer No. + 62 (Fig. 7) and other similar isopach maps have been used to conclude (Andersen 1937), that the volcanoes were situated in Skagerrak (see Fig. 1). Geophysical anomalies (Åm 1973) and basalt dredged from the area (Noe-Nygaard 1967) support this hypothesis.

Fig. 8 is a profile from NW to SE in Denmark, showing grain sizes and thicknesses from NW to SE in Denmark for ash layer No. + 62 and for

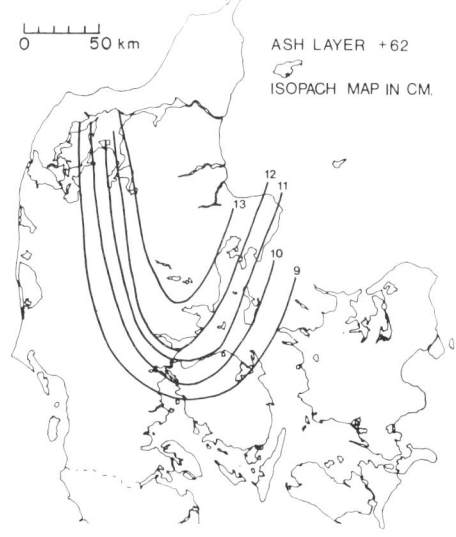

FIG. 7. Isopach map of ash layer No. + 62 (modified from Andersen 1937).

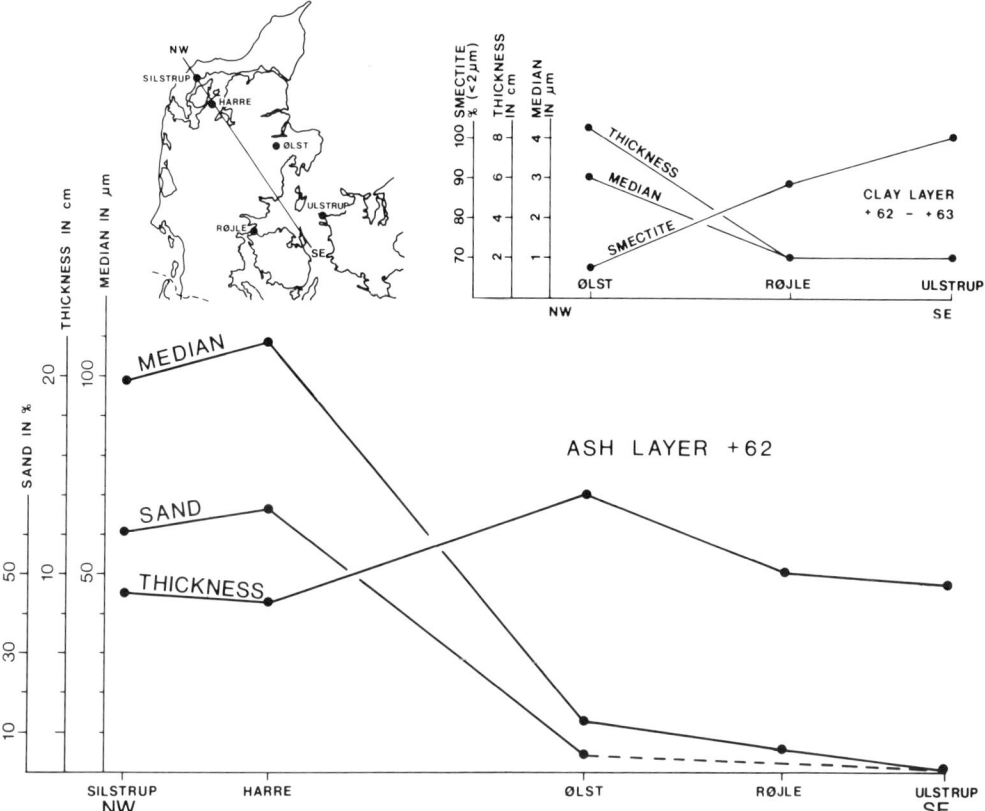

Fɪɢ. 8. NW–SE profile of ash layer No. +62 and clay layer between ash layers Nos. +62 and +63, showing the variation in grain size, thickness and smectite content.

the clay layer between ash layer Nos +62 and +63. Fig. 4 shows the same profile, in which the total thicknesses and average grain-size parameters and smectite contents (only for clay layers) are plotted for all ash and clay layers identified at selected localities.

In the following, the regional variations in thickness and grain size of ash layers and interbedded clay layers will be compared. The thickness patterns outside the Limfjord area and their relations to the Limfjord area as a whole will be discussed first, and afterwards the detailed distribution within the Limfjord area will be described and interpreted.

Outside the Limfjord area

It is evident from Fig. 8 that the ash layer No. +62 is coarsest in the Limfjord area and much finer grained to the SE. The thickness is greatest at Ølst and decreases to the S or SE. In the clay layer between ash layers Nos +62 and +63 the same decrease in grain size and thickness is seen. The decrease in the grain size is expressed as an

increase in smectite content because of the smaller size of smectite particles compared to other clay minerals. The pattern for ash layer No. +62 and for the clay layer between ash layers Nos +62 and +63, seems to be characteristic of most of the ash and clay layers. As seen from Fig. 4 the thickness, grain size and smectite patterns for all ash and clay layers identified at four localities (see Fig. 6) are similar from Ølst towards the S and SE. The clay layers were deposited by settling in the sea water and the patterns of parameters seem to indicate a supply from the N and NW and a dispersal by water currents to the S and SE. As the ash layers show the same trends outside the Limfjord area, it is suggested that the supply and dispersal mechanism for the ash layers S and SE of the Limfjord area were the same, i.e. a transport from N and NW to S and SE by a current.

The Limfjord area

The maximum grain size of the ashes in the Limfjord area is ca. 0.5 mm, with an average size of 0.1 mm. The relatively coarse grain size in the Limfjord area (mean = 100 µm) and its rapid decrease to Ølst (mean = 13 µm) indicates a relatively short distance to the source of the volcanic material. It is suggested that the main transport mechanism from the volcanic centre(s) to the Limfjord area was the generally northerly winds, overprinted by a water current transport becoming successively more dominant with decreasing grain size. The increase in thickness from Silstrup to Ølst is interpreted as indicating that the original ash contained a considerable amount of silt-sized fractions relative to the sand. The silt-sized ash particles were more subject to water-current transport than the sand, and are therefore deposited further away from the volcano(es). The very good degree of sorting in most ashes supports the hypothesis of silt winnowing.

Within the Limfjord area the grain-size and thickness distribution of the ashes show characteristic trends (Bøggild 1918; Andersen 1937; Pedersen *et al.* 1975). According to these authors the thickest and coarsest layers are generally found on Fur. Both parameters decrease towards the NW, S and E, indicating that the windborne supply of coarse ash particles was located to the N and NNE (see Pedersen *et al.* 1975).

As indicated on Figure 1, there may have been more volcanic centres in the Skagerrak. Malm *et al.* (1984) also suggested a source in the Skagerrak region for the main part of the basaltic ashes in Denmark. It is most likely, however, that some of the acidic and more fine-grained ashes, as for example No. +19, have had another source, possibly the British volcanic province.

In the North Sea Basin the clay mineralogy of the Palaeogene section from the Norwegian Wells 30/5-1, 15/6-2 and 2/8-2, located in the central parts of the Viking and Central Grabens, have been analyzed (Sørensen & Nielsen 1981a, b, c). In the Danish offshore sector the Wells B-1, C-1, D-1, E-1, F-1, M-2x (Nielsen 1980), Lulu-1 and Inez-1 were analyzed in this study. Onshore the borings Harre, Viborg-1 (Thiede *et al.* 1980), LB 38, DGI 83101 (Nielsen & Heilmann–Clausen 1986), the wells Tønder-3 and Plön in W Germany, besides several outcrops (Nielsen 1974; Heilmann–Clausen *et al.* 1985) have been included in the project.

Smectite is a dominant, but variable, component in the clay fraction in the Palaeocene and in parts of the Eocene sediments. This variation is due in part to changes in the supply of other minerals, i.e. the dilution effect. In order to find an absolute measure for smectite deposition the sedimentation rates for smectite have been calculated.

Comparing these sedimentation rates for smec-

tites in different regions of the North Sea (Fig. 9) quite distinct differences are revealed. In the Central Graben, the Norwegian–Danish Basin, on the Ringkøbing-Fyn High and in the N German Basin, the sedimentation rates were small and with only small variations, although there is an overall southward decrease in thickness. In the Viking Graben the sedimentation rate was much greater. This might have been caused by a shorter distance to main volcanic centres in Scotland and in the opening N Atlantic.

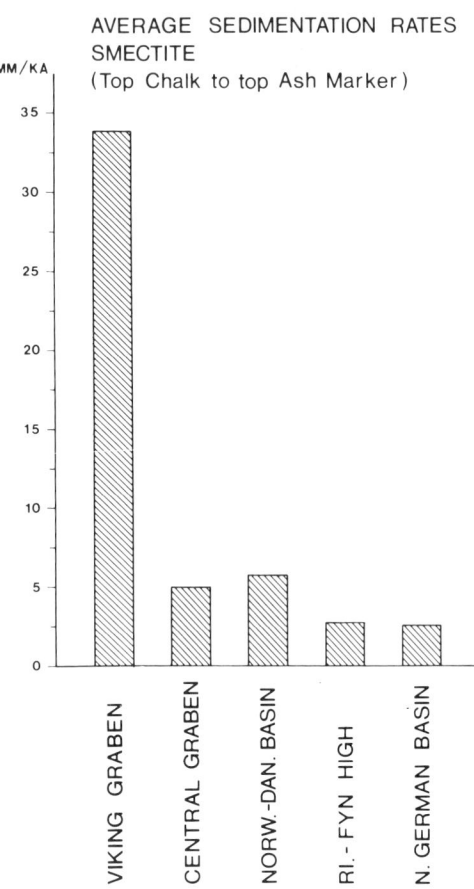

FIG. 9. Average sedimentation rates in mm/ka for smectites in different regions of the North Sea from top Ekofisk Formation/top Danian Limestone to top Balder Formation/top Fur and Ølst Formations.

Conclusions

(1) The *evidence* for the Palaeogene volcanism is provided by:

a) The presence of more than 200 ash layers.

b) The mineralogy of the sediments, and especially the similarity in mineralogy of argillized ash layers and interbedded clays, i.e. the dominance of smectite and presence of zeolites and cristobalite.

c) The almost identical chemical composition of smectites from argillized ash layers and the interbedded clays and the clays of the Holmehus Formation indicates that a volcanic origin for the Palaeogene smectites is probable.

(2) The proposed *ages* for the Danish Palaeogene pyroclastic phases are:

a) The earliest recorded volcanic ash layers are of late Palaeocene age, belonging to the Danish dinoflagellate Zones 5 and 6 (probably equivalent to the NP9 Zone). The ashes are present in the Holmehus Formation and in the lower part of the Ølst Formation.

b) The age of the main phase in the Fur Formation and upper part of the Ølst Formation is still unknown, but there is some evidence for suggesting an early Eocene (NP10 Zone) age.

c) The latest phase is of Early Eocene age, and terminated near the Early–Middle

Eocene boundary (i.e. during NP13 or NP14).

(3) The *provenance and transport mechanism* of the volcanic sediments are deduced as follows:

a) The grain-size and thickness variations suggest a relatively short distance to the eruption centres, and that they were situated N of the Limfjord area.

b) In southern Denmark both the ashes and the interbedded clays show evidence of transport by water currents from NW to SE.

c) The deposition rate of smectite in the Viking Graben is six times greater than in the remaining part of the North Sea, indicating a supply from a volcanic source to the N or W, closer to the region than the Skagerrak volcano.

ACKNOWLEDGEMENTS: Samples for this study were kindly placed at our disposal by Norwegian Petroleum Directorate, Geological Survey of Denmark, Mærsk Olie & Gas A/S and Deutsche Texaco A/G. Financial support has been given by the Danish Natural Science Research Council.

References

ANDERSEN, S. A. 1937. De vulkanske Askelag i Vejgennemskæringen ved Ølst og deres Udbredelse i Danmark. *Danmarks Geologiske Undersøgelse*, **59**, 50 pp.

BERGGREN, W. A., KENT, D. V. & FLYNN, J. J. 1985. Jurassic to Paleogene: Part 2. Paleogene geochronology and chronostratigraphy. *In*: SNELLING, N. J. (ed) *The Chronology of the Geological Record*. Memoir of the Geological Society of London, **10**, pp. 141–186.

BØGGILD, O. B. 1918. Den vulkanske Aske i Moleret samt en Oversigt over Danmarks ældre Tertiærbjærgarter. *Danmarks Geologiske Undersøgelse, Raekke 2*, **33**, 159 pp.

BRINDLEY, G. W. & BROWN, G. 1980. *Crystal Structures of Clay Minerals and their X-ray Identification*. Mineralogical Society (Monograph), **5**, 495 pp.

COSTA, L. I. & DOWNIE, C. 1976. The distribution of the dinoflagellate *Wetzeliella* in the Palaeogene of north-western Europe. *Palaeontology*, **19**, 591–614.

—— & MÜLLER, C. 1978. Correlation of Cenozoic dinoflagellate and nannoplankton zones from the NE Atlantic and NW Europe. *Newsletters on Stratigraphy*, **7**, 65–72.

——, DENISON, C. & DOWNIE, C. 1978. The Paleocene/

Eocene boundary in the Anglo–Paris Basin. *Journal of the Geological Society of London*, **135**, 261–264.

DE CONINCK, J. 1975. Microfossiles a paroi Organique de l'Ypresien du Bassin Belge. *Service Geologique de Belge, Professional Paper 1975*, **12**, 165 pp.

—— 1977. Organic walled microfossils from the Eocene of the Woensdrecht borehole, southern Netherlands. *Mededelingen Rijks Geologische Dienst, Nieuwe Serie*, **28**, 33–64.

—— 1981. Especes indicatrices de microfossiles a paroi organique des depots de l'Ypresien superieur et du Lutetien dans le sondage de Kallo. *Bulletin de la Societé Belge de Géologie*, **89**, 309–317.

——, DE DECKLER, M., DE HEINZELIN, J. & WILLEMS, W. 1981. L'age des faunes d'Erquelinnes. *Bulletin de la Societé Belge de Géologie*, **90**, 121–154.

DEEGAN, C. E. & SCULL, B. J. 1977. A Standard Lithostratigraphic Nomenclature for the Central and Northern North Sea. *Report of the Institute of Geological Sciences*, **77/25**, 36 pp.

FORCHHAMMER, G. 1835. *Danmarks geognostiske Forhold, forsaavidt som de ere afhængige af Dannelser, der ere sluttede*. Hostrup Schulz, Copenhagen, 112 pp.

GERSNER, F. 1980. *En lithostratigrafisk, kemisk og mineralogisk undersøgelse af Røsnæsleret og en del af Lillebæltleret i Danmark.* Thesis (unpublished). Geologisk Institut, Aarhus Universitet.

GIBBS, R. J. 1971. X-ray diffraction mounts. *In*: CARVER, R. (ed) *Procedures in Sedimentary Petrology.* Wiley–Interscience New York, pp 531–539.

HANSEN, J. M. 1979. Age of the Mo-Clay Formation. *Bulletin of the Geological Society of Denmark,* **27**, 89–91.

HEILMANN–CLAUSEN, C. 1982. The Paleocene–Eocene boundary in Denmark. *Newsletters on Stratigraphy,* **11**, 55–63.

—— 1983. *Dinoflagellate zonation and lithostratigraphy of Palaeocene and Eocene sediments from Denmark.* PhD thesis (unpublished). Aarhus Universitet.

—— 1985. Dinoflagellate stratigraphy of the uppermost Danian to Ypresian in the Viborg 1 borehole, central Jylland, Denmark. *Danmarks Geologiske Undersøgelse,* Series **A7**, 69 pp.

——, NIELSEN, O. B. & GERSNER, F. 1985. Lithostratigraphy and depositional environments in the Upper Paleocene and Eocene of Denmark. *Bulletin of the Geological Society of Denmark,* **33**, 287–323.

KNOX, R. W. O'B. 1984. Nannoplankton zonation and the Palaeocene/Eocene boundary beds of NW Europe: an indirect correlation by means of volcanic ash layers. *Journal of the Geological Society of London,* **141**, 993–999.

—— & HARLAND, R. 1979. Stratigraphical relationships of the early Palaeogene ash-series of NW Europe. *Journal of the Geological Society of London,* **136**, 463–470.

——, MORTON, A. C. & HARLAND, R. 1981. Stratigraphical relationships of Palaeocene sands in the UK sector of the central North Sea. *In*: ILLING, L. V. & HOBSON, G. D. (eds) *Petroleum Geology of the Continental Shelf of North-West Europe.* Institute of Petroleum, London, 267–281.

MADIRAZZA, I. & FREGERSLEV, S. 1969. Lower Eocene tuffs at Mønsted, North Jutland. *Bulletin of the Geological Society of Denmark,* **19**, 283–318.

MALM, O. A., BRUNN CHRISTENSEN, O., FURNES, H., LØVLIE, R., RUESLÅTTEN, H. & LORANGE ØSTBY, K. 1984. The Lower Tertiary Balder Formation: an organogenic and tuffaceous deposit in the North Sea region. *In*: SPENCER, A. M. *et al.* (eds) *Petroleum Geology of the North European Margin.* Graham & Trotman, London, pp. 149–170.

MARTINI, E. & MÜLLER, C. 1986. Current Tertiary and Quaternary calcareous nannoplankton stratigraphy and correlations. *Newsletters on Stratigraphy,* **16**, 99–112.

MORTON, A. C., BACKMAN, J. & HARLAND, R. 1983. A reassessment of the stratigraphy of DSDP Hole 117A, Rockall Plateau: implications for the Palaeocene–Eocene boundary in NW Europe. *Newsletters on Stratigraphy,* **12**, 104–111.

MÜLLER, G. 1967. Diagenesis in argillaceous sediments. *In*: LARSEN, G. & CHILINGAR, G. V. (eds) *Diagenesis in sediments.* Developments in Sedimentology. **8**, pp. 128–177.

NIELSEN, O. B. 1974. Sedimentation and diagenesis of Lower Eocene sediments at Ölst, Denmark. *Sedimentary Geology,* **12**, 25–44.

—— 1980. A sedimentological mineralogical investigation of the Tertiary sediments from the borehole M-2X in Central Trough, North Sea. *Danmarks Geologiske Undersøgelse,* Årbog, **1979**, 41–50.

—— & HEILMANN–CLAUSEN, C. 1986. Lithology and stratigraphy of the Tertiary section. *In*: MØLLER, J. T. (ed) *Twenty-five years of Geology in Aarhus. Geoskrifter,* **24**, 235–253.

NOE-NYGAARD, A. 1967. Dredged Basalts from Skagerrak. *Meddelelser fra Dansk Geologisk Forening,* **17**, 285–287.

NORIN, R. 1940. Problems concerning the volcanic ash layers of the lower Tertiary of Denmark. *Geologiska Föreningens i Stockholm, Förhandlingar,* **62**, 31–44.

PEDERSEN, A. K., ENGELL, J. & RØNSBO, J. G. 1975. Early Tertiary volcanism in the Skagerrak: New chemical evidence from ash-layers in the Mo-clay of northern Denmark. *Lithos,* **8**, 255–268.

PEDERSEN, G. K. 1981. Anoxic events during sedimentation of a Palaeogene diatomite in Denmark. *Sedimentology,* **28**, 487–504.

—— & SURLYK, F. 1983. The Fur Formation, a late Paleocene ash-bearing diatomite from northern Denmark. *Bulletin of the Geological Society of Denmark,* **32**, 43–65.

PERCH-NIELSEN, K. 1967. Nannofossilien aus dem Eozän von Dänemark. *Eclogae geologicae Helvetiae,* **60**, 19–32.

—— 1971. Elektronenmikroskopische Untersuchungen an Coccolithen und verwandten Formen aus dem Eozän von Dänemark. *Det Kongelige Danske Videnskabernes Selskab, Biologiske Skrifter,* **18(3)**, 76 pp.

—— 1976. New silicoflagellates and a silicoflagellate zonation in north European Palaeocene and Eocene diatomites. *Bulletin of the Geological Society of Denmark,* **25**, 27–40.

PRINZ, W. & VAN ERMENGEM, E. 1883. Recherches sur la structure de quelques diatomées contenues dans le "Cementstein" du Jutland. *Annales de la Societé Belge de Microscopie,* **8**, 7–65.

SCHULTZ, L. G. 1960. Quantitative interpretation of mineralogical composition from X-ray and chemical data for the Pierre shale. *United States Geological Survey, Professional Paper,* **391-C**, 31 pp.

SØRENSEN, S. & NIELSEN, O. B. 1981a. Biostratigrafi, lithostratigrafi og sedimentpetrografi av tertiærsedimenter fra den norske kontinentalsokkel. Brønn 2/8-2. *Institutt for Geologi, Universitetet i Oslo, Intern Skriftserie,* **33**, 10 pp.

—— & —— 1981b. Biostratigrafi, lithostratigrafi og sedimentpetrografi av tertiærsedimenter fra den norske kontinentalsokkel. Brønn 15/6-2. *Institutt for Geologi, Universitetet i Oslo, Intern Skriftserie,* No. **34**, 10 pp.

——, —— 1981c. Biostratigrafi, lithostratigrafi og sedimentpetrografi av tertiærsedimenter fra den norske kontinentalsokkel. Brønn 30/5-1. *Institutt*

for Geologi, Universitetet i Oslo, Intern Skriftserie,
35, 11 pp.

THIEDE, J., NIELSEN, O. B. & PERCH–NIELSEN, K. 1980.
Lithofacies, mineralogy and biostratigraphy of
Eocene sediments in Northern Denmark (Deep

Test Viborg-1). *Neues Jahrbuch für Geologie und
Paläontologie, Abhandlungen,* **160**, 149–172.

ÅM, K. 1973. Geophysical indications of Permian and
Tertiary igneous activity in the Skagerrak. *Norges
Geologiske Undersøkelse,* **287**, 1–25.

O. B. NIELSEN & C. HEILMANN–CLAUSEN, Department of Geology, Aarhus University, DK
8000 Aarhus C, Denmark.

The record of early Tertiary N Atlantic volcanism in sediments of the North Sea Basin

R. W. O'B. Knox & A. C. Morton

SUMMARY: Volcaniclastic deposits in sedimentary sequences of the North Sea Basin and adjacent areas indicate that two phases of early Palaeogene explosive volcanism took place in the north-eastern Atlantic region. The earlier, late Palaeocene (NP5–NP6) phase involved significant activity along a N–S trend that included both the British and Faeroe–Greenland Tertiary volcanic provinces. The later phase spanned the latest Palaeocene and early Eocene (NP9 to NP13), with much or all of the activity taking place in the Faeroe–Greenland Province. Early ashfalls of mixed basaltic to silicic compositions may have included contributions from the final phase of British volcanism, but were followed by a series of 200 or more tholeiitic ashfalls of Faeroe–Greenland provenance. These tholeiitic eruptions appear to have marked the onset of separation of Greenland from Europe in mid NP10 times. A subsequent return to pyroclastic activity of more variable compositions appears to have marked the re-establishment of stresses within the E Greenland crust that continued throughout the early Eocene (mid NP10 to end NP13). The mechanism of eruption of the tholeiitic ashes, which are equivalent to a magma volume of several thousand cubic kilometres, is uncertain, but they would appear to involve hydrovolcanic processes.

A major problem in dating and correlating the early Palaeogene igneous events in the north-eastern Atlantic region is the limited biostratigraphical control. Interbedded sediments, where present, are of little assistance in regional correlation since they generally lack biostratigraphically useful fossil assemblages.

An alternative means of relating igneous events to standard stratigraphical schemes is to look for expressions of volcanic activity in adjacent sedimentary basins. The most direct information is provided by pyroclastic ashfalls, but reworking of terrestrial igneous rocks may locally provide useful information. Additionally, less direct information may be obtained from the record of epeirogenic events within the sedimentary basin, since such events may well be closely associated with igneous activity. For the Tertiary of the north-eastern Atlantic region, the best opportunity for such an approach is provided by the North Sea Basin, where volcanic contributions to the sedimentary sequence can be identified both onshore and in numerous offshore boreholes. Accordingly, it is the sequences of the North Sea Basin that receive most attention in this paper, although brief reference is also made to pyroclastic occurrences in the Faeroe–Shetland Basin and the eastern Atlantic Basin.

North Sea stratigraphy

A composite sequence for the central North Sea area is shown in Fig. 1, which includes both the local lithostratigraphical divisions of Deegan & Scull (1977) and the regional Palaeocene 'Units A–E' of Knox et al. (1981). Regional studies have identified a series of basin-wide events reflected in successive regressions and transgressions (Knox et al. 1981; Stewart 1987). The principal tectonic/sedimentary events were: (1) end-Cretaceous hiatus (base Unit A); (2) top Ekofisk regression (base Unit B); (3) late Andrew event (base Unit C); (4) base Forties sand regression (base Unit D); (5) top Forties transgression (base Unit E); (6) base Balder hiatus; and (7) top Balder hiatus. These events are most evident in marginal sequences, where the transgressive/regressive cycles are reflected in hiatuses and facies changes. Volcaniclastic sediments are encountered at several levels within the sequence. With one exception they are of undoubted pyroclastic origin, and two main phases of pyroclastic activity can be recognized (Phases 1 & 2 of Knox & Morton 1983). These are described in the next section.

Biostratigraphical control is mainly provided by dinoflagellate cyst assemblages, for which a basin-wide zonation has been established (Costa et al. 1978; Knox et al. 1981). Indigenous calcareous nannofossils are absent from most of the early Palaeogene sequence of the North Sea area, but Zones NP1–5 have been identified in Unit A (Ekofisk Formation and overlying marl) and Zone NP12 in the early Eocene variegated clays that overlie the Balder Formation (see Knox et al. 1981). Calcareous nannofossils are slightly more common in onshore sequences marginal to the North Sea Basin (Aubry 1985; Siesser et al. 1987) and are abundant in the sequences of the

From MORTON, A. C. & PARSON, L. M. (eds), 1988, *Early Tertiary Volcanism and the Opening of the NE Atlantic*, Geological Society Special Publication No. 39, pp. 407–419.

407

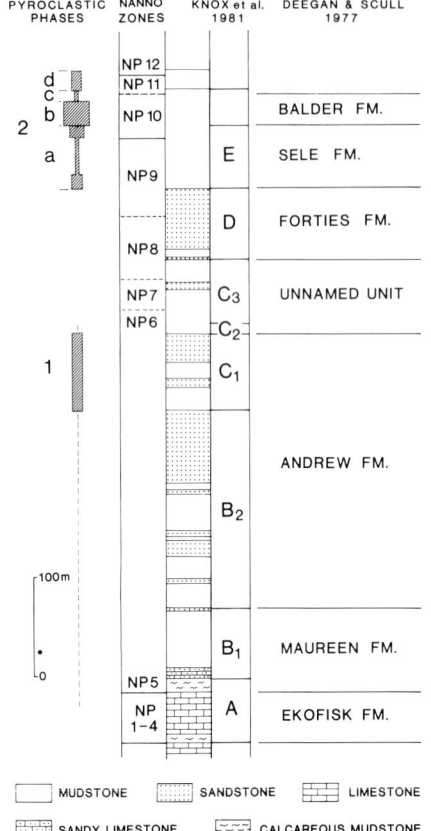

FIG. 1. Representative early Palaeogene sequence for the central North Sea Basin.

eastern Atlantic (Müller 1985). Indirect correlation between the Atlantic sequences and those of the North Sea (Knox 1984) allows an approximate nannoplankton zonation to be applied to the central North Sea sequence (see Fig. 1). The nanno-plankton data are complemented to some extent by magnetostratigraphical data (Townsend & Hailwood 1985). Of particular importance is the presence of the Thanet normal polarity zone, which has been identified in the lower part of the Thanet sands of Kent and in an approximately equivalent clay sequence in Norfolk (Cox et al. 1985). The clay sequence is the onshore representative of the Unit C3 clays of the North Sea sequence. The Thanet normal polarity zone is correlated with chron 26N (Townsend & Hailwood 1985; Cox et al. 1985). Chron 26N straddles the NP6/NP7 boundary, indicating that the base of Unit C3 is of late NP6 age; biostratigraphical information from equivalent onshore sections (Siesser et al. 1987) indicates

that Unit C3 extends upwards into zone NP8. On this basis, Units B1, B2, C1 and C2 would occupy the interval mid NP5 to late NP6 (equivalent to the bulk of chron 26R) and Unit D (Forties sand) would be no older than NP8. The NP8/NP9 boundary probably lies within Unit D.

The NP9/NP10 boundary was considered by Knox (1984) to fall within the lower part of the Sele Formation. However, the dating of the ashes of Unit C1/C2 as no younger than NP7 indicates that the early NP9 ash of Hole 549 should be correlated with those of the basal part of the Sele Formation, rather than with those of Unit C. The base of the Sele Formation may therefore lie close to the base of NP9, and the NP9/NP10 boundary may correspond to a level in the upper part of the Sele Formation. Unit D (Forties sand) may thus span the interval mid/late NP8 to earliest NP9. The zonation of the remainder of the early Eocene sequence is unchanged from that of Knox (1984).

The volcaniclastic sediments

As mentioned above, two discrete phases of pyroclastic sedimentation can be identified in the North Sea sequences (Fig. 1). Phase 1 spans Unit B (Maureen and Andrew Formations) to Unit C2 ('undifferentiated' unit). Phase 2 spans the Sele Formation (Unit E), the Balder Formation, and the immediately overlying early Eocene variegated clay unit. Phase 2 is further divisible into Subphases 2a, 2b, 2c and 2d. The volcaniclastic deposits that represent these phases and subphases are described below.

Preservation of the volcaniclastic material studied is highly variable. Fresh volcanic glass is present only exceptionally; more often, partial or complete alteration to smectitic clay has taken place. Routine petrology thus provides only a broad assessment of the original ash compositions, and in the more substantially argillized layers the primary composition has been deduced mainly from residual phenocryst minerals (feldspars, amphiboles, etc.). More detailed information on magma chemistry should be obtained from current investigations into trace element and rare-earth element compositions.

Phase 1

This phase is characterized by sporadic graded airfall ash layers, but also contains one or more units of volcaniclastic sand. Because of their very different character, the volcaniclastic sands are described separately from the graded ash layers.

Volcaniclastic sand

Description. The volcaniclastic sand occurs in units of up to 30 m (Fig. 2), often with little or no contamination by terrigenous sediments. The sand is composed of irregular to rounded vitric particles. In thin section the particles generally appear structureless, although some display vesicular or amygdaloidal texture; they are green in colour, and have undergone partial alteration to smectite.

Distribution. The volcaniclastic sands are best developed in the southern part of the Outer Moray Firth Basin, off eastern Scotland (Fig. 3A); thin representatives extend eastwards as far as the Forties area.

Composition. Incipient alteration of the volcaniclastic sands precludes accurate whole-rock chemical analysis, but a basaltic composition is indicated by the brown to olive-green colour of the glass and by low SiO_2 and high MgO contents,

as determined by microprobe analysis. No phenocryst minerals have been observed; microlitic feldspar is the only optically identifiable mineral phase.

Derivation. The distribution of the basaltic sands within the central North Sea area clearly points to derivation from the W, the only reasonable source being the Hebridean region. The purity and relative freshness of many of the deposits suggests rapid redistribution with a minimum of terrestrial transport. These features, together with the dominant vitric character of the particles, favours penecontemporaneous reworking of pyroclastic deposits that accumulated close to the basin margin.

Graded ash layers

Description. The airfall ash layers of Phase 1 are poorly represented in cores and at outcrop. They typically consist of argillized layers up to a few centimetres in thickness.

Distribution. The phase 1 ashes have only rarely been cored in the offshore area, but examination of cuttings samples indicates that they extend throughout the UK sector (see Fig. 3B), being most abundant in northern and central parts. Knox (1979) reported disseminated ash particles in the basal Thanet sands of Kent (SE England). From equivalent sediments in the adjacent offshore area, Morton (1982) reported euhedral aegirine, arfvedsonite, and Mg-kataphorite in association with disseminated lithic volcaniclastic grains. More recently Cox *et al.* (1985) have reported a single discrete tuff layer from the basal part of Unit C3 (Lista clay facies) in Norfolk. The latter is the youngest confirmed representative of Phase 1, at least in the southern North Sea area, and is late chron 26R in age. Bentonitic ash layers have also been reported by Heilmann-Clausen *et al.* (1985) from the broadly equivalent sequence in Denmark (Holmehus Formation).

A single ash layer recorded in an offshore BGS borehole (82/12) W of the Shetland Islands (Morton *et al.* 1988) is also ascribed to this phase.

Composition. The few mineralogical analyses available indicate that the North Sea and onshore UK ashes include basaltic and more silicic types; the presence in some layers of aegirine, arfvedsonite, and magnesio-kataphorite indicate peralkaline compositions. The ash layer from W of the Shetlands is composed of relatively fresh glass, and microprobe analysis indicates a Fe-Ti tholeiite composition comparable to that of basalts

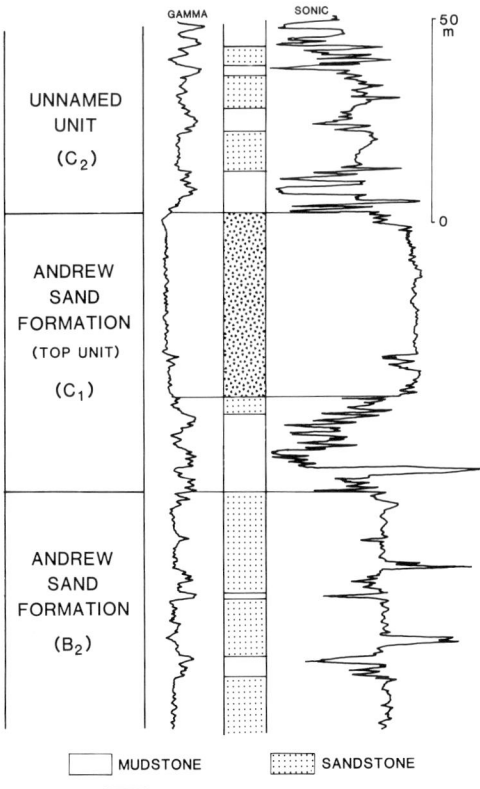

FIG. 2. Representative Phase 1 volcaniclastic sand unit, showing typical wireline log responses (UK Well 21/2-1).

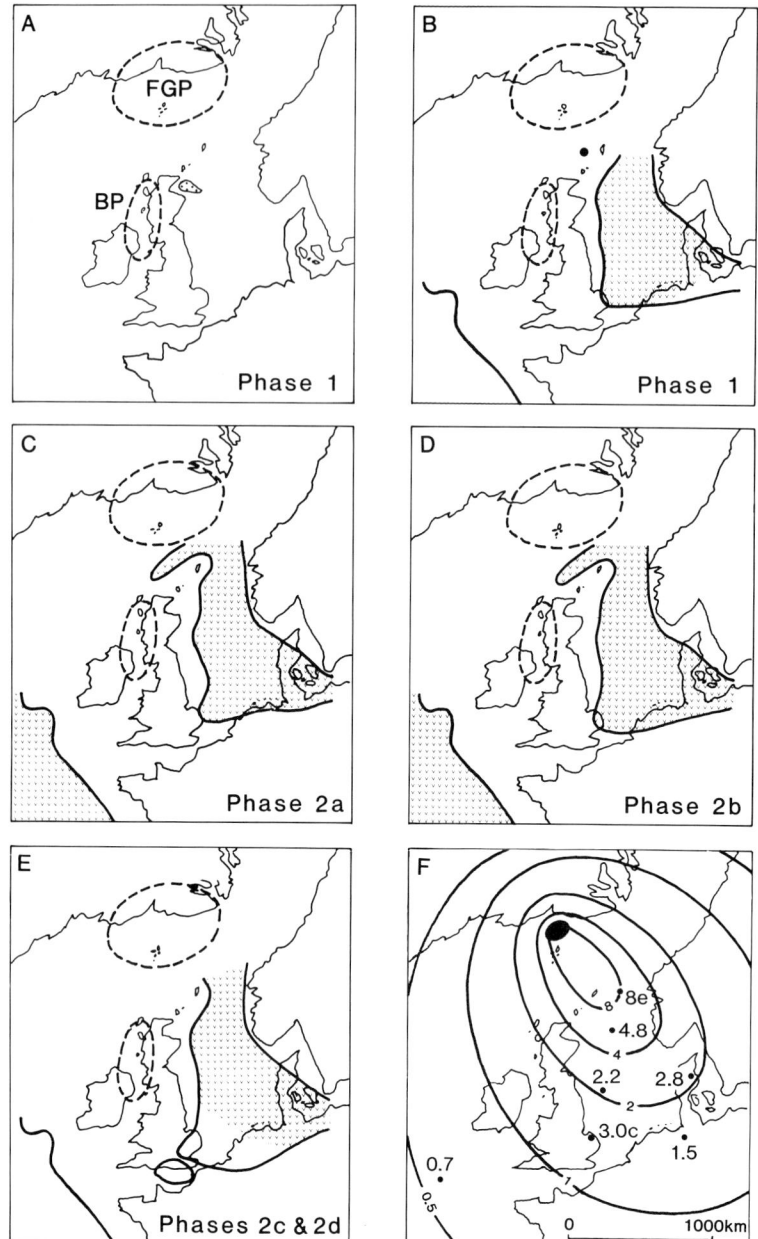

FIG. 3. Distribution of early Palaeogene volcaniclastic sediments in NW Europe. Dashed lines mark generalized volcanic provinces: BP: British Province; FGP: Faeroe–Greenland Province. Continuous lines mark limits of facies suitable for the preservation of airfall ash layers. Ornamented areas show known distribution of airfall ash layers: (A) Phase 1 volcaniclastic sands. (B) Phase 1 airfall ashes. The absence of airfall ashes in cores from the eastern Atlantic Basin may result from hiatus rather than non-deposition. The black circle marks the site of the Selandian ash of BGS Borehole 82/12. (C) Subphase 2a airfall ashes. (D) Subphase 2b airfall ashes. (E) Subphases 2c and 2d airfall ashes. (F) Subphase 2b total ash thickness. Larger figures refer to ash thicknesses (in metres) in individual measured sections: suffix 'e' indicates thickness estimated from an incomplete sequence, suffix 'c' indicates contamination by detrital sediment. Contours show postulated thickness distribution with supposed effects of contamination and subaqueous redistribution removed. Measured ash sequences are: northern North Sea: Norwegian Well 30/2-1 (thicknesses taken from Malm *et al.* 1984, but excluding the redeposited layer); central North Sea: UK Well 16/7-2; south-western North Sea: BGS Borehole 81/46A (Lott *et al.* 1983); onshore UK: BGS Ormesby Borehole (Cox *et al.* 1985); onshore Denmark: Ølst (thicknesses taken from Andersen 1937); eastern Atlantic Basin: DSDP Hole 550 (Knox 1985). Black area represents possible location of source volcanoes.

of the Faeroe–Greenland province (Morton *et al.* 1988).

Derivation. Little is known of the detailed distribution of the airfall ashes of Phase 1, except that they appear to be most abundant in the N of the North Sea Basin. A source in the Hebridean province is possible, this being supported by the close association of the eruptive phase with a well-defined period of uplift and erosion of central Scotland (Knox & Morton 1983). A notable feature of these ashes is the presence of aegirine, which may provide a clue as to their origin. In the British Tertiary Volcanic Province, aegirine occurs as a minor constituent of the Rockall granite (Hawkes *et al.* 1975) and of at least two Skye granites (J. D. Bell, written communication, 1987). It is, however, relatively common in E Greenland (Nielsen 1980). It is thus possible that both provinces contributed to the Phase 1 ash sequence.

Subphase 2a

Description. The pyroclastic deposits of Subphase 2a occur as a series of graded airfall ash layers that are best preserved in the laminated mudstone facies of the Sele Formation (Fig. 4). The layers are generally less than 1 cm thick, and rarely exceed 3 cm, although one layer of 27 cm has been recorded.

Distribution. Subphase 2a ashes have been identified throughout the North Sea Basin, except in extreme marginal areas (see Fig. 3C), where non-marine facies preclude their preservation (Knox & Morton 1983). They have recently been identified in Norfolk, in strata equivalent to the Woolwich & Reading Beds (Cox *et al.* 1985). Although the Subphase 2a ashes have been encountered in several offshore cores, variable recovery and local facies changes do not allow the assessment of regional trends in number and thickness of the layers. Distal representatives of the Subphase 2a ashes have been identified in Deep Sea Drilling Project (DSDP) cores from the Bay of Biscay and Goban Spur regions of the N Atlantic (Knox 1985). No consistent trend in distribution can be detected, except that the ashes become attenuated and scarcer in the extreme S of the region.

Composition. Only in Denmark are some of the ashes sufficiently fresh for accurate whole-rock analysis; Pedersen *et al.* (1975) have shown that ashes of tholeiitic, rhyolitic, and peralkaline compositions are represented.

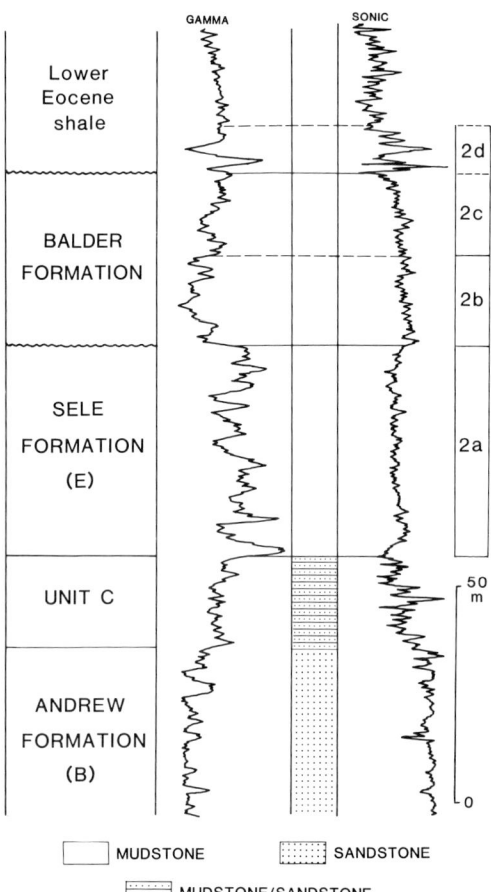

MUDSTONE SANDSTONE

MUDSTONE/SANDSTONE

FIG. 4. Representative late Palaeocene to early Eocene sequence of the central North Sea Basin, showing the distribution of Phase 2 ashes and typical wireline log responses for the associated sediments (UK Well 16/28-4).

For the remainder of the ash layers, some indication of the chemical composition can be obtained from the residual phenocryst phases. In particular, the feldspars display compositions ranging from labradorite to sanidine (Fig. 5), indicating compositions ranging from basaltic to silicic. Additional minerals identified from certain Subphase 2a ashes from Denmark include titaniferous aegirine, magnesio-kataphorite, kaersutite, Ti-augite, salite, aenigmatite, sphene, ilmenite, pseudobrookite, and brookite (Pedersen *et al.* 1975), indicating peralkaline parent magmas.

Derivation. Since no consistent trend in number and thickness of the ashes can be detected, their distribution does not provide direct evidence of

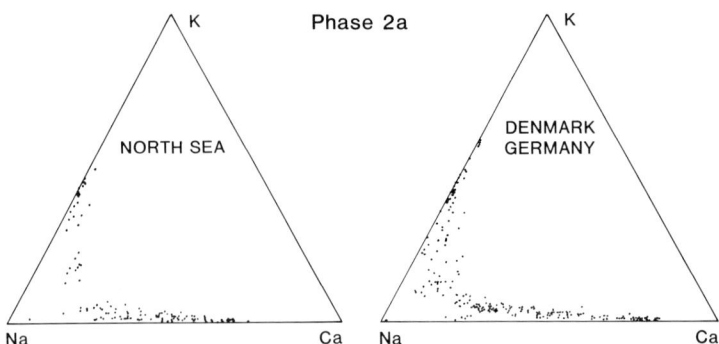

FIG. 5. Feldspar compositions of Subphase 2a ashes from the North Sea and Denmark/Germany, showing the close similarity between the two areas.

derivation; however, the gross pattern is sufficiently similar to that of the Subphase 2b ash distribution (see below) as to indicate that they were substantially of the same derivation. The occurrence of titaniferous aegirine is of particular significance, since it is a scarce mineral whose only reported occurrence in the north-eastern Atlantic region is in E Greenland (Nielsen 1979). It thus seems certain that at least some layers were derived from the Faeroe–Greenland province.

Subphase 2b

Description. The volcaniclastic sediments of Subphase 2b occur as closely-spaced graded ash layers that are best preserved in the laminated mudstone of the Balder Formation (see Fig. 4), whose indurated character produces the 'ash-marker' seismic reflector. Individual layers are generally less than 3 cm in thickness, although individual layers of up to 28 cm have been recorded (Malm *et al.* 1984). They are generally less altered than those of Subphase 2a, especially where they are associated with siliceous (diatomaceous) mudstones. In some marginal areas, however, they are bentonitic.

Distribution. Subphase 2b ashes have been identified throughout the North Sea area and beyond (see Fig. 3D). They occur in the lower part of the Balder Formation of offshore areas, and are well exposed in Denmark, where they constitute the positive-numbered ash series of Bøggild (1918). They have also been identified in north-western Germany (Andersen 1938), in the Netherlands (Pannekoek 1956), and in SE England (Knox & Ellison 1979). In more marginal parts of the basin they are absent due to hiatus.

The Subphase 2b ashes extend into the Faeroe–Shetland Basin (Ridd 1983), and distal representatives have been identified in the Bay of Biscay and Goban Spur areas of the N Atlantic (Knox 1984, 1985), indicating that the ashes must have covered an area of more than six million square kilometres. The existence of cores and outcrops that display more or less complete sequences through the Subphase 2b ashes allows regional assessment of thickness trends, in which the ashes show an overall northward increase in total ash thickness (see Fig. 3F). However, substantial local thickness variations occur in the region of Denmark and north-western Germany, where both the total ash thickness and the thickness of certain individual ash layers show a marked northward increase (Andersen 1937; Nielsen & Heilmann–Clausen 1988). The significance of these local thickness patterns is discussed below. The excessive thickness of one ash layer (81 cm) in Norwegian Well 30/2-1 has been ascribed by Malm *et al.* (1984) to reworking by gravity flow.

Composition. In the Fur Formation ('mo-clay') diatomite facies of Denmark, many of the Subphase 2b ash layers ('positive series') occur as virtually unaltered tuffs. As indicated by Bøggild (1918) the Danish Subphase 2b tuffs are, with one exception, basaltic. Microprobe analysis by Pedersen *et al.* (1975) confirmed this, and indicated that they are of Fe-Ti tholeiite parentage. Feldspars are of calcic plagioclase composition, except for one acidic layer (equivalent to +19 of Denmark), which contains anorthoclase.

Malm *et al.* (1984) presented major element analyses, plus a limited suite of trace element data, for the tuffs of the Balder Formation of Well 30/2-1 (Norwegian sector). These results indicated that the North Sea Subphase 2b ashes are of basaltic origin and have 'within-plate' basalt geochemistry when plotted on discrimination diagrams such as those of Pearce & Cann (1973).

Derivation. As shown in Fig. 3F, the trend of northward thickening of the Subphase 2b ashes continues to the N of mainland Britain, indicating that most if not all, of the ashes were derived from a distant northerly source. However, the northward increase in thickness of ash in and around Denmark, as noted above, has long been regarded as indicating derivation from a nearby source in the Skagerrak region (Bøggild 1918; Andersen 1937; Nielsen & Heilmann–Clausen 1988). It is, however, possible to explain local variations in ash thickness as resulting from the drifting of suspended ash by prevailing water currents. Such a process could account for the excessive thickness displayed by some of the Danish ash layers; the association of the greatest ash thicknesses with the 'mo-clay' diatomite facies may in part be a reflection of reduced compaction, but it could also be interpreted as reflecting close hydraulic equivalence of the ash particles and diatom frustules. The accumulation of buoyant particles in the Danish area may have been related to a long-term south-eastward surface water flow induced by prevailing north-westerly winds.

A separate source for the Danish ashes was also proposed by Malm *et al.* (1984), on the basis of contrasting plagioclase compositions, as determined by X-ray diffraction. However, microprobe plagioclase analyses carried out by us on tuffs from both the south-western North Sea (BGS Borehole 81/46A, Lott *et al.* 1983) and Denmark reveal essentially similar compositions (Fig. 6). Also, unaltered glass inclusions within plagioclase phenocrysts from the same tuffs have been shown by microprobe analysis to possess major-element compositions virtually identical to those of glass shards in the Danish sequence (Morton, unpublished data). Furthermore, trace element and rare earth element patterns of the Balder tuffs of UK Well 16/7a-2 are remarkably similar to those of the Danish ashes.

Chemical evidence thus appears to favour the concept of a common source for the Subphase 2b ashes, as indicated by their uniform stratigraphical relationships throughout the North Sea Basin and beyond (Knox 1984, 1985). Information on the probable location of the source area is provided by the trace element characteristics of the tuffs, which indicate a parent magma of Ti-tholeiite type (Morton, unpublished data). This composition would seem to preclude the Rockall Plateau and the Vøring Plateau as possible source areas, since the tholeiites of both areas lack titanium enrichment. Fe-Ti tholeiites are, however, typical of the Greenland–Iceland–Faeroes trend, whose geographical location fits well with the observed trends of ash distribution. Nothing is known of the basalts in the Møre Basin, so a source within the broader Faeroe–Møre region cannot be ruled out. Nevertheless, the exceptional association of basaltic magmas with highly explosive eruptions suggests special, and hence perhaps highly localized, magmatic processes; these are most likely to have been provided by the Faeroe–Iceland–Greenland mantle plume. On this basis, a schematic reconstruction of airfall ash thickness is shown in Fig. 3F as emanating from a Faeroe–Greenland source.

Detailed discussion of the mechanism of eruption of the basaltic ashes is beyond the scope of this paper. It may, however, be pointed out that this sequence of up to 200 closely spaced basaltic ash layers appears to be unique. The mean total ash thickness for the estimated minimum areal coverage is ca. 2 m. Assuming a maximum post-compactional interparticle and intraparticle porosity of 50%, the total volume of magma erupted amounts to a minimum of ca. 6000 km^3. With an average total of ca. 150 individual layers, each layer would thus represent 40 km^3 of magma. The magnitude and wide distribution of the eruptions indicates that they were produced by violently explosive eruptions.

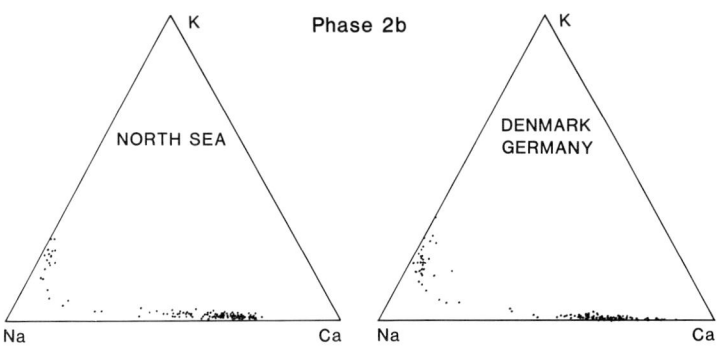

FIG. 6. Feldspar compositions of Subphase 2b ashes from the North Sea and Denmark/Germany, showing the close similarity between the two areas.

A textural study of Subphase 2b basaltic tephras from Denmark led Pedersen & Jørgensen (1981) to propose a Surtseyan-type eruptive mechanism. We would agree that the textures indicate a hydroclastic mechanism, but the widespread distribution of the tephra implies a violence of eruption that cannot be explained by the Surtseyan model. If it is assumed that the violence of a hydrovolcanic eruption is related to the surface area of contact between magma and water, then the most likely mechanism for a violent eruption is the intrusion of a dyke into water-bearing sediments. This process seems to have accounted for the violence of the Tawawera eruption of 1886 (Walker et al. 1984), in which basalt interacted with groundwater held within proximal pyroclastic deposits. A similar setting may account for the Faeroe–Greenland eruptions, although they would appear to have been of a more violent nature. Alternatively, basalt dykes may have been intruded into an earlier sedimentary sequence, with groundwater/magma interaction taking place at depth. However, the restriction of violently explosive eruptions to the Faeroe–Greenland plume region suggests that some intrinsic feature of the magma itself, such as anomalously high temperature, may also have contributed.

Subphases 2c and 2d

Description. The Subphase 2c ashes are similar to those of Subphase 2b, but are thin and very sporadic. The Subphase 2d ashes occur as thin, graded airfall layers, and are invariably bentonitized. The recognition of two separate subphases is based more on their contrasting tectonic and stratigraphical setting than on their intrinsic character; further work on their mineralogy and chemistry should provide a firmer basis for categorizing these ashes.

Distribution. Subphase 2c ashes have been identified in the central and northern North Sea areas (Knox & Morton 1983), in Denmark, and in NW Germany (see Fig. 3E). Subphase 2d ashes are well represented in the central and northern North Sea areas and in Denmark (Røsnæs Clay). Isolated layers have been recorded from the southern North Sea and NW Germany. Subphase 2c and Subphase 2d ashes are absent or extremely attenuated in southern England and the eastern Atlantic (Bay of Biscay and Goban Spur areas).

Composition. Little is known of the compositions of the Subphase 2c and Subphase 2d ashes; they include some basaltic layers, but more intermediate varieties are also present, as indicated by feldspar compositions. Aegirine is present in some Subphase 2d ashes.

Derivation. The presence of aegirine in Subphase 2d ashes indicates a continued source in the Faeroe–Greenland Province. Since the distribution of these ashes corresponds to the thicker, central zone of distribution of the Subphase 2b ashes, their scarcity or absence from the SW of the region and their relative thinness compared with those of Subphase 2b may reflect lower intensity of the source eruptions. However, the scarcity of Subphase 2c and 2d ashes in the western North Sea compared with Denmark indicates a possible eastward shift in their distribution pattern. Such a shift could be explained by: (1) a shift in the focus of volcanism within the Faeroe–Greenland province; (2) a shift in the position of E Greenland sources relative to the North Sea as a result of seafloor spreading; or (3) a shift in the prevailing wind direction as a result of the development of the north-eastern Atlantic seaway.

Discussion

The record of volcaniclastic sedimentation in the early Palaeogene sequence of the North Sea Basin indicates at least two distinct phases of explosive volcanism in adjacent areas. Deposits of the first phase appear to represent activity in both the British and Faeroe–Greenland provinces, whereas those of the second phase appear to represent activity that took place largely, if not exclusively, in the Faeroe–Greenland province. If correct, these conclusions place definite constraints on the timing of major explosive volcanism in the two provinces. Through the correlations shown in Fig. 1, this timing may be expressed in terms of standard biozones and magnetozones (Fig. 7). It should also be possible to make a direct comparison between the absolute timing indicated by radiometric dating of the igneous rocks themselves and that obtained from correlation of radiometric dates with the biozones and magnetozones. Unfortunately, however, there is no universal agreement on the latter correlation, so two contrasting schemes are presented in Fig. 7 (Curry & Odin 1982; Berggren et al. 1985) for comparison.

British volcanism

Although the history of individual areas within the British province may be rather complex in detail, the sequence of activity (at least for the

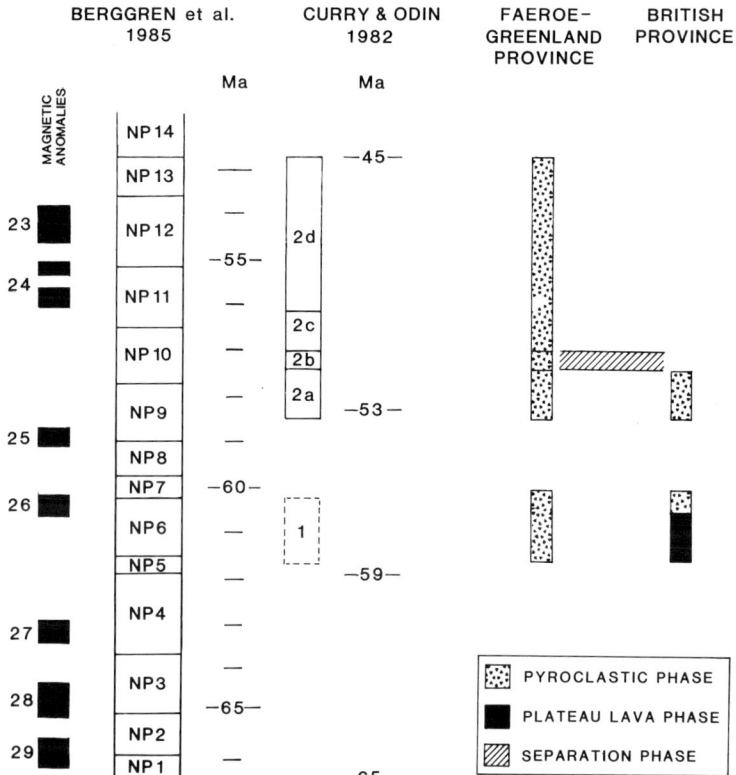

FIG. 7. Summary of stratigraphical relationships of pyroclastic deposits in the Palaeogene of the north-eastern Atlantic region. Inferred phases of pyroclastic activity are shown for the Faeroe–Greenland province and for the British province (basaltic sand phase only). The postulated phase of British province plateau lava eruption is also shown.

Hebridean region) can generally be considered in terms of two more or less distinct phases, with widespread eruption of basaltic lavas ('plateau lava' phase) being followed by volcanism and intrusion of more centralized type ('central volcanic' phase), involving silicic as well as basaltic magmas. The lavas, of both plateau and central type, are exclusively of reversed polarity, whereas the central intrusives commonly show a trend from reversed to normal with decreasing age (Mussett *et al.* 1988). The late intrusions are mostly reversed. Mussett *et al.* conclude that this sequence of polarity reversals most probably spans the interval 26R–26N–25R, with much of the activity dated as in the interval 60–58 Ma.

The Hebridean sequences include significant proportions of pyroclastic deposits, including both basaltic and acidic types (Bell & Emeleus 1988), but how much of the pyroclastic activity was sufficiently violent to be recorded in adjacent sedimentary basins is not clear.

Unfortunately, in the absence of more detailed

mineralogical and chemical analyses, it is not possible to identify any of the Phase 1 airfall ash layers as being specifically of British provenance. Only the volcaniclastic sands (Unit C1) of the Moray Firth area can be ascribed with certainty to a British source. Unit C1 probably falls within the upper part of chron 26R, corresponding perhaps to the middle part of NP6. This would correspond to an age of ca. 59 Ma on the Berggren *et al.* time-scale and ca. 57.5 Ma on the Curry & Odin time-scale (see Fig. 7). Both ages are within the range indicated by Mussett *et al.* (1988) for the central volcanic phase, so that the Unit C1 volcaniclastic sands may well be related to the Palaeocene phase of central-type volcanism in the British province.

Whether any of the more acidic Phase 1 ashes represent Hebridean activity is not certain, but their restriction to the latest part of chron 26R (latest NP6) would preclude correlation with any of the normal-polarity Hebridean igneous phases. Thompson (1980) considered that Plinian eruptions probably took place during the emplacement of the Western Red Hills granites of Skye, and it is reasonable to suppose that any such eruptions would be represented in the North Sea sequence. However, the Western Red Hills granites are mostly of normal polarity (Mussett *et al.* 1988), with reversed polarity being restricted to the Loch Ainort epigranite, which is placed in the middle of the eruptive sequence (and dated at 58.7 ± 0.9 Ma by Dickin 1981). Since all of the Phase 1 pyroclastics predate chron 26N, it is not possible to relate the Western Red Hills volcanic phase (dominantly normal polarity) to the Phase 1 pyroclastics. It is thus possible that the Hebridean contribution to the Phase 1 pyroclastics was more or less limited to the Unit C1 volcaniclastic sand, with the airfall ashes being derived from elsewhere.

Dates obtained from the Skye intrusives indicate that a significant time interval separated the intrusion of the Western Red Hills granites (ca. 59 Ma) and Eastern Red Hills granites (ca. 54 Ma) (Dickin 1981; Mussett *et al.* 1988). On the time-scale of Berggren *et al.* the later phase would correspond to the later part of Subphase 2d (see Fig. 6). A more plausible correlation, with Subphase 2a, is obtained with the Odin time-scale, in which case some of the Unit E ashes could be of British province provenance. Application of the Odin time-scale to the youngest British province radiometric dates (ca. 52 Ma, according to Mussett *et al.* 1988), indicates that British province volcanism had more or less ceased by the end of Subphase 2a. This fits well with the conclusion reached below, that the British province could not have produced the tholeiitic ashes of Subphase 2b.

If it is assumed that any British contribution to the Phase 1 ashes is related to the later, central volcanic phase in the British province, then the plateau lavas must have been erupted during sedimentation of Units A and B of the North Sea sequence. Since the onset of lava eruptions obviously took place in response to a major change in crustal stresses, it seems most likely that it coincided with the marked change in the patterns of tectonism and sedimentation that took place in the North Sea Basin at the Unit A/Unit B boundary. Since this new pattern of basin development continued up to the end of Unit B times, it is reasonable to conclude that Unit B represents the plateau basalt phase. This would date the plateau basalts as ranging from mid NP5 to mid NP6, corresponding to an interval of about one million years, in the region of 61.5 Ma on the Berggren *et al.* time-scale, and 58 Ma on the Curry & Odin time-scale (see Fig. 7). Of these two dates, the former compares better with the

age range of ca. 60–63 Ma given by Mussett *et al.* (1988) for the early lavas.

N Atlantic volcanism

As discussed earlier, some of the Phase 1 airfall ashes contain minerals, such as aegirine, that are indicative of peralkaline parent magmas. No obvious source can be found in the British province (apart from Rockall Island), but aegirine-bearing peralkaline rocks do occur in E Greenland (Nielsen 1980). Although mineralogical and geochemical analysis of the ashes is required before any firm conclusions can be drawn, we believe that the existing sketchy data argue strongly for a source in the Faeroe–Greenland province. On the Faeroe Islands themselves, the tholeiitic Lower Lava Series displays a magnetic stratigraphy that is interpreted by Waagstein (1988) as representing chrons 26R to 24R. This supports the concept of Phase 1 volcanism within the province, as does the tholeiitic tuff recorded by Morton *et al.* (1988).

The Gardiner complex, near Kangerdlugssuaq, bears aegirine (Ti-rich, Nielsen 1979) and is overlain by plateau basalts (Nielsen 1987). It was, therefore, active early in the Tertiary volcanic history of E Greenland. The aegirines in Phase 1 could well have been sourced from this or a similar centre. Several lines of evidence therefore indicate that the Faeroe–Greenland province was volcanically active during Phase 1.

The evidence for Faeroe–Greenland contributions to the Phase 2 ashes is more compelling. In particular, no other source can be found for the titaniferous aegirines that occur in some Subphase 2a ash layers.

A Faeroe–Greenland contribution to the Phase 2 ashes is also supported by the distinctive Fe-Ti tholeiitic nature of the succeeding Subphase 2b ashes, since such compositions appear to be restricted to the Faeroe–Greenland section of the north-eastern Atlantic rift zone. The tholeiitic ashes are remarkably voluminous and, as discussed above, must have been produced by extremely violent eruptions. They are also significant in that they are associated with a major change in the patterns of subsidence and sedimentation in the North Sea Basin and other parts of the NW European continental shelf. In the North Sea Basin, the change from ashes of mixed composition (Subphase 2a) to ashes of more or less purely basaltic composition (Subphase 2b) was accompanied by the cessation of established patterns of relative subsidence within the North Sea Basin and by the cessation of source uplift and sand influx (Knox *et al.* 1981). The Phase 2b ashes were thus erupted at a time of remarkably uniform subsidence and sedimentation.

We propose that the abrupt reduction of tectonic activity took place in response to relief of crustal tension along the north-eastern Atlantic rift zone, manifested also by extrusion of the considerable amounts of basaltic lavas that form the dipping reflector sequences of chron 24R age along the NE Atlantic margins. Another result of the change in regional stress pattern appears to have been the cessation of volcanic activity in the British province, as discussed in the preceding section.

The origin of the Subphase 2c and Subphase 2d ashes is not known, but the presence of aegirine also favours a Faeroe–Greenland source. Since they appear to postdate the onset of seafloor spreading, these ashes presumably represent continued, if sporadic, activity under a crustal stress regime that must have been somewhat different from that of Subphase 2a, which was of broadly similar composition. The end of Subphase 2d pyroclastic activity at about the NP13/NP14 boundary (ca. 45 Ma on the Curry & Odin time-scale) presumably marks yet another significant crustal event, as a result of which the early Eocene phase of explosive volcanism was finally brought to a close.

Dating of the various subphases of Phase 2 is relatively straightforward (see Fig. 1). Subphase 2a ranges from ?mid NP9 to early NP10, Subphase 2b from early NP10 to mid NP10, Subphase 2c from mid NP10 to late NP10/early NP11, and Subphase 2d from early NP11 to late NP13. Work in progress on the magnetostratigraphy of BGS Borehole 81/46A (Hailwood & Knox, in prep.) indicates that Subphases 2a, 2b, and 2c fall within chron 24R, and Subphase 2d within chron 24N. The potential thus exists for a relatively refined correlation between the North Sea ash sequence and volcanic sources in the Faeroe–Greenland province.

Conclusions

The volcaniclastic sediments of the North Sea and adjacent areas provide a record of two distinct phases of explosive volcanism, the first taking place within the late Palaeocene and the second within the period latest Palaeocene to latest early Eocene. Direct correlation between the igneous provinces and the pyroclastic sequences is somewhat hampered by uncertainties in the correlation between radiometric dates and the standard biostratigraphical and magnetostratigraphical zones. Our 'best-fit' correlations indicate that of the two scales discussed (see Fig. 7),

that of Berggren *et al.* (1985) is the better for the Palaeocene interval, whereas that of Curry & Odin (1982) is better for the early Eocene interval.

Two main centres of pyroclastic activity appear to be represented, the British province and the Faeroe–Greenland province, with the latter dominating volumetrically. It is clear from the above descriptions and discussion that the eruptions that took place in the Faeroe–Greenland province were remarkable both in their magnitude and in the nature of their parent magma. Indeed, the sequence of up to 200 violently explosive basaltic eruptions appears to be unique. As discussed on p. 414, some form of hydrovolcanic process appears to have been involved.

Although it is not possible to say whether individual phases of activity in the British province and in the Faeroe–Greenland province were precisely synchronous, it is clear that the volcanism in both areas developed in response to crustal stresses that preceded rifting and eventual seafloor spreading between Greenland and Europe. In pre-separation times, the Faeroe–Greenland province lay on the northward projection of the British province lineation (see Fig. 3A). We consider that the evidence for more or less synchronous Phase 1 activity in the two provinces suggests that their common alignment is genetic rather than coincidental, and that volcanism was initiated in both regions in response to a single phase of regional E–W crustal tension. Indeed, the volcanism may have been linked with that of the W Greenland province, whose activity was more or less coeval with that of the British province (Upton 1988).

The Faeroe–Greenland province, and its associated mantle plume, may thus be seen as having been located on the intersection of the earlier (late Palaeocene) N–S lineation and the later (latest Palaeocene to early Eocene) NE–SW lineation of the north-eastern Atlantic rift zone (Fig. 8). The two phases of pyroclastic activity identified in the sedimentary record thus mark a shift in the location of highly explosive volcanism from one lineation to the other, presumably in response to a fundamental change in the crustal stress pattern that led to the separation of Greenland from Europe. The N–S lineation appears to have become inactive with the onset of Subphase 2b basaltic volcanism, as indicated by the termination of volcanism in the British province (and perhaps also in the W Greenland province). However, its point of intersection with the NE Atlantic rift zone continued to be the site of anomalous explosive activity, probably until the end of early Eocene times. The resurgence of more acidic pyroclastic events (Subphases 2c & 2d) after the tholeiitic phase of Subphase 2b

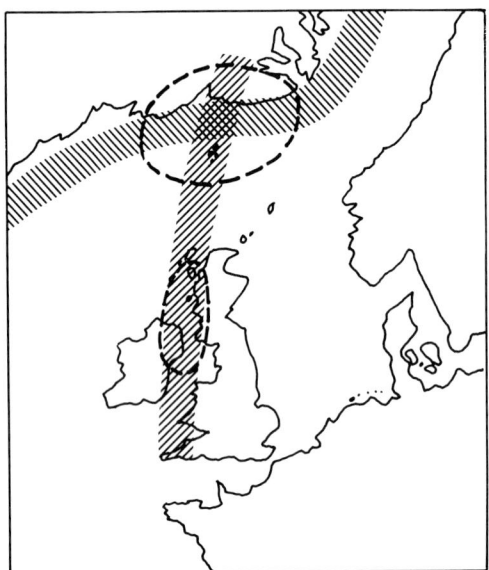

FIG. 8. Alignments of volcanic activity in the north-eastern Atlantic region.

suggests that the Greenland plate experienced renewed internal stresses that were not directly associated with the Faeroe–Greenland separation, and that may therefore have been related to the impending north-eastward extension of the zone of separation.

As a final comment, it may be noted that the observed distribution of ash not only provides information on the location of the volcanic sources, but also provides an indication of the prevailing wind direction at the time of the eruptions. It would seem that during the Phase 2 eruptions, at least, the prevailing wind direction was from the NW; since there was no fundamental change in the configuration of the eastern Atlantic ocean between early Palaeocene and early Eocene times, this interpretation of the prevailing wind direction probably applies to the entire interval.

ACKNOWLEDGEMENTS: This paper is published with the approval of the Director, British Geological Survey (NERC).

References

ANDERSEN, S. A. 1937. De vulkanske Askelag i Vejgennemskæringen ved Ølst og deres Udbredelse i Danmark. *Danmarks Geologisk Undersøgelse, Række 2*, **59**, 50pp.

—— 1938. Die Verbreitung der eozänen vulkanischen Ascheschichten in Dänemark und Nordwestdeutschland. *Zeitschrift der Geschieberforschungen Flachlandsgeologie*, **14**, 179–207.

AUBRY, M-P. 1985. Northwestern European Paleogene magnetostratigraphy, biostratigraphy, and paleogeography: Calcareous nannofossil evidence. *Geology*, **13**, 198–202.

BELL, B. R. & EMELEUS, C. H. 1988. Pyroclastic rocks of the British Tertiary Volcanic Province. *In*: MORTON, A. C. & PARSON, L. M. (eds) *Early Tertiary Volcanism and the Opening of the NE Atlantic*. Geological Society of London, Special Publication, **39**, pp. 365–379.

BERGGREN, W. A., KENT, D. V., FLYNN, J. J. & VAN COUVERING, J. A. 1985. Cenozoic geochronology. *Bulletin of the Geological Society of America*, **96**, 1407–1418.

BØGGILD, O. B. 1918. Den vulkanse Aske i Moleret. *Danmarks Geologisk Undersøgelse, Række 2*, **33**, 159pp.

COSTA, L., DENNISON, C. & DOWNIE, C. 1978. The Paleocene/Eocene boundary in the Anglo–Paris Basin. *Journal of the Geological Society of London*, **135**, 261–264.

COX, F., HAILWOOD, E. A., HARLAND, R., HUGHES, M. J., JOHNSTON, N. & KNOX, R. W. O'B. 1985. Palaeocene sedimentation and stratigraphy in Norfolk, England. *Newsletters on Stratigraphy*, **14**, 168–185.

CURRY, D. & ODIN, G. S. 1982. Dating of the Palaeogene. *In*: ODIN, G. S. (ed) *Numerical Dating in Stratigraphy*. John Wiley & Sons, Chichester, 607–629.

DEEGAN, C. E. & SCULL, B. J. (compilers) 1977. A standard lithostratigraphic nomenclature for the Central and Northern North Sea. *Institute of Geological Sciences Report, 77/25; Norwegian Petroleum Directorate Bulletin*, **1**.

DICKIN, A. P. 1981. Isotope geochemistry of Tertiary igneous rocks from the Isle of Skye. *Journal of Petrology*, **22**, 155–189.

HAWKES, J., MERRIMAN, R. J., HARDING, R. R., & DARBYSHIRE, D. P. F. 1975. Rockall Island: new geological, petrological, chemical and Rb-Sr age data. *In*: *Expeditions to Rockall 1971–72. Institute of Geological Sciences Report*, **75/1**, pp. 11–51.

HEILMANN-CLAUSEN, C., NIELSEN, O. B. & GERSNER, F. 1985. Lithostratigraphy and depositional environments in the Upper Paleocene and Eocene of Denmark. *Bulletin of the Geological Society of Denmark*, **33**, 287–323.

KNOX, R. W. O'B. 1979. Igneous grains associated with zeolites in the Thanet Beds of Pegwell Bay, northeast Kent. *Proceedings of the Geologists' Association*, **90**, 55–59.

—— 1984. Nannoplankton zonation and the Palaeocene/Eocene boundary beds of NW Europe: an indirect correlation by means of volcanic ash layers. *Journal of the Geological Society of London*, **141**, 993–999.

—— 1985. Stratigraphic significance of volcanic ash in Paleocene and Eocene sediments at Sites 549 and 550. *In*: GRACIANSKY, P. C. DE, POAG, C. W. *et al. Initial Reports of the Deep Sea Drilling Project*. US Government Printing Office, Washington, **80**, pp. 845–850.

—— & ELLISON, R. A. 1979. A Lower Eocene ash sequence in south-eastern England. *Journal of the Geological Society of London*, **136**, 251–254.

—— & MORTON, A. C. 1983. Stratigraphical distribution of early Palaeogene pyroclastic deposits in the North Sea Basin. *Proceedings of the Yorkshire Geological Society*, **44**, 355–363.

——, MORTON, A. C. & HARLAND, R. 1981. Stratigraphical relationships of Palaeocene sands in the UK sector of the central North Sea. *In*: ILLING, L. V. & HOBSON, G. D. (eds) *Petroleum Geology of the Continental Shelf of North-West Europe*. Heyden & Sons, London, pp. 267–281.

LOTT, G. K., KNOX, R. W. O'B., HARLAND, R. & HUGHES, M. J. 1983. The stratigraphy of Palaeogene sediments in a cored borehole off the coast of north-east Yorkshire. *Institute of Geological Sciences Report*, **83/9**.

MALM, O. A., CHRISTENSEN, O. B., FURNES, H. LØVLIE, R. RUSELÅTTEN, H. & ØSTBY, K. L. 1984. The Lower Tertiary Balder Formation: an organogenic and tuffaceous deposit in the North Sea region. *In*: A. M. SPENCER *et al.* (eds) *Petroleum Geology of the North European Margin*. Graham & Trotman, London, pp. 149–170.

MORTON, A. C. 1982. Provenance and diagenesis of Palaeogene sandstones of southeast England as indicated by heavy mineral analysis. *Proceedings of the Geologists' Association*, **93**, 263–274.

——, EVANS, D., HARLAND, R., KING, C. & RITCHIE, D. 1988. Volcanic ash in a cored borehole W of the Shetland Islands: evidence for Selandian (late Palaeocene) volcanism in the Faeroes region. *In*: MORTON, A. C. & PARSON, L. M. (eds) *Early Tertiary Volcanism and the Opening of the NE Atlantic*. Geological Society of London, Special Publication, **39**, pp. 263–269.

MÜLLER, C. 1985. Biostratigraphic and paleoenvironmental interpretation of the Goban Spur region based on a study of calcareous nannoplankton. *In*: GRACIANSKY, P. C. DE, POAG, C. W. *et al. Initial Reports of the Deep Sea Drilling Project*. US Government Printing Office, Washington, **80**, pp. 573–599.

MUSSETT, A. E., DAGLEY, P. & SKELHORN, R. R. 1988. Time and duration of activity in the British Tertiary igneous province. *In*: MORTON, A. C. & PARSON, L. M. (eds) *Early Tertiary Volcanism and the Opening of the NE Atlantic*. Geological Society of London, Special Publication, **39**, pp. 337–348.

NIELSEN, O. B. & HEILMANN-CLAUSEN, C. 1988. Palaeogene volcanism: the sedimentary record in Denmark. *In*: MORTON, A. C. & PARSON, L. M. (eds) *Early Tertiary Volcanism and the Opening of the NE Atlantic*. Geological Society of London, Special Publication, **39**, pp. 395–405.

NIELSEN, T. F. D. 1979. The occurrence and formation of Ti-aegirines in peralkaline syenites. An example from the Tertiary ultramafic alkaline Gardiner complex, East Greenland. *Contributions to Mineralogy and Petrology*, **69**, 235–244.

—— 1980. The petrology of a melilitolite, melteigite, carbonatite and syenite ring dyke system, in the Gardiner complex, East Greenland. *Lithos*, **13**, 181–197.

—— 1987. Tertiary alkaline magmatism in East Greenland: a review. *In*: FITTON, J. G. & UPTON, B. G. J. (eds) *Alkaline Igneous Rocks*. Geological Society of London, Special Publication, **30**, pp. 489–515.

PANNEKOEK, A. J. 1956. Eocene. *In*: PANNEKOEK, A. J. (ed) *Geological History of the Netherlands*. The Hague, pp. 57–59.

PEARCE, J. A. & CANN, J. R. 1973. Tectonic setting of basic volcanic rocks determined using trace element analyses. *Earth Science and Planetary Letters*, **19**, 290–300.

PEDERSEN, A. K. & JØRGENSEN, K. A. 1981. A textural study of basaltic tephras from lower Tertiary diatomites in Northern Denmark. *In*: SELF, S. & SPARKS, R. S. J. (eds) *Tephra Studies*. Reidel, Dordrecht, pp. 213–218.

——, ENGELL, J. & RØNSBO, J. G. 1975. Early Tertiary volcanism in the Skagerrak: new chemical evidence from ash layers in the mo-clay of northern Denmark. *Lithos*, **8**, 255–268.

RIDD, M. F. 1983. Aspects of the Tertiary geology of the Faeroe–Shetland Channel. *In*: BOTT, M. H. P., SAXOV, S., TALWANI, M. & THIEDE, J. (eds) *Structure and Development of the Greenland–Scotland Ridge*. Plenum Press, New York, pp. 91–108.

SIESSER, W. G., WARD, D. J. & LORD, A. R. 1987. Calcareous nannoplankton biozonation of the Thanetian Stage (Palaeocene) in the type area. *Journal of Micropalaeontology*, **6**, 85–102.

STEWART, I. J. 1987. A revised stratigraphic interpretation of the early Palaeogene of the central North Sea. *In*: BROOKS, J. & GLENNIE, K. W. (eds) *Petroleum Geology of North West Europe*. Graham & Trotman, London, pp. 557–576.

THOMPSON, R. N. 1980. Askja 1875, Skye 56 Ma: basalt-triggered, Plinian, mixed-magma eruptions during the emplacement of the Western Redhills granites, Isle-of-Skye, Scotland. *Geologische Rundschau*, **69**, 245–262.

TOWNSEND, H. A. & HAILWOOD, E. A. 1985. Magnetostratigraphic correlation of Palaeogene sediments in the Hampshire and London Basins, southern U.K. *Journal of the Geological Society of London*, **142**, 957–982.

UPTON, B. G. J. 1988. History of Tertiary igneous activity in the N Atlantic borderlands. *In*: MORTON, A. C. & PARSON, L. M. (eds) *Early Tertiary Volcanism and the Opening of the NE Atlantic*. Geological Society of London, Special Publication, **39**, pp. 429–453.

WAAGSTEIN, R. 1988. Structure, composition and age of the Faeroe basalt plateau. *In*: MORTON, A. C. & PARSON, L. M. (eds) *Early Tertiary Volcanism and the Opening of the NE Atlantic*. Geological Society of London, Special Publication, **39**, pp. 225–238.

WALKER, G. P. L., SELF, S. & WILSON, L. 1984. Tarawera 1886, New Zealand—a basaltic Plinian fissure eruption. *Journal of Volcanology and Geothermal Research*, **21**, 61–78.

R. W. O'B. KNOX & A. C. MORTON, British Geological Survey, Keyworth, Nottingham NG12 5GG, UK.

Some aspects of Tertiary tectonics and sedimentation along the western Barents Shelf

A. Nøttvedt, L. T. Berglund, E. Rasmussen & R. J. Steel

Recent marine geophysical research has suggested that the north-easternmost Atlantic Ocean opened by a three-stage series of events involving seafloor spreading: (1) S of the Senja fracture zone from 60 Ma, (2) between the Senja fracture zone and Hornsund fault zone from 48 Ma, and (3) opposite the Hornsund fault zone from 37 Ma. This history involved early large-scale shear movement and later rifting.

These major tectonic events which produced the new continental margin had a marked effect, both structural and sedimentational, on the adjacent Barents Sea platform. Examples of these effects from Svalbard and the southwestern Barents Sea include a range of strike-slip, compressional and extensional tectonic features as well as a marked changing character to the style of Tertiary sedimentation.

Fig. 1 outlines the main Tertiary structures along the NE Atlantic margin. East of the Hornsund fault zone, the W Spitsbergen foldbelt (Eocene–early Oligocene) extends along the W coast of Spitsbergen for more than 300 km. The fold belt is flanked to the E by the Spitsbergen Central Tertiary Basin (Palaeocene–mid-Eocene) and cut by the late orogenic strike slip/extensional Forlandsundet Graben (mid-Eocene–early Oligocene) (Steel *et al.* 1985). The fold belt terminates somewhere N of Bjørnøya, where only extensional post-Palaeozoic tectonism has been recorded (Horn & Orvin 1928; Gjelberg 1981). A similar, extensional regime is indicated for the entire Stappen High (Faleide *et al.* 1984). No pre-rift (early Oligocene) sediments are present on the high. To the W of the Stappen High, between the Senja fracture zone and Hornsund fault zone, Palaeocene–Eocene extensional shear, and also contemporaneous compressional shear, have recently been speculated (Boldreel, pers. comm.). The inferred presence of an underlying oceanic basement related to an intermediate stage of opening from 48 Ma (Myhre *et al.* 1982) is rejected, and a complete pre-rift (Palaeocene–early Oligocene) sequence is suggested to rest uncomformably on older Mesozoic rocks in this area (Boldreel, pers. comm.). Contiguous with the Stappen High lies the Bjørnøya Basin, where late Cretaceous or early Tertiary compression and the presence of a thin Tertiary pre-rift sequence (Palaeocene–? early Eocene/Oligocene) is reported (Faleide *et al.* 1984; Spencer *et al.*

1984). From our seismic data, however, it is possible that the deformation noted relates to mobilization of deeper salt layers. The bordering Senja Ridge, to the S, has been assigned to both compression (Rønnevik 1984; Faleide *et al.* 1984) and pure extension (Spencer *et al.* 1984) in Tertiary times. The latter hypothesis seems likely for at least part of the ridge. A Tertiary pre-rift sequence (late Palaeocene–early Eocene) is present, which expands (late Palaeocene–mid-Oligocene) into the bordering, extensional Tromsø Basin (Spencer *et al.* 1984). To the far S, W of the Troms–Finnmark Platform, indications of early Tertiary compressional shear have been noted. To the W, the rifted Western Barents Sea Basin (early Oligocene–present) is continuous along the entire W Barents Sea margin.

Fig. 2 shows some key sections across the NE Atlantic margin, from Spitsbergen southward to the Senja Ridge/Tromsø Basin. In late Cretaceous–early Tertiary times there were local and subregional truncation and planation of faulted Cretaceous strata along the Greenland–Barents Sea boundary (see Fig. 2, lines a & d), followed by broad late Palaeocene subsidence and sedimentation in an epicontinental sea area. On Spitsbergen, extensional or transtensional subsidence commenced in early Palaeocene times (Steel *et al.* 1985; Manum & Throndsen 1986) (see Fig. 2, line a′–unit I′) prior to the initial seafloor spreading. Early Tertiary subsidence is reported also in the Bjørnøya Basin (Faleide *et al.* 1984) (see Fig. 2, line c–unit I) and, questionably, in the area between the Senja fracture zone and Hornsund fault zone (Boldreel, pers. comm.) (see Fig. 2, line b–unit II). Gentle warping and subsidence continued and spread to the Senja Ridge/Tromsø Basin (see Fig. 2, line d–unit I) during the initial seafloor spreading in the Norwegian–Greenland Sea and onset of major shear along the Greenland–Barents Sea boundary in late Palaeocene times. A draping style of sedimentary infill supports a general uniform shear or weak extensional shear/transtension during this period. A thick Palaeocene infill and evolving pull-apart basin conditions between the Senja fracture zone and Hornsund fault zone (Boldreel, pers. comm.) do not contradict this general picture.

Early Eocene to early Oligocene times brought northwards propagating differential shear and

From MORTON, A. C. & PARSON, L. M. (eds), 1988, *Early Tertiary Volcanism and the Opening of the NE Atlantic*, Geological Society Special Publication No. 39, pp. 421–425.

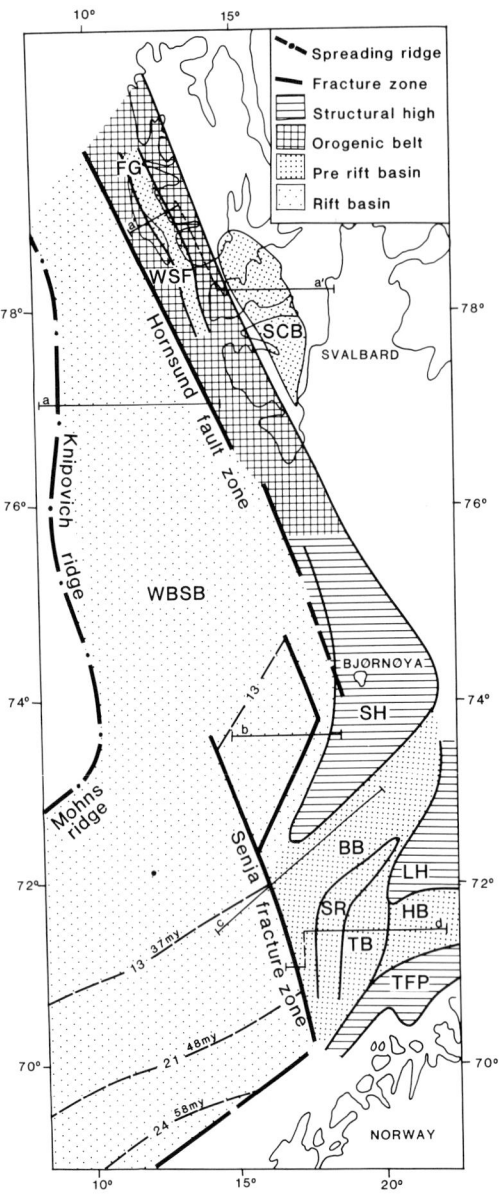

Tertiary structural elements

FIG. 1. Schematic illustration of the key structural elements of the western Barents Sea Margin. TFP: Troms-Finnmark Platform; TB: Tromsø Basin; SR: Senja Ridge; HB: Hammerfest Basin; LH: Loppa High; BB: Bjørnøya Basin; SH: Stappen High; WBSB: Western Barents Sea Basin; WSF: West Spitsbergen Fold Belt; SCB: Spitsbergen Central Basin; FG: Forlandsundet Graben. Modified from Faleide et al. (1984) and Spencer et al. (1984).

the accompanying break-up of the earlier broad basins. Major orogenesis and deformation along the W Spitsbergen Fold Belt (see Fig. 2, line a') was accompanied to the S by weak uplift of the Senja Ridge (Spencer et al. 1984) (see Fig. 2, line d) and the NW Bjørnøya Basin/Stappen High (Faleide et al. 1984) (see Fig. 2, line c). Continued subsidence is noted in the Spitsbergen Central Basin, between the Senja fracture zone and Hornsund fault zone, S Bjørnøya Basin and Tromsø Basin (see Fig. 2). The change in overall tectonic regime corresponds roughly to the initiation of seafloor spreading along the Senja fracture zone from about 53 Ma (see Fig. 1). In great contrast to the broad sedimentation basins of the Palaeocene, the smaller Eocene basins show rapid infilling by clinoforms from areas rising due to local and regional transpression (see Fig. 2, lines a' and d–unit II).

The differential shear noted during Eocene to mid-Oligocene times raises some very important questions with respect to the opening history. In the W Spitsbergen Fold Belt, high-angle reverse faults, thrust faults and asymmetric folds within the core of the orogen couple eastward into the flanking Spitsbergen Central Basin to extensive decollement and thrust-ramp deformation (Rasmussen & Nøttvedt, in prep.). Adding at least 5 km of shortening within the Spitsbergen Central Basin to an estimated shortening of 10–15 km within the foldbelt (Birkenmajer 1981), gives a minimum of 15–20 km of total compressive shortening during the Eocene to early Oligocene period on Spitsbergen. Although early Tertiary compressional shear has also been noted further to the S (see above; Bjørnøya Basin, ? Senja Ridge, Troms–Finnmark Platform), evidence for substantial shortening along the Senja fracture zone is lacking. Assuming an orthogonal relationship to bounding spreading ridges and overall parallel shear along the Senja fracture zone, and given a lateral displacement of some 500 km along this margin (anomaly 24 to 13), a 20 km shortening across the W Spitsbergen Foldbelt would imply a non-orthogonal relationship of this segment of the Hornsund fault zone to the bounding spreading ridges at an angle of 2–3° (anticlockwise) to the orthogon. A similar oblique relationship of the Hornsund fault zone to the Senja fracture zone is also concluded by Faleide et al. (1988), on the basis of geometric reconstructions of the opening of the NE Atlantic margin.

It is presently not clear whether a 2–3° angle of relative compression is sufficient to create regional detachment and thrusting on a scale of that observed on Spitsbergen. Experimental work by Lamons et al. (1987) seems to indicate the occurrence of plate boundary-parallel, but fairly

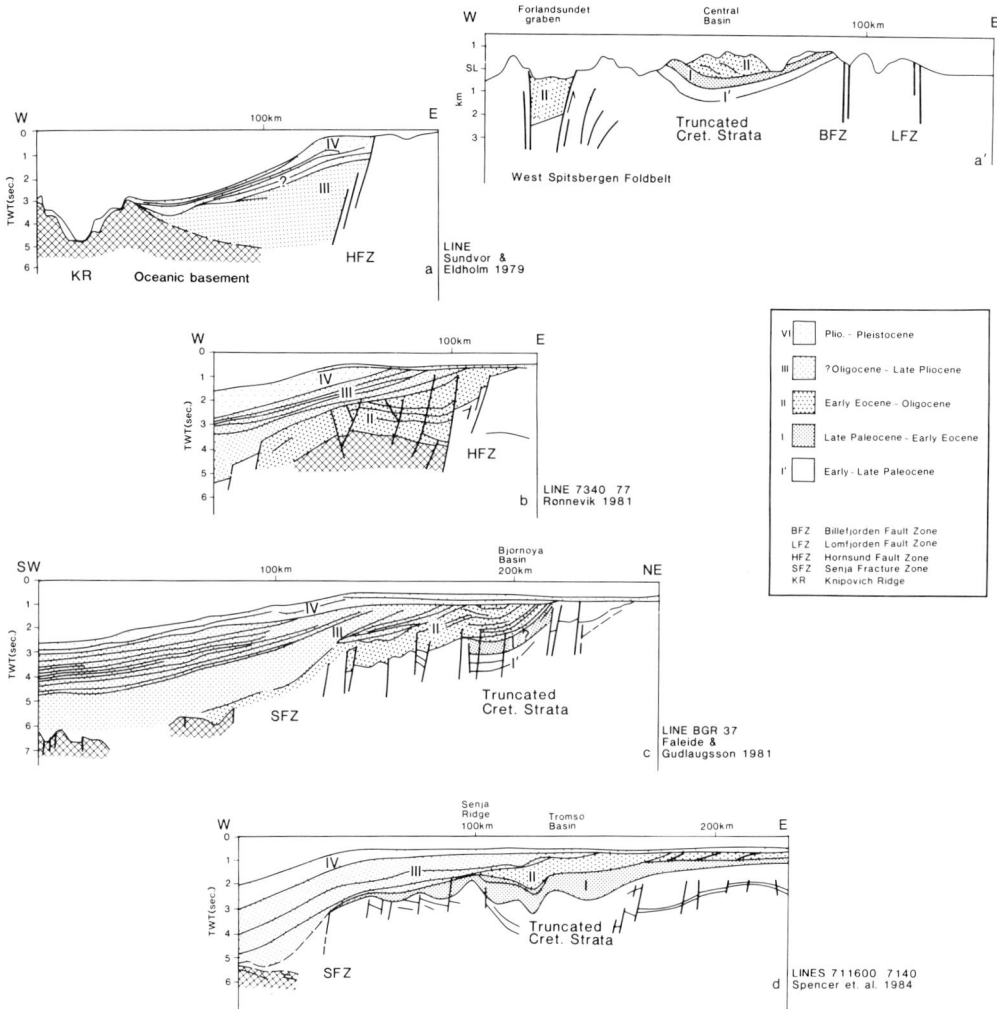

FIG. 2. Geoseismic profiles and geotraverses of the continental margin from 71° to 79°N. See Fig. 1 for locations. Line a is from Sundvor *et al.* 1979, Line a′ from Steel *et al.* 1985 and lines b, c and d are modified from Spencer *et al.* (1984).

steeply dipping, major thrusts at such low angles. It is expected, however, that continued weak compressional shear would produce proper compression shortly away from a shear zone as the growth of the shear zone welt becomes isostatically abandoned.

From early Oligocene times, the formation of a new rifted Barents Sea margin caused the main Tertiary depocentre to shift westwards, into the newly forming Western Barents Sea Basin (Fig. 3). Continued subsidence is observed across the Senja fracture zone (see Fig. 2, lines c & d–units II & III), as well as the Hornsund fault zone (see Fig. 3), and up to 7 8 km of sediments accumu-

lated typically by shelf/slope progradation and accretion across this subsided margin.

The isopach map (see Fig. 3) clearly shows that the maximum thickness of this post-rift sequence is located between latitudes 72° and 75°N, indicating that the main provenance area was the Barents Shelf itself, and not the surrounding mainlands (i.e. Norway & Svalbard). Volumetric calculations of these post-early Oligocene deposits suggest widespread uplift and erosion eastwards across a large area of the Barents Shelf (see Fig. 3). From the Barents Shelf bathymetric map two major drainage systems seem to be present on the shelf, which correspond to two local

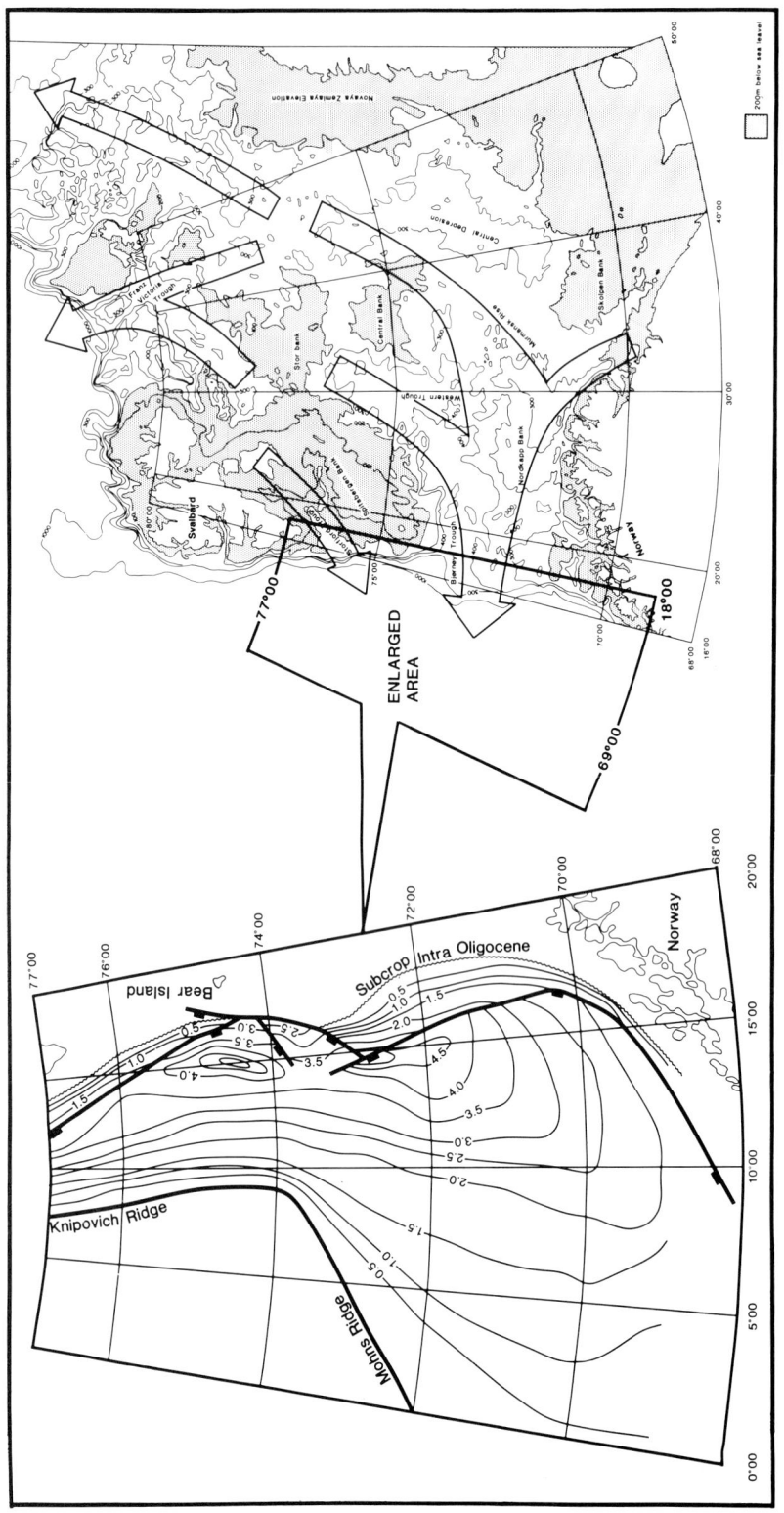

**Post Mid. - Oligocene Isopach map
(in sec. T.W.T.)**

Barents sea physiographic map

FIG. 3. Isopach map of the post mid-Oligocene sedimentary wedge and its relationship to the Barents Sea bathymetric map. Arrows on bathymetric map indicate major Tertiary drainage routes on the Barents Shelf.

thickness maxima on the isopach map (see Fig. 3): the Storfjorden Trough draining the Spitsbergen Bank, and the Bjørnøya Trough draining most of Barents Shelf eastwards to Novaja Semlya. Assuming that the total drainage area has not changed significantly during the Neogene, an estimated removal of more than 1 km of strata, on average, is deduced. This has important consequences for the prospectivity and current exploration of parts of the Barents Shelf.

References

BIRKENMAJER, K. 1981. The geology of Svalbard, the western part of the Barents Sea and the continental margin of Scandinavia. *In*: NAIRN, A. E. M., CHURKIN, M. & STEHLI, F. C. (eds) *The Ocean Basins and Margins (Vol. 5, The Arctic Ocean)*. Plenum Press, New York, pp. 265–329.

FALEIDE, J. I. & GUDLAUGSSON, S. T. 1981. *Geology of the western Barents Sea—a regional study based on marine geophysical data*. Cand. real thesis, University of Oslo, 160 pp.

——, —— & JACQUART, G. 1984. Evolution of the western Barents Sea. *Marine and Petroleum Geology*, **1**, 123–150.

——, MYHRE, A. M. & ELDHOLM, O. 1988. Early Tertiary volcanism at the western Barents Sea margin. *In*: MORTON, A. C. & PARSON, L. M. (eds) *Early Tertiary Volcanism and the Opening of the NE Atlantic*. Geological Society of London, Special Publication, **39**, pp. 135–146.

GJELBERG, J. 1981. Upper Devonian (Famennian) to Middle Carboniferous succession of Bjørnøya. *Norsk Polarinstitutts Skrifter*, **174**, 67 pp.

HORN, G. & ORVIN, A. K. 1928. Geology of Bear Island with special reference to the coal deposits and an account of the history of the island. *Skrifter om Svalbard og Ishavet*, **15**, 152 pp.

LAMONS, R., BRUN, J-P. & VAN DEN DRIESSCHE, J. 1987. Physical models of oblique convergence (abstract). *Geological kinematics and dynamics meeting, April 22–24 1987, Uppsala.*

MANUM, S. B. & THRONDSEN, T. 1986. Age of Tertiary formations on Spitsbergen. *Polar Research*, **4**, 103–131.

MYHRE, A. M., ELDHOLM, O. & SUNDVOR, E. 1982. The margin between Senja and Spitsbergen fracture zones: implications from plate tectonics. *Tectonophysics*, **89**, 33–50.

RØNNEVIK, H. C. 1981. Geology of the Barents Sea. *In*: ILLING, L. V. & HOBSON, G. D. (eds) *Petroleum Geology of the Continental Shelf of North-West Europe*. Heyden & Son, London, pp. 395–406.

SPENCER, A. M., HOME, P. C. & BERGLUND, L. T. 1984. Tertiary structural development of the western Barents Shelf: Troms to Svalbard. *In*: SPENCER, A. M. *et al.* (eds) *Petroleum Geology of the North European Margin*. Graham & Trotman, London, pp. 199–209.

STEEL, R J., GJELBERG, J., HELLAND-HANSEN, W., KLEINSPEHN, K., NØTTVEDT, A. & RYE-LARSEN, M. 1979. The Tertiary strike-slip basins and orogenic belt of Spitsbergen. *In*: BIDDLE, K. T. & CHRISTIE-BLICK, N. (eds) *Strike-slip Deformation, Basin Formation and Sedimentation*. Society of Economic Paleontologists and Mineralogists Special Publication No. 37, pp. 339–359.

—— & WORSLEY, D. 1984. Svalbard's post-Caledonian strata—an atlas of sedimentational patterns and paleogeographic evolution. *In*: SPENCER, A. M. *et al.* (eds) *Petroleum Geology of the North European Margin*. Graham & Trotman, London, pp. 109–135.

SUNDVOR, E. & ELDHOLM, O. 1979. The western and northern margin off Svalbard. *Tectonophysics*, **59**, 239–250.

A. NØTTVEDT, Norsk Hydro Research Centre, PO Box 4313, N-5013, Bergen, Norway.

L. T. BERGLUND & E. RASMUSSEN, Norsk Hydro Exploration, PO Box 31, N-9401, Harstad, Norway.

R. J. STEEL, Norsk Hydro Exploration, PO Box 200, N-1301, Stabekk, Norway.

Review of Igneous Activity

History of Tertiary igneous activity in the N Atlantic borderlands

B. G. J. Upton

SUMMARY: Stretching and thinning of the Laurasian continental lithosphere, which had proceeded intermittently from late Palaeozoic through the Mesozoic, reached a climax in early Tertiary times with copious generation of basalt magma, invasion of the attenuated crust as dyke swarms and surface eruption, principally by fissure volcanism. Continued basin subsidence allowed basaltic lavas, erupted at near sea-level, to accumulate to thicknesses which in places (e.g. W Greenland, E Greenland, Faeroes) attained several kilometres. In some zones dyke-swarm injection and crustal attenuation led to sea floor spreading and generation of ocean crust. Fissure swarms not infrequently changed position so that there could be one or more 'unsuccessful' rifting events prior to establishment of a 'successful' spreading axis.

While some uncertainty still exists regarding the precise timing of the onset of magmatism in the various zones of failure, the bulk occurred between 60 and 50 Ma. Across the British Isles basalt magmatism occurred early (ca. 60–59 Ma) whereas the great volumes of the E Greenland basalts appear to have erupted rapidly over the interval 54–52 Ma. The onset of magmatism in the Faeroes may have preceded that in E Greenland and possibly that of W Greenland–Baffin Island.

Major sill swarms developed, commonly within the Mesozoic sedimentary strata. At favoured channels for magma ascent, commonly controlled by major faults, longer-lived central-vent volcanoes developed. Slow cooling of large magma bodies at such foci produced mafic and ultramafic cumulates; production of salic magmas was mainly confined to these central complexes, the majority of which are located close to the E Greenland coast between ca. 66° and 74°N and a N–S zone through the British Isles from ca. 58°–51°N. Although the salic magmas were predominantly silicic (rhyolitic), feldspathoidal (phonolitic) magmas were important in some of the E Greenland centres. Generation of salic magmas was generally late with respect to the main phases of basalt eruption in the various sectors of activity. However, notable sequences of peraluminous silicic lavas may have preceded basaltic eruption in the Rockall Trough and Vøring Plateau areas. It is probable that magmatic activity in some of the sectors that had experienced the most intense early-phase volcanism persisted, on a small-scale, long after the sea floor spreading had been initiated and activity on the passive margins of both W and E Greenland may not have finally terminated until ca. 30 Ma.

The total compositional range of early Tertiary magmas associated with the N Atlantic marginal regions was extensive. Whereas tholeiites, varying from FeTi-rich to N-type MORB, were erupted in greatest volume, compositions varied from picrites, through tholeiites to ferro-basalts, icelandites, dacites and rhyolites. More alkalic mafic magmas included alkali olivine-basalt (and its differentiated products), nephelinites, lamprophyres and melilitites, with some associated carbonatites. Salic magmas ranged widely from silicic to feldspathoidal and from peraluminous to peralkaline.

The concept of a Thulean volcanic province was formulated by Holmes (1918) to embrace the igneous rocks of E Greenland, Iceland, Faeroes and the Hebridean–Northern Ireland region (Fig. 1). It was subsequently shown that the oldest rocks of Iceland, seen in the NW and E, are of Miocene age and not older than 16 Ma, whereas those of E Greenland–Faeroes–Hebrides were principally erupted in the Palaeocene and Eocene. A broad review of the whole Thulean province was provided by Noe-Nygaard (1974). This account attempts to present an updated *synthesis* of events in the Palaeocene, Eocene and Oligocene epochs, and to draw together an overall perspective of the igneous activity that immediately preceded, accompanied and postdated the

parting of the northern continents in the early part of the Tertiary period.

A very extensive literature relating to the early Tertiary volcanism around the N Atlantic has come into being, with an expansion that shows no sign of abating. The sequence of events, particularly in E Greenland, is complex and much detail has had to be passed over in order to present a generalized account of acceptable length.

Some of the activity formerly regarded as early Tertiary is now regarded as of Cretaceous age, dating from ca. 80–70 Ma. Thus, the submarine occurrences on the continental shelf to the N and W of Britain (including the Anton Dohrn seamount, the Blackstones Complex, etc.) and

From MORTON, A. C. & PARSON, L. M. (eds), 1988, *Early Tertiary Volcanism and the Opening of the NE Atlantic*, Geological Society Special Publication No. 39, pp, 429–453.

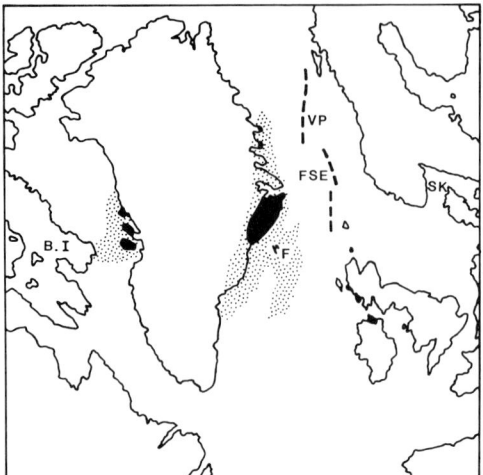

FIG. 1. Sketch map showing pre-drift reconstruction of N Atlantic region. Black: principal extant, on-shore outcrops of early Tertiary volcanic rocks; Stippled: likely off-shore and former on-shore extent of early Tertiary volcanic rocks; B.I.: Baffin Island; F: Faeroes; VP: Vøring Plateau 'dipping reflectors'; FSE: Faeroe–Shetland Escarpment; SK: Skagerrak.

the ash horizons in the chalk of eastern England probably date from this time (Macintyre *et al.* 1975; Durant *et al.* 1976; Pacey 1984).

Furthermore, it is now reasonably firmly established that the commencement of volcanism in the western part of the British Isles (the British Tertiary Volcanic Province, BTVP) may have pre-dated the start of activity in E Greenland, Faeroes and Skagerrak by several million years. Activity appears to have been concentrated in a number of relatively intense phases in the late Cretaceous and early Tertiary, often affecting wide areas of what were to become the N Atlantic borderlands, with these phases correlatable with the initiation or modification of plate motions (Macintyre *et al.* 1975; Fitch *et al.* 1978).

Whereas the various regions bordering the new ocean basins underwent distinct tectono-magmatic histories, a common pattern emerges from their study so that a generalized sequence of events can be construed. Thus, it was general for continental failure to take place within a relatively narrow (<100 km?) rift zone within which normal faulting leading to graben or half-graben development had been occurring for tens or even hundreds of million years prior to genesis of new oceanic crust. The positioning of these continental rift zones was typically controlled by older (Palaeozoic or Precambrian) lineaments. Subsidence within the rifts permitted accumulation of considerable sequences of both marine and non-

marine sediments, principally within the Mesozoic era. Typically, these early (pre-Tertiary) stages were not accompanied by volcanism. Accounts of the distribution and evolution of these Mesozoic rifts are given by Ziegler (1978), Kent (1977), Hallam (1971), Surlyk (1977, 1978), Henderson (1973) and Henderson *et al.* (1976).

Following widespread regression, tectonism and erosion, in the late Cretaceous, lower Tertiary strata generally lie with a marked disconformity or unconformity upon the older formations. Further faulting in the early Tertiary saw rejuvenation of the rift basins with shallow-water (typically non-marine) sedimentation and the onset of volcanism so that, in the axial parts of the basins, the first evidence of volcanism is generally of ashy sediments, often overlain by hyaloclastite breccias and/or pillow lavas. By this stage the continental lithosphere beneath the rifts is inferred to have been thinned through tensional stress with brittle deformation at shallow depths and plastic deformation at deeper levels, and to be progressively penetrated by dykes as episodic dyke (or fissure) swarms developed. With rapid attainment of maximum rates of magma production and ascent, accumulation of extrusive products became fast enough to establish subaerial lavas, i.e. lava accumulation outpaced basin subsidence. Highly mobile basaltic lavas flowed laterally from fissure volcanoes to produce quasi-horizontal and often extensive lava fields in which the younger lavas progressively overstepped bounding fault scarps to overlie older formations. Faulting and down-warping permitted axial subsidence to compensate for lava accumulation so that the lava successions rarely built up to heights very much above sea level. Following initial, high energy phases diminishing rates of magma supply gave rise to increasing time lapses between successive flows so that subaerial weathering of lavas and intercalation of fluvial or lacustrine sediments commonly becomes more prominent up the succession. As the patterns of structural failure changed, fissure swarms became relocated and major disconformities and, occasionally, unconformities occurred within the lava sequences.

Sills developed within the strata beneath the developing lava plateaux and, in places, formed important sill-swarms in the late Palaeozoic, Mesozoic and/or early Tertiary bedded successions.

Along particular rifted sectors which experienced the maximum tensional stresses, magma ascent became increasingly concentrated with repeated fissuring, becoming more and more focused upon a particular plane of weakness so that there was progression from dispersed dyke swarms in continental crust to sheeted dyke complexes in which the pre-dyke component diminished to zero. By the time this critical stage of proto-ocean development commenced the degree of partial melting in the mantle is believed to have increased and to have been taking place at increasingly shallow depths so that, concomitantly with the structural changes, the magmas undergo changes in composition to MORB type.

Clearly, different parts of the rifted basins of the N Atlantic region did not all experience the idealized sequence of events listed above. Thus, in the North Sea there was no Tertiary volcanic culmination to the Mesozoic rifting, just as in the Sea of the Hebrides and Skagerrak fissuring stopped short of initiation of oceanic floor.

Preferential channelling of magma along specific sites of crustal weakness (e.g. at intersection of major lineaments) permitted formation of large crustal magma chambers above which complex and relatively long-lived central volcanoes could develop. As a generalization, such structures formed relatively late in the magmatic developments in the individual 'provinces', apparently after the relief of much of the extensional stress. Hence, the major central volcanoes tended to grow upon the lava fields that had principally formed from fissure eruptions. Development of large crustal reservoirs of basaltic magma allowed a variety of phenomena to take place, more or less concurrently. Crystal fractionation permitted growth of layered ultrabasic and basic cumulates and, in essentially 'closed systems', the genesis of highly evolved salic residua. Heat losses through cooling of basic magma and latent heat of crystallization caused melting of wall rocks. Magma mixing events and contamination of mantle-derived basic magmas by anatectic crustal magmas undoubtedly occurred on both large and small scales.

A model involving early ascent of granitic diapirs above rising basaltic magma, with effects including early doming, cone-sheet formation and late-stage caldera subsidence was proposed by Walker (1975) for the British Tertiary Volcanic Province central-type complexes and may be of relevance elsewhere in the N Atlantic region.

On the margins of the newly created Greenland continent magmatism did not cease when the proto-ocean spreading centres were generated. Large vertical displacements occurred with relative uplift (on the km scale) of the 'passive margins' and subsidence of the new-formed oceanic lithosphere as it cooled and thickened. Step-faulted or monoclinal zones developed along, or close to, the continental margin with strata tilted or flexed towards the ocean, (as in E

Greenland S of Scoresby Sund and in W Greenland in the Svartenhuk–Ubekendt area).

In such zones, with magma continuing to ascend along near vertical fissures as the tectonic compensation occurred, early dykes acquired strong dips toward the continent while later dykes are tilted to a lesser (or zero) degree. Early lavas were correspondingly more steeply inclined towards the developing rift axis than later ones. Furthermore, there may be a thickening of the lava pile in the area of downwarping, as has been noted in the W Greenland province (Rosenkranz & Pulvertaft 1969; Henderson 1973). Thus a situation developed analogous to that envisaged by Bodvarsson & Walker (1964) for the neovolcanic zone in Iceland.

The early Tertiary volcanic rocks now seen as eroded relics around the margins of the fragmented continents will be considered from W to E. Whereas Baffin Bay and the Skagerrak are geographically remote from the N Atlantic, their evolution has so many parallels to those of the N Atlantic borderlands and is undoubtedly so closely related in terms of plate tectonics to the early Tertiary igneous history of E Greenland and the British Isles, that they will be covered in this review account.

Baffin Island—W Greenland

The activity that gave rise to the bulk of the volcanic suites preserved along, or close to, the coasts of Baffin Island and W Greenland appears to have been roughly contemporaneous with the main phase of activity in the Hebridean–Irish region. Thus, most is believed to have erupted between 63 and 56 Ma, i.e. somewhat earlier than the onset of the intense East Greenland volcanism. General reviews covering the magmatic history have been presented by Clarke & Pedersen (1976) and Clarke (1977).

According to Srivastava (1983), active sea floor spreading in the S Labrador Sea commenced in the late Cretaceous, causing severe tension in the region of Davis Strait, with production of many N–S faults, forming grabens and half-grabens that later filled with sediments. By the time sea floor spreading started in the northern part of the Labrador Sea (in Danian times), minor separation commenced in the Davis Strait area accompanied by onset of basaltic volcanism. The first indications of this are provided by tuffs in Danian sediments.

By early mid-Palaeocene, a considerable portion of the S Davis Strait or northern Labrador Sea had evolved and it was at this stage that copious eruption of tholeiitic basalt commenced, accompanying the attenuation of crust between Baffin Island and that part of W Greenland lying between 69 and 73°N, immediately preceding true ocean-crust formation in Baffin Bay.

It is clear that a complex graben system was already developed between Baffin Island and the interior highlands of W Greenland in the Mesozoic (Henderson 1973; Henderson *et al.* 1976). Whereas this graben system saw accumulation of thick sequences of marine and non-marine Cretaceous sediments, with basin subsidence and sedimentation continuing into the Palaeocene, the sedimentation may have begun in the Jurassic (Henderson 1973) and the structure may possibly have had Palaeozoic antecedents (Fahrig *et al.* 1971; Bridgwater *et al.* 1973). Profound subsidence continued after the start of volcanism allowing accumulation of up to 10 km of lavas, as on the Svartenhuk peninsula (Noe-Nygaard 1942). The accumulation rate of volcanic extrusives was sufficiently high for sediments to be reduced to a subordinate role in the stratigraphy after commencement of volcanism.

The volcanic rocks on Baffin Island (at ca. 67°N) are confined to a narrow fringe along some 90 km of nearly linear NW–SE-trending coast from Cape Dyer northwards (Fig. 2). They comprise a few hundred metres thickness of olivine-rich tholeiitic basalts and picrites, commencing with subaqueous hyaloclastite and pillow-breccias, passing abruptly up-sequence to sub-aerial lava-flows, when the rate of accumulation sufficiently exceeded subsidence rates for the exclusion of water from the basin. The lavas became thinner, with increasingly well-defined reddened (oxidized) surfaces up-sequence as eruptive rates declined and time intervals between one flow and the next lengthened (Clarke & Upton 1971). The original succession on Baffin Island is unlikely to have been substantially thicker, or to have had significantly greater inland extent than indicated by the existing outcrops.

By contrast, to the NW in western Greenland, the volcanic rocks exhibit much greater thickness, lithological variety and areal extent (Clarke & Pedersen 1976). Thus, the outcrop area (Fig. 3) is some 370 km (N–S) by ca. 125 km across, covering (with offshore continuations) some 55 000 km². As on Baffin Island, subaqueous successions were followed by subaerial piles of olivine-rich tholeiitic (picritic) extrusives. These are succeeded by less-magnesian aphyric and plagioclase-phyric basalts which are locally overlain by transitional, to alkaline olivine basalts (on Hareøen). A variety of products were erupted at a late stage across what is now Svartenhuk peninsula and Ubekendt Ejland, including trachybasalts, trachytes and rhyolites, as lavas and tuffs. On Ubekendt Ejland salic extrusives appear to be overlain by the largely alkaline basaltic Erquâ Formation (Larsen 1977), with a NNW-trending swarm of lamprophyre dykes representing the latest stage in the magmatic history (Clarke *et al.* 1983). A central-type intrusive complex, composed of gabbro and granophyre was developed at Sarqatâ qáqâ on Ubekendt Ejland.

The early magnesian tholeiites in W Greenland are regarded as contemporaneous, and originally contiguous, with those of Baffin Island (Clarke 1977). In most pre-drift continental reconstructions Nugssuaq is the point in Greenland closest to Cape Dyer on Baffin Island (Rosenkranz & Pulvertaft 1969). The early extrusives, and particularly those of Baffin Island, include very primitive and probably unmodified primary magma products (Clarke 1970). These magnesian tholeiites, with MORB-like geochemistry (low contents of large-ion lithophile elements (LILE) and

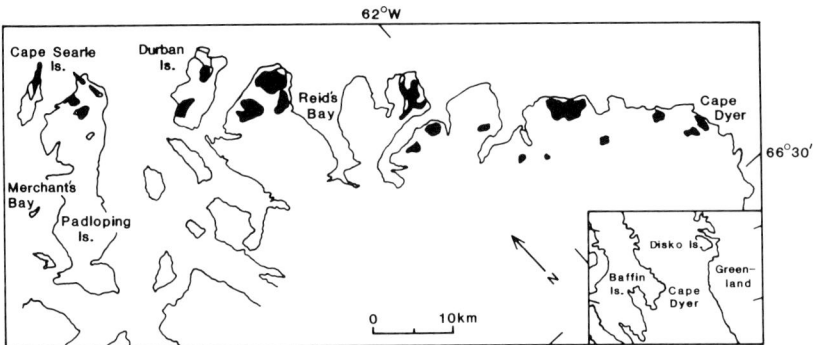

FIG. 2. Sketch map showing outcrop areas of early Tertiary volcanic rocks and associated sediments on the coast of Baffin Island in the vicinity of Cape Dyer.

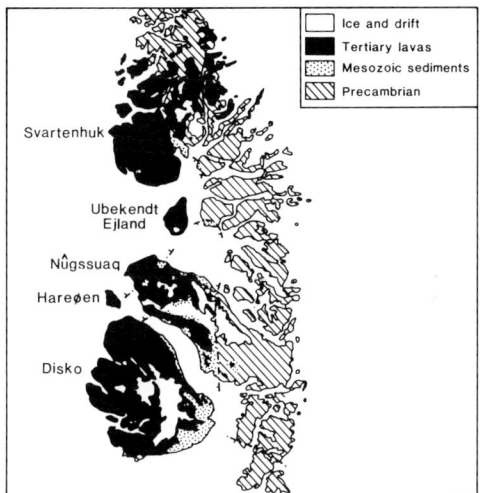

☐ Ice and drift
■ Tertiary lavas
▨ Mesozoic sediments
▧ Precambrian

Svartenhuk

Ubekendt
Ejland

Nûgssuaq

Hareøen

Disko

FIG. 3. Sketch map of part of the W Greenland coast, showing outcrop areas.

low $^{87}Sr/^{86}Sr_i$ and relatively light-REE depleted patterns; O'Nions & Clarke 1972) imply high degrees of partial melting and rapid, unimpeded ascent from the mantle during the early stages of magmatism. The overall evolution of the W Greenland succession indicates decreasing magma ascent rates with time (increasing fractionation, increasing time intervals between events, and increasing opportunities for crustal interaction) and late eruption of small volumes of magma arising from smaller degrees of partial melting at increasingly greater mantle depths or as high-pressure fractionation residues.

The voluminous, early, high-temperature and primitive basalts probably erupted from fissures (?NW–SE-trending) which were, at first, rather narrowly restricted within the rift axis. The indications are that the topography was vigorous, with steep fault-defined escarpments in the Precambrian highland regions to the E and to the W. A boundary fault system that defined the eastern limits of Cretaceous–Tertiary sedimentation in W Greenland (Rosenkranz & Pulvertaft 1969) also confined the earlier basaltic extrusives. Consequently, the thickest volcanic successions occur within the graben (Henderson 1973). Early picritic breccias in W Greenland filled topographic hollows and locally (as in the Nugssuaq peninsula) attained thicknesses of up to 700 m (Clarke & Pedersen 1976). Whereas most of the foreset bedding features in the W Greenland breccias indicate eruptive sources to the W of the present coastline, those on Baffin Island point to eruptive sites to the NE. The manner in which

the subhorizontal flows continually expanded the area of the lava fields as bounding fault-scarps were progressively overstepped, also points to eruptive sites which lay to the seaward side of the respective Baffin Island and W Greenland outcrop areas.

In W Greenland the early picritic tholeiite series was followed by a sharp change to relatively Mg-poor tholeiites, regarded as products of relatively low-pressure fractionation derived from the picritic tholeiite parental magmas. The change of lava type coincided, at least locally, with a prolonged quiescence during which (on Svartenhuk) several tens of metres of unconsolidated sands with thin coals accumulated. The later Mg-poor tholeiite flows tend to be thicker (averaging 25–30 m) and more widespread (occasionally traceable for more than 50 km) than the underlying picritic series. Red boles tend to be well-developed and the younger lavas transgress onto the Precambrian basement. Whereas the higher lavas were probably mainly erupted through fissures, there is some evidence for central vent eruption (Münther 1973). On Hareøen arenaceous sediments (with coals) testify to continued subsidence after the main volcanic phase (Hald 1971).

Dates on the main-phase volcanic rocks remain somewhat uncertain. Beckinsale *et al.* (1974) obtained a Rb-Sr isochron age of 65 ± 5 Ma on the granite–gabbro Sarqâta qáqâ complex (Ubekendt Ejland) and inferred that the W Greenland basalts erupted before 65 Ma. On the basis of palaeomagnetic data, Athavale & Sharma (1975) considered the main activity to be at ca. 63 Ma. However, using dates from $^{40}Ar/^{39}Ar$ studies, Parrott (1976) (quoted in Clarke *et al.* 1983) ascribed the bulk of activity to the bracket 60–56 Ma, which would make it essentially synchronous with the main-phase activity in the Hebridean–Northern Irish region (see below). However, Hansen & Pedersen (1985), taking into account the work of Athavale & Sharma (1975) and a revised magnetic polarity time-scale for the Palaeocene–Eocene, concluded that eruption of the lower part of the lava succession took place between magnetic anomalies 25 and 24 at approximately 56–52 Ma. A fission-track age of ca. 45 Ma, on rhyolitic glass within the upper part of the (mainly) basaltic upper sequence of Disko–Nugssuaq region suggests prolonged (but declining) activity (Hansen & Pedersen 1985). $^{40}Ar/^{39}Ar$ data on three of the late lamprophyre dykes from Ubekendt Ejland suggests an age of 32.6 ± 4.4 Ma for the most recent magmatism (Parrott & Reynolds 1975).

Thus, a highly productive early eruptive stage (pre-ocean crust genesis), was followed in the W

Greenland region by a lessening of eruption frequency. Reduction in magma supply rates was associated with a tendency for increasingly fractionated magmas to erupt. The later magma batches may have supplied volcanoes on the passive (western) margin of the Greenland continent well after initiation of oceanic spreading in Baffin Bay. By the time the very late (Oligocene) lamprophyres on Ubekendt Ejland were intruded, the Baffin Bay spreading centre was remote and perhaps inactive. Clarke *et al.* (1983) suggested that the late-stage lamprophyric magmas may have arisen as very delayed fractionates from magmas residual from the earlier tholeiitic activity.

Whereas the lavas on the landward (eastern) side of the W Greenland province tend to be sub-horizontal, faulting and down-flexing to the W takes the lavas below sea-level. Thus, SW of a flexure line across Svartenhuk the lavas dip SW at 12–37°. Westward tilting and step-wise down-faulting to the W is also seen on the western parts of Nugssuaq peninsula and Disko Island. Clarke (1977) noted that the Ubekendt Ejland lamprophyre dykes are nearly vertical, suggesting that the lava dip (15–35° W) had been acquired prior to dyke emplacement.

Despite the fact that the W Greenland and E Greenland provinces developed wholly independently, the two provinces show many features in common. Mesozoic rifting on both sides of Greenland was a prelude to massive tholeiite basalt volcanism immediately prior to plate separation. Both passive margins saw continued magmatism after continental severance with a tendency for late products to be more alkaline and/or salic.

Northern E Greenland: Scoresby Sund to Shannon Island

In this part of the E Greenland coast, from ca. 70°–ca. 76°N, plate separation was largely coincident with the trends of the Caledonian orogeny. It is clear that the final act of continental lithosphere failure from 70°N to at least 76°N had been preceded by perhaps as much as 250 Ma of repeated extension, with faulting, basin formation and sedimentation. A structural element that played a key role in the history of the region, from the late Palaeozoic onwards, was the post-Devonian main fault (Vischer 1943). This NNE-SSW lineament acted as a boundary separating a stable cratonic region to the W from a zone that experienced progressive fragmentation and rifting, to the E (Surlyk 1977, 1978; Larsen 1980).

The region W of the fault was rarely transgressed by the sedimentary accumulations (Permian, Triassic, Jurassic & Cretaceous) to the E, and it also formed a topographic barrier to the westward spread of lavas in the early Tertiary. The outcrop distribution of Mesozoic and Tertiary strata is shown in Fig. 4.

The zone of rifting experienced major tectonic disturbance around the Jurassic–Cretaceous time boundary at ca. 135 Ma (Surlyk 1978). The faulting at this time has been regarded as heralding the opening of the new ocean basin (Hallam 1971). However, the Cretaceous was generally tectonically quiescent until relatively late when coarse (often conglomeratic) sediments made an abrupt appearance. These conglomerates may be inferred to relate to rejuvenation of rift-marginal faults with uplift of the stable block to the W that was to become the continental interior. This phase may have triggered the onset of mantle melting, magma ascent and surface volcanism. Ashy sediments, hyaloclastites and pillow lavas mark the onset of Tertiary volcanism. The arkosic conglomerates have been presumed to be of Lower Tertiary age on the geological maps of Koch & Haller (1971) and Mayne (1942). However, at least those on Wollaston Forland are now regarded as being of late Cretaceous age (Surlyk, *in* Hald 1978).

In some areas, e.g. on the SE of the Hold With Hope peninsula, subaerial basalt lavas directly overlie the arkosic conglomerates (Upton *et al.* 1980). However, further W subsidence produced a shallow water gulf in which ca. 50 m of water-sorted tuffs, hyaloclastites, shales and thin coals accumulated, whereas near Giesecke Bjerge over 300 m of hyaloclastites (with lava intercalations), underlie the main subaerial lava succession.

Erosional remnants of a formerly extensive basalt lava field can be found from Shannon (75°30′N) in the N to Kejser Franz Josephs Fjord (ca. 73°N) in the S. The thickest preserved succession is seen on the Hold With Hope peninsula where it reaches 800 m thickness. The sequence on Hold With Hope has been divided into an upper and a lower series (Upton *et al.* 1980). The lower series comprises tholeiitic basalt lava flows, typically 10–30 m thick (Noe-Nygaard 1976). With the exception of rare hyaloclastitite pillow lavas these are subaerial. The lack of obvious volcanic feeders and the banking of lavas against bounding faults to the west has led to the supposition that they were mainly erupted from fissure volcanoes situated east of the present coast line. The early lavas spread over a roughly levelled surface, across formations ranging in age from Devonian to late Cretaceous and early Tertiary.

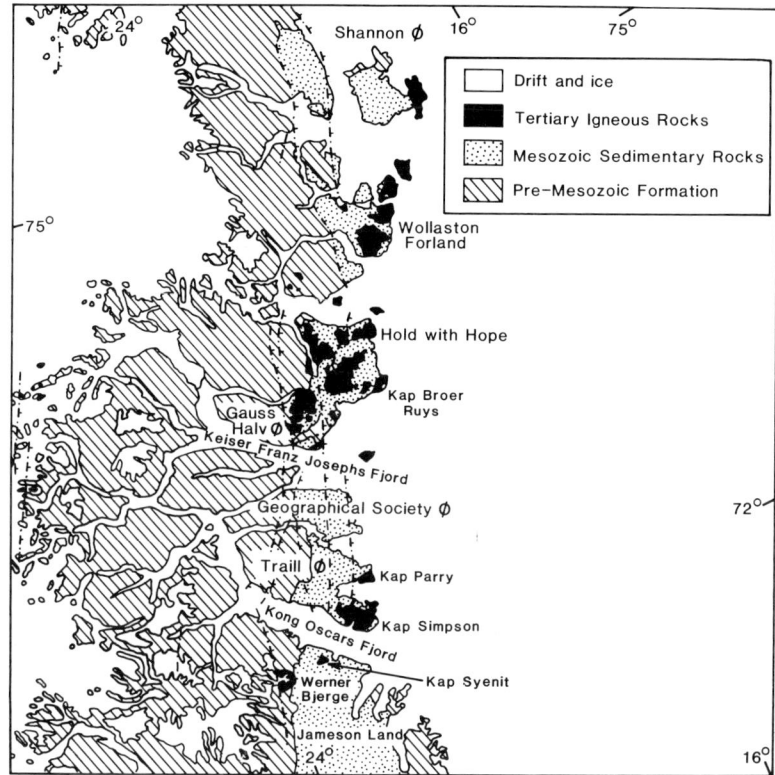

FIG 4. Sketch map of part of the East Greenland coast north of Scoresby Sund showing outcrop areas.

Tholeiitic sills are abundant in the Mesozoic sediments and early Tertiary sediments and volcanic breccias, throughout Hold With Hope, Gauss Halvø, Traill Ø and Geographical Society Ø, between ca. 72° and 73°40'N (Noe-Nygaard 1976). Sills total 200 m S of Giesecke Bjerge (Hald 1978). Some, possibly most, may relate to the Lower Series basalt lavas. Traill Ø sills thicken to the W and possibly the bounding fault zones acted as conduits (Surlyk 1977). The early lavas and sills are dominantly quartz tholeiites and, unlike the early products in the Baffin Bay region, are relatively Mg-poor. There is no reason to consider that the lava pile was ever very much thicker than that now preserved on Hold with Hope and Gauss Halvø. Basin subsidence with accompanying volcanic infill was less dramatic than that in W Greenland and central E Greenland (see below).

The precise timing of the onset of volcanism in this northern province is not known. The sequence of events is very similar to that attending the early stages of volcanism in the Kangerdlugssuaq area in central E Greenland (see below) and

a generalized contemporaneity may have pertained.

About two-thirds up the lava sequence in Gauss Halvø and Hold With Hope the basalts show an abrupt change of character. The lower basalts are relatively uniform microphyric quartz tholeiites with low contents of K and other LIL elements. By contrast the upper basalts are notably enriched in LILE, range from quartz tholeiite to basanitic and are very variable in phenocryst content. Many of the flows are thin and eruptive rates are thought to have been much lower than in the lower basalts. Textural, geochemical and volcanological features suggest that the upper units of the lava sequence erupted from a central-vent volcano to build a large, low-angled shield, rather than from predominantly fissure volcanoes as was the case earlier in the volcanic history. The Upper Plateau lavas were followed by emplacement of a NNE–NE-trending basaltic dyke swarm, in turn cut by the Myggbukta caldera complex. The Myggbukta Complex comprises extrusives and intrusives ranging from picrite to rhyolite. According to Upton *et al.*

(1984) the Upper Plateau lavas were produced in the early growth of the volcano. A rifting event followed in which the dyke swarm was emplaced prior to caldera collapse over a relatively shallow-level crustal magma chamber. REE patterns suggest that the basaltic magmas of the dyke swarm and the Myggbukta Complex reflect a greater degree of mantle melting than that involved in the genesis of the upper plateau lavas. The Myggbukta volcano and its related fissure swarm may have grown within a 'lateral rift' at least 100 km W of the contemporary 'axial rifts' (Mohns and Aegir) along which new oceanic lithosphere was forming (Upton *et al.* 1984). If so, the lower basalts and sills may have immediately preceded plate separation whereas the upper lavas and later events may have erupted through thinned, faulted passive margins while eruption of MORB was taking place further E.

Myggbukta is but one of some half-dozen central-type complexes in the northern region. These include the Werner Bjerge in Scoresby Land (Bearth 1959; Kapp 1960; Brooks *et al.* 1982) and the Kap Syenit, Kap Simpson and Kap Parry complexes on Traill Ø (Rex *et al.* 1979). At each location suites of alkaline intrusive rocks are exposed; general review accounts of these complexes have been presented by Noe-Nygaard (1976) and Nielsen (1987). The largest of these complexes is the Werner Bjerge Complex, adjacent to the post-Devonian fault line and consisting of three centres: (1) an early SE 'basic alkali' centre containing agglomerates and breccias as well as alkali pyroxenites and gabbros, syenogabbros and syenites; (2) a northern alkali syenite and granite centre; and (3) (the youngest?) nepheline syenite centre. Whole-rock isochron ages for the Kap Parry, Kap Simpson and Werner Bjerge Complexes are 40.3 ± 1.2, 38.5 ± 5 and 30 ± 2 Ma respectively (Rex *et al.* 1979).

The Myggbukta Centre is demonstrably the culminating magmatic event at the end of a complex magmatic evolution. It is likely that the various alkaline and/or salic centres from Werner Bjerge through Kap Parry to Kap Broer Ruys also had a late-stage genesis and were emplaced by stoping and/or ring faulting in the passive margin well after commencement of oceanic growth.

Although the bulk of volcanism in the northern province occurred within the rifted sector in proximity to the new spreading axis, some activity took place within the stable block, well to the W of the post-Devonian fault line. This activity produced a variety of highly silica-undersaturated basic lavas, now seen as cappings to nunataks at around 74°N. They include nephelinites, basanites and nepheline-hawaiites that have yielded

K-Ar dates of from 58–44 Ma (Brooks *et al.* 1979). It is probable that these erupted from a number of small central-vent volcanoes, and that they may denote small volumes of primitive magmas which originated as small-scale partial melts at relatively deep levels in the mantle (Nielsen 1987).

Central E Greenland: Angmagssalik–Scoresby Sund

South of ca. 70°, what is now the E Greenland coastal region was the region in which the greatest volume of magma was erupted in the period immediately preceding the separation of Greenland and Norway (Fig. 5). The trend of the coastline S of Scoresby Sund shows a marked deflection from that to the N and for some 500 km the coastline follows a relatively smooth bowed course, convex to the SE. The coastline shows a marked inflection or re-entrant at Kangerdlugssuaq, a large fjord at ca. 68°N. South and W of Kangerdlugssuaq the coastline is again essentially smooth and little indented for ca. 300 km to ca. 66°N. These two sections of Atlantic coast have been considered to represent the 'successful' arms and the Kangerdlugssuaq lineament the 'unsuccessful' arm of a *rrr* triple junction (Burke & Dewey 1973; Brooks 1973a). The smooth coastlines to the NE and SW of Kangerdlugssuaq are defined by a narrow zone which experienced intense faulting, flexuring and multiple intrusion (as coast-marginal dyke swarms and central-type intrusive complexes) around the time of plate separation. The faulting and flexing (the latter producing a profound monoclinal feature) of this sector of the E Greenland continental margin have been likened to the features produced in SE Africa (affecting Swaziland, Mozambique, Zimbabwe), including the Limpopo lineament and the Lebombo monocline associated with the disruption of Gondwanaland and the Karoo magmatism (Wager 1947; Cox *et al.* 1965).

A broad region of some 80 000 km² between Kangerdlugssuaq and Scoresby Sund extending some 200 km inland from the Blosseville Kyst, preserves the greatest quantity of early Tertiary igneous rocks in the N Atlantic marginal areas and ranks among the world's larger Phanerozoic continental flood-basalt provinces. The total volume of lavas is estimated at ca. 2×10^5 km³ (L. M. Larsen, pers. comm.).

Although to the S of Kangerdlugssuaq, coastal dyke swarm(s) are intense (Wager & Deer 1939; Bridgwater *et al.* 1978; Larsen 1978; Myers 1980) their presumed extrusive products have been

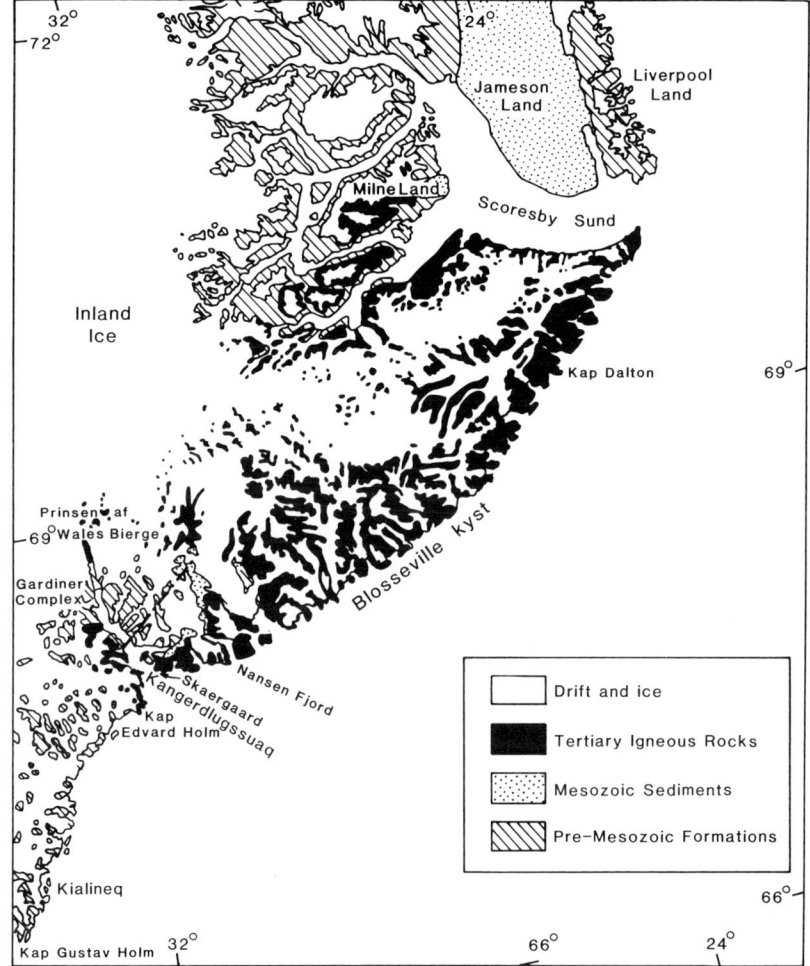

FIG. 5. Sketch map of central E Greenland showing outcrop areas.

(almost completely) lost by erosion. The flood basalts inland of the Blosseville Kyst (the Blosseville Group) are generally thickest towards the coast (Larsen & Watt 1985). Former estimates for maximum thickness of the lavas in the Kangerdlugssuaq–Scoresby Sund region of up to 10 km (Wager 1947; Soper *et al*. 1976) have been revised downwards and, probably, nowhere does the actual thickness exceed 3.5 km (Nielsen & Brooks 1981). The bulk of the lavas appear (as in the subordinate northern province) to have been erupted from fissure volcanoes within a rift valley of low relief and to have been restrained or dammed to the W by highlands of Precambrian gneiss. Thus the lavas, with near-horizontal attitude (regional dip ca. 1°SE, Larsen & Watt

1985) lap onto an elevated continental interior to the N and W and, in the S and E are dropped below sea-level by faulting and/or flexing along the Atlantic coast. According to Bott (1987) the lava field may have extended eastwards, across and beyond the Faeroe Block, with the main eastern edge forming the Faeroe–Shetland Escarpment.

The base of the lava succession in the coast-marginal area is exposed near Kangerdlugssuaq. In this region some 200 m of Cretaceous and lower Tertiary sediments intercede between the Precambrian basement and the Tertiary volcanic succession. These shallow water sediments accumulated in an embayment of the pre-Atlantic Sea prior to the start of volcanism (Higgins & Soper

1981). Facies changes in sediments and volcanoes indicate deeper water to the E (Brooks & Nielsen 1982). Although the base of the volcanic succession is not seen over most of the Blosseville Kyst region, Mesozoic sediments probably underlie the entire eastern half of the lava pile (Larsen & Watt 1985). Sedimentological data indicate shallowing of the marine basin or gulf in the Danian (Soper *et al.* 1976), followed by an unconformity with a basal conglomerate (basal Sparnacian; Higgins & Soper 1981), passing up into coarse sands and volcanogenic sediments. The latter are thus the first indication of the intense basaltic eruptive event that was to follow. At this stage magma generation may have just commenced and the first penetrative fissures through the attenuating continental lithosphere may be inferred to have reached surface levels.

There is some evidence that the direction of flow in the lower lavas of the Kangerdlugssuaq area was from E to W, suggesting eruptive sites within the present-day shelf area (Wager 1934; Soper *et al.* 1976). Foreset beds in the pillow breccias of the same area indicate a southerly source, i.e. again offshore from the modern coastline (Brooks & Nielsen 1982).

The duration of the activity that produced the Blosseville Group has been established biostratigraphically to have been short-lived, and to have occupied a few million years only (Brooks 1973b; Soper & Costa 1976; Soper *et al.* 1976). This conclusion is supported by the apparent restriction of almost all of the lavas to a single (reversed) magnetic polarity epoch, and the bulk of the Blosseville Group must have accumulated rapidly within some 2 or 3 million years corresponding to magnetic anomaly interval C24R (Soper *et al.* 1976; Brooks & Nielsen 1982; Larsen & Watt 1985). Due to lack of consensus regarding time-scale calibrations (cf. Odin & Mitchell 1983; Berggren *et al.* 1985) there remains uncertainty concerning the true date of this major eruptive phase. According to Bott (1987), the most satisfactory estimate is that it occurred (mainly) between 54 and 52 Ma. The biostratigraphic age is well established as straddling the Palaeocene–Eocene boundary, extending from highest Thanetian to lowest Ypresian (Soper & Costa 1976; Soper *et al.* 1976).

The lower lavas of the Blosseville Group in the Kangerdlugssuaq area (Vandfaldsdalen & Mikis Formations) are rather variable, ranging from picrites and ankaramites to basalts and basaltic andesites (Brooks & Nielsen 1982). New data on the lower lavas and discussion of their genesis are presented by Gill *et al.* (1988). Although the bulk of the Blosseville lavas were thought to be relatively uniform FeTi-rich oversaturated tholeiites (Brooks *et al.* 1976), recent work in the Scoresby Sund area has shown considerable variation among the tholeiites.

Detailed investigations of the lavas in the Scoresby Sund region have revealed two major and one minor volcanic episode (Larsen & Watt 1985; Larsen *et al.* 1988). The lavas of the first episode have been divided into an initial phase (Magga Dan Formation) followed by: (1) the Milne Land and (2) the Geikie Plateau Formations. These lavas were erupted from inland areas to the N and W of Scoresby Sund. After a period of quiescence, marked by a sandstone unit up to 5 m thick and the production of a nephelinitic tuff, the second major episode commenced, with production of: (1) the Rømer Fjord and (2) Skraenterne Formations.

The earlier stages of both of these major eruptive episodes (viz. Milne Land and Rømer Fjord Formations) produced tholeiitic lavas of variable composition, followed by main stages (viz. Geikie Plateau and Skraenterne Formations), which both show systematic upward trends towards more Ti-poor and generally more primitive compositions but with closing phases of more evolved, (Ti-rich) tholeiites, i.e. there is a cyclical compositional variation shown by the lavas of the two major episodes (Larsen & Watt 1985; Larsen *et al.* 1988).

In the second episode eruptive centres lay to the E of those that fed the first. The Rømer Fjord lavas erupted from sites mainly towards the Atlantic coast, whereas the Skraenterne Formation is derived from sources S and E of the present coastline.

Larsen *et al.* (1988) address the problems of correlating the lavas of the Scoresby Sund region with those of the Kangerdlugssuaq region to the S. They conclude that the Scoresby Sund upper basalts (Rømer Fjord and Skraenterne Formations) may be equivalent to the upper plateau basalts of the Kangerdlugssuaq region. Although the lower basalts of Scoresby Sund may be time-equivalents of the 2 km thick lower basalts of Kangerdlugssuaq, the former, as stated above, originated from sites N and W of Scoresby Sund, whereas the Kangerdlugssuaq lower basalts were erupted S and E of the present coast.

The plateau basalts of the Faeroe Islands are discussed in a later section. In a pre-drift reconstruction the Faeroes lay immediately S of the Kangerdlugssuaq area. Larsen *et al.* (1988) note that the Faeroese basalts were probably erupted from separate volcanic systems and that no strict stratigraphic correlation with either the Kangerdlugssuaq or Scoresby Sund regions of E Greenland should be anticipated.

The third (minor) volcanic episode recognized

in the Scoresby Sund region involved normal faulting and an intense coast-parallel dyke swarm which fed lava flows of the Igtertivâ Formation, now only preserved in graben-like downfaulted occurrences near the coast. The Igtertivâ lavas and dykes, now volumetrically insignificant, are normally magnetized and are correlated with anomaly 24 (C24N) (Larsen & Watt 1985). The third event (dykes and Igtertivâ lavas) involve compositions ranging from low-Ti to high-Ti tholeiites as well as some alkaline (hawaiitic) types. The Igtertivâ lavas are often interbedded with sediments of basal Ypresian age (Soper & Costa 1976; Larsen *et al*. 1988).

At Kap Dalton volcanogenic conglomerates and sandstones were deposited upon the upper-most lavas. These sediments, of middle Ypresian age, are thought to have succeeded from the lavas after only a short time interval and indicate continued subsidence after cessation of volcanism (Soper & Costa 1976; Brooks & Nielsen 1982).

Coast-marginal dyke swarms constitute per-haps the single most remarkable phenomenon associated with volcanism in the N Atlantic marginal areas. The swarms can be traced for some 350 km along the coastal strip from Kap Wendel in the S to Nansen Fjord in the N, with an offshore continuation (detected aeromagneti-cally), for a total of 780 km between 63°N and 69°45′N (Larsen 1978; Myers 1980).

The intensity of the coast-marginal swarms has led some authors (Brooks 1973b; Nielsen 1978) to suggest comparison with the sheeted complexes of ophiolites; Larsen (1978) described the coast-parallel swarm(s) as constituting a continental spreading centre. The E Greenland coast mar-ginal swarm(s) thus appear to exhibit features transitional from those of a continental basic dyke swarm developed during tensile stress to those of an ophiolite-like sheeted complex result-ing from repetitive injection along a young spreading centre. In an account of the opening of the Norwegian–Greenland Sea, Bott (1987) pro-posed that sea floor spreading commenced at least 100 km E of the flexure zone and coast-marginal dyke-swarm, and that the intervening shelf area is underlain by thinned continental crust. On this interpretation the crustal thinning was due to a short-lived rifting event prior to active spreading, with the flexure marking the inner edge of the thinned, subsided crust. How-ever, the continental severance may have oc-curred as close as 15 km to the present coastline between Kap Gustav Holm and the Kangerdlugs-suaq area (cf. Gill *et al*., 1988).

Vertical movements accompanied lithospheric stretching during emplacement of the swarm(s) so that intrusion of the dykes took place while flexuring and normal faulting were occurring. The flexuring and faulting, with predominant downthrow to the SE, gave rise to the asymmetric monoclinal feature, first recognized by Wager (1935), that defines the actual coastline of E Greenland over several hundred kilometres and which presumably overlies a fundamental zone of fracturing in the deeper lithosphere.

With relative uplift on the western (continental interior) side and down-throw towards the eastern side, crustal blocks along the hinge zone (coastal monocline) were rotated by flexing and/or antith-etic faulting to acquire an oceanward dip (Nielsen 1975; Nielsen & Brooks 1981). The dykes are believed to have been emplaced as essentially vertical intrusions. Since emplacement of the swarms embraced a considerable time period, the earliest dykes were rotated the most and the youngest, the least. The relationship, first pointed out in the Kangerdlugssuaq region by Nielsen (1975), that the younger the dyke, the less it will have been affected by tilting, has since been shown to hold true as a general rule to as far S as 66°30′ (Bridgwater *et al*. 1978; Nielsen 1978). In the Kangerdlugssuaq area coast-parallel dykes have been subdivided into an intense pre-flexure generation and a less intense swarm intruded when flexuring was almost at a close (Nielsen 1978; Brooks & Nielsen 1982; Gill *et al*. 1988). In addition a still younger coast-parallel swarm, involving MORB-like tholeiites, has been recog-nized in the vicinity of I.C. Jakobsen Fjord and regarded as following closely the termination of flexuring (Brooks & Nielsen 1982).

Whereas in the early stages of rifting fissure volcanism was paramount, with the passage of time and reduction of tensional stress, ascending basaltic magmas were increasingly liable to become arrested and ponded to form crustal magma chambers. Thus, a number of central complexes developed along the coast marginal zone, at least some of which fed central-type volcanoes. These (relatively early) central com-plexes are dominantly of gabbroic compositions and include strikingly layered cumulate succes-sions. They tend to be funnel-shaped and may be associated with cone-sheet emplacement. The best known of these funnel-shaped layered gab-bros is the Skaergaard intrusion close to the Kangerdlugssuaq 'triple junction' (Wager & Deer 1939; Wager & Brown 1968; McBirney 1975). The Skaergaard intrusion, like the nearby gabbro intrusion at Kap Edvard Holm, appears to have developed where basaltic magma exploited the flat-lying upper limb of the developing flexure, with intrusion largely controlled by the uncon-formity between the Precambrian basement and the superjacent sediment and lava pile (Myers

1980). A zircon fission-track age of 54.6 ± 1.7 Ma for Skaergaard (Brooks & Gleadow 1977) accords reasonably well with geological inferences concerning the age of intrusion (Brooks & Nielsen 1982). The Skaergaard was itself intruded by a massive sill of tholeiitic gabbro (Basistoppen sill). An extensive sill complex with tholeiitic sills up to 500 m thick, intruded into the sediments and lavas east of Kangerdlugssuaq, may be essentially contemporaneous with the Skaergaard intrusion (Brooks & Nielsen 1982). The late MORB-like coast-parallel dyke swarm (referred to above) cuts this sill complex.

Emplacement of the large tholeiitic gabbro intrusions tended to occur towards the end of the period of flexuring and coast-parallel dyke intrusion. They are commonly (like Skaergaard and the Basistoppen sill) tilted themselves by late flexuring and some (e.g. the Kap Gustav Holm and Imilik gabbros) may have been undergoing tilting during their crystallization (Myers 1980; Brooks & Nielsen 1982). The Kruuse Fjord layered gabbro, however, is exceptional in having been emplaced inland of the coastal flexure and is consequently unaffected by flexuring (Myers 1980).

With separation of the Greenland and Norwegian continental lithosphere the E Greenland flexure zone became progressively distanced from the evolving spreading centre. Although magmatism continued, a marked temporal change in style of intrusion and composition of magmas can be discerned. Intrusive complexes (in some cases demonstrably subvolcanic) tended to become increasingly dominated by salic magmas (mainly forming quartz syenites and syenites).

The relatively low-density salic magmas arose by diapirism, stoping or ring-faulting and are thought to have been generated through processes of crystal fractionation combined with crustal anatexis and contamination, and to have been formed from bodies of basic magma trapped at depth and residual from the early intense basaltic activity phases (Myers 1980; Nielsen 1987). Near the supposed triple-junction of Kangerdlugssuaq Fjord lies the Kangerdlugssuaq syenite complex (Kempe *et al.* 1970; Brooks & Gill 1982; Deer *et al.* 1984). This complex, intruded at ca. 50 Ma (Pankhurst *et al.* 1976), is composed of a variety of quartz syenites, syenites and nepheline syenites. With a diameter of ca. 33 km it is the largest of all the central complexes associated with the early Tertiary volcanism around the N Atlantic.

The syenitic intrusions were generally emplaced some 3–5 Ma after the gabbros (Gleadow & Brooks 1979; Brooks & Nielsen 1982) at a time when the early phase of fissuring, faulting and flexuring had terminated and supply of (rela-

tively) primitive tholeiitic magma from the mantle had also ceased. The late-stage magmatism close to the E Greenland coast saw eruption of a wide range of magma types, including mildly alkaline to nephelinitic mafic magmas, as well as salic magmas, many of which were notably alkaline (Nielsen 1987). This activity, geographically related to the 'initial magmatic lineament' (Nielsen 1987) persisted long after the termination of early Tertiary tholeiitic magmatism. Dykes cutting the Kangerdlugssuaq syenite complex (i.e. post ca. 50 Ma) are of lamprophyric character (Nielsen 1987). A further example of late intrusion of lamprophyres comes from the coastal region S of Kangerdlugssuaq, between Kap Japetus Steenstrup and Tasilaq, where olivine- and clinopyroxene-phyric lamprophyre dykes are widespread (Bridgwater *et al.* 1978). In this latter region the latest manifestation of volcanic activity takes the form of hydrothermal alteration associated with coast-parallel fissures.

Clasts of strongly alkaline rocks in the basal conglomerates overlying tholeiitic plateau basalts at Kap Dalton show that here too the last eruptions were of alkalic magma (Wager 1935).

Much remains to be done to establish more precise dates for the late alkaline magmatism associated with the E Greenland coast. However, at Kialineq (one of the central complexes on the coast S of Kangerdlugssuaq) a whole-rock Rb-Sr isochron date of 35 ± 2 Ma was obtained for the emplacement of the quartz syenites and granites (Brown *et al.* 1977). Thus, as in W Greenland and NE Greenland, a brief and intense phase of tholeiitic basalt magmatism heralding the start of continental separation was followed by a period lasting many millions of years during which alkaline and/or salic magmas were intermittently erupted. Total cessation of volcanism along the passive margins of the 'new' continent may not have been finally attained until late Oligocene times.

Although tholeiitic activity dominated the early stages of separation along what was to become the coastal zone, alkaline magmas were erupted inland to the W through what is likely to have been thick, relatively cold, continental lithosphere (Nielsen & Brooks 1981). Thus, inland of the Kangerdlugssuaq, an area of ca. 500 km^2 encompassing Prinsen af Wales Bjerge was covered by alkali basaltic, hawaiitic, basanitic and nephelinitic lavas. These appear to have constituted a volcanic field above the tholeiitic plateau lavas with interdigitation of lavas erupted from a large number of central-type volcanoes (Fawcett *et al.* 1982; Nielsen 1987). The Gardiner Complex is a remarkable ring-complex involving both plutonic and volcanic rocks, within the

Kangerdlugssuaq lineament and ca. 40 km S of Prinsen af Wales Bjerge. It is composed of a variety of ultramafic cumulates (including melilite-rich rocks), agpaitic syenites, carbonatites, apatite and magnetite-rich rocks (Nielsen 1980, 1981; Brooks & Nielsen 1982). According to Nielsen (1987) these rocks formed within a nephelinitic volcano that could have fed lavas similar to those of Prinsen af Wales Bjerge. The Gardiner Complex has been dated at ca. 50 Ma (Gleadow & Brooks 1979). The inland alkaline volcanism may have occurred within the interval 55–50 Ma, i.e. roughly contemporaneously to the main tholeiite-dominated activity along the evolving continental margin (Nielsen 1987).

After plate separation and the consequent coastal flexuring, a profound domal uplift commenced at ca. 50 Ma, centred on the Kangerdlugssuaq region (Brooks 1973a, 1979; Brooks & Nielsen 1982). Thus a major change in tectonic environment attended the transition from dominantly tholeiitic basalt magmatism to the more alkaline (and commonly salic) magmatism. A further regional uplift occurred in the mid-Oligocene at ca. 35 Ma, i.e. at much the same time as magmatic activity ended. Brooks & Nielsen (1982) noted that these events may have been causally connected with a major re-organization of plate geometry in the N Atlantic at anomaly 13 time, i.e. at ca. 36 Ma.

The cause of the post-separation uplifts remains conjectural. However, in view of the fact that the Kangerdlugssuaq region may have been the site at which the greatest volume of basalt was extruded, it is possible that the epeirogenic elevation was related to massive continental under-plating by basic intrusions in the deep crust or along the crust–mantle boundary (Cox 1980; McKenzie 1984; Furlong & Fountain 1986).

The Faeroes

The Faeroe Islands are composed almost entirely of subaerial tholeiitic basalt lavas. Extending over an area of some 1400 km² and with generalized easterly to south-easterly dips of 1–3°, a sequence some 3 km thick is exposed (Fig. 6). The base, neither seen nor reached in drilling, is thought to lie some 2–4 km below sea-level (Hald & Waagstein 1984) and the entire sequence may exceed 5 km thickness (Waagstein 1988). The lavas are believed to overlie continental crust (Casten 1973; Bott et al. 1974, 1976; Hald & Waagstein 1984), with the boundary towards Tertiary oceanic crust located along an escarpment W of the islands. Mesozoic sediments and

sills may intervene between Precambrian basement and the lava pile (Gibb et al. 1986; Gibb & Kanaris-Sotiriou 1988).

Pre-drift reconstructions place the Faeroe basalts relatively close to the Blosseville Kyst of Greenland and it is likely that the Faeroes represent a portion of the early Tertiary lava plateau generated shortly before ocean crust formation (Brooks 1973a). The Faeroes have been interpreted as a continental fragment created during the N Atlantic opening (Casten & Nielsen 1975; Bott 1987).

The visible succession was subdivided into a Lower, Middle and Upper Series on the basis of field-characteristics of the flows and weak unconformities (Noe-Nygaard & Rasmussen 1968, 1984). Slight easterly tilting, followed by dyke intrusion, succeeded the Lower Series eruptions; further tilting followed the Middle Series and yet more tilting took place after eruption of the Upper Series lavas. These movements were due to uplift to the W, near the crest of the present Wyville-Thomson Ridge (Noe-Nygaard 1974). There is some evidence that the Lower Series lavas, which may exceed 3 km thickness in the

FIG. 6 Geological sketch map of the Faeroe Islands.

southern and central part of the Faeroes (Waag-stein 1988) were erupted from sites to the W. Interflow clay or ash-rich sediments increase up sequence indicating temporal decline in magma supply rates (Noe-Nygaard & Rasmussen 1968, 1984; Noe-Nygaard 1974).

A long period of quiescence followed the Lower Series activity, during which time a ca. 10 m horizon of clays and coals was deposited. This is overlain by tuff–agglomerate deposits. Botanical evidence correlates the coal-bearing horizon with the late Palaeocene (Lund 1983) whereas the explosive activity has been correlated with the main 'ash-marker' in the North Sea (Smythe *et al.* 1983; Jacque & Thouvenin 1975).

The Middle Series is composed of very thin ropy lava units. In contrast to the nearly aphyric Lower Series lavas those of the Middle Series are typically porphyritic. Whereas the Lower Series lavas are mainly silica-oversaturated tholeiites, the Middle and Upper Series lavas are typically olivine tholeiites, with increasing predominance of MORB-type compositions. A detailed account of compositional variation within the Faeroese sequence is given by Waagstein (1988).

Although the Middle and Upper Series are magnetically reversed, the Lower Series contains at least two zones showing normal polarity (Nielsen 1983). Problems relating to the precise ages and correlation of the magnetic stratigraphy of the Faeroese succession with those of the oceanic magnetic anomaly patterns are addressed by Waagstein (1988). He suggests that the Lower Series activity commenced near the start of C26R and ended at the beginning of C24R, in which case the Series may pre-date the onset of E Greenland basalt eruption which probably did not occur before the start of C24R. Waagstein proposes that the zone of active rifting that gave the Lower Series may have switched to E Greenland at the start of C24R. The younger, MORB-like, Ti-poor Faeroese tholeiites, which appear to be derived from sources N of the Faeroes, may relate to the still later eastward shift of the zone of active rifting in E Greenland (Larsen & Watt 1985; Waagstein 1988).

Rockall Plateau and Rockall Trough

The Rockall Plateau is composed of Precambrian basement, overlain by Palaeocene volcanics, in turn overlain by Upper Palaeocene, Lower Eocene and Oligocene sediments (Roberts 1975). The early, fissure-fed basalts are thought to have been subaerial: they now compose the 'dipping (seismic) reflector sequence' around the margins

of the Rockall microcontinent and data from recent drilling programmes indicates composi-tions resembling N-type MORB (viz. Merriman *et al.* 1988). The only part of the Rockall Plateau exposed above sea-level is the Rockall sea-stack, composed of varieties of aegirine granite dated at 52 ± 9 Ma (Sabine 1960).

The nature of the Rockall Trough, which separates the Rockall Plateau from the British Isles, remains controversial. It may be underlain by oceanic crust, but could be floored by atten-uated continental crust (cf. Roberts 1975; Morton *et al.* 1988). Within the Trough there are Mesozoic sediments, Palaeocene volcanics and a younger sedimentary cover. A recent drilling in the Trough encountered a thick lava sequence beneath upper Palaeocene sediments; the lava sequence comprises 689 m of basalts overlying at least 356 m of peraluminous dacites. The higher basalts are olivine tholeiites with picritic and alkaline tendencies while the lower basalts are olivine tholeiites whose compositions resemble N-type MORB. The dacites are fine-grained to glassy cordierite–hypersthene–phyric rocks thought to have originated from melting of aluminous sediments (?Cretaceous black shales). The necessary heat is speculated to have been derived from crystallization of a very large basaltic magma chamber (Morton *et al.* 1988).

Early Eocene (Ypresian) volcanogenic sedi-ments, involving within-plate tholeiitic basalt ashes are present in substantial thickness in the northern part of the Rockall Trough and may have originated from explosive activity in the Faeroese province (Jones & Ramsay 1982).

Vøring Plateau and the Faeroe–Shetland escarpment

Whereas the E coast of Greenland shows occur-rences of early Tertiary volcanic rocks along a reach of some 1000 km, such rocks are absent on the opposing Norwegian coastal regions. Com-parable subaerial or shallow-water extrusive sequences of lower Tertiary age, overlying rifted, attenuated continental crust may, however, be present offshore on the continental shelf. Atten-tion has been focused specifically upon two NE–SW trending escarpments, the Vøring Plateau escarpment and the Faeroe–Shetland escarp-ment, features which may mark the ocean–continent boundary on the eastern side of the Norwegian Sea (Talwani & Eldholm 1972; Eldholm 1978) (Fig. 7).

According to Talwani & Eldholm (1972) the E

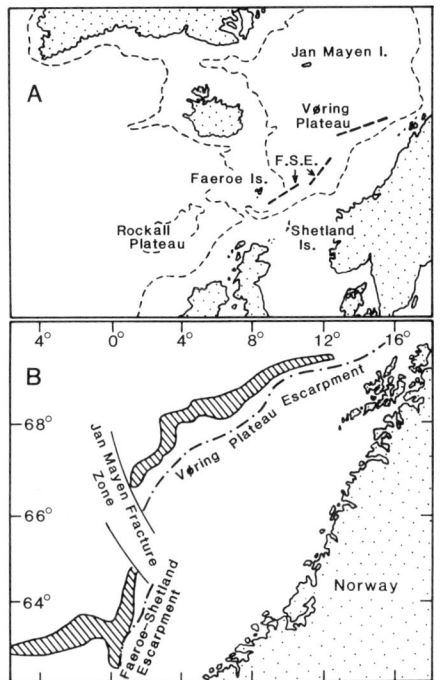

FIG. 7. Sketch maps showing: (**A**) relationships of Vøring Plateau, Faeroe–Shetland Escarpment (F.S.E.), Faeroe Islands and the Rockall Plateau; (**B**) positions of the dipping seismic reflectors in the vicinity of the Vøring Plateau and Faeroe–Shetland Escarpments (after Hinz 1981).

part of the Vøring Plateau consists of Mesozoic rocks (continental or oceanic) covered by subaerial Tertiary lavas and volcaniclastic horizons. Hole 642E, ODP Leg 104, on the Vøring Plateau 'Marginal High' penetrated over 300 m of sediment overlying a sequence of lavas and intercalated volcaniclastic sediments a little more than 900 m thick (Eldholm 1978; Eldholm *et al.* 1987). The lava sequence was divided into a Lower Series of generally silicic, and commonly glassy, flows (some of which are peraluminous, with cordierite, hypersthene and plagioclase phenocrysts) and an Upper Series (ca. 760 m thick) composed of olivine tholeiites of N-MORB type. A 7 m succession of volcaniclastic and mudstone material separates the two Series. The Lower Series appears to have been formed from two magma types: (1) derived by partial fusion of sediments (or metasediments) and (2) MORB-tholeiites, like those of the Upper Series, but strongly contaminated by crustal melts (Eldholm *et al.* 1987). There are clearly close comparisons to be drawn between this succession and that of the Rockall Trough, discussed above.

South of the Jan Mayen Fracture Zone, as a pair of offset linear features trending NE–SW, (with the southern sector roughly midway between Faeroe and Shetland) lies the Faeroe–Shetland escarpment. The southern sector lies well E of the Faeroes, with Tertiary lavas (and Mesozoic sediments?) overlying a Precambrian basement. The Faeroe–Shetland escarpment is built up of a pile of lower Tertiary basalts, which have been correlated with the Lower and Middle Series of Faeroese lavas, overlying 5–6 km of Mesozoic sediments. Whether these overlie stretched and subsided continental crust or pre-mid-Cretaceous oceanic crust is, however, uncertain (Smythe *et al.* 1983).

Within the Lower Series basalts the presence of two partly eroded central volcanoes has recently been detected by seismic techniques (Ridd 1983; Gatliff *et al.* 1984). These two, the Erlend and W Erlend volcanic complexes, which are ca. 30 km apart, are thought to be underlain by quasi-cylindrical mafic or ultramafic plutons. The Erlend volcano appears to consist of extrusives dipping radially out from a central vent. The complexes are of substantial size, with overall dimensions comparable to those of the principal Hebridean volcanic centres (see below) (Gatliff *et al.* 1984).

Britain and Ireland

Volcanism associated with early Tertiary rifting events in the region of the British Isles was concentrated in the N and W and the products are mainly seen around the Hebrides and Northern Ireland (Fig. 8). Magma ascent in this British Tertiary Volcanic Province (BTVP) was principally through dykes in the earlier stages, with major central-type volcanic centres generally becoming established and important towards the end of the activity. The literature dealing with the BTVP is very extensive (cf. Thompson 1982) and continues to grow. Recent review accounts include those of Emeleus (1983) for the Hebridean region and Preston (1981) for the Irish region. A series of reviews (Brown; Preston; Emeleus; Wadsworth; Bell; King; Speight *et al.*; Thompson) on specific topics within the BTVP is presented in Sutherland (1982).

The dykes, dominantly basaltic, have generalized NW–SE to N–S trends, with more nearly E–W trends pertaining in western Ireland. The majority belong to what have been termed regional linear swarms (Speight *et al.* 1982). The swarms strike at a large angle to the continental margin and developed through fissuring that was approximately normal to the developing oceanic

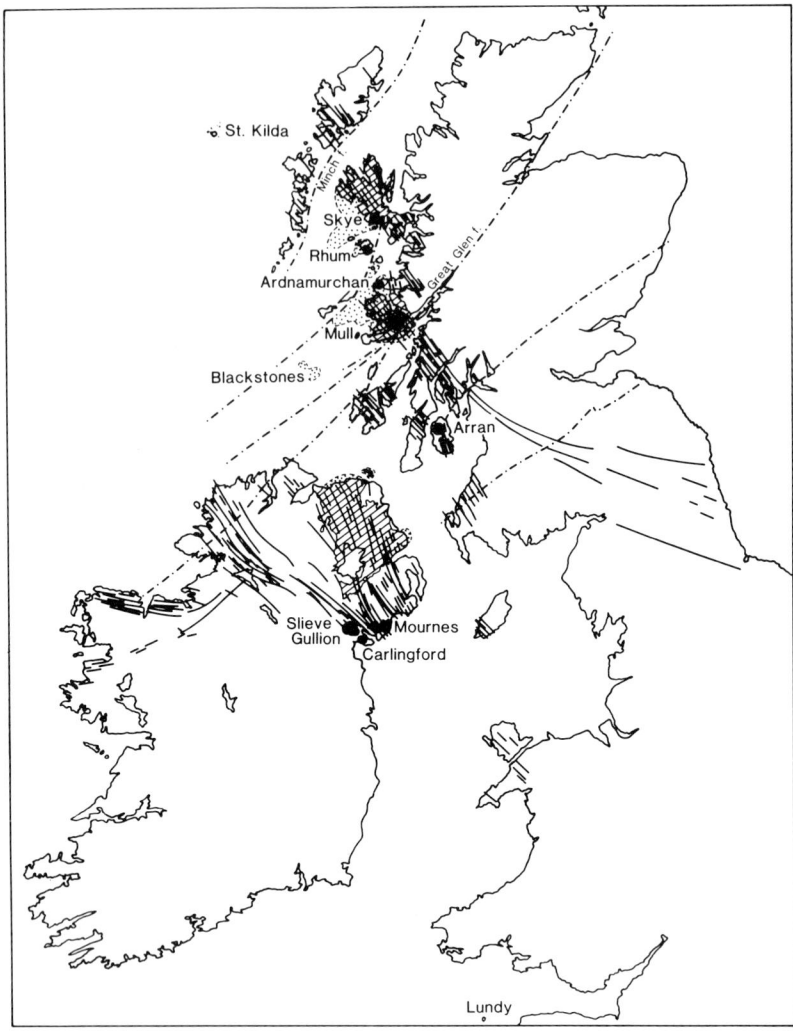

FIG. 8. Sketch map of the British Tertiary Volcanic Province showing: principal on- and off-shore outcrops of early Tertiary lavas (hatched and stippled respectively); main central volcanic complexes (black), and (diagrammatic) distribution of the dyke swarms.

ridge to the N and W. Although the bulk of the volcanic rocks occur in the Hebridean–Northern Irish area, early Tertiary dykes occur over a very wide area, extending from the Outer Hebrides through the western Scottish Highlands to NE England, and from the coasts of western Ireland to N Wales and central England. Among the most southerly representatives are the dykes of the Dingle Peninsula (SW Ireland) and the granites and basic dykes of Lundy Island in the Bristol Channel at ca. 51°N.

Of the principal central volcanic complexes, eight occupy a roughly N–S tract, some 40 km

broad by 250 km long extending from Skye in the N, through Rhum, Ardnamurchan, Mull and Arran to the three complexes in Northern Ireland (Mournes, Slieve Gullion and Carlingford), and which may be extrapolated southwards to include the Lundy complex. Whereas intersection of the dyke swarms with major fracture zones (e.g. the Camasunary–Skerryvore, Great Glen and High-land Boundary faults) almost certainly influenced the siting of some, the reason for the general N–S disposition remains enigmatic. Macintyre *et al.* (1975) have suggested that this N–S lineament may, before opening of the N Atlantic, have been a southerly continuation of the lineament responsible for the Kangerdlugssuaq 'rift' in E Greenland. Among recent discussions of the structural controls of the main BTVP complexes the reader is referred to analyses by Walker (1975), Vann (1978), Preston (1981, 1982) and Emeleus (1983). Whereas the northernmost complex in this zone (Skye) lay some 300 km E of the eventual site of ocean ridge initiation, the southernmost complex (Lundy) was separated from it by approximately 500 km. The St. Kilda intrusive complex, generally similar to the Hebridean and Northern Irish central complexes in composition, structural style and age, is remote from the Hebridean centres, lying some 50–60 km W of the Outer Hebrides on the continental shelf.

Although the principal eruptive sites lay to the N and W of the British Isles, ash horizons within Lower Tertiary sedimentary strata occur over an area of some 400 000 km^2 in the North Sea (Rhys 1974; Jacque & Thouvenin 1975). The ash layers occur low in the Eocene or in the uppermost Palaeocene and may largely represent down-wind fall-out from explosive eruptions from the BTVP, with some admixture of material erupted from volcanoes in the Skagerrak and the N Atlantic rift. Thus, of the two main phases of ash deposition, one at 58–57 Ma may reflect the Hebridean volcanism, with a second phase at 55–52 Ma mainly due to the E Greenland–Faeroe–Skagerrak (etc.) activity immediately preceding ocean crust genesis (Knox & Morton 1983). Similar late Palaeocene–early Eocene ash horizons are also reported from the W Shetland Basin (Ridd 1983) and the Rockall Trough (Jones & Ramsay 1982).

Whereas some eruption of basalt may have taken place over the whole broad region affected by dyke intrusion, three principal areas of basalt accumulation remain, which almost certainly denote the basinal areas in which the most persistent volcanism occurred. These include: (1) the Skye lavas; (2) those of the Ardnamurchan–Mull region; and (3) those of the Antrim Plateau (Northern Ireland). The two northern areas were largely controlled by three roughly NE–NNE-trending faults inherited from Caledonian structures, viz. the Minch, Camasunary–Skerryvore and Great Glen Faults. These faults bound asymmetric Mesozoic sedimentary basins (McQuillin & Binns 1973) which formed part of the zone of Mesozoic grabens and half-grabens which, further N, were to control the separation of Greenland from Norway. To the S of the Hebridean basins, the Antrim lavas largely occupy a much older (Upper Palaeozoic) sedimentary basin, the SW extension of the Midland Valley of Scotland. Basin subsidence continued

during the Mesozoic and the period of lava accumulation. Preserved sequences are generally less than 1 km thick (viz. Emeleus 1983), except on Mull where some 1.8 km thickness is preserved. In Mull the upper part of the lava sequence may be related mainly to central vent formation, rather than fissure eruption. Subsidence continued over the main fissure swarms after cessation of volcanism producing synclinal basins, as in Mull (Speight *et al.* 1982) and Antrim (Preston 1981), with further faulting and tilting of fault blocks (McQuillin & Binns 1973).

The lava successions in the Hebridean areas tend to be somewhat more alkalic than the typical basalts erupted early in the Baffin Bay, E Greenland and Faeroe regions. Flow-top reddening, production of lateritic palaeosols, interflow fluviatile and lacustrine sediments are relatively common features. As a broad generalization, in comparison with the principal early Tertiary sequences of Baffin Bay, E Greenland and the Faeroes, it would appear that much of the basaltic magmas of the BTVP resulted from relatively small-scale partial melting events and that the magma supply rates were more subdued.

In the Skye succession (Anderson & Dunham 1966; Esson *et al.* 1975) the main lava series is largely composed of alkali basalts and basalts of transitional type, with more fractionated compositions becoming commoner upwards. They are overlain, however, by tholeiitic basalts (Preshal–Mhor type) poor in LILE and with LREE depletion, which may have been erupted in substantial volume. Whereas the Preshal–Mhor type lavas have largely been eroded, some 70% of the dyke swarm is composed of this basalt type (Mattey *et al.* 1977).

In Muck, Eigg and Canna the lavas are mostly alkaline basalts. The Muck and Eigg lavas, overlying Mesozoic sediments, probably extended across Rhum and pre-dated the development of the Rhum central igneous complex (Emeleus *et al.* 1985). The siting of the Rhum complex appears to have been determined by the N–S Long Loch Fault. This fault, which probably had a significant pre-Tertiary history (as was the case for the Camasunary–Skerryvore Fault) was active during development of the Rhum Complex with fault movement probably responsible for the emplacement of the Central Series ultrabasic rocks in this complex (Emeleus *et al.* 1985). Deep erosion of the (inferred) Rhum central volcano was followed by deposition of fluviatile sediments and lavas across its western flanks. Tholeiitic basalt pebbles occur in the conglomerates immediately underlying alkali basalts and hawaiitic lavas. The latter are succeeded, after an erosional interlude, by tholeiitic andesites and, after another erosional interval, by icelandite flows. Hawaiite lavas, after a further interval, represent the youngest eruptive products (Ridley 1973; Emeleus & Forster 1979; Emeleus 1985). Emeleus (1983) concluded that, in each of the Hebridean lava fields, a general progression from early alkali basalts to alkali basalt differentiates, can be discerned, with tholeiitic flows appearing later. In the Mull area alkali olivine basalts, succeeded by the Porphyritic Central Type (consisting of alkali basalts and their derivatives) were followed by tholeiitic basalts of the 'Non-Porphyritic Central Type'.

In the Antrim plateau, however, magmatism was dominantly tholeiitic. The lava pile, on the basis of two principal erosional breaks, has been divided into three, the Lower, Middle and Upper Series, with the Middle Series having very limited distribution. The lower basalts, described by Preston (1981) as primitive olivine tholeiites, are overlain in the Giant's Causeway region by silica-oversaturated tholeiitic basalts. In mid-Antrim some 60 m of rhyolitic extrusives accumulated around a central vent (Sandy Braes), and are believed to be rough time-equivalents of the Middle Series basalts. Following a period of quiescence that produced deep lateritic weathering and fluviatile deposits, renewed fissure volcanism gave rise to the Upper Series basalts. They differ from the lower basalts, which are relatively LREE enriched, in being LREE depleted and there are compositional similarities between the Upper Series and the Preshal–Mhor type basalts (Lyle 1985). The latter may have resulted from an increase in the degree of partial melting in the source areas. In both the Skye and Antrim areas it appears that the degree of partial melting reached its climax late in the activity, but nonetheless the processes of extension and magma genesis failed to produce new oceanic crust.

Accurate dating of the various stages of volcanism in the BTVP has proved difficult and contentious, many of the published data being inconsistent with the stratigraphy (Mussett 1980; Mitchell & Mohr 1986). It is possible that the Antrim basalts commenced extrusion at ca. 66 Ma, with the Upper basalts erupted at ca. 61–58.5 Ma (Fitch et al. 1978). The Antrim lavas are directly overlain by Oligocene sediments (Curry et al. 1978). Such a broad time span envisaged for the development of the Antrim lava plateau is, however, at variance with the conclusion of Wilson (in Preston 1981) based on palaeomagnetic data, that the whole lava sequence may have grown in less than 1 My, possibly within a single (reversed) magnetic polarity epoch.

There is, however, now general consensus that the main Hebridean activity dates from ca. 60–57 Ma, and produced the lava plateaux of Skye–Mull–Antrim, and the ash horizons in the North Sea (Macintyre et al. 1975, 1979; Mussett 1980, 1984, 1986; Knox & Morton 1983). Macintyre et al. (1975, 1979) and Macintyre (1977), while concluding that the activity was mainly encompassed between 60 and 50 Ma, interpreted the data as indicative of a major phase at ca. 60–59 Ma and a later, subordinate, phase at 53–52 Ma correlative with the onset of intense activity further N and W that produced the (main) E Greenland succession.

Some of the most recent and precise $^{39}Ar/^{40}Ar$ dating on Hebridean volcanic rocks has been presented by Mussett (1984, 1986). Within the Rhum–Canna–Skye region, the Rhum complex was emplaced, unroofed and partly covered by younger lavas within a single reversed magnetic polarity interval (26R), occupying some 2.8 My at ca. 59 Ma, but with some activity persisting into the ensuing normal polarity epoch. Thus, a complex sequence of eruptive, tectonic and erosional events was encompassed in less than 3 My (Mussett 1980). In the Mull district, extrusion commenced at ca. 60 Ma, continuing to ca. 57 Ma (again probably within chron 26R) with central complex evolution persisting into chron 26N and the later Mull dykes intruded during 25R. On this basis, volcanism in the Mull area was confined to between 0.7 and 4.7 My.

The younger activity appears to have been mainly silicic; thus, the Conachair granite (St. Kilda) is dated at 55 ± 1 Ma (Brooks 1984) and the Eastern Red Hills of Skye, at 53.5 ± 0.4 Ma (Dickin 1981). The Lundy granite has recently been assigned an age of 53 ± 5 Ma (Hampton & Taylor 1983). Firm evidence for still younger activity has been obtained from the Mourne granites, Northern Ireland, and from Eigg in the 'Small Isles' of the Hebrides. Granite emplacement in the Mournes is now thought to have taken place from 56–52 Ma (Gibson et al. 1987; Thompson et al. 1987) and eruption of the Sgurr of Eigg pitchstone at 52.1 ± 0.5 Ma (Dickin & Jones 1983).

The youngest ages reported for Tertiary igneous rocks in the British Isles are the (K-Ar) dates of 45 and 25 Ma from dolerite dykes in the Dingle Peninsula, SW Ireland (Horne & Macintyre 1975). These ages were taken as evidence for exceptionally late magmatism associated with reactivation of a NNW-trending fault zone and, if substantiated, indicate that small-scale localized activity persisted many millions of years beyond the initial, highly productive, eruptive phases as appears to have been demonstrated for the W and E Greenland provinces.

The Skagerrak

The Skagerrak, between southern Norway and Jutland has been regarded by Pedersen *et al.* (1975) as having acted as a 'failed arm' during the early Tertiary rifting events attendant upon the N Atlantic opening. It may well represent a faulted basin marginal to the eastern side of the Atlantic which saw accumulation of Mesozoic sediments, followed by lower Tertiary volcanism.

Boulders of Mesozoic sediment in Jutland glacial deposits are almost certainly derived from the Skagerrak and blocks of tholeiitic basalt have been dredged from the Skagerrak itself (Noe-Nygaard 1967). Much of the evidence for Skagerrak volcanism comes from the ash layers within the lower Eocene sediments in Jutland, N Germany, Netherlands, Belgium, SE England and the Forties Field in the North Sea, where isopach data indicate the Skagerrak as the source area (Elliott 1971; Pedersen *et al.* 1975).

In Jutland, ash layers are overlain by fossiliferous clays, correlated with Atlantic DSDP cores dated at 52 Ma. The ash horizons themselves are considered to date from ca. 53–54 Ma, i.e. essentially contemporaneous with much of the volcanism in E Greenland and some of that in the British Hebridean and N Irish province. Whereas palaeomagnetic data suggest a duration of activity of ca. 3 My, varve evidence indicates that it may have been as short as ca. 60 000 years (Pedersen *et al.* 1975). Pedersen *et al.* discerned four stages: *Stage 1* involving subalkaline rhyolitic ashes (from one or more central volcanoes?) followed by tholeiitic basalt eruptions; *Stage 2* showing waning activity with alkaline and peralkaline magmas from one or more isolated volcanoes; *Stage 3* saw an increase in activity, during which some basaltic and more evolved ashes were deposited. The latter suggest that central-vent volcanoes were involved; *Stage 4* was the most intense and gave rise to 110 layers of monotonous FeTi-rich tholeiitic ashes.

Gravity and magnetic anomalies at approximately the same location in the north-western part of the Skagerrak have been interpreted as indicative of a volcano buried beneath ca. 250 m of sediments (Sharma 1970). There remains, however, doubt as to whether this structure represents the eruptive site for the ash horizons described above.

Concluding remarks

Early Tertiary igneous activity preceding, accompanying and following the attempted or successful break-up of the northern continents in the N Atlantic (*sensu lato*) region clearly embraced a very wide range of magmatic phenomena and produced a broad spectrum of lithologies. Volcanic sequences comprise rocks as diverse as picrites, nephelinites and rhyolites while plutonic suites include peridotites, gabbros, diorites, granites, syenites and foyaites, with rare development of agpaites, melilitolites and carbonatites.

Detailed stratigraphic and geochemical study of what were, until quite recently, regarded as monotonous sequences of tholeiitic plateau basalts, has demonstrated significant variability in time and place. The basalts have been shown to vary widely in terms of Mg-number, silica-saturation and incompatible element contents, with implications for significant variability in mantle source-rock compositions and the degree and conditions of source-rock melting, as well as for interaction with lithospheric wall-rocks at all levels, crystal-fractionation and magma mixing.

Alkali basalts occur in W and E Greenland (albeit subordinate to the tholeiites) as well as in the Hebrides. Si-deficient alkaline mafic magmas, including lamprophyres and nephelinites, were generated in small volumes. Some, e.g. the Ubekendt lamprophyres may be high-pressure fractionation residues (Clarke *et al.* 1983), whereas others may represent very small-scale mantle melt extracts. Some were erupted in initial phases immediately preceding voluminous production of tholeiites (as in Scoresby Sund) (Larsen *et al.* 1988) and at Hold With Hope (Upton *et al.* 1984). Others (e.g. the Ubekendt dykes) were emplaced as the ultimate phase, following extensive long-lived volcanism. Still others (e.g. Prinsen af Wales Bjerge and the 74°N E Greenland nunataks) were erupted through colder, thicker crust, remote from the principal theatre of contemporary basalt extrusion. A thin lamprophyre dyke containing mantle xenoliths, on the Isle of Lewis, dated at ca. 50 Ma (Menzies *et al.* 1987) may be a comparable example in the Hebrides.

Salic magmas, typically confined to the longer lived central-type volcanic centres, also displayed great variability from silica-rich (granites and rhyolites) through silica-saturated to silica-deficient feldspathoidal syenites. Granitic magmas, produced on a subordinate scale in W Greenland, were prominently developed near the E Greenland coast from S of Kangerdlugssuaq to Hold With Hope, as well as in virtually all of the principal British centres from Lundy to St. Kilda. Compositions of the granites (and rhyolites) grade from peraluminous to peralkaline, with Rockall exemplifying the peralkaline extreme. Feldspathoidal salic magmas were generated in

E Greenland from Kangerdlugssuaq N to Werner Bjerge. In the Hebridean province they are insignificant and generally confined to late fractionation residues in the larger alkali dolerite sills.

Highly peraluminous eruptives (andesites, dacites and rhyolites) are notable in the offshore volcanic successions in the Vøring Plateau and Rockall Trough. Peraluminous rhyolite glasses are also known from W Greenland (Hansen & Pedersen 1985). These occurrences may represent larger-scale analogues of the buchites commonly developed in the Hebrides where basic intrusions cut argillaceous sediments, and are probably derived from crustal fusion of aluminous sedimentary rocks.

In conclusion, after well over 100 years of intensive investigation, the early Tertiary igneous provinces around the N Atlantic continue to reveal histories of ever more diversity and complexity as increasingly sophisticated levels of field, petrological, geochemical and geophysical techniques are brought to bear.

ACKNOWLEDGEMENTS: Of the various colleagues who have offered advice and criticism I would particularly thank C. H. Emeleus, L. M. Larsen, A. C. Morton, T. F. D. Nielsen, and an anonymous reviewer. My thanks go also to L. Begg and D. Baty for typing and draughting.

References

ABRAHAMSEN, N., SCHOENHARTING, G. & HEINESEN, M. 1984. Palaeomagnetism of the Vestmanna core and magnetic age and evolution of the Faeroe Islands. *In:* BERTHELSEN, O., NOE-NYGAARD, A. & RASMUSSEN, J. (eds). *The deep drilling project 1980–1981 in the Faeroe Islands.* Føroya Frodskaparfelag, Torshavn, pp 93–108.

ANDERSON, F. W. & DUNHAM, K. C. 1966. *The geology of Northern Skye.* Memoir of the Geological Survey of Great Britain.

ATHAVALE, R. N. & SHARMA, P. V. 1975. Palaeomagnetic results on early Tertiary lava flows from West Greenland and their bearing on the evolution history of the Baffin Bay–Labrador Sea region. *Canadian Journal of Earth Sciences*, **12**, 1–18.

BEARTH, P. 1959. On the alkali massif of the Werner Bjerge in East Greenland. *Meddelelser om Grønland*, **153(4)**, 62 pp.

BECKINSALE, R. D., THOMPSON, R. N. & DURHAM, J. J. 1974. Petrogenetic significance of initial $^{87}Sr/^{86}Sr$ Ratios in the North Atlantic Tertiary Igneous Province in the light of Rb-Sr, K-Ar and ^{18}O abundance studies of the Sarqâta qáqâ Intrusive Complex, Ubekendt Island, West Greenland. *Journal of Petrology*, **15**, 525–538.

——, PANKHURST, R. J., SKELHORN, R. R. & WALSH, J. N. 1978. Geochemistry and petrogenesis of the early Tertiary lava pile of the Isle of Mull, Scotland. *Contributions to Mineralogy and Petrology*, **66**, 415–427.

BELL, J. D. 1982. Acid intrusions. *In:* SUTHERLAND, D. S. (ed) *Igneous Rocks of the British Isles.* Wiley & Sons, New York, 427–440.

BERGGREN, W. A., KENT, D. V. & FLYNN, J. J. 1985. Jurassic to Paleogene: Part 2 Paleogene geochronology and chronostratigraphy. *In:* SNELLING, N. J. (ed). *The chronology of the geological record.* Memoir of the Geological Society of London, **10**, 141–195.

BODVARSSON, G. & WALKER, G. P. L. 1964. Crustal drift in Iceland. *Geophysical Journal of the Royal Astronomical Society*, **8**, 285–300.

BOTT, M. H. P. 1987. The continental margin of central East Greenland in relation to North Atlantic plate tectonic evolution. *Journal of the Geological Society of London*, **144**, 561–568.

BOTT, M. H. P., SUNDERLAND, J., SMITH, P. J., CASTEN, U. & SAXOV, S. 1974. Evidence for continental crust beneath the Faeroe Islands. *Nature*, **248**, 202–204.

——, NIELSEN, P. H. & SUTHERLAND, J. 1976. Converted P-waves originating at the continental margin between the Iceland–Faeroe Ridge and the Faeroe Block. *Geophysical Journal of the Royal Astronomical Society*, **44**, 229–238.

BRIDGWATER, D., DAVIES, F. B., GILL, R. C. O., GORMAN, B. E., MYERS, J. S., PEDERSEN, S. & TAYLOR, P. 1978. Precambrian and Tertiary geology between Kangerdlugssuaq and Angmagssalik, East Greenland. A preliminary report. *Grønlands Geologiske Undersøgelse Rapport*, **83**, 17 pp.

——, ESCHER, A., JACKSON, G. D., TAYLOR, F. C. & WINDLEY, B. F. Development of the Precambrian Shield in West Greenland, Labrador and Baffin Island. *Memoir of the American Association of Petroleum Geologists*, **19**, 99–116.

BROOKS, C. K. 1973(a). Rifting and doming in southern East Greenland. *Nature, Physical Sciences*, **244**, 23–24.

—— 1973(b). The Tertiary of Greenland—a volcanic and plutonic record of continental breakup. *Memoir of the American Association of Petroleum Geologists*, **19**, 150–160.

—— 1979. Geomorphological observations at Kangerdlugssuaq, East Greenland. *Meddelelser om Grønland, Geoscience*, **4**, 24 pp.

—— & GLEADOW, A. J. W. 1977. A fission-track age for the Skaergaard intrusion and the age of the East Greenland basalts. *Geology*, **5**, 539–540.

—— & GILL, R. C. O. 1982. Compositional variation in the pyroxenes and amphiboles of the Kangerdlugssuaq intrusion, East Greenland: Further evidence for the crustal contamination of a syenitic magma. *Mineralogical Magazine*, **45**, 1–9.

—— & NIELSEN, T. F. D. 1982. The Phanerozoic development of the Kangerdlugssuaq area, East Greenland. *Meddelelser om Grønland, Geoscience*, **9**, 30 pp.

—— & —— 1982. The E. Greenland continental margin: a transition between oceanic and continental magmatism. *Journal of the Geological Society of London*, **139**, 265–275.

——, —— & PETERSEN, T. S. 1976. The Blosseville Coast Basalts of East Greenland: their occurrence, composition and temporal variations. *Contributions to Mineralogy and Petrology*, **58**, 279–292.

——, PEDERSEN, A. K. & REX, D. C. 1979. The petrology and age of alkaline mafic lavas from the nunatak zone of central East Greenland. *Bulletin of the Grønlands Geologiske Undersøgelse*, **133**, 28 pp.

——, ——, LARSEN, L. M. & ENGELL, J. 1982. The mineralogy of the Werner Bjerge Complex, East Greenland. *Meddelelser om Grønland, Geoscience*, **7**.

BROOKS, M. 1984. The age of the Conachair Granite. *In:* HARDING, R. R., MERRIMAN, R. J. & NANCARROW, R. H. A. (eds). St. Kilda: an illustrated account of the geology. *Report of the British Geological Survey*, **16(7)**, 40–1.

BROWN, G. M. 1982. Introduction to Part 7: An appraisal of the igneous history. *In:* SUTHERLAND, D. S. (ed). *Igneous Rocks of the British Isles*. Wiley & Sons, New York, 345–350.

BROWN, P. E., VAN BREEMEN, O., NOBLE, R. H. & MACINTYRE, R. M. 1977. Mid-Tertiary igneous activity in East Greenland—the Kialineq Complex. *Contributions to Mineralogy and Petrology*, **64**, 109–122.

BURKE, K. & DEWEY, J. F. 1973. Plume-generated triple junctions: key indicators in applying plate tectonics to old rocks. *Journal of Geology*, **81**, 406–433.

CASTEN, U. 1973. The crust beneath the Faeroe Islands. *Nature*, **241**, 83–84.

—— & NIELSEN, P. H. 1975. Faeroe Islands—a microcontinental fragment? *Journal of Geophysics*, **41**, 357–366.

CLARKE, D. B. 1970. Tertiary Basalts of Baffin Bay: possible primary magma from the mantle. *Contributions to Mineralogy and Petrology*, **25**, 203–224.

—— 1977. The Tertiary Volcanic Province of Baffin Bay. *Geological Association of Canada, Special Paper*, **16**, 445–460.

—— & PEDERSEN, A. K. 1976. Tertiary volcanic province of West Greenland. *In:* ESCHER, A. & WATT, W. S. (eds). *Geology of Greenland*. Geological Survey of Greenland, Copenhagen, pp. 365–385.

——, MUECKE, G. K. & PE-PIPER, G. 1983. The lamprophyres of Ubekendt Ejland, West Greenland: products of renewed partial melting or

extreme differentiation? *Contributions to Mineralogy and Petrology*, **83**, 117–127.

—— & UPTON, B. G. J. 1971. Tertiary Basalts of Baffin Island: field relations and tectonic setting. *Canadian Journal of Earth Science*, **8**, 248–258.

COX, K. G. 1980. A model for flood basalt vulcanism. *Journal of Petrology*, **21**, 629–650.

——, JOHNSON, R. L., MONKMAN, L. J., STILLMAN, C. J., VAIL, J. R. & WOOD, D. N. 1965. The geology of the Nuanetsi Igneous Province. *Philosophical Transactions of the Royal Society of London*, **A257**, 71–218.

CURRY, D., ADAMS, C. G., BOULTER, M. C., DILLEY, F. C., EAMES, F. E., FUNNELL, B. M. & WELLS, M. K. 1978. *A correlation of Tertiary rocks in the British Isles*. Geological Society of London, Special Report, **12**, 72 pp.

DEER, W. A. 1976. Tertiary igneous rocks between Scoresby Sund and Kap Gustav Holm, East Greenland. *In:* ESCHER, A. & WATT, W. S. (eds). *Geology of Greenland*. Geological Survey of Greenland, Copenhagen, 405–429.

——, KEMPE, D. R. C. & JONES, G. C. 1984. Syenitic and associated intrusions of the Kap Edvard Holm region, Kangerdlugssuaq, East Greenland. *Meddelelser om Grønland, Geoscience*, **12**, 26 pp.

DICKIN, A. P. 1981. Isotope geochemistry of Tertiary igneous rocks from the Isle of Skye, N.W. Scotland. *Journal of Petrology*, **22**, 155–189.

—— & JONES, N. W. 1983. Isotopic evidence for the age and origin of pitchstones and felsites, Isle of Eigg, N.W. Scotland. *Journal of the Geological Society of London*, **140**, 690–700.

DURANT, G. P., DOBSON, M. R., KOKELAAR, B. P., MACINTYRE, R. M. & REA, W. J. 1976. Preliminary report on the nature and age of the Blackstones Bank Igneous Centre, western Scotland. *Journal of the Geological Society of London*, **132**, 319–326.

ELDHOLM, O. 1978. Observations on the margin off Norway (66–70°N) and the history of early Cenozoic rifting. *In:* RAMBERG, I. B. & NEUMANN, E. R. (eds). *Tectonics and Geophysics of Continental Rifts*, D. Reidel, Dordrecht, pp. 229–239.

——, THIEDE, J., TAYLOR, E. *et al.* 1987. Site 642. *Initial Reports of the Ocean Drilling Program (Part A)*. US Government Printing Office, Washington, **104**, 53–213.

ELLIOTT, G. F. 1971. Eocene volcanics in south-east England. *Nature, Physical Science*, **230**, 9.

EMELEUS, C. H. 1982. The central complexes. *In:* SUTHERLAND, D. S. (ed). *Igneous Rocks of the British Isles*. Wiley & Sons, New York, pp. 369–414.

—— 1983. Tertiary igneous activity. *In:* CRAIG, G. Y. (ed). *Geology of Scotland*. Scottish Academic Press, Edinburgh, 357–397.

—— 1985. The Tertiary lavas and sediments of northwest Rhum, Inner Hebrides. *Geological Magazine*, **122**, 419–437.

—— & FORSTER, R. M. 1979. *Tertiary Igneous Rocks of Rhum—Field Guide*. Nature Conservancy Council, Newbury.

——, WADSWORTH, W. J. & SMITH, N. J. 1985. The early igneous and tectonic history of the Rhum

Tertiary volcanic centre. *Geological Magazine*, **12**, 451–457.

ESSON, J., DUNHAM, A. C. & THOMPSON, R. N. 1975. Low alkali, high calcium olivine tholeiite lavas from the Isle of Skye, Scotland. *Journal of Petrology*, **16**, 488–497.

FAHRIG, W. F., IRVING, E. & JACKSON, G. D. 1971. Palaeomagnetism of the Franklin diabases. *Canadian Journal of Earth Science*, **8**, 455–467.

FAWCETT, J. J., GITTINS, J. RUCKLIDGE. 1982. Petrology of Tertiary lavas from the western Kangerdlugssuaq area, East Greenland. *Mineralogical Magazine*, **45**, 211–218.

FITCH, F. J., HOOKER, P. J., MILLER, J. A. & BRERETON, N. R. 1978. Glauconite dating of Palaeocene—Eocene rocks from East Kent and the time-scale of Palaeocene volcanism in the North Atlantic region. *Journal of the Geological Society of London*, **135**, 499–512.

FURLONG, K. P. & FOUNTAIN, D. M. 1986. Continental crustal underplating: thermal considerations and seismic-petrological consequences. *Journal of Geophysical Research*, **90**, 8285–8294.

GATLIFF, R. W., HITCHEN, K., RITCHIE, J. D. & SMYTHE, D. K. 1984. Internal structure of the Erlend volcanic complex, north of Shetland, revealed by seismic reflection. *Journal of the Geological Society of London*, **141**, 555–562.

GIBB, F. G. F. & KANARIS-SOTIRIOU, R. 1988. The geochemistry and origin of the Faeroe–Shetland sill complex. *In*: MORTON, A. C. & PARSON, L. M. (eds) *Early Tertiary Volcanism and the Opening of the NE Atlantic*. Geological Society of London, Special Publication, **39**, pp. 241–252.

——, —— & NEVES, R. 1986. A new Tertiary sill complex of mid-ocean ridge basalt type off the Shetland Isles: a preliminary report. *Transactions of the Royal Society of Edinburgh (Earth Science)*, **77**, 223–30.

GIBSON, D., McCORMICK, A. G., MEIGHAN, I. G. & HALLIDAY, A. N. 1987. The British Tertiary Igneous Province: Young Rb-Sr ages for the Mourne Mountains Granites. *Scottish Journal of Geology*, **23**, 221–5.

GILL, R. C. O., NIELSEN, T. F. D., BROOKS, C. K. & INGRAM, G. A. 1988. Tertiary volcanism in the Kangerdlugssuaq region, E Greenland: trace element geochemistry of the Lower Basalts and tholeiitic dyke swarms. *In*: MORTON, A. C. & PARSON, L. M. (eds) *Early Tertiary Volcanism and the Opening of the NE Atlantic*. Geological Society of London Special Publication, **39**, pp. 161–179.

GLEADOW, A. J. W. & BROOKS, C. K. 1979. Fission track dating, thermal histories and tectonics of igneous intrusions in East Greenland. *Contributions to Mineralogy and Petrology*, **71**, 45–60.

HALD, N. 1971. An investigation of the igneous rocks on Hareøen and Western Nugssuaq, West Greenland. *Grønlands Geologiske Undersøgelse Rapport*, **35**.

—— 1976. Early Tertiary flood basalts from Hareøen and western Nugssuaq, West Greenland. *Grønlands Geologiske Undersøgelse Bulletin*, **120**, 36 pp.

—— 1978. Tertiary igneous activity at Giesecke Bjerge,

northern East Greenland. *Danske Geologiske Forening, Special Issue*, **27**, 109–115.

—— & WAAGSTEIN, R. 1983. Silicic basalts from the Faeroe Islands: Evidence of crustal contamination. *In*: BOTT, M. H. P., SAXOV, S., TALWANI, M. & THIEDE, J. (eds). *Structure and development of the Greenland–Scotland Ridge*. Plenum Press, New York, pp. 343–348.

—— & WAAGSTEIN, R. 1984. Lithology and chemistry of a 2 km sequence of Lower Tertiary tholeiitic lavas drilled on Suduroy, Faeroe Islands (Lopra-1). *In*: BERTHELSEN, O., NOE-NYGAARD, A. & RASMUSSEN, J. (eds) *The deep drilling project 1980–1981 in the Faeroe Islands*. Føroya Frodskaparfelag, Torshavn, pp. 15–38.

HALLAM, A. 1971. Mesozoic geology and the opening of the North Atlantic. *Journal of Geology*, **79**, 129–157.

—— 1983. Jurassic, Cretaceous and Tertiary sediments. *In*: CRAIG, G. Y. (ed). *Geology of Scotland*. Scottish Academic Press, Edinburgh, pp. 343–356.

HAMPTON, C. M. & TAYLOR, P. N. 1983. The age and nature of the basement of southern Britain: evidence from Sr and Pb isotopes in granites. *Journal of the Geological Society of London*, **140**, 499–509.

HANSEN, K. & PEDERSEN, A. K. 1985. Fission track dating of lower Tertiary rhyolitic glass rocks from Disko. *Grønlands Geologiske Undersøgelse Rapport, Report of Activities, 1984*, **125**, 28–30.

HENDERSON, G. 1973. The geological setting of the West Greenland Basin in the Baffin Bay region. Earth science symposium on offshore eastern Canada, *Geological Survey of Canada Paper*, 71–23, 521–544.

——, ROSENKRANTZ, A. & SCHIENER, E. J. 1976. Cretaceous—Tertiary sedimentary rocks of West Greenland. In: ESCHER, A. & WATT, W S (eds) *Geology of Greenland*. Geological Survey of Greenland, Copenhagen, pp. 340–362.

HIGGINS, A. C. & SOPER, N. J. 1981. Cretaceous–Palaeogene sub-basaltic and intrabasaltic sediments of the Kangerdlugssuaq area, Central East Greenland. *Geological Magazine*, **118**, 337–448.

HINZ, K. 1981. A hypothesis on terrestrial catastrophes. Wedges of very thick oceanward dipping layers beneath passive continental margins—their origin and palaeoenvironmental significance. *Geologisches Jahrbuch*, **E22**, 3–28.

HOLMES, A. 1918. The basaltic rocks of the Arctic region. *Mineralogical Magazine*, **18**, 180–223.

HORNE, R. R. & MACINTYRE, R. M. 1975. Apparent age and significance of Tertiary dykes in the Dingle Peninsula, S.W. Ireland. *Scientific Proceedings of the Royal Dublin Society*, **A5**, 293–299.

JACQUE, M. & THOUVENIN, J. 1975. Lower Tertiary tuffs and volcanic activity in the North Sea. *In*: WOODLAND, A. W. (ed). *Petroleum and the continental shelf of North West Europe, (Vol. 1: Geology)*. Applied Science Publishers, London, pp. 455–465.

JONES, E. J. W. & RAMSAY, A. T. S. 1982. Volcanic ash deposits of early Eocene age from the Rockall Trough. *Nature*, **299**, 342–344.

KAPP, H. 1960. Zur petrologie der subvolkane zwischen

Mesters Vig und Antarctic Havn (Ost–Grönland). *Meddelelser om Grønland*, **153**.

KEMPE, D. R. C., DEER, W. A. & WAGER, L. R. 1970. Geological investigations in East Greenland, Part VII. The petrology of the Kangerdlugssuaq alkaline intrusion, East Greenland. *Meddelelser om Grønland*, **190(2)**, 49 pp.

KENT, P. E. 1977. The Mesozoic development of aseismic continental margins. *Journal of the Geological Society of London*, **134**, 1–18.

KHARIN, G. N. 1974. The petrology of magmatic rocks, DSDP Leg 38. *In:* TALWANI, M., UDINTSEV, G. *et al. Initial Reports of the Deep Sea Drilling Project*. US Government Printing Office, Washington, **38**, 685–702.

KING, B. C. 1982. Composite intrusions: association of acid and basic magmas. *In:* SUTHERLAND, D. S. (ed). *Igneous Rocks of the British Isles*. Wiley & Sons, New York, pp. 441–447.

KOCH, L. & HALLER, J. 1971. Geological map of East Greenland 72°N–76°N lat (1:250,000). *Meddelelser om Grønland*, **183**.

KNOX, R. W. O'B. & MORTON, A. C. 1983. Stratigraphical distribution of early Palaeogene pyroclastic deposits in the North Sea Basin. *Proceedings of the Yorkshire Geological Society*, **44**, 355–363.

LARSEN, H. C. 1978. Offshore continuation of East Greenland dyke swarm and North Atlantic Ocean formation. *Nature*, **274**, 220–223.

—— 1980. Geological perspectives of the East Greenland continental margin. *Bulletin of the Geological Society of Denmark*, **29**, 77–101.

LARSEN, J. G. 1977. Transition from low potassium olivine tholeiites to alkali basalts on Ubekendt Ejland. *Meddelelser om Grønland*, **200(1)**, 1–42.

LARSEN, L. M. & WATT, W. S. 1985. Episodic volcanism during break-up of the North Atlantic: evidence from the East Greenland plateau basalts. *Earth and Planetary Science Letters*, **73**, 105–116.

——, —— & WATT, M. 1988. Geology and petrology of the Lower Tertiary plateau basalts of the Scoresby Sund region, East Greenland. *Bulletin of the Grønlands Geologiske Undersøgelse*. In press.

LUND, J. 1983. Biostratigraphy of interbasaltic coals. *In:* BOTT, M. H. P., SAXOV, S., TALWANI, M. & THIEDE, J. (eds). *Structure and development of the Greenland–Scotland Ridge*. Plenum Press, New York. 417–423.

LYLE, P. 1985. The petrogenesis of the Tertiary basaltic and intermediate lavas of northeast Ireland. *Scottish Journal of Geology*, **21**, 71–84.

MCBIRNEY, A. R. 1975. Differentiation of the Skaergaard Intrusion. *Nature*, **253**, 691–694.

MACINTYRE, R. M. 1977. Anorogenic magmatism, plate motion and Atlantic evolution. *Journal of the Geological Society of London*, **133**, 375–384.

——, MCMENAMIN, T. & PRESTON, J. 1975. K-Ar results from Western Ireland and their bearing on the timing and siting of Thulean magmatism. *Scottish Journal of Geology*, **11**, 227–249.

——, —— & —— 1979. K-Ar results from Western Ireland and their bearing on the timing and siting of Thulean magmatism. (reply to Mussett). *Scottish Journal of Geology*, **15**, 251–254.

MCKENZIE, D. 1984. A possible mechanism for epeirogenic uplift. *Nature*, **307**, 616–618.

MCQUILLIN, R. & BINNS, P. E. 1973. Geological structure in the Sea of the Hebrides. *Nature, Physical Science*, **241**, 2–4.

MATTEY, D. P., GIBSON, I. L., MARRINER, G. F. & THOMPSON, R. N. 1977. The diagnostic geochemistry, relative abundance, and spatial distribution of high-calcium, low-alkali olivine tholeiite dykes in the Lower Tertiary regional swarm of the Isle of Skye, N.W. Scotland. *Mineralogical Magazine*, **41**, 273–286.

MAYNC, W. 1942. Stratigraphie und Faziesverhältnisse der oberpermischen Ablagerungen Ostgrönlands (olim "Oberkarbon–Unterperm") zwischen Wollaston Forland und dem Kejser Franz Josephs Fjord. *Meddelelser om Grønland*, **115**, 128 pp.

MENZIES, M., HALLIDAY, A., HUNTER, R. N., MACINTYRE, R. M. & UPTON, B. G. J. 1987. The age, composition and significance of a xenolith-bearing monchiquite dike, Lewis, Scotland. *Australian Journal of Earth Sciences Special Publication*, IKC IV. In press.

MERRIMAN, R. J., TAYLOR, P. N. & MORTON, A. C. 1988. Petrochemistry and isotope chemistry of early Palaeogene basalts forming the dipping reflector sequence SW of Rockall Plateau, NE Atlantic. *In:* MORTON, A. C. & PARSON, L. M. (eds) *Early Tertiary Volcanism and the Opening of the NE Atlantic*. Geological Society of London, Special Publication, **39**, pp. 123–134.

MITCHELL, J. G. & MOHR, P. 1986. K-Ar systematics in Tertiary dolerites from West Connacht, Ireland. *Scottish Journal of Geology*, **22**, 225–240.

MORTON, A. C., DIXON, J. E., FITTON, J. G., MACINTYRE, R. M., SMYTHE, D. K. & TAYLOR, P. N. 1988. Early Tertiary volcanic rocks in Well 163/6–1A, Rockall Trough. *In:* MORTON, A. C. & PARSON, L. M. (eds) *Early Tertiary Volcanism and the Opening of the NE Atlantic*. Geological Society of London Special Publication, **39**, pp. 293–308.

MÜNTHER, V. 1973. Results from a geological reconnaissance around Svartenhuk Halvø, West Greenland. *Grønlands Geologiske Undersøgelse Rapport*, **50**, 26 pp.

MUSSETT, A. E. 1980. British Tertiary Igneous Province probably not associated with East Greenland lavas. *Nature*, **284**, 376–377.

—— 1984. Time and duration of Tertiary igneous activity of Rhum and adjacent areas. *Scottish Journal of Geology*, **20**, 273–279.

—— 1986. ^{40}Ar–^{39}Ar step-heating ages of the Tertiary igneous rocks of Mull, Scotland. *Journal of the Geological Society of London*, **143**, 887–896.

MYERS, J. S. 1980. Structure of the coastal dyke swarm and associated plutonic intrusions of East Greenland. *Earth and Planetary Science Letters*, **46**, 407–418.

NIELSEN, P. H. 1983. Geology and crustal structure of the Faeroe Island—a review. *In:* BOTT, M. H. P., SAXOV, S., TALWANI, M. & THIEDE, J. (eds). *Structure and Development of the Greenland–Scotland Ridge*. Plenum Press, New York, pp. 77–86.

NIELSEN, T. F. D. 1975. Possible mechanism of

continental break-up in the North Atlantic. *Nature*, **253**, 182–184.

—— 1978. The Tertiary dike swarms of the Kangerdlugssuaq area, East Greenland. An example of magmatic development during continental break-up. *Contributions to Mineralogy and Petrology*, **67**, 63–78.

—— 1980. The petrology of a melilitolite, melteigite, ijolite, nepheline, syenite and carbonatite ring dike system in the Gardiner Complex, East Greenland. *Lithos*, **13**, 181–197.

—— 1981. The ultramafic cumulate series, Gardiner Complex, East Greenland. Cumulates in a shallow level magma chamber of a nephelinitic volcano. *Contributions to Mineralogy and Petrology*, **76**, 60–72.

—— 1987. Tertiary alkaline magmatism in East Greenland: a review. *In:* FITTON, J. G. & UPTON, B. G. J. (eds) *Alkaline Igneous Rocks*. Geological Society of London, Special Publication, **30**, 489–515.

—— & BROOKS, C. K. 1981. The E. Greenland rifted continental margin: an examination of the coastal flexure. *Journal of the Geological Society of London*, **138**, 559–568.

——, SOPER, N. J., BROOKS, C. K., FALLER, A. M., HIGGINS, A. C. & MATTHEWS, D. W. 1981. The pre-basaltic sediments and the Lower Basalts at Kangerdlugssuaq, East Greenland: their stratigraphy, lithology, palaeomagnetism and petrology. *Meddelelser om Grønland, Geoscience*, **6**, 25 pp.

NOE-NYGAARD, A. 1942. On the geology and petrography of the West Greenland Basalt Province, Part III. The Plateau Basalts of Svartenhuk Peninsula. *Meddelelser om Grønland*, **137(3)**, 78 pp.

—— 1967. Dredged basalts from Skagerrak. *Meddelelser Dansk Geologisk Forening*, **17**, 285–287.

—— 1974. Cenozoic to Recent Volcanism in and around the North Atlantic Basin. *In:* NAIRN, A. E. M. & STEHLI, F. G. (eds). *The Ocean Basins and Margins (Vol. 2 The North Atlantic)*. Plenum Press, New York, pp. 391–443.

—— 1976. Tertiary igneous rocks between Shannon and Scoresby Sund, East Greenland. *In:* ESCHER, A. & WATT, W. S. (eds). *Geology of Greenland*. Geological Survey of Greenland, Copenhagen, pp. 386–402.

—— & RASMUSSEN, J. 1968. Petrology of a 3,000 metre sequence of basaltic lavas in the Faeroe Islands. *Lithos*, **1**, 286–304.

——, —— 1984. Introduction: Geological review and choice of drilling sites. *In:* BERTHELSEN, O., NOE-NYGAARD, A. & RASMUSSEN, J. (eds). *The deep drilling project 1980–1981 in the Faeroe Island*. Føroya Frodskaparfelag, Torshavn, pp. 9–12.

NUNNS, A. G. 1983. Plate tectonic evolution of the Greenland–Scotland Ridge and surrounding regions. *In:* BOTT, M. H. P., SAXOV, S., TALWANI, M. & THIEDE, J. (eds) *Structure and development of the Greenland–Scotland Ridge*. Plenum Press, New York, pp. 11–28.

ODIN, G. S. & MITCHELL, J. G. 1983. Dating of the Palaeocene–Eocene Blosseville Group basalts, Scoresby Sund, East Greenland: a review. *Newsletters on Stratigraphy*, **12**, 112–121.

O'NIONS, R. K. & CLARKE, D. B. 1972. Comparative trace element geochemistry of Tertiary basalts from Baffin Bay. *Earth and Planetary Science Letters*, **15**, 436–446.

PACEY, N. R. 1984. Bentonites in the chalk of central eastern England and their relation to the opening of the northeast Atlantic. *Earth and Planetary Science Letters*, **67**, 48–60.

PANKHURST, R. J., BECKINSALE, R. D. & BROOKS, C. K. 1976. Strontium and oxygen isotopic evidence relating to the petrogenesis of the Kangerdlugssuaq alkaline intrusion, East Greenland. *Contributions to Mineralogy and Petrology*, **54**, 17–42.

PARROTT, R. J. E. 1976. $^{40}Ar/^{39}Ar$ dating on Labrador Sea volcanics and their relation to sea-floor spreading. M.Sc. thesis, Dalhousie University, Halifax.

—— & REYNOLDS, P. H. 1975. Argon 40/argon 39 geochronology: age determinations of basalts from the Labrador Sea area. *Geological Society of America, Abstracts with Programs*, **7**, 835.

PEDERSEN, A. K., ENGELL, J. & RØNSBO, J. G. 1975. Early Tertiary volcanism in the Skagerrak: new chemical evidence from ash-layers in the mo-clay of northern Denmark. *Lithos*, **8**, 255–268.

PRESTON, J. 1981. Tertiary igneous activity. *In:* HOLLAND, C. H. (ed) *A Geology of Ireland*. Scottish Academic Press, Edinburgh. 213–223.

—— 1982. Eruptive volcanism. *In:* SUTHERLAND, D. S. (ed). *Igneous Rocks of the British Isles*. Wiley & Sons, New York, pp. 351–368.

REX, D. C., GLEDHILL, A. R., BROOKS, C. K. & STEENFELT, A. 1979. Radiometric ages of Tertiary salic intrusions near Kong Oscars Fjord, East Greenland. *Grønlands Geologiske Undersøgelse Rapport*, **95**, 106–109.

RHYS, G. H. 1974. A proposed standard lithostratigraphic nomenclature for the southern North Sea and an outline structure nomenclature for the whole of the (U.K.) North Sea. *Report of the Institute of Geological Sciences*, **74/8**.

RIDD, M. F. 1983. Aspects of the Tertiary geology of the Faeroe–Shetland Channel. *In:* BOTT, M. H. P., SAXOV, S., TALWANI, M. & THIEDE, J. (eds). *Structure and Development of the Greenland–Scotland Ridge*. Plenum Press, New York, pp. 91–108.

RIDLEY, W. I. 1973. The petrology of volcanic rocks from the Small Isles of Inverness-shire. *Report of the Institute of Geological Sciences*, **73/10**.

ROBERTS, D. G. 1975. Tectonic and stratigraphic evolution of Rockall Plateau and Trough. *In:* WOODLAND, A. W. (ed) *Petroleum and the Continental Shelf of North West Europe (Vol I: Geology)*. Applied Science Publishers, London, pp. 77–89.

ROSENKRANTZ, A. & PULVERTAFT, T. C. R. 1969. Cretaceous–Tertiary stratigraphy and tectonics in Northern West Greenland. *In:* KAY, M. (ed). *North Atlantic Geology and Continental Drift*. Memoir of the American Association of Petroleum Geologists, **12**, 883–898.

SABINE, P. A. 1960. The geology of Rockall, North Atlantic. *Bulletin of the Geological Survey of Great Britain*, **16**, 156–178.

SHARMA, P. V. 1970. Geophysical evidence for a buried volcanic mount in the Skagerrak. *Bulletin of the Geological Society of Denmark*, **19**, 368–377.

SMYTHE, D. K., CHALMERS, J. A., SKUCE, A. G., DOBINSON, A. & MOULD, A. S. 1983. Early opening history of the North Atlantic—I. Structure and origin of the Faeroe–Shetland Escarpment. *Geophysical Journal of the Royal Astronomical Society*, **72**, 373–398.

SOPER, N. J. & COSTA, L. I. 1976. Palynological evidence for the age of Tertiary basalts and post-basaltic sediments at Kap Dalton, central East Greenland. *Grønlands Geologiske Undersøgelse Rapport*, **80**, 123–127.

——, HIGGINS, A. C., DOWNIE, C., MATTHEWS, D. W. & BROWN, P. E. 1976. Late Cretaceous–early Tertiary stratigraphy of the Kangerdlugssuaq area, East Greenland, and the age of opening of the north-east Atlantic. *Journal of the Geological Society of London*, **132**, 85–104.

SPEIGHT, J. M., SKELHORN, R. R., SLOAN, T. & KNAAP, R. J. 1982. The dyke swarms of Scotland. *In:* SUTHERLAND, D. S. (ed). *Igneous Rocks of the British Isles*. Wiley & Sons, New York, pp. 449–459.

SRIVASTAVA, S. P. 1983. Davis Strait: Structures, origin and evolution. *In:* BOTT, M. H. P., SAXOV, S., TALWANI, M. & THIEDE, J. (eds). *Structure and development of the Greenland–Scotland Ridge*. Plenum Press, New York, pp. 159–189.

SURLYK, F. 1977. Mesozoic faulting in East Greenland. *Geologie en Mijnbouw*, **56**, 311–327.

—— 1978. Jurassic basin evolution in East Greenland. *Nature*, **274**, 130–133.

SUTHERLAND, D. S. (ed) 1982. *Igneous Rocks of the British Isles*. Wiley & Sons, New York.

TALWANI, M. & ELDHOLM, O. 1972. Continental margin off Norway: A geophysical study. *Bulletin of the Geological Society of America*, **83**, 3575–3606.

—— & UDINTSEV, G. 1974. Tectonic synthesis. *In:* TALWANI, M., UDINTSEV, G. *et al. Initial Reports of the Deep Sea Drilling Project*, US Government Printing Office, Washington, **38**, pp. 1213–1242.

THOMPSON, P., MUSSETT, A. E. & DAGLEY, P. 1987. Revised ^{40}Ar-^{39}Ar age for granites of the Mourne Mountains, Ireland. *Scottish Journal of Geology*, **23**, 215–220.

THOMPSON, R. N. 1982. Geochemistry and magma genesis. *In:* SUTHERLAND, D. S. (ed). *Igneous Rocks of the British Isles*. Wiley & Sons, New York, pp. 461–477.

—— 1982. Magmatism of the British Tertiary Volcanic Province. *Scottish Journal of Geology*, **18**, 49–107.

UPTON, B. G. J., EMELEUS, C. H. & HALD, N. 1980. Tertiary volcanism in northern E. Greenland: Gauss Halvø and Hold with Hope. *Journal of the Geological Society of London*, **137**, 491–508.

——, —— & BECKINSALE, R. D. 1984. Petrology of the northern East Greenland Tertiary flood basalts: Evidence from Hold with Hope and Wollaston Forland. *Journal of Petrology*, **25**, 151–184.

——, ——, —— & MACINTYRE, R. M. 1984. Myggbukta and Kap Broer Ruys: the most northerly of the East Greenland Tertiary igneous centres (?). *Mineralogical Magazine*, **48**, 323–343.

VANN, I. R. 1978. The siting of Tertiary vulcanicity. *In:* BOWES, D. R. & LEAKE, B. E. (eds). *Crustal Evolution in North-western Britain and Adjacent Regions*. Geological Journal Special Issue, **10**, 393–414.

VISCHER, A. 1943. Die postdevonische Tektonik von östgrönland zwischen 74° und 75°N. Br. Kuhn Ø, Wollaston Forland, Clavering Ø und angrenzende Gebiete. *Meddelelser om Grønland*, **133**, 1–94.

WAAGSTEIN, R. 1988. Structure, composition and age of the Faeroe basalt plateau. *In:* MORTON, A. C. & PARSON, L. M. (eds) *Early Tertiary Volcanism and the Opening of the NE Atlantic*. Geological Society of London, Special Publication, **39**, pp. 225–238.

WADSWORTH, W. J. 1982. The major basic intrusions. *In:* SUTHERLAND, D. S. (ed). *Igneous Rocks of the British Isles*. Wiley & Sons, New York, pp. 415–425.

WAGER, L. R. 1934. Geological investigations in East Greenland. Part I. General geology from Anmagssalik to Kap Dalton. *Meddelelser om Grønland*, **105(2)**, 46 pp.

—— 1935. Geological investigations in East Greenland. Part II. Geology of Kap Dalton. *Meddelelser om Grønland*, **105(3)**, 32 pp.

—— 1947. Geological investigations in East Greenland. Part IV. The stratigraphy and tectonics of Knud Rasmussens Land and the Kangerdlugssuaq region. *Meddelelser om Grønland*, **134**, 1–64.

—— & DEER, W. A. 1939. Geological investigation in East Greenland. Part III. The petrology of the Skaergaard intrusion, Kangerdlugssuaq. *Meddelelser om Grønland*, **105**, 1–352.

—— & BROWN, G. M. 1968. *Layered Igneous Rocks*. Oliver & Boyd, London.

WALKER, G. P. L. 1975. A new concept of the evolution of the British Tertiary intrusive centres. *Journal of the Geological Society of London*, **131**, 121–141.

ZIEGLER, P. A. 1978. North-western Europe: tectonics and basin development. *Geologie en Mijnbouw*, **57**, 589–626.

B. G. J. UPTON, Grant Institute of Geology, University of Edinburgh, West Mains Road, Edinburgh EH9 3JW, UK.

Index

Page numbers in italic type refer to pages on which illustrations or tables appear.

Index